Corrosion Behavior and Mechanical Properties of Metallic Materials

Corrosion Behavior and Mechanical Properties of Metallic Materials

Guest Editor

Xiaogang Li

MDPI

Basel • Beijing • Wuhan • Barcelona • Belgrade • Novi Sad • Cluj • Manchester

Guest Editor
Xiaogang Li
Institute of Nuclear and New
Energy Technology
Key Laboratory of Advanced
Reactor Engineering and
Safety of Ministry of
Education
Tsinghua University
Beijing
China

Editorial Office
MDPI AG
Grosspeteranlage 5
4052 Basel, Switzerland

This is a reprint of the Special Issue, published open access by the journal *Materials* (ISSN 1996-1944), freely accessible at: www.mdpi.com/journal/materials/special_issues/7O27S5A2M9.

For citation purposes, cite each article independently as indicated on the article page online and as indicated below:

Lastname, A.A.; Lastname, B.B. Article Title. *Journal Name* **Year**, *Volume Number*, Page Range.

ISBN 978-3-7258-3748-9 (Hbk)
ISBN 978-3-7258-3747-2 (PDF)
https://doi.org/10.3390/books978-3-7258-3747-2

Contents

About the Editor

Xiaogang Li

Xiaogang Li is an assistant professor of Tsinghua University. Xiaogang Li focused on the design of fuel-handling systems in high-temperature gas-cooled reactors, reactor equipment design, the microstructure and structural integrity assessment of high-temperature alloys, failure analysis, life evaluation and life extension of metal components, and welding. Xiaogang Li currently serves as a Member of the Early Career Editorial Board for the *Chinese Journal of Mechanical Engineering, International Journal of Structural Integrity, Rare Metals, Materials, Exploration, Results in Engineering, Journal of Iron and Steel Research International, Additive Manufacturing Frontiers, Journal of Materials Informatics, Corrosion Communications, Complex Engineering Systems, Energy Storage and Applications, Microstructures, Advanced Manufacturing, Journal of Electrochemistry*, and other journals. Xiaogang Li was invited to serve as a long-term reviewer of the {Journal of Magnesium and Alloys, Journal of Materials Science & Technology, Journal of Power Sources, Materials Science & Engineering A, Applied Surface Science, International Journal of Hydrogen Energy, Journal of Alloys and Compounds, Journal of Materials Research and Technology, Chinese Journal of Mechanical Engineering, Advanced Engineering Materials, Computational Materials Science, Steel Research International, Journal of Pipeline Systems Engineering and Practice, International Journal of Electrochemical Science}, and other journals.

materials

Editorial

Corrosion Behavior and Mechanical Properties of Metallic Materials

Xiaogang Li [1,*], Xu Zhang [2], Xuelan Yan [3] and Haiquan Zhang [1]

1 Institute of Nuclear and New Energy Technology, Tsinghua University, Beijing 100084, China; haiquanzh@mail.tsinghua.edu.cn
2 Beijing Institute of Astronautical Systems Engineering, Beijing 100071, China; thuzhangxu@outlook.com
3 Gansu Academy of Sciences, Lanzhou 730030, China; yanxuelan-01@163.com
* Correspondence: xiaogangli@mail.tsinghua.edu.cn

Worldwide, more than USD 400 million/year is spent on corrosion protection and productivity losses due to corrosion. Corrosion, a process associated with chemical/electrochemical reaction, often has a deleterious consequence on mechanical properties, ultimately resulting in the degradation of a material. Metallic components, widely employed in various industries, i.e., oil, gas, marine, nuclear, fuel cells, medicine and generating electricity, often suffer from severe corrosion, which might be detrimental to service life and even cause serious accidents. Meanwhile, some extreme corrosive environments are also the main restriction on the applications of advanced metallic materials with excellent mechanical properties. Thus, understanding corrosion behavior and its effect on mechanical properties will always be of great practical significance to the development and application of metallic materials. Corrosion behavior leads to the degradation of mechanical properties, and both are affected by the service environment and the physical and chemical properties of the material itself.

The purpose of this Special Issue is to provide a research forum to report corrosion behavior and the related mechanical properties, chemical composition, and microstructure for metallic materials to address existing corrosion challenges and assist in the development of super-corrosion-resistant materials. Topics of interest include, but are not limited to, the studies mentioned above. Other relevant studies, such as hydrogen embrittlement, characterization of the corroded microstructure, the corrosion mechanisms of advanced materials, the method of surface treatment to improve corrosion resistance, the evolution mechanism of mechanical properties in corrosion environment, and the design of novel corrosion-resistant material, will also be considered, which could enhance our knowledge of corrosion protection. Research articles and reviews in this area of study are welcome.

Zhang et al. [1] contributed a review entitled "Surface-Modification Strategy to Produce Highly Anticorrosive $Ti_3C_2T_x$ MXene-Based Polymer Composite Coatings: A Mini-Review" which reviewed $Ti_3C_2T_x$ MXene/epoxy resin and MXene/waterborne polyurethane polymer anticorrosive composite coatings. It introduced the structure, characteristics, and main synthesis methods of MXenes and summarized the surface-modification strategies to improve the dispersion, compatibility, stability, and anti-aggregation properties of MXenes. This paper outlined the current challenges and future opportunities regarding MXene-based corrosion-resistant organic composite coatings.

Lin et al. [2] contributed a paper entitled "Molecular Dynamics Simulations of Displacement Cascades in BCC-Fe: Effects of Dislocation, Dislocation Loop and Grain Boundary" in which the interactions between displacement cascades and dislocations, dislocation loops, and grain boundaries in BCC-Fe were investigated. The atomic structural changes

Received: 10 February 2025
Revised: 21 February 2025
Accepted: 21 February 2025
Published: 25 February 2025

Citation: Li, X.; Zhang, X.; Yan, X.; Zhang, H. Corrosion Behavior and Mechanical Properties of Metallic Materials. *Materials* **2025**, *18*, 1009. https://doi.org/10.3390/ma18051009

induced by displacement cascades were clarified to understand how irradiation damage affects dislocations, dislocation loops, and grain boundaries, revealing irradiation damage at the microscale.

Muthu et al. [3] contributed a paper entitled "Accelerated Formation of Oxide Layers on Zircaloy-4 Utilizing Air Oxidation and Comparison with Water-Corroded Oxide Layers" which examined the formation of an oxide layer on a Zircaloy-4 cladding tube by means of a combination of high-temperature water corrosion and air oxidation. Morphology and oxide phases by the above accelerated method were comparable to those formed through high-temperature water corrosion, which is of great engineering value in Canada Deuterium Uranium fuel integrity analyses.

Zurnadzhy et al. [4] contributed a paper entitled "Enhancing the Tensile Properties and Ductile-Brittle Transition Behavior of the EN S355 Grade Rolled Steel via Cost-Saving Processing Routes" which reports on the structural characterization and mechanical property evaluations of low-carbon micro-alloyed structural steel subjected to different technological processes at the final stage of steel sheet manufacturing. The work focused on a detailed comparison of the effect of different technologies on the properties of the rolled steel produced from the same melt and paid attention to the evaluation of the effect of structural factors influencing the strength of steel in regard to the variations in its ductile–brittle transition behavior.

Zhao et al. [5] contributed a paper entitled "Effect of Boron Content in LiOH Solutions on the Corrosion Behavior of Zr-Sn-Nb Alloy" in which the effects of boron (through the injection of boric acid) on improving the corrosion resistance of the Zr-Sn-Nb alloy in LiOH solutions were clarified. It was noted that B^{3+} was added into oxide films and, thus, the oxygen vacancies caused by Li^{+} aggregation were reduced, slowing down the oxidation of Zr-Sn-Nb alloys.

Cheng et al. [6] contributed a paper entitled "Investigation of Precipitation Behavior of a Novel Ni-Fe-Based Superalloy during High-Temperature Aging Treatment" which investigated the precipitation behaviors in a novel Ni-Fe-based superalloy developed for advanced ultra-supercritical coal-fired power plant applications. The abnormal coarsening behavior of precipitations in the novel alloy was noted, and its effects on hardness and crack were discussed.

Łosiewicz et al. [7] contributed a paper entitled "Effect of Artificial Saliva Modification on Corrosion Resistance of Metal Oxide Coatings on Co-Cr-Mo Dental Alloy" which prepared the single-layer TiO_2-ZrO_2 sol–gel coatings on the Co-Cr-Mo dental alloy using the method of dip-coating. The TiO_2-ZrO_2 coatings significantly reduced pitting corrosion susceptibility at physiological and acidic pH, with the 500 °C sintered coating showing better protective properties, which showed the potential of TiO_2-ZrO_2 coatings in enhancing the performance of Co-Cr-Mo dental alloys.

Faraji et al. [8] contributed a paper entitled "Effect of Natural Inhibitors on the Corrosion Properties of Grade 2 Titanium Alloy" which investigated the corrosion inhibition capabilities of natural extracts such as pomegranate juice, tomato and algae extract on the corrosion of grade 2 Titanium in basic solutions or in solutions containing chlorides. The use of the above compounds as natural inhibitors could be very important in the near future due to the fact that a large amount of more commonly diffused inhibitors is generally toxic and pollutant and the use of new natural compounds could be important in regard to global sustainability.

Chen et al. [9] contributed a paper entitled "Phase, Microstructure and Corrosion Behaviour of $Al_{0.3}FeCoNiCr_x$ High-Entropy Alloys via Cr Addition" which investigated the microstructure evolution of $Al_{0.3}FeCoNiCr_x$ high-entropy alloys and their short-term and long-term corrosion behaviors. The corrosion resistance of $Al_{0.3}FeCoNiCr_x$ increased with the increase in Cr content, exhibiting excellent corrosion resistance compared to the existing high-entropy alloys in the AlFeCoNiCr composition system.

Zhang et al. [10] contributed a paper entitled "Steel Catenary Riser Fatigue Assessment: Fracture Mechanics Approach Versus S–N Curve Method" which investigated the fatigue resistance of a full-scale Steel Catenary Riser girth weld by the Strength–Number of cycles curve method based on weld formation quality and fracture mechanic approaches, and the test results are superior to the design curve E in BS 7608. The method provided a rational estimation of full-scale girth welds, and it was noted that the short crack growth phase is crucial to improving the accuracy and that the geometries of the weld cap are the predominant factors of fatigue life.

Yang et al. [11] contributed a paper entitled "Corrosion Behavior of Copper Foil on PCB Substrates Under Atmospheric Environment in Sichuan-Tibet Region of China" which investigated the corrosion behavior of copper foil on printed circuit boards exposed for one year in a closed atmospheric environment across 22 different sites in the Sichuan-Tibet region. It was noted that copper foils in areas with a large temperature change, higher humidity and more rainfall exhibit more severe corrosion, and copper sulfides are more likely to grow in humid areas.

Sulima et al. [12] contributed a paper entitled "Effect of ZrB_2 Content on the Properties of Copper Matrix Composite" which prepared the Cu-xZrB_2 composites reinforced with zirconium diboride (ZrB_2) to achieve higher hardness, compressive strength and wear resistance with lower relative densities. It should be noted that there is disproportionate decrease in the corrosion resistance of this composite with an increase in the concentration of the ceramic phase in sulfuric acid environment.

Hempel et al. [13] contributed a paper entitled "Hydrogen Diffusion in Deformed Austenitic TRIP Steel—A Study of Mathematical Prediction and Experimental Validation" which investigated the hydrogen embrittlement behavior of TRIP steel. The effects of pre-deformation on hydrogen diffusion and hydrogen embrittlement were clarified, and a good agreement between the theoretical calculations and the fracture surfaces was obtained.

Dobkowska and Kubasek [14] contributed a paper entitled "On the Possibility of the Deformation of Mg and Alloys Without Preheating of Initial Billets: Understanding Their Corrosion Performance via Electrochemical Tests" which focused on the possibility of KoBo extrusion of hcp-structured Mg alloys with different chemical compositions; it also investigated the influence of major microstructural components on the corrosion behavior of Mg and its alloys. A comparison of various groups of Mg that were extruded using KoBo was created to ascertain the most corrosion resistive alloys.

Charazińska et al. [15] contributed a paper entitled "Reconstruction of the Passive Layer of AISI 304 and 316 Steel After Scratching" which investigated the reconstruction of the passive film of austenitic stainless steels after scratching. A new approach in the conducted research was to scratch the test samples under controlled conditions and then evaluate the dynamics of the passive layer reconstruction using an atomic force microscope.

Liu et al. [16] contributed a paper entitled "Numerical Simulation of Galvanic Corrosion and Electrical Insulation for TC4/304 Galvanic Couple" which investigated the galvanic corrosion and electrical insulation between TC4 Ti-alloy and 304 stainless steel coupled in pipe joints. A finite element model was utilized to examine the influence of geometric and environmental factors on the galvanic corrosion process, which showed that electrical insulation performance significantly affects the corrosion process.

Feng et al. [17] contributed a paper entitled "The Evolution of the Tensile Properties of MoS_2-Coated Titanium Alloy Bolts Under the Synergistic Damage of NaCl Corrosion and Preloading" in which titanium alloy bolts coated with MoS_2 (TC4+MoS_2) and bolts treated with a composite treatment of anodizing oxidation and MoS_2 coating (TC4+AO+MoS_2)

were corroded while the tensile properties of them were evaluated. The coatings could provide excellent corrosion resistance, and the tensile strength and fracture behavior of titanium alloy bolts after coating were slightly affected under the synergistic damage of sea water corrosion and preloading.

Donnerbauer et al. [18] contributed a paper entitled "Methodology for Hydrogen-Assisted Fatigue Testing Using In Situ Cathodic Charging" in which a setup was developed utilizing a customized electrochemical charging cell built into a dynamic testing system. The methodology for hydrogen-assisted fatigue testing was achieved using in situ cathodic charging, which could represent hydrogen charging under superimposed mechanical loading.

Gu et al. [19] contributed a paper entitled "The Effect of Low-Temperature Short-Term Annealing on the Microstructure and Properties of Ultrafine-Grained Pure Titanium, in which industrial pure titanium was processed by equal-channel angular pressing and short-term annealed to achieve better mechanical properties for industrial applications. The strengthening of industrial pure titanium after the above processes was likely due to the improved basal texture and the refined grain size.

Vertal' et al. [20] contributed a paper entitled "Improving Biodegradable Mg-Zn(-Ca) Alloys by Surface Treatment via Plasma Electrolytic Oxidation" which investigated the influence of plasma electrolytic oxidation preparation time on the degradation resistance of Mg-1Zn and Mg-1Zn-0.4Ca alloys. The paper noted that low-alloyed, biocompatible Mg-Zn(-Ca) alloys could achieve corrosion resistance comparable to high-performance Mg-4Y-3RE-0.4Zr alloys through appropriate surface treatment.

This Special Issue highlights the novel approach and view, which is committed to clarifying corrosion behavior and the mechanical properties of metallic materials. We sincerely thank the authors who submitted manuscript to our Special Issue and also extend our sincere gratitude to the readers. We hope that our Special Issue will serve as a platform for dialog between scientists in the field of metallic materials. In addition, in view of our continued intake of high-quality articles on the topic, we have begun work on the Special Issue "Corrosion Behavior and Mechanical Properties of Metallic Materials (Second Edition)", and submissions are now welcome. The submission address is https://www.mdpi.com/journal/materials/special_issues/GOO9WH0316 (accessed on 20 February 2025). The topics of interest include, but are not limited to, corrosion behavior, mechanical properties, chemical composition, microstructures, hydrogen embrittlement, characterization of the corroded microstructure, corrosion mechanism of advanced materials, method of surface treatment to improve corrosion resistance, the evolution mechanism of mechanical properties in corrosion environments, and the design of novel corrosion-resistant material. Research articles and reviews regarding this area of study are welcome.

Funding: We appreciate the financial support by the National Natural Science Foundation of China [No. 52405385].

Acknowledgments: The author sincerely appreciates the invaluable support of our department members.

Conflicts of Interest: The authors declare no conflict of interest.

References

1. Zhang, S.; Zhang, G.; Fang, L.; Wang, Z.; Wu, F.; Liu, G.; Wang, Q.; Nian, H. Surface-Modification Strategy to Produce Highly Anticorrosive $Ti_3C_2T_x$ MXene-Based Polymer Composite Coatings: A Mini-Review. *Materials* **2025**, *18*, 653. [CrossRef]
2. Lin, P.; Cui, S.; Nie, J.; He, L.; Cui, W. Molecular Dynamics Simulations of Displacement Cascades in BCC-Fe: Effects of Dislocation, Dislocation Loop and Grain Boundary. *Materials* **2023**, *16*, 7497. [CrossRef]
3. Muthu, S.M.; Lee, H.-B.; Okonkwo, B.O.; Wang, D.; Jang, C.; Na, T. Accelerated Formation of Oxide Layers on Zircaloy-4 Utilizing Air Oxidation and Comparison with Water-Corroded Oxide Layers. *Materials* **2023**, *16*, 7589. [CrossRef]

4. Zurnadzhy, V.; Stavrovskaia, V.; Chabak, Y.; Petryshynets, I.; Efremenko, B.; Wu, K.; Efremenko, V.; Brykov, M. Enhancing the Tensile Properties and Ductile-Brittle Transition Behavior of the EN S355 Grade Rolled Steel via Cost-Saving Processing Routes. *Materials* **2024**, *17*, 1958. [CrossRef] [PubMed]
5. Zhao, Y.; Wu, Z.; Chen, Z.; Yin, Z.; Tang, M.; Xiong, J.; Deng, P. Effect of Boron Content in LiOH Solutions on the Corrosion Behavior of Zr-Sn-Nb Alloy. *Materials* **2024**, *17*, 2373. [CrossRef] [PubMed]
6. Cheng, J.; Li, K.; Yang, Z.; Huo, X.; Fan, M.; Li, S.; Li, S.; Liu, Q.; Ma, Q.; Cai, Z. Investigation of Precipitation Behavior of a Novel Ni-Fe-Based Superalloy during High-Temperature Aging Treatment. *Materials* **2024**, *17*, 4875. [CrossRef] [PubMed]
7. Łosiewicz, B.; Osak, P.; Górka-Kulikowska, K.; Maszybrocka, J. Effect of Artificial Saliva Modification on Corrosion Resistance of Metal Oxide Coatings on Co-Cr-Mo Dental Alloy. *Materials* **2024**, *17*, 5166. [CrossRef] [PubMed]
8. Faraji, M.; Pezzato, L.; Yazdanpanah, A.; Nardi, G.; Esmailzadeh, M.; Calliari, I. Effect of Natural Inhibitors on the Corrosion Properties of Grade 2 Titanium Alloy. *Materials* **2024**, *17*, 5202. [CrossRef] [PubMed]
9. Chen, M.; Shen, H.; Wen, C.; Wang, N.; Tian, Y.; Zhong, W.; Huang, H. Phase, Microstructure and Corrosion Behaviour of $Al_{0.3}FeCoNiCr_x$ High-Entropy Alloys via Cr Addition. *Materials* **2024**, *17*, 5259. [CrossRef] [PubMed]
10. Zhang, N.; Deng, C.; Zhang, W.; Li, S.; Gong, B. Steel Catenary Riser Fatigue Assessment: Fracture Mechanics Approach Versus S–N Curve Method. *Materials* **2024**, *17*, 5677. [CrossRef] [PubMed]
11. Yang, X.; Jin, H.; Jia, D.; Wang, J.; Chen, Z.; Tian, R.; Du, S.; Ma, L. Corrosion Behavior of Copper Foil on PCB Substrates Under Atmospheric Environment in Sichuan-Tibet Region of China. *Materials* **2024**, *17*, 6039. [CrossRef] [PubMed]
12. Sulima, I.; Kowalik, R.; Stępień, M.; Hyjek, P. Effect of ZrB_2 Content on the Properties of Copper Matrix Composite. *Materials* **2024**, *17*, 6105. [CrossRef] [PubMed]
13. Hempel, C.; Mandel, M.; Quitzke, C.; Wendler, M.; Kreschel, T.; Volkova, O.; Krüger, L. Hydrogen Diffusion in Deformed Austenitic TRIP Steel—A Study of Mathematical Prediction and Experimental Validation. *Materials* **2024**, *17*, 6114. [CrossRef] [PubMed]
14. Dobkowska, A.; Kubasek, J. On the Possibility of the Deformation of Mg and Alloys Without Preheating of Initial Billets: Understanding Their Corrosion Performance via Electrochemical Tests. *Materials* **2024**, *17*, 6182. [CrossRef] [PubMed]
15. Charazińska, S.; Sikora, A.; Malczewska, B.; Lochyński, P. Reconstruction of the Passive Layer of AISI 304 and 316 Steel After Scratching. *Materials* **2024**, *17*, 6238. [CrossRef] [PubMed]
16. Liu, K.; Wu, Y.; Liu, H.; Chen, H.; Li, W. Numerical Simulation of Galvanic Corrosion and Electrical Insulation for TC4/304 Galvanic Couple. *Materials* **2025**, *18*, 38. [CrossRef] [PubMed]
17. Feng, D.; Xie, M.; Yu, W.; Li, C.; Guo, R.; Hu, Y.; Ming, Q.; Wan, Q. The Evolution of the Tensile Properties of MoS_2-Coated Titanium Alloy Bolts Under the Synergistic Damage of NaCl Corrosion and Preloading. *Materials* **2025**, *18*, 123. [CrossRef] [PubMed]
18. Donnerbauer, K.; Nickel, T.; von Pavel, M.; Otto, J.L.; Gerdes, L.; Vasquez, J.R.; Lingnau, L.A.; Koch, A.; Walther, F. Methodology for Hydrogen-Assisted Fatigue Testing Using In Situ Cathodic Charging. *Materials* **2025**, *18*, 339. [CrossRef] [PubMed]
19. Gu, Y.; Jiang, J.; Ma, A.; Gu, Y.; Yang, Z.; Wu, H.; Song, C.; Yang, Q.; Ni, C. The Effect of Low-Temperature Short-Term Annealing on the Microstructure and Properties of Ultrafine-Grained Pure Titanium. *Materials* **2025**, *18*, 517. [CrossRef] [PubMed]
20. Vertal', J.; Kajánek, D.; Kubásek, J.; Minárik, P. Improving Biodegradable Mg-Zn(-Ca) Alloys by Surface Treatment via Plasma Electrolytic Oxidation. *Materials* **2025**, *18*, 747. [CrossRef]

materials

MDPI

Article

Effect of Bulk Phase Composition on the Growth of PEO Coatings on the Biomedical Ti-6Al-4V Alloy

José Roberto Ferreira Neto [1], Rafael Parra Ribeiro [1], Nilson Cristino da Cruz [1], Elidiane Cipriano Rangel [1], Bruna de Oliveira Pinto [2], Jhuliene Elen Muro Torrento [2], Carlos Roberto Grandini [2], Ulisses Ferreira Kaneko [3] and Diego Rafael Nespeque Correa [2,*]

[1] Laboratory of Technological Plasma, Institute of Science and Technology, Campus Sorocaba, São Paulo State University (UNESP), Sorocaba 18087-180, SP, Brazil; jose.roberto-ferreira@unesp.br (J.R.F.N.); rafael-parra.ribeiro@unesp.br (R.P.R.); nilson.cruz@unesp.br (N.C.d.C.); elidiane.rangel@unesp.br (E.C.R.)

[2] Laboratory of Anelasticity and Biomaterials, School of Sciences, Campus Bauru, São Paulo State University (UNESP), Bauru 17033-360, SP, Brazil; bruna.oliveira-pinto@unesp.br (B.d.O.P.); jhuliene.torrento@unesp.br (J.E.M.T.); carlos.r.grandini@unesp.br (C.R.G.)

[3] Institute of Geosciences and Exact Sciences, Campus Rio Claro, São Paulo State University (UNESP), Rio Claro 13506-900, SP, Brazil; ulisses.kaneko@unesp.br

* Correspondence: diego.correa@unesp.br

Abstract: This study investigated the effects of plasma electrolytic oxidation (PEO) treatment in a Ca- and P-rich electrolyte on the surface of the Ti-6Al-4V alloy with distinct α/β phase proportions previously induced by heat treatments. The results revealed that the α/β phase proportions were successfully altered by the heat treatment temperatures, forming α phase plates surrounded by β phase precipitates. PEO-treated samples exhibited a thick and microsized porous TiO_2 coating in the anatase and rutile crystalline forms. The oxide layer was depleted by Al and V atoms, while Ca and P were gradually enriched along the coatings. Chemical analysis also indicated the absorption of water and organic molecules into the outer layer. PEO-treated samples had microscale roughness and thickness, hydrophilic behavior, and surface energy mainly formed by the dispersive component. The bulk's elastic modulus decreased with β phase precipitation, while the alloying elements directly influenced the Vickers microhardness. The corrosion tests indicated a stable and protective layer in the PEO-treated samples, showing better corrosion resistance than untreated ones. Overall, the findings indicated that the α and β phase proportion significantly impacts the mechanical properties, while the PEO treatment acts in the corrosion protection and surface aspects, suggesting that combining both approaches could be a powerful tool in biomedical applications.

Keywords: biomaterials; Ti-6Al-4V; heat treatment; phase composition; PEO; corrosion

Academic Editor: Xiaogang Li

Received: 19 January 2025
Revised: 10 February 2025
Accepted: 18 February 2025
Published: 21 February 2025

Citation: Neto, J.R.F.; Ribeiro, R.P.; Cruz, N.C.d.; Rangel, E.C.; Pinto, B.d.O.; Torrento, J.E.M.; Grandini, C.R.; Kaneko, U.F.; Correa, D.R.N. Effect of Bulk Phase Composition on the Growth of PEO Coatings on the Biomedical Ti-6Al-4V Alloy. *Materials* **2025**, *18*, 955. https://doi.org/10.3390/ma18050955

1. Introduction

Titanium (Ti) and its alloys have been extensively used as biomaterials due to their lower elastic modulus, recognized biocompatibility, and better corrosion resistance than conventional stainless steel and cobalt alloys [1]. Titanium's biocompatibility is directly related to the chemical stability of the passive TiO_2 layer naturally formed on its surface [2]. Moreover, the Ti-6Al-4V alloy (wt%, ASTM F136 [3], or commercially pure Ti grade 5) is currently the most popular Ti alloy worldwide, with versatile and tunable properties due to the coexistence of the α and β phases in the microstructure [4]. Nowadays, the material is still largely employed in manufacturing load-bearing orthopedic implants, despite the remaining concerns about the ion-releasing of harmful and toxic aluminum (Al) and vanadium (V) ions in long-term implantation [5]. Therefore, novel strategies for the surface modification of the Ti-6Al-4V alloy targeting attenuating corrosion in the human body are in demand.

Among the diverse methods for modifying metallic surfaces, plasma electrolytic oxidation (PEO), or micro-arc oxidation (MAO), is the most attractive for Ti and its alloys as it can produce a porous, thick, and firmly adherent TiO_2 coating [6]. Based on the anodic oxidative reactions and the plasma discharges promoted by the dielectric breakdown, PEO treatment can significantly modify the surface topography, chemical composition, crystallinity, hydrophilicity, and roughness, which are key factors to consider in implant–tissue interaction and osseointegration [7]. Furthermore, the PEO treatment also permits the surface enrichment of bioactive elements, such as calcium (Ca) and phosphorous (P), through the driven force provided by the anodic voltage [8]. As is well known, calcium phosphate deposition on titanium surfaces has been a strategy used to make the implant's surface friendlier to human bone tissues, enabling fast bone growth and recovery [9]. Even though there is vast amounts of research focused on the PEO treatment of Ti alloys [10,11], there is still a lack of a profound understanding of the role of the α and β phases on the coating's growth mechanisms.

For the Ti-6Al-4V alloy, the alloying elements can decay the β-transus temperature (initially at 882 °C), making the control of the amount and distribution of α and β phases by proper heat treatments possible [12]. Previous studies have demonstrated their effectiveness for diverse industrial applications [13]. However, the role of the α and β phases on the oxide growth mechanisms during surface modification, especially PEO treatment, is scarcely reported. For example, Luz et al. [14] studied TiO_2 nanotubes grown on the Ti-10Nb alloy using anodic oxidation treatment. The coating grown on the α phase depicted self-organized nanotubes, whereas that grown on the β phase presented some transversal holes. The α and β phases also affected wettability and P enrichment. Even though the applied voltage was not high enough to produce plasma discharges, the study provided new insights into developing biofunctional surfaces. In contrast, Yavari et al. [15] studied the effect of PEO treatment on the commercially pure Ti grade 2 (stable α), Ti-6Al-7Nb (α + β), Ti-35Zr-10Nb (metastable β), and Ti-45Nb (stable β) alloys in an electrolyte composed of calcium glycerophosphate ($C_3H_7CaO_6P$) and calcium acetate ($C_4H_6CaO_4$). The authors found that the pore characteristics (size and density) were larger for the β-type alloys, with the anatase TiO_2 phase preferentially forming in the stable α- and β-type alloys in detriment of the metastable ones. However, the findings involving the growth mechanisms depended not exclusively on the phase composition but also on the alloying elements in solid solution.

Given that, the present study aimed to evaluate the effect of the α/β proportion of the Ti-6Al-4V alloy on the oxide layer growth by PEO treatment. The substrate underwent heat treatments at distinct temperatures and was then subjected to PEO treatment in a Ca- and P-rich electrolyte. After that, the chemical and phase composition, roughness, wettability, and mechanical and corrosion properties were evaluated for potential use in biomedical implants.

2. Materials and Methods

2.1. Sample Processing

Square-shaped Ti-6Al-4V samples (20 × 20 × 3 mm) were used as substrates [3]. The samples were previously polished with waterproof SiC emery papers, from #220 to #1200 mesh, ultrasonically cleaned in acetone for 5 min, and dried in an air stream. The samples were subjected to an initial heat treatment for microstructural homogenization at 1000 °C for 12 h, with subsequent slow cooling in a turned-off furnace. Then, the samples were separated into three groups, which were heat-treated at 600 °C, 800 °C, and 1000 °C for 30 min, followed by fast water cooling (Figure 1a). All the heat treatments were performed at a vacuum of 10^{-6} Torr. The temperatures were set up to produce distinct amounts of α and β phases in the microstructure, as indicated in Figure 1b. The corresponding sample nomenclature is indicated in Table 1.

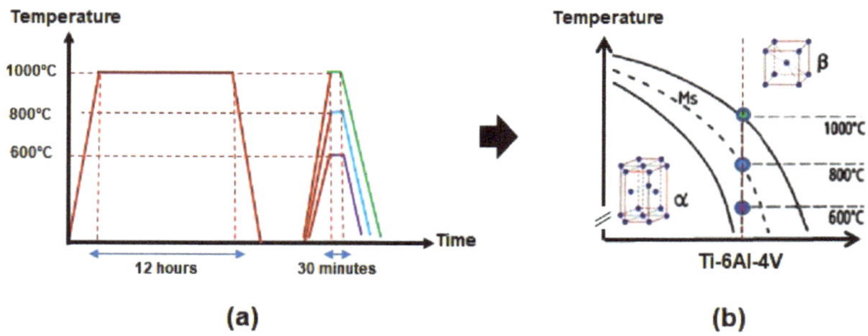

Figure 1. Processing route of the samples: (**a**) time–temperature plot and (**b**) the corresponding pseudo-binary phase diagram (adapted from [16]).

Table 1. Sample labeling.

Sample	Condition
As-received	Raw metal
HT 600	Heat-treated at 600 °C
HT 800	Heat-treated at 800 °C
HT 1000	Heat-treated at 1000 °C
As-received PEO	PEO-treated raw metal
HT 600 PEO	Heat-treated at 600 °C and PEO-treated
HT 800 PEO	Heat-treated at 800 °C and PEO-treated
HT 1000 PEO	Heat-treated at 1000 °C and PEO-treated

2.2. Surface Modification

The PEO treatment was conducted in the same experimental apparatus reported by Marcuz et al. [17]. In short, the surface treatment was carried out in a two-liter stainless steel tank, surrounded by a refrigeration system that maintained the electrolyte under constant temperature. The sample was connected to the positive terminal (anode) using a clamp, and the other terminal (cathode) was connected to the metal container. Then, the PEO treatments were carried out using a pulsed power source (MAO-30, Plasma Technology Ltd., Hong Kong, China) operated with positive pulses of 480 V, frequency of 100 Hz, and duty cycle of 60% for 120 s. The electrolyte was composed of 0.2 mol·L^{-1} of calcium acetate ($C_4H_6CaO_4$) and 0.02 mol·L^{-1} of sodium glycerophosphate ($C_3H_7CaO_6P$) in deionized water, being previously mixed by a magnetic stirrer. After that, the samples were cleaned with distilled water and dried in an air stream.

2.3. Sample Characterization

Microstructural and topographical characteristics were examined by optical microscopy (OM; BX51M microscope, Olympus Ltd., Westborough, MA, USA) and scanning electron microscopy (SEM; JSM-6010LA, JEOL Ltd., Peabody, MA, USA), using the secondary electron beam mode (SE), with a spot size of 30 μm, at a voltage of 3 kV. For the microstructural analysis, the selected part of the samples underwent conventional metallographic preparation based on grinding in waterproof SiC papers (#360 to #1500 mesh), polishing in alumina suspension (0.25 μm), and etching in Kroll's solution. The phase composition was evaluated by X-ray diffraction measurement (XRD; Panalytical X'Pert PRO, Westborough, MA, USA), with monochromatic Cu-Kα radiation (λ = 0.1544 nm), at a voltage of 45 kV, a current of 40 mA, in the θ-2θ mode, with a scan speed of 0.02° per second, and a step size of 0.02°. The diffracted peaks were indexed by the Highscore Plus® software version 3.0 and crystallographic datasheets from the ICDD (International Centre for Diffraction Data).

Semi-quantitative chemical microanalysis and elemental mapping were performed using X-ray dispersive energy spectroscopy (EDS) in a detector (JSM-6010LA, JEOL Ltd., USA) coupled to the SEM equipment operating at 10 kV. In the transmittance mode, the vibrational characteristics of molecular groups absorbed in the surfaces were analyzed using Fourier-transform infrared spectroscopy (FTIR; Jasco Corp., Tokyo, Japan) at room temperature, with 120 scans and a resolution of 4 cm^{-1}. The vibrational characteristics of molecular absorbed groups in the surfaces were analyzed by Raman spectroscopy through an optical spectrometer (HRS300 model, Princeton Instruments Inc., Trenton, NJ, USA) set to a 1200 grooves/mm grating and equipped with a contact Peltier module (PIXIS100BR, Teledyne Princeton Instruments Inc., Trenton, NJ, USA). All spectra were collected in a confocal configuration with a 10× objective and a numerical aperture of 0.28 (Mitutoyo Inc., Aurora, IL, USA). The excitation source (Verdi 6V, Coherent Laser Corp., Saxonburg, PA, USA) was adjusted to 532 nm, with a power of 50 mW, and an accumulation time of 50 s. The surface chemical aspects were further investigated by X-ray photoelectron spectroscopy (XPS; K-Alpha model from Thermo Scientific Inc., Waltham, MA, USA), operating with monochromatic radiation source of Al-Kα (1486 eV), spot size of 400 µm, energy step of 200 eV for long-scan (survey) and 50 eV for short-scan spectra (high-resolution), with a resolution of 1 eV for long- and 0.01 eV for short-scan spectra. The CasaXPS® software version 2.3.24 was used to evaluate the XPS spectra, with a previous energy calibration with the C1s peak.

Arithmetic roughness (Ra) was measured by optical profilometry (Dektak 150 profilometer, Veeco Metrology, Tucson, AZ, USA), with a stylus radius of 12.5 µm, scan length of 2000 µm, resolution of 0.16 µm, during 20 s, and applied load of 3 mgf. The surface wettability was checked by contact angle measurements (Ramé-hart Instrument Corp., Succasunna, NJ, USA) using droplets (30 µL) of distilled water, diiodomethane, and saline 0.9% NaCl solution at room temperature. The surface energy was calculated using Young's equation [18]. Thickness values were estimated by the eddy current method (ME240 equipment, Instrutherm Inc., Sao Paulo, Brazil).

2.4. Sample Testing

The Vickers micro-hardness test was conducted in hardener equipment (Shimadzu Ltd., model HMV-G, Tokyo, Japan) operated at a load of 0.300 kg (2.942 N) and a dwell time of 15 s. The impulse excitation method acquired the elastic modulus using Sonelastic equipment (ATCP Physical Engineering Inc., Ribeirão Preto, SP, Brazil). The tests were repeated 10 times to collect reliable average values for each sample. The electrochemical tests were conducted in a three-electrode system, with the sample set up as a working electrode, a platinum rod (Pt) as a counter electrode, and an Ag/AgCl wire as a reference electrode. The experiments were carried out in an Autolab PGSAT128N potentiostat/galvanostat (Metrohm Ltd., Riverview, FL, USA). The electrolyte was composed of an aqueous 0.9% NaCl solution at room temperature, which is conventionally used to simulate the concentration of sodium chloride in human blood plasma [19]. The sample contact area with the electrolyte was limited by an O-ring (Ø = 1 cm). The open-circuit potential (OCP) was measured for 3600 s, while the potentiodynamic polarization (PDP) test was measured in the potential range of −1 to 2 V vs. OCP at a scan rate of 1 mV·s^{-1} and step size of

1 mV. Electrochemical impedance spectroscopy (EIS) measurements were performed by applying a perturbation of 10 mV concerning the OCP value in a frequency range from 10^{-2} to 10^5 Hz, with 10 points per decade. The results were evaluated using NOVA® software version 2.1.5, which also fitted the equivalent electrical circuit (EEC) of the curves. The corrosion potential (E_{corr}) and corrosion current density (I_{corr}) were calculated from the PDP curves using Tafel's extrapolation method, with the polarization resistance (R_p) obtained using the Butler–Volmer equation [20].

3. Results

Figure 2 shows the decay of the current density with the time for the samples during the PEO treatment. The current density sharply declined in the first seconds in all samples, starting at around 0.9–1.4 A·cm^{-2}, followed by an exponential drop with some fluctuations after around 60 s, and ending between 0.4 and 0.5 A·cm^{-2}. As is well known, the initial drop is related to the anodic growth of a dense oxide layer, which blocks the passage of electrical charges. The smooth decay and fluctuations are the result of establishing plasma discharges along the surface through the micro-arcs. Overall, the α and β phase proportions did not significantly affect the curves, suggesting that the alloying elements affect the growth kinetics more than the phase composition. Our previous study detected the same trend with the Ti-6Al-4V alloy when subjected to PEO treatment in a ZrO_2-rich electrolyte and under the same electrical conditions [21].

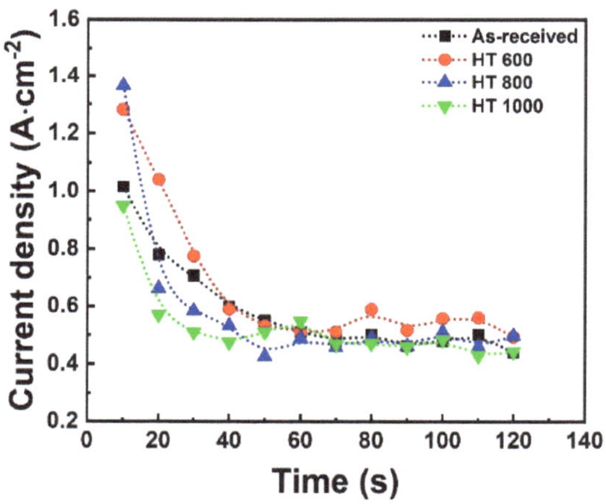

Figure 2. Current density vs. time plot for the Ti-6Al-4V samples in each condition.

Figure 3 shows the bulk microstructure and corresponding topography of the untreated and PEO-treated samples, respectively. The as-received sample possessed an irregular microstructure, with some β phase precipitates immersed in the α phase matrix. This microstructure is a result of previous plastic deformation performed by the supplier in the raw metal. The HT 600 sample depicted a partially recrystallized microstructure composed of typical α phase plates and some amount of β phase in the boundaries. As expected, the HT 800 sample presented a refined microstructure with thin α phase plates, indicating an augmenting of the β phase in the boundaries. Lastly, the HT 1000 sample possessed thinner α phase plates and coarse β phase precipitates, which the water cooling retained from the high temperature. Overall, the microstructures followed the expected α and β phase proportions by the distinct heat treatment temperatures, as illustrated in Figure 1. For the PEO-treated samples, a thick oxide layer with microsized porous structures without cracks or spallation can be seen, indicating that the α and β phases did not affect the global aspect of the topography, e.g., pore distribution and size. In our earlier study [22], Ti-15Zr-xMo (x = 0, 5, 10, and 15 wt%) alloys were subjected to a similar PEO treatment in a DC power source. The chemical and phase composition played a significant role in bioactivity, cell viability, mineralization, and differentiation ability, even though the topography remained almost the same.

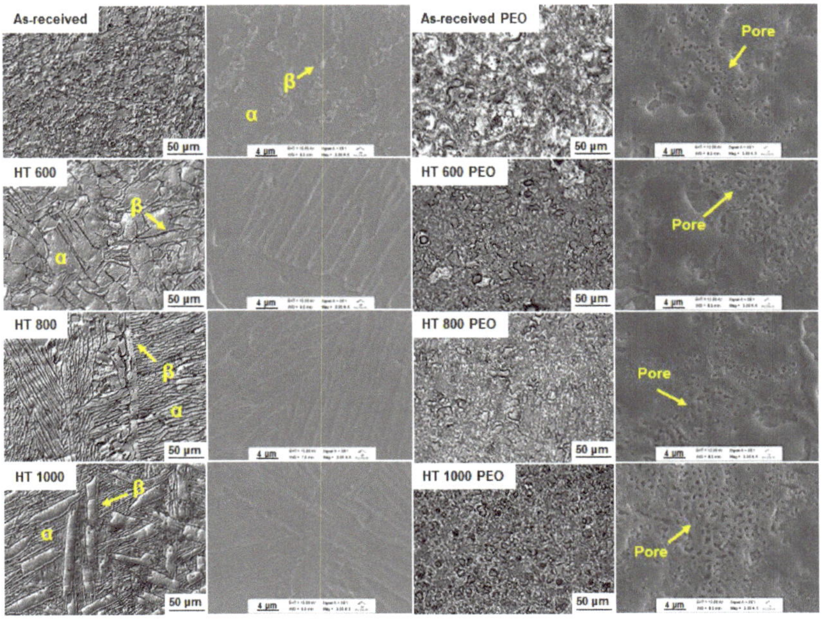

Figure 3. Microstructure and the corresponding surface topography of the samples: optical (**left**) and SEM (**right**) images.

The XRD profiles in Figure 4 provide details about the phase composition of the bulk and surface. The untreated samples (Figure 4a) possessed a strong preferred orientation of some peaks from the α and β phases due to the previous plastic deformation of the raw samples. However, it is still possible to note that the as-received and HT 600 samples depicted major α and minor β phases, while the HT 800 and HT 1000 samples possessed a predominance of β phase. This result supports the microstructural modification previously shown in Figure 3. In the PEO-treated samples (Figure 4b), the peaks from the bulk phases were attenuated, indicating the growth of a thick oxide layer. In addition, it is possible to note some peaks from TiO_2 as anatase (A) and rutile (R) phases. The A(101) and R(110) peaks remained at similar intensity in all sample conditions (Figure 4c), indicating that the bulk phase proportion did not present a major influence on the crystallinity of the coating. Similar results were found by Sobolev, Kossenko, and Borodianskiy [23] when studying the effect of current pulse frequency on the Ti-6Al-4V alloy subjected to PEO treatment in an AC power source and electrolyte enriched with Na_2CO_3 and Na_2SiO_3.

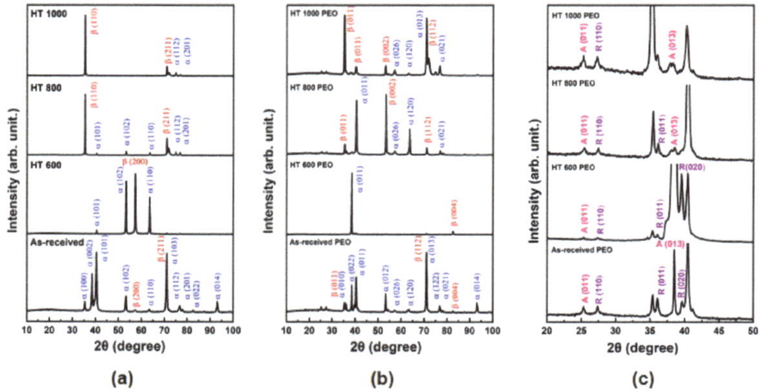

Figure 4. XRD profiles of the untreated (**a**) and PEO-treated (**b**) Ti-6Al-4V samples, with the corresponding zoomed-in view (**c**).

Figure 5 shows the elemental mapping of the PEO-treated samples. The signal for Ti and O was more intense than that for Al and V, indicating a positive abundance of these elements in the coatings. The uniform distribution of Ti and O confirms a uniform coating after the PEO treatment. Ca and P atoms were detected in all samples without clear differences between the heat treatment conditions. Ca retained a more intense signal than P, suggesting a Ca/P ratio higher than 1.0, approaching that expected for hydroxyapatite (1.67). As already reported by previous studies, Ca- and P-rich PEO coatings on Ti surfaces can provide proper bioactivity through the calcium phosphate formation capability when immersed in body fluids [24].

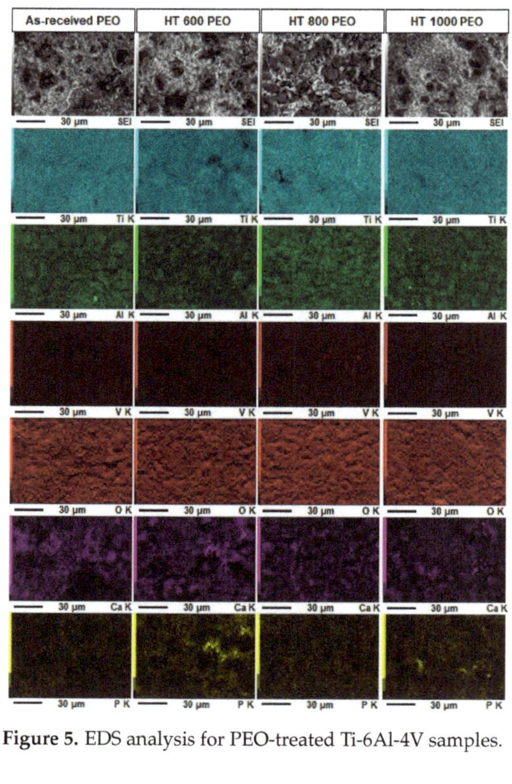

Figure 5. EDS analysis for PEO-treated Ti-6Al-4V samples.

Figure 6 shows the corresponding semi-quantitative EDS analysis for the whole area of Figure 5 compared with the untreated samples. The PEO-treated samples possessed considerable oxygen due to the anodic oxide layer growth and a visible depletion of Al and V atoms compared to the bulk. This can positively avoid the release of harmful ions into the human blood. Ca atoms were detected in equal amounts and P remained as a trace, while the Ca/P ratio remained between 3.0 and 8.0. Overall, all heat-treated samples depicted the same chemical composition after PEO treatment when detected by EDS. However, considering the typical microsized thickness of the coatings and the considerable penetration depth of the electron beam, it can be presumed that the acquired chemical composition originates from bulk and surface signals [17].

Figure 6. Semi-quantitative EDS analysis for Ti-6Al-4V samples in all studied conditions.

The FTIR and Raman spectra of the PEO-treated samples are presented in Figure 7. The FTIR spectra (Figure 7a,b) showed characteristic bands produced by stretching (ν) and bending (δ) of radical groups related to absorbed water (O-H), and organic (C-O, C=C, and C-H) and inorganic (N-H) molecules, and some related to phosphate (P-O) and calcium-based compounds (Ca-O). The Raman spectra depicted active vibration bands from TiO_2 (anatase and rutile phases), with some distinct signal-to-noise ratio with the α and β phases. The bands in the Raman spectra were similar to those found in other PEO-treated Ti surfaces [23]. The presence of vibrational bands originating from Ca and P (FTIR) and TiO_2 (Raman) shows that the PEO treatment successfully provided the growth of an anodic oxide layer rich in bioactive elements independently of the α and β phase proportions, which can guarantee favorable bioactive responses under implantation [25].

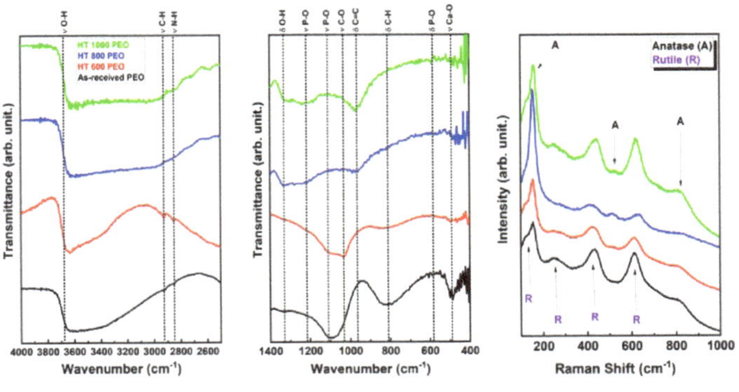

Figure 7. FTIR results at high (**a**) and low (**b**) wavenumbers and Raman spectra (**c**) for PEO-treated Ti-6Al-4V samples.

Figure 8a shows the survey spectra of all samples, where the main elements of interest are labeled. The semi-quantitative analysis (Figure 8b) presented a distinct chemical proportion compared to the EDS (Figure 6) due to the lower penetration depth of the X-ray beam. Thus, the outer layer comprised C from absorbed organic molecules and O from the oxide layer, as indicated in the FTIR (Figure 7a,b) and Raman spectra (Figure 7c), respectively. The PEO treatment also produced Ca enrichment of the outer layer, with a minor amount of P. The Ca/P ratio remained between 2.0 and 3.0, distinct from the EDS analysis, showing that the Ca and P atoms were not uniformly distributed into the coatings. The result also indicated a depletion of Al and a complete absence of V atoms in the outer layer, reinforcing the positive effect of the PEO treatment on the surface, as already highlighted in previous studies [26].

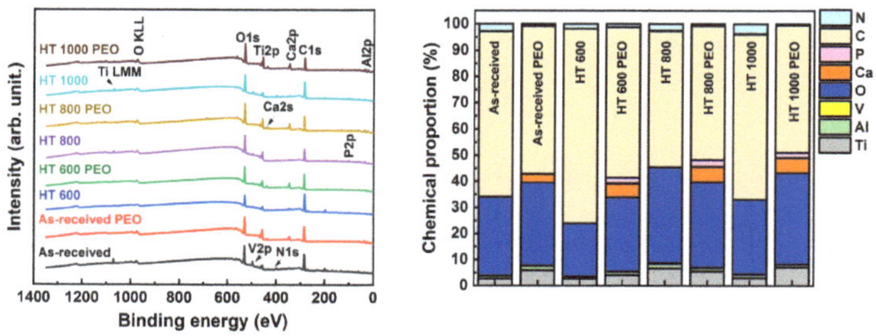

(a) (b)

Figure 8. XPS semi-quantitative results: survey spectrum (**a**) and corresponding chemical composition (**b**) for Ti-6Al-4V in all studied conditions.

The semi-quantitative results from the high-resolution (HR) spectra of the C1s, O1s, and Ti2p peaks are illustrated in Figure 9. The fitted spectra of all elements of interest are found in the Supplementary Materials. The C1s spectrum (Figure 9a) exhibited a convolution of three peaks related to major C-C, C-O, and minor O=C-O bonding. The PEO treatment tended to increase the C-C bonding and decay that of C-O. The O1s spectrum (Figure 9b) also possessed three main components related to major O-H and some amounts of O-C and O-Metal bonding. Overall, the PEO treatment augmented the O-Metal component to the detriment of the O-C. Lastly, the Ti2p spectrum (Figure 9c) depicted mainly the Ti^{4+} chemical state (TiO_2), the presence of Ti^{3+}, which is related to sub-oxides (e.g., Ti_2O_3), and Ti^0 (metallic bonding). Overall, the PEO-treated samples were mainly formed by the TiO_2 bonded with organic and hydroxyl radical groups in the outer layer. According to Parfenov et al. [27], the PEO treatment is a useful tool to biofunctionalize the surface of Ti and its alloys, improving their chemical activity with peptides and organic molecules that trigger the cell mechanisms.

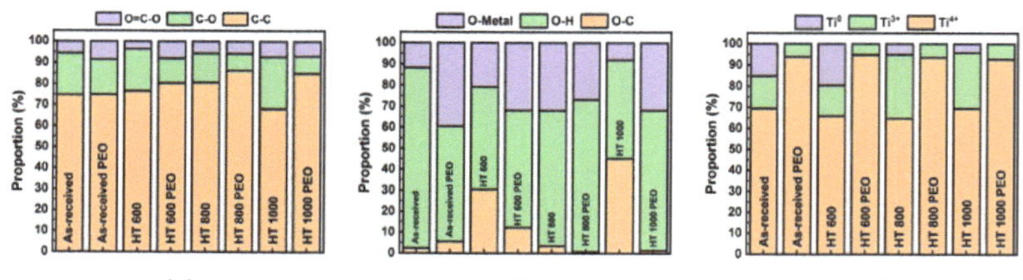

(a) (b) (c)

Figure 9. XPS HR results for C1s (**a**), O1s (**b**), and Ti2p (**c**) for Ti-6Al-4V samples in all studied conditions.

Figure 10 presents the PEO-treated samples' roughness (Ra and Rq) and thickness values. The roughness remained in the order of some micrometers, between 4 and 5 μm for Ra and 5 and 7 μm for Rq, while the thickness exhibited typical values of around 6–7 μm. A sensible difference in the values between the samples was undetected, indicating that the α and β phases did not play a role in these parameters. As is well known, micrometric surface roughness favors cell adhesion, growth, and proliferation, an important biological parameter to consider during implantation [28].

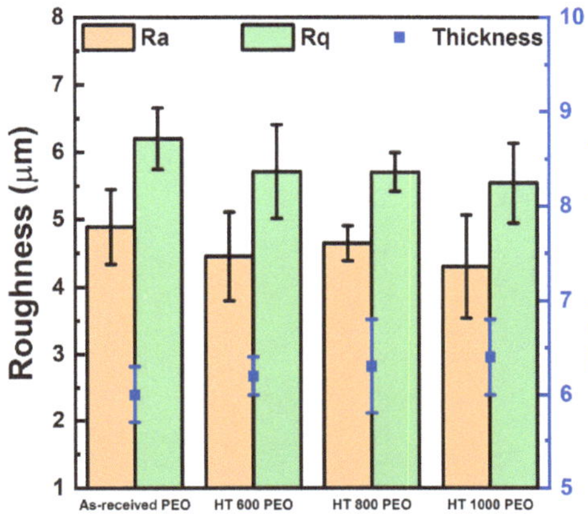

Figure 10. Roughness and thickness for PEO-treated Ti-6Al-4V samples.

Figure 11a compares the contact angle values collected in distilled water, diiodomethane, and 0.9% NaCl with the corresponding surface energy values. Overall, the samples depicted hydrophilic behavior (contact angle lower than 90°) in all droplets except for the HT 600 PEO sample. The contact angle possessed the smallest values in diiodomethane without a clear tendency to vary with the α and β phases or the PEO treatment. Regarding the surface energy (Figure 11b), all samples depicted most of the dispersive component instead of the polar one, with the PEO treatment tending to diminish the total surface energy compared to the untreated material. According to Ponsonnet et al. [29], the dispersive component is related to the force interactions produced by non-polar radical groups, guaranteeing better cell proliferation along the surface.

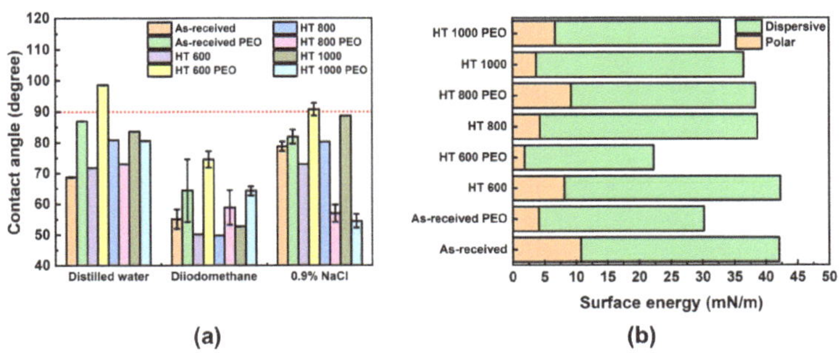

Figure 11. Wettability test: contact angle (**a**) and surface energy (**b**) for Ti-6Al-4V samples in all studied conditions. The dotted red line refers to the limit between hydrophilic and hydrophobic surfaces.

Figure 12 compares the Vickers microhardness and elastic modulus only for the untreated samples since it was not possible to acquire the values for the PEO-treated samples precisely. The elastic modulus depicted a gradual decay with the heat treatments, resulting directly from the retained β phase formation to the detriment of the α phase. Regarding the Vickers microhardness, the values were higher than those of the commercially pure Ti grade 2 (~185 HV). However, there was no clear tendency for variation with the heat treatments, indicating that the solid-solution-strengthening and work-hardening mechanisms could be the main influence instead of the phase precipitation hardening. Overall, the combination of low elastic modulus and high hardness is useful for implants as it can attenuate the stress shielding effect and provide support for biomechanical loads [30]. Further evaluation of the mechanical properties by tensile tests can provide an in-depth view of the influence of the α/β phases and PEO treatment.

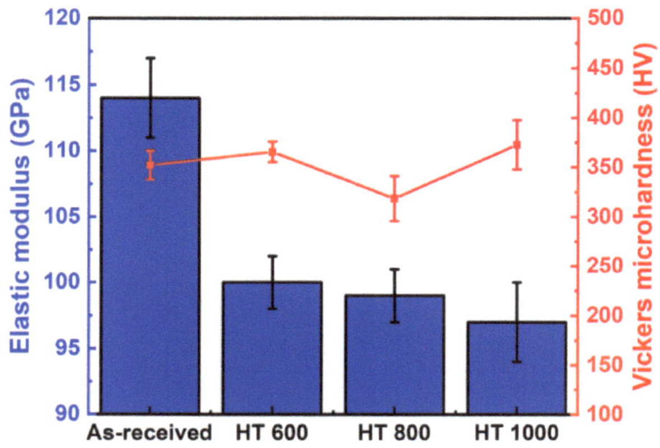

Figure 12. Selected mechanical properties for untreated Ti-6Al-4V samples.

Figure 13 presents the OCP curves and a comparison of the average values. All samples presented a stable OCP curve during the evaluated time (Figure 13a), with a small tendency to increase, indicating the presence of a stable oxide layer. Furthermore, the PEO-treated samples showed more positive values than the untreated ones, confirming the favorable protective effect of the oxide layer grown on the surface. The average values (Figure 13b) remained positive for the PEO-treated samples, showing a remarkable anticorrosion ability of the PEO treatment compared to the heat treatments. The obtained values were similar to those of Garcia-Cabezón et al. [26], who investigated the anticorrosion properties of the Ti-6Al-4V alloy prepared by powder metallurgy and subjected to the same PEO treatment.

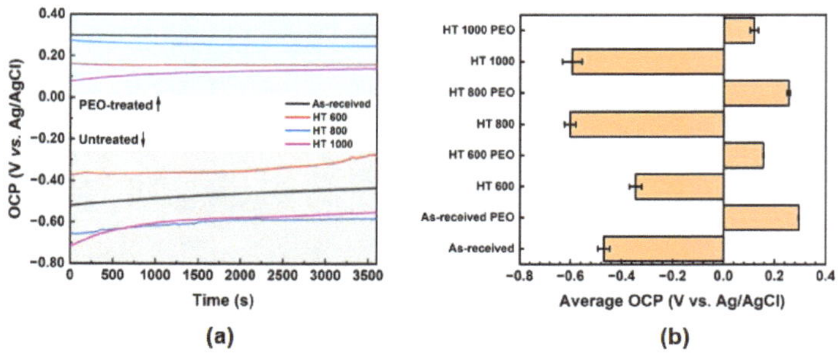

Figure 13. OCP results: OCP vs. time plot (**a**) and average OCP value (**b**) for Ti-6Al-4V samples in all studied conditions.

Figure 14 and Table 2 present the PDP curves and the calculated quantitative results. Overall, the PEO-treated samples, besides a similar corrosion current, exhibited a more positive corrosion potential than the untreated ones. The corresponding Rp value remained in the order of dozens of kΩ, with the HT 600 PEO sample possessing the highest value. In the same way, the PEO treatment acted more directly against the corrosion mechanisms than the heat treatments. According to Sobolev, Zinigrad, and Borodianskiy [31], the superior anticorrosion properties of the PEO-treated surfaces are due to a double electric layer formed in the coating, with the outer layer being porous and enriched with bioactive species and the inner layer dense and responsible for protecting the substrate.

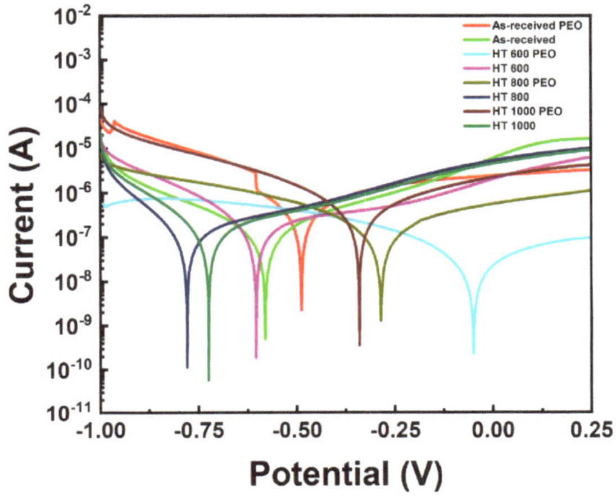

Figure 14. PDP results for Ti-6Al-4V samples in all studied conditions.

Table 2. Quantitative results from the PDP curves.

Sample	E_{corr} (mV)	j_{corr} (nA·cm^{-2})	R_p (kΩ)
As-received	−580	120	544
As-received PEO	−490	120	98
HT 600	−600	180	252
HT 600 PEO	−50	70	2380
HT 800	−780	120	381
HT 800 PEO	−290	260	565
HT 1000	−720	140	402
HT 1000 PEO	−340	640	164

The EIS curves, Bode and Nyquist plots are indicated in Figure 15. The phase angles for the untreated and PEO-treated samples (Figure 15a,b) were distinct, indicating different oxide layers on the surface. Furthermore, the impedance modulus (|Z|) at low frequencies (10^{-1} Hz) remained in the order of 10^9 Ω·cm^2 for the untreated samples and 10^{10}–10^{12} Ω·cm^2 for the PEO-treated ones, suggesting a better protective effect of the PEO coatings compared to the naturally formed passive layer. The Nyquist plot (Figure 15c) for the untreated sample depicted a clear semi-circle corresponding to one capacitive phase element, gradually increasing the radius with the augmenting of the β phase. In contrast, the PEO-treated samples (Figure 15d) depicted two semi-circles, suggesting two distinct capacitive phase elements with a larger radius than the previous one. Previous reports [32–34] indicated the presence of multilayers in the PEO coatings, which results in distinct elements in the electrical equivalent circuit in EIS spectra, giving beneficial responses to the anticorrosion properties.

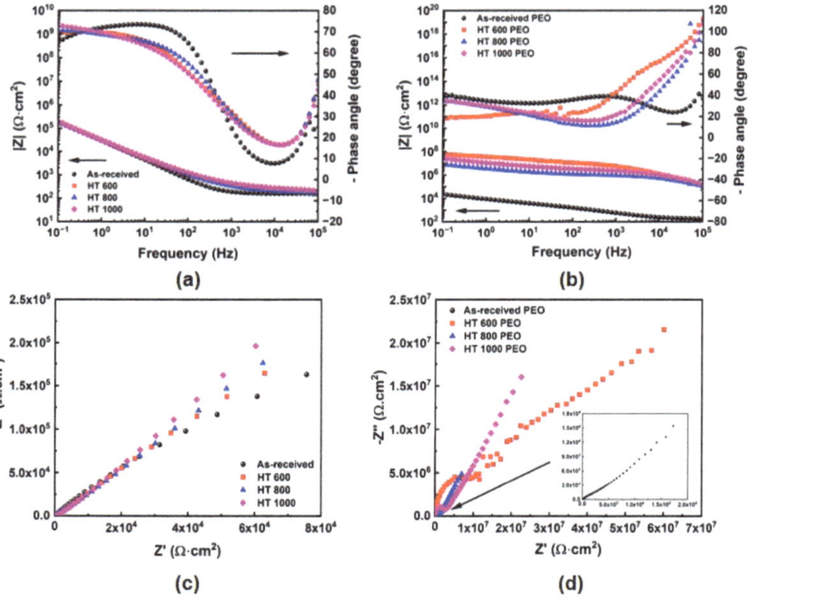

Figure 15. EIS results: Bode plots for the untreated (**a**) and PEO-treatment (**b**) Ti-6Al-4V samples and the corresponding Nyquist plots in (**c**,**d**).

Figure 16 summarizes the EECs diagrams used to fit the EIS curves, and Tables 3 and 4 show the corresponding quantitative values. A goodness-of-fitness value (χ^2) below 1.0 was considered for a successful fitting. Overall, the untreated samples depicted a simple connection between the solution resistance (Rs) and a constant phase element (CPE) related to the metallic surface. The CPE component is characterized by a parallel configuration of the polarization resistance (Rp) and an imperfect capacitor (C_{dl}) related to the protective capability and charge accumulation on the surface, respectively. The heat-treated samples also depicted a Warburg element (W), which is associated with diffusion processes into the surface. Thus, for the untreated samples, it is possible to conclude that the corrosion resistance came from a single passive layer naturally formed on the surface, with the α/β phases having minor effects on the surface chemical activity. On the other hand, the PEO-treated samples possessed a more complex EEC, composed of three CPE components. The CPEs components were related to the outer porous layer, inner dense layer, and the metallic surface, respectively. The corresponding resistive and capacitive elements for the porous (R_{pore} and C_{pore}), dense (R_{ox} and C_{ox}) oxide layers, and substrate (R_p and C_{dl}) guarantee superior protection against the corrosion of the PEO-treated samples when compared to the untreated samples. It is worth pointing out that the capacitive element for the porous and dense oxide layer was a pure capacitor, which is related to a uniform and defect-free layer. Furthermore, the n/p value for the PEO-treated samples depicted is not so close to 1.0 as the untreated ones, indicating more resistive behavior, which is positive in avoiding surface degradation. According to the literature [35,36], the combination of the oxidative reactions and plasma discharges of the PEO treatment are the origins of the multi-layered EEC, being beneficial for protection against the corrosive environment.

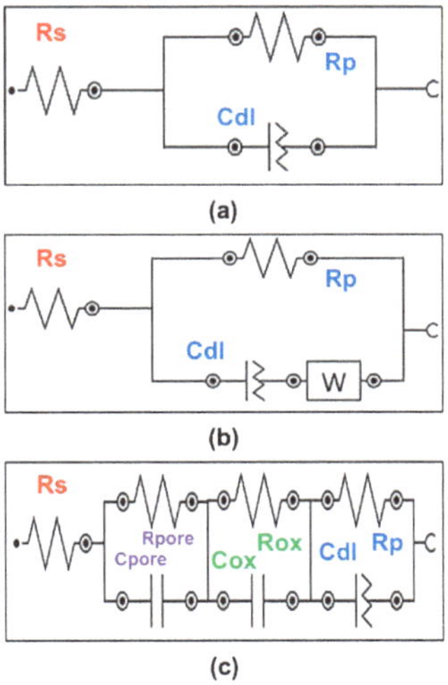

Figure 16. EEC diagrams of the EIS curves: (**a**) as-received sample; (**b**) HT 600, HT 800, HT 1000, and as-received PEO samples; (**c**) HT 600 PEO, HT 800, and HT 1000 PEO samples.

Table 3. Quantitative results from fitting the EIS curves for the untreated samples.

Parameters	As-Received	HT 600	HT 800	HT 1000
Rs (MΩ·cm^2)	1.52×10^{-4}	1.55×10^{-4}	1.68×10^{-4}	2.18×10^{-4}
Rp (MΩ·cm^2)	7.63×10^{-3}	1.29	4.69	7.63
C_{dl} (μMho·s^n)	7.31	8.65	8.26	7.88
n/p	0.842	0.87	0.821	0.856
W (μMho·s$^{1/2}$)	-	69.7	99.0	57.8
χ^2	0.053511	0.025978	0.03358	0.021966

Table 4. Quantitative results from fitting the EIS curves for the PEO-treated samples.

Parameters	As-Received PEO	HT 600 PEO	HT 800 PEO	HT 1000 PEO
Rs (MΩ·cm^2)	1.2×10^{-4}	3.58×10^{-2}	3.75×10^{-2}	9.36×10^{-3}
Rpore (MΩ·cm^2)	-	6.59	8.44×10^{-1}	2.06
Cpore (pF)	-	18.5	22.6	13.9
Rox (MΩ·cm^2)	-	31.9	8.86	27.3
Cox (nF)	-	32.4	233	58.7
Rp (MΩ·cm^2)	5.03×10^{-3}	36.6	4.07	11.2
C_{dl} (μMho·s^n)	14.7	3.49×10^{-3}	75.2×10^{-3}	17.7×10^{-3}
n/p	0.557	0.600	0.559	0.559
W (μMho·s$^{1/2}$)	46.6	-	-	-
χ^2	0.027026	0.41719	0.29647	0.46814

4. Conclusions

This study evaluated the effect of distinct α/β phase proportions of the Ti-6Al-4V alloy on the growth mechanisms of the PEO treatment in a Ca- and P-rich electrolyte. From the results, it is possible to summarize the findings below:

- The α/β phase proportion changed with the temperature of the heat treatments, leading to α phase plates and β phase precipitate formation in the boundaries;
- All PEO-treated samples possessed a microsized porous surface of TiO_2 with equal amounts of anatase and rutile;

- The amounts of Al and V were depleted in the oxide layer, while Ca and P were enriched. The chemical bonding detected in the outer layer gave evidence for the incorporation of calcium, phosphorus, absorbed water, and organic molecules;
- The PEO-treated samples possessed a microsized roughness and thickness, with hydrophilic contact angles and surface energy composed mostly of the dispersive component;
- The elastic modulus showed a gradual decay with the β phase precipitation, while the Vickers microhardness was exclusively affected by the alloying elements and the work-hardening effect;
- The OCP curves indicated a more stable protective layer for the PEO-treated samples than the bulk, while the PDP curves highlighted their superior corrosion resistance. The EIS indicated the presence of multiple protective elements in the oxide layer of the PEO-treated samples, which beneficially contributed to corrosion resistance. Further biological testing can provide a better view of the protective ability of the PEO treatment;
- The findings show that the α/β phase proportion considerably affected the mechanical properties of the Ti-6Al-4V alloy, while the PEO treatment had an effect on the anticorrosion properties and surface aspects. Thus, combining heat and PEO treatments makes it possible to handle the surface and bulk properties of this commercial biomaterial to provide superior performance for usage as biomedical implants.

Supplementary Materials: The following supporting information can be downloaded at: https://www.mdpi.com/article/10.3390/ma18050955/s1, Figure S1: High-resolution XPS spectra for aluminum; Figure S2: High-resolution XPS spectra for vanadium; Figure S3: High-resolution XPS spectra for calcium; Figure S4: High-resolution XPS spectra for phosphorus; Figure S5: High-resolution XPS spectra for carbon; Figure S6: High-resolution XPS spectra for oxygen. Auger peak (Na KLL) is an adventitious impurity; Figure S7: High-resolution XPS spectra for titanium; Table S1: Fitting details for aluminum; Table S2: Fitting details for calcium; Table S3: Fitting details for phosphorus; Table S4: Fitting details for carbon; Table S5: Fitting details for oxygen; Table S6: Fitting details for titanium.

Author Contributions: Conceptualization, J.R.F.N. and D.R.N.C.; methodology, J.R.F.N., R.P.R., J.E.M.T. and D.R.N.C.; validation, J.R.F.N., R.P.R., J.E.M.T., U.F.K. and D.R.N.C.; formal analysis, J.R.F.N., R.P.R., B.d.O.P., J.E.M.T., U.F.K. and D.R.N.C.; investigation, J.R.F.N., R.P.R., B.d.O.P. and J.E.M.T.; resources, E.C.R., N.C.d.C., C.R.G. and D.R.N.C.; data curation, J.R.F.N., R.P.R., B.d.O.P., U.F.K. and D.R.N.C.; writing—original draft preparation, J.R.F.N. and D.R.N.C.; writing—review and editing, J.R.F.N., R.P.R., E.C.R., N.C.d.C., B.d.O.P., J.E.M.T., U.F.K., C.R.G. and D.R.N.C.; visualization, J.R.F.N., R.P.R., J.E.M.T., U.F.K. and D.R.N.C.; supervision, D.R.N.C.; project administration, D.R.N.C.; funding acquisition, E.C.R., N.C.d.C., C.R.G. and D.R.N.C. All authors have read and agreed to the published version of the manuscript.

Funding: This study was financially supported by the Coordination of Superior Level Staff Improvement (CAPES, Finance Code 001), National Council for Scientific and Technological Research (CNPq grants #407251/2018-9, #314810/2021-8, #404020/2023-2, and #421677/2023-6), and São Paulo Research Foundation (FAPESP grants #2018/24931-7, #2021/13921-3, and #2024/03148-3) funding agencies.

Institutional Review Board Statement: Not applicable.

Informed Consent Statement: Not applicable.

Data Availability Statement: The original contributions presented in this study are included in the article and Supplementary Materials. Further inquiries can be directed to the corresponding author.

Acknowledgments: The authors thank the National Laboratory of Nanotechnology (LNNano) for the XPS experiment (Proposal #20232420) and the undergraduate student Beatriz Lahóz Ribeiro Pagliuso Almada from UNESP—Campus Bauru for the support with the metallographic preparation.

Conflicts of Interest: The authors declare no conflicts of interest.

Abbreviations

The following abbreviations are used in this manuscript:

PEO	Plasma electrolytic oxidation
MAO	Micro-arc oxidation
XRD	X-ray diffraction
OM	Optical microscopy
SEM	Scanning electron microscopy
FTIR	Fourier-transform infrared spectroscopy
XPS	X-ray photoelectron spectroscopy
OCP	Open-circuit potential
PDP	Potentiodynamic polarization
EIS	Electrochemical impedance spectroscopy
EEC	Equivalent electrical circuit
$\lvert Z \rvert$	Impedance modulus
Z'	Real component of the impedance
Z''	Imaginary component of the impedance
CPE	Constant phase element
Rs	Solution resistance
Rp	Polarization resistance
Cdl	Imperfect capacitor
Rpore	Polarization resistance of the porous layer
Cpore	Pure capacitor of the porous layer
Rox	Polarization resistance of the inner oxide layer
Cox	Pure capacitor of the inner oxide layer
W	Warburg element
n/p	Exponent of the imperfect capacitor
$\chi 2$	Goodness-of-fitness factor

References

1. Elhadad, A.A.; Romero-Resendiz, L.; Rossi, M.C.; Rodríguez-Albelo, L.M.; Lascano, S.; Afonso, C.R.M.; Alcudia, A.; Amigó, V.; Torres, Y. Findings and Perspectives of β-Ti Alloys with Biomedical Applications: Exploring beyond Biomechanical and Biofunctional Behaviour. *J. Mater. Res. Technol.* **2024**, *33*, 3550–3618. [CrossRef]
2. Yang, J.; Song, Y.; Dong, K.; Han, E.-H. Research Progress on the Corrosion Behavior of Titanium Alloys. *Corros. Rev.* **2023**, *41*, 5–20. [CrossRef]
3. *ASTM F136-13*; Specification for Wrought Titanium-6Aluminum-4Vanadium ELI (Extra Low Interstitial) Alloy for Surgical Implant Applications (UNS R56401). ASTM International: West Conshohocken, PA, USA, 2013.
4. Liu, S.; Shin, Y.C. Additive Manufacturing of Ti6Al4V Alloy: A Review. *Mater. Des.* **2019**, *164*, 107552. [CrossRef]
5. Pinto, B.O.; Torrento, J.E.; Grandini, C.R.; Galindo, E.L.; Pintão, C.A.F.; Santos, A.A.; Lisboa-Filho, P.N.; Pontes, F.M.L.; Correa, D.R.N. Development of Ti–Al–V Alloys for Usage as Single-Axis Knee Prostheses: Evaluation of Mechanical, Corrosion, and Tribocorrosion Behaviors. *Sci. Rep.* **2023**, *13*, 4349. [CrossRef] [PubMed]
6. Ming, X.; Wu, Y.; Zhang, Z.; Li, Y. Micro-Arc Oxidation in Titanium and Its Alloys: Development and Potential of Implants. *Coatings* **2023**, *13*, 2064. [CrossRef]
7. Aliofkhazraei, M.; Macdonald, D.D.; Matykina, E.; Parfenov, E.V.; Egorkin, V.S.; Curran, J.A.; Troughton, S.C.; Sinebryukhov, S.L.; Gnedenkov, S.V.; Lampke, T.; et al. Review of Plasma Electrolytic Oxidation of Titanium Substrates: Mechanism, Properties, Applications and Limitations. *Appl. Surf. Sci. Adv.* **2021**, *5*, 100121. [CrossRef]
8. Rokosz, K.; Hryniewicz, T.; Dudek, Ł. Phosphate Porous Coatings Enriched with Selected Elements via PEO Treatment on Titanium and Its Alloys: A Review. *Materials* **2020**, *13*, 2468. [CrossRef] [PubMed]
9. Wu, T.; Blawert, C.; Serdechnova, M.; Karlova, P.; Dovzhenko, G.; Wieland, D.C.F.; Zheludkevich, M.L. Role of Polymorph Microstructure of Ti6Al4V Alloy on PEO Coating Formation in Phosphate Electrolyte. *Surf. Coat. Technol.* **2021**, *428*, 127890. [CrossRef]

10. Cardoso, G.C.; Grandini, C.R.; Rau, J.V. Comprehensive Review of PEO Coatings on Titanium Alloys for Biomedical Implants. *J. Mater. Res. Technol.* **2024**, *31*, 311–328. [CrossRef]
11. Pesode, P.; Barve, S. Surface Modification of Titanium and Titanium Alloy by Plasma Electrolytic Oxidation Process for Biomedical Applications: A Review. *Mater. Today Proc.* **2021**, *46*, 594–602. [CrossRef]
12. Najafizadeh, M.; Yazdi, S.; Bozorg, M.; Ghasempour-Mouziraji, M.; Hosseinzadeh, M.; Zarrabian, M.; Cavaliere, P. Classification and Applications of Titanium and Its Alloys: A Review. *J. Alloys Compd. Commun.* **2024**, *3*, 100019. [CrossRef]
13. Sarma, J.; Kumar, R.; Sahoo, A.K.; Panda, A. Enhancement of Material Properties of Titanium Alloys through Heat Treatment Process: A Brief Review. *Mater. Today Proc.* **2020**, *23*, 561–564. [CrossRef]
14. Luz, A.R.; Santos, L.S.; Lepienski, C.M.; Kuroda, P.B.; Kuromoto, N.K. Characterization of the Morphology, Structure and Wettability of Phase Dependent Lamellar and Nanotube Oxides on Anodized Ti-10Nb Alloy. *Appl. Surf. Sci.* **2018**, *448*, 30–40. [CrossRef]
15. Yavari, S.A.; Necula, B.S.; Fratila-Apachitei, L.E.; Duszczyk, J.; Apachitei, I. Biofunctional Surfaces by Plasma Electrolytic Oxidation on Titanium Biomedical Alloys. *Surf. Eng.* **2016**, *32*, 411–417. [CrossRef]
16. Sibisi, P.N.; Popoola, A.P.I.; Arthur, N.K.K.; Pityana, S.L. Review on Direct Metal Laser Deposition Manufacturing Technology for the Ti-6Al-4V Alloy. *Int. J. Adv. Manuf. Technol.* **2020**, *107*, 1163–1178. [CrossRef]
17. Marcuz, N.; Ribeiro, R.P.; Rangel, E.C.; Cruz, N.C.; Possato, L.G.; Coan, K.S.; Grandini, C.R.; Correa, D.R.N. Exploiting the Effect of PEO Parameters on the Surface of AISI 1020 Low-Carbon Steel Treated in a TaOH-Rich Electrolyte. *Surf. Coat. Technol.* **2024**, *477*, 130374. [CrossRef]
18. Jothi Prakash, C.G.; Prasanth, R. Approaches to Design a Surface with Tunable Wettability: A Review on Surface Properties. *J. Mater. Sci.* **2021**, *56*, 108–135. [CrossRef]
19. Yilmaz, B.; Engin, A.; Tezcaner, A.; Evis, Z. Historical Development of Simulated Body Fluids Used in Biomedical Applications: A Review. *Microchem. J.* **2020**, *155*, 104713. [CrossRef]
20. Lukács, Z.; Kristóf, T. Determination of Kinetic Parameters from a New Quadratic Approximation of the Butler-Volmer Equation. *J. Electroanal. Chem.* **2022**, *918*, 116443. [CrossRef]
21. da Silva, A.C.N.; Ribeiro, R.P.; Rangel, E.C.; da Cruz, N.C.; Correa, D.R.N. Production of Porous ZrO_2–TiO_2 Ceramic Coatings on the Biomedical Ti-6Al-4V Alloy via AC PEO Treatment and Their Effects on the Corrosion Behavior in 0.9% NaCl. *Coatings* **2024**, *14*, 866. [CrossRef]
22. Correa, D.R.N.; Rocha, L.A.; Ribeiro, A.R.; Gemini-Piperni, S.; Archanjo, B.S.; Achete, C.A.; Werckmann, J.; Afonso, C.R.M.; Shimabukuro, M.; Doi, H.; et al. Growth Mechanisms of Ca- and P-Rich MAO Films in Ti-15Zr-XMo Alloys for Osseointegrative Implants. *Surf. Coat. Technol.* **2018**, *344*, 373–382. [CrossRef]
23. Sobolev, A.; Kossenko, A.; Borodianskiy, K. Study of the Effect of Current Pulse Frequency on Ti-6Al-4V Alloy Coating Formation by Micro Arc Oxidation. *Materials* **2019**, *12*, 3983. [CrossRef] [PubMed]
24. Wang, R.; Ni, S.; Ma, L.; Li, M. Porous Construction and Surface Modification of Titanium-Based Materials for Osteogenesis: A Review. *Front. Bioeng. Biotechnol.* **2022**, *10*, 973297. [CrossRef] [PubMed]
25. López-Valverde, N.; López-Valverde, A.; Aragoneses, J.M.; Macedo de Sousa, B.; Rodrigues, M.J.; Ramírez, J.M. Systematic Review and Meta-Analysis of the Effectiveness of Calcium-Phosphate Coating on the Osseointegration of Titanium Implants. *Materials* **2021**, *14*, 3015. [CrossRef] [PubMed]
26. Garcia-Cabezón, C.; Rodriguez-Mendez, M.L.; Amigo Borrás, V.; Raquel, B.; Rodriguez Cabello, J.C.; Ibañez Fonseca, A.; Martin-Pedrosa, F. Application of Plasma Electrolytic Oxidation Coating on Powder Metallurgy Ti-6Al-4V for Dental Implants. *Metals* **2020**, *10*, 1167. [CrossRef]
27. Parfenov, E.; Parfenova, L.; Mukaeva, V.; Farrakhov, R.; Stotskiy, A.; Raab, A.; Danilko, K.; Rameshbabu, N.; Valiev, R. Biofunctionalization of PEO Coatings on Titanium Implants with Inorganic and Organic Substances. *Surf. Coat. Technol.* **2020**, *404*, 126486. [CrossRef]
28. Majhy, B.; Priyadarshini, P.; Sen, A.K. Effect of Surface Energy and Roughness on Cell Adhesion and Growth—Facile Surface Modification for Enhanced Cell Culture. *RSC Adv.* **2021**, *11*, 15467–15476. [CrossRef] [PubMed]
29. Ponsonnet, L.; Reybier, K.; Jaffrezic, N.; Comte, V.; Lagneau, C.; Lissac, M.; Martelet, C. Relationship between Surface Properties (Roughness, Wettability) of Titanium and Titanium Alloys and Cell Behaviour. *Mater. Sci. Eng. C* **2003**, *23*, 551–560. [CrossRef]
30. Sarraf, M.; Rezvani Ghomi, E.; Alipour, S.; Ramakrishna, S.; Liana Sukiman, N. A State-of-the-Art Review of the Fabrication and Characteristics of Titanium and Its Alloys for Biomedical Applications. *Bio Des. Manuf.* **2022**, *5*, 371–395. [CrossRef]
31. Sobolev, A.; Zinigrad, M.; Borodianskiy, K. Ceramic Coating on Ti-6Al-4V by Plasma Electrolytic Oxidation in Molten Salt: Development and Characterization. *Surf. Coat. Technol.* **2021**, *408*, 126847. [CrossRef]
32. Park, M.G.; Choe, H.C. Corrosion Behaviors of Bioactive Element Coatings on PEO-Treated Ti-6Al-4V Alloys. *Surf. Coat. Technol.* **2019**, *376*, 44–51. [CrossRef]

33. Garcia-Cabezón, C.; Rodríguez-Méndez, M.L.; Borrás, V.A.; Bayón, R.; Salvo-Comino, C.; Garcia-Hernandez, C.; Martin-Pedrosa, F. Improvements in Tribological and Anticorrosion Performance of Porous Ti-6Al-4V via PEO Coating. *Friction* **2021**, *9*, 1303–1318. [CrossRef]
34. Hwang, I.-J.; Choe, H.-C.; Brantley, W.A. Electrochemical Characteristics of Ti-6Al-4V after Plasma Electrolytic Oxidation in Solutions Containing Ca, P, and Zn Ions. *Surf. Coat. Technol.* **2017**, *320*, 458–466. [CrossRef]
35. Dilimon, V.S.; Shibli, S.M.A. A Review on the Application-Focused Assessment of Plasma Electrolytic Oxidation (PEO) Coatings Using Electrochemical Impedance Spectroscopy. *Adv. Eng. Mater.* **2023**, *25*, 2201796. [CrossRef]
36. Lazanas, A.C.; Prodromidis, M.I. Electrochemical Impedance Spectroscopy—A Tutorial. *ACS Meas. Sci. Au* **2023**, *3*, 162–193. [CrossRef]

materials

Article

Improving Biodegradable Mg-Zn(-Ca) Alloys by Surface Treatment via Plasma Electrolytic Oxidation

Jakub Vertaľ [1], Daniel Kajánek [2], Jiří Kubásek [3] and Peter Minárik [1,2,*]

1 Charles University, Department of Physics of Materials, Ke Karlovu 5, 121 16 Prague, Czech Republic; kubovertal@gmail.com
2 University of Žilina, Research Centre, Univerzitná 8215/1, 01026 Žilina, Slovakia; daniel.kajanek@uniza.sk
3 University of Chemistry and Technology, Department of Metals and Corrosion Engineering, Technická 5, 166 28 Prague, Czech Republic; jiri.kubasek@vscht.cz
* Correspondence: peter.minarik@matfyz.cuni.cz

Abstract: This study investigated the influence of plasma electrolytic oxidation (PEO) preparation time on the degradation resistance of Mg-1Zn (Z1) and Mg-1Zn-0.4Ca (ZX10) alloys, with comparisons to pure Mg and commercial Mg-4Y-3RE-0.4Zr (WE43). PEO layers were formed with varying preparation times (5, 10, and 15 min) and analyzed for microstructure, morphology, and corrosion resistance. The results indicated that PEO layers with a 10 min preparation time had the most homogeneous structure and optimal corrosion resistance. Prolonged PEO preparation times increased pore density, crack formation, and layer thickness while also promoting layer degradation during extended immersion in 0.9% NaCl corrosive media. The dissolution of phosphates from PEO layers contributes to the formation of a protective corrosion layer, enhancing long-term resistance. These findings demonstrate that low-alloyed, biocompatible Mg-Zn(-Ca) alloys can achieve corrosion resistance comparable to high-performance WE43 alloys through appropriate surface treatment.

Keywords: magnesium; plasma electrolytic oxidation; microstructure; corrosion

Academic Editor: Xiaogang Li

Received: 29 December 2024
Revised: 30 January 2025
Accepted: 5 February 2025
Published: 8 February 2025

Citation: Vertaľ, J.; Kajánek, D.; Kubásek, J.; Minárik, P. Improving Biodegradable Mg-Zn(-Ca) Alloys by Surface Treatment via Plasma Electrolytic Oxidation. *Materials* **2025**, *18*, 747. https://doi.org/10.3390/ma18040747

1. Introduction

The development of magnesium (Mg) alloys for medical applications has been the focus of intensive research in recent years, yielding significant advancements, including the commercialization of the first biodegradable implants [1,2]. However, a critical challenge limiting their application, particularly in the production of larger biodegradable medical devices such as bone fixation systems, is the combination of inadequate corrosion resistance during the initial implantation period and the unpredictable nature of corrosion attack. To mitigate these limitations, implants are typically designed with increased thickness to prevent mechanical failure during the healing process.

One promising approach to address this issue is the formation of a surface layer that provides high corrosion resistance while remaining biodegradable. This layer is intended to delay the onset of substantial corrosion until the surrounding tissue has healed, rendering the material's mechanical strength less critical [3].

In this study, special attention was devoted to plasma electrolytic oxidation (PEO), a surface treatment that generates a ceramic-like oxide coating on the material's surface. This layer, typically with a thickness of up to tens of microns, exhibits excellent corrosion resistance and high strength, with strong adhesion to the substrate due to the chemical bonding between the oxide layer and the metal [4]. The PEO process is similar to anodizing, but operates at higher voltages—up to 600 V—resulting in dielectric breakdown and plasma discharges on the sample surface. During the process, molten oxide is expelled through channels in the material, solidifying it into a hard, porous layer upon contact with the cold electrolyte [5]. However, the PEO layer's porous and cracked structure is a key drawback, as these defects compromise its corrosion resistance. The composition of the electrolytic solution, which may include aluminate-, phosphate-, or silicate-based components, also influences the layer's properties [6]. This surface treatment does not depend on the sample shape, is affordable, and has a scalability that depends mainly on the capacity of the power source. Therefore, it is very interesting in terms of samples of complex shapes, such as fixation implants.

The composition of the PEO layer strongly depends on the composition of the substrate and the electrolyte [7]. Therefore, there are strong limitations for biodegradable applications, which pose strict requirements for both composition and biocompatibility. In this regard, the phosphate electrolyte is a very good choice for biodegradable magnesium alloys and was repeatedly used in different in vivo [8,9] and in vitro studies [10–13] with very promising results.

Among the Mg alloys frequently employed in biomedical applications, Mg-Zn-Ca alloys stand out due to the biocompatibility of both zinc and calcium [14,15]. Zinc enhances the alloy's mechanical properties through solid solution strengthening and grain boundary strengthening [16], while calcium contributes to improved strength and serves as a vital component of bone and teeth, playing essential roles in blood clotting and nerve function [17]. A representative example of such alloys is Mg-1Zn-0.4Ca (ZX10, wt.%), which has been extensively studied for biodegradable applications across various microstructural states, including ultrafine-grained conditions [18]. ZX10 exhibits favorable mechanical properties, high strength, and a controlled, homogeneous degradation process with minimal hydrogen evolution [17,19]. The PEO formation on ZX00 (<1Zn, <1Ca, wt%) [9] and ZX20 (2.1Zn, 0.6Ca, wt%) [13] already showed a significant increase in the corrosion resistance in this class of magnesium alloys. However, the first study focused primarily on in vivo degradation with only limited characterization of the PEO layer, and the second one focused on the incorporation of Mn_3O_4 nanoparticles into the PEO layer. Therefore, a comprehensive characterization of PEO layer formation and its effect on the corrosion resistance of the well-known ZX10 alloy is still missing.

This work aimed to investigate the influence of PEO preparation time on the degradation resistance of Mg-1Zn (Z1) and Mg-1Zn-0.4Ca (ZX10) alloys. The comparison of Z1 and ZX10 alloys serves to elucidate the role of calcium during PEO formation and resulting corrosion resistance in the case of this class of magnesium alloys. For comparative analysis, pure Mg and commercial Mg-4Y-3RE-0.4Zr (WE43) alloy were subjected to similar preparation and testing conditions.

2. Materials and Methods

2.1. Materials and Preparation of PEO

This study investigated three magnesium alloys—Mg-1Zn (Z1), Mg-1Zn-0.4Ca (ZX10), and Mg-4Y-3RE-0.4Zr (WE43)—and pure magnesium (Mg, 99.96%). The Z1 and ZX10 alloys were fabricated through conventional casting, and their chemical composition was analyzed using energy-dispersive X-ray spectroscopy (EDS). The WE43 alloy and pure Mg were obtained commercially in extruded and cast forms, respectively. Prior to experimentation, all alloys underwent annealing for 16 h at 530 °C in an argon atmosphere to homogenize their chemical composition and dissolve secondary phases.

The PEO process was performed with the samples serving as the anode and a stainless-steel sheet acting as the cathode using a N8762A DC power source (Keysight, Santa Rosa,

CA, USA). To ensure homogeneity of the electrolyte, a rotating stirrer was employed. The electrolyte consisted of 12 g/L $Na_3PO_4 \cdot 12H_2O$ and 1 g/L KOH, and a constant current density of 50 mA/cm^2 was applied to the anode. In our previous study, these preparation parameters were found to be the most beneficial for PEO thickness and compactness [20]. Samples with dimensions of $15 \times 15 \times 5$ mm^3 were exposed to the PEO process for three different durations—5, 10, and 15 min—to optimize coating morphology and corrosion resistance. Prior to PEO, all samples were mechanically ground and polished using a 3 µm diamond suspension, followed by electro-polishing with a Lectropol device (Struers, Ballerup, Denmark) and Struers AC2 solution.

2.2. Microstructure

The microstructure of the alloys was examined using optical microscopy. Sample preparation for microstructural analysis followed the same protocol as for pre-PEO treatment. Samples were etched with a solution comprising 10 mL acetic acid, 4.2 g picric acid, 10 mL H_2O, and 70 mL ethanol. Microstructural observations were conducted using a LSM 700 optical microscope (ZEISS, Oberkochen, Germany).

Morphological, cross-sectional, and compositional analyses of the PEO layers were carried out using a APREO 2 scanning electron microscope (SEM; ThermoFischer, Waltham, MA, USA) equipped with energy-dispersive X-ray spectroscopy (EDS; EDAX, Pleasanton, CA, USA). Samples for cross-sectional observations were embedded in conductive epoxy resin, ground, and polished with a 50 nm alumina suspension.

2.3. Corrosion

Corrosion resistance was evaluated using electrochemical impedance spectroscopy (EIS) with a VSP-300 potentiostat (Biologic, Seyssinet-Pariset, France) in a three-electrode configuration. The frequency range was set between 100 kHz and 10 mHz, with frequency steps of 10 per decade and a potential amplitude of 10 mV. A 0.9% NaCl solution at room temperature served as the corrosive medium. Samples were exposed for up to 168 h to monitor degradation processes. Additionally, the corrosion resistance of PEO-free samples was measured after 2 h of exposure for each alloy.

3. Results

3.1. Initial Microstructure

The initial microstructure of the investigated alloys was analyzed using optical microscopy on etched surfaces. As shown in Figure 1, no secondary phases comprising alloying elements are observed. This absence is attributed to the solution annealing process performed during production, which ensured the dissolution of alloying elements into the material matrix.

The microstructures of the ZX10 and Z1 alloys, as well as pure magnesium, were characterized by irregularly shaped, large grains. The grains in the ZX10 and Z1 alloys were comparable in both shape and size. The average grain size evaluated via the linear interception method was ~0.49 mm and ~0.32 mm for Z1 and ZX10, respectively. In contrast, the grains in pure magnesium were significantly larger, with an average grain size of ~1.35 mm. The WE43 alloy represented an exception, as it was supplied in an as-extruded condition. Following homogenization treatment, the WE43 alloy exhibited a microstructure composed of polyhedral grains. Notably, the WE43 alloy had the smallest average grains among the tested materials, approximately ~0.15 mm, making it distinct from the other alloys.

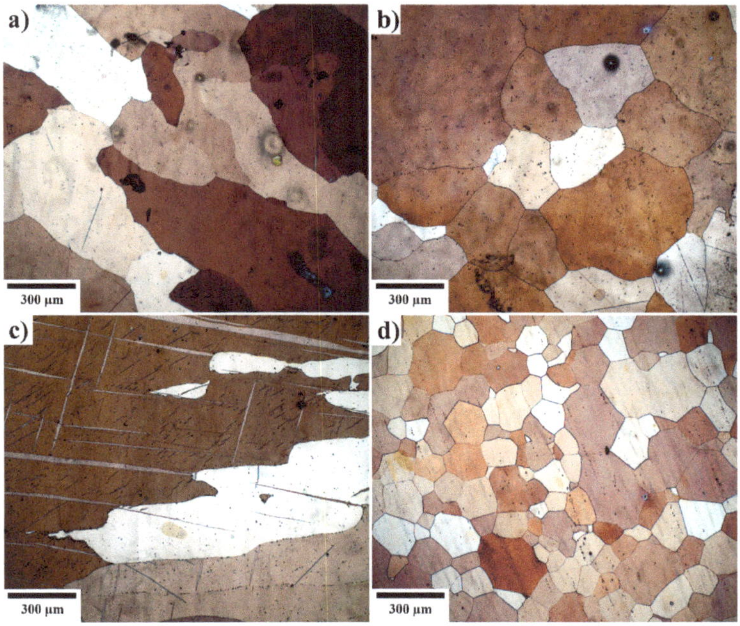

Figure 1. Optical micrographs of the initial microstructure before PEO preparation. (**a**) Z1, (**b**) ZX10, (**c**) pure Mg, and (**d**) WE43.

3.2. PEO Layers

As anticipated, the plasma electrolytic oxidation (PEO) layers of all samples exhibited a porous structure. These pores are formed due to the rapid cooling of molten oxide when it contacts the cold electrolyte. Simultaneously, this interaction generates an accumulation of voltage, resulting in the formation of microcracks within the layer structure.

Figure 2 presents the topography of the PEO layers developed on the studied alloys for three different preparation times. After 5 min of preparation, the PEO layers were predominantly characterized by a high surface density of small pores (<1 μm). Notably, the Z1 alloy was the only sample to exhibit larger pores (>10 μm) surrounded by thicker ridges in the PEO layer. Cross-sectional measurements indicated an average PEO layer thickness of approximately 12 μm, with the exception of the WE43 alloy, which exhibited a thinner layer of ~7 μm (see Table 1). These measurements were conducted in pore-free regions of the layers.

Table 1. Average thickness of the PEO layers *d* evaluated from cross sections in the pore-free areas.

	d (μm)			
	Z1	**ZX10**	**Mg**	**WE43**
5 min	12 ± 1	13 ± 1	12 ± 1	7 ± 1
10 min	18 ± 2	14 ± 1	17 ± 2	14 ± 1
15 min	25 ± 2	23 ± 2	22 ± 2	27 ± 2

Increasing the preparation time to 10 min resulted in a marked rise in the surface density of large pores, particularly in the Z1 and Mg samples. In the Z1 alloy, large pores became the dominant feature, covering most of the surface. In other samples, large pores formed clusters interspersed with regions of smaller pores. The WE43 alloy displayed the lowest surface density of large pores and did not develop pronounced ridges around them, as observed in other materials. A higher surface density of large pores was found in the ZX10 alloy, followed by the Mg sample, where these pores covered more than three-quarters of the surface area. This extended preparation time also increased the thickness of the PEO layers. As shown in Table 1, the Z1 and Mg samples exhibited thicker layers (~18 μm) compared to the ZX10 and WE43 alloys (~14 μm). The greater thickness in Z1 and Mg samples is attributable to the higher surface density of large pores and more prominent ridges.

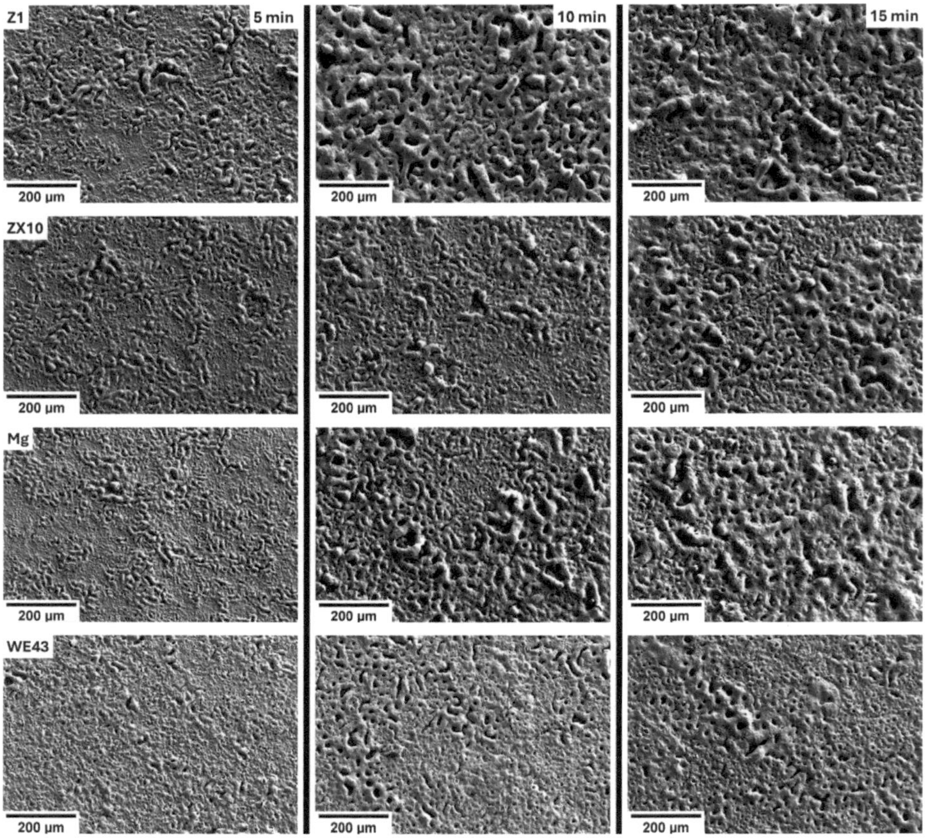

Figure 2. The topography of the PEO layers formed on the investigated materials after three preparation times.

Further, increasing the preparation time to 15 min did not induce substantial qualitative changes in the topography of the PEO layers for individual samples. The Z1 and Mg samples continued to exhibit the highest surface density of large pores, which, along with high ridges, covered nearly the entire surface. The ZX10 alloy displayed a modest increase in large pore density, with its surface appearing similar to that of the Mg sample after 10 min of preparation. Notably, the PEO layer of the WE43 alloy remained largely unchanged even after 15 min, with the surface maintaining a highly homogeneous distribution of predominantly small pores. Occasional clusters of large pores were observed, but these were not surrounded by the prominent ridges seen in the other materials. The thickness of the PEO layers increased further with the extended preparation time, reaching comparable levels (~23 μm) across all alloys, except for the WE43 alloy, which developed the thickest layer at approximately 27 μm.

Cracks were observed in the PEO layers of all investigated samples. Figure 3 provides a detailed depiction of crack development in the Z1 alloy. The formation and progression of cracks showed a systematic relationship with the preparation time, both in terms of their quantity and thickness. After 5 min of preparation, cracks were predominantly localized near larger pores and appeared unconnected. However, as the preparation time increased, these cracks began to interconnect, and their thickness expanded progressively. By 15 min, the PEO layer exhibited a relatively dense network of interconnected cracks, as illustrated in Figure 3. This trend of crack development, characterized by increasing connectivity and thickness with longer preparation times, was consistently observed across all studied materials.

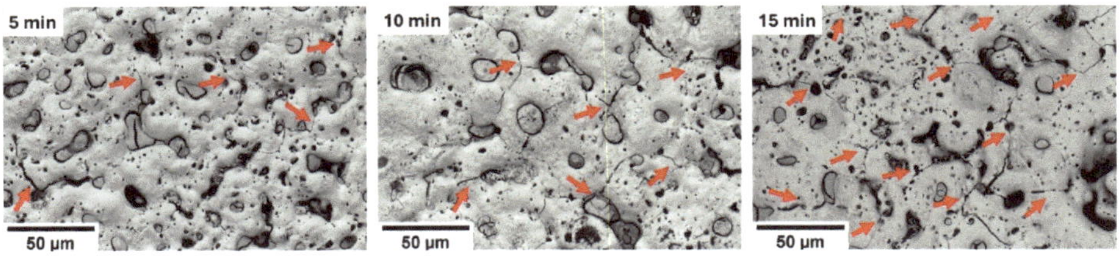

Figure 3. Development of the cracks, marked by red arrows, in the PEO layer with increasing preparation time for the Z1 alloy (BSE signal).

The investigation of PEO cross sections revealed significant heterogeneity within the layers. Large pores frequently extended almost to the substrate surface, contributing to the formation of cavities, as demonstrated in Figure 4 for the ZX10 alloy prepared with a 10 min treatment time. Notably, layers produced with a 10 min preparation time were systematically the most compact across all the investigated alloys. Figure 4 also presents EDS elemental maps of the PEO layer cross section, indicating that the layer primarily consists of magnesium, oxygen, and phosphorus. This is a typical composition of PEO formed in a phosphate-rich solution, resulting in the formation of $Mg_3(PO_4)_2$ and MgO [7]. Their formation follows the equations:

$$Mg \rightarrow Mg^{2+} + 2e^- \quad (1)$$

$$Mg^{2+} + 2OH^- \rightarrow Mg(OH)_2 \quad (2)$$

$$Mg(OH)_2 \rightarrow MgO + H_2O \quad (3)$$

$$Mg^{2+} + 2PO_4^{3-} \rightarrow Mg_3(PO_4)_2 \quad (4)$$

in which (1) represents the magnesium dissolution reaction, (2) and (3) represent the formation process of MgO, and (4) represents the formation of $Mg_3(PO_4)_2$.

Small amounts of zinc and calcium were detected, indicating that their concentrations were considerably lower compared to those in the substrate. Note that this relative decrease in alloying elements concentration is primarily due to the PEO layer enrichment by oxygen and phosphorus. The amount of zinc and calcium was comparable to the matrix composition when only ratios between magnesium and alloying elements were considered. These findings were consistent across all studied alloys, irrespective of their composition, as shown in Table 2. Interestingly, in the ZX10 alloy, a small PEO layer enrichment by calcium was observed, indicating a possible formation of calcium oxide or calcium phosphate. However, no direct observation of such phases was achieved.

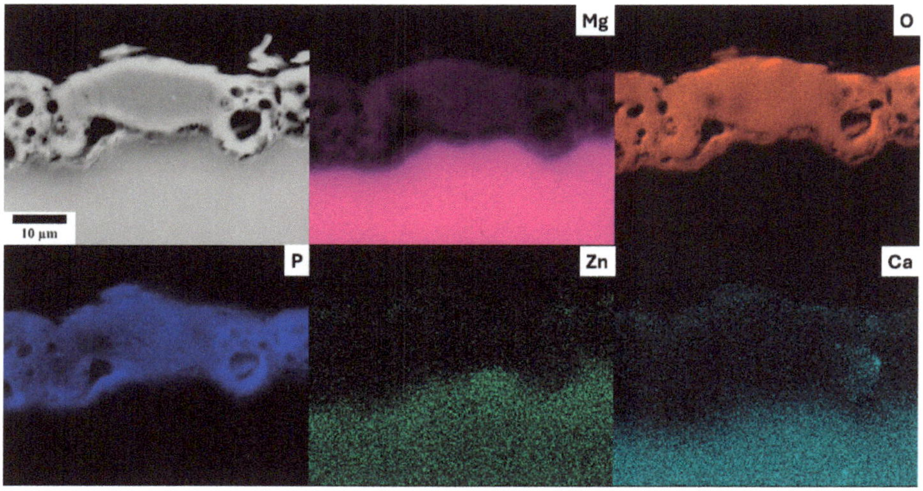

Figure 4. Cross section of the PEO layer formed on the ZX10-10 min sample with the chemical composition analysis.

Table 2. Average EDS composition of PEO layer formed on the 10 min samples excluding oxygen and phosphorous.

	Mg	Zn	Ca	Y	Nd
Z1	99.6 ± 0.1	0.4 ± 0.1	-	-	-
ZX10	98.8 ± 0.1	0.3 ± 0.1	1.0 ± 0.1	-	-
WE43	93.7 ± 0.1	-	-	3.3 ± 0.1	3.0 ± 0.1

3.3. Corrosion Resistance

The protective performance of the PEO layers was evaluated in a 0.9% NaCl solution over a one-week exposure period. The corrosion resistance of the alloys with PEO coatings was monitored periodically by electrochemical impedance spectroscopy (EIS) at intervals of 2 h, 4 h, 8 h, 12 h, 24 h, 48 h, and 168 h. The representative Nyquist plots are shown in Figure 5 for all 10 min samples. The experimental data were fitted using the equivalent circuit shown in Figure 6, which most effectively represents the physical phenomena occurring at the PEO layer. Two capacitive loops represent the occurrence of areas with different electrochemical behavior, usually referring to the coating/corrosion product layer (first loop) and the existence of a sub-layer or the presence of a double layer on a surface (second loop). In this model, the parameter R_S denotes the solution resistance, while R_1 and R_2 represent the charge transfer resistance through distinct regions of the layer. A detailed description of the physical meaning of the individual parameters of the equivalent circuit is given elsewhere [21,22]. Nevertheless, from the corrosion resistance point of view, the most important parameter is the total impedance $R_{sum} = R_1 + R_2$, which represents the total corrosion resistance of the material. Note that during certain measurements (see Figure 5), the Nyquist plot displayed only a single arc. In these instances, the R_2 and CPE_2 elements were excluded from the equivalent circuit.

The polarization resistance of the uncoated samples measured after 2 h of exposure was (3.6 ± 0.2) kΩ·cm^2 for Z1, (3.6 ± 0.1) kΩ·cm^2 for ZX10, (0.82 ± 0.02) kΩ·cm^2 for Mg, and (4.7 ± 0.2) kΩ·cm^2 for WE43. Clearly, the formation of the PEO layer significantly affected the corrosion resistance of the materials, as shown in Figure 7. Generally, samples with PEO layers formed for longer durations (10 and 15 min) exhibited higher corrosion resistance than those formed for 5 min. The polarization resistance after 2 h of exposure was approximately 20 kΩ·cm^2 for the Z1 alloy (10 and 15 min samples), whereas it was only around 14 kΩ·cm^2 for the ZX10 alloy (10 and 15 min samples). For the Mg sample, the highest resistance was observed in the Mg-15 min sample (~30 kΩ·cm^2), and for the WE43 alloy, the highest value was approximately 80 kΩ·cm^2 for the WE43-15 min sample. However, due to considerable data scatter, the WE43-15 min sample can be considered statistically similar to the WE43-10 min sample, which had a resistance of ~60 kΩ·cm^2. As the exposure time increased, all samples exhibited a continuous decrease in corrosion resistance, reaching approximately 5 kΩ·cm^2 after 24 h. The only exception was the Z1-5 min sample, which initially showed a relative decrease in polarization resistance after PEO formation (~2.3 kΩ·cm^2), followed by a gradual increase in resistance over time, reaching ~5.3 kΩ·cm^2 after 24 h.

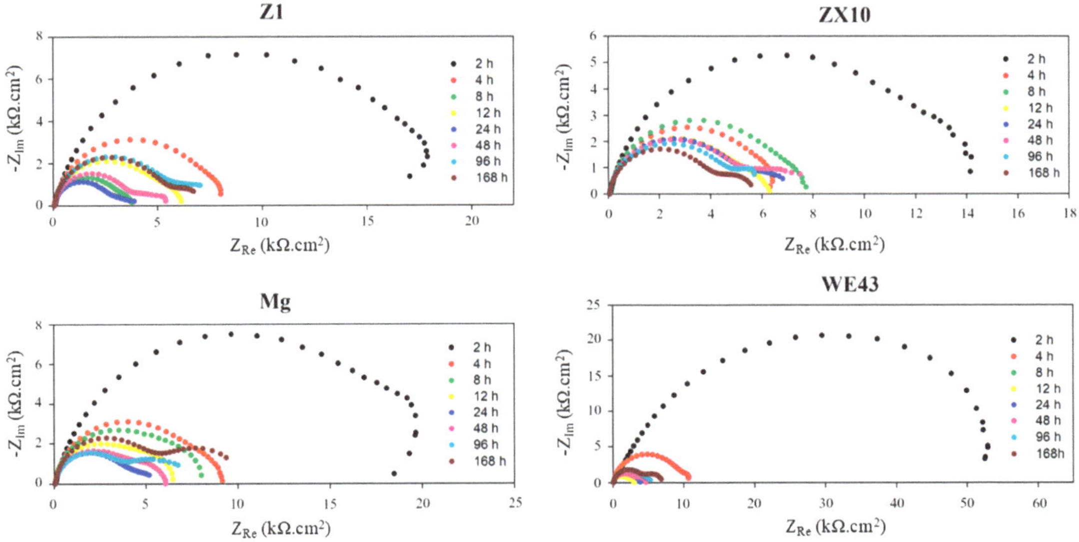

Figure 5. Representative Nyquist plots of the investigated 10 min samples.

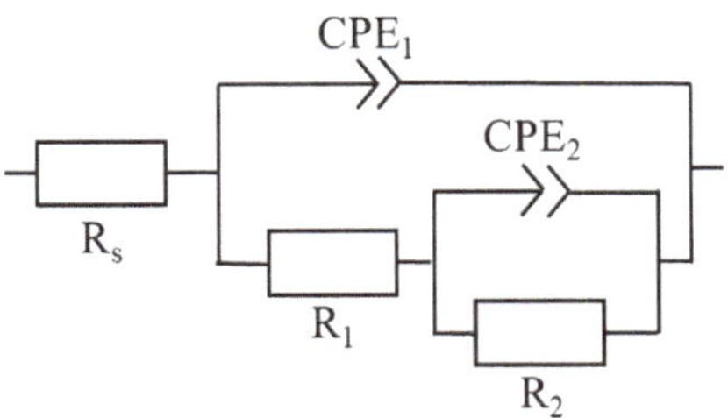

Figure 6. Equivalent circuit used for the EIS data analysis.

After 24 h of immersion, the corrosion resistance of nearly all samples began to increase, as shown in Figure 7. The notable exceptions were the Z1-5 min and ZX10-15 min samples, which displayed a continuous decrease in resistance with prolonged immersion time. After 168 h of immersion, the highest polarization resistance among the Mg-Zn(-Ca) samples was observed in the Z1-10 min ((6.1 ± 1.5) kΩ·cm^2) and ZX10-10 min ((5.5 ± 0.8) kΩ·cm^2) samples. The Mg and WE43 samples exhibited resistances of (8.4 ± 2.8) kΩ·cm^2 and (7.1 ± 2.0) kΩ·cm^2, respectively, across all PEO preparation times. Although there were relatively significant differences in the average corrosion resistance values, the high data scatter after one week of immersion observed in all samples indicates that these results are statistically comparable within the measurement error.

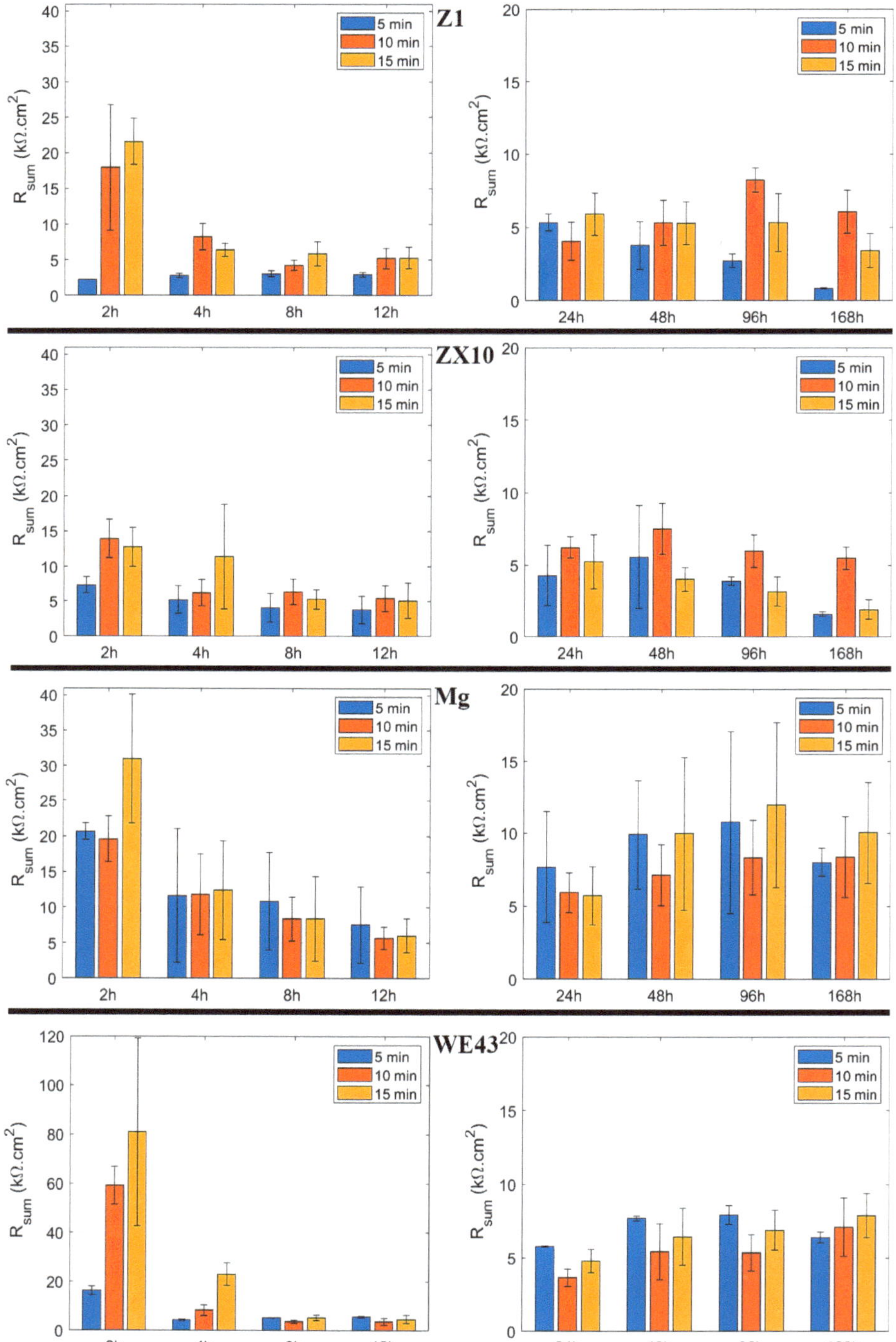

Figure 7. Graphical representation of R_{sum} values of the PEO layers with different preparation times.

The examination of the cross section through the surface layer after 168 h of immersion revealed significant dissolution of the PEO layer during the degradation process in 0.9% M NaCl solution, with the formation of a compact corrosion layer beneath it. This phenomenon was observed for all samples investigated. Figure 8 illustrates this degradation in the ZX10-10 min sample, where the average thickness of the PEO layer decreased to approximately 8 μm and was severely shattered. EDS elemental mapping indicated that phosphorus from the dissolved PEO layer was incorporated into the corrosion layer, which also contained zinc from the substrate material. Note that phosphorous was not present in the corrosion solution, and the PEO layer was the only possible source. Additionally, zinc was found to segregate at the interface between the substrate and the corrosion layer. In contrast, no calcium enrichment was detected in the corrosion layer. Elemental mapping of the Z1 alloy yielded similar results (excluding calcium), while in the case of the WE43 alloy, an enrichment of the corrosion layer with rare earth elements was observed.

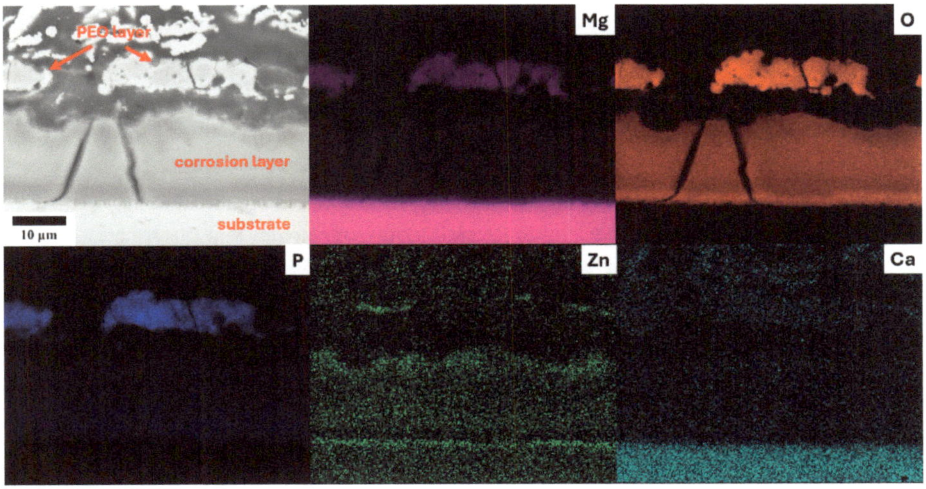

Figure 8. Cross section of the PEO layer formed on the ZX10-10 min sample after 168 h of immersion in 0.9% NaCl solution, including the chemical composition analysis.

4. Discussion

The formation of PEO is associated with the flow of molten oxide inside the electric discharges, which forms on the surface during the preparation. The molten oxide gradually lies over the previous layers, and the thickness of the coating grows. The formation of pores and cracks inside the PEO is inseparable from this process, and a lot of work has been done to minimize their size and density [23]. Most of the processing parameters (the solution composition and the current density) were chosen according to a previous study on commercial AZ31 [20], and the only free variable used in this study was the preparation time. The results showed that the gradual increase in PEO layer thickness with time was relatively comparable between the studied materials and was in the range of 1.4–1.7 μm/min. These values are significantly higher than the measured 0.9 μm/min for the polished AZ31 prepared with the same parameters [24]. Nevertheless, a growth rate of 0.5–1.6 μm/min is commonly observed for magnesium alloys [13,25], and the PEO thicknesses achieved in this study are comparable to other work [6,26]. For a very short preparation time (1 min), a growth rate of 3 μm/min can be achieved [10], but the further increase in thickness is evidently slowed by the formation of a non-conductive surface [12]. Therefore, the measured results are in accordance with the literature. In general, the direct comparison to other studies is problematic because of high variation in the processing parameters.

Figure 2 shows that the more alloyed materials (ZX10 and WE43) develop higher surface density of large pores later than the Mg and Z1 materials. In addition, the WE43 alloy, having the highest content of alloying elements, showed the lowest porosity even after 15 min of preparation. It was found that the decrease in porosity may be achieved by the increase in the surface defect density and topography prior to the PEO preparation [24]. However, this is not the reason for the smoother PEO surface of the WE43 alloy, as the grain size was on one hand lower than in the case of the other investigated materials, but on the other hand, it was significantly higher than the pore size and average distance between the pores. A small decrease in porosity with increasing content of alloying elements has already been reported [27], but the general reason was not revealed. In this study, the lower porosity was observed in the alloys with a higher content of alloying elements—WE43 and ZX10. This is clearly visible in Figure 2, particularly in the 10 and 15 min samples. The chemical analysis revealed that the PEO layer had a comparable ratio between the substrate alloying elements: Mg/Zn(/Ca) for Mg-1Zn(-0.4Ca) and Mg/Y/Nd for WE43. Therefore, the alloying elements were built into the oxide layer during the PEO process, which had some effect on discharges and consequently porosity. However, no particles rich in alloying elements were observed in this study. This statement also holds for the calcium in the ZX10 alloy, which was the only element observed with increased concentration in the PEO layer compared to the substrate. Note that the formation of calcium phosphates is convenient when an excess amount of calcium and phosphates is available during anodization [28]. However, no direct observation of this phase was achieved. It was shown that with increasing preparation time of the AZ-type alloys as a substrate, a $MgAl_2O_4$ spinel formed in the inner PEO layer [29,30]. Nevertheless, more alloyed AZ91 exhibited a higher surface roughness than AZ31 with the same preparation parameters [30]. Therefore, further work regarding the effect of alloying elements on porosity needs to be conducted to understand this phenomenon fully.

The formation of the PEO layer clearly had a significant impact on the corrosion resistance of the investigated materials. After 2 h of immersion in the corrosive media, the overall corrosion resistance manifested by the polarization resistance R_{sum} in all samples exhibited a significant increase. The only exception was the Z1-5 min sample, which exhibited the highest porosity of the coating of all 5 min samples. Therefore, the combination of relatively thin and highly porous layers did not positively impact the corrosion resistance, and faster local degradation occurred. Otherwise, the highest corrosion resistance was measured for the WE43-15 min sample, which showed the best combination of high PEO layer thickness and low porosity.

The PEO layer is commonly described to be composed of three parts—the outer porous layer, the intermediate dense layer, and the inner dense layer [26]. The intermediate dense layer is responsible for the good corrosion resistance of the material, and the inner dense layer for a good attachment to the surface. The cross-section analysis showed that the inner dense layer was not compact through the whole surface of any of the studied samples and often contained pores encapsulated in the layer (red arrows in Figure 9) and columnar pores reaching the inner dense layer (yellow arrows in Figure 9). The rapid decrease in R_{sum} within the first 24 h of immersion observed for all investigated samples can be associated with the dissolution of the inner dense layer at the bottom of the pores and the onset of degradation of the substrate. The further increase in the immersion time resulted in a continuous increase in the corrosion resistance in most of the samples because of the formation of a corrosion layer below the PEO layer on the coating–substrate interface (cf. Figures 7 and 8). The formation of primarily $Mg(OH)_2$ hydroxide [31] occurred at the beginning in the points of PEO layer failure. These products tend to fill the faults and defects of the PEO coating, thereby locally improving its integrity and also increasing the barrier effect of the coating [24]. Subsequently, a uniform layer of the corrosion products formed on the whole surface together with a continuous dissolution of the PEO layer, which became significantly thinner and highly fragmented (Figure 8). As a result of the PEO layer dissolution, the phosphates released from the dissolved PEO layer partially built into the newly created corrosion layer in the case of all samples (see Figure 8), making it more compact and protective. Phosphates are well known for inhibiting the corrosion process in biological media, which are rich in these elements [32]. The results showed that the most convenient PEO preparation time for all studied materials was 10 min (Figure 7). This result is in good agreement with the literature, where it has been repeatedly shown that the barrier effect of the PEO layer may be damaged by prolonged preparation time due to the formation of higher porosity and crack density [20]. The same effect was observed also in this study (cf. Figures 3 and 6).

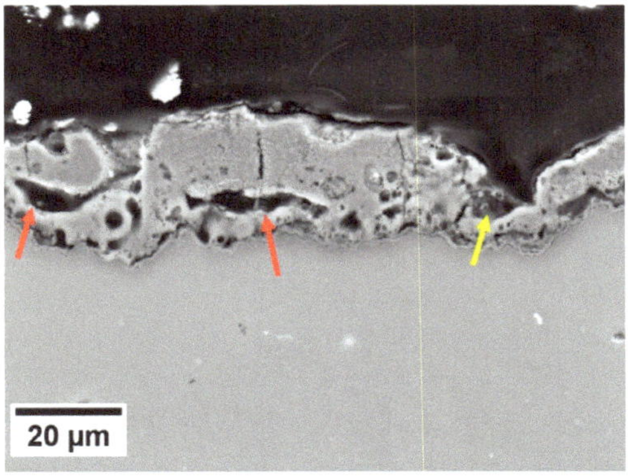

Figure 9. Cross section of the PEO layer formed on the ZX10-10 min sample with highlighted pores.

Finally, the investigated alloys Z1 and ZX10 showed slightly lower average corrosion resistance after 168 h of immersion in 0.9% NaCl solution than the reference ones (Mg and WE43). Nevertheless, note that the corrosion resistance of all investigated alloys was comparable after 168 h of immersion, considering the statistical error. This result is very important regarding the development of biodegradable magnesium-based materials because it shows that the low-alloyed alloys with high biocompatibility, like Mg-Zn(-Ca), can achieve comparable corrosion resistance to a highly corrosion-resistant WE43 alloy by a proper surface treatment. This result is of particular interest because of the recent successful in vivo test of WE43-based magnesium implants with PEO surface modification [8]. In addition, it is shown that the formation of the PEO layer is not beneficial only before its dissolution, but a transfer of phosphate into the newly created magnesium hydroxide improves the corrosion resistance of the material, even for a longer time.

5. Conclusions

The subject of this study was the analysis of the influence of the preparation time of the PEO process on the degradation resistance of Mg-1Zn (Z1) and Mg-1Zn-0.4Ca (ZX10) alloys in comparison to the pure Mg and commercial Mg-4Y-3RE-0.4Zr (WE43) prepared similarly. The following conclusions may be drawn from this study.

- The morphology of the PEO layers showed a typical pore-like nature. A longer preparation time caused an increase in the size and coverage of pores on the surface of all materials, as well as a growth in microcracks' size and density. The cross-section observation showed that the PEO layers prepared for 10 min were the most homogeneous ones.

- The corrosion resistance measured by EIS showed that the best results were achieved after 10 min of PEO preparation. Overall, the corrosion resistance of all investigated materials was higher than that of their PEO-free counterparts, as shown in the results measured after two hours of immersion.
- The corrosion resistance of all investigated alloys was comparable within the statistical error after 168 h of immersion. This result shows that the low-alloyed alloys with high biocompatibility, like Mg-Zn(-Ca), can achieve comparable long-term corrosion resistance to the highly corrosion-resistant WE43 alloy by proper surface treatment.
- The cross-sectional and chemical analysis of the samples after 168 h of immersion showed that the PEO layer degrades during the immersion and the corrosion layer develops beneath it. The dissolved phosphates build into the newly formed corrosion layer and improve its properties. The formation of the PEO layer is not beneficial only before its dissolution, but a transfer of phosphate into the newly created magnesium hydroxide improves the corrosion resistance of the material, even for a longer time.

Author Contributions: Conceptualization, D.K. and P.M.; methodology, P.M.; formal analysis, J.V., D.K. and P.M.; investigation, J.V. and P.M.; resources, J.K. and D.K.; writing—original draft preparation, J.V. and P.M.; writing—review and editing, D.K., J.K. and P.M.; visualization, J.V. and P.M.; supervision, P.M.; project administration, P.M. and D.K. All authors have read and agreed to the published version of the manuscript.

Funding: Financial support from the Czech Science Foundation project GA24-10626S is gratefully acknowledged. Financial support from the EU NextGenerationEU through the Recovery and Resilience Plan for Slovakia under the project No. 09I03-03-V04-00315 is acknowledged as well.

Institutional Review Board Statement: Not applicable.

Informed Consent Statement: Not applicable.

Data Availability Statement: The original contributions presented in the study are included in the article, further inquiries can be directed to the corresponding author.

Conflicts of Interest: The authors declare no conflicts of interest.

References

1. Tsakiris, V.; Tardei, C.; Clicinschi, F.M. Biodegradable Mg Alloys for Orthopedic Implants—A Review. *J. Magnes. Alloys* **2021**, *9*, 1884–1905. [CrossRef]
2. Biber, R.; Pauser, J.; Geßlein, M.; Bail, H.J. Magnesium-Based Absorbable Metal Screws for Intra-Articular Fracture Fixation. *Case Rep. Orthop.* **2016**, *2016*, 9673174. [CrossRef]
3. Shi, B.; Xu, J.; Zou, J.; Li, Y.R.; Zhou, Z.; Liu, K.; Jia, Q.; Jiang, H.B. Advances in Amelioration of Plasma Electrolytic Oxidation Coatings on Biodegradable Magnesium and Alloys. *Heliyon* **2024**, *10*, e24348. [CrossRef]
4. Barati Darband, G.; Aliofkhazraei, M.; Hamghalam, P.; Valizade, N. Plasma Electrolytic Oxidation of Magnesium and Its Alloys: Mechanism, Properties and Applications. *J. Magnes. Alloys* **2017**, *5*, 74–132. [CrossRef]
5. Shishir, R.; Lokeshkumar, E.; Manojkumar, P.; Nasiruddin, U.; Premchand, C.; Ponnilavan, V.; Rameshbabu, N. Development of Biocompatible and Corrosion-Resistant Plasma Electrolytic Oxidation Coating over Zinc for Orthopedic Implant Applications. *Surf. Coat. Technol.* **2022**, *450*, 128990. [CrossRef]
6. Toulabifard, A.; Rahmati, M.; Raeissi, K.; Hakimizad, A.; Santamaria, M. The Effect of Electrolytic Solution Composition on the Structure, Corrosion, and Wear Resistance of PEO Coatings on AZ31 Magnesium Alloy. *Coatings* **2020**, *10*, 937. [CrossRef]
7. Durdu, S.; Aytaç, A.; Usta, M. Characterization and Corrosion Behavior of Ceramic Coating on Magnesium by Micro-Arc Oxidation. *J. Alloys Compd.* **2011**, *509*, 8601–8606. [CrossRef]
8. Herzog, P.; Rendenbach, C.; Turostowski, M.; Ellinghaus, A.; Prates Soares, A.; Heiland, M.; Duda, G.N.; Schmidt-Bleek, K.; Fischer, H. Titanium versus Plasma Electrolytic Oxidation Surface-Modified Magnesium Miniplates in a Forehead Secondary Fracture Healing Model in Sheep. *Acta Biomater.* **2024**, *185*, 98–110. [CrossRef]

9. Kopp, A.; Fischer, H.; Soares, A.P.; Schmidt-Bleek, K.; Leber, C.; Kreiker, H.; Duda, G.; Kröger, N.; van Gaalen, K.; Hanken, H.; et al. Long-Term In Vivo Observations Show Biocompatibility and Performance of ZX00 Magnesium Screws Surface-Modified by Plasma-Electrolytic Oxidation in Göttingen Miniature Pigs. *Acta Biomater.* **2023**, *157*, 720–733. [CrossRef]
10. Muñoz, M.; Fernández, J.P.; Torres, B.; Pulido, N.; Zhang, G.; Shanov, V.; Moreno, L.; Matykina, E.; Rams, J. Plasma Electrolytic Oxidation of AZ31 Magnesium Stents for Degradation Rate Control. *Surf. Coat. Technol.* **2024**, *479*, 130539. [CrossRef]
11. Rahmati, M.; Mohammadi Zahrani, E.; Atapour, M.; Noorbakhsh Nezhad, A.H.; Hakimizad, A.; Alfantazi, A.M. In Situ Synthesis and Electrochemical Corrosion Behavior of Plasma Electrolytic Oxidation Coating Containing an Osteoporosis Drug on AZ31 Magnesium Alloy. *Mater. Chem. Phys.* **2024**, *315*, 128983. [CrossRef]
12. Azzahra, N.G.; Widakdo, J.; Sudiro, T.; Hung, W.-S.; Anawati, A. Characteristics of Plasma Electrolytic Oxide Coating on Magnesium-Hydroxyapatite Composites. *Mater. Lett.* **2024**, *356*, 135601. [CrossRef]
13. Bahrampour, S.; Bordbar-Khiabani, A.; Hossein Siadati, M.; Gasik, M.; Mozafari, M. Improving the Inflammatory-Associated Corrosion Behavior of Magnesium Alloys by Mn3O4 Incorporated Plasma Electrolytic Oxidation Coatings. *Chem. Eng. J.* **2024**, *483*, 149016. [CrossRef]
14. Roh, H.-J.; Park, J.; Lee, S.-H.; Kim, D.-H.; Lee, G.-C.; Jeon, H.; Chae, M.; Lee, K.-S.; Sun, J.-Y.; Lee, D.-H.; et al. Optimization of the Clinically Approved Mg-Zn Alloy System through the Addition of Ca. *Biomater. Res.* **2022**, *26*, 41. [CrossRef]
15. Xu, Z.; Smith, C.; Chen, S.; Sankar, J. Development and Microstructural Characterizations of Mg–Zn–Ca Alloys for Biomedical Applications. *Mater. Sci. Eng. B* **2011**, *176*, 1660–1665. [CrossRef]
16. Cáceres, C.H.; Blake, A. The Strength of Concentrated Mg–Zn Solid Solutions. *Phys. Status Solidi A* **2002**, *194*, 147–158. [CrossRef]
17. Malvestiti, L.; Tano de la Hoz, M.F.; Usach, V.; Pastore, J.I.; Setton, P.; Katunar, M.R.; Ceré, S. Mg ZX10 Alloy for Biomedical Applications: Surface Characterization and Comparative Corrosion Studies under Simulated Biological Conditions. *Mater. Chem. Phys.* **2024**, *315*, 128933. [CrossRef]
18. Merson, E.D.; Poluyanov, V.A.; Myagkikh, P.N.; Merson, D.L.; Vinogradov, A.Y. Improving Mechanical Performance of Biodegradable ZX10 Alloy in Hanks' Solution through Thermo-Mechanical Treatment. *Lett. Mater.* **2024**, *14*, 223–230. [CrossRef]
19. Jafari, S.; Raman, R.K.S.; Davies, C.H.J.; Hofstetter, J.; Uggowitzer, P.J.; Löffler, J.F. Stress Corrosion Cracking and Corrosion Fatigue Characterisation of MgZn1Ca0.3 (ZX10) in a Simulated Physiological Environment. *J. Mech. Behav. Biomed. Mater.* **2017**, *65*, 634–643. [CrossRef]
20. Hadzima, B.; Kajánek, D.; Jambor, M.; Drábiková, J.; Březina, M.; Buhagiar, J.; Pastorková, J.; Jacková, M. PEO of AZ31 Mg Alloy: Effect of Electrolyte Phosphate Content and Current Density. *Metals* **2020**, *10*, 1521. [CrossRef]
21. Amirudin, A.; Thieny, D. Application of Electrochemical Impedance Spectroscopy to Study the Degradation of Polymer-Coated Metals. *Prog. Org. Coat.* **1995**, *26*, 1–28. [CrossRef]
22. Gu, Y.; Bandopadhyay, S.; Chen, C.; Ning, C.; Guo, Y. Long-Term Corrosion Inhibition Mechanism of Microarc Oxidation Coated AZ31 Mg Alloys for Biomedical Applications. *Mater. Des.* **2013**, *46*, 66–75. [CrossRef]
23. Yang, C.; Chen, P.; Wu, W.; Sheng, L.; Zheng, Y.; Chu, P.K. A Review of Corrosion-Resistant PEO Coating on Mg Alloy. *Coatings* **2024**, *14*, 451. [CrossRef]
24. Kajánek, D.; Pastorek, F.; Hadzima, B.; Bagherifard, S.; Jambor, M.; Belány, P.; Minárik, P. Impact of Shot Peening on Corrosion Performance of AZ31 Magnesium Alloy Coated by PEO: Comparison with Conventional Surface Pre-Treatments. *Surf. Coat. Technol.* **2022**, *446*, 128773. [CrossRef]
25. Anawati; Asoh, H.; Ono, S. Enhanced Uniformity of Apatite Coating on a PEO Film Formed on AZ31 Mg Alloy by an Alkali Pretreatment. *Surf. Coat. Technol.* **2015**, *272*, 182–189. [CrossRef]
26. Hussein, R.O.; Nie, X.; Northwood, D.O. An Investigation of Ceramic Coating Growth Mechanisms in Plasma Electrolytic Oxidation (PEO) Processing. *Electrochimica Acta* **2013**, *112*, 111–119. [CrossRef]
27. Comparison of 2D and 3D Plasma Electrolytic Oxidation (PEO)-Based Coating Porosity Data Obtained by X-Ray Tomography Rendering and a Classical Metallographic Approach. Available online: https://www.mdpi.com/1996-1944/15/18/6315 (accessed on 29 December 2024).
28. Hiromoto, S.; Shishido, T.; Yamamoto, A.; Maruyama, N.; Somekawa, H.; Mukai, T. Precipitation Control of Calcium Phosphate on Pure Magnesium by Anodization. *Corros. Sci.* **2008**, *50*, 2906–2913. [CrossRef]
29. Chen, Y.; Yang, Y.; Zhang, W.; Zhang, T.; Wang, F. Influence of Second Phase on Corrosion Performance and Formation Mechanism of PEO Coating on AZ91 Mg Alloy. *J. Alloys Compd.* **2017**, *718*, 92–103. [CrossRef]
30. Rakoch, A.G.; Monakhova, E.P.; Khabibullina, Z.V.; Serdechnova, M.; Blawert, C.; Zheludkevich, M.L.; Gladkova, A.A. Plasma Electrolytic Oxidation of AZ31 and AZ91 Magnesium Alloys: Comparison of Coatings Formation Mechanism. *J. Magnes. Alloys* **2020**, *8*, 587–600. [CrossRef]

31. Pardo, A.; Merino, M.C.; Coy, A.E.; Arrabal, R.; Viejo, F.; Matykina, E. Corrosion Behaviour of Magnesium/Aluminium Alloys in 3.5 Wt.% NaCl. *Corros. Sci.* **2008**, *50*, 823–834. [CrossRef]
32. Mei, D.; Lamaka, S.V.; Lu, X.; Zheludkevich, M.L. Selecting Medium for Corrosion Testing of Bioabsorbable Magnesium and Other Metals—A Critical Review. *Corros. Sci.* **2020**, *171*, 108722. [CrossRef]

materials

MDPI

Review

Surface-Modification Strategy to Produce Highly Anticorrosive $Ti_3C_2T_x$ MXene-Based Polymer Composite Coatings: A Mini-Review

Shufang Zhang [1,2,3], Guoqin Zhang [2,4], Liang Fang [1,2,5,*], Zhiheng Wang [2], Fang Wu [2], Gaobin Liu [2], Qirui Wang [1] and Hongen Nian [1,*]

1 Key Laboratory of Green and High-End Utilization of Salt Lake Resources, Qinghai Institute of Salt Lake, Chinese Academy of Sciences, Xining 810008, China; roseymcn.cn2000@foxmail.com (S.Z.); qrwang@isl.ac.cn (Q.W.)
2 Chongqing Key Laboratory of Interface Physics in Energy Conversion, College of Physics, Chongqing University, Chongqing 400044, China; zgq@cqyu.edu.cn (G.Z.); 202027021033@cqu.edu.cn (Z.W.); wufang@cqu.edu.cn (F.W.); gbl@cqu.edu.cn (G.L.)
3 College of AI and BigData, Chongqing Polytechnic University of Electronic Technology, Chongqing 401331, China
4 Aviation and Automobile School, Chongqing Youth Vocational & Technical College, Chongqing 400712, China
5 Center of Modern Physics, Institute for Smart City of Chongqing University in Liyang, Liyang 213300, China
* Correspondence: lfang@cqu.edu.cn (L.F.); nhe@isl.ac.cn (H.N.)

Abstract: MXenes are a group of novel two-dimensional (2D) materials with merits such as large specific surface area, abundant surface-functional groups, high chemical activity, excellent mechanical properties, high hydrophilicity, and good compatibility with various polymers. In recent years, many novel high-performance organic anticorrosion coatings using MXenes as nanofillers have been reported and have attracted widespread attention. As the first successfully prepared MXene material, $Ti_3C_2T_x$ is the most extensively studied and typical member of the MXene family. Therefore, it is taken as the representative of its family, and the status of $Ti_3C_2T_x$ MXene/epoxy resin (EP) and MXene/waterborne polyurethane (WPU) polymer anticorrosive composite coatings is reviewed. Firstly, the structure, characteristics, and main synthesis methods of MXenes are briefly introduced. Then, the latest progress of four surface-modification strategies to improve the dispersion, compatibility, stability, and anti-aggregation properties of MXenes, namely functionalization grafting, orientation regulation, heterostructure nanocomposite design, and stabilization and greening treatment, are analyzed and summarized. Finally, the current challenges and future opportunities regarding MXene-based corrosion-resistant organic composite coatings are discussed prospectively.

Keywords: MXene composites; $Ti_3C_2T_x$; organic composite coatings; anticorrosion coatings; 2D materials; surface modification

Academic Editors: Antonio Polimeni and Csaba Balázsi

Received: 1 December 2024
Revised: 10 January 2025
Accepted: 21 January 2025
Published: 1 February 2025

Citation: Zhang, S.; Zhang, G.; Fang, L.; Wang, Z.; Wu, F.; Liu, G.; Wang, Q.; Nian, H. Surface-Modification Strategy to Produce Highly Anticorrosive $Ti_3C_2T_x$ MXene-Based Polymer Composite Coatings: A Mini-Review. *Materials* **2025**, *18*, 653. https://doi.org/10.3390/ma18030653

1. Introduction

Corrosion is a ubiquitous and costly issue for diverse industries [1–8]. The economic losses caused by corrosion are around USD 2.5 trillion annually for the whole world, as evaluated by the National Association for Corrosion Engineers (NACE) [1]. Additionally, as early as 2017, corrosion and related destruction were estimated to account for a loss of approximately RMB 2128 billion (≈USD 310 billion) in China, which is nearly 3.34% of the value of the gross domestic product (GDP) [2]. Effective corrosion control is important not only for economic savings but also for the safety, longevity, and sustainability of metal assets.

In order to minimize its adverse effects, various methods [9–11], such as protective coatings, surface treatments, and corrosion inhibitors, have been adopted to prevent corrosion. Among these, organic anticorrosion coatings [12–16] have the advantages of simplicity, low cost, high efficiency, and environmental friendliness and are widely used in fields such as construction, automobiles, aerospace, and ocean engineering. Organic anticorrosion coatings are usually based on epoxy resin (EP), polyurethane (PU), acrylic, or another polymer matrix [17–20] and commonly suffer from defects such as coating shrinkage, internal pores, cracks, and peeling caused by solvent evaporation. These defects introduce channels for corrosive media (such as water molecules and Cl^- ions) to diffuse into the surface of the metal substrate, weakening the protective effect of the coating. Therefore, there is an urgent need to develop protective coatings with as few internal defects (such as microholes) as possible.

The incorporation of fillers can enhance corrosion resistance by reducing defects such as pores in organic coatings, increasing coating density, and prolonging the diffusion path of corrosive substances [17–20]. Among various optional fillers, due to their large specific surface area, adjustable electronic structure, good flexibility, and excellent mechanical properties, two-dimension (2D) layered materials have great potential to improve the performance of protective coatings in blocking the penetration of corrosive ions. A variety of 2D materials such as graphene (Gr), graphene oxide (GO), hexagonal boron nitride (h-BN), molybdenum disulfide (MoS_2), and layered double hydroxides (LDHs) have been successively attempted [21–25]. However, the cumbersome and costly preparation of Gr and h-BN and their inherent chemical inertness result in weak interactions with organic substrates and easy aggregation, leading to poor corrosion resistance.

In contrast, MXenes (2D transition metal carbides/nitrides) have abundant surface chemical terminal groups, which easily form bonds with other materials or undergo surface modification or functionalization to improve the corrosion resistance of organic coatings [26–30]. Therefore, MXenes have attracted widespread attention in recent years [31–36]. For example, in 2019, H. Yan et al. first introduced an MXene ($Ti_3C_2T_x$) in the field of corrosion prevention [30]. Subsequently, research on new organic anticorrosion coatings based on MXenes has grown rapidly. However, actual MXenes may have many structural defects, and the characteristics of easy stacking, good conductivity, and proneness to oxidation are adverse to corrosion protection. Therefore, there is controversy about whether MXenes can be applied for corrosion protection or not. Thus, a few basic questions need to be answered, such as: if it can be applied, what are its advantages and disadvantages? How can it be used appropriately? Is there any special attention essential? Therefore, it is necessary to summarize and analyze the characteristics, performance, and development trends of MXene-based anticorrosion coatings.

$Ti_3C_2T_x$ was the first successfully synthesized 2D material in the MXene class; as a result, it has been the most studied and has garnered constant interest from academia and industry. The MXene $Ti_3C_2T_x$ possesses the highest electrical conductivity (up to 24,000 $S{\cdot}cm^{-1}$) and one of the highest stiffness values, with a Young's modulus of ~334 GPa, among water-dispersible conductive 2D materials [3]. The negative surface charge of MXenes is conducive to their effective dispersion in aqueous and other polar solvents. There are many reviews on $Ti_3C_2T_x$ polymer composites for catalysis, flexible electronics, and energy storage, but most of them are focused on the whole 2D family, not especially on MXenes or $Ti_3C_2T_x$, resulting in insufficient depth and specificity in analysis and discussion. To the best of our knowledge, only one review particularly discussed $Ti_3C_2T_x$/polymer composites for anticorrosion applications has been published, by I. Amin et al. in 2022 [3].

In order to report the current progress, especially after 2022, this mini-review first briefly introduces the structural characteristics and synthesis methods of MXene. Then, focusing on how to improve the dispersibility, compatibility, stability, and anti-agglomeration of MXene to enhance the corrosion resistance of organic composite coatings, four surface-modification methods on $Ti_3C_2T_x$ MXene, namely functional grafting, orientation arrangement, heterostructure nanocomposite design, and stabilization and greening treatment, are discussed in detail. Finally, the current challenges and future development opportunities of MXene-based coatings in the field of anticorrosion are prospected.

The main reviews related to recent advances in the synthesis and application of corrosion-resistant MXene-based polymer composite materials published after 2022 are compared in the present review. The title, focus, and published year of these review papers are summarized in Table 1, which illustrates that the novelty and uniqueness of this review lie in summarizing and proposing four surface-modification strategies (functionalization grafting, orientation regulation, heterostructure hybrid, and stabilization and greening) to achieve high-performance anticorrosive $Ti_3C_2T_x$ MXene-based polymer composite coating.

Table 1. Comparison of the main reviews related to the synthesis and application of the anticorrosive MXene-based polymer composites published after 2022, and the present review.

Title	Focus	Year	Ref.
$Ti_3C_2T_x$ MXene polymer composites for anticorrosion: an overview and perspective	In this brief review, the current progress of $Ti_3C_2T_x$ polymer composites for corrosion protection is summarized, and the possible ways to exploit $Ti_3C_2T_x$ polymer anticorrosive composites more effectively are discussed, and a perspective beyond $Ti_3C_2T_x$ MXene composition for the development of future anticorrosion coatings is provided.	2022	[3]
Emerging layered materials and their applications in the corrosion protection of metals and alloys	Focus on the emerging 2D materials beyond Gr, TMD such as MoS2, MXenes, LDHs, h-BN, and graphitic carbon nitride in the formulation of effective and protective films and coatings is described and discussed.	2022	[5]
Beyond graphene and boron nitride: why MXene can be used in composite for corrosion protection on metals?	The intrinsic surface protection ability and advantages of MXene beyond Gr and BN for application in corrosion protection are analyzed. The recent progress of experimental results and theoretical calculations are summarized to explore the roles of MXene in anticorrosion. The challenges and perspectives of MXene-based anticorrosion coatings are put forward.	2024	[4]
2D materials for marine corrosion protection: A review	Recent advancements in the development and potential to apply for marine corrosion protection of 2D material (including Gr, LDH, BN, MXenes, and MoS_2)-enhanced coatings through improved performance metrics and sustainability are reviewed and emphasized. The challenges and further research to overcome application hurdles are highlighted.	2024	[1]
Two-dimensional nanomaterials reinforced organic coatings for marine corrosion protection: state of the art, challenges, and future prospectives	The preparation strategies, properties of 2D NMs, and diverse protection models based on composite coatings for marine corrosion protection, including a physical barrier, self-healing, and cathodic protection, are illustrated. Furthermore, computational simulations and critical factors on the anticorrosion of composite coatings are clarified. Finally, the remaining challenges and prospects for marine corrosion protection based on 2D NMs reinforced organic coatings are highlighted.	2024	[2]

Table 1. *Cont.*

Title	Focus	Year	Ref.
Research progress on MXene-based organic composite anticorrosion coatings	The current research status, mainly before 2023, and challenges and prospects of anticorrosion coatings based on MXene $Ti_3C_2T_x$ MXene-based epoxy resin organic composite are reviewed and discussed in Chinese.	2024	[6]
Research on two-dimensional layered materials for metal corrosion protection: Advances and challenges	According to material attributes, 2D layered materials (2DLM) are sorted into natural, conventional and emerging three categories. Their recent progress in anticorrosion applications is reviewed with emphasis. The advantages and limitations of various 2DLM-based anticorrosion coatings are compared, and the challenges for future preparation and application in the field of corrosion protection are outlined.	2024	[7]
Toward self-healing two-dimensional MXene coatings for corrosion protection on metals: design strategies and mechanisms	A comprehensive review of the current findings on the anticorrosive and self-healing of MXene-based coating on metals is presented. The review focuses on the design, properties, and mechanisms of four types of self-healing MXene coatings (synergy with Ce^{3+}, LDH, inhibitors, and self-healing polymer).	2025	[8]
Surface-modification strategy towards highly anticorrosive $Ti_3C_2T_x$ MXene-based polymer composite coating: a mini-review	The latest progress in the anticorrosion performance of $Ti_3C_2T_x$/EP and $Ti_3C_2T_x$/WPU polymer composite coatings is reviewed. Four surface-modification strategies (functionalization grafting, orientation arrangement, heterostructure hybrid, and stabilization and greening) to improve the dispersion, compatibility, stability, and anti-aggregation properties of MXene are summarized and emphasized. The challenges and opportunities of anticorrosive MXene-based organic composite coatings are prospected.	2025	This work

2. Brief Introduction of MXene and Overview of $Ti_3C_2T_x$'s Anticorrosion

2.1. Structure and Main Preparation Methods of MXene

MXene is a new category of 2D lamellar material discovered in 2011, derived from pre-transition metal carbides or nitrides of MAX phase [37–40], which can be written as $M_{n+1}AX_n$, where M denotes the pre-transition metal containing Sc, Ti, V, Cr, Zr, Nb, Mo, Hf, Ta, etc., A represents the third/fourth main group elements, such as Al, Ga, Ln, Tl, Si, Ge, Sn, Pb, etc., and X refers to C or N; n = 1, 2, or 3. MAX phase consists of layered ternary carbides and nitrides, with over 70 types reported so far. In the MAX crystal structure, the M layer is densely packed, with X atoms filling its octahedral position and A atoms alternating with $M_{n+1}X_n$ layers to form a layered structure. Removing the A element from MAX results in 2D MXene, commonly abbreviated as $M_{n+1}X_nT_x$, where T_x indicates surface-functional groups such as =O, –OH, and –F left after etching, and n represents the number of atomic layers of the elements occurring in a sandwich-like layered morphology. In the periodic table, the elements used to build MXenes represented by different colors are shown in Figure 1a, and various MXenes (M_2XT_x, $M_3X_2T_x$, $M_4X_3T_x$, and $M_5X_4T_x$) correspondingly obtained from MAX phases (M_2AX, M_3AX_2, M_4AX_3, and M_5AX_4) with the general formula $M_{n+1}AX_n$ (n = 1–4) are demonstrated in Figure 1b [39].

Compared with other 2D materials, such as Gr, GO, BN, and MoS_2, MXenes have the following main unique advantages [4]: (1) The intact structure gives it a better conductivity (1.1–2.4 MS/m) than GO and BN (ranges from 0.032 to 0.119 MS/m for GO). (2) The presence of various surface-functional groups, such as =O, –OH, and –F, gives it an intrinsic negative charge characteristic and hydrophilicity, which is beneficial to combine with the positively charged NMs, such as LDH, to prevent self-agglomeration. In contrast to GO with –COOH, the –OH on MXene causes it to be hydrophilic, being conducive to dispersing well in water-based solvents or polymers. (3) The formation of TiO_2 on Ti-based MXene ($Ti_3C_2T_x$, Ti_2CT_x, etc.) can further enhance the corrosion resistance of composite coatings because the formed nanoparticles can serve as filler in the coating. (4) Distinct from Gr or BN, MXene is a large family. Owing to the tunable compositions, multiple types of MXenes offer various kinds of functions and practical applications. Ti-, Cr-, V-, and Mo-based MXenes, like $Ti_3C_2T_x$, $Cr_3C_2T_x$, V_2CT_x, and Mo_2CT_x, and others, as well as the amorphous MXene and emerging high-entropy MXene (HE-MXene), provide the possibility for multi-functional coatings and present promising potential for various applications.

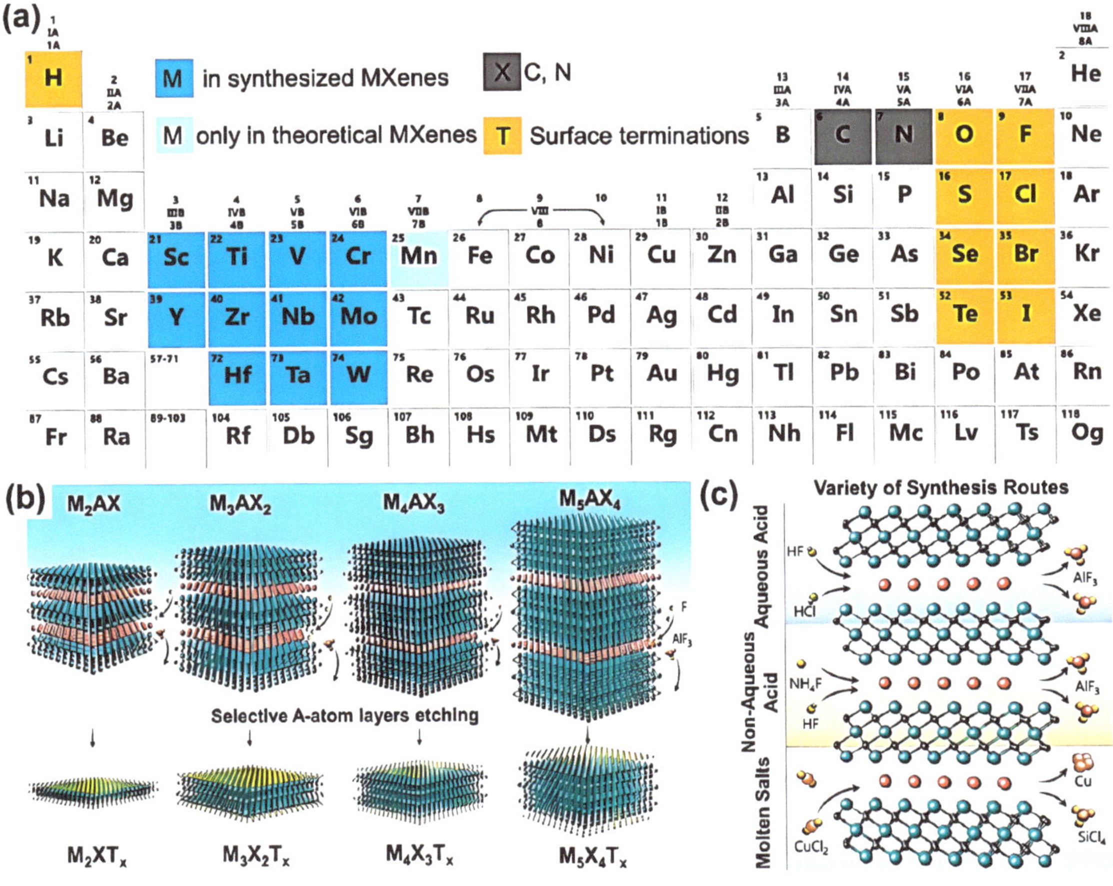

Figure 1. (**a**) Periodic table with color-coded elements that are adopted to construct MXenes. (**b**) Schematic representation of various MXenes (M2XTx, M3X2Tx, M4X3Tx, and M5X4Tx) obtained from the corresponding MAX phases (M2AX, M3AX2, M4AX3, and M5AX4) with the general formula Mn+1AXn (n = 1–4), where X is C and/or N; Tx is –H, =O, –F, and –OH in MXenes. (**c**) Schematic illustration of the three most employed synthesis routes for MXenes via aqueous acids (mixture of HCl/HF), nonaqueous acids (NH4F/HF), and molten salts [39]. Reprinted with permission from [39], copyright 2022, American Chemical Society.

As for the preparation of MXenes, there are generally two strategies [41–46]: "bottom-up" or "top-down". The former is based on templates and synthesized through chemical vapor deposition (CVD), while the latter is through peeling off MAX bulk crystals into several layers or single-layer sheets. At present, the more commonly used approach is "top-down", which includes two main steps: etching and exfoliation. In the etching process, a strong corrosive agent is usually employed to selectively remove the A layer in the MAX material to obtain an "accordion" shaped material. Then, intercalation agents or mechanical vibrations such as ultrasound and ball milling are adopted for exfoliation (stripping) treatment, therefore achieving multi-, few-, or single-layer MXene. The three most employed methods to produce MXenes are though aqueous acid (HCl/HF mixture), nonaqueous acid (NH_4F/HF), or molten salt. The corresponding synthetic routes are illustrated in Figure 1c [39].

Generally speaking, MXenes can be fabricated through diverse etching methods, but the MXenes obtained by various preparation approaches will possess different characteristics, such as different structures, surface chemical states, and physical or chemical properties. Therefore, it is necessary to select appropriate preparation and etching methods according to the special target application scenario and performance requirements of MXenes. For example, if high conductivity, large sheet size, wide interlayer spacing, and high mechanical performance are required, mild etching conditions (typically LiF + HCl) are recommended, but if MXenes with small sheet size and rich defects are wanted, HF etching may be the better choice.

Similar to Gr, owing to the large surface area, good chemical and mechanical stability, excellent conductivity, and high flexibility, extensive work and exploration on the application of MXene nanosheets have been carried out in various fields such as energy storage [47], electromagnetic shielding [48,49], adsorption, flexible devices, catalysis, superconductivity [45], and corrosion prevention.

So far, over 50 varieties of MXene have been reported to have been synthetized, and more than 100 have been explored theoretically [32]. Since the synthesis and properties of MXene and $Ti_3C_2T_x$ have been discussed in numerous reviews elsewhere [50–55], they are not within the scope of this review.

2.2. Overview of $Ti_3C_2T_x$ and Its Anticorrosion

As the first 2D MXene material obtained from the Ti_3AlC_2 MAX phase [37], $Ti_3C_2T_x$ is the most studied one and has gained the greatest attention in the MXene class since its first synthesis in 2011. $Ti_3C_2T_x$ is composed of three layers of Ti atoms and two layers of C atoms placed in the layers of Ti–C–Ti–C–Ti. The Tx component in the formula denotes the surface terminations (typically –OH, =O, –F, –Cl) existing on the outer planes of Ti. The negatively charged surface of $Ti_3C_2T_x$ is beneficial to its excellent dispersion in aqueous and other polar solvents. The lamellar structure of $Ti_3C_2T_x$, combined with its good solubility across a wide range of solvents, tunable surface functionality, excellent interface interaction, and stability with other organic/polymeric materials, makes it a potential material for anticorrosive coatings. In addition, the "top-down" synthesis route through wet chemical selective etching from its precursor, the Ti_3AlC_2 MAX phase, causes it to have quite good scalability for industrial production. Due to these superior features and their feasibility for solution process, scalability, and surface functionality, various applications of $Ti_3C_2T_x$, such as in energy storage [56–59], flexible electronics and biosensors [60,61], adsorption [61–66], electromagnetic shielding [67], and anticorrosion [68], have been reported. Therefore, in the review, $Ti_3C_2T_x$ is taken as a typical representative of the MXene family, and its corrosion-resistant properties for organic composite coatings are discussed.

Four groups of topics, namely "MXene", "$Ti_3C_2T_x$", "$Ti_3C_2T_x$ and Corrosion", and "$Ti_3C_2T_x$ and Anticorrosion/Anticorrosive" were selected, and the total and annual number of published papers in the past decade (from 2015 to November 2024, the date of this article) were queried through Web of Science. The results are shown in Figure 2. The annual publication of papers on "MXene", "$Ti_3C_2T_x$", "$Ti_3C_2T_x$", and "$Ti_3C_2T_x$ and Corrosion" is shown in Figures 2a and 2b, respectively. They show that the number of papers on these three topics has grown year by year, indicating that research on MXene and its anticorrosion is receiving increased attention. By November 2024, a total of approximately 417 papers have been published on the topic of "$Ti_3C_2T_x$ and Corrosion", with 95 and 110 papers published in 2023 and 2024, respectively. There are 79, 27, and 20 papers, respectively, on the topic of "$Ti_3C_2T_x$ and Anticorrosion/Anticorrosion", indicating that research on MXene corrosion and anticorrosion is in the early stage of rapid growth and it is, therefore, worth paying attention.

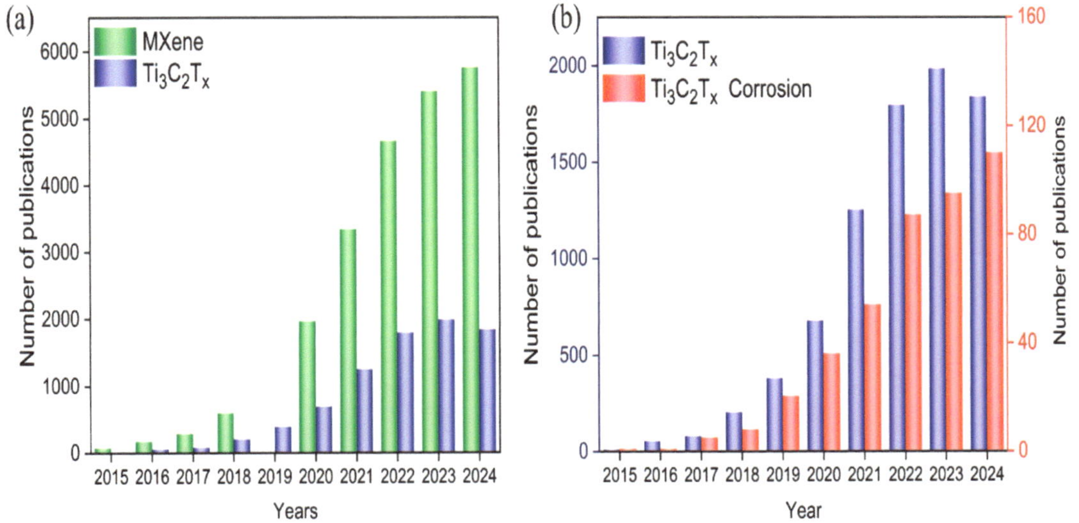

Figure 2. Annual publication of papers on MXene, $Ti_3C_2T_x$, or "$Ti_3C_2T_x$ and Corrosion" as a topic over the past decade: (**a**) MXene and $Ti_3C_2T_x$, (**b**) $Ti_3C_2T_x$ and "$Ti_3C_2T_x$ and Corrosion". Source: Web of Science Search index: [topic = MXene or $Ti_3C_2T_x$ or "$Ti_3C_2T_x$ and Corrosion"].

3. Recent Progress of $Ti_3C_2T_x$-Based Polymer Composite Anticorrosion Coatings

To exploit the corrosion resistance of MXene nanosheets (NSs), the most popular way to prepare the anticorrosion coating is the physical mixing (such as magnetic stirring) of $Ti_3C_2T_x$ with waterborne epoxy (WEP) or waterborne polyurethane (WPU). In order to prevent agglomeration, various surface-treatment methods have been adopted to improve the dispersibility, compatibility, stability, and orientation properties of MXene. Thus, the discussion in the section is organized as follows: (i) pristine MXenes/polymer matrix composites, (ii) surface functionalization grafting methods, (iii) orientation arrangement, (iv) heterostructure and hybrid composite, and (v) some comparison and summary of the corrosion performance.

3.1. Pristine $Ti_3C_2T_x$

The first work about the anticorrosive properties of pristine $Ti_3C_2T_x$ was made in 2019 by Yan et al. [30], in which few-layer $Ti_3C_2T_x$ NSs were incorporated into EP with an amine curing agent. The effect of $Ti_3C_2T_x$ content on the morphology, internal structure, and corrosion resistance of the $Ti_3C_2T_x$/EP composite coating was studied. It is reported that the addition of MXene can prevent the micropores in the cured EP coating from becoming coated on carbon steel Q345 (Figure 3a–d). The Tafel plots (Figure 3e) of four coating samples after 96 h immersion in 3.5% NaCl solution show that the 1.0 wt.% $Ti_3C_2T_x$ containing composite coating has the highest corrosion potential (E_{corr}) and the lowest corrosion current density (I_{corr}), which is 3.39×10^{-8} A·cm^{-2}, in contrast to that of the pure EP counterpart (1.00×10^{-6} A·cm^{-2}). As illustrated in Figure 3f, the results of the salt spray test after 15 days indicate that 1.0 wt.% $Ti_3C_2T_x$ offers the strongest protection. The improved anticorrosion performance is attributed to the barrier effect of MXene flakes as a thin film to inhibit the diffusion of electrolytes, thus providing corrosion protection to the substrate. However, excessive MXenes (such as 2.0 wt.%) will agglomerate in the EP, reducing the anticorrosion performance of composite coatings.

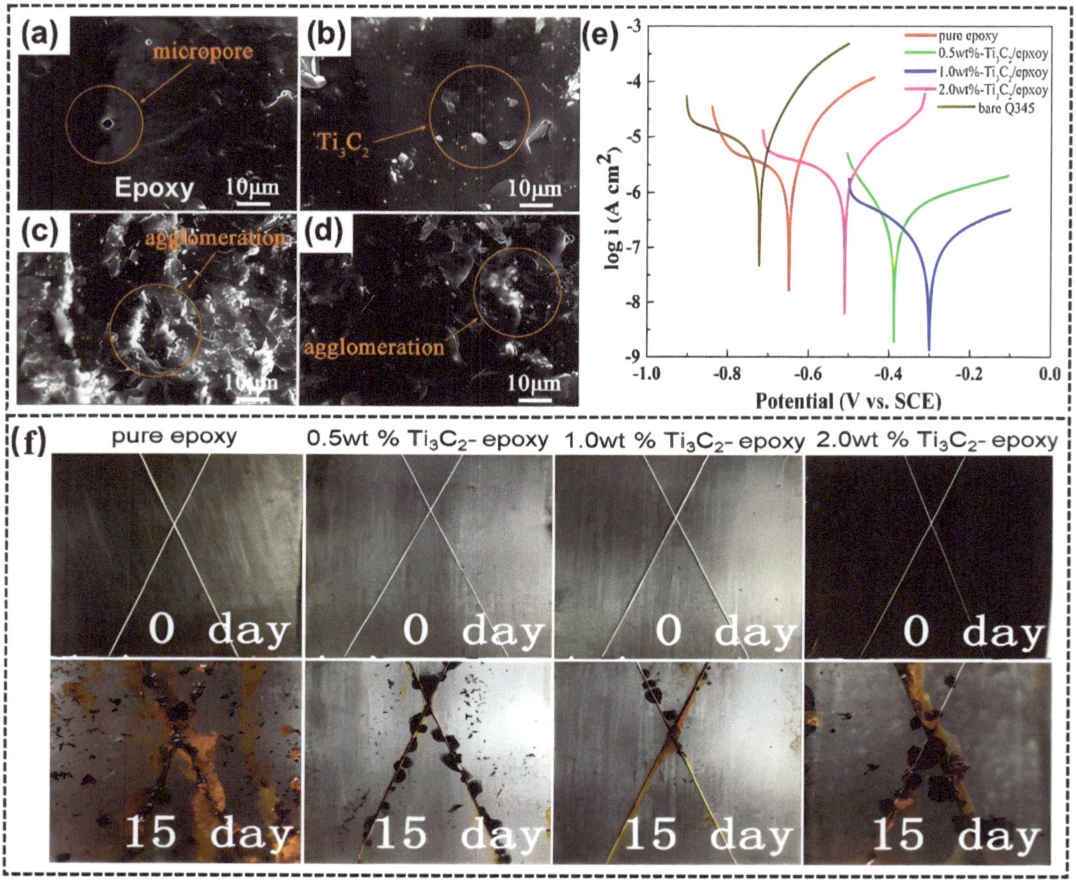

Figure 3. SEM images of surface morphology of MXene/epoxy coating with different $Ti_3C_2T_x$ content: (**a**) bare EP, (**b**) 0.5 wt.% $Ti_3C_2T_x$/EP, (**c**) 1.0 wt.% $Ti_3C_2T_x$/EP, (**d**) 2.0 wt.% $Ti_3C_2T_x$/EP; (**e**) the polarization curves of uncoated and coated $Ti_3C_2T_x$/EP samples after 96 h of immersion in 3.5% NaCl, (**f**) photographs of four kinds of coating samples before and after the 15-day salt spray test [30]. Reprinted with permission from [30], copyright 2019, Elsevier.

In addition to EP, silane has also been used as a polymer matrix. Polymer composites of $Ti_3C_2T_x$ and sulfhydryl silane were prepared on copper substrate by Cao et al. in 2022 [69]. The optimal loading amount of $Ti_3C_2T_x$ is 0.25 Mg/mL. Compared with bare silane coating, the impedance modulus at the lowest frequency ($|Z|_{0.01Hz}$) of the silane composite coating is enhanced by 1.8 orders of magnitude, and the I_{corr} decreases by about 5 times [69].

Besides organic coating, MXene can be adopted to reinforce the corrosion protection of metal matrix composite and metallic coating. The $Ti_3C_2T_x$/Cu [70] metallic matrix composite and the $Ti_3C_2T_x$–Ce [71] and Ni–W–$Ti_3C_2T_x$ [72] composite coating with enhanced corrosion resistance were reported.

3.2. Surface Functionalization of $Ti_3C_2T_x$ MXene

Although MXene has the potential to enlarge the anticorrosion performance of coatings, its enhancement efficiency is lower than expected. The reason for this is that, similar to other 2D materials, MXene also exhibits a strong irreversible stacking and aggregation trend due to interlayer van der Waals forces (vdW) and hydrogen-bonding interactions. Aggregated MXene NSs cannot provide a sufficient protective network and, in the worst case, may affect the continuity of the coating and exacerbate corrosion. In addition, the weak interaction and compatibility between MXene and the coating matrix limit the loading capacity. Therefore, good dispersibility, high compatibility, and moderate content are the primary requirements for preparing high-performance MXene-based composite coatings. Dispersion and compatibility are key to breaking off the MXene conductive network and achieving its maximum blocking effect. Therefore, surface modification of MXene is required to improve its dispersion and compatibility.

Surfactant modification, covalent functionalization, and tuning of surface-functional groups are three effective surface-treatment methods for a nanofiller to disperse homogenously into a polymer matrix [27,29]. The surface terminations, i.e., surface-functional groups (typically –OH, =O, –F, –Cl) existing on the surface of MXene render MXenes hydrophilic. Therefore, it is relatively easy to obtain stable MXene-containing suspensions in water or polar solvents and composites of MXene and polymer due to these groups allowing the reactions and interactions with hydrophilic polymers such as polyacrylic acid (PAA), polydiallyldimethylammonium chloride (PDDA), polyvinyl alcohol (PVA), and many biopolymers. The first MXene polymer nanocomposites (NCs) were reported in 2014 when $Ti_3C_2T_x$ was introduced to hydrophilic PDDA and PVA polymers [15].

Since this first report, there has been increased interest in the research of MXene polymer NCs. In 2015, Chen et al. prepared one kind of polymer brush structure by grafting poly(2-(dimethylamino) ethyl methacrylate) (PDMAEMA) on vanadium carbide (V_2CT_x) via the self-initiated photo-grafting and photo-polymerization of DMAEMA [16]. This NC presented dual CO_2 and temperature response, showing promise as a stimulus-responsive smart hybrid material that could be applied in biological sensing. In 2017, four functional groups, $–NH_2$, –COOR, $–C_6H_6$, and $–C_{12}H_{26}$, were grafted onto the surface of $Ti_3C_2T_x$ by Hao et al. [17] through the introduction of either (3-aminopropyl)-triethoxy-silane (APTES), (γ-methacryl-oxypropyl)-trimethoxysilane (MAPTES), (methyl aniline)-triethoxy-silane (AMTES), or (dodecyl)-triethoxy-silane (DTES). In 2019, Si et al. prepared polystyrene (PS) NCs with colloidal $Ti_3C_2T_x$, respectively, modified by three types of cationic surfactants [18], namely decyltrimethylammonium bromide (DTAB), octadecyl trimethylammonium bromide (OTAB), and didodecyldimethylammonium bromide (DDAB). It is observed that the basal spacing of $Ti_3C_2T_x$ increases with the length of the cationic surfactant.

To date, most work has focused on the use of $Ti_3C_2T_x$. In addition, multiple polymer hosts have been reported, including hydrophilic polymers such as PVA, PAA, silicones, and epoxies.

Functionalizing MXene through grafting modification with coupling agents is one of the most convenient methods to prevent its aggregation and stabilize within the polymer matrix. Certain functional groups grafting onto the active oxidation sites on the surface of MXene will form covalent bonds. Many functional groups can be used to achieve chemical covalent modification, such as organic small molecules, organic polymers, inorganic nano oxides, etc. The most commonly used organic small molecules are various coupling agents, such as silane and titanate.

3.2.1. Functionalization Through Silane-Based Coupling Agents

Owing to the rich reactive surface groups, silane coupling agents have been widely adopted as grafting agents to improve interfacial binding. Surface silanization reaction is beneficial to inhibit its structural degradation, thus effectively stabilizing the MXene and enhancing the surface properties with adjustable hydrophilicity [73–83].

Aminopropyltriethoxysilane (APTES) is the most used silane coupling agent for surface functionalization. The primary amino functional groups of APTES provide a few functionalization feasibilities from bio-conjugation to nanoparticle impregnation. Ji et al. reported that APTES-functionalized $Ti_3C_2T_x$ demonstrates improved anti-oxidation stability and regulable hydrophilicity [73].

The first work to investigate the corrosion resistance of amino-modified $Ti_3C_2T_x$ MXene was conducted by Yan et al. in 2020 [74]. The synthesis process of $Ti_3C_2T_x$ and its surface functionalization through APTES is illustrated in Figure 4a. The photographs of the 0.5 wt.% l-M (pristine $Ti_3C_2T_x$) and f-M (APTES-modified $Ti_3C_2T_x$) samples dispersed in EP after 0–30 days are displayed in Figure 4b, where the f-M film remains stable and evenly distributed in the EP slurry for 30 days. After soaking in a 3.5 wt.% NaCl solution for 4 weeks, the f-M 0.5 wt.% composite coating on Al alloy has the highest $|Z|_{0.01Hz}$ value of (1.02×10^7 $\Omega \cdot cm^2$), which was two orders of magnitude higher than that of the bare EP coating. The significant improvement in corrosion resistance is mainly attributed to two points: (a) the good dispersibility and compatibility of f-M flakes in EP, and (b) the interaction between the amino groups of APTES and the hydroxyl groups on the metal surface, which improves the adhesion between the coating and the metal substrate. The work indicates the importance of the ligand functionalization of $Ti_3C_2T_x$ in enhancing its dispersibility, keeping its chemical stability, and slowing down degradation owing to oxidation in the polymer matrix [74–77].

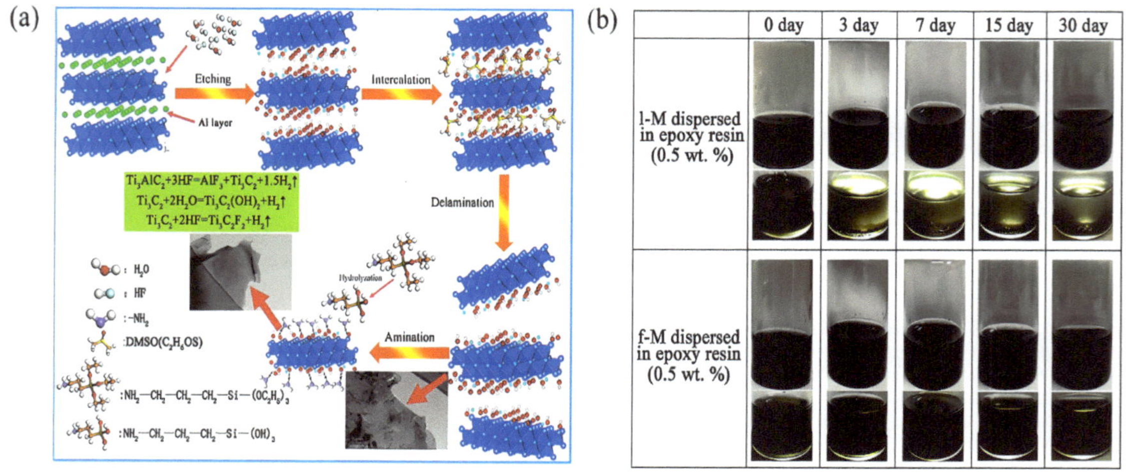

Figure 4. Preparation and stability of amino-functionalized $Ti_3C_2T_x$ [74]. (**a**) Schematic illustration of synthesis and surface APTES-functionalized $Ti_3C_2T_x$, (**b**) settlement of 0.5 wt.% l-M and f-M dispersed in EP after 0–30 days of static placement. Reprinted with permission from [74], copyright 2020 Elsevier.

Similarly, [3-(2-aminoethyl) aminopropyl]-trimethoxysilane (AEAPTES) was employed to surface-functionalize $Ti_3C_2T_x$ by Zhang et al. in 2021 [75]. The AEAPTES-modified $Ti_3C_2T_x$ (named Ti_3C_2@Si) was then added into WPU with $Ti_3C_2T_x$ ratios of 0.05, 0.1, and 0.15 wt.%. The introduction of amino functional groups on $Ti_3C_2T_x$ facilitates intercalation with isocyanate groups in WPU, causing a strong and compact structure and stable dispersions of Ti_3C_2@Si into WPU. Therefore, the formed network of an effective barrier creates complex diffusion paths, slowing the diffusion rates of a corrodent [75]. Importantly, the functionalized Ti_3C_2@Si enhanced the hydrophobicity of WPU, which decreased the absorption of water and increased the corrosion performance of Ti_3C_2/WPU composite coatings. These findings are similar to those of Gr-based materials, where covalently functionalized Gr showed better corrosion protection than pristine Gr.

In addition to aminosilane, epoxy silane coupling agents were also attempted. In 2022, Li et al. prepared epoxy-functionalized multilayer $Ti_3C_2T_x$ nanosheets (GPS-$Ti_3C_2T_x$) by grafting (3-glycidoxypropyl) trimethoxysilane (GPTMS) [78]. They found that the scratch resistance, adhesion, and anticorrosion of the composite coating were all improved, but the flexibility was reduced. The optimal content of GPS-$Ti_3C_2T_x$ was 0.5 wt.% [78], and the impedance of the corresponding organic coating was nearly three and one orders of magnitude higher than that of bare EP and unmodified $Ti_3C_2T_x$ coatings, respectively. This article suggests that the performance of GPS-$Ti_3C_2T_x$ functionalized with epoxy groups should be better than that of APTES-$Ti_3C_2T_x$ functionalized with amino groups. However, Pourhashem et al. [79] found that amino-functionalized APTES-GO has better dispersion and anticorrosion than epoxy-functionalized GPS-GO, indicating that the modification effects of the same type of silane on MXene and Gr are different.

In 2022, Wang et al. realized silane-grafted $Ti_3C_2T_x$ by APTES [80]. Due to the chemical reaction between the hydroxyl groups on the MXene and silane molecules, some black dots were observed on the sheets, thus decreasing the transparency of the modified MXene NSs coating. After a 120-day immersion test, the coating resistance was $1.4755 \times 10^6\ \Omega\cdot cm^2$ [80].

The silane@fresh MXene/WEP coating obtained by Zhou et al. in 2020 demonstrated a high $|Z|_{0.01Hz}$ value of $5.5 \times 10^8\ \Omega\cdot cm^2$ after 40 days' immersion test [81]. Meanwhile, it kept a larger coating resistance than the oxidized MXene/EP coating after a 60-day immersion test [81].

In 2023, Yin et al. fabricated a novel superhydrophobic $Ti_3C_2T_x$/epoxy/cerium conversion composite (STECC) coating with a sandwich structure (Figure 5a) [82]. The water-contact angle (WCA) of the composite coating was $155.8 \pm 2°$ after the fluorosilane modification (Figure 5b). The $|Z|_{0.01Hz}$ of the STECC coating was up to $7.49 \times 10^6\ \Omega\cdot cm^2$, which was larger by four orders of magnitude than the bare Mg alloy (Figure 5c). The corrosion-protection mechanism of the STECC coating is shown in Figure 5d. The superhydrophobic $Ti_3C_2T_x$ coating on the top avoids the penetration of corrosive mediums and prevents their direct contact with the substrate, solving the problem that the conductive $Ti_3C_2T_x$ tends to bring out galvanic coupling corrosion. The middle EP coating acts as a corrosion mitigation of the substrate. Furthermore, the cerium conversion coating at the bottom not only increases adhesion between the intermediate layer and the Mg alloy but also forms a passivation layer to protect the substrate. This work provides an effective new strategy for MXene-based composite anticorrosion coatings.

Additionally, MXene NSs-enhanced vinyl ester resin (VER) coatings modified by 3-methacryloxypropyl-trimethoxysilane (γ-MPS@MXene) were developed by Chen et al. in 2023 [83]. The composite coating achieved a remarkable anticorrosion. The 0.02 wt.% γ-MPS@MXene modified coating exhibited significantly lower I_{corr} than the unmodified counterpart coatings. Furthermore, it shows that the 0.08 wt.% γ-MPS@MXene is the most effective at hindering corrosive media infiltration and safeguarding the carbon steel substrate [83].

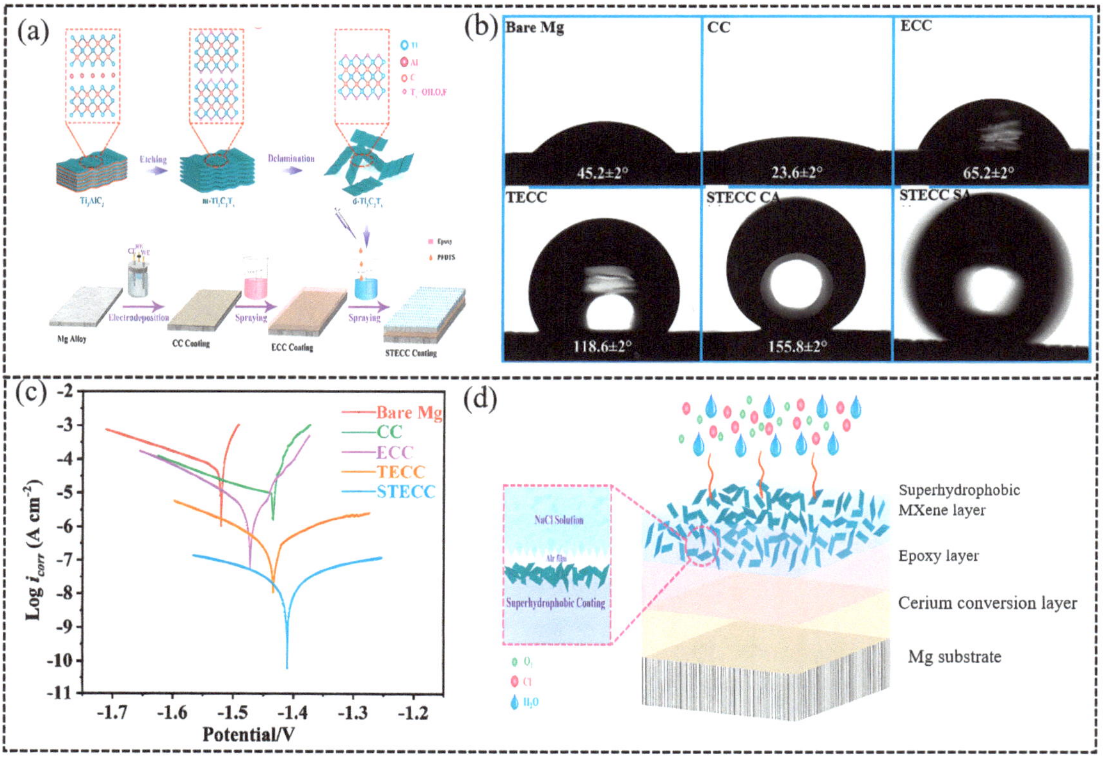

Figure 5. Preparation and performance of functionalized superhydrophobic $Ti_3C_2T_x$/epoxy/cerium conversion (STECC) composite coating [82]. (**a**) Fabrication process of STECC composite coating on the surface of Mg alloy; (**b**) The WCA of the bare Mg alloy, CC, ECC, TECC, STECC coating, and SA of the STECC coating; (**c**) The polarization curves of as-fabricated specimens after 1 day of immersion in 3.5 wt.% NaCl solution; (**d**) Schematic diagram of the STECC coating corrosion-protection mechanism. Reprinted with permission from [82], copyright 2023, John Wiley and Sons.

3.2.2. Functionalization Through Polymer Molecule

The polymeric layer can be used for surface modification to improve the stability of the MXene [84–91]. Conductive polymers (CPs) such as polythiophene (PT), polypyrrole (PPy), and polyaniline (PANI) exhibit anticorrosion properties. Among them, PANI has received widespread attention for anticorrosion owing to its ease of synthesis, good thermal stability, and reversible acid/base doping/dedoping. As early as 2016, Jafari et al. used PANI-modified Gr as a corrosion-protection layer for Cu and found that the obtained coating had low electrical conductivity and excellent barrier properties, which could prevent water and oxygen molecules from reaching the coating/metal interface [84]. Similar to Gr, the polymerization treatment of MXene is also expected to enhance its corrosion resistance.

In 2021, Cai et al. studied the interfacial interaction between $Ti_3C_2T_x$ and Fe and found that the high conductivity of $Ti_3C_2T_x$ promotes an oxygen-reduction reaction, accelerating the corrosion of low-carbon steel Q345 [85]. Therefore, it is necessary to perform polymerization treatment on MXene to reduce its conductivity. In the same year, they first applied $Ti_3C_2T_x$@PANI composite materials (TPCs) for the corrosion protection of low-carbon steel [86]. As shown in Figure 6a,b, the multilayer TPCs (MTPCs) with epoxy at the bottom and top and TPCs in the middle were prepared by layer-by-layer (LBL) deposition. By adjusting the PANI content, the conductivity of the composite coating was reduced from 1307 S·cm^{-1} (pristine $Ti_3C_2T_x$) to 0.5 S·cm^{-1} (TPCs). The photographs of the samples before and after the 30-day salt spray test are displayed in Figure 6c,d, which indicates that the MTPCs12 coating ($Ti_3C_2T_x$ and aniline mass ratio of 1:2) has the best anticorrosion performance. The corresponding anticorrosion mechanism is illustrated in Figure 6e.

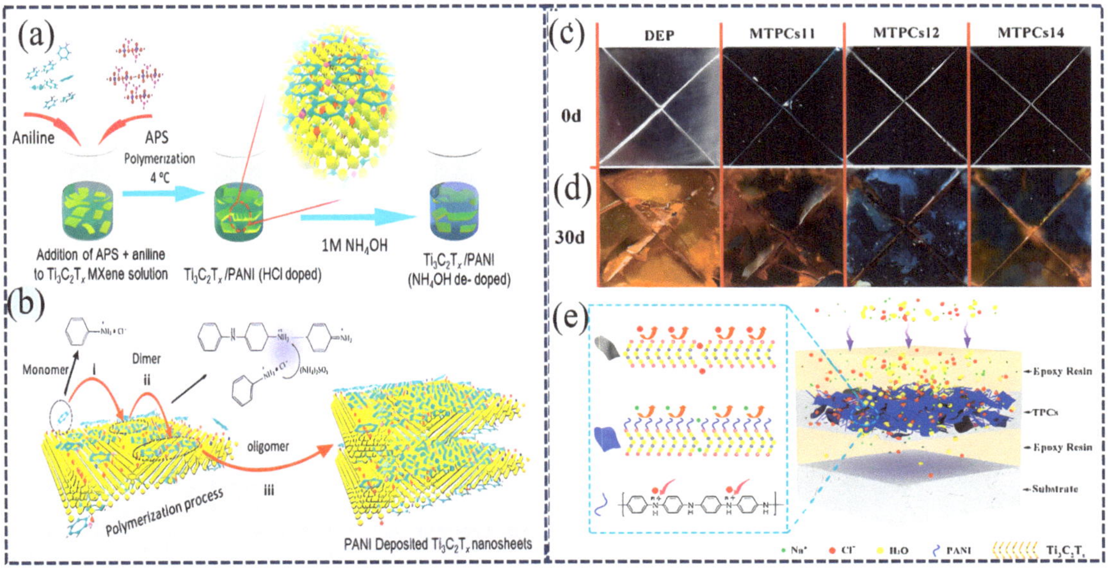

Figure 6. Schematic illustration of the preparation of Ti_3C_2/PANI composites [86]. (**a**) Preparation of Ti_3C_2 NSs. (**b**) Synthesis of Ti_3C_2/PANI composites (TPCs) with the mechanism of oxidative polymerization of aniline on Ti_3C_2. (**c**,**d**) Photograph of the samples before and after the salt spray test for 30 days. (**e**) Diagram explaining the corrosion-resistant mechanism of multilayer thin polymer coatings (MTPCs). Reprinted with permission from [86], copyright 2021, Elsevier.

In 2022, Li et al. synthesized a Ti_3C_2@PANI composite through in situ intercalation polymerization [87]. It was found that, owing to the synergistic effect of Ti_3C_2 NSs' barrier effect and PANI's passivation effect, the $|Z|_{0.01Hz}$ and charge-transfer resistance of 0.3 wt.% Ti_3C_2@PANI/EP coating in 3.5 wt.% NaCl is increased by 1–2 orders of magnitude compared to the bare EP and Ti_3C_2/EP. Specifically, the author proposes that Ti_3C_2 treated with LiF and HCl accelerates the failure of EP coatings due to the insertion of Li^+, which results in higher conductivity than the MXene material obtained by HF etching [87].

In 2022, Kaewsaneha et al. explored the corrosion resistance of Zn–EP coatings boosted by $Ti_3C_2T_x$@PANI NSs hybrid [88]. It shows that, compared with standard EP and Zn–EP coatings, the 2.0 wt.% $Ti_3C_2T_x$@PANI significantly decreased corrosion products, water absorption, and iron leakage. The superior anticorrosion capabilities are ascribed to the fact that the PANI/@$Ti_3C_2T_x$ hybrid provided a more effective barrier against corrosion and enhanced the electrical connectivity for reliable cathodic protection.

In 2023, the PANI@Ti_3C_2 MXene/polyester-epoxy coating fabricated by Nazarlou et al. [89] exhibited a lower I_{corr} (9.15×10^{-7} A·cm^{-2}) than that without PANI-modified coating (2.05×10^{-5} A·cm^{-2}), and the $|Z|_{0.01Hz}$ value increased from ~4×10^6 Ω·cm^2 to ~8.5×10^7 Ω·cm^2 [89].

The aforementioned works indicate that the Ti_3C_2@PANI composite with lower conductivity can achieve efficient corrosion protection in the EP coating on Q235 steel.

In addition to PANI, chitosan (CS) and polydimethylsiloxane (PDMS) have also been tried [90,91]. In 2022, He et al. successfully grafted MXene with phosphoric acid-modified chitosan (P-CS) due to the strong bond (Ti–O–P) formed through the chemical reaction between the hydroxyl groups of MXene and P-CS [90]. The functionalized $Ti_3C_2T_x$ MXene that was used as a composite nanofiller for WEP not only improved the dispersion state and compatibility of MXene in WEP but also effectively inhibited galvanic corrosion. After a 50-day immersion test, the $|Z|_{0.01Hz}$ of CS@MXene/EP coating was 4.73×10^7 Ω·cm^2, 4 times larger than that of the MXene/EP coating (~10^7 Ω·cm^2) [90].

In 2023, Zhang et al. explored the PDMS-modified $Ti_3C_2T_x$ MXene to improve the dispersion and stability [91]. Through a silylation reaction, the hydroxyl-terminated PDMS was covalently grafted on the $Ti_3C_2T_x$ NSs to form the Ti–O–Si bond. The agglomeration of $Ti_3C_2T_x$ NSs was prevented by the spatial site resistance effect. The I_{corr} reduces from 3.93×10^{-7} (MXene/PU) to 1.49×10^{-9} A·cm^{-2} (PDMS@MXene/PU), while the coating resistance increases from 3.02×10^4 to 1.06×10^7 Ω·cm^2.

3.2.3. Functionalization Through Amine-Based Molecule

With richly hydrophilic groups, polydopamine (PDA) and p-phenylenediamine (PPD) were used to modify the $Ti_3C_2T_x$ MXene [92–97]. Chemical bonds formed between DA and MXene will cause the self-polymerization of DA, which leads to DAs becoming a benzoquinone structure and the MXene NSs being wrapped by PDAs. The PDA molecule in the outer layer of MXene can avoid the further oxidation of MXenes. The interaction between the existing –OH of MXene NSs and the benzoquinone structure of DAs improved the stability of MXene [94]. The PDA@MXene/EP coating kept a higher $|Z|_{0.01Hz}$ value at ~10^8 Ω·cm^2 than the MXene/EP coating after 30 days' immersion test [94]. It was found that, compared with the pristine MXene, the DA modification did not damage the 2D structure of MXenes, and the modified MXene nanosheet became thicker [95]. After a 25-day immersion test, the DA-modified MXene/EP coating exhibited a high coating resistance of 6.78×10^7 Ω·cm^2, and its $|Z|_{0.01Hz}$ value (7.94×10^7 Ω·cm^2) was higher than that of unmodified MXene/EP coating (2.75×10^5 Ω·cm^2) after a 30-day immersion test [95]. The I_{corr} value of the modified MXene coating (6.452×10^{-8} A·cm^{-2}) was lower than that of the unmodified MXene coating (7.004×10^{-7} A·cm^{-2}). In another work, it shows that after the modification of PDA, the coating resistance and charge-transfer resistance increased from 4250 to 6857 Ω·cm^2 and from 4.674×10^4 to 2.383×10^5 Ω·cm^2, respectively [93].

In 2023, Cai et al. reported that PDA-modified $Ti_3C_2T_x$ significantly enhanced the protective properties of zinc-rich epoxy (ZRE) coating, which contributes to the superior physical barrier and electrical contact effect of ZRE-$Ti_3C_2T_x$@PDA [96]. After immersion, the corrosion of Zn particles in ZRE resulted in electrical disconnection between Zn particles. By bringing into $Ti_3C_2T_x$@PDA, the electrical contact between Zn particles and substrate was significantly improved, thereby increasing the cathodic protection of ZRE under long-term immersion. Therefore, due to the physical barrier and conductive properties of $Ti_3C_2T_x$, effective cathodic protection can be achieved for a longer period of time. The schematic anticorrosion mechanisms of ZRE-$Ti_3C_2T_x$ @PDA/WEP coating and the polymerization mechanisms of DAs and binding on $Ti_3C_2T_x$ NSs are displayed in Figure 7a, Figure 7b, and Figure 7c, respectively.

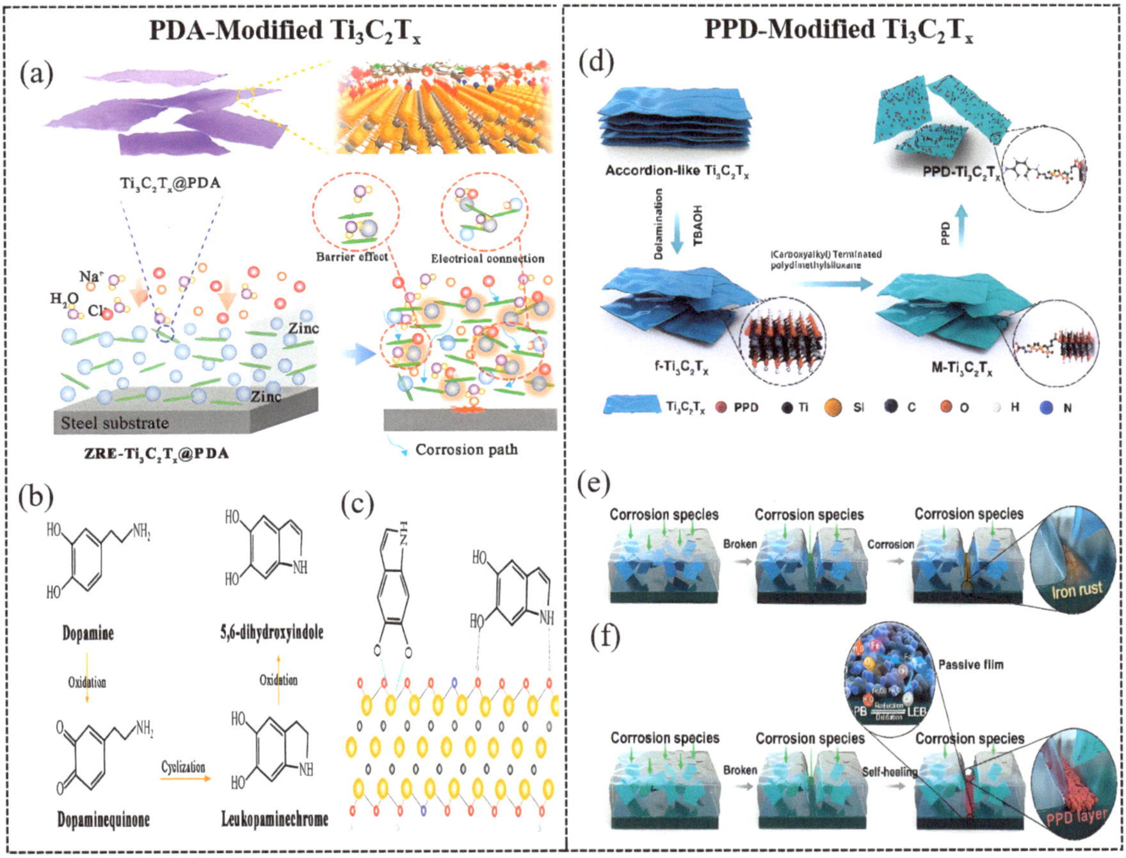

Figure 7. (**a**) The schematic anticorrosion mechanisms of ZRE-$Ti_3C_2T_x$ @PDA/WEP coating; the mechanisms of (**b**) polymerization of dopamine and (**c**) binding on $Ti_3C_2T_x$ NSs [96]. Reprinted with permission from [96], copyright 2023 Elsevier. (**d**) The synthesis process for PPD@$Ti_3C_2T_x$ hybrids. Schematic representation of anticorrosion and self-healing protection mechanism of (**e**) 0.5f-$Ti_3C_2T_x$/EP, (**f**) 0.5PPD@$Ti_3C_2T_x$/EP [97]. Reprinted with permission from [97], copyright 2023 Elsevier.

As well as PDA, p-phenylenediamine (PPD) was utilized to functionalize $Ti_3C_2T_x$ (PPD@$Ti_3C_2T_x$) via the hydrothermal method by Shi et al. in 2023 [97]. The synthesis process for PPD@$Ti_3C_2T_x$ hybrids is shown in Figure 7d. The PPD@$Ti_3C_2T_x$ nanohybrid was adopted to strengthen the EP matrix. Under a 20 MPa alternating hydrostatic pressure (AHP) environment, the $|Z|_{0.01Hz}$ of the composite coating enhanced from 3.72×10^7 to 3.74×10^8 $\Omega \cdot cm^2$ with immersion time increasing from 2 to 6 days. Furthermore, the composite coating presented superior self-healing and great physical barrier behaviors. The anticorrosion and self-healing protection mechanism of 0.5f-$Ti_3C_2T_x$/EP and 0.5PPD@$Ti_3C_2T_x$/EP are schematically illustrated in Figures 7e and 7f, respectively.

3.2.4. Functionalization Through Other Organic Compounds

Other compounds have also been explored to functionalize the MXene NSs to enhance their anti-oxidation properties [98–102]. It is known that, despite its superior intrinsic properties, $Ti_3C_2T_x$ flakes are prone to hydrolysis and oxidation in hydrated environments, causing them to potentially transform into TiO_x and TiO_2. The degradation is a multistep process, usually initiated at the defect and edge sites of MXenes, and the defect sites undergo systematic hydrolysis with TiO_2 as the outcome of the final formed degraded product. This chemical degradation may impede the anticorrosion behavior of MXene in the long term. Therefore, strategies to mitigate oxidation are requisite.

Recently, it has been discovered that $Ti_3C_2T_x$ can be passivated by imidazolium salt ionic liquids (ILs), which can quench reactive oxygen species. In 2021, Zhao et al. developed a stable MXene through the noncovalent modification of 1-(3-Aminopropyl)-3-methylimidazolium bromide (APMImBr) on $Ti_3C_2T_x$ MXene [98]. The tight adhesion of IL loaded on the $Ti_3C_2T_x$ surface via noncovalent interaction endows the EP composite coating with a good barrier and self-healing properties caused by IL passive films. The IL @MXene NSs structure, corrosion mechanism of the neat WEP, the IL@MXene/WEP composite coating, and digital photographs of different coatings-protected steels after NSST for 500 h are schematically illustrated in Figure 8a.

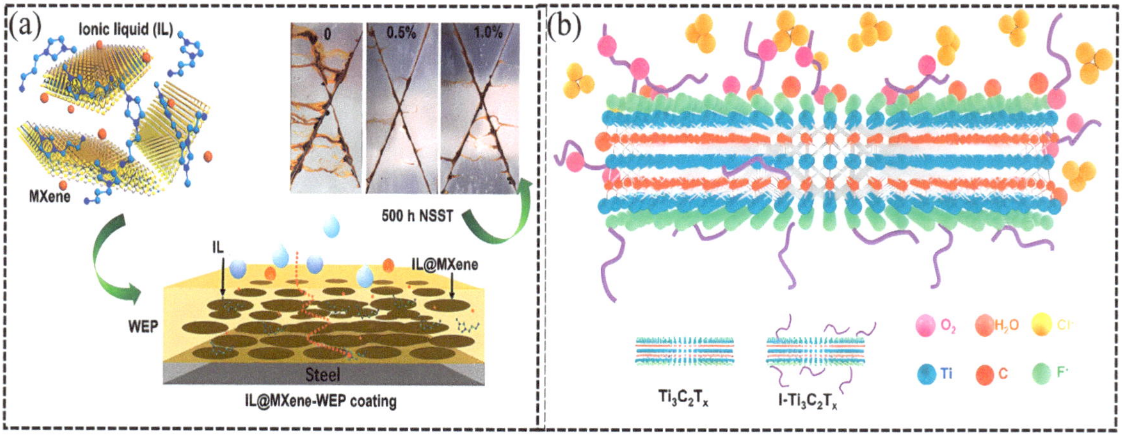

Figure 8. (**a**) Schematic illustration of IL@MXene nanosheets structure, corrosion mechanism of the neat WEP, the IL@MXene/WEP composite coating, and digital photographs of steels protected by different coatings after NSST for 500 h [98]. Reprinted with permission from [98], copyright 2021, American Chemical Society. (**b**) Schematic representation of the corrosion process with I-$Ti_3C_2T_x$ contained in epoxy resin [99]. Reprinted with permission from [99], copyright 2023 Elsevier.

In 2023, imidazolium salt (1-allyl-3-methylimidazolium iodide) ([AMIm]I) was used to modify the surface of MXene by Ning et al. [99]. The treated $Ti_3C_2T_x$ was stable for 90 days in an aqueous solution, showing excellent antioxidant stability. The imidazolium ions on $Ti_3C_2T_x$ MXene inhibit its reaction with O_2. The adhesion between the coatings and the substrate was significantly enhanced, owing to the strong compatibility between the modified MXene and EP. The anticorrosion behavior of IL-modified $Ti_3C_2T_x$ is schematically represented in Figure 8b.

Given the increasing awareness of environmental protection, there is an urgent need for environmentally friendly, easily accessible, and corrosion-resistant biological reagents for research on anticorrosion performance, so organic and green modifiers, such as amino acids, are proposed.

In 2021, Li et al. successfully prepared functionalized $Ti_3C_2T_x$ through a covalent reaction between amino acid (L-cysteine) and $Ti_3C_2T_x$ surface [101]. Due to the rich groups ($-NH_2$, –SH, and –OH) of the L-cysteine, N–Ti and S–Ti bonds will come into being after the condensation reaction between the L-Cys and the MXene. Thus, L-Cys can improve the dispersion and compatibility of MXene in EP and inhibit galvanic corrosion.

Moreover, in 2023, Zhao et al. employed tannic acid (TA) to modify $Ti_3C_2T_x$ and successfully prepared a parallel arranged P-TA@$Ti_3C_2T_x$/EP coating via the blade-coating method. The corrosion resistance under alternating hydrostatic pressure (AHP) was investigated [102].

The characteristics of the above-mentioned functionalization grafting methods are compared and summarized in Table 2.

Table 2. Summary of characteristics of functionalization grafting methods.

Functional Groups	Bonding	Typical Examples	Aim	Mechanism	Ref.
Silane-based coupling agent	Covalent	APTES, AEAPTES, GPTMS, MPS	Surface silanization reaction is used to effectively stabilize the MXene against structural degradation and improve the surface properties with adjustable hydrophilicity.	Ti–O–Si bonds and the primary amino functional group make MXene positively charged.	[73–83]
Polymer molecule	Covalent	PANI, PDMS, P-CS	Wrap a layer of polymer insulation on the surface of MXene to reduce its conductivity and passivate it.	Ti–O–Si or Ti–O–P bonds.	[84–91]
Amine-based	Covalent	PDA, PPD	DA undergoes self-polymerization under the action of MXene's hydroxyl group, forming an insulating polymer that protects MXene from oxidation.	Interaction between the –OH of MXene NSs and DA's benzoquinone structure improved the stability of MXene.	[92,97]
Other organic compounds	Noncovalent	IL, such as [APMIm]Br, [APMIm]I	Prevent oxidation and improve stability.	ILs isolate oxygen and inhibit oxidation and degradation of MXene.	[98,99]
Other organic compounds	Covalent	Glycine, L-Cys, TA,	Using organic reagents such as amino acids and increasing environmental friendliness.	N–Ti and S–Ti bonds.	[100–103]

3.3. Orientation Regulation

It is known that the incorporation of 2D nanomaterials (NMs) into polymer coatings can significantly boost the physical barrier ability of the composite, as these nanofillers have a sufficiently high aspect ratio to block the diffusion path of a corrosion-related species, forcing corrosion ions to follow a longer, more tortuous path through the coating matrix. However, the anticorrosive performance of 2D NM reinforced-composite coatings usually fails to fulfill the theoretical prediction. This is not only related to the dispersion and compatibility of NMs into the matrix, but more importantly, it is also closely related to their alignment and orientation [2]. In most previous studies, 2D NMs are usually distributed randomly instead of in an orderly mode, which would not achieve optimal barrier performance. The physical barrier of the 2D MXene NSs can be maximized by achieving highly oriented NSs in composite coating. The assembly strategies to obtain high-orientation 2D NM reinforced polymer composite coatings mainly include layer-by-layer (LbL) deposition, magnetic field, electrodeposition, and spin-coating [2].

To date, electrodeposition, magnetic field, and self-assembly caused by airflow or blade-coating approaches have been proposed to obtain highly oriented MXenes for anticorrosive coatings [92,102–107]. In 2022, Fan et al. obtained well-oriented $Ti_3C_2T_x$ in EP coating through electrodeposition [103]. The $Ti_3C_2T_x$ became protonated with a positive charge (f^+-$Ti_3C_2T_x$) via amino-functionalization (Figure 9a). The parallel arranged mechanism under the electric field is as follows: the positive f^+-$Ti_3C_2T_x$ flakes move towards the cathode under the action of electric field force and rotate in the direction perpendicular to the electric field; meanwhile, the f^+-$Ti_3C_2T_x$ flakes with the same charge repel each other, causing the flakes to quickly align parallel to the coating in a short period of time. The EP coating with 1 wt.% f+-$Ti_3C_2T_x$ exhibits an almost perfect internal aligned structure, and its corrosion resistance is four orders of magnitude higher than that of randomly arranged $Ti_3C_2T_x$ coatings. After a 4-week immersion test, the $|Z|_{0.01Hz}$ of the aligned MXene/EP coating reduced from 6.96×10^8 to 1.05×10^8 $\Omega \cdot cm^2$, but became 2.63×10^4 $\Omega \cdot cm^2$ for the randomly distributed coating, while the coating resistances of the aligned and the random MXene/EP coating were, respectively, 1.04×10^8 $\Omega \cdot cm^2$ and 2.83×10^4 $\Omega \cdot cm^2$ after the immersion test [92]. The preparation process, the cross-sectional SEM images, and the anticorrosion mechanism of the parallel $Ti_3C_2T_x$ MXene/EP coating via electrodeposition are displayed in Figure 9a.

In 2022, Ding et al. prepared the carbon dot (CD)-functionalized $Ti_3C_2T_x$ MXene and realized its alignment in EP coating through an airflow-induced self-assembly method (Figure 9b) [104]. It is found that the oriented distributed NSs can suppress galvanic corrosion, but the random distribution will accelerate the galvanic corrosion on the coating defects. After a 35-day immersion test, the $|Z|_{0.01Hz}$ of the oriented CD@ $Ti_3C_2T_x$/EP coating reduced from 4.1×10^{10} to 5.1×10^9 $\Omega \cdot cm^2$, and that of the random MXene/EP coating decreased from 3.4×10^9 to 1.1×10^7 $\Omega \cdot cm^2$. The coating resistance of the oriented MXene/EP coating (3.3×10^9 $\Omega \cdot cm^2$) was higher than that of the randomly distributed one (1.3×10^7 $\Omega \cdot cm^2$) after the immersion test [104]. The improvement of the stability and durability of the coating is attributed to the elimination of connections of MXene/MXene and MXene/metal perpendicular to the diffusion direction of corrosive media.

In 2024, Fan et al. synthesized an MXene/PDDA/Fe_3O_4 (MPF) hybrid via self-assembly induced by a magnetic field (Figure 9c) [107].

Moreover, inspired by the structure of a mille crepe cake, Tan et al. prepared tannic acid (TA)@$Ti_3C_2T_x$/EP coatings by incorporating TA-modified MXene using a facile blade-coating approach (Figure 9d) in 2023 [102]. The $Ti_3C_2T_x$ fillers demonstrated near-perfect parallel orientation within the TA@$Ti_3C_2T_x$/EP composite coating. The $|Z|_{0.01Hz}$ of the highly aligned TA@ $Ti_3C_2T_x$ coating is 1–3 orders of magnitude larger than that of the randomly arranged one. In addition, molecular dynamics (MD) simulation results show that TA molecules can strongly adhere to the substrate in 3.5 wt.% NaCl solution both under a 1 atm and 20 MPa environment, indicating a synergistic protective effect of TA and MXene NSs.

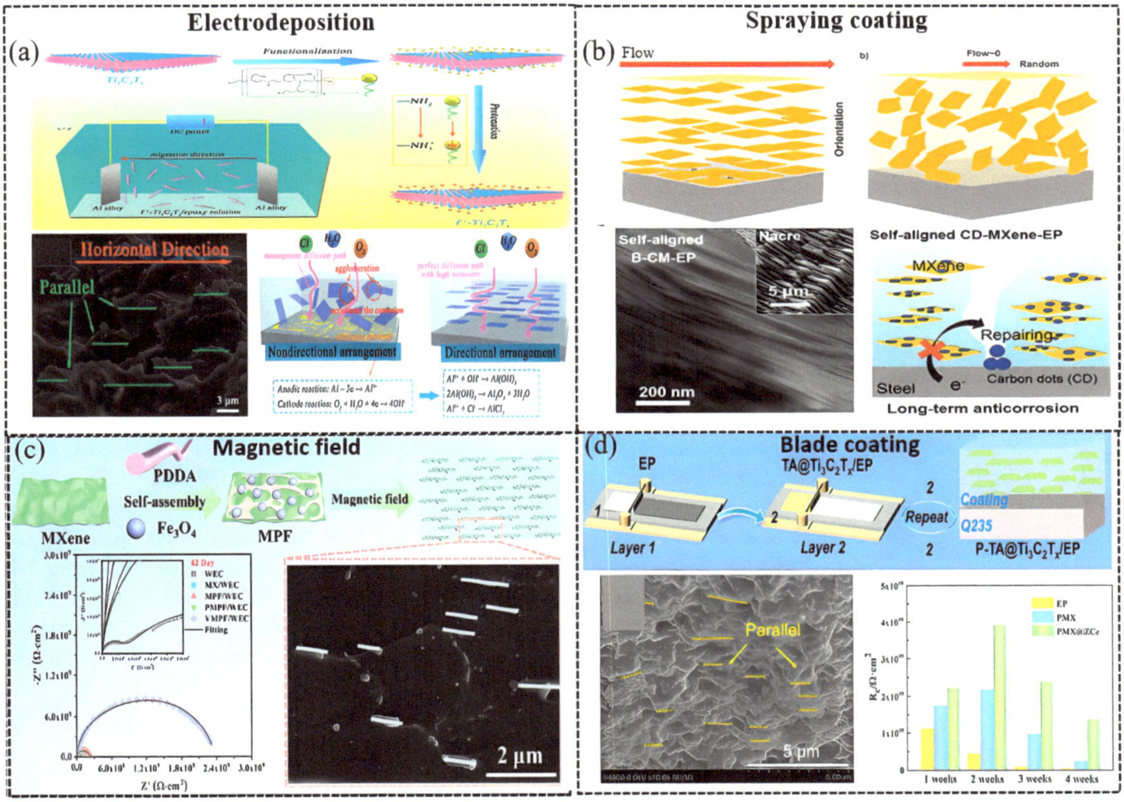

Figure 9. (**a**) Preparation process of orientation aligned $Ti_3C_2T_x$ MXene/EP coating through electrodeposition [92], including the cross-sectional SEM images and the anticorrosion mechanism. Reprinted with permission from [92], copyright 2022 Elsevier. (**b**) Preparation of oriented CD functionalized $Ti_3C_2T_x$ MXene via spraying method [104], including the schematic process of airflow-induced self-assembly, the cross-sectional SEM images, and the anticorrosion mechanism of oriented CDs@MXene/EP coating. Reprinted with permission from [104], copyright 2022, Elsevier. (**c**) The MXene/PDDA/Fe_3O_4 (MPF) hybrid synthesized via self-assembly induced by magnetic field [107]. Reprinted with permission from [107], copyright 2024, American Chemical Society. (**d**) The preparation process of oriented TA@$Ti_3C_2T_x$/EP coatings by incorporating tannic acid (TA)-modified MXene into EP using a facile blade-coating method [102]. Reprinted with permission from [102], copyright 2023 Elsevier.

The comparison of the above-discussed methods to obtain highly oriented $Ti_3C_2T_x$ MXenes NSs in EP composite coatings is listed in Table 3. Each method has advantages and weaknesses. Thus, a strategy to produce highly aligned MXene 2D NSs in composite coatings economically, reliably, and suitable for large-scale preparation is still desired.

Table 3. Summary of the methods to achieve oriented MXenes NSs in composite coatings.

Methods	Advantages	Disadvantages	Ref.
Electrodeposition	Easy to precisely control the structure and suitable for large-scale production	External force required, expensive, high energy consumption	[92,103,106]
Magnetic field	Easy to precisely control the structure	External force required, expensive	[105]
Spraying	Airflow-induced self-assembly, simple process, low cost	Low repeatability, hard to precisely control the structure	[104]
Blade coating	Low cost, alternate	Low repeatability, hard to precisely control the structure	[102]

3.4. Heterostructure Composite

MXenes are prone to agglomeration and oxidation. A heterogeneous structure composite is another strategy to maintain its structural stability and conductivity. Zero-dimensional (nanoparticle (NP), quantum dots (QD)), one-dimensional (nanofiber (NF), nanotube (NT), nanowire (NW)), and two-dimensional (Gr, GO, LDH, h-BN, MoS_2, etc.) NMs can be adopted to hybridize with two-dimensional MXene to create nanocomposites or heterostructures. By now, carbon-based NMs (typical, CD, CNTs, and Gr and GO) and LDH are two kinds of main NMs used to integrate with MXenes to produce heterostructures [108–128].

Herein, this section is arranged based on the hybridization between MXene and the following NMs: (i) Carbon-based NMs, (ii) LDH, (iii) 0D NPs, (iv) 1D NMs (CNT, CNF), and (v) other 2D NMs, such as h-BN, MoS_2, and covalent organic frameworks (COFs) [128] or MOF. Some typical heterogeneous composites formed between multiple NMs and MXenes are shown in Figure 10 [4].

3.4.1. Hybridization Between MXene and Carbon-Based NMs

Carbon-based NMs (Gr [108], GO [109–111], and CNTs [80,111,112]) have been adopted to complexes with MXene. Gr was first integrated with MXene in 2020 by Yan et al. [108]. It is found that $Ti_3C_2T_x$ was wrapped by Gr. The $Ti_3C_2T_x$/@Gr/EP composite coating involving hetero-structured $Ti_3C_2T_x$@Gr into EP demonstrated excellent anticorrosion properties, exhibiting a corrosion-resistance modulus of 2.14×10^9 $\Omega \cdot cm^2$, much larger than those of bare EP (1.06×10^8 $\Omega \cdot cm^2$), $Ti_3C_2T_x$/EP (1.51×10^9 $\Omega \cdot cm^2$), and Gr-EP (1.53×10^9 $\Omega \cdot cm^2$). The boosted performance of the $Ti_3C_2T_x$/@Gr/EP composite was ascribed to three aspects: (i) protective films formed owing to dual hybrid surfaces, (ii) excellent thermal conductivity and lubricant properties of $Ti_3C_2T_x$ and Gr, and (iii) the synergistic effects of $Ti_3C_2T_x$@Gr-interweaved structures that greatly enhanced the corrosion-resistant performance of organic coatings.

In 2021, Shen et al. synthesized GO@$Ti_3C_2T_x$ heterojunction with the conductivity of 1380 ± 100 $S \cdot cm^{-1}$ and incorporated this nanohybrid into zinc-rich coatings (ZRC) [109]. The ZRC loaded with 0.5 wt.% GO@$Ti_3C_2T_x$ displayed the highest coating-resistance value (3.047×10^4 $\Omega \cdot cm^2$) after a 50-day immersion test in 3.5 wt.% NaCl solution, which was one order of magnitude higher than that of other samples.

GO-modified $Ti_3C_2T_x$ flakes were also prepared by Qiang et al. in 2023 [110]. It shows that the hybridized GO@$Ti_3C_2T_x$ flake is ~3.9 nm thick and C, O, and Ti elements are distributed uniformly. The $|Z|_{0.01Hz}$ of GO@$Ti_3C_2T_x$/EP coating retained about 1.84×10^8 $\Omega \cdot cm^2$ after a 8-day immersion test [110].

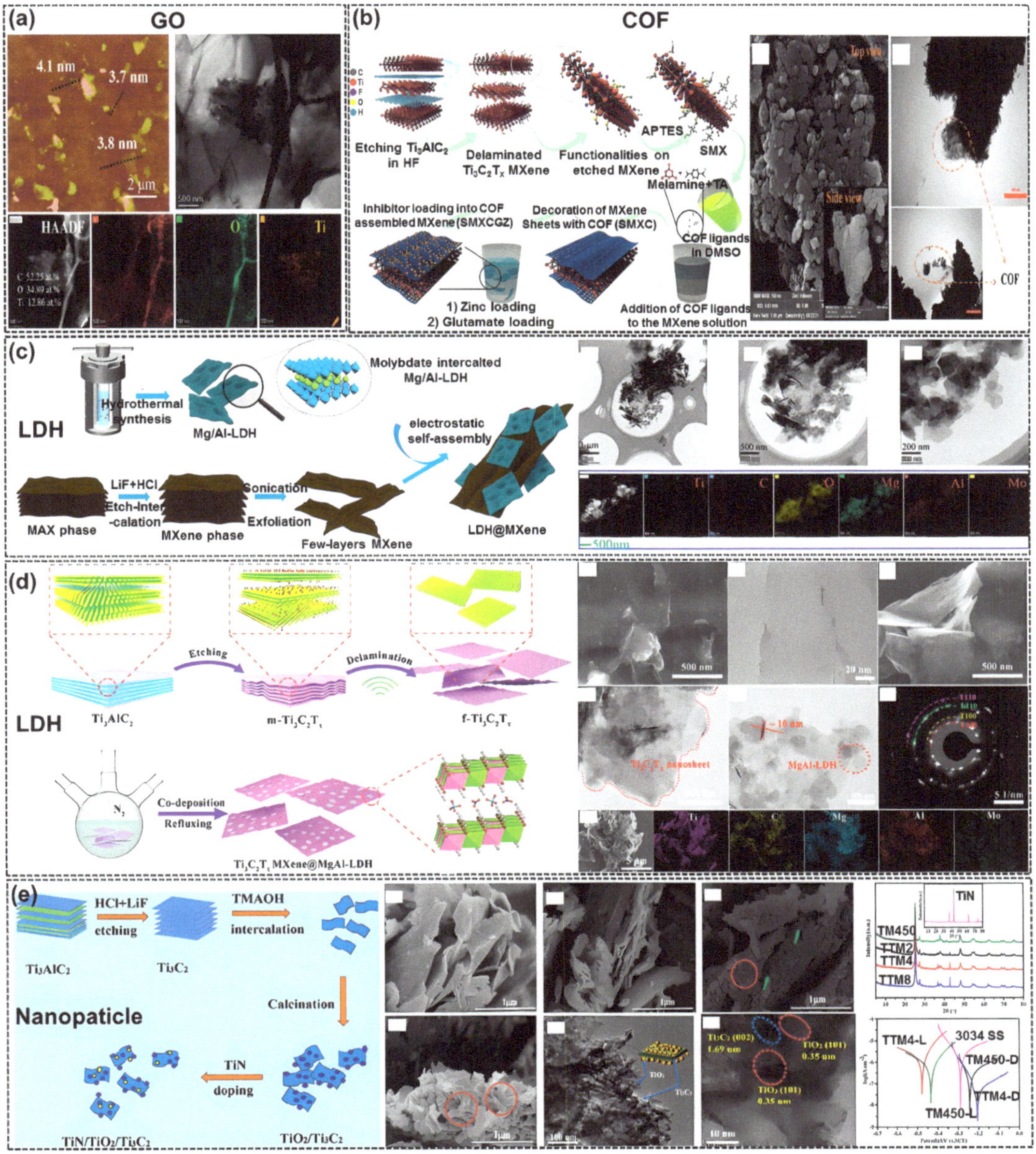

Figure 10. Heterogeneous composite materials formed between multiple NMs and MXenes [4]. Reprinted with permission from [4], copyright 2024 Elsevier. (**a**) AFM and TEM images of GO-$Ti_3C_2T_x$ NSs, including the HAADF image and EDS mapping of GO-$Ti_3C_2T_x$ [110]. Reprinted with permission from [110], copyright 2023 Elsevier. (**b**) Schematic illustration of the fabrication process of COF-modified MXene, including FE-SEM and TEM images of synthesized MXene [128]. Reprinted with permission from [128], copyright 2023 Elsevier. (**c**) Synthesis process of $Ti_3C_2T_x$@MgAl-LDH by the hydrothermal method, including the SEM and TEM images and EDS elemental mappings of the Ti3C2Tx@MgAl-LDH [115]. Reprinted with permission from [115], copyright 2023 Elsevier. (**d**) The

co-deposition of $Ti_3C_2T_x$ MXene@MgAl-LDH heterostructure, including SEM, HRTEM images, SAED pattern, and EDS mapping of $Ti_3C_2T_x$ MXene@MgAl-LDH [114]. Reprinted with permission from [114], copyright 2021 Elsevier. (**e**) Preparation process of MXene modified by the TiN/TiO_2 nanoparticles, including the SEM and TEM images of the modified MXene nanosheets, XRD patterns, and polarization curve of the samples [121]. Reprinted with permission from [121], copyright 2022 Elsevier.

In 2022, the CDs were used for the surface modification of $Ti_3C_2T_x$ MXene for corrosion protection by Ding et al. [104]. After a 35-day immersion test, the $|Z|_{0.01Hz}$ of the CDs@$Ti_3C_2T_x$/EP coating decreased from 4.1×10^{10} to 5.1×10^9 $\Omega \cdot cm^2$ [104].

In 2022, Wang et al. synthesized the A-$Ti_3C_2T_x$@S-CNTs/WPU composite coating [80], in which the A-$Ti_3C_2T_x$ (AEAPTMS-modified $Ti_3C_2T_x$) became positively charged, and S-CNTs (sodium dodecyl-benzene-sulfonate (SDBS)-treated CNTs) became negatively charged. Therefore, they were able to assemble through electrostatic interaction, preventing their self-aggregation. The optimal content of A-$Ti_3C_2T_x$@S-CNTs in WPU coatings is found to be 0.2 wt.%. The reaction of $Ti_3C_2T_x$ flakes and AEAPTMS and electrostatic attraction between A-$Ti_3C_2T_x$ and S-CNTs are displayed in Figures 11a and 11b, respectively.

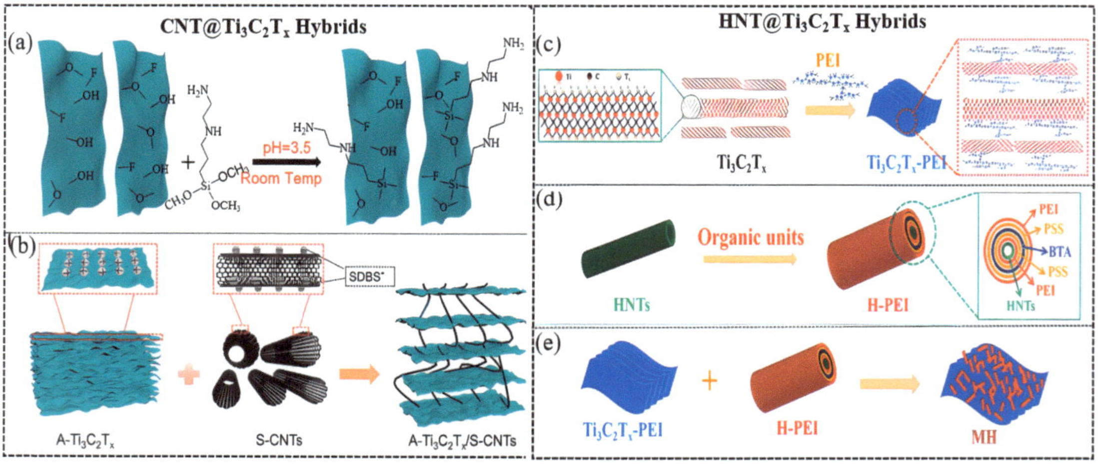

Figure 11. (**a**,**b**) Silanized MXene/carbon nanotube (A-$Ti_3C_2T_x$/S-CNT) composite structure [80]. (**a**) Schematic diagram of the reaction between $Ti_3C_2T_x$ flakes and AEAPTMS. (**b**) Schematic electrostatic attraction between A-$Ti_3C_2T_x$ and S-CNTs. Reprinted with permission from [80], copyright 2022, American Chemical Society. (**c**–**e**) Schematic illustration of the fabrication of MXene@HNTs (MH) and PEI composite [113]: (**c**) H-PEI, (**d**) MXene-PEI, and (**e**) MH. Reprinted with permission from [113], copyright 2022, Elsevier.

In addition, halloysite nanotubes (HNTs) were used to functionalize $Ti_3C_2T_x$ in 2022 by Deng et al. [113]. The synergistic effects of HNTs and $Ti_3C_2T_x$ improved the corrosion resistance of the composite coating. The fabrication of MXene@HNTs/PEI composite is shown in Figure 11c–e, respectively.

3.4.2. Hybridization Between MXene and LDH

Due to the inherent positively charged surface of LDH and the negatively charged surface of MXene, a heterostructure of MXene/LDH is easy to obtain via electrostatic interaction [68,114–118]. For instance, MgAl-LDH has been composited with MXene through in situ growing on $Ti_3C_2T_x$ via the hydrothermal method [114] (Figure 10c) or co-deposition [115] (Figure 10d). After the surface modification of LDH, the $Ti_3C_2T_x$ MXene retained a sheet-like morphology, but its surface became rougher. After a 21-day immersion test in 3.5 wt.% NaCl solution, the $|Z|_{0.01Hz}$ of MXene@LDH was 3.9×10^6 $\Omega \cdot cm^2$, two times higher than that of EP (2.0×10^6 $\Omega \cdot cm^2$) [114]. The assembly of these two types of NSs solved their inherent problem of restacking, and the composite still exhibited excellent dispersion and good compatibility with EP after 160 days of standing.

In 2023, our group prepared a novel $Ti_3C_2T_x$@LDH/EP hybrid coating using a simple in situ assembled method on AZ31 magnesium alloy [68]. The environmentally friendly organic acid, L-cysteine acid (L-Cys), was encapsulated into MgAl-LDH as a corrosion inhibitor. The $Ti_3C_2T_x$@MgAl-LDH (TML) heterostructure NSs demonstrate excellent corrosion resistance and self-healing. Its I_{corr} and $|Z|_{0.01\ Hz}$ are 1.4×10^{-9} $A \cdot cm^{-2}$ and 1.66×10^7 $A \cdot cm^{-2}$, which is 4 orders of magnitude lower than that of the uncoated Mg substrate and 1–2 orders of magnitude larger than that of EP (8.7×10^{-5} $A \cdot cm^{-2}$), respectively. The great enhancement of the corrosion-resistant and self-healing performance of the TML/EP composite coating is mainly attributed to the synergistic effect of the "barrier effect" of TML and the corrosion inhibition effect of the L-cysteine acid intercalated in MgAl-LDH. This work proposes valuable insight to widen the functional application of $Ti_3C_2T_x$ and the design of highly anticorrosive and self-healing organic coatings for Mg alloys. The preparation, corrosion resistance, and mechanism of self-healing of L-Cys-intercalated $Ti_3C_2T_x$@MgAl-LDH/EP hybrid coating are shown in Figure 12 [68].

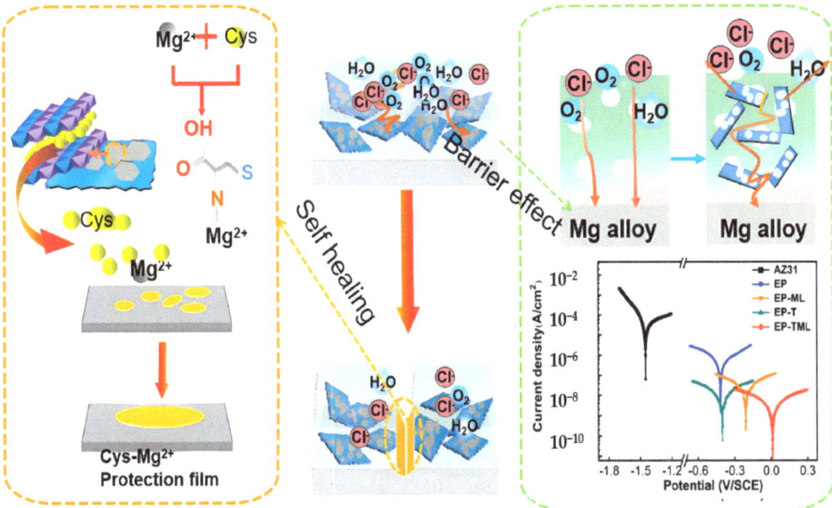

Figure 12. The preparation, corrosion resistance, and mechanism of self-healing of L-Cys-intercalated $Ti_3C_2T_x$@MgAl-LDH/EP hybrid coating [68]. Reprinted with permission from [68], copyright 2023, Springer Nature.

3.4.3. 0D NPs/2D MXene Hybrid Structure

Except for the aforementioned carbon-based and LDH NMS, 0D NPs have also been introduced to integrate with 2D MXene [77,104,119–123]. For example, CDs have been used to surface-modify $Ti_3C_2T_x$ MXene for corrosion protection by Ding et al. in 2022 [104]. The TiN@TiO_2 NPs were synthesized on the MXene via the calcination and doping process [121]. The formed TiO_2 NPs were dispersed on both sides of the MXene sheets and between the Ti_3C_2 MXene layers. The calcined MXene was observed to possess a rougher surface, a heterostructure of TiN@TiO_2, and an improved anticorrosion property of the coating.

In 2022, the aminated silica grown in situ on $Ti_3C_2T_x$ MXene NSs was adopted as the functional filler by He et al. [122]. The $|Z|_{0.01Hz}$ of SiO_2@ $Ti_3C_2T_x$/WEP coating was 1.33×10^9 $\Omega \cdot cm^2$. After a 21-day immersion test in 3.5 wt.% NaCl, the coating resistance was $4.456 \times 10^8\ \Omega \cdot cm^2$ [122]. The fabrication process, structure, and antiwear and anticorrosion performance of aminated $SiO_2@Ti_3C_2T_x$/WEP composite coating are given in Figure 13.

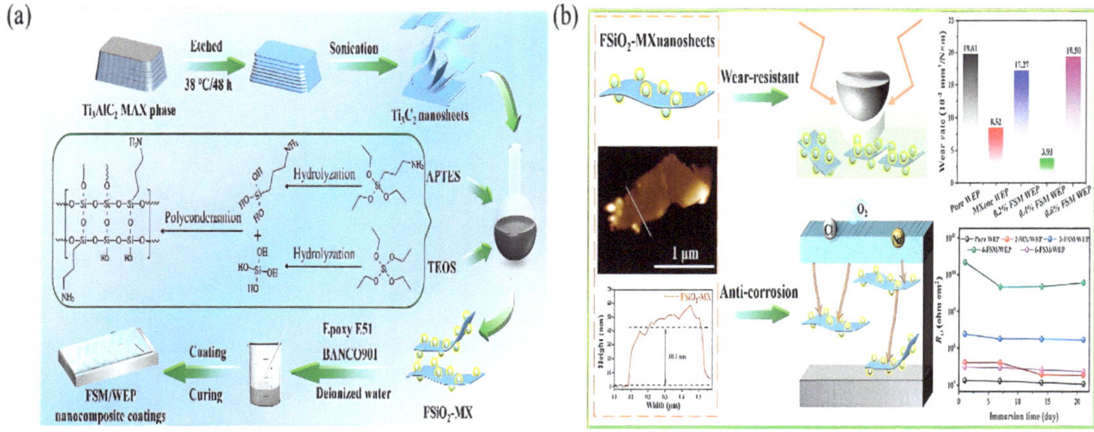

Figure 13. The 0D/2D hybrid structure of in situ growth of aminated silica on MXene nanosheets: (**a**) The fabrication process; (**b**) the structure and antiwear and anticorrosion performance in waterborne epoxy composite coatings [122]. Reprinted with permission from [122], copyright 2022, Elsevier.

3.4.4. 1D Nanofiber/2D MXene Hybrid Structure

Silk fiber (SF) and cellulose nanofiber (CNF) are two typical cases of 1D nanofiber to complex with MXene [124,125]. In 2021, Chen et al. synthesized a novel silk fibroin-$Ti_3C_2T_x$ (SF-$Ti_3C_2T_x$) composite filler [124]. The preparation procedure of the SF-$Ti_3C_2T_x$ is schematic illustrated in Figure 14a. After SF fiber modification, the surface of MXene NSs became rougher and covered by network materials (Figure 14b), enhancing the adhesive strength between $Ti_3C_2T_x$ NSs and EP, which is beneficial for obtaining composites with higher density and better mechanical properties. The anticorrosion performance of SF-$Ti_3C_2T_x$/EP composite coating under atmospheric pressure (AP) and a simulated deep-sea environment was studied, and it was found that the SF-$Ti_3C_2T_x$-augmented organic coating also has excellent corrosion resistance under high hydrostatic pressure. As shown in Figure 14c, after soaking the 0.5SF-$Ti_3C_2T_x$/EP composite coating under 20 MPa hydrostatic pressure for 240 h, the interface between the composite coating and steel substrate was tightly integral, without delamination and still had high adhesion strength (3.16 MPa), much larger than that of bare EP (only 0.88 MPa) (Figure 14d).

Furthermore, Wu et al. also designed an MXene-reinforced EP coating (Figure 14e–h) using cellulose nanofiber (CNF)-modified $Ti_3C_2T_x$ in 2022 [125]. As shown in SEM images (Figure 14g), before immersion in AHP, the three coatings, namely bare EP, $Ti_3C_2T_x$/EP, and $Ti_3C_2T_x$@CNF/EP, all presented smooth and intact morphologies without visible microdefects. However, numerous micropores were observed on bare EP coating after a 10-day immersion test under the AHP environment, but the $Ti_3C_2T_x$@CNF/EP still retained a compact morphology (Figure 14g). Figure 14h shows that the $Ti_3C_2T_x$@CNF/EP coating has the greatest adhesion strength, regardless of as-prepared, standard AP, or AHP conditions.

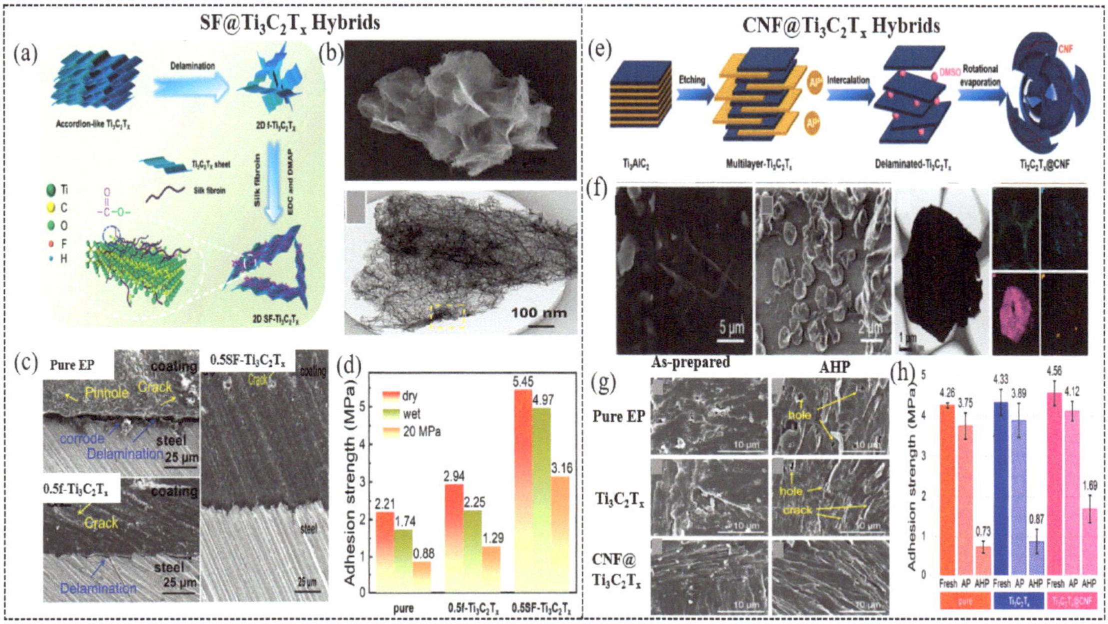

Figure 14. (**a–d**) SF@$Ti_3C_2T_x$/EP hybrid structure [124]: (**a**) Schematic illustration of the preparation procedure of the SF-$Ti_3C_2T_x$; (**b**) SEM and TEM image of SF-$Ti_3C_2T_x$; (**c**) SEM images of coating/steel interface for pure EP, 0.5f-$Ti_3C_2T_x$/EP, and 0.5SF-$Ti_3C_2T_x$/EP after a 240-h immersion under 20 MPa hydrostatic pressure; (**d**) the adhesion strength for different coatings under dry, wet, and 20 MPa environments. Reprinted with permission from [124], copyright 2021, Elsevier. (**e–h**) CNF@$Ti_3C_2T_x$/EP hybrid structure [125]: (**e**) Schematic representation of the synthesis procedure of the delaminated $Ti_3C_2T_x$ and CNF@$Ti_3C_2T_x$ via a wet etching method; (**f**) SEM images of surface and the cross-sectional of $Ti_3C_2T_x$@CNF film, TEM image of $Ti_3C_2T_x$@CNF topological structure and its elemental mapping; (**g**) cross-sectional SEM images of bare EP, $Ti_3C_2T_x$/EP, and $Ti_3C_2T_x$@CNF/EP coatings before and after under AHP environment; (**h**) adhesion strength of three coatings under freshly as-prepared AP and AHP environments. Reprinted with permission from [125], copyright 2022, Elsevier.

3.4.5. Other 2D NMs@2D MXene Hybrid Structure

Other 2D lamellar NMs, such as BN, MoS_2, MOF, and COF, have also been explored [126–130]. In 2022, Zhou et al. prepared $Ti_3C_2T_x$-modified h-BN NMs [126]. They show that the $Ti_3C_2T_x$@BN/EP coating could maintain a time constant after 21 days of soaking in a 3.5 wt.% NaCl solution and its $|Z|_{0.01Hz}$ value was 1.5 orders of magnitude higher than that of the bare EP coating.

In 2022, Cai et al. fabricated a $Ti_3C_2T_x$@MoS_2 layered hybrid structure through a simple one-step hydrothermal method [127]. The MoS_2 NSs vertically anchored on $Ti_3C_2T_x$ NSs took the shape of a unique layered 2D structure that effectively inhibits the restacking of $Ti_3C_2T_x$ and MoS_2. The 0.1 wt.% $Ti_3C_2T_x$@MoS_2/EP composite coating exhibits the best corrosion-resistant performance.

The COF-modified MXene NSs were synthesized through the in situ growth method by Najmi et al. in 2023 [128]. Owing to the high chemical stability of COFs with imine linkages, the COF-treated $Ti_3C_2T_x$ MXene presented a stable anticorrosion. After the incorporation of COFs@MXene in the EP coating, the I_{corr} reduced from 6.701 to 0.815 $\mu A \cdot cm^{-2}$ [128].

The characteristics of hybridization methods to obtain 0D-,1D-2D NMs/MXene heterostructure are summarized in Table 4.

Table 4. Summary of characteristics of heterostructure hybridization methods.

Hybridization	Major Objective	Typical Instances	Ref.
0D/MXene	Simultaneously improve corrosion resistance and wear resistance	CDs, TiO_2, SiO_2, TiO_2/TiN	[77,104,119–123]
1D/MXene	Improve the adhesion between the coating and the substrate	CNT, HNT, CNF	[124,125]
2D/MXene	Preventing agglomeration, provide nanocontainers for corrosion inhibitors to improve self-healing performance	Gr, GO	[108,110]
		LDH	[68,114,115]
		BN, MoS_2, etc.	[126–128]

3.5. Comparison of Corrosion Performance of $Ti_3C_2T_x$-Based Organic Composite Coatings

The comparison of preparation conditions and corrosion resistance of MXene-based organic composite coatings is listed in Table 5. It can be seen that:

(1) The most popular method to prepare anticorrosion coatings is through physical mixing of MXene with polymer. Layered $Ti_3C_2T_x$ was employed as a filler to disperse in an organic matrix, among which WEP or WPU are the two most commonly used substrates. The loaded amount of $Ti_3C_2T_x$ in EP-based composite is generally below 2 wt.%, with major cases being 0.5 wt.%. The content of $Ti_3C_2T_x$ added in WPU-based composite is usually less than 0.2 wt.%, lower than that of WEP-based composite coating.

(2) The anticorrosion performance of organic coatings expressed by the impedance modulus ($|Z|_{0.01Hz}$) is approximately on the order of 10^8 $\Omega \cdot cm^2$. Among them, the CD-$Ti_3C_2T_x$/EP coated on Q235 steel has the highest $|Z|_{0.01Hz}$ (4.1×10^{11} $\Omega \cdot cm^2$) [104], while that of the APTES-modified A-$Ti_3C_2T_x$-Ce^{3+}/high solid phase EP coated on mild steel is the lowest (4.8×10^4 $\Omega \cdot cm^2$) [77].

(3) To increase its dispersibility, compatibility, and stability, the $Ti_3C_2T_x$ MXene is usually required to undergo surface treatment prior to being compounded into a polymer matrix. Surface treatment mainly includes functionalized grafting, surface modification, or hybridization composite with other 2D NMs.

(4) Currently, carbon steel and aluminum alloys are the two most commonly employed metal substrates for MXene-based organic anticorrosive coatings. Only a few studies have been carried out on magnesium [68], and there has been almost no work on titanium alloys, which provides a lot of research space.

In summary, in order to achieve a high anticorrosion performance of MXene-based polymer composite coating, four surface-modification strategies—surface functionalization grafting, orientation arrangement, heterostructure composite, and antioxidant stabilization and environmentally friendly modification (greening)—were discussed and analyzed. The basic principles and typical cases of four surface-modification strategies are summarized in Figure 15, and the advantages and limitations of these surface-modification methods are given in Table 6.

Table 5. Comparison of corrosion performance of $Ti_3C_2T_x$-based organic composite coatings.

Metal Substrate	MXene	Polymer Matrix	C_{MX}/ wt./%	Corrosion Performance				Ref.
				E_{corr}/ V	I_{corr}/ 10^{-9}A·cm^{-2}	R_{corr}/ mm·year^{-1}	$\|Z\|_{0.01Hz}$/ $10^8 \Omega \cdot cm^2$	
Steel/Q345	$Ti_3C_2T_x$	EP (E44)	1.0	−0.30	33.90	3.9×10^{-4}	6.23	[30]
Cu	$Ti_3C_2T_x$	Silane	0.25	−0.39	4.04	--	0.19	[69]
Al alloy/6082	APTES-$Ti_3C_2T_x$	WEP (H228A)	0.5	0.04	1.01	1.10×10^{-5}	7.94	[74]
Steel/Q235	AEAPTES-$Ti_3C_2T_x$	WPU	0.1	−0.05	2.67	3.31×10^{-5}	0.16	[75]
Sn plate	GPTES-$Ti_3C_2T_x$	EP (E44)	0.5	--	--	--	0.02	[78]
Mild steel	L-Cysteine-$Ti_3C_2T_x$	EP (E51)	0.2	−0.22	0.12	7.95×10^{-6}	12.1	[101]
Steel/Q235	P-CS-$Ti_3C_2T_x$	EP (E51)	0.2	−0.36	5.44	3.53×10^{-4}	19.8	[90]
Steel/Q235	$Ti_3C_2T_x$ and TiO_2	WPU	0.1	−0.04	2.79	3.25×10^{-5}	0.825	[119]
Steel/Q235	SF-$Ti_3C_2T_x$	WEP	0.5	--	--	--	8.30	[124]
Steel/Q235	IL-$Ti_3C_2T_x$	WEP	0.5	−0.81	14		0.016	[98]
Steel/Q235	CD-$Ti_3C_2T_x$	WEP	0.5	−0.08	--	--	4100	[104]
Al alloy/6082	Parallelized f$^+$-$Ti_3C_2T_x$	WEP	1	--	--	--	3.78 (1 week)	[92]
Steel/Q345	$Ti_3C_2T_x$@MoS_2	WEP	0.1	--	--	--	0.018	[127]
Steel/Q345	$Ti_3C_2T_x$@MgAl-LDH	WEP (H228A)	1.5	--	--	--	0.15	[114]
Steel/Q235	$Ti_3C_2T_x$/h-BN	WEP	0.5	--	--	--	1.2	[126]
Steel/Q235	AEAPTES-$Ti_3C_2T_x$/SDBS-CNTs	WPU	0.2	0.72	0.22	--	31.2	[80]
Steel/Q345	$Ti_3C_2T_x$/PANI	WEP (H228A)	--	−0.11	--	--	7.00	[86]
Steel/Q235	$Ti_3C_2T_x$@PANI	EP (WH53)	0.3	−0.58	47	3.63×10^{-4}	6.7×10^{-4}	[87]
Al alloy/6082	f-$Ti_3C_2T_x$-ZB($Ti_3C_2T_x$/PEI/ZIF-8/BTA)	WEP (H228A)	0.5	--	--	--	4.37×10^{-2}	[76]
Mild steel	APTES-$Ti_3C_2T_x$-Ce^{3+}	EP (High solid phase)	2.0	−0.67	3610	4.2×10^{-2}	4.8×10^{-4}	[77]
Mg alloy/AZ31	$Ti_3C_2T_x$@MgAl-LDH	WEP (H228A)	0.16	0.01	1.4	--	0.166 (as prepared) 0.00252 (3 weeks)	[68]

Note: C_{MX}: Content of MXene composite; E_{corr}: Corrosion potential; I_{corr}: Corrosion current density; R_{corr}: Corrosion rate, $|Z|_{0.01Hz}$: Impedance modulus. The corrosion performance was measured by immersing in a 3.5 wt.% NaCl solution without scratches for 30 min. Abbreviation: EP: Epoxy resin; WEP: Waterborne epoxy resin; PU: Polyurethane; WPU: Waterborne polyurethane; APTES: 3-aminopropyltriethoxysilane; AEAPTES: 3-(2-aminoethylamino)-propyl] trimethoxysilane; P-CS: Phosphorylated chitosan functionalized; GPTES: (3-glycidoxypropyl) trimethoxysilane; SF: Silk fibroin protein; IL: Ionic liquid; CD: Carbon quantum dots; SDBS: Sodium dodecylbenzenesulfonate; PANI: Polyaniline; PEI: Polyethyleneimine; ZIF-8: 2-methylimidazole; BTA: Benzotriazole.

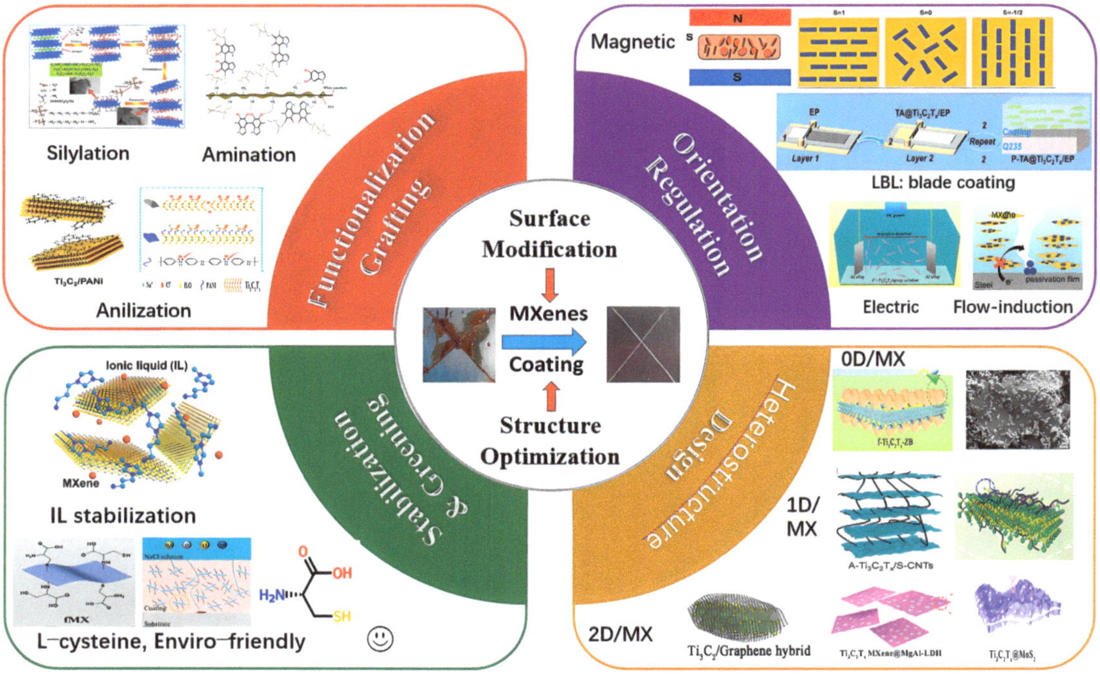

Figure 15. Summary of four surface-modification strategies to improve anticorrosion performance of MXene-based polymer composite coating.

Table 6. Comparison of advantages and weaknesses of the surface-modification methods to improve corrosion performance of $Ti_3C_2T_x$-based organic composite coatings.

Modification	Advantage	Disadvantage
Functionalization grafting	Versatile coupling agent; wide range of function adjustment; from non-conductive to conductive, from hydrophilic to hydrophobic	Design at atomic/molecular scale; complex preparation process; cost; reagents may not be environmentally friendly
Orientation regulation	Simple design idea; without new additive	Limited to flat surface; difficulty in uniform parallel arrangement; poor adhesion between coating and substrate
Heterostructure Nanocomposite	Simple design; versatile nanomaterials to choose from; easy-to-obtain multifunctionality (self-healing, etc.)	Complex preparation process; difficult to obtain precise layer-to-layer assembly; cost; unclear interaction between the heterostructure and the coating
Stabilization	Ionic liquids isolate oxygen, prevent oxidation, and improve stability	Limited types of reagents; cost; unclear interaction between the heterostructure and the coating
Greening	Using organic reagents such as amino acids; environmentally friendly	Limited available types of organic reagents; poor adhesion between coating and substrate

However, it should be noted that in practical applications, in addition to corrosion-protection performance, other properties, such as self-healing, anti-wear, and weather resistance, are also required. A single protective coating is difficult to meet both the requirements of antiwear and anticorrosion applications in harsh environments [131–133]. Therefore, multi-functional composite coatings are needed. To obtain MXene-based organic coatings with self-healing function, the groups of H. Yan (2021) [76], S.A. Haddadi (2021) [77], Z.H. Wang (2023) [68], X. Li (2023) [134], X. Huang (2024) [135], and other researchers have studied the self-healing performance of EP coatings by functionalizing

$Ti_3C_2T_x$ with amino groups and loading-corrosion inhibitors. A few months ago, the self-healing performance and mechanisms of MXene-based coatings for corrosion protection on metals were comprehensively reviewed by Cao's group [8].

In addition, an inorganic–organic hybridization multilayer protective system composed of MXene is being attempted and deserves attention.

4. Conclusions and Prospect

Based on its large specific surface area, abundant surface-functional groups, excellent hydrophilicity, mechanical properties, and ion impermeability, MXene has broad application prospects in the field of anticorrosion. This review focuses on the analysis of organic coatings based on $Ti_3C_2T_x$ nanostructures. MXene-based composite coating is a holistic system, and its corrosion-resistance behavior mainly depends on the interaction between MXene/polymer and the existing state of MXene in the internal space of the coating. By surface functionalization and composite treatment of MXene, the interface interaction between it and organic substrates can be improved by achieving better dispersion and compatibility. The uniform distribution and good orientation alignment of MXene are important for boosting the shielding effect of MXene on corrosive ions in organic coatings. Methods such as flow induction [104] and electrophoretic deposition [92] can achieve parallel alignment of MXene. Conductive MXene and metal substrates may undergo galvanic corrosion, which can accelerate the local corrosion of the metal. Therefore, it is necessary to reduce the conductivity of MXene, and insulation treatment is one of the feasible solutions.

In a critical review in 2024, H. Cao identified the challenges and opportunities to be addressed for MXene-based anticorrosion coatings from four aspects, namely preparation, structure, application and mechanism, and proposed the potential future jobs, which is shown in Figure 16. For more details, please refer to reference [4].

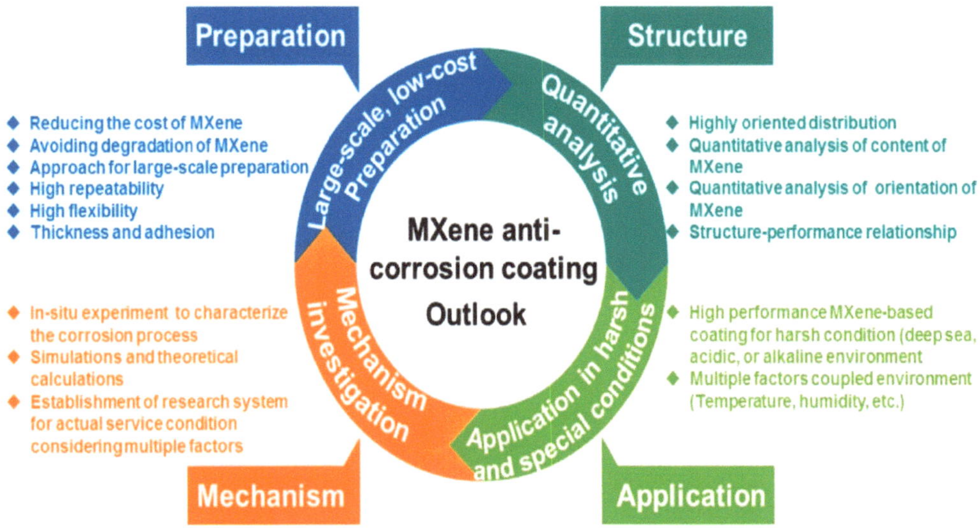

Figure 16. Challenges and opportunities in MXene-based anticorrosion coatings [4]. Reprinted with permission from [4], copyright 2024 Elsevier.

In our opinion, the following main issues still need to be addressed:

(1) The mass production of high-quality, ultra-thin, high-yield, and low-cost single- or few-layer MXenes nanosheets remains a significant challenge. Obtaining customized MXene-based materials with the required wettability, mechanical, chemical, or elec-

trical properties is key to achieving excellent protective performance in applications where the large-scale synthesis of fluorine-free, inexpensive, low conductivity, and highly stable MXene is the foundation. Currently, the exfoliation processes for MXene nanosheets are marked by high costs and low yields, predominantly confined to the laboratory scale. This necessitates the pursuit of cost-effective synthesis methods to make MXene nanosheets more accessible for broader applications.

(2) In terms of the preparation of organic coatings, the uniform dispersion and orientation arrangement of MXene nanosheets are the key factors in determining the anticorrosion performance of composite coatings. Addressing the restacking tendency of these nanosheets, various surface-modification techniques have been employed to prevent their aggregation. However, although covalent modifications are effective, they may disrupt the structural integrity of nanosheets and entail utilizing dangerous chemicals, which pose risks to both the environment and health. Hence, there is a growing call for greener, more sustainable, noncovalent modification methods, such as electrostatic and π–π interactions and hydrogen bonding, which preserve the nanosheet structure while enhancing dispersion and functional integration into coatings. Meanwhile, a simple, efficient, and low-cost coating-processing technology is crucial for industrial applications.

(3) At present, most of the research is conducted on the short-term corrosion behavior of simple components in laboratory environments. To achieve practical applications, it is necessary to conduct in-depth research on the failure mechanism of MXene-based organic anticorrosion coatings in long-term and complex corrosive environments. In addition, the understanding of the mechanism of uniform distribution of MXene and MXene/organic matrix interface is still insufficient, and more advanced experimental tools and/or theoretical methods should be developed to conduct in-depth research to better reveal the relationship between complex interfaces and the corrosion resistance of composite coatings, paving the way to optimize the internal structure of coatings.

(4) At present, the most studied MXene material is $Ti_3C_2T_x$, so it is necessary to expand the research object to other members of the MXene family. In addition to carbides, it should also be expanded to nitrides and carbonitrides. Meanwhile, apart from Ti-based MXenes, MXenes based on other metals, such as V-, Mo-, W-, etc., are also worth considering. The large family of MXenes will undoubtedly bring more choices for the design of MXene-based anticorrosion coatings.

(5) In practical applications, protection coatings should possess multiple functions. How to apply MXene to obtain composite coatings with synergistic functions such as corrosion resistance, self-healing, self-cleaning, anti-pollution, and antiwear is a new topic worthy of attention. However, to obtain more functionality and higher performance, the difficulty, complexity, and cost of the manufacturing process will undoubtedly increase. Currently, the production process faces hurdles such as high costs, both in the procurement of nanosheets and in the extensive use of organic solvents needed for optimal dispersion, pointing to an urgent need for a balance between synthesis, functionality, cost, and environmental impact.

In short, as a new member of a large class of 2D material, MXene has broad application prospects in the field of anticorrosion. Research at home and abroad is still in its nascent stages, and there is a lot of work worth carrying out. This article aims to provide some useful references and inspirations for researchers engaged in MXene preparation, performance, and MXene-based anticorrosion coatings.

Finally, in order to let readers conveniently understand the many abbreviations that occur in this review, a glossary of acronyms is given below.

Author Contributions: Conceptualization, S.Z. and L.F.; resources, G.Z., G.L. and F.W.; writing—original draft preparation, S.Z., G.Z., Z.W. and Q.W.; writing—review and editing, S.Z. and L.F.; supervision, L.F. and F.W.; project administration, L.F. and H.N.; funding acquisition, L.F. and H.N. All authors have read and agreed to the published version of the manuscript.

Funding: This work was supported by National Natural Science Foundation of China (Grant No. NSFC U22A2078), Open Fund of Key Laboratory of Green and High-end Utilization of Salt Lake Resources (Grant No. ISL2024-02), the Chongqing Natural Science Foundation Innovation and Development Joint Fund Project (Grant No. CSTB2022NSCQ-LZX0032), the Fundamental Research Funds for the Central Universities (No. 2022CDJQY-007, 2024CDJCYJ-001, 2024CDJXY022), Open Fund of Chongqing Key Laboratory of Interface Physics in Energy Conversion, Chongqing University (Grant No. CQSM202502), the Sharing Fund of Large-scale Equipment of Chongqing University (202303150164, 202303150169, 202303150178). We also thank the Analytical and Testing Center of Chongqing University for performing XRD, FT-IR, and SEM tests.

Institutional Review Board Statement: Not applicable.

Informed Consent Statement: Not applicable.

Data Availability Statement: No new data were created or analyzed in this study.

Conflicts of Interest: The authors declare no conflicts of interest.

List of Acronyms

0D	Zero-dimensional
1D	One-dimensional
2D	Two-dimensional
3D	Three-dimensional
AEAPTES	3-(2-amino-ethylamino)-propyl] trimethoxysilane, $C_8H_{22}O_3N_2Si$, KH-792
AHP	Alternating hydrostatic pressure
[AMIm]Br	1-allyl-3-methylimidazolium bromide
[AMIm]I	1-allyl-3-methylimidazolium iodide
AMTES	Anilino-methyl-triethoxy-silane, $C_{13}H_{23}NO_3Si$, KH-ND42
AP	Atmospheric pressure
APTES	3-aminopropyl-triethoxy-silane, $C_9H_{23}NO_3Si$, KH-550
[APMIm]Br	1-aminopropyl-3-methylimidazolium bromide
BTA	Benzotriazole
CD	Carbon quantum dots
CNT	Carbon nanotube
COF	Covalent organic frameworks
CP	Conductive polymers
CS	Chitosan
CVD	Chemical vapor deposition
DA	Dopamine, C8H11NO2
DDAB	didodecyldimethylammonium bromide
DTAB	decyltrimethylammonium bromide
DTES	Dodecyltriethoxysilane, $C_{18}H_{40}O_3Si$
E_{corr}	Corrosion potential
EIS	Electrochemical impedance spectroscopy
EP	Epoxy resin
GDP	Gross domestic product
GPTES	(3-glycidoxypropyl) trimethoxysilane
GO	Grapheme oxide
Gr	Grapheme

h-BN	Hexagonal boron nitride
HNTs	Halloysite nanotubes
I_{corr}	Corrosion current density
IL	Ionic liquid
LbL	Layer-by-layer deposition
L-Cys	L-cysteine amino acid
LDH	Layered double hydroxide
LEIS	Local electrochemical impedance spectroscopy
MAPTES	γ-methacryloxy-propyl-trimethoxysilane, $C_{10}H_{20}O_5Si$, KH-570
MOF	Metal organic framework
MPS	3-methacryloxy-propyltrimethoxy-silane
MTPCs	Multilayer $Ti_3C_2T_x$/PANI composite
MXene	2D transition metal carbides/nitrides
NF	Nanofiber
NP	Nanoparticle
NMs	Nanomaterials
NSs	Nanosheets
NT	Nanotube
NW	Nanowire
OTAB	Octadecyl trimethylammonium bromide
PAA	Polyacrylic acid
PANI	Polyaniline
PBZ	Polybenzoxazine
PDA	Polydopamine
PDDA	Poly(diallyldimethylammonium chloride)
PDMS	Polydimethylsiloxane
PDMAEMA	Poly2-dimethylaminoethyl metchacrylate
PEI	Polyethyleneimine
PPD	Phenylenediamine
PPy	Polypyrrole
PS	Polystyrene
PT	Polythiophene
PU	Polyurethane
PVA	Polyvinyl alcohol
QD	Quantum dots
SDBS	Sodium dodecylbenzenesulfonate
SF	Silk fiber
STECC	Superhydrophobic $Ti_3C_2T_x$/epoxy/cerium conversion composite coating
TA	Tannic acid
TML	$Ti_3C_2T_x$/MgAl-LDH heterostructure
TPCs	$Ti_3C_2T_x$/PANI composite materials
vdW	van der Waals
VER	Vinyl ester resin
WCA	Water-contact angle
WEP	Waterborne epoxy
WPU	Waterborne polyurethane
$\vert Z_{0.01Hz}\vert$	Impedance modulus corresponding to 0.01 Hz
ZIF-8	Zeolitic imidazolate framework-8
ZRC	Zinc-rich coatings
ZRE	Zinc-rich epoxy coating

References

1. Shang, C.; Sang, D.; Li, C.; Zou, L.; Wu, J.; Wang, Q. 2D materials for marine corrosion protection: A review. *APL Mater.* **2024**, *12*, 060601. [CrossRef]

2. Wu, Y.; Sun, Y.; Zhao, W.; Wang, L. Two-dimensional nanomaterials reinforced organic coatings for marine corrosion protection: State of the art, challenges, and future prospectives. *Adv. Mater.* **2024**, *36*, 2312460. [CrossRef] [PubMed]
3. Amin, I.; Brekel, H.V.D.; Nemani, K.; Batyrev, E.; de Vooys, A.; van der Weijde, H.; Anasori, B.; Shiju, N.R. $Ti_3C_2T_x$ MXene polymer composites for anticorrosion: An overview and perspective. *ACS Appl. Mater. Interfaces* **2022**, *14*, 43749–43758. [CrossRef]
4. Cao, H. Beyond graphene and boron nitride: Why MXene can be used in composite for corrosion protection on metals? *Compos. Part B Eng.* **2023**, *271*, 111168. [CrossRef]
5. Sukanya, R.; Barwa, T.N.; Luo, Y.; Dempsey, E.; Breslin, C.B. Emerging layered materials and their applications in the corrosion protection of metals and alloys. *Sustainability* **2022**, *14*, 4079. [CrossRef]
6. Fang, L.; Zhang, S.F.; Wang, Z.H.; Wu, F.; Xiao, L.L. Research progress on MXene-based organic composite anti-corrosion coatings. *J. Sichuan Univ. Sci. Eng. (Nat. Sci. Ed.)* **2024**, *37*, 1–14.
7. Zhang, W.; Yan, W.; Zhao, X.; Huang, K.; Li, J. Research on two-dimensional layered materials for metal corrosion protection: Advances and challenges. *Prog. Org. Coat.* **2024**, *194*, 08589. [CrossRef]
8. Ma, X.; Gong, B.; Wang, T.; Hou, J.; Ji, S.; Xu, Q.; Cao, H. Toward self-healing two dimensional MXene coatings for corrosion protection on metals: Design strategies and mechanisms. *Adv. Colloid Interface Sci.* **2025**, *335*, 103340. [CrossRef]
9. Shi, L.; Hu, J.; Lin, X.; Fang, L.; Wu, F.; Xie, J.; Meng, F. A robust superhydrophobic PPS-PTFE/SiO_2 composite coating on AZ31 Mg alloy with excellent wear and corrosion resistance properties. *J. Alloys Compd.* **2017**, *721*, 157–163. [CrossRef]
10. Xie, J.; Hu, J.; Lin, X.; Fang, L.; Wu, F.; Liao, X.; Luo, H.; Shi, L. Robust and anti-corrosive PDMS/SiO_2 superhydrophobic coatings fabricated on magnesium alloys with different-sized SiO_2 nanoparticles. *Appl. Surf. Sci.* **2018**, *457*, 870–880. [CrossRef]
11. Tang, Y.; Wu, F.; Fang, L.; Ruan, H.; Hu, J.; Zeng, X.; Zhang, S.; Luo, H.; Zhou, M. Effect of deposition sequence of MgAl-LDH and SiO_2@ PDMS layers on the corrosion resistance of robust superhydrophobic/self-healing multifunctional coatings on magnesium alloy. *Prog. Org. Coat.* **2023**, *174*, 107299. [CrossRef]
12. Lyon, S.B.; Bingham, R.; Mills, D.J. Advances in corrosion protection by organic coatings: What we know and what we would like to know. *Prog. Org. Coat.* **2017**, *102*, 2–7. [CrossRef]
13. Ou, B.; Wang, Y.; Lu, Y. A review on fundamentals and strategy of epoxy-resin-based anticorrosive coating materials. *Polym.-Plast. Technol. Mater.* **2021**, *60*, 601–625. [CrossRef]
14. Jiang, W.; Yang, L.; Zheng, Q.; Dong, J.; Zhong, C. Research progress on enhancing corrosion resistance of organic polymeric coatings by MOFs. *Surf. Technol.* **2024**, *53*, 1–14.
15. Ling, Z.; Ren, C.E.; Zhao, M.Q.; Yang, J.; Giammarco, J.M.; Qiu, J.; Barsoum, M.W.; Gogotsi, Y. Flexible and conductive MXene films and nanocomposites with high capacitance. *Proc. Natl. Acad. Sci. USA* **2014**, *111*, 16676–16681. [CrossRef] [PubMed]
16. Chen, J.; Chen, K.; Tong, D.; Huang, Y.; Zhang, J.; Xue, J.; Huang, Q.; Chen, T. CO_2 and temperature dual responsive "Smart" MXene phases. *Chem. Commun.* **2015**, *51*, 314–317. [CrossRef]
17. Hao, L.; Zhang, H.; Wu, X.; Zhang, J.; Wang, J.; Li, Y. Novel thin-film nanocomposite membranes filled with multi-functional $Ti_3C_2T_x$ nanosheets for task-specific solvent transport. *Compos. Part A Appl. Sci. Manuf.* **2017**, *100*, 139–149. [CrossRef]
18. Si, J.Y.; Tawiah, B.; Sun, W.L.; Lin, B.; Wang, C.; Yuen, A.C.Y.; Yu, B.; Li, A.; Yang, W.; Lu, H.D. Functionalization of MXene nanosheets for polystyrene towards high thermal stability and flame retardant properties. *Polymers* **2019**, *11*, 976. [CrossRef] [PubMed]
19. Cai, M.; Yan, H.; Song, S.; He, D.; Lin, Q.; Li, W.; Fan, X.; Zhu, M. State-of-the-art progresses for $Ti_3C_2T_x$ MXene reinforced polymer composites in corrosion and tribology aspects. *Adv. Colloid Interface Sci.* **2022**, *309*, 102790. [CrossRef]
20. He, X.; Cui, C.; Chen, Y.; Zhang, L.; Sheng, X.; Xie, D. MXene and polymer collision: Sparking the future of high-performance multifunctional coatings. *Adv. Funct. Mater.* **2024**, *34*, 2409675. [CrossRef]
21. Tang, Y.; Wu, F.; Fang, L.; Guan, T.; Hu, J.; Zhang, S. A comparative study and optimization of corrosion resistance of ZnAl layered double hydroxides films intercalated with different anions on AZ31Mg alloys. *Surf. Coat. Technol.* **2019**, *358*, 594–603. [CrossRef]
22. Chen, J.; Fang, L.; Wu, F.; Xie, J.; Hu, J.; Jiang, B.; Luo, H. Corrosion resistance of a self-healing rose-like MgAl-LDH coating intercalated with aspartic acid on AZ31 Mg alloy. *Prog. Org. Coat.* **2019**, *136*, 105234. [CrossRef]
23. Chen, J.L.; Fang, L.; Wu, F.; Hu, J.; Zhang, S.F.; Jiang, B.; Luo, H.J. Comparison of corrosion resistance of MgAl-LDHs and ZnAl-LDHs films intercalated with organic anions ASP on AZ31 Mg alloys. *Trans. Nonferrous Met. Soc. China* **2020**, *30*, 2424–2434. [CrossRef]
24. Yan, H.; Wang, J.; Meng, C.; Wang, X.; Song, S.; Fan, X.; Zhang, L.; Li, H.; Li, W.; Zhu, M. Towards long-term corrosion and wear protection of Al alloy: Synergy of $Ti_3C_2T_x$ flake and micro-arc oxidation coating. *Corros. Sci.* **2020**, *174*, 108813. [CrossRef]
25. Song, Y.H.; Tang, Y.; Fang, L.; Wu, F.; Zeng, X.; Hu, J.; Zhang, S.F.; Jiang, B.; Luo, H.J. Enhancement of corrosion resistance of AZ31 Mg alloys by one-step in situ synthesis of ZnAl-LDH films intercalated with organic anions (ASP, La). *J Magnes. Alloys* **2021**, *9*, 658–667. [CrossRef]
26. Gong, K.; Zhou, K.; Qian, X.; Shi, C.; Yu, B. MXene as emerging nanofillers for high-performance polymer composites: A review. *Compos. Part B Eng.* **2021**, *217*, 108867. [CrossRef]

27. Riazi, H.; Nemani, S.K.; Grady, M.C.; Anasori, B.; Soroush, M. Ti_3C_2 MXene-Polymer nanocomposites and their applications. *J. Mater. Chem. A.* **2021**, *9*, 8051–8098. [CrossRef]
28. Cao, H.; Neal, N.N.; Pas, S.; Radovic, M.; Lutkenhaus, J.L.; Green, M.J.; Pentzer, E.B. Architecting MXenes in polymer composites. *Prog. Polym. Sci.* **2024**, *153*, 101830. [CrossRef]
29. Carey, M.; Barsoum, M.W. MXene polymer nanocomposites: A review. *Mater. Today Adv.* **2021**, *9*, 100120. [CrossRef]
30. Yan, H.; Li, W.; Li, H.; Fan, X.; Zhu, M. Ti_3C_2 MXene nanosheets toward high-performance corrosion inhibitor for epoxy coating. *Prog. Org. Coat.* **2019**, *135*, 156–167. [CrossRef]
31. Firouzjaei, M.D.; Karimiziarani, M.; Moradkhani, H.; Elliott, M.; Anasori, B. MXenes: The two-dimensional influencers. *Mater. Today Adv.* **2022**, *13*, 100202. [CrossRef]
32. Khan, K.; Tareen, A.K.; Iqbal, M.; Hussain, I.; Mahmood, A.; Khan, U.; Khan, M.F.; Zhang, H.; Xie, Z. Recent advances in MXenes: A future of nanotechnologies. *J. Mater. Chem. A* **2023**, *11*, 19764–19811. [CrossRef]
33. Rostami, M.; Badiei, A.; Ziarani, G.M. A review of recent progress in the synthesis of 2D Ti_3C_2Tx MXenes and their multifunctional applications. *Inorg. Chem. Commun.* **2024**, *163*, 112362. [CrossRef]
34. Wang, P.; Wang, B.Q.; Wang, R. Progress in the synthesis process and electrocatalytic application of Mxene materials. *Materials* **2023**, *16*, 6816. [CrossRef] [PubMed]
35. Abid, Z.; Rafiq, K.; Aslam, A.; Xie, R.; Hussain, E. Scope, evaluation and current perspectives of MXene synthesis strategies for state of the art applications. *J. Mater. Chem. A.* **2024**, *12*, 7351–7395. [CrossRef]
36. Zhang, C.J. Interfacial assembly of two-dimensional MXenes. *J. Energy Chem.* **2021**, *60*, 417–434. [CrossRef]
37. Naguib, M.; Kurtoglu, M.; Presser, V.; Lu, J.; Niu, J.; Heon, M.; Hultman, L.; Gogotsi, Y.; Barsoum, M.W. Two-dimensional nanocrystals produced by exfoliation of Ti_3AlC_2. *Adv. Mater.* **2011**, *23*, 4248–4253. [CrossRef]
38. Naguib, M.; Mashtalir, O.; Carle, J.; Presser, V.; Lu, J.; Hultman, L.; Gogotsi, Y.; Barsoum, M.W. Two-dimensional transition metal carbides. *ACS Nano* **2012**, *6*, 1322–1331. [CrossRef]
39. Murali, G.; Reddy Modigunta, J.K.; Park, Y.H.; Lee, J.H.; Rawal, J.; Lee, S.Y.; In, I.; Park, S.J. A review on MXene synthesis, stability, and photocatalytic applications. *ACS Nano* **2022**, *16*, 13370–13429. [CrossRef]
40. Liu, F.; Zhou, A.; Chen, J.; Jia, J.; Zhou, W.; Wang, L.; Hu, Q. Preparation of Ti_3C_2 and Ti_2C MXenes by fluoride salts etching and methane adsorptive properties. *Appl. Surf. Sci.* **2017**, *416*, 781–789. [CrossRef]
41. Ghidiu, M.; Lukatskaya, M.R.; Zhao, M.Q.; Gogotsi, Y.; Barsoum, M.W. Conductive two-dimensional titanium carbide 'clay'with high volumetric capacitance. *Nature* **2014**, *516*, 78–81. [CrossRef] [PubMed]
42. Li, T.; Yao, L.; Liu, Q.; Gu, J.; Luo, R.; Li, J.; Yan, X.; Wang, W.; Liu, P.; Chen, B.; et al. Fluorine-free synthesis of high-purity $Ti_3C_2T_x$ (T= OH, O) via alkali treatment. *Angew. Chem. Int. Ed.* **2018**, *57*, 6115–6119. [CrossRef] [PubMed]
43. Yang, S.; Zhang, P.; Wang, F.; Ricciardulli, A.G.; Lohe, M.R.; Blom, P.W.; Feng, X. Fluoride-free synthesis of two-dimensional titanium carbide (MXene) using a binary aqueous system. *Angew. Chem. Int. Ed.* **2018**, *130*, 15717–15721. [CrossRef]
44. Ding, H.; Li, Y.; Li, M.; Chen, K.; Liang, K.; Chen, G.; Lu, J.; Palisaitis, J.; Persson, P.O.; Eklund, P.; et al. Chemical scissor–mediated structural editing of layered transition metal carbides. *Science* **2023**, *379*, 1130–1135. [CrossRef] [PubMed]
45. Kamysbayev, V.; Filatov, A.S.; Hu, H.; Rui, X.; Lagunas, F.; Wang, D.; Klie, R.F.; Talapin, D.V. Covalent surface modifications and superconductivity of two-dimensional metal carbide MXenes. *Science* **2020**, *369*, 979–983. [CrossRef] [PubMed]
46. Naguib, M.; Barsoum, M.W.; Gogotsi, Y. Ten years of progress in the synthesis and development of MXenes. *Adv. Mater.* **2021**, *33*, 2103393. [CrossRef]
47. Lukatskaya, M.R.; Mashtalir, O.; Ren, C.E.; Dall'Agnese, Y.; Rozier, P.; Taberna, P.L.; Naguib, M.; Simon, P.; Barsoum, M.W.; Gogotsi, Y. Cation intercalation and high volumetric capacitance of two-dimensional titaniumcarbide. *Science* **2013**, *341*, 1502–1505. [CrossRef] [PubMed]
48. Shahzad, F.; Alhabeb, M.; Hatter, C.B.; Anasori, B.; Man Hong, S.; Koo, C.M.; Gogotsi, Y. Electromagnetic interference shielding with 2D transition metal carbides (MXenes). *Science* **2016**, *353*, 1137–1140. [CrossRef] [PubMed]
49. Iqbal, A.; Shahzad, F.; Hantanasirisakul, K.; Kim, M.K.; Kwon, J.; Hong, J.; Kim, H.; Kim, D.; Gogotsi, Y.; Koo, C.M. Anomalous absorption of electromagnetic waves by 2D transition metal carbonitride Ti_3CNT_x (MXene). *Science* **2020**, *369*, 446–450. [CrossRef] [PubMed]
50. Yury, G.; Anasori, B. The rise of MXenes. *ACS Nano* **2019**, *13*, 8491–8494. [CrossRef]
51. Lei, J.C.; Zhang, X.; Zhou, Z. Recent advances in MXene: Preparation, properties, and applications. *Front. Phys.* **2015**, *10*, 276–286. [CrossRef]
52. Alhabeb, M.; Maleski, K.; Anasori, B.; Lelyukh, P.; Clark, L.; Sin, S.; Gogotsi, Y. Guidelines for synthesis and processing of two-dimensional titanium carbide ($Ti_3C_2T_x$ MXene). *Chem. Mater.* **2017**, *29*, 7633–7644. [CrossRef]
53. Ng, V.M.H.; Huang, H.; Zhou, K.; Lee, P.S.; Que, W.; Xu, J.Z.; Kong, L.B. Recent progress in layered transition metal carbides and/or nitrides (MXenes) and their composites: Synthesis and applications. *J. Mater. Chem. A* **2017**, *5*, 3039–3068. [CrossRef]
54. Verger, L.; Xu, C.; Natu, V.; Cheng, H.M.; Ren, W.; Barsoum, M.W. Overview of the synthesis of MXenes and other ultrathin 2D transition metal carbides and nitrides. *Curr. Opin. Solid State Mater. Sci.* **2019**, *23*, 149–163. [CrossRef]

55. Jiang, X.; Kuklin, A.V.; Baev, A.; Ge, Y.; Ågren, H.; Zhang, H.; Prasad, P.N. Two-dimensional MXenes: From morphological to optical, electric, and magnetic properties and applications. *Phys. Rep.* **2020**, *848*, 1–58. [CrossRef]
56. He, L.; Wu, F.; Bai, M.; Ruan, H.; Fang, L.; Hu, J.; Luo, H.; Zhang, S.; Hu, B.; Zhou, M. NiCo-S nanosheets grown on $Ti_3C_2T_x$-wrapped carbon cloth for high- performance flexible supercapacitor. *J. Alloys Compd.* **2022**, *908*, 164493. [CrossRef]
57. He, L.; Wu, F.; Fang, L.; Hu, J.; Luo, H.; Zhang, S.; Hu, B.; Zhou, M. 3D Self-supported $NiS_2/Ti_3C_2T_x$-CC composite electrode for high-performance flexible supercapacitors. *Integr. Ferroelectr.* **2022**, *226*, 172–184. [CrossRef]
58. Zhou, H.; Wu, F.; Fang, L.; Hu, J.; Luo, H.J.; Guan, T.; Hu, B.S.; Zhou, M. Layered NiFe-LDH/MXene nanocomposite electrode for high-performance supercapacitor. *Int. J. Hydrogen Energy* **2020**, *45*, 13080–13089. [CrossRef]
59. Zhou, H.; Lu, Y.; Wu, F.; Fang, L.; Luo, H.; Zhang, Y.; Zhou, M. MnO_2 nanorods/MXene/CC composite electrode for flexible supercapacitors with enhanced electrochemical performance. *J. Alloys Compd.* **2019**, *802*, 259–268. [CrossRef]
60. Li, M.; Fang, L.; Zhou, H.; Wu, F.; Lu, Y.; Luo, H.; Zhang, Y.; Hu, B. Three-dimensional porous MXene/NiCo-LDH composite for high performance non-enzymatic glucose sensor. *Appl. Surf. Sci.* **2019**, *495*, 143554. [CrossRef]
61. Xiang, Y.; Fang, L.; Wu, F.; Zhang, S.; Ruan, H.; Luo, H.; Zhang, H.; Li, W.; Long, X.; Hu, B.; et al. 3D crinkled Alk-Ti_3C_2 MXene based flexible piezoresistive sensors with ultra-high sensitivity and ultra-wide pressure range. *Sci. Technol. Adv. Mat.* **2021**, *6*, 2001157. [CrossRef]
62. Wang, H.; Wu, F.; Wang, Z.; Wang, Y.; Zhang, S.; Luo, H.; Zheng, Z.; Fang, L. Ultra-fast and ultra-efficient removal of Cr (VI) by the aqueous solutions of monolayer MXene ($Ti_3C_2T_x$). *Chemosphere* **2022**, *308*, 136573. [CrossRef] [PubMed]
63. Luo, X.; Wang, H.; Ren, X.; Liu, G.; Luo, H.; Zheng, Z.; Wu, F. Effective adsorptive removal of heavy metal anion Cr(VI) and dye cation methylene blue by the novel 2D material $TiVCT_x$ MXene. *J. Alloys Compd.* **2024**, *1010*, 177409. [CrossRef]
64. Yi, L.; Wang, H.; Ren, X.; Liu, G.; Nian, H.; Zheng, Z.; Wu, F. Enhancing Cr(VI) removal performance of $Ti_3C_2T_x$ through structural modification by using a spray freezing method. *RSC Adv.* **2024**, *14*, 28320. [CrossRef]
65. Wang, H.; Ren, X.; Wu, F.; Liu, G.; Zhang, S.; Luo, H.; Zheng, Z.; Fang, L. Rational design of OH-functionalized Ti_2CT_x nanosheets and composite aerogel spheres: A highly efficient approach for U(VI) removal from simulated seawater. *J. Water. Process. Eng.* **2024**, *67*, 106091. [CrossRef]
66. Wang, H.; Ren, X.; Wu, F.; Zhang, S.; Fang, L. Investigation on high EMI shielding effectiveness and shielding mechanism of spherical $Ti_3C_2T_x$ microfilm prepared by spray-freezing. *J. Alloys Compd.* **2024**, *976*, 173317. [CrossRef]
67. Hu, Y.; Wang, H.; Ren, X.; Wu, F.; Liu, G.B.; Zhang, S.F.; Fang, L. Enhancing methylene blue adsorption performance of $Ti_3C_2T_x$@sodium alginate foam through pore structure regulation. *Nanomaterials* **2024**, *14*, 1925. [CrossRef] [PubMed]
68. Wang, Z.; Fang, L.; Wu, F.; Ruan, H.; Tang, Y.; Hu, J.; Zeng, X.; Zhang, S.; Luo, H. Anti-corrosion, self-healing and environmental-friendly $Ti_3C_2T_x$/MgAl-LDH@epoxy composite organic coating for Mg alloy protection. *J. Mater. Sci.* **2023**, *58*, 3283–3306. [CrossRef]
69. Cao, H.; Fang, M.; Jia, W.; Liu, X.; Xu, Q. Remarkable improvement of corrosion resistance of silane composite coating with $Ti_3C_2T_x$ MXene on copper. *Compos. Part B Eng.* **2022**, *228*, 109427. [CrossRef]
70. Song, G.; Deng, Q.; Wang, B.; Liu, Z.; Ye, C.; Wei, X.; Dai, D.; Pan, Z.; Li, W.; Song, S.; et al. Thermal and corrosion behavior of Ti_3C_2/ copper composites. *Compos. Commun.* **2020**, *22*, 100498. [CrossRef]
71. Zhang, L.; Huang, S.; Weng, Y.; Li, J.; Han, P.; Ye, S.; Zhang, X. Preparation of Ni-P-$Ti_3C_2T_x$-Ce composite coating with enhanced wear resistance and electrochemical corrosion behavior on the surface of low manganese steel. *Surf. Coat. Technol.* **2022**, *441*, 128508. [CrossRef]
72. Xiao, Q.; Zou, J.; Chen, B.; Luo, J.; Lu, Y.; Chen, W.; Zhang, X.; Liang, T. Influence of $Ti_3C_2T_x$ addition on the corrosion behavior and tribological properties of electrodeposited Ni-W-$Ti_3C_2T_x$ nanocomposite coatings. *Mater. Sci. Technol.* **2022**, *38*, 703–715. [CrossRef]
73. Ji, J.; Zhao, L.; Shen, Y.; Liu, S.; Zhang, Y. Covalent stabilization and functionalization of Mxene via silylation reactions with improved surface properties. *FlatChem* **2019**, *17*, 100128. [CrossRef]
74. Yan, H.; Cai, M.; Li, W.; Fan, X.; Zhu, M. Amino-functionalized $Ti_3C_2T_x$ with anti-corrosive/wear function for waterborne epoxy coating. *J. Mater. Sci. Technol.* **2020**, *54*, 144–159. [CrossRef]
75. Zhang, F.; Liu, W.; Wang, S.; Liu, C.; Shi, H.; Liang, L.; Pi, K. Surface functionalization of $Ti_3C_2T_x$ and its application in aqueous polymer nanocomposites for reinforcing corrosion protection. *Compos. Part B Eng.* **2021**, *217*, 108900. [CrossRef]
76. Yan, H.; Fan, X.; Cai, M.; Song, S.; Zhu, M. Amino-functionalized $Ti_3C_2T_x$ loading ZIF-8 nanocontainer@ benzotriazole as multifunctional composite filler towards self-healing epoxy coating. *J. Colloid Interface Sci.* **2021**, *602*, 131–145. [CrossRef] [PubMed]
77. Haddadi, S.A.; Hu, S.; Ghaderi, S.; Ghanbari, A.; Ahmadipour, M.; Pung, S.Y.; Li, S.; Feilizadeh, M.; Arjmand, M. Amino-functionalized MXene nanosheets doped with Ce (III) as potent nanocontainers toward self-healing epoxy nanocomposite coating for corrosion protection of mild steel. *ACS Appl. Mater. Inter.* **2021**, *13*, 42074–42093. [CrossRef] [PubMed]
78. Li, X.; Zhou, S. Epoxy-functionalized Ti_3C_2 nanosheet for epoxy coatings with prominent anticorrosion performance. *Prog. Org. Coat.* **2022**, *162*, 106559. [CrossRef]

79. Pourhashem, S.; Rashidi, A.; Vaezi, M.R.; Bagherzadeh, M.R. Excellent corrosion protection performance of epoxy composite coatings filled with amino-silane functionalized graphene oxide. *Surf. Coat. Technol.* **2017**, *317*, 1–9. [CrossRef]
80. Wang, Y.; Chang, W.; Wang, Z.; Ma, J.; Yu, X.; Li, Y.; Zhang, L. Silanized MXene/carbon nanotube composites as a shielding layer of polyurethane coatings for anticorrosion. *ACS Appl. Nano Mater.* **2022**, *5*, 1408–1418. [CrossRef]
81. Zhou, Z.; Pourhashem, S.; Wang, Z.; Sun, J.; Ji, X.; Zhai, X.; Duan, J.; Hou, B. Mxene structure: A key parameter in corrosion barrier performance of organic coatings. *J. Ind. Eng. Chem.* **2022**, *116*, 310–320. [CrossRef]
82. Yin, X.; Li, B.; Xue, S.; Mu, P.; Li, J. Fabrication of $Ti_3C_2T_x$-based superhydrophobic composite coating for reinforcement corrosion resistance of AZ31B Mg alloy. *Adv. Eng. Mater.* **2023**, *25*, 2201573. [CrossRef]
83. Chen, W.; Wu, Z.; He, X.; Su, Y.; Zheng, G.; Oh, S.K.; Mei, C. Achieving superior anti-corrosion properties of vinyl ester resin coatings via compositing with 3-methacryloxy propyl trimethoxysilane functionalized MXene nanosheets. *Polym. Test.* **2023**, *127*, 108203. [CrossRef]
84. Jafari, Y.; Ghoreishi, S.M.; Shabani-Nooshabadi, M.J.S.M. Polyaniline/graphene nanocomposite coatings on copper: Electropolymerization, characterization, and evaluation of corrosion protection performance. *Synthetic Met.* **2016**, *217*, 220–230. [CrossRef]
85. Cai, M.; Yan, H.; Li, Y.; Li, W.; Fan, X.; Zhu, M. Elucidating the electrochemical mechanism for enhanced corrosion of $Ti_3C_2T_x$-coated mild steel. *Surf. Topogr. Metrol. Prop.* **2021**, *9*, 035033. [CrossRef]
86. Cai, M.; Yan, H.; Li, Y.; Li, W.; Li, H.; Fan, X.; Zhu, M. $Ti_3C_2T_x$/PANI composites with tunable conductivity towards anticorrosion application. *Chem. Eng. J.* **2021**, *410*, 128310. [CrossRef]
87. Li, C.; Xu, J.; Xu, Q.; Xue, G.; Yu, H.; Wang, X.; Lu, J.; Cui, G.; Gu, G. Synthesis of $Ti_3C_2T_x$ MXene@PANI composites for excellent anticorrosion performance of waterborne epoxy coating. *Prog. Org. Coat.* **2022**, *165*, 106673. [CrossRef]
88. Kaewsaneha, C.; Thananukul, K.; Opaprakasit, P.; Tangboriboonrat, P.; Sreearunothai, P. Hybrid MXene ($Ti_3C_2T_x$)/polyaniline nanosheets as additives for enhancing anticorrosion properties of Zn-epoxy coating. *Prog. Org. Coat.* **2022**, *173*, 107173. [CrossRef]
89. Nazarlou, Z.; Hosseini, S.F.; Seyed Dorraji, M.S.; Rasoulifard, M.H.; Aydemir, U. Ti_3C_2 MXene/polyaniline/montmorillonite nanostructures toward solvent-free powder coatings with enhanced corrosion resistance and mechanical properties. *ACS Appl. Nano Mater.* **2023**, *6*, 8804–8818. [CrossRef]
90. He, X.; Li, S.; Shen, R.; Ma, Y.; Zhang, L.; Sheng, X.; Chen, Y.; Xie, D.; Huang, J. A high-performance waterborne polymeric composite coating with long-term anti-corrosive property based on phosphorylation of chitosan-functionalized $Ti_3C_2T_x$ MXene. *Adv. Compos. Hybrid Mater.* **2022**, *5*, 1699–1711. [CrossRef]
91. Zhang, Y.; Chen, C.; Chen, Z.; Zhang, T.; Wang, Y.; Cao, S.; Ma, J. Superior anticorrosion performance of well-dispersed MXene-polymer composite coatings enabled by covalent modification and ambient electron-beam curing. *ACS Appl. Mater. Interfaces* **2023**, *15*, 11099–11110. [CrossRef] [PubMed]
92. Fan, X.; Yan, H.; Cai, M.; Song, S.; Huang, Y.; Zhu, M. Achieving parallelly-arranged $Ti_3C_2T_x$ in epoxy coating for anti-corrosive/wear high- efficiency protection. *Compos. Part B Eng.* **2022**, *231*, 109581. [CrossRef]
93. He, X.; Wu, J.; Chen, Y.; Zhang, L.; Sheng, X. A trace amount of MXene@PDA nanosheets for low-temperature zinc phosphating coatings with superb corrosion resistance. *Appl. Surf. Sci.* **2022**, *603*, 154455. [CrossRef]
94. Chen, Y.; Zhan, Y.; Dong, H.; Li, Y.; Yang, X.; Sun, A.; Chen, X.; Zhu, F.; Jia, H. Two-dimensional lamellar MXene nanosheets/waterborne epoxy composite coating: Dopamine triggered surface modification and long-term anticorrosion performance. *Colloids Surf. A* **2023**, *674*, 131865. [CrossRef]
95. Li, X.; Zhou, S. Dopamine monomer functionalized Ti_3C_2 nanosheets and their anticorrosion improvement for waterborne epoxy coatings. *J. Mater. Res. Technol.* **2022**, *20*, 210–220. [CrossRef]
96. Cai, M.; He, C.; Li, W.; Fan, X.; Wang, J.; Zhu, M. Polydopamine-modified $Ti_3C_2T_x$ to enhance anticorrosion of waterborne zinc-rich epoxy coating. *Carbon* **2023**, *215*, 118467. [CrossRef]
97. Shi, Y.; Chen, C.; Li, Y.; Zhao, W. Achieving dual functional corrosion resistance for epoxy coatings under alternating hydrostatic pressure via constructing P-phenylenediamine/$Ti_3C_2T_x$ hybrids. *Carbon* **2023**, *201*, 1048. [CrossRef]
98. Zhao, H.; Ding, J.; Zhou, M.; Yu, H. Air-stable titanium carbide Mxene nanosheets for corrosion protection. *ACS Appl. Nano Mater.* **2021**, *4*, 3075–3086. [CrossRef]
99. Ning, Y.; Jian, D.; Liu, S.; Chen, F.; Song, Y.; Li, S.; Liu, B. Designing a $Ti_3C_2T_x$ MXene with long-term antioxidant stability for high-performance anti-corrosion coatings. *Carbon* **2023**, *202*, 20–30. [CrossRef]
100. Mao, X.; He, Y.; Li, C.; Li, H.; Gou, R.; Liu, H.; Zhao, Y.; Chen, W.; Yan, J.; Yuan, X. Glycine-$Ti_3C_2T_x$ hybrid material improves the electrochemical corrosion resistance of a water-borne epoxy coating. *Langmuir* **2024**, *40*, 11817–11827. [CrossRef]
101. Li, S.; Huang, H.; Chen, F.; He, X.; Ma, Y.; Zhang, L.; Sheng, X.; Chen, Y.; Shchukina, E.; Shchukin, D. Reinforced anticorrosion performance of waterborne epoxy coating with eco-friendly L-cysteine modified $Ti_3C_2T_x$ MXene nanosheets. *Prog. Org. Coat.* **2021**, *161*, 106478. [CrossRef]
102. Tan, W.; Zhao, W. Mille crepe cake structure-inspired fabricating parallelly-arranged TA@$Ti_3C_2T_x$ composite coating for long-term corrosion resistance under alternating hydrostatic pressure. *Chem. Eng. J.* **2023**, *454*, 140334. [CrossRef]

103. Huang, L.; Du, Z.; Tian, Y.; Sun, J.; Li, J.; Xing, Y.; Liang, X.; Zhang, X. L-Cysteine-encapsulated MXene nanosheet possessing ultra-antioxidation in aqueous suspension for electrophoretic deposition assisted carbon-fiber-surface modification. *Adv. Mater. Interfaces* **2024**, *11*, 2300851. [CrossRef]
104. Ding, J.; Zhao, H.; Yu, H. Structure and performance insights in carbon dots-functionalized MXene-epoxy ultrathin anticorrosion coatings. *Chem. Eng. J.* **2022**, *430*, 132838. [CrossRef]
105. Wang, T.; Cao, H. Orientation control and characterization strategies towards developing 2D materials-based composite coatings for corrosion protection on metals: Progresses and challenges. *Prog. Org. Coat.* **2023**, *182*, 107715. [CrossRef]
106. He, C.; Cai, M.; Huang, Y.; Fan, X.; Zhu, M. Self-alignment of amino-functionalized $Ti_3C_2T_x$ modified with cerium-doped ZIF-8 nanocontainer towards anti-corrosive/ wear and self-healing application. *Compos. Part B Eng.* **2023**, *271*, 111144. [CrossRef]
107. Fan, Y.; Wu, J.; Zhang, L.; Sheng, X. Self-oriented functionalized MXene with excellent antioxidant properties for enhancing anticorrosion properties of waterborne epoxy resin. *Ind. Eng. Chem. Res.* **2024**, *63*, 10637–10650. [CrossRef]
108. Yan, H.; Zhang, L.; Li, H.; Fan, X.; Zhu, M. Towards High Performance Additive of Ti_3C_2/Graphene Hybrid with a Novel Wrapping Structure in Epoxy Coating. *Carbon* **2020**, *157*, 217–233. [CrossRef]
109. Shen, L.; Zhao, W.; Wang, K.; Xu, J. GO-Ti_3C_2 two-dimensional heterojunction nanomaterial for anticorrosion enhancement of epoxy zinc-rich coatings. *J. Hazard. Mater.* **2021**, *417*, 126048. [CrossRef] [PubMed]
110. Qiang, Y.; Ran, B.; Li, M.; Xu, Q.; Peng, J. GO-functionalized MXene towards superior anti-corrosion coating. *J. Colloid Interface Sci.* **2023**, *642*, 595–603. [CrossRef] [PubMed]
111. Verma, C.; Barsoum, I.; Alfantazi, A.; Quraishi, M.A. Carbon allotropes and MXenes composites as superdurable nanofillers for polymer coatings: Emerging materials. *Mater. Lett.* **2023**, *337*, 133892. [CrossRef]
112. Lou, D.; Chen, H.; Liu, J.; Wang, D.; Wang, C.; Jasthi, B.K.; Zhu, Z.; Younes, H.; Hong, H. Improved anticorrosion properties of polyurethane nanocomposites by $Ti_3C_2T_x$ MXene/functionalized carbon nanotubes for corrosion protection coatings. *ACS Appl. Nano Mater.* **2023**, *6*, 12515. [CrossRef]
113. Deng, Y.; Song, G.L.; Zhang, T.; Lang, Z.; Wu, P.; Zheng, D. Loading halloysite nanotubes on MXene as functional composite filler towards a polybenzoxazine anticorrosion coating. *Colloids Surf. A* **2022**, *650*, 129498. [CrossRef]
114. Cai, M.; Fan, X.; Yan, H.; Li, Y.; Song, S.; Li, W.; Li, H.; Lu, Z.; Zhu, M. In situ assemble $Ti_3C_2T_x$ MXene@MgAl-LDH heterostructure towards anticorrosion and antiwear application. *Chem. Eng. J.* **2021**, *419*, 130050. [CrossRef]
115. Li, C.; He, Y.; Zhao, Y.; Li, Z.; Sun, D.; Li, H.; Chen, W.; Yan, J.; Wu, G.; Yuan, X. Cataphoretic deposition of an epoxy coating with the incorporation of $Ti_3C_2T_x$@Mg-Al layered double hydroxide for long-term active corrosion protection effect. *Prog. Org. Coat.* **2023**, *175*, 107333. [CrossRef]
116. Wu, Y.; Wu, L.; Yao, W.; Chen, Y.; Zhou, Y.; Wang, L.; Wu, T.; Jiang, B.; Atrens, A.; Pan, F. ZnAl-LDH@MXene modified by inhibitor as a nanofiller applied on Mg alloys coating protection. *Carbon* **2023**, *204*, 36. [CrossRef]
117. Xing, X.; Sui, Y.; Li, Q.; Yuan, M.; Zhao, H.; Zhou, D.; Chu, X.; Liu, S.; Tang, E. Multifunctional ZnAl-MoO_4 LDH assembled $Ti_3C_2T_x$ MXene composite foractive/passive corrosion protection behavior of epoxy coatings. *Appl. Surf. Sci.* **2023**, *623*, 157092. [CrossRef]
118. Zhang, X.; Yin, C.; Liu, S.; Cong, D.; Li, Z.; Wang, J.; Sun, T.; Liu, X.; Liang, Z.; Jiang, S.; et al. Doubly doped Mg-$AlV_2O_7{}^{4-}$ layered double hydroxide/Mo_2CT_x MXene nanosheet composites for wear-and corrosion-resistant coatings. *ACS Appl. Nano Mater.* **2023**, *6*, 14308–14321. [CrossRef]
119. Zhang, F.; Liu, W.; Wang, S.; Shi, H.; Liu, C.; Liang, L.; Pi, K. Engineering MXenes ($Ti_3C_2T_x$) surface with TiO_2 for enhancing anti-corrosion performance of coatings. *Polymers* **2021**, *230*, 124086. [CrossRef]
120. Yao, Y.; Li, D.; Jin, S.; Wang, Z.; Wang, D.; Shu, Q. Improved corrosion resistance and thermal stability of insensitive NTO explosives by MXene modification in the presence of non-covalent bonds. *New J. Chem.* **2022**, *46*, 9389–9396. [CrossRef]
121. Dong, Y.; Wu, K.; Yin, Y.; Du, X.; Zhou, Q. TiN/TiO_2/MXene ternary composite with photocathodic protection for 304 stainless steel. *Chem. Eng. J.* **2022**, *442*, 136390. [CrossRef]
122. He, X.; Wu, J.; Li, S.; Chen, Y.; Zhang, L.; Sheng, X. In situ growth of aminated silica on MXene nanosheets: A novel 0D/2D hybrid structure for multifunctional waterborne epoxy composite coatings. *Prog. Org. Coat.* **2022**, *171*, 107042. [CrossRef]
123. An, H.; Jiang, C.; Yin, X.; Liu, K.; Liang, S.; Wang, X.; Xiao, J.; Zhao, X.; Sun, Z. Polyaniline/TiO_2/MXene ternary composites for enhancing corrosion resistance of waterborne epoxy coatings. *ACS Appl. Nano Mat.* **2025**, *8*, 340–350. [CrossRef]
124. Chen, J.; Zhao, W. Silk fibroin-$Ti_3C_2T_x$ hybrid nanofiller enhance corrosion protection for waterborne epoxy coatings under deep sea environment. *Chem. Eng. J.* **2021**, *423*, 130195. [CrossRef]
125. Wu, Y.; Zhao, W.; Liu, S.; Wang, L. $Ti_3C_2T_x$@CNF nanohybrid improving the corrosion resistance of waterborne epoxy coating under 15 MPa alternating hydrostatic pressure. *Chem. Eng. J.* **2022**, *438*, 135483. [CrossRef]
126. Zhou, M.; Zhao, C.; Liu, P.; Yu, H. Adsorption behavior of $Ti_3C_2T_x$ with h-BN nanosheet and their application in waterborne epoxy anti-corrosion coating. *Appl. Surf. Sci.* **2022**, *586*, 152778. [CrossRef]
127. Cai, M.; Feng, P.; Yan, H.; Li, Y.; Song, S.; Li, W.; Li, H.; Fan, X.; Zhu, M. Hierarchical $Ti_3C_2T_x$@MoS_2 heterostructures: A first principles calculation and application in corrosion/wear protection. *J. Mater. Sci. Technol.* **2022**, *116*, 151–160. [CrossRef]

128. Najmi, P.; Keshmiri, N.; Ramezanzadeh, M.; Ramezanzadeh, B.; Arjmand, M. Porous 2D Ti_3C_2 MXene nanosheets sandwiched between imine-based covalent organic frameworks (COFs) for excellent corrosion protective coatings. *Chem. Eng. J.* **2023**, *456*, 141001. [CrossRef]
129. Wang, P.; He, B.; Wang, B.; Liu, S.; Ye, Q.; Zhou, F.; Liu, W. MXene/metal–organic framework based composite coating with photothermal self-healing performances for antifouling application. *Chem. Eng. J.* **2023**, *474*, 145835. [CrossRef]
130. Xu, Y.; Wang, S.; Liu, Z.; Lin, S.; Cheng, X.; Wang, H. Fabrication of organic/inorganic hybrid waterborne polyurethane coating based on CeNPs@m$Ti_3C_2T_x$ composites for anti-corrosion applications. *Prog. Org. Coat.* **2023**, *182*, 107668. [CrossRef]
131. Yue, Z.; Fan, X.Q.; Zhu, M.H. Review of fretting wear and surface protection technology. *Surf. Sci. Technol.* **2024**, *53*, 1–20.
132. Wang, P.; He, B.; Du, Y.; Wang, B.; Gao, J.; Liu, S.; Ye, Q.; Zhou, F. Functionalized $Ti_3C_2T_x$-based nanocomposite coatings for anticorrosion and antifouling applications. *Chem. Eng. J.* **2022**, *448*, 137668. [CrossRef]
133. An, D.; Wang, Z.; Qin, L.; Wu, Y.; Lu, S.; Yang, H.; Ma, Z.; Mawignon, F.J.; Liu, J.; Hao, L.; et al. Preparation of MXene/EP coating for promising anticorrosion and superlow friction properties. *Prog. Org. Coat.* **2023**, *183*, 107779. [CrossRef]
134. Li, X.; Xue, Z.; Sun, W.; Chu, J.; Wang, Q.; Tong, L.; Wang, K. Bio-inspired self-healing MXene/polyurethane coating with superior active/passive anticorrosion performance for Mg alloy. *Chem. Eng. J.* **2023**, *454*, 140187. [CrossRef]
135. Huang, X.; Luo, J.; Zhang, L.; Xie, D.; Sheng, X. MXene@curcumin nanosheets doped with Ce(iii) toward water-borne epoxy nanocomposite coating for long-term corrosion protection and wear resistance. *Ind. Eng. Chem. Res.* **2024**, *63*, 15508–15523. [CrossRef]

materials

Communication

The Effect of Low-Temperature Short-Term Annealing on the Microstructure and Properties of Ultrafine-Grained Pure Titanium

Yanxia Gu [1,2,*], Jinghua Jiang [2,3,*], Aibin Ma [2,3], Yaxiao Gu [2], Zhenquan Yang [2], Haoran Wu [4], Chenlong Song [2], Qingping Yang [2] and Chaobing Ni [2]

1 College of Marine and Electrical Engineering, Jiangsu Maritime Institute, Nanjing 211100, China
2 College of Materials Science and Engineering, Hohai University, Nanjing 210098, China; aibin-ma@hhu.edu.cn (A.M.); 230208020001@hhu.edu.cn (Y.G.); yang-zhen-quan@hhu.edu.cn (Z.Y.); 231625010026@hhu.edu.cn (C.S.); 231625010039@hhu.edu.cn (Q.Y.); 231625010021@hhu.edu.cn (C.N.)
3 Suqian Institute, Hohai University, Suqian 223800, China
4 Institute of Materials, Henan Academy of Sciences, Zhengzhou 450046, China; haoranwuted@hnas.ac.cn
* Correspondence: guyanxiaj@126.com (Y.G.); jinghua-jiang@hhu.edu.cn (J.J.)

Abstract: Industrial pure titanium was processed through 1–4 passes by equal-channel angular pressing (ECAP), and the processed samples were subsequently short-term annealed for 15 min at 300 °C, to achieve better mechanical properties for industrial applications. The microstructure was analyzed using TEM, EBSD, and XRD observations. The mechanical properties were studied through tensile testing. The TEM and EBSD results showed that the grain size of industrial pure titanium was refined to approximately 420 nm after four passes of ECAP processing, with very little grain growth after annealing. The XRD analysis proved the enhanced basal texture in the subsequent annealed samples. Tensile tests indicated that the strength of the processed sample increased with more ECAP passes and was improved by 39% after four passes compared with the as-received state; in addition, the low-temperature short-term annealing resulted in a further strengthening phenomenon. It was concluded that the strengthening after annealing in industrial pure titanium was likely due to the improved basal texture, resulting in texture strengthening.

Keywords: titanium; annealing; texture strengthening

Academic Editor: Yong-Cheng Lin

Received: 6 December 2024
Revised: 9 January 2025
Accepted: 21 January 2025
Published: 23 January 2025

Citation: Gu, Y.; Jiang, J.; Ma, A.; Gu, Y.; Yang, Z.; Wu, H.; Song, C.; Yang, Q.; Ni, C. The Effect of Low-Temperature Short-Term Annealing on the Microstructure and Properties of Ultrafine-Grained Pure Titanium. *Materials* **2025**, *18*, 517. https://doi.org/10.3390/ma18030517

1. Introduction

In recent years, ultrafine-grained (UFG) materials have received much attention owing to their remarkable mechanical properties, and extensive research has been carried out on the associated preparation methods, microstructure observation, and properties of ultrafine crystalline materials [1,2]. The implementation of severe plastic deformation (SPD) methods has facilitated the fabrication of ultrafine crystalline microstructures [3–5]. Among various deformation methods, equal-channel angular pressing (ECAP) is a potential method for SPD, which enables the production of ultrafine crystalline bulk metals with uniform microstructures and large dimensions [6,7]. The sample is forced through two channels with an identical cross-section in the die, which are orientated at an angle relative to each other. This process induces shear deformation at the channel corners, resulting in substantial grain refinement. Titanium alloys are characterized by the light weight nature, high strength, low elasticity, and desirable corrosion resistance and have considerable potential in various fields, including aerospace, marine engineering, and biomedicine [8–12]. Existing research has demonstrated that the mechanical properties [13–18] and corrosion resistance [19–21] of titanium alloys can be further enhanced via ECAP, which considerably broadens the potential applications of these materials and meets the high demands [22,23]. In dental surgery, the Ti-6Al-4V alloy is often used as a medical implant due to its higher strength compared to pure Ti. However, the Al and V atoms are harmful to human health. It was reported that the strength of UFG pure Ti processed by SPD was comparable to that of Ti-6Al-4V, indicating its potential for use in long-term dental implants [11,24–26].

Although SPD exhibits considerable advantages, the ultrafine crystalline materials tend to exhibit unstable microstructures owing to the accumulation of numerous dislocations and internal stresses, which can adversely affect their performance and reliability in subsequent applications [27,28]. To address this issue, appropriate annealing heat treatment can be used. Several researchers have reported that reasonable annealing of ultrafine-grained materials could result in the further improvement of their strength, ductility, electrical conductivity, and surface properties [29–31]. Mavlyutov et al. [29] demonstrated that the annealing of UFG Al at 90–200 °C led to an increase in microhardness and electrical conductivity. Orlova et al. [30] reported that annealing a UFG Al-Zr alloy at 230 °C for 60 min led to an increase in strength and electrical conductivity simultaneously, which was attributed to the rearrangement of grain boundaries. Kim et al. [32] reported that annealing at 300 °C for 60 min brought about an increased strength in nanoscale high Mn steel, which was associated with C clusters driven by annealing. Chen et al. [33] demonstrated that the strength of a Cu-Zn alloy was further enhanced by annealing at 300 °C for 30 min. Sotniczuk et al. [31,34] found that annealing UFG pure Ti at 250 °C for 15 min led to an enhancement of microhardness and corrosion resistance. Majchrowicz et al. [35] reported that a strength enhancement of UFG pure Ti was obtained by annealing at 25 °C for 15 min. The above unexpected phenomenon is primarily attributed to the recovery of the ultrafine structure [35,36], or the segregation of impurity atoms [37]. Because texture evolution occurs during annealing [38], texture likely plays a significant role in enhancing mechanical properties and corrosion resistance. Although several studies have been carried out on the preparation methods and properties of ultrafine crystalline materials, comparatively fewer studies have explored the influences of annealing on the microstructure and mechanical properties of these materials. In particular, the role of texture in the strength enhancement of UFG materials after annealing has not been comprehensively investigated.

In this research, the effect of low-temperature short-term annealing treatment on industrial pure titanium processed via ECAP for 1–4 passes was examined. The modified microstructures and properties of the specimens after each pass were systematically discussed to investigate further improvements in the properties of pure titanium subjected to ECAP.

2. Materials and Methods

Experiments were performed using industrial pure titanium (TA2, Baimtec Material Co., Ltd., Beijing, China); the chemical composition of this material is provided in Table 1. The initial grain size was ~10 μm. The samples subjected to ECAP (Wuxi Haofei Machinery Factory, Wuxi, China) were sectioned into dimensions of 20 mm × 20 mm × 45 mm. As illustrated in Figure 1, the samples were processed through a rotary die at a temperature of 420 °C for 1–4 passes. There were two orthogonal channels in the die. The right, left, and bottom punches were loaded in the channels, and then the sample and the top punch were placed into the die. The plunger pressed the sample out of the right channel. It was unnecessary to remove the sample from the die during the pressing passes; instead, the die was simply rotated anticlockwise to facilitate the subsequent pass—the detailed operation steps were described in our early work [7]. After ECAP processing, the samples underwent annealing at 300 °C for 15 min, and the mechanical and microstructurally tested specimens were selected from the longitudinal section of the sample. For simplicity, the samples investigated after these procedures are summarized in Table 2.

Table 1. Chemical composition of TA2 pure Ti (wt.%).

Fe	C	N	O	Ti
0.06	0.02	0.01	0.06	Balance

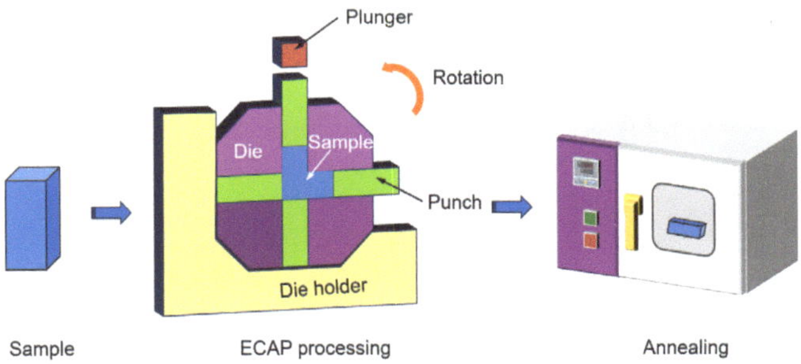

Figure 1. The schematic of the rotary-die ECAP and annealing treatment.

Table 2. Corresponding numbers and specific processes of different states.

Number	Specific Treatment
0P	As-received
1P	ECAP for 1 pass
2P	ECAP for 2 passes
3P	ECAP for 3 passes
4P	ECAP for 4 passes
1P + A	ECAP for 1 pass + 300 °C × 15 min annealing treatment
2P + A	ECAP for 2 passes + 300 °C × 15 min annealing treatment
3P + A	ECAP for 3 passes + 300 °C × 15 min annealing treatment
4P + A	ECAP for 4 passes + 300 °C × 15 min annealing treatment

For room-temperature tensile tests, the spacing dimensions of each specimen were 6 mm × 2 mm × 2 mm (length × width × height). The tests were conducted at a temperature and tensile rate of 25 °C and 1 mm/min, respectively. Each test was performed in triplicate to ensure satisfactory data reproducibility and accuracy.

Transmission electron microscopy (TEM, FEI Tecnai G T20, Hillsboro, OR, USA) foils were prepared for microstructure analysis by mechanically grinding them to 0.1 mm, followed by electrolytic double-jet thinning using a perchloric acid–alcohol mixture as the electrolyte. The crystalline structures were analyzed using a X-ray diffractometer (D8 Discover, Bruker, Billerica, MA, USA) with Cu Kα radiation from 30° to 80°.

Electron Backscattered Diffraction (EBSD) tests of samples were performed using a scanning electron microscope (SEM, Gemini 300, Carl Zeiss, Oberkochen, Germany) equipped with an Oxford EBSD detector. The working voltage was at 20 kV. The specimens had dimensions of 10 mm × 5 mm × 2 mm. The specimens were grinded using metallographic sandpapers, and electrochemically polished in the mixed solution of perchloric acid and methanol (1:9 by volume). The electrolytic polishing was performed at −30 °C with a voltage of 30 V and an electric current of 0.8 A. The scanning area was 57 μm × 42 μm and the step size was 0.09 μm for the 2P and 2P + A samples. The scanning area was

30 μm × 20 μm and the step size was 0.045 μm for the 4P and 4P + A samples. The kernel average misorientation (KAM) mapping was calculated according to the following formula: $KAM = \sum(\Delta\varphi i)/n$, where $\Delta\varphi i$ represents the grain boundary angle of each grain boundary, and n represents the total number of grain boundaries.

3. Results and Discussion

3.1. Microstructure

Figure 2 displays the TEM microstructure after each ECAP pass and subsequent low-temperature short-term annealing treatment. The deformation-twinning structure and considerably coarse undeformed grain boundaries can be identified, as illustrated in Figure 2a. Furthermore, the microstructure exhibits pronounced dislocation entanglement, but clear sub-grain boundaries have not been formed, indicating that the structure remains predominantly deformed after a single pass. After two passes (Figure 2b), numerous dislocation tangle zones are distributed within the material and coarse slats with uneven widths are visible. As shown in Figure 2c, an increased number of pressing passes generates enhanced grain deformation, accompanied by dislocation annihilation or rearrangement, resulting in the gradual formation of small dislocation cells. As illustrated in Figure 2d, the dislocation walls forming the cell wall can continuously absorb surrounding dislocations to form sub-grain boundaries. This process results in more distinct grain boundaries and the development of finer grains with an average crystallite size of ~420 nm. The grain size was similar to the previous work of pure Ti after ECAP reported by Stolyarov et al. [39]. During annealing, it was reported that the grain growth of UFG pure Ti started at 350 °C and higher [40]. Valiev et al. [40] found that the annealing strengthening effect was optimum at a temperature range of 250–300 °C, and there was little annealing strengthening effect below 250 °C. Therefore, this research focused on the microstructure after annealing at 300 °C. Our previous work found that annealing at 300 °C with a duration of 30 min or longer caused visible grain growth and a decrease in strength [41]. So, the microstructure evolution of each-pass samples after heating at 300 °C for 15 min was investigated in this work. As shown in Figure 2e–h, when the subsequent annealing treatment was applied after various pressing passes, minimal growth in grain size was observed due to low annealing temperatures and short annealing duration [42]. Furthermore, the recovery process reduced the dislocation density and led to the development of distinct grain boundaries.

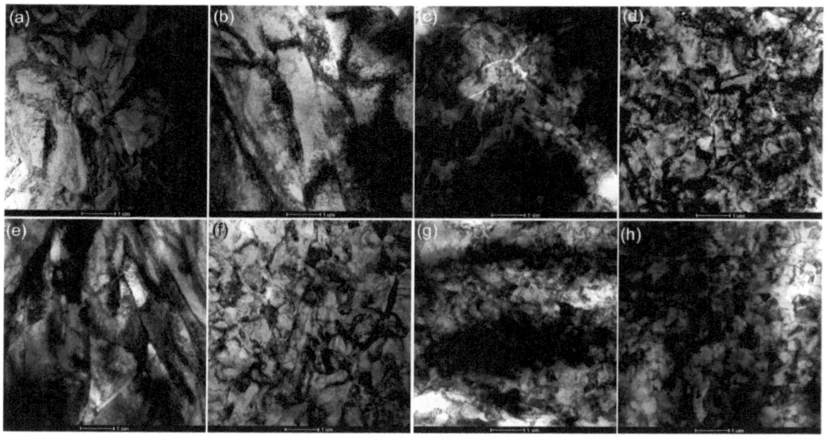

Figure 2. TEM images of pure titanium processed via various ECAP passes and annealing: (**a**) 1P; (**b**) 2P; (**c**) 3P; (**d**) 4P; (**e**) 1P + A; (**f**) 2P + A; (**g**) 3P + A; (**h**) 4P + A.

Figure 3 depicts the EBSD inverse-pole figures (IPF) and corresponding grain size distributions of the 2P, 2P + A, 4P, and 4P + A samples. The 2P and 2P + A samples displayed a nonuniform structure; the grains were composed of fine grains and coarse grains. The average grain size (AGS) was about 0.71 μm after two passes. After annealing, although the proportion of fine grains decreased, the AGS remained 0.71 μm for the 2P + A sample. The 4P sample possesses a uniform grain structure and the grains were refined to about 0.42 μm. Annealing treatment resulted in a decrease in the proportion of fine grains and an increase in the AGS to about 0.43 μm for the 4P + A sample. The EBSD analysis confirmed very little grain growth during the annealing treatment.

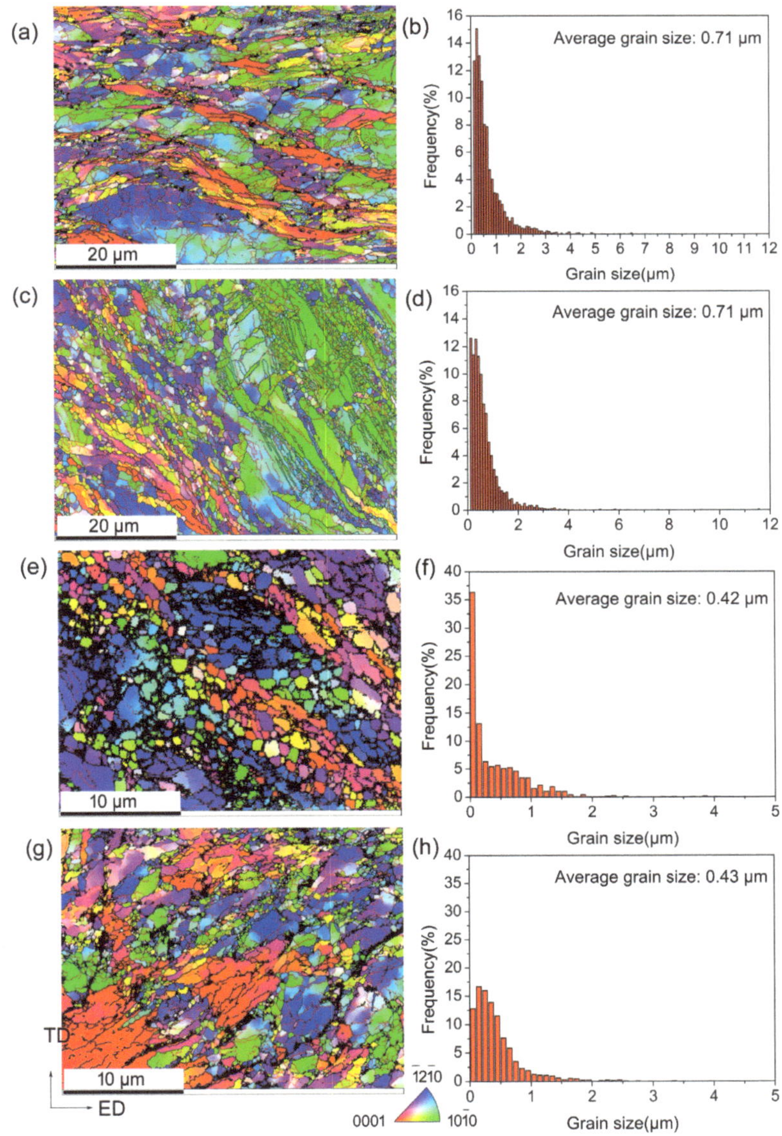

Figure 3. EBSD IPF maps (**a**,**c**,**e**,**f**) and grain size statistics (**b**,**d**,**f**,**h**) of (**a**,**b**) 2P; (**c**,**d**) 2P + A; (**e**,**f**) 4P; and (**g**,**h**) 4P + A samples.

Figure 4 shows the EBSD kernel average misorientation (KAM) maps and the corresponding data of the 2P, 2P + A, 4P, and 4P + A samples. The dislocation density can be determined by KAM values according to the following formula [43]:

$$\rho = \frac{2\theta}{ub}$$

where ρ is dislocation density; θ stands for the average KAM value, and the θ values are presented in Figure 4; u refers to the step size in the EBSD observation; and b denotes the Burgers vector, which is 0.286 nm for pure Ti. According to the above formula, the KAM value can indicate the dislocation density. The larger KAM value implies higher dislocation density. The step size, namely, the u, is 0.09 μm for 2P and 2P + A samples, and 0.045 μm for 4P and 4P + A samples. Thus, the dislocation density in the 2P sample was 6.3×10^{16} m^{-2}, and it decreased to 5.5×10^{16} m^{-2} after short-term annealing. The dislocation density in the 4P sample was the highest (1.1×10^{17} m^{-2}), and it dropped a little after annealing.

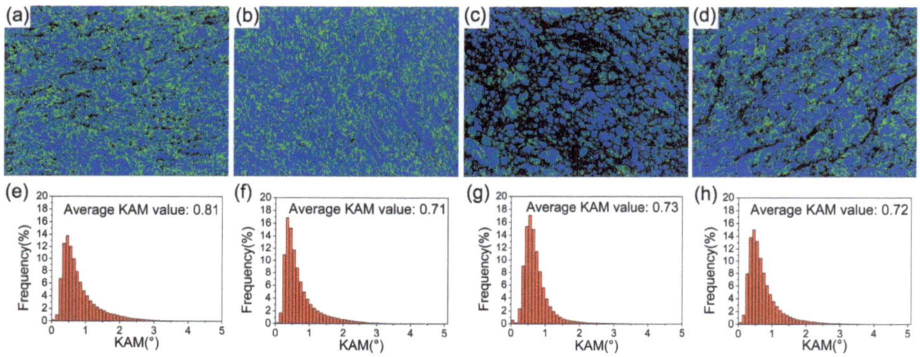

Figure 4. KAM maps of the (**a**) 2P, (**b**) 2P + A, (**c**) 4P, and (**d**) 4P + A samples. (**e–h**) The corresponding local misorientation statistics, respectively.

Figure 5 shows the misorientation angle distributions of the 2P, 2P + A, 4P, and 4P + A samples. It is considered that high-angle grain boundaries (HAGBs) refer to grain boundaries where the misorientation between adjacent grains is greater than 15°, and the rest are low-angle grain boundaries (LAGBs). The HAGBs accounted for 41.7% in the 2P sample and the proportion increased to 53.2% after annealing. In the 4P sample, the ratio of HAGBs was higher (44.9%) than that in the 2P sample. After annealing, the percentage of HAGBs increased to 53.5%. The evolution of grain boundaries for 2 and 4 passes illustrated that increasing the pass number of ECAP promoted grain refinement, accompanied by the increase in the grain boundary angle. The grain refinement mechanism pathway of 1–4 passes during ECAP is as illustrated in Figure 6. Dring ECAP, in 1–2 passes, the grains were elongated from equiaxed crystals, and the dislocation accelerated as the pass number increased. As deformation continues, dislocations accumulated and rearranged, forming structures such as dislocation walls, which further evolved into LAGBs. Then, the LAGBs partially transformed into HAGBs and divided the grains into units, thereby achieving grain refinement.

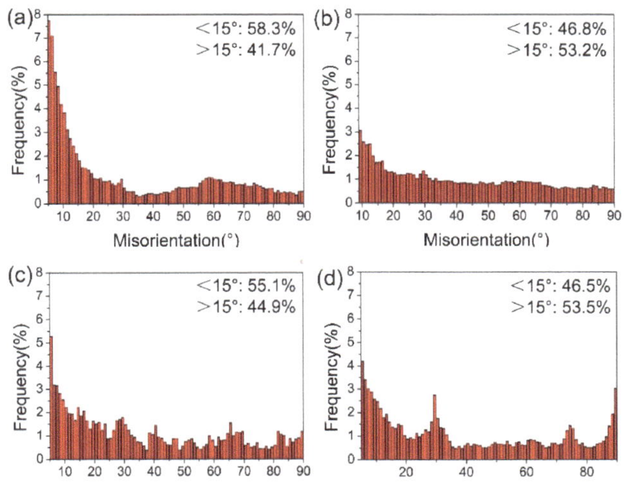

Figure 5. Misorientation angle distributions of (**a**) 2P, (**b**) 2P + A, (**c**) 4P, and (**d**) 4P + A samples.

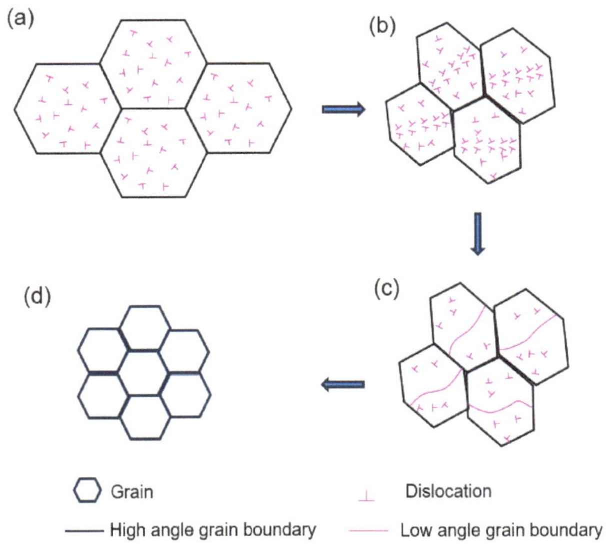

Figure 6. The refining mechanism pathway during ECAP. (**a**) Dislocation formation, (**b**) Dislocation rearrangement, (**c**) Formation of low angle grain boundaries, (**d**) Fine grains with high angle grain boundaries.

3.2. Texture Evolution

Figure 7 depicts the X-ray diffraction (XRD) patterns of industrial pure titanium after ECAP processing and low-temperature short-term annealing treatment for each pass. The strongest peaks of the specimens processed via ECAP for 1–4 passes are the diffraction peaks corresponding to $(10\bar{1}1)$ crystal plane. As the number of pressing passes increased, the integrated intensity of (0002) peaks progressively increased with accumulated deformation. After low-temperature short-duration annealing treatment, the intensity of the diffraction peaks corresponding to the (0002) plane increased compared with the pre-annealing state. The strongest peak observed was the (0002) peak for the samples subjected to 3-pass or 4-pass pressing followed by annealing treatment.

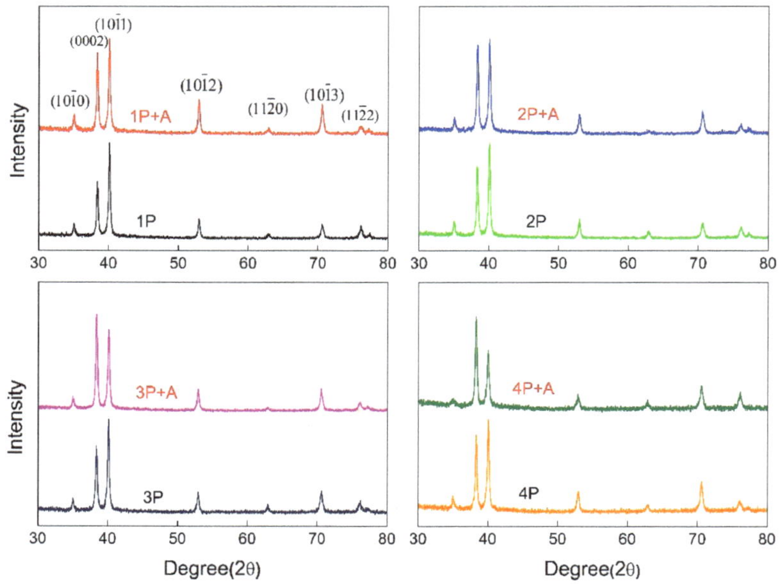

Figure 7. X-ray diffraction patterns of pure titanium after different processing.

The texture coefficient (TC) was adopted to better evaluate the preferred orientation of the grains and analyze the evolution of the texture. The texture coefficient was calculated according to Equation (1) [44,45]:

$$TC_{(hkil)} = \frac{I_{(hkil)}}{I_{0(hkil)}} \left[\frac{1}{n} \sum_{n=1}^{n} \frac{I_{(hkil)}}{I_{0(hkil)}} \right]^{-1} \quad (1)$$

where $I_{(hkil)}$ and $I_{0(hkil)}$ represent the experimentally measured diffraction intensity of the crystal plane and the standard diffraction intensity in the PDF card, respectively, and n is the number of diffraction planes. The TC value represents the degree of preferred grain orientation. A TC value of >1 indicates that grains have developed a preferred orientation. A higher TC value indicates a more pronounced preferred orientation. Based on Equation (1), the relative texture coefficients of the five crystal planes of industrial pure titanium in each state were computed; these coefficients are listed in Table 3. The texture coefficient of the (0002) crystal plane of industrial pure titanium after various processing steps is shown in Figure 8. As shown in Table 3 and Figure 8, the texture coefficients for the (0002) crystalline planes of the specimens processed via ECAP in each pass exceed 1, indicating a preferred orientation of the (0002) crystal planes parallel to sample surface. After low-temperature short-duration annealing treatment, the texture coefficients for the (0002) planes of the specimens across all passes increased, indicating a more pronounced selective orientation. The annealing process resulted in an increased alignment of the (0002) planes parallel to the specimen surfaces. The pole figures in our previous work also confirmed that the (0002) texture in 4P sample was strengthened after short-term annealing [41].

Table 3. Texture coefficients of pure titanium after different processes.

Sample	Texture Coefficient (*TC*)				
	$(10\bar{1}0)$	(0002)	$(10\bar{1}1)$	$(10\bar{1}2)$	$(11\bar{2}0)$
1P	0.52	1.55	0.76	1.29	0.49
2P	0.55	1.73	0.69	1.17	0.63
3P	0.52	1.66	0.68	1.31	0.68
4P	0.49	1.55	0.55	1.11	0.60
1P + A	0.51	1.59	0.55	1.62	0.45
2P + A	0.51	1.98	0.63	1.07	0.34
3P + A	0.43	2.22	0.56	1.23	0.30
4P + A	0.36	1.78	0.35	0.90	0.73

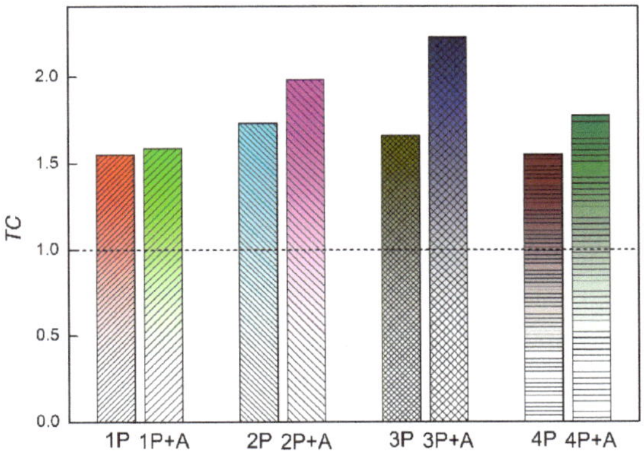

Figure 8. Texture coefficient of (0002) crystal planes of pure titanium after different processing.

3.3. Mechanical Properties

Figure 9 and Table 4 illustrate tensile strengths, yield strengths, and elongations of pure titanium processed by ECAP for different passes and annealing. The initial tensile strength and elongation of the specimen was 450 MPa and 48%, respectively. After ECAP, the tensile strengths and yield strengths increased gradually as the pass number increased from 1 to 4. In particular, the tensile strength of specimens subjected to four passes increased to 627 MPa, reflecting a 39% improvement compared with the initial counterpart. The tensile strength of 627 MPa was close to the reported value (~630 MPa) of pure Ti after four passes of ECAP [46]. The EBSD observations proved that the grain size of the 4P sample was smaller than that of 2P sample, so the 4P sample exhibited a higher strength than the 2P sample, arising from grain refinement. Our previous work found that after 8 passes, there was almost no further improvement in strength compared to the 4-pass counterpart, because there was a dynamic balance between grain refinement and recovery [47]. Therefore, the mechanical properties after 1–4 passes were tested in the present study. The elongations of samples decreased to about 30% after ECAP. In addition, the strength of the specimens in each pass after low-temperature short-duration annealing exhibited an enhancement relative to that in their pre-annealing state. As the number of pass increased, the strengths of the annealed samples increased progressively. The tensile strength of the specimen processed through four passes and subjected to annealing treatment reached 663 MPa, indicating a 5% increase compared with its strength in the ECAP-processed condition before annealing. The annealing strengthening level of 5% was similar to the reported work [13,34]. Regarding elongation after annealing, in general, the elongation can increase after an annealing treatment. However, there were also investigations found that strength was increased and elongation was decreased after annealing in UFG Al [48–50], and this phenomenon was attributed to dislocation source-limited strengthening. In our present study, the elongations of annealed samples remained almost unchanged compared to their pre-annealing state. The specific underlying reasons are not clear and still need further in-depth exploration. This is the limitation of this study. We will try to address this issue in our future research.

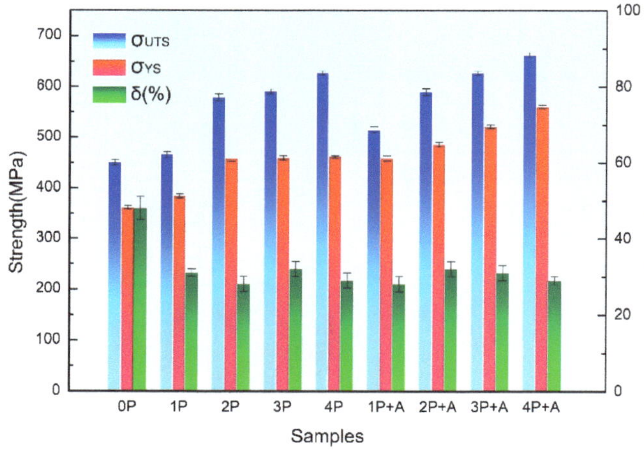

Figure 9. Tensile strengths, yield strengths, and elongations of pure titanium processed by ECAP for different passes and annealing.

Table 4. Corresponding data of tensile strengths, yield strengths, and elongations of pure titanium in each state.

Samples	Tensile Strength σb (MPa)	Yield Strength σs (MPa)	Elongations δ (%)
0P	450 ± 5	362 ± 3	48 ± 3
1P	466 ± 6	384 ± 4	31 ± 1
2P	579 ± 7	458 ± 5	28 ± 2
3P	591 ± 5	460 ± 4	32 ± 2
4P	627 ± 4	462 ± 3	29 ± 2
1P + A1	515 ± 6	459 ± 5	28 ± 2
2P + A1	590 ± 7	486 ± 4	32 ± 2
3P + A1	627 ± 4	522 ± 3	31 ± 2
4P + A1	663 ± 4	561 ± 3	29 ± 1

Industrial pure titanium subjected to multiple ECAP passes after low-temperature short-duration annealing treatment exhibited a strength improvement similar to that observed for ultrafine-grained pure Cu, Al, and Mg [51–53]. Typically, the annealing of processed materials is beneficial for recovery and recrystallization, which results in a reduction in the defect density and an increase in the grain size, thereby decreasing the final strength. However, research has shown that applying annealing treatment to ultrafine and nanocrystalline crystals can enhance their strength, which is contrary to the conventional softening effect observed in coarse crystalline materials during annealing. Annealing strengthening exhibits different behaviors in various metals, resulting in considerable differences in the mechanistic explanations [54]. For instance, Akbari et al. [51] observed that the tensile strength of pure Cu processed increased after low-temperature short-duration annealing due to the generation of small grains during the annealing process. Zhao et al. [52] conducted annealing treatment on ultrafine crystalline aluminum at 60 °C for 30 min and ascribed the improved strength to the pinning of dislocations by the grain boundaries and the obstruction of the dislocation motion. Furthermore, Volkov et al. [53] demonstrated that the increased strength of ultrafine pure magnesium following low-temperature annealing was due to the self-locking of dislocations. Chen et al. [33] reported the unusual annealing strengthening observed in the Cu-Zn alloy and attributed this phenomenon to the impurity segregation. Semenova et al. [37] discovered that the remarkable enhancement in the strength of UFG Ti (Grade 4) after annealing was in connection with the recovery of nonequilibrium boundaries and impurity segregation.

Although various researchers have provided different explanations for the annealing strengthening phenomenon, relatively few studies have examined the influence of annealing on the texture of ultrafine crystalline materials. Compared with the cubic structure, industrial pure titanium, which has fewer slip systems, is more susceptible to texture-strengthening effects [41,55–57]. Texture strengthening in Al and Mg alloys has also been reported [58–60]. The impurity content in the industrial pure titanium used in this study was considerably low, indicating that the possibility of strengthening due to impurity segregation was substantially minimal. In this study, microstructure observation validated that the grain sizes of the annealed samples were similar to those observed before annealing treatment, indicating that the possibility of fine-grain strengthening resulting from annealing was also considerably low. The literature reported that the bimodal grain size distribution can provide additional strengthening through the hetero-deformation-induced strengthening [61]. The grain size distribution was unimodal before and after annealing in the present study, so the possibility of an extra strengthening effect by grain size distribution was slim. Based on the XRD results and texture coefficient values, the texture of the basal (0002) plane of industrial pure titanium was enhanced after annealing, which resulted in strength enhancement. In summary, the improved strength of ECAP-processed industrial pure titanium may be attributed to the following factors: (1) the low annealing temperature and short duration prevented considerable grain growth; (2) low-temperature short-duration annealing enhanced the (0002) basal texture, resulting in a small Schmid factor for the basal plane, thereby impeding dislocation movement and leading to texture strengthening, which contributed to the increased strength.

4. Conclusions

This study systematically investigated the microstructure and mechanical properties of industrial pure titanium subjected to 1–4 ECAP passes, followed by low-temperature short-term annealing after each pass. The results suggested a new strategy of ECAP combined low-temperature short-term annealing aimed at further increasing the strength of UFG pure titanium. In addition, this study innovatively proposed the mechanism of texture strengthening caused by annealing in UFG pure titanium. The conclusions are as follows:

(1) After ECAP processing, the microstructure of industrial pure titanium became increasingly refined with the increasing number of passes, resulting in an average grain size of ~420 nm after four passes. Furthermore, the strength of industrial pure titanium was enhanced via ECAP processing, and the tensile strength of the specimens after four passes was improved by 39% compared with that of the sample in the initial state.

(2) Microstructure observation revealed that the grain growth of ECAP-processed samples was very little during short-term annealing. The calculated results of the EBSD analysis

demonstrated that the dislocation density was decreased after annealing. Moreover, the proportion of HAGBs was increased by the aging treatment. The XRD observations showed that the basal texture was strengthened after short-term annealing.

(3) The ECAP-processed specimens subjected to annealing treatment at 300 °C for 15 min exhibited increased strength compared with the pre-annealed specimens. This strengthening effect was likely attributable to the fact that the low annealing temperature and short duration prevented substantial grain growth and enhanced basal texture, thereby impeding dislocation movement.

Author Contributions: Conceptualization, Y.G. (Yanxia Gu) and J.J.; methodology, A.M.; validation, C.S., Q.Y. and C.N.; formal analysis, Y.G. (Yanxia Gu); investigation, Y.G. (Yanxia Gu), Y.G. (Yaxiao Gu), Z.Y., H.W., C.S., Q.Y. and C.N.; resources, J.J.; data curation, Y.G. (Yaxiao Gu), Z.Y. and H.W.; writing—original draft preparation, Y.G. (Yanxia Gu), Y.G. (Yaxiao Gu) and Z.Y.; writing—review and editing, J.J.; supervision, A.M.; project administration, Y.G. (Yanxia Gu) and J.J.; funding acquisition, Y.G. (Yanxia Gu), J.J. and A.M. All authors have read and agreed to the published version of the manuscript.

Funding: This research was funded by the National Natural Science Foundation of China, grant number 52305365, Postdoctoral Science Foundation of China, grant number 2020M681474, and Jiangsu Qinglan Project.

Institutional Review Board Statement: Not applicable.

Informed Consent Statement: Not applicable.

Data Availability Statement: The original contributions presented in this study are included in the article. Further inquiries can be directed to the corresponding author.

Conflicts of Interest: The authors declare no conflicts of interest.

References

1. Edalati, K.; Bachmaier, A.; Beloshenko, V.A.; Beygelzimer, Y.; Blank, V.D.; Botta, W.J.; Bryła, K.; Čížek, J.; Divinski, S.; Enikeev, N.A.; et al. Nanomaterials by severe plastic deformation: Review of historical developments and recent advances. *Mater. Res. Lett.* **2022**, *10*, 163–256. [CrossRef]
2. Ovid'ko, I.A.; Valiev, R.Z.; Zhu, Y.T. Review on superior strength and enhanced ductility of metallic nanomaterials. *Prog. Mater. Sci.* **2018**, *94*, 462–540. [CrossRef]
3. Lin, H.-K.; Li, G.Y.; Mortier, S.; Bazarnik, P.; Huang, Y.; Lewandowska, M.; Langdon, T.G. Processing of CP-Ti by high-pressure torsion and the effect of surface modification using a post-HPT laser treatment. *J. Alloys Compd.* **2019**, *784*, 653–659. [CrossRef]
4. Edalati, K.; Ahmed, A.Q.; Akrami, S.; Ameyama, K.; Aptukov, V.; Asfandiyarov, R.N.; Ashida, M.; Astanin, V.; Bachmaier, A.; Beloshenko, V.; et al. Severe plastic deformation for producing superfunctional ultrafine-grained and heterostructured materials: An interdisciplinary review. *J. Alloys Compd.* **2024**, *1002*, 174667. [CrossRef]
5. Valiev, R.Z.; Estrin, Y.; Horita, Z.; Langdon, T.G.; Zehetbauer, M.J.; Zhu, Y. Producing Bulk Ultrafine-Grained Materials by Severe Plastic Deformation: Ten Years Later. *JOM* **2016**, *68*, 1216–1226. [CrossRef]
6. Dyakonov, G.S.; Mironov, S.; Enikeev, N.; Semenova, I.P.; Valiev, R.Z.; Semiatin, S.L. Annealing behavior of severely-deformed titanium Grade 4. *Mater. Sci. Eng. A* **2019**, *742*, 89–101. [CrossRef]
7. Gu, Y.; Jiang, J.; Ma, A.; Lu, L.; Wu, H. Corrosion behavior of pure titanium processed by rotary-die ECAP. *J. Mater. Res. Technol.* **2021**, *15*, 1873–1880. [CrossRef]
8. Mishnaevsky, L.; Levashov, E.; Valiev, R.Z.; Segurado, J.; Sabirov, I.; Enikeev, N.; Prokoshkin, S.; Solov'yov, A.V.; Korotitskiy, A.; Gutmanas, E.; et al. Nanostructured titanium-based materials for medical implants: Modeling and development. *Mater. Sci. Eng. R* **2014**, *81*, 1–19. [CrossRef]
9. Wang, J.L.; Liu, R.L.; Majumdar, T.; Mantri, S.A.; Ravi, V.A.; Banerjee, R.; Birbilis, N. A closer look at the in vitro electrochemical characterisation of titanium alloys for biomedical applications using in-situ methods. *Acta Biomater.* **2017**, *54*, 469–478. [CrossRef]
10. Sotniczuk, A.; Kuczyńska-Zemła, D.; Kwaśniak, P.; Thomas, M.; Garbacz, H. Corrosion behavior of Ti-29Nb-13Ta-4.6Zr and commercially pure Ti under simulated inflammatory conditions—Comparative effect of grain refinement and non-toxic β phase stabilizers. *Electrochim. Acta* **2019**, *312*, 369–379. [CrossRef]

11. Elias, C.N.; Fernandes, D.J.; Souza, F.M.d.; Monteiro, E.d.S.; Biasi, R.S.d. Mechanical and clinical properties of titanium and titanium-based alloys (Ti G2, Ti G4 cold worked nanostructured and Ti G5) for biomedical applications. *J. Mater. Res. Technol.* **2019**, *8*, 1060–1069. [CrossRef]
12. Kaur, M.; Singh, K. Review on titanium and titanium based alloys as biomaterials for orthopaedic applications. *Mater. Sci. Eng. C* **2019**, *102*, 844–862. [CrossRef] [PubMed]
13. Li, Z.; Fu, L.; Fu, B.; Shan, A. Effects of annealing on microstructure and mechanical properties of nano-grained titanium produced by combination of asymmetric and symmetric rolling. *Mater. Sci. Eng. A* **2012**, *558*, 309–318. [CrossRef]
14. Sheremetyev, V.; Churakova, A.; Derkach, M.; Gunderov, D.; Raab, G.; Prokoshkin, S. Effect of ECAP and annealing on structure and mechanical properties of metastable beta Ti-18Zr-15Nb (at.%) alloy. *Mater. Lett.* **2021**, *305*, 130760. [CrossRef]
15. Raab, G.I.; Kodirov, I.S.; Aksenov, D.A.; Valiev, R.Z. The formation of a high-strength state in martensitic Ti Grade 4 by ECAP. *J. Alloys Compd.* **2022**, *922*, 166205. [CrossRef]
16. Xu, H.Q.; Wei, K.X.; Wei, W.; Dzugan, J.; Alexandrov, I.V.; An, X.L.; Wang, D.D.; Liu, X.K.; Daniel, M.; Hradil, D.; et al. Microstructure and mechanical properties evolution of Ti-13Nb-13Zr alloy processed by ECAP-Conform and rotary swaging. *J. Alloys Compd.* **2023**, *969*, 172351. [CrossRef]
17. Tesař, K.; Koller, M.; Vokoun, D.; Tyc, O.; Čech, J.; Sedlák, P. Texture, elastic anisotropy and thermal stability of commercially pure titanium prepared by room temperature ECAP. *Mater. Des.* **2023**, *226*, 111678. [CrossRef]
18. Yang, X.-R.; Wang, Z.-L.; Dai, Y.; Li, B.-W.; Jin, Z.; Luo, L.; Liu, X.-Y.; Wang, J.-Z. Evolution of microstructure, macrotexture and mechanical properties of high strength biomedical TA4 pure titanium during multi-pass ECAP. *J. Mater. Res. Technol.* **2024**, *28*, 3976–3987. [CrossRef]
19. Fattah-Alhosseini, A.; Imantalab, O.; Ansari, G. The role of grain refinement and film formation potential on the electrochemical behavior of commercial pure titanium in Hank's physiological solution. *Mater. Sci. Eng. C* **2017**, *71*, 827–834. [CrossRef] [PubMed]
20. Lu, J.; Zhang, Y.; Huo, W.; Zhang, W.; Zhao, Y.; Zhang, Y. Electrochemical corrosion characteristics and biocompatibility of nanostructured titanium for implants. *Appl. Surf. Sci.* **2018**, *434*, 63–72. [CrossRef]
21. Sotniczuk, A.; Chromiński, W.; Adamczyk-Cieślak, B.; Pisarek, M.; Garbacz, H. Corrosion behaviour of biomedical Ti under simulated inflammation: Exploring the relevance of grain refinement and crystallographic texture. *Corros. Sci.* **2022**, *200*, 110238. [CrossRef]
22. Gu, Y.; Ma, A.; Jiang, J.; Yuan, Y.; Wu, H. Microstructure and tensile anisotropy of pure Ti processed by up-scaled RD-ECAP. *Mater. Charact.* **2020**, *168*, 110513. [CrossRef]
23. Gu, Y.; Jiang, J.; Ma, A.; Wu, H.; Lu, L. Enhanced corrosion behavior of ultrafine-grained pure titanium in simulated high-temperature seawater. *J. Mater. Res. Technol.* **2022**, *19*, 1–6. [CrossRef]
24. Agarwal, K.M.; Tyagi, R.K.; Singhal, A.; Bhatia, D. Effect of ECAP on the mechanical properties of titanium and its alloys for biomedical applications. *Mater. Sci. Energy. Technol.* **2020**, *3*, 921–927. [CrossRef]
25. Fintova, S.; Dlhy, P.; Mertova, K.; Chlup, Z.; Duchek, M.; Prochazka, R.; Hutar, P. Fatigue properties of UFG Ti grade 2 dental implant vs. conventionally tested smooth specimens. *J. Mech. Behav. Biomed. Mater.* **2021**, *123*, 104715. [CrossRef]
26. Rezyapova, L.R.; Safargalina, Z.A.; Usmanov, E.I.; Valiev, R.R.; Minasov, T.B.; Valiev, R.Z. Nanostructured Titanium with Ultrafine-Grained Structure as Advanced Engineering Material for Biomedical Application. *Adv. Eng. Mater.* **2024**, *26*, 2400394. [CrossRef]
27. Jiang, Q.-W.; Li, X.-W. Effect of pre-annealing treatment on the compressive deformation and damage behavior of ultrafine-grained copper. *Mater. Sci. Eng. A* **2012**, *546*, 59–67. [CrossRef]
28. Edalati, K.; Hashiguchi, Y.; Iwaoka, H.; Matsunaga, H.; Valiev, R.Z.; Horita, Z. Long-time stability of metals after severe plastic deformation: Softening and hardening by self-annealing versus thermal stability. *Mater. Sci. Eng. A* **2018**, *729*, 340–348. [CrossRef]
29. Mavlyutov, A.M.; Bondarenko, A.S.; Murashkin, M.Y.; Boltynjuk, E.V.; Valiev, R.Z.; Orlova, T.S. Effect of annealing on microhardness and electrical resistivity of nanostructured SPD aluminium. *J. Alloys Compd.* **2017**, *698*, 539–546. [CrossRef]
30. Orlova, T.S.; Latynina, T.A.; Mavlyutov, A.M.; Murashkin, M.Y.; Valiev, R.Z. Effect of annealing on microstructure, strength and electrical conductivity of the pre-aged and HPT-processed Al-0.4Zr alloy. *J. Alloys Compd.* **2019**, *784*, 41–48. [CrossRef]
31. Sotniczuk, A.; Kuczyńska-Zemła, D.; Królikowski, A.; Garbacz, H. Enhancement of the corrosion resistance and mechanical properties of nanocrystalline titanium by low-temperature annealing. *Corros. Sci.* **2019**, *147*, 342–349. [CrossRef]
32. Kim, J.G.; Seol, J.B.; Bae, J.W.; Kim, H.S. On the mechanistic understanding of annealing-induced strength enhancement of ultrafine-grained high-Mn steel. *Materialia* **2020**, *13*, 100837. [CrossRef]
33. Chen, J.; Su, Y.; Zhang, Q.; Sun, J.; Yang, D.; Jiang, J.; Song, D.; Ma, A. Enhancement of strength-ductility synergy in ultrafine-grained Cu-Zn alloy prepared by ECAP and subsequent annealing. *J. Mater. Res. Technol.* **2022**, *17*, 433–440. [CrossRef]
34. Sotniczuk, A.; Kuczyńska-Zemła, D.; Majchrowicz, K.; Kijeńska-Gawrońska, E.; Kruszewski, M.; Nikiforow, K.; Pisarek, M.; Swieszkowski, W.; Garbacz, H. Tailoring mechanical and surface properties of UFG CP-Ti by the low-temperature annealing. *Appl. Surf. Sci.* **2023**, *607*, 155038. [CrossRef]

35. Majchrowicz, K.; Sotniczuk, A.; Adamczyk-Cieślak, B.; Chromiński, W.; Jóźwik, P.; Pakieła, Z.; Garbacz, H. The influence of microstructure and texture on the hardening by annealing effect in cold-rolled titanium. *J. Alloys Compd.* **2023**, *948*, 169791. [CrossRef]
36. Renk, O.; Maier-Kiener, V.; Issa, I.; Li, J.H.; Kiener, D.; Pippan, R. Anneal hardening and elevated temperature strain rate sensitivity of nanostructured metals: Their relation to intergranular dislocation accommodation. *Acta Mater.* **2019**, *165*, 409–419. [CrossRef]
37. Semenova, I.; Salimgareeva, G.; Da Costa, G.; Lefebvre, W.; Valiev, R. Enhanced Strength and Ductility of Ultrafine-Grained Ti Processed by Severe Plastic Deformation. *Adv. Eng. Mater.* **2010**, *12*, 803–807. [CrossRef]
38. Huang, X.; Xin, Y.; Cao, Y.; Li, W.; Huang, G.; Zhao, X.; Liu, Q.; Wu, P. Understanding the mechanisms of texture evolution in an Mg-2Zn-1Ca alloy during cold rolling and annealing. *Int. J. Plast.* **2022**, *158*, 103412. [CrossRef]
39. Stolyarov, V.V.; Zeipper, L.; Mingler, B.; Zehetbauer, M. Influence of post-deformation on CP-Ti processed by equal channel angular pressing. *Mater. Sci. Eng. A* **2008**, *476*, 98–105. [CrossRef]
40. Valiev, R.Z.; Sergueeva, A.V.; Mukherjee, A.K. The effect of annealing on tensile deformation behavior of nanostructured SPD titanium. *Scr. Mater.* **2003**, *49*, 669–674. [CrossRef]
41. Gu, Y.; Ma, A.; Jiang, J.; Li, H.; Song, D.; Wu, H.; Yuan, Y. Simultaneously improving mechanical properties and corrosion resistance of pure Ti by continuous ECAP plus short-duration annealing. *Mater. Charact.* **2018**, *138*, 38–47. [CrossRef]
42. Singh, A.V.; Vyas, V.; Salve, T.S.; Cortelli, D.; Dellasega, D.; Podesta, A.; Milani, P.; Gade, W.N. Biofilm formation on nanostructured titanium oxide surfaces and a micro/nanofabrication-based preventive strategy using colloidal lithography. *Biofabrication* **2012**, *4*, 025001. [CrossRef]
43. Yang, Z.; Ma, A.; Xu, B.; Jiang, J.; Wu, H.; Sun, J. Tailoring the microstructure and texture to achieve high strength in a high-Al content Mg alloy by stepless-temperature-reduction ECAP. *Mater. Sci. Eng. A* **2022**, *841*, 143023. [CrossRef]
44. Stylianou, R.; Tkadletz, M.; Schalk, N.; Penoy, M.; Czettl, C.; Mitterer, C. Effects of reference materials on texture coefficients determined for a CVD α-Al_2O_3 coating. *Surf. Coat. Technol.* **2019**, *359*, 314–322. [CrossRef]
45. Wang, Y.; Tang, W.; Zhang, L. Crystalline Size Effects on Texture Coefficient, Electrical and Optical Properties of Sputter-deposited Ga-doped ZnO Thin Films. *J. Mater. Sci. Technol.* **2015**, *31*, 175–181. [CrossRef]
46. Latysh, V.; Krallics, G.; Alexandrov, I.; Fodor, A. Application of bulk nanostructured materials in medicine. *Curr. Appl. Phys.* **2006**, *6*, 262–266. [CrossRef]
47. Wu, H.; Jiang, J.; Liu, H.; Sun, J.; Gu, Y.; Tang, R.; Zhao, X.; Ma, A. Fabrication of an Ultra-Fine Grained Pure Titanium with High Strength and Good Ductility via ECAP plus Cold Rolling. *Metals* **2017**, *7*, 563. [CrossRef]
48. Huang, X.; Hansen, N.; Tsuji, N. Hardening by annealing and softening by deformation in nanostructured metals. *Science* **2006**, *312*, 249–251. [CrossRef]
49. Huang, X.; Kamikawa, N.; Hansen, N. Increasing the ductility of nanostructured Al and Fe by deformation. *Mater. Sci. Eng. A* **2008**, *493*, 184–189. [CrossRef]
50. Huang, X. Tailoring dislocation structures and mechanical properties of nanostructured metals produced by plastic deformation. *Scr. Mater.* **2009**, *60*, 1078–1082. [CrossRef]
51. Akbari Mousavi, S.A.A.; Bahadori, S.R. The effects of post annealing on the mechanical properties, microstructure and texture evolutions of pure copper deformed by twist extrusion process. *Mater. Sci. Eng. A* **2011**, *528*, 1242–1246. [CrossRef]
52. Zhao, F.X.; Xu, X.C.; Liu, H.Q.; Wang, Y.L. Effect of annealing treatment on the microstructure and mechanical properties of ultrafine-grained aluminum. *Mater. Des.* **2014**, *53*, 262–268. [CrossRef]
53. Volkov, A.Y.; Kliukin, I.V. Improving the mechanical properties of pure magnesium through cold hydrostatic extrusion and low-temperature annealing. *Mater. Sci. Eng. A* **2015**, *627*, 56–60. [CrossRef]
54. Singh, A.V.; Galluzzi, M.; Borghi, F.; Indrieri, M.; Vyas, V.; Podesta, A.; Gade, W.N. Interaction of bacterial cells with cluster-assembled nanostructured titania surfaces: An atomic force microscopy study. *J. Nanosci. Nanotechnol.* **2013**, *13*, 77–85. [CrossRef]
55. Nasiri-Abarbekoh, H.; Abbasi, R.; Ekrami, A.; Ziaei-Moayyed, A.A. Notch-texture strengthening mechanism in commercially pure titanium thin sheets. *Mater. Des.* **2014**, *55*, 683–689. [CrossRef]
56. Liu, G.; Yang, Y.; Luo, X.; Huang, B.; Kou, Z.; Li, P. Improving the mechanical properties of titanium films by texture strengthening. *Mater. Charact.* **2017**, *127*, 365–370. [CrossRef]
57. Srinivasan, N.; Velmurugan, R.; Singh, S.K.; Pant, B.; Kumar, R. Texture strengthening and anisotropic hardening of mill annealed Ti-6Al-4V alloy under equi-biaxial tension. *Mater. Charact.* **2020**, *164*, 110349. [CrossRef]
58. Li, R.G.; Li, H.R.; Pan, H.C.; Xie, D.S.; Zhang, J.H.; Fang, D.Q.; Dai, Y.Q.; Zhao, D.Y.; Zhang, H. Achieving exceptionally high strength in binary Mg-13Gd alloy by strong texture and substantial precipitates. *Scr. Mater.* **2021**, *193*, 142–146. [CrossRef]
59. Liu, F.; Liu, Z.Y.; He, G.Y.; Ou, L.N. Dislocation ordering and texture strengthening of naturally aged Al-Cu-Mg alloy. *J. Mater. Sci. Technol.* **2022**, *118*, 1–14. [CrossRef]

60. Wang, S.; Hou, J.P.; Zhang, Z.J.; Gong, B.S.; Qu, Z.; Wang, H.; Zhou, X.H.; Jiang, H.C.; Wang, Q.; Li, X.W.; et al. Grain design in ultra-fine Al wire with remarkable combination of strength and conductivity: Ultra-fine-long grains with super strong <111> texture. *Scr. Mater.* **2024**, *238*, 115746. [CrossRef]
61. Sun, J.; Xu, B.; Yang, Z.; Zhuo, X.; Han, J.; Wu, Y.; Song, D.; Liu, H.; Jiang, J.; Ma, A. Developing an industrial-scale ECAP Mg-Al-Zn alloy with multi-heterostructure for synchronously high strength and good ductility. *Mater. Charact.* **2020**, *164*, 110341. [CrossRef]

materials

MDPI

Article

Methodology for Hydrogen-Assisted Fatigue Testing Using In Situ Cathodic Charging

Kai Donnerbauer [1,*], Timo Nickel [1], Matthias von Pavel [1,2], Johannes L. Otto [1], Lars Gerdes [1], Julian Rozo Vasquez [1], Lars A. Lingnau [1], Alexander Koch [1] and Frank Walther [1,*]

1 Chair of Materials Test Engineering (WPT), TU Dortmund University, Baroper Str. 303, D-44227 Dortmund, Germany; lars.lingnau@tu-dortmund.de (L.A.L.)
2 Institute for Research and Transfer e.V. (RIF), Joseph-von-Fraunhofer-Str. 20, D-44227 Dortmund, Germany
* Correspondence: kai.donnerbauer@tu-dortmund.de (K.D.); frank.walther@tu-dortmund.de (F.W.); Tel.: +49-231-755-8416 (K.D.); +49-231-755-8028 (F.W.)

Abstract: With hydrogen being a promising candidate for many future and current energy applications, there is a need for material-testing solutions, which can represent hydrogen charging under superimposed mechanical loading. Usage of high purity gaseous hydrogen under high pressure in commercial solutions entails huge costs and also potential safety concerns. Therefore, a setup was developed utilizing a customized electrochemical charging cell built into a dynamic testing system. With this setup, two heat treatment states of AISI 4140 (DIN 1.7225, 42CrMo4) with varying yield and ultimate tensile strength were characterized in constant amplitude tests. S-N (Woehler) curves differ between heat-treated states, and when comparing testing in air with in situ cathodic hydrogen-charged specimens, hydrogen proves to be detrimental to the material properties. For both states considered, the presence of hydrogen leads to a reduction in fatigue life. Fractographic analyses by scanning electron microscopy reveals that for in situ cathodic hydrogen-charged specimens, the crack initiation mechanisms change for the higher strength heat treatment state.

Keywords: fatigue engineering; AISI 4140; in situ cathodic hydrogen charging; electrochemical hydrogen charging; hydrogen embrittlement; fractography

Academic Editor: Yongfu Zhu

Received: 25 November 2024
Revised: 2 January 2025
Accepted: 8 January 2025
Published: 14 January 2025

Citation: Donnerbauer, K.; Nickel, T.; von Pavel, M.; Otto, J.L.; Gerdes, L.; Vasquez, J.R.; Lingnau, L.A.; Koch, A.; Walther, F. Methodology for Hydrogen-Assisted Fatigue Testing Using In Situ Cathodic Charging. *Materials* **2025**, *18*, 339. https://doi.org/10.3390/ma18020339

1. Introduction

With the increasing role of hydrogen in various industries, particularly in the context of clean energy and environmental sustainability [1], there is an emerging and significant need for testing structural materials under the presence of hydrogen [2]. With this trend towards a hydrogen economy, the infrastructure for production, storage and distribution will expand significantly. Hence follows, that many structural materials will be used in applications where mechanical loading and hydrogen charging are superimposed [3]. However, hydrogen is detrimental to the mechanical properties of many metallic materials [4]. Steels, in general, but especially medium- and high-strength quenched and tempered martensitic steels are endangered [5,6]. An uptake of hydrogen is known to reduce ductility up to the possibility of brittle fractures even at concentrations of around only 1 mass ppm [7,8]. Also, under tensile stress the hydrogen desorption is increased [9]. Especially, the uptake of dislocation-trapped hydrogen is increased according to diffusion-coupled finite element analysis [10]. Hydrogen also leads to reduced fatigue performance [11–15]. In general, uniaxial fatigue data are recorded by constant amplitude testing in servohydraulic, electrodynamic, or resonance testing systems, whereby resonance testing systems are used in the very high cycle-fatigue regime since they can reach much higher testing frequencies. Though many test methods have been developed to measure decreasing mechanical properties under hydrogen [16], currently available standards (ASTM G129, ASTM F1624 and ISO 16573 [17–19]) merely cover incremental step test [20], constant load test and slow strain rate test [21]. In all tests, the mechanical load is applied as quasi-static or static, but not cyclic, which is often present under service in relevant applications. Fracture surfaces on tempered martensitic steel caused by hydrogen embrittlement can exhibit transgranular quasi-cleavage (QC) features but also intergranular (IG) features [22–25]. The underlying mechanisms are still discussed and can occur simultaneously [22,26–28], but brittle mechanisms of IG fractures such as the lowering of cohesive energies [29] and ductile mechanisms of QC fractures like hydrogen-enhanced localized plasticity [30] have been proposed. QC fractures can show river-like patterns on the facets of transgranular fracture portions [31]. IG fractures occur along prior austenite grain boundaries (PAGB) and are commonly accompanied by secondary cracks [32,33]. The influence of PAGB has been researched extensively. For martensitic steels, it has been reported that more hydrogen will be accumulated around PAGB with increasing initial dislocation density [34]. Mechanical testing can be performed using specimens pre-charged electrochemically in a liquid electrolyte or under a high-purity high-pressure gaseous hydrogen atmosphere [35]. Pre-charged specimens do not fully represent service conditions, since mechanical loading and hydrogen charging happen simultaneously during service. There are already existing testing setups to apply mechanical loading during hydrogen charging under gaseous hydrogen atmospheres, but these are very cost-intensive, hardly available and developed for specific-use cases, which is due to the properties and dangers of pressurized hydrogen [36,37]. If specimens are pre-charged, hydrogen concentration in lattice decreases greatly with increasing testing frequency. This effect is much less pronounced in an open system, where hydrogen is provided constantly [38]. Therefore, in situ testing methodologies enabling broad frequency ranges and testing conditions similar to servohydraulic testing systems would be desirable. With an increase in the hydrogen economy as it is currently occurring, a need for efficient testing setups for the determination of material properties during simultaneous hydrogen charging and mechanical loading arises. Electrochemical hydrogen charging is commonly applied followed by mechanical testing, but due to its relatively easy experimental setup, it has also been applied during quasi-static mechanical testing [39,40]. In this work, an efficient setup is presented allowing electrochemical hydrogen charging inside a universal fatigue testing system under dynamic loading conditions realizing testing frequencies of 10 Hz. However, the limitation in terms of testing frequencies depends only on the servohydraulic testing system and is not decreased due to the charging cell. Due to the very low hydrogen volume in comparison to testing under high-pressure gaseous hydrogen, no additional safety equipment was necessary. The components are also widely available without high costs. Therefore, the method can significantly improve the accessibility of dynamic mechanical testing under the influence of hydrogen.

2. Materials and Methods

2.1. Material and Heat Treatment

Steel AISI 4140 (DIN 1.7225, 42CrMo4), manufactured from RIVA Stahl GmbH, Henningsdorf, Germany was chosen as a model material. Chemical composition determined by optical emission spectroscopy of the material was in accordance with DIN EN ISO 683-2 and is given in Table 1 [41].

To investigate the influence of strength on hydrogen susceptibility under fatigue loading, two different material states obtained by means of heat treatment were studied. The austenitizing temperature was 840 °C under a nitrogen atmosphere and kept for 30 min, followed by quenching in water for both conditions. Two different tempering temperatures of 650 °C and 400 °C were used for 30 min in air. These temperatures are subsequently referred to as medium strength (MS) and high strength (HS). Optical microscopy and analysis by electron backscatter diffraction (EBSD) in horizontal cross-sections are provided in Figure 1. Specimens for EBSD underwent metallographic preparation including cutting, embedding, grinding, and polishing up to a 0.05 µm colloidal silica suspension. For optical microscopy, alcoholic nitric acid was used as an etchant. EBSD images were acquired in a field emission gun scanning electron microscope Tescan Mira III (Tescan, Brno, Brno, Czech Republic) equipped with APEX EDAX Velocity Pro detector (AMETEK Inc., Berwyn, PA, USA) at 25 kV acceleration voltage and a step size of 50 nm.

Table 1. Chemical composition of AISI 4140 (DIN 1.7225, 42CrMo4) batch measured by optical emission spectroscopy.

			C	Si	Mn	P	S	Cr	Mo	Cu
Measured		[wt.%]	0.41	0.28	0.79	0.010	0.020	1.06	0.20	0.16
Standard	min		0.38	0.10	0.60	-	-	0.90	0.15	-
	max		0.45	0.40	0.90	0.025	0.035	1.20	0.30	0.40

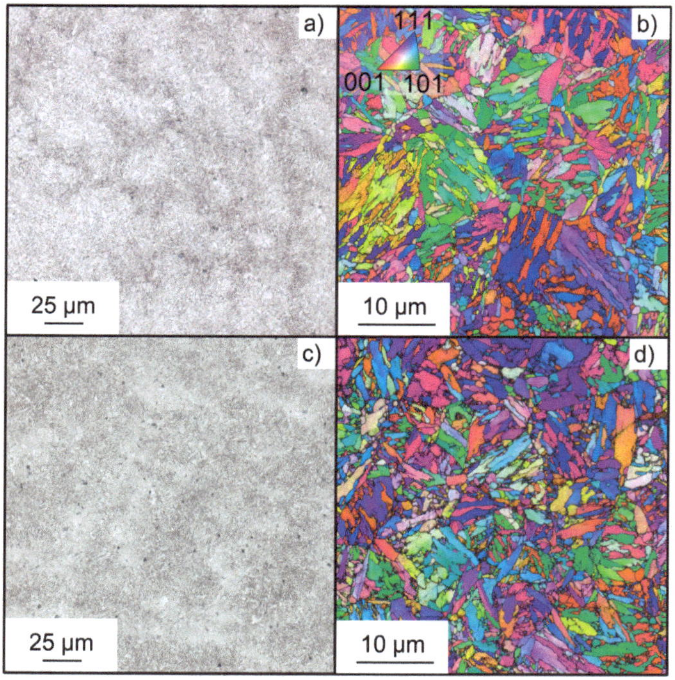

Figure 1. Optical microscopy images and orientation mappings by electron backscatter diffraction of (**a**,**b**) medium strength (MS) and (**c**,**d**) high strength (HS) heat treatment states.

Both material states exhibit a tempered martensitic microstructure with fine needle-like grains and prior austenite grains being recognizable by orientation contrast. The quasi-static properties in air were determined by tensile testing and are summarized as a reference for further fatigue testing in Table 2. Also, the Vickers hardness is given.

After heat treatment, material states differed greatly in quasi-static strength and hardness with values of HS being approximately 60% higher than for MS.

Table 2. Hardness and quasi-static material properties from testing in air at room temperature as a reference.

Heat Treatment State	Hardness HV10	Yield Strength R_e [MPa]	Ultimate Strength R_m [MPa]
Medium strength (MS), 650 °C	320	877	969
High strength (HS), 400 °C	496	1455	1619

2.2. Fatigue Specimens and Preparation

Fatigue tests were carried out using unnotched specimens with a diameter of 10 mm and a length of 12 mm in the gauge section. The specimen geometry is shown in Figure 2. After turning, specimens were polished mechanically. To define a constant area for electrochemical charging, specimens were then coated with the polyurethane varnish Urethan 71 by CRC Industries (CRC Industries Inc., Horsham, PA, USA). However, the gauge length was masked during coating in order to ensure a metallic surface in the smallest cross-section for hydrogen entry.

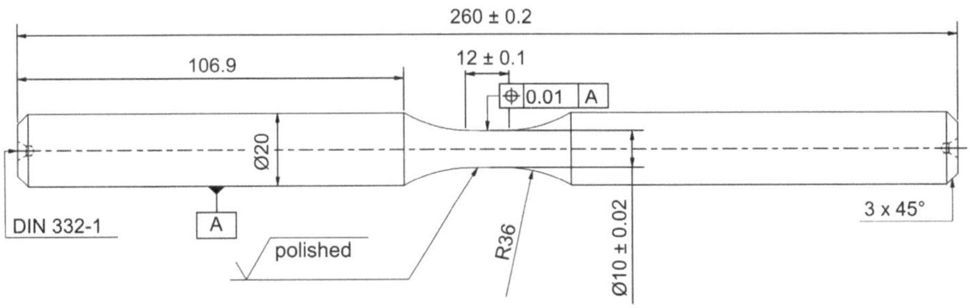

Figure 2. Unnotched fatigue testing specimen geometry.

2.3. Experimental Procedure

All fatigue tests were performed with the servohydraulic fatigue testing system Instron 8802 (Instron, Darmstadt, Germany) with a maximum load capacity of 250 kN. R = 0.1 was chosen as the stress ratio, to ensure that the material is constantly under tensile stress because of the already-investigated stress dependence of hydrogen diffusion. The testing frequency f was constant at 10 Hz and runouts were defined at 3×10^6 cycles, because fracture occurred beyond the common limit of 2×10^6 cycles. A two-electrode electrochemical system in a custom build polymethyl methacrylate (PMMA) cell connected to galvanostat Gamry 1000 A (Gamry Instruments, Warminster, PA, USA) was mounted around the fatigue specimen, as shown in Figure 3.

The electrolyte, which is a 3.5 wt.% NaCl solution, was prepared using deionized water and analytical grade NaCl. Usage of solutions of NaCl are in accordance with ISO 16573 and have been applied by various authors before [42–45]. Due to the cathodic corrosion protection, no corrosive effects such as pitting were observed macroscopically or fractographically in this work. The electrolyte was circulated and tempered to 20 °C by a thermostat. The susceptibility to hydrogen-assisted fatigue was evaluated by in situ cathodic hydrogen charging with a constant current density of 0.075 A/mm^2 with a platinum mesh electrode surrounding the gauge length as the counter and the fatigue specimen as the working electrode. The charging time is dependent on the time until the specimen fracture, since hydrogen charging was only applied during mechanical loading. The charging process during the fatigue loading is schematically depicted in more detail across length scales in Figure 4.

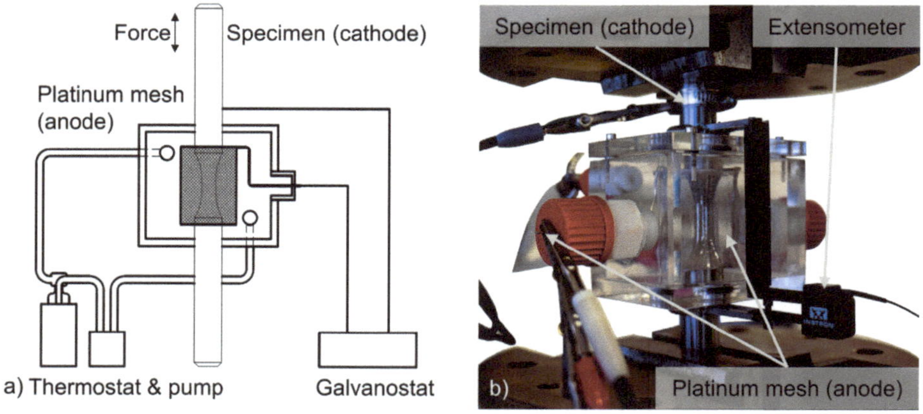

Figure 3. In situ cathodic hydrogen charging inside a fatigue testing system. (**a**) Schematic sketch, (**b**) real experimental setup.

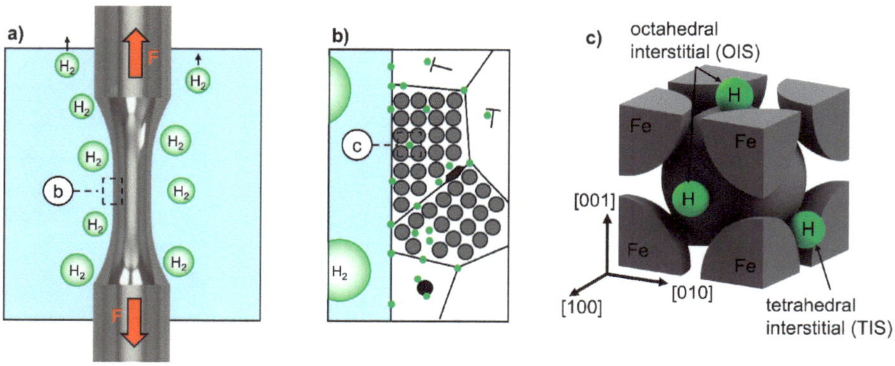

Figure 4. Scale-bridging schematic representation of in situ electrochemical hydrogen charging: (**a**) Macroscopic scale with specimen under cyclic tension stress and molecular hydrogen in form of gas bubbles; (**b**) detail of polycrystalline specimen surface with adsorbed atomic hydrogen and diffusion into the metallic lattice with trapped hydrogen at typical defects; and (**c**) 3D representation of an iron bcc unit cell with hydrogen in interstitial tetrahedral and octahedral sites.

Most of the hydrogen provided by the electrochemical charging recombines into molecular hydrogen and escapes in the form of gas bubbles, Figure 4a. However, some of the hydrogen is adsorbed on the specimen surface as atomic hydrogen, which can diffuse into the polycrystalline metal lattice. However, it can be trapped at metallic defects such as pores, grain boundaries, triple grain boundaries, dislocations, stacking faults and precipitates, as shown in Figure 4b. Integration into the body-centered cubic (bcc) metal lattice as mobile hydrogen can occur interstitially in tetrahedral (TIS) or octahedral (OIS) form, as shown in Figure 4c.

3. Results and Discussion

3.1. Constant Amplitude Testing

Figure 5 illustrates the results of constant amplitude tests in S-N curves representing stress amplitude σ_a over the number of cycles to failure N_f. For both heat treatment states, testing in air leads to higher N_f than testing with in situ cathodic hydrogen charging. On average, comparing power law fits, the S-N curve of HS state in air is 15% higher compared to the in situ cathodic hydrogen charging in the region where experimental data were recorded. On the other side, the fit for the MS state in air is only 7.2% higher, meaning that the hydrogen-assisted fatigue was more severe for the higher strength material. However, it has to be stated that fit curves for the HS state show a similar slope, thus, being independent on the number of cycles to failure. In the MS state, the distance between curves increases slightly with a higher number of cycles to failure, which indicates an increasing effect of hydrogen in higher cycle fatigue regimes. This can be interpreted as a higher hydrogen concentration being needed to negatively influence the lower strength heat treatment. This is in accordance with the known correlation between strength and susceptibility to hydrogen. Also, there is no decline of the S-N curve with an increasing number of cycles to failure, as would be expected, if anodic corrosion processes decreased the fatigue life. The constant slope supports the fact that material degradation, due to presence of hydrogen, is the reason for the reduction in fatigue life.

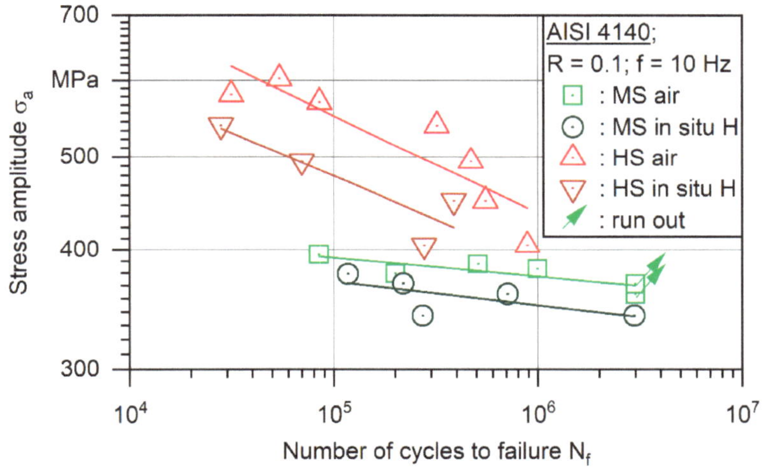

Figure 5. S-N data of heat-treated steel AISI 4140 (DIN 1.7225, 42CrMo4) in medium strength (MS) and high strength (HS) states, tested in air and in situ cathodic hydrogen charging conditions.

For the HS state, the S-N curve is much steeper in general. However, even at $\sigma_{max} = 0.62 \cdot R_e$ no runouts were achieved, whereas at the MS state, a runout occurred at maximum stress $\sigma_{max} = 0.935 \cdot R_e$. All crack initiations in air for both states and in situ cathodic hydrogen charging for MS were detected to occur at inclusions. For the HS state, crack initiation happened at specimen surfaces and at inclusions close to the surface under hydrogen charging. Due to the nature of cathodic charging, this corresponds to the highest local hydrogen concentrations and also underlines the increasing susceptibility with increasing strength. Since most cracks were initiated at inclusions, the scatter of data can be explained by the varying size of those. Interestingly, fracture occurred at 2.96×10^6 at $\sigma_a = 340$ MPa for the MS state under cathodic hydrogen charging beyond the commonly used runout limit of 2×10^6. In contrast, two runouts were recorded at higher stress levels for testing in air. This implies that even in a very high cycle-fatigue regime, fracture will occur, if hydrogen is offered.

3.2. Fractographic Investigations

In order to obtain information about the failure mechanisms involved, the specimens were fractographically examined in the scanning electron microscope Tescan Mira III (Tescan, Brno, Brno, Czech Republic). Figure 6 shows typical fracture surfaces for the medium strength heat treatment. Fracture initiation always occurred at inclusions, which can be identified as aluminum oxide or magnesium oxide and can be found either as shown inside the specimen bulk or close to the surface. When comparing the fracture surfaces for MS, no systematic differences between testing in air and testing under in situ cathodic hydrogen charging regarding crack initiation at inclusions could be observed. Nevertheless, the morphology of fracture surfaces around the inclusions changed. Both show transgranular QC morphologies, but under hydrogen charging, the surface is more faceted and shows river-like patterns. The most likely mechanism for the change in morphology of the transgranular fracture surfaces is hydrogen-enhanced localized plasticity. If crack initiation takes place at an inclusion in bulk, a corrosive influence can be ruled out. Therefore, material degradation due to hydrogen diffusion towards the stress concentration at the inclusion is the explanation for the observed reduction in fatigue lifetime, even for steel with an ultimate tensile strength below 1000 MPa.

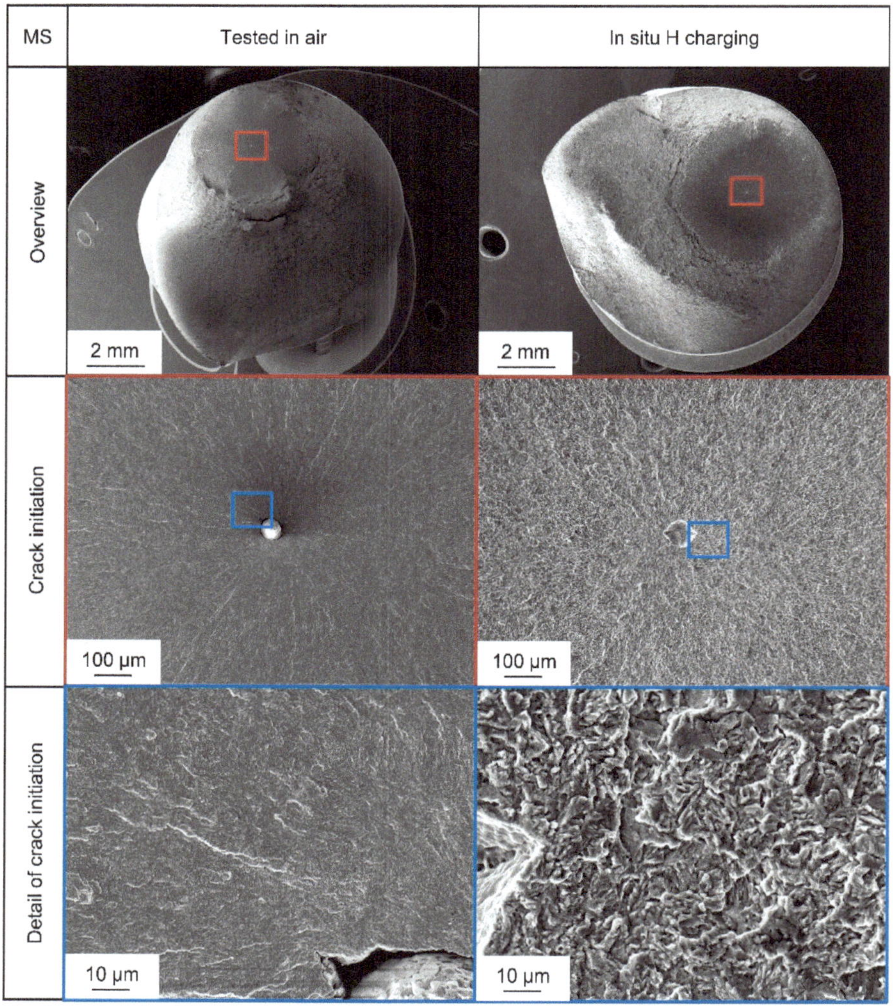

Figure 6. Comparison of fracture surfaces of two representative specimens of heat-treated steel AISI 4140 (DIN 1.7225, 42CrMo4) in medium strength (MS) state, tested in air (on the left) at σ_a = 380 MPa and in situ cathodic hydrogen charging condition (on the right) at σ_a = 360 MPa.

Figure 7 shows typical fracture surfaces for the HS state, where significant differences to medium strength and also between testing in air and testing under in situ cathodic hydrogen charging are observable. Fracture initiation in air again always occurred at inclusions. The fatigue fracture surfaces show a comparable fracture surface to Figure 6 for testing in air with transgranular fracture initiated at an inclusion. When hydrogen is provided, electrochemically fractured surfaces change and show IG fracture portions along PAGB, together with QC transgranular regions. Since for the HS state an increased uptake of hydrogen around PAGB can be expected because of higher initial dislocation density [34], brittle IG fracture portions only for this state are a coherent result. Crack initiation for the specimen in Figure 7 occurred in bulk at an inclusion even though it occurred close to the surface. Since a crack initiation due to corrosion would have to start at the surface, it can be concluded that the local stress concentration of the inclusion, together with the presence of hydrogen, led to hydrogen-assisted failure.

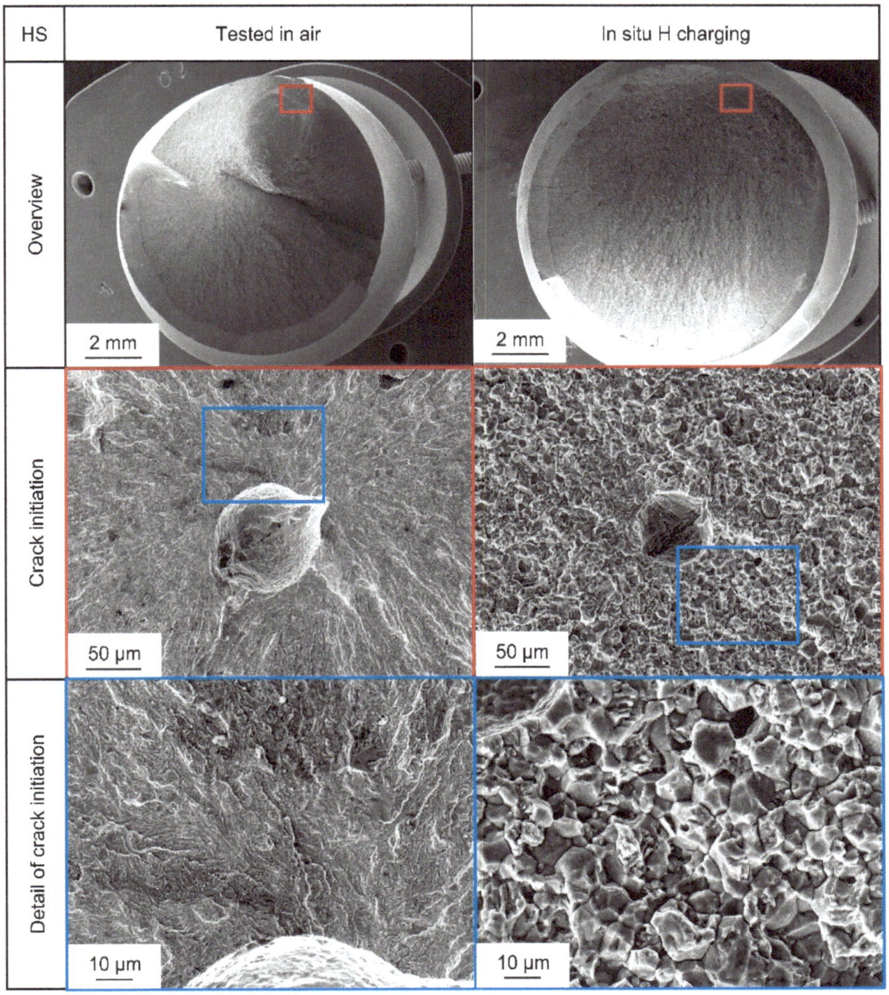

Figure 7. Comparison of fracture surfaces of two representative specimens of heat-treated steel AISI 4140 (DIN 1.7225, 42CrMo4) in a high strength (HS) state, tested in air (on the left), at σ_a = 450 MPa, and in situ cathodic hydrogen charging condition (on the right) at σ_a = 450 MPa.

These transgranular QC fracture portions also change in morphology, as it is typical for hydrogen-induced fracture due to hydrogen-enhanced localized plasticity [23,24]. This is very visible at higher magnifications. Moreover, secondary cracks along grain boundaries become visible, which is, together with brittle intergranular fracture surfaces, another characteristic feature for hydrogen-induced fracture by a hydrogen-enhanced decohesion mechanism, as shown in Figure 8. Therefore, both mechanisms seem to be active at the same time for the HS state. Synergistic activity of both mechanisms has been reported before [46].

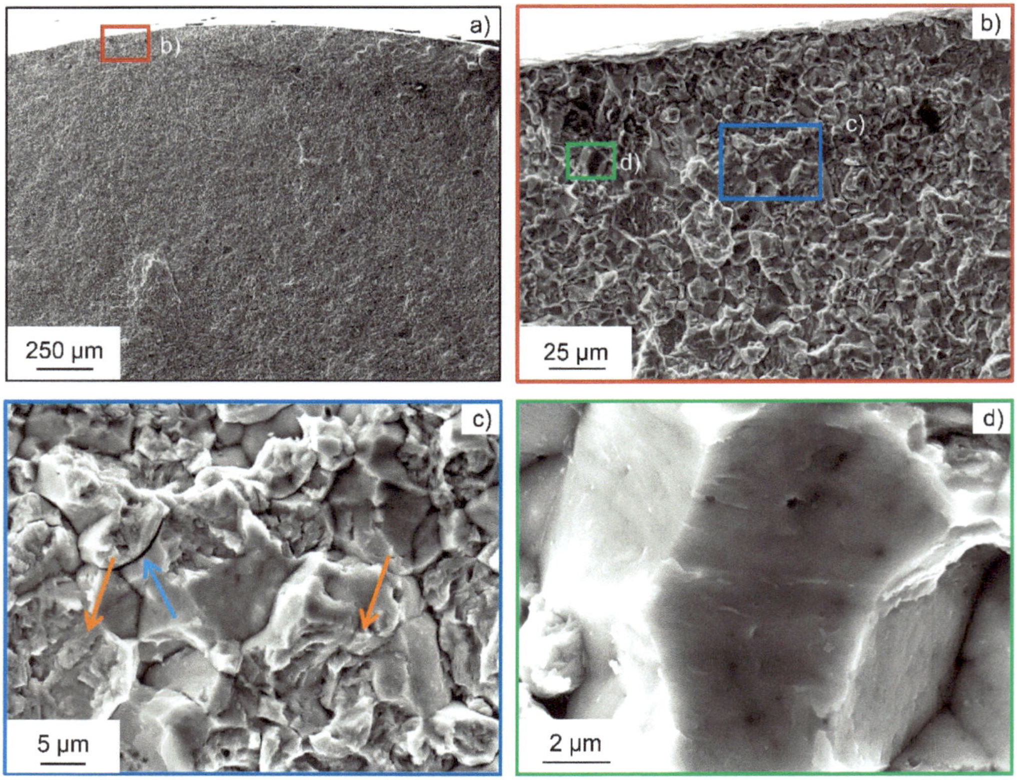

Figure 8. Representative fracture surface of the heat-treated steel AISI 4140 (DIN 1.7225, 42CrMo4) in high-strength state subjected to in situ cathodic hydrogen charging at σ_a = 540 MPa: (**a**) Overview of the crack initiation area close to the specimen surface; (**b**) crack initiation area showing intergranular and transgranular portions; (**c**) evidence of hydrogen-induced cracks by facets (orange arrows) and secondary cracks along grain boundaries (blue arrows); and (**d**) detail of an intergranular crack.

In summary, regardless of the location of crack initiation (surface or inclusion), for the HS treatment, the fracture initiation mechanism changes from transgranular to a mixture of transgranular QC, with high IG portions. As with the MS specimens, the transgranular portions change to a more faceted morphology. Figure 9 shows insights about the remaining fracture. There is a transient area, where intergranular features are not present anymore, but secondary cracks remain, and facets become less present. Figure 9b shows a partly ductile dimple fracture surface in the specimen center. Since hydrogen concentration is expected to be the lowest in the specimen center, it is consistent that the most ductile fracture portion could be detected there. Quite prominent secondary cracks were found again towards the specimen surface.

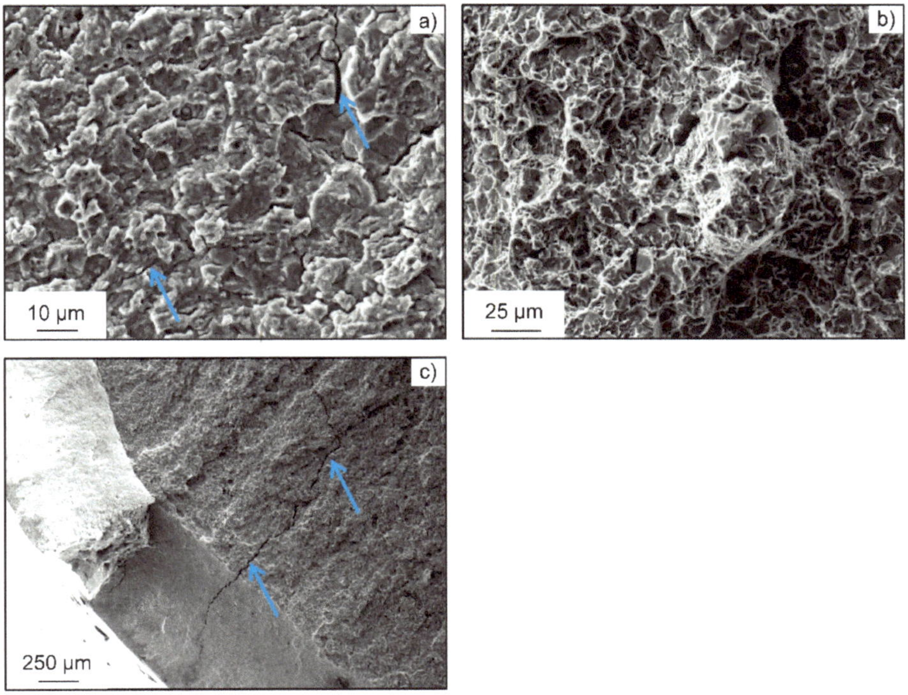

Figure 9. Representative fracture surface of the heat-treated steel AISI 4140 (DIN 1.7225, 42CrMo4) in high-strength state subjected to in situ cathodic hydrogen charging at σ_a = 540 MPa: (**a**) transient area with secondary cracks (blue arrows) and quasi cleavage fracture surface; (**b**) partly ductile dimple fracture in the center of the specimen; and (**c**) secondary cracks on the opposite side of the crack initiation area near the specimen surface (blue arrows).

4. Discussion, Conclusions and Outlook

This work presents a methodology for hydrogen-assisted fatigue testing using in situ cathodic charging. Results agree with the expected behavior, showing a more severe degradation for the high-strength heat treatment compared to the lower strength one, when specimens are charged with hydrogen under simultaneous fatigue loading. The fractographic analysis revealed a significant change in fracture morphology. Only for specimens with high-strength heat treatment, intergranular fractures are dominant, due to the hydrogen-enhanced decohesion mechanism, while no intergranular fracture occurred for specimens tested in air. Crack initiation happened mostly at inclusions, ruling out an influence of corrosion, but a reduction in fatigue lifetime was still observed even at a testing frequency of 10 Hz for a material with an ultimate tensile strength below 1000 MPa. This means that even comparatively low strength steel grades should be investigated in detail, if they are exposed to a possible hydrogen uptake. If the results are to be transferred to another grade, a comparable chemical composition should be present, but most importantly, yield and ultimate strength should not differ, since a clear influence was shown.

The developed setup is highly versatile, allowing easy implementation into other testing systems without the necessity to fulfill additional conditions, aside from those inherent to the fatigue testing system itself. This flexibility underscores the potential for broad application across various materials and testing conditions. In comparison to testing with gaseous hydrogen under high pressure, the setup is easier to handle, safer, cheaper, and therefore more accessible.

Future work should expand the current S-N and fractographic data, allowing a more statistically robust quantitative assessment. More fractographic data could allow quantitative separation of intergranular and transgranular features by image segmentation. The method should be applied to a wider array of electrolytes and materials to further validate the methodology. Steels with a face-centered cubic crystal structure like AISI 316, which is used in hydrogen storage and transport applications, would be interesting candidates. Implementation into a resonance fatigue testing system would allow insights into very high cycle-fatigue regime under hydrogen loading. The influence of low loading levels at a very high number of cycles to failure and, therefore, comparatively long in situ charging times, could be studied in resonance testing systems. Additionally, a comprehensive comparison of data from tests with gaseous hydrogen would provide valuable insights and a deeper understanding of hydrogen embrittlement mechanisms across different testing environments. By conducting finite element predictions of lattice hydrogen concentration, the influence of hydrostatic stress could be investigated to better compare the experimental results between different material strengths.

Author Contributions: Conceptualization, K.D.; data curation, K.D. and M.v.P.; formal analysis, M.v.P., L.G., L.A.L., J.R.V. and A.K.; investigation, K.D., T.N. and M.v.P.; methodology, K.D., J.L.O., L.G., L.A.L. and A.K.; project administration, F.W.; validation, K.D.; visualization, K.D., T.N., M.v.P. and J.L.O.; writing—original draft, K.D. and M.v.P.; writing—review and editing, J.R.V., A.K. and F.W. All authors have read and agreed to the published version of the manuscript.

Funding: The authors gratefully acknowledge the funding by Germany's Federal Ministry for Economic Affairs and Climate Action (BMWK), managed by AiF Projekt GmbH within the Central Innovation Program for SMEs (ZIM, project no. KK5120203SY2 and KK5072229JP2) of the projects "HμMag—Development of a method for the non-destructive analysis of steel components for the prediction of hydrogen-induced failure by local hydrogen analysis and micromagnetic measurements" and "H2Life—Development of a cryogenic test chamber system for testing and qualifying materials and components for the hydrogen economy at high and very high fatigue lifetimes of up to 500 million cycles in contact with liquid hydrogen at −253 °C".

Institutional Review Board Statement: Not applicable.

Informed Consent Statement: Not applicable.

Data Availability Statement: The raw data supporting the conclusions of this article will be made available by the authors on request due to an ongoing study.

Conflicts of Interest: The authors declare no conflicts of interest.

References

1. Noyan, O.F.; Hasan, M.M.; Pala, N. A global review of the hydrogen energy eco-system. *Energies* **2023**, *16*, 1484. [CrossRef]
2. Ahad, M.T.; Bhuiyan, M.M.H.; Sakib, A.N.; Becerril Corral, A.; Siddique, Z. An overview of challenges for the future of hydrogen. *Materials* **2023**, *16*, 6680. [CrossRef]
3. Laureys, A.; Depraetere, R.; Cauwels, M.; Depover, T.; Hertelé, S.; Verbeken, K. Use of existing steel pipeline infrastructure for gaseous hydrogen storage and transport: A review of factors affecting hydrogen induced degradation. *J. Nat. Gas Sci. Eng.* **2022**, *101*, 104534. [CrossRef]
4. Gavriljuk, V.G.; Shyvaniuk, V.M.; Teus, S.M. Hydrogen in metallic alloys—Embrittlement and enhanced plasticity: A review. *Corros. Rev.* **2024**, *42*, 267–301. [CrossRef]
5. Liu, Q.; Atrens, A. A critical review of the influence of hydrogen on the mechanical properties of medium-strength steels. *Corros. Rev.* **2013**, *31*, 85–103. [CrossRef]
6. Oriani, R.A. Whitney award lecture—1987: Hydrogen—The versatile embrittler. *Corrosion* **1987**, *43*, 390–397. [CrossRef]
7. Koyama, M.; Tasan, C.C.; Akiyama, E.; Tsuzaki, K.; Raabe, D. Hydrogen-assisted decohesion and localized plasticity in dual-phase steel. *Acta Mater.* **2014**, *70*, 174–187. [CrossRef]
8. Han, J.; Nam, J.-H.; Lee, Y.-K. The mechanism of hydrogen embrittlement in intercritically annealed medium Mn TRIP steel. *Acta Mater.* **2016**, *113*, 1–10. [CrossRef]

9. Matsumoto, Y.; Miyashita, T.; Takai, K. Hydrogen behavior in high strength steels during various stress applica-tions corresponding to different hydrogen embrittlement testing methods. *Mater. Sci. Eng. A* **2018**, *735*, 61–72. [CrossRef]
10. Hussein, A.; Krom, A.H.; Dey, P.; Sunnardianto, G.K.; Moultos, O.A.; Walters, C.L. The effect of hydrogen content and yield strength on the distribution of hydrogen in steel: A diffusion coupled micromechanical FEM study. *Acta Mater.* **2021**, *209*, 116799. [CrossRef]
11. Wu, X.; Teng, M.; Jia, W.; Cai, J. Study on the mechanical properties of X80 pipeline steel under pre-charged high-pressure gaseous hydrogen. *Int. J. Hydrogen Energy* **2024**, *84*, 39–52. [CrossRef]
12. Dwivedi, S.K.; Vishwakarma, M. Hydrogen embrittlement in different materials: A review. *Int. J. Hydrogen Energy* **2018**, *43*, 21603–21616. [CrossRef]
13. Kanezaki, T.; Narazaki, C.; Mine, Y.; Matsuoka, S.; Murakami, Y. Effects of hydrogen on fatigue crack growth behavior of austenitic stainless steels. *Int. J. Hydrogen Energy* **2008**, *33*, 2604–2619. [CrossRef]
14. Stalheim, D.; Boggess, T.; San Marchi, C.; Jansto, S.; Somerday, B.; Muralidharan, G.; Sofronis, P. Microstructure and mechanical property performance of commercial grade API pipeline steels in high pressure gaseous hydrogen. In Proceedings of the 2010 8th International Pipeline Conference, Calgary, AB, Canada, 27 September–1 October 2010; ASME: New York, NY, USA, 2010; Volume 2, pp. 529–537; ISBN 978-0-7918-4421-2.
15. Nguyen, T.T.; Park, J.; Nahm, S.H.; Tak, N.; Baek, U.B. Ductility and fatigue properties of low nickel content type 316L austenitic stainless steel after gaseous thermal pre-charging with hydrogen. *Int. J. Hydrogen Energy* **2019**, *44*, 28031–28043. [CrossRef]
16. Correa Marques, S.; Lima Molter, D.; Almeida, L.d.S.; dos Santos, D.S. The influence of the experimental methodology on evaluating the hydrogen embrittlement susceptibility of AISI 4340 steel manufactured by different routes. *Eng. Fail. Anal.* **2024**, *162*, 108361. [CrossRef]
17. *ASTM G129*; Standard Practice for Slow Strain Rate Testing to Evaluate the Susceptibility of Metallic Materials to Environmentally Assisted Cracking. ASTM: New York, NY, USA, 2021.
18. *ASTM F1624*; Standard Test Method for Measurement of Hydrogen Embrittlement Threshold in Steel by the Incremental Step Loading Technique. ASTM: New York, NY, USA, 2024.
19. *ISO 16573*; Measurement Method for the Evaluation of Hydrogen Embrittlement Resistance of High Strength Steels. ISO: Vernier, Switzerland, 2022.
20. Das, T.; Rajagopalan, S.K.; Brahimi, S.V.; Wang, X.; Yue, S. A study on the susceptibility of high strength tem-pered martensite steels to hydrogen embrittlement (HE) based on incremental step load (ISL) testing methodology. *Mater. Sci. Eng. A* **2018**, *716*, 189–207. [CrossRef]
21. Martínez-Pañeda, E.; Harris, Z.D.; Fuentes-Alonso, S.; Scully, J.R.; Burns, J.T. On the suitability of slow strain rate tensile testing for assessing hydrogen embrittlement susceptibility. *Corros. Sci.* **2020**, *163*, 108291. [CrossRef]
22. Lynch, S. Hydrogen embrittlement phenomena and mechanisms. *Corros. Rev.* **2012**, *30*, 105–123. [CrossRef]
23. Djukic, M.B.; Sijacki Zeravcic, V.; Bakic, G.M.; Sedmak, A.; Rajicic, B. Hydrogen damage of steels: A case study and hydrogen embrittlement model. *Eng. Fail. Anal.* **2015**, *58*, 485–498. [CrossRef]
24. Ogawa, K.; Matsumoto, Y.; Suzuki, H.; Takai, K. Hydrogen embrittlement susceptibility evaluation of tempered martensitic steels showing different fracture surface morphologies. *ISIJ Int.* **2019**, *59*, 1705–1714. [CrossRef]
25. Nagao, A.; Smith, C.D.; Dadfarnia, M.; Sofronis, P.; Robertson, I.M. Interpretation of hydrogen-induced fracture surface morphologies for lath martensitic steel. *Procedia Mater. Sci.* **2014**, *3*, 1700–1705. [CrossRef]
26. Djukic, M.B.; Bakic, G.M.; Sijacki Zeravcic, V.; Sedmak, A.; Rajicic, B. The synergistic action and interplay of hydrogen embrittlement mechanisms in steels and iron: Localized plasticity and decohesion. *Eng. Fract. Mech.* **2019**, *216*, 106528. [CrossRef]
27. Gangloff, R.P. Hydrogen assisted cracking of high strength alloys. In *Comprehensive Structural Inteqrity*; Milne, I., Ritchie, R.O., Karihaloo, B., Petit, J., Scott, P., Eds.; Elsevier Science: New York, NY, USA, 2003; Volume 6, pp. 31–101.
28. Nagumo, M. Advances in physical metallurgy and processing of steels. Function of hydrogen in embrittlement of high-strength steels. *ISIJ Int.* **2001**, *41*, 590–598. [CrossRef]
29. Troiano, A.R. The role of hydrogen and other interstitials in the mechanical behavior of metals. *Trans. ASM* **1960**, *52*, 54–80. [CrossRef]
30. Birnbaum, H.K.; Sofronis, P. Hydrogen-enhanced localized plasticity—A mechanism for hydrogen-related fracture. *Mater. Sci. Eng. A* **1994**, *176*, 191–202. [CrossRef]
31. Chen, T.; Chiba, T.; Koyama, M.; Shibata, A.; Akiyama, E.; Takai, K. Hierarchical characteristics of hydro-gen-assisted crack growth and microstructural strain evolution in tempered martensitic steels: Case of qua-si-cleavage fracture. *Met. Mater. Trans. A* **2021**, *52*, 4703–4713. [CrossRef]
32. Latypova, R.; Seppälä, O.; Tun Nyo, T.; Kauppi, T.; Mehtonen, S.; Hänninen, H.; Kömi, J.; Pallaspuro, S. Influence of prior austenite grain structure on hydrogen-induced fracture in as-quenched martensitic steels. *Eng. Fract. Mech.* **2023**, *281*, 109090. [CrossRef]
33. Nie, Y.; Kimura, Y.; Inoue, T.; Yin, F.; Akiyama, E.; Tsuzaki, K. Hydrogen embrittlement of a 1500-MPa tensile strength level steel with an ultrafine elongated grain structure. *Met. Mater. Trans. A* **2012**, *43*, 1670–1687. [CrossRef]

34. Momotani, Y.; Shibata, A.; Yonemura, T.; Bai, Y.; Tsuji, N. Effect of initial dislocation density on hydrogen ac-cumulation behavior in martensitic steel. *Scr. Mater.* **2020**, *178*, 318–323. [CrossRef]
35. Koren, E.; Hagen, C.M.H.; Wang, D.; Lu, X.; Johnsen, R.; Yamabe, J. Experimental comparison of gaseous and electrochemical hydrogen charging in X65 pipeline steel using the permeation technique. *Corros. Sci.* **2023**, *215*, 111025. [CrossRef]
36. Zafra, A.; Álvarez, G.; Benoit, G.; Henaff, G.; Martinez-Pañeda, E.; Rodríguez, C.; Belzunce, J. Hydrogen-assisted fatigue crack growth: Pre-charging vs in-situ testing in gaseous environments. *Mater. Sci. Eng. A* **2023**, *871*, 144885. [CrossRef]
37. Briottet, L.; Moro, I.; Escot, M.; Furtado, J.; Bortot, P.; Tamponi, G.M.; Solin, J.; Odemer, G.; Blanc, C.; Andrieu, E. Fatigue crack initiation and growth in a CrMo steel under hydrogen pressure. *Int. J. Hydrogen Energy* **2015**, *40*, 17021–17030. [CrossRef]
38. Fernández-Sousa, R.; Betegón, C.; Martínez-Pañeda, E. Analysis of the influence of microstructural traps on hydrogen assisted fatigue. *Acta Mater.* **2020**, *199*, 253–263. [CrossRef]
39. Imdad, A.; Arniella, V.; Zafra, A.; Belzunce, J. Tensile behaviour of 42CrMo4 steel submitted to annealed, nor-malized, and quench and tempering heat treatments with in-situ hydrogen charging. *Int. J. Hydrogen Energy* **2024**, *50*, 270–280. [CrossRef]
40. Arniella, V.; Zafra, A.; Álvarez, G.; Belzunce, J.; Rodríguez, C. Comparative study of embrittlement of quenched and tempered steels in hydrogen environments. *Int. J. Hydrog. Energy* **2022**, *47*, 17056–17068. [CrossRef]
41. *DIN EN ISO 683-2*; Heat-Treatable Steels, Alloy Steels and Free-Cutting Steels—Part 2: Alloy Steels for Quenching and Tempering. DIN EN ISO: Berlin, Germany, 2018.
42. Tavares, S.S.; Bastos, I.N.; Pardal, J.M.; Montenegro, T.R.; Da Silva, M.R. Slow strain rate tensile test results of new multiphase 17%Cr stainless steel under hydrogen cathodic charging. *Int. J. Hydrogen Energy* **2015**, *40*, 16992–16999. [CrossRef]
43. Thomas, S.; Ott, N.; Schaller, R.F.; Yuwono, J.A.; Volovitch, P.; Sundararajan, G.; Medhekar, N.V.; Ogle, K.; Scully, J.R.; Birbilis, N. The effect of absorbed hydrogen on the dissolution of steel. *Heliyon* **2016**, *2*, e00209. [CrossRef] [PubMed]
44. Venezuela, J.; Gray, E.; Liu, Q.; Zhou, Q.; Tapia-Bastidas, C.; Zhang, M.; Atrens, A. Equivalent hydrogen fugacity during electrochemical charging of some martensitic advanced high-strength steels. *Corros. Sci.* **2017**, *127*, 45–58. [CrossRef]
45. Venezuela, J.; Lim, F.Y.; Liu, L.; James, S.; Zhou, Q.; Knibbe, R.; Zhang, M.; Li, H.; Dong, F.; Dargusch, M.S.; et al. Hydrogen embrittlement of an automotive 1700 MPa martensitic advanced high-strength steel. *Corros. Sci.* **2020**, *171*, 108726. [CrossRef]
46. Wasim, M.; Djukic, M.B.; Ngo, T.D. Influence of hydrogen-enhanced plasticity and decohesion mechanisms of hydrogen embrittlement on the fracture resistance of steel. *Eng. Fail. Anal.* **2021**, *123*, 105312. [CrossRef]

materials

MDPI

Article

The Evolution of the Tensile Properties of MoS_2-Coated Titanium Alloy Bolts Under the Synergistic Damage of NaCl Corrosion and Preloading

Derong Feng [1,2], Maoyang Xie [1,2], Weilin Yu [1], Chao Li [1], Raolong Guo [1,2], Yunpeng Hu [1,2], Quanyuan Ming [3] and Qiang Wan [3,*]

1 Henan Aerospace Precision Manufacturing Co., Ltd., Xinyang 464000, China; xyfdrong@126.com (D.F.); xiemy369@163.com (M.X.); yuweilin.hi@163.com (W.Y.); chaogto@163.com (C.L.); 13383768531@163.com (R.G.); hyp1378293@163.com (Y.H.)
2 Henan Key Laboratory of Fastening and Connection Technology, Xinyang 464000, China
3 College of Engineering, Huazhong Agricultural University, Wuhan 430070, China; mingquanyuan1023@163.com
* Correspondence: wanqiang0915@mail.hzau.edu.cn; Tel.: +86-1329-658-4057

Abstract: MoS_2 coating is a newly developed method to prevent bolt corrosion and the seizure of bolts used in equipment in sea areas. It is of great significance to investigate the evolution of the tensile properties and intact coatings for the maintenance of coated bolts. To evaluate the tensile properties of MoS_2-coated titanium alloy bolts, titanium alloy bolts coated with MoS_2 (TC4+MoS_2) and bolts treated with a composite treatment of anodizing oxidation and MoS_2 coating (TC4+AO+MoS_2) were corroded in salt spray tests for 4300 h. The MoS_2 coating significantly enhanced the bolts' corrosion resistance, demonstrating exceptional protective performance by only experiencing minor peeling due to oxidation-induced cracking of the coating during the extensive 4300 h salt spray test. The tensile strengths of the TC4+MoS_2 and TC4+AO+MoS_2 bolts both decreased as compared with the original bolts. The bolts pretreated with anodic oxidation revealed lighter coating peeling and maintained a higher tensile strength after corrosion. Therefore, it can be concluded that the coatings provided excellent corrosion resistance, leading to a minor impact on the bolts' tensile strength and fracture behavior under the synergistic damage of sea water corrosion and preloading.

Keywords: assembly TC4 bolt; MoS_2 coating; anodic oxidation, corrosion; tensile property

Academic Editor: Young Gun Ko

Received: 10 November 2024
Revised: 6 December 2024
Accepted: 11 December 2024
Published: 31 December 2024

Citation: Feng, D.; Xie, M.; Yu, W.; Li, C.; Guo, R.; Hu, Y.; Ming, Q.; Wan, Q. The Evolution of the Tensile Properties of MoS_2-Coated Titanium Alloy Bolts Under the Synergistic Damage of NaCl Corrosion and Preloading. *Materials* **2025**, *18*, 123. https://doi.org/10.3390/ma18010123

1. Introduction

TC4 (Ti6Al4V) bolts are widely used in aerospace, shipbuilding, biomedical, and other fields due to their excellent mechanical properties, corrosion resistance, and low density [1,2]. Seizing is an important failure mode in the application of titanium alloy bolts [3]. The fundamental reason for seizing is enhanced galling, which is a form of wear caused by metal surfaces in gliding contact with one another. Enhanced galling is attributed to increased friction [4]. There are three reasons for the increased wear of titanium bolts. First, the high stress induced by preloading greatly increases the friction coefficient [5]. Second, the smooth surface brings about a large contact area between threads, which intensifies the friction [6]. Third, a lack of lubrication can also lead to an increase in the friction coefficient [6]. Currently, coatings with a low friction coefficient are being developed to avoid the increased wear and prevent seizing [7,8].

The coatings used for bolts mainly include hot-dip galvanizing, zinc–nickel alloy coating, Dacromet coating, and fluorocarbon coating [9]. However, hot-dip galvanizing coatings with poor corrosion resistance and a high friction coefficient of over 0.15 are not suitable for applications in titanium bolts. Dacromet coating and fluorocarbon coating, with a friction coefficient of 0.08–0.12, present poor adhesion with the bolt substrate and are limited as they providing a low friction coefficient over a long duration. Also, another limiting factor in the application of the above two coatings is aging failure. Obvious exfoliation was observed after 1200 h of corrosion in a salt spray corrosion test.

Molybdenum disulfide (MoS_2) is a layered structure that permits the easily sliding of layers under the action of shear force, thereby significantly reducing the friction coefficient [10]. Meanwhile, it possesses good corrosion resistance with doping metal particles and multi-functional structures [11,12]. Therefore, it has been widely used to coat titanium alloy bolts in recent years [13]. TC4 bolts coated with molybdenum disulfide show excellent tribological properties, which can effectively reduce the friction and wear of bolts. Under the same pre-tightening torque conditions, molybdenum-disulfide-coated bolts present the best anti-loosening performance due to their high pre-tightening force [14] and are currently the most promising bolt protective coating [15].

The main problem of MoS_2 coating is the peeling of the coating [16–18], especially under preloading. The stress resulting from preloading can accelerate crack propagation in the coating and the passivation film formed during corrosion, accelerating the corrosion of the bolt substrate. Further improving the coating's adhesion and ensuring the service life of the coating is key to ensuring the long-term effectiveness of the coating. Pulse anodizing technology can form micropores on the surface of titanium alloy, which, in turn, enhances the adhesion of coatings by providing binding sites [19,20]. The combination of pulse anodizing and MoS_2 coating has become a new strategy in the surface treatment of TC4 bolts.

Many studies on the friction properties and anti-seizing properties of MoS_2 coating on bolts have been carried out. Unlike fluid lubrication, MoS_2 is suitable for use in vertically arranged components because it lacks migration. For this reason, sputtered MoS_2 coating is primarily considered as a low-friction/anti-seizing coating method of the ITER blanket central bolt insert [21]. Zhu attributed the best anti-loosening performance to MoS_2 coating on a bolt compared to PTFE and TiN coatings [15,22]. Usually, titanium alloy bolts are subjected to NaCl corrosion during service. Corrosion can induce the failure of MoS_2 coatings and can even lead to the corrosion of the substrate when the coating fails [23,24]. This will greatly decrease the tensile bearing capacity of the bolt under the action of salt spraying and tension [25,26]. Wei et al. revealed that the ultimate tensile strength and load-bearing capacity of M24 $\times$ 130 mm 20 MnTiB bolts corroded after 10 years in air decreased by 2.5–6.8% and 1.25–5.5%, respectively [27]. Notably, the load at 75% displacement dropped from 390 kN to 340 kN, a reduction of over 15%. Corrosion reduces the clamping force, slip coefficient, initial slip load (or friction shear strength), full slip load, yield load, ultimate load, and ductility, all of which are closely related to the corrosion loss of the entire specimen [28]. The slip resistance coefficient (μ) of specimens after corrosion is approximately linear with respect to the increase in the steel mass loss rate [29]. Therefore, exploring the mechanical property evolution and coating stability of TC4 under the synergistic action of preloading and salt spraying to evaluate the service duration of the surface coating is key to the application of coated bolts used in hot and humid marine environments.

In this study, corrosion tests were carried out on titanium alloy bolts coated with MoS_2 and bolts treated through a composite treatment of anodizing and MoS_2 coating via a salt spray test and coastal hanging corrosion for 4300 h. After corrosion, the bearing capacity of the bolts was tested via a tensile test. The surface morphology and tensile fracture morphology of the bolts were observed in this study in order to explore the underlying mechanisms for the failure of the coating in the salt spray environment and to confirm the protective aging effect of the MoS_2 coating on the bolts.

2. Materials and Methods

2.1. Corrosion Tests

M12-TC4 bolts, provided by Henan Aerospace Precision Manufacturing Co., Ltd., Xinyang, China, were either coated with MoS_2 or treated with a composite treatment of anodizing and MoS_2 coating, and the samples were named TC4+MoS_2 and TC4+AO+MoS_2. The MoS_2 coating process is as follows: first, the bolts were degreased and then cleaned in hot water, acid, and hot pure water in order; next, the bolts were dried and coated with MoS_2 via immersion; third, the MoS_2 was allowed to solidify. The bolts were installed in an aluminum fixture with a pre-tightening force of 40 kN, which is the installation requirement for M12 bolts in applications. The corrosion test of the coated bolts was carried out using a salt spray corrosion test. The salt spray corrosion test was conducted in a salt spray corrosion test chamber according to the GB/T 10125-2021 standard. During the test, a 3.5% NaCl neutral aqueous solution with a pH value between 6.5 and 7.2 was used as the corrosive solution at 35 °C. During the test, the uniformity and stability of the spray were strictly controlled to ensure the consistency and repeatability of the corrosive environment. The salt spray corrosion times were set at 96, 192, 1100, 2100, and 4300 h. The times for the coastal hanging corrosion test were 2100 h and 4300 h.

2.2. Friction Tests

The friction coefficient of the MoS_2-coated TC4 plates with and without pulse anodic oxidation treatment was tested by a sliding friction tester (RTEC MFT-5000). Plate specimens with a diameter of 20 mm were polished to a mirror finish. The thread friction coefficient test was conducted on the coated bolts in accordance with the ISO 16047-2005 "fasteners—torque/clamp force testing" standard after the salt spray corrosion, as shown in Figure 1. The axial force of the bolt was set at 40 kN. The experiment was performed on a SCHATZ threaded fastener analysis system (5416-2777/04/CS). Each set of experiments was repeated three times, and the average value of the friction coefficients was taken as the final result.

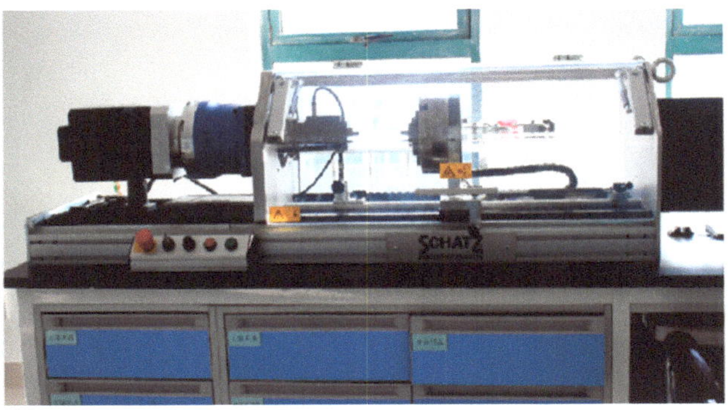

Figure 1. SCHATZ threaded fastener analysis system.

2.3. Tensile Test of Assembly Bolts After Corrosion

The tensile testing was conducted on multiple pairs of assembled bolts using a CMT5305 electronic universal testing machine in accordance with the Chinese mechanical testing standards for fasteners, specifically GB/T 3098.1-2010. The stress and elongation were recorded for each test. Three specimens were tested for each corrosion time to obtain the average values of the yield strength and tensile strength. The yield strength was determined at the turning point between linear and nonlinear changes from the force–displacement curves. The tensile strength was calculated from the maximum force of the tensile curve by dividing by the area of the fracture cross-section. Bolts exposed at the Sanya coastal area for 2100 and 4300 h were also tensioned to identify the effectiveness of the salt spray corrosion.

2.4. Microstructure Characterization

The microstructure and chemical composition of the treated bolts were characterized with an FEI scanning electron microscope (SEM) system equipped with energy-dispersive spectrometry (EDS) at the surface and cross-section. Additionally, the surface morphology of the corroded specimens and the fracture surfaces of the tensile failures were characterized and analyzed to identify the corrosion products and fracture morphology. Specimens for cross-sectional observation were cut from the bolts. Then, they were ground and polished to a mirror-like surface. The characterization of the corroded MoS_2 coating and the fracture morphology was conducted from the original state. The working distance of the SEM was 13 mm, and the working voltage was 20 kV.

3. Results and Discussion

3.1. Microstructure of the Coating on Bolts

Figure 2 shows the cross-sectional morphology and compositions of the MoS_2-coated bolts. EDS mapping profiles and line scans were carried out to reveal the elemental distribution in the coatings. According to the elemental distribution of the mapping profiles, the upper layer was mainly composed of Mo and S, while the lower layer was rich in Ti, V, Al, and O. The results suggested that the upper layer was the MoS_2 coating, and the lower layer was the TC4 substrate. The line scanning distribution also confirmed the above conclusion. The width of the Mo and S peaks in the element scan lines along the cross-section was the thickness of the MoS_2 coating. Combined with the thickness calculated from the morphology figures, the MoS_2 coating was about 42.55 μm thick. The observed O element was attributed to pollution during the specimen production process. Figure 3 shows the cross-sectional morphology and compositions of the TC4+AO+MoS_2 bolt. Similarly, the top MoS_2 coating and the TC4 substrate were confirmed by the aggregation of S and Ti in the top surface and the bottom layer. An obvious O peak was found between the MoS_2 coating and the TC4 substrate. According to the process, this should be the anodic oxidation layer, which mainly consisted of Ti and O.

3.2. Friction Coefficient and Thread Friction Coefficient Test

Figure 4 reveals the friction curves of the plates treated with the MoS_2 coating and with the composite treatment of anodic oxidation and MoS_2 coating. The friction coefficient (COF) of the plate coated with MoS_2 only increased with the slide time and reached a maximum value. The average friction coefficient was calculated to be 0.166. In comparison, the friction coefficient of the plate treated with the composite treatment of anodic oxidation and MoS_2 coating revealed a similar trend but had a higher average friction coefficient of 0.192. The higher friction coefficient was related to the higher roughness induced by anodic oxidation [30].

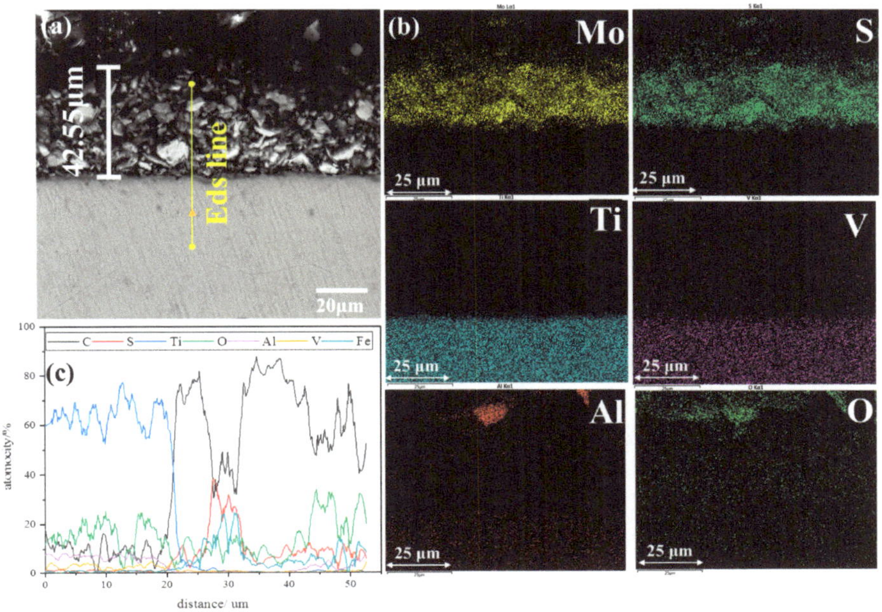

Figure 2. Cross-sectional morphology of TC4+MoS_2 bolts: (**a**) cross-sectional morphology, (**b**) EDS mapping profiles, (**c**) EDS line scanning.

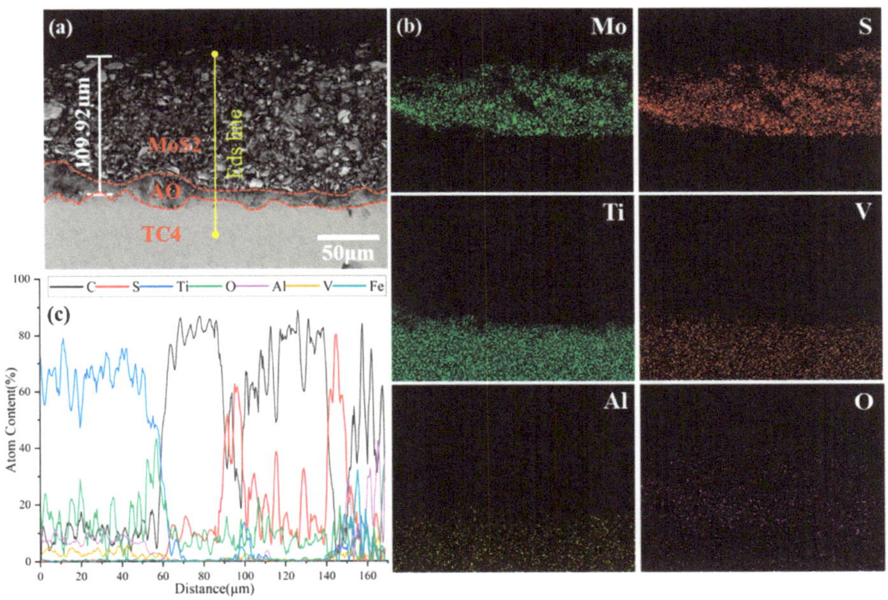

Figure 3. Cross-sectional morphology of TC4+AO+MoS_2 bolts: (**a**) cross-sectional morphology, (**b**) EDS mapping profiles, (**c**) EDS line scanning.

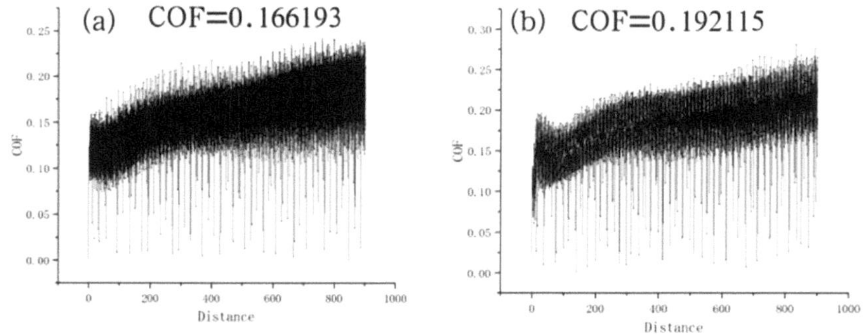

Figure 4. Friction curves of (**a**) TC4+MoS_2 bolts and (**b**) TC4+AO+MoS_2 bolts.

The thread friction coefficient of the two treated bolts was calculated by measuring the tightening torque and thread torque according to Formulas (1) and (2).

$$\mu_{tot} = \frac{\frac{T}{F} - \frac{P}{2\pi}}{0.5777d_2 + 0.5D_b} \tag{1}$$

$$T = T_b + T_{th} \tag{2}$$

μ_{tot} is the thread friction coefficient, d_2 is the pitch diameter, d_b is the friction diameter of the nut bearing surface, p is the pitch, t is the tightening torque, t_b is the torque of the end-face friction, T_{th} is thread torque, and f is the axial force.

The thread friction coefficient of the bolt coated with MoS_2 was 0.062, while that of the TC4+AO+MoS_2 bolt was 0.065. The thread friction coefficient was consistent with the sliding friction coefficient obtained from plates [31]. However, the friction coefficient of the bearing surface of the bolt coated only with MoS_2 was 0.056, and that of the bolt additionally treated with anodic oxidation was 0.049. A lower friction coefficient could lead to more tightening torque being converted into the axial preloading of the bolt and decrease the wear during assembly or disassembly [32]. Also, the lower friction coefficient of the bearing surface of the TC4+AO+MoS_2 bolt resulted in a slightly smaller total friction coefficient of 0.055.

3.3. *Surface Morphologies of Corroded Bolts*

Figures 5 and 6 show the macroscopic morphology of the bolts after surface passivation and molybdenum disulfide coating under long-term 3.5% NaCl salt solution spraying. The salt spray corrosion times were 0 h, 192 h, 1100 h, 2100 h, and 4300 h. As can be seen from the figures, after 4300 h of salt spray corrosion, all the bolts basically showed no macroscopic rust spots, indicating good corrosion resistance. However, the coating on both the bolt head and the screw rod had peeled off, exposing the TC4 bolt substrate. Among these, the peeling area ratio of the bolt head was less than that of the screw rod. The reason is that during the bolt removal process, the screw rod rubbed against the clamping nut, causing the surface coating to peel off. As the corrosion time increased, the peeling area of the coating on the screw rod surface of the bolts treated with the two methods increased significantly. It can be seen that the corrosion seriously reduced the stability of the coating, and there was no significant difference in the peeling area between the screw rod of the bolt with anodic oxidation treatment and the screw rod without anodic oxidation treatment.

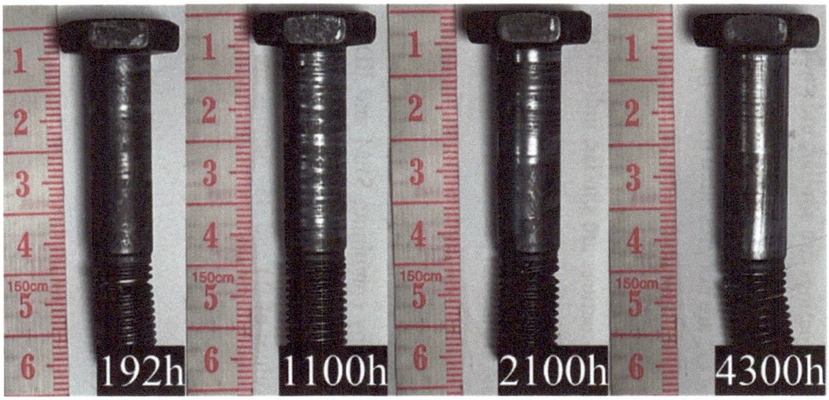

Figure 5. Macroscopic morphology of TC4+MoS_2 bolts with different corrosion times.

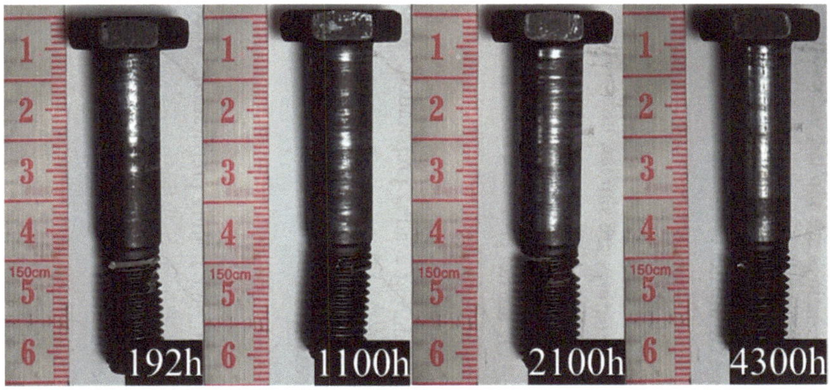

Figure 6. Macroscopic morphology of TC4+AO+MoS_2 bolts with different corrosion times.

Figures 7–9 show the microscopic morphologies and compositions of the TC4 bolts coated with MoS_2 coating and with the composite treatment of anodic oxidation and MoS_2 coating after the corrosion test. Figure 7 shows that significant coating delamination occurred on the surface of the MoS_2-coated bolt after 192 h of salt spray corrosion. The EDS mapping profiles indicated that the exposed areas were primarily composed of Ti, Al, and V elements, which are the main components of the TC4 substrate. It is suggested that delamination of the coating occurred. The surface layer mainly consisted of Mo and S elements, indicating that the surface coating was MoS_2. A high density of O elements about 1.6 at. % was observed on the specimen surface in the mapping profiles, indicating oxidation of the MoS_2. Due to the delamination of the surface coating, the TC4 alloy substrate also oxidized with a 15.3% O content. This suggests that the delamination of the MoS_2 led to a loss of corrosion protection. As the salt spray corrosion time increased, the O content in the MoS_2 coating further increased [23,33]. For the specimen corroded for 2100 h, the O content in the MoS_2 coating increased to 22.2%. Correspondingly, the TC4 substrate material also underwent more severe corrosion, with the surface O content increasing to over 44%. After 4300 h, the O content in the coating surface further increased to 44.6%, accompanied by significant cracks on the coating surface. However, the O content in the oxide layer on the TC4 surface did not increase further. The results indicated that the MoS_2 coating oxidized in the salt spray corrosion environment, leading to the cracking and subsequent delamination of the coating due to the stress growth during the oxidation process, which resulted in the exposed TC4 substrate undergoing metal corrosion.

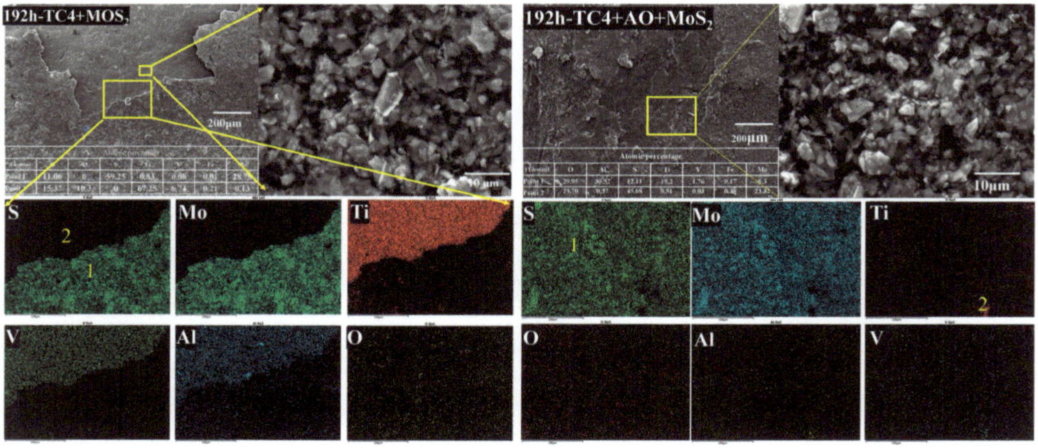

Figure 7. Microscopic morphology and elemental distribution of two bolts after 192 h of salt spray corrosion.

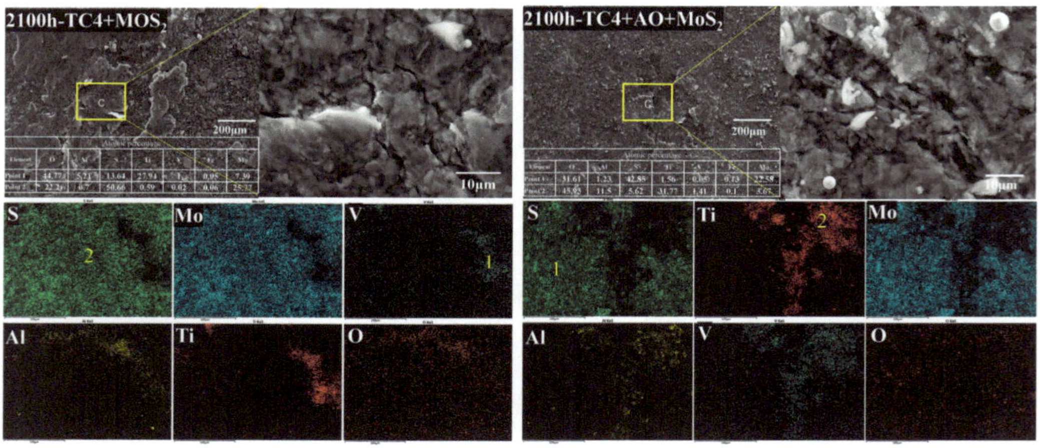

Figure 8. Microscopic morphology and elemental distribution of two bolts after 2100 h of salt spray corrosion.

The surface of the bolts with the composite treatment showed a uniform distribution of Mo and S. No obvious presence of Ti, Al, or V elements was observed after 192 h of salt spray corrosion. The results indicated that there was no MoS_2 delamination on the surface of the bolts. The EDS spectrum analysis revealed that the surface oxygen content was consistently around 29%, which was higher than that of the bolts without anodic oxidation treatment. However, the surface MoS_2 coating on the bolts treated with pulse anodic oxidation was more complete, even after severe oxidation. This suggested that the anodic oxidation enhanced the adhesion of the coating in the salt spray environment, ensuring the effective protective performance of the MoS_2 coating [19]. As the salt spray corrosion test was prolonged to 2100 h, the coating on the bolt surface started to exfoliate, with the presentation of Ti elements, as shown in Figure 8. Spot analysis revealed that the oxygen content in the exposed substrate surface reached 45%, which was significantly higher than the 22.2% observed on the corroded substrate of the MoS_2-coated bolts. The high oxygen content was not derived from corrosion of the exposed substrate but from the oxide layer formed by anodic oxidation. Since the MoS_2 coating had not completely exfoliated, the measured oxygen content was lower than that in the anodic oxide layer. Therefore, when the corrosion time was increased to 4300 h, more of the MoS_2 coating delaminated, and the measured oxygen content on the surface of the bolt substrate further increased to 59.85%.

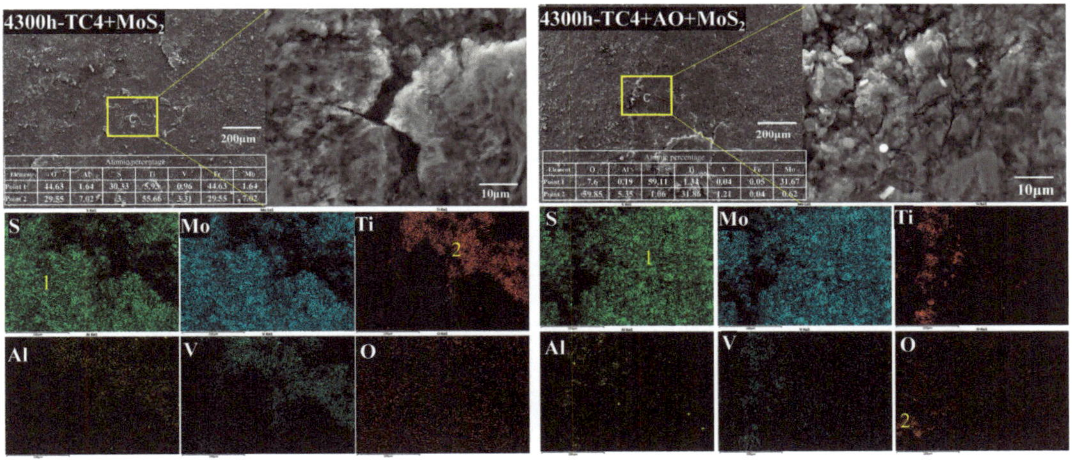

Figure 9. Microscopic morphology and elemental distribution of two bolts after 4300 h of salt spray corrosion.

It is evident that these MoS_2 and anodic oxidation treatments can improve the corrosion resistance of bolts, but the bonding strength between the coating and the substrate may decrease during long-term exposure to corrosive environments. A decrease in the bonding strength can lead to easy exfoliation of the coating, exposing the substrate material and further exacerbating corrosion [34]. Pulsed anodic oxidation can improve the adhesion of the coating, slowing down the process of coating exfoliation during the oxidation of MoS_2.

3.4. Tensile Properties of Bolts After Corrosion

Figure 10 shows the tension curves of the two different types of treated bolts after salt spray corrosion. The yield strength and tensile strength are shown in Tables 1 and 2. The tension process indicates that the bolts underwent two stages: elastic deformation and plastic deformation. During the elastic deformation stage, the load on the bolt increased linearly with increasing displacement. When the load reached the yield strength of the material, the curve began to deviate from linearity and entered the plastic deformation stage. Further increasing the load, the tensile strength point was reached. Then, local necking appeared, resulting in a reduction in the actual load-bearing area and a rapid decrease in the load. According to the standard for the mechanical testing of fasteners (ISO 898-1:2013, fitting formulas were developed on the basis of the line segment corresponding to the elastic deformation of the bolts. The yield strength and tensile strength of the bolts after corrosion were obtained.

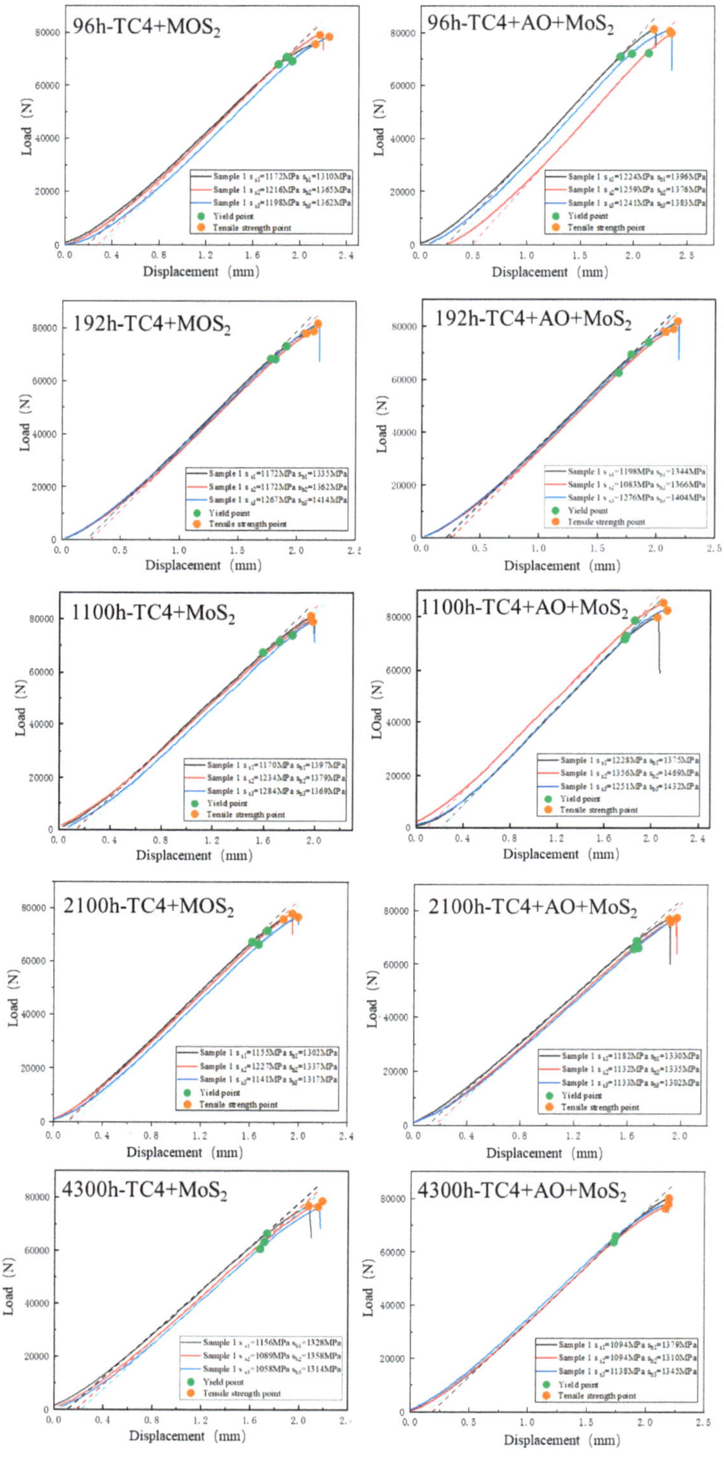

Figure 10. Tensile curves of TC4+MoS_2 and TC4+AO+MoS_2 bolts after different times of salt spray corrosion.

Table 1. Yield strength of TC4 bolts treated differently at different times.

	96 h	192 h	1100 h	2100 h	4300 h
TC4+MoS_2 (salt spray)	1195.3 ± 20.7	1203.7 ± 63.3	1185.7 ± 54.7	1174.3 ± 52.7	1101.0 ± 55.0
TC4+AO+MoS_2 (salt spray)	1241.3 ± 17.7	1185.7 ± 90.3	1229.3 ± 77.7	1149.0 ± 33.0	1108.7 ± 29.3
TC4+MoS_2 (costal hanging)	-	-	-	1114.5 ± 90.5	1186.3 ± 20.3
TC4+AO+MoS_2 (costal hanging)	-	-	-	1157.5 ± 49.5	1178.3 ± 22.3

Table 2. Tensile strength of TC4 bolts treated differently at different times.

	96 h	192h	1100 h	2100 h	4300 h
TC4+MoS_2 (salt spray)	1345.7 ± 25.7	1370.3 ± 43.7	1381.7 ± 15.3	1318.7 ± 18.3	1333.3 ± 24.7
TC4+AO+MoS_2 (salt spray)	1385.0 ± 11.0	1371.3 ± 32.7	1424.94 ± 50.3	1322.3 ± 20.3	1344.7 ± 34.7
TC4+MoS_2 (costal hanging)	-	-	-	1372.0 ± 20.0	1344.3 ± 12.3
TC4+AO+MoS_2 (costal hanging)	-	-	-	1387.0 ± 15.0	1346.0 ± 28.0

Figure 11 shows tensile curves of the TC4+MoS_2 and TC4+AO+MoS_2 bolts after different times of coastal hanging corrosion, respectively. Figure 12 shows the yield strength and tensile strength of TC4+MoS_2 and TC4+AO+MoS_2 bolts after different times of salt spray corrosion and coastal hanging corrosion. When the corrosion time was shorter than 1100 h, both the yield strength and tensile strength exhibited slight fluctuations within a certain range without a significant decrease. This indicated that short-term corrosion revealed little impact on the tensile mechanical properties of the MoS_2-protected bolts. When the corrosion time increased from 1100 h to 2100 h, the yield strength of both types of bolts showed a significant decrease, dropping from above 1200 MPa to around 1150 MPa. With the further increase in the spray corrosion time, the yield strength of the bolts continued to decrease, reaching 1000 MPa after 4300 h of corrosion. The tensile strength of both types of bolts also followed a similar trend. Within the first 1100 h of corrosion, there was almost no impact on the tensile strength of the bolts. As the corrosion time increased to over 2100 h, the tensile strength of the bolts decreased. After 4300 h, the tensile strengths of the TC4+MoS_2 and TC4+AO+MoS_2 bolts were 1332 and 1344 MPa, respectively. The decrements were 38 and 26 MPa compared with those of the original bolts. The TC4+AO+MoS_2 bolt revealed a smaller decrement compared with the TC4+MoS_2 bolt. The main reason for the decreased strength was the increased number of crack sources, the reduced effective bearing area, and the shortened crack propagation path of the bolt, resulting from the surface corrosion layer of the TC4 substrate [28]. Under the synergistic effect of preloading and corrosion, the corrosion sites on the surface became the weak points of the bolt, which were prone to cracking, thereby increasing the number of crack sources and decreasing the bearing area [35]. As the corrosion time increased, more corrosion sites formed and more crack sources appeared, leading to a continuous decrease in the tensile strength. Furthermore, the bolts hanging in the coastal area revealed a higher tensile strength as compared with those subjected to the salt spray test after the same corrosion time. This indicated that the damage caused by salt spraying was more severe than that caused by hanging in coastal areas. In addition, it should be pointed out that the tensile strength of the bolts treated with pulse anodic oxidation was higher than that of the untreated bolts under different corrosion times. It can be seen that the pulse anodic oxidation could reduce the tensile strength degradation of the bolts under preloading and corrosion damage by enhancing the adhesion of the MoS_2 coating and ensuring the long-term anti-corrosion protection of the MoS_2 coating. Therefore, the composite treatment of anodic oxidation and MoS_2 coating is a better method for protecting TC4 bolts in corrosion environments.

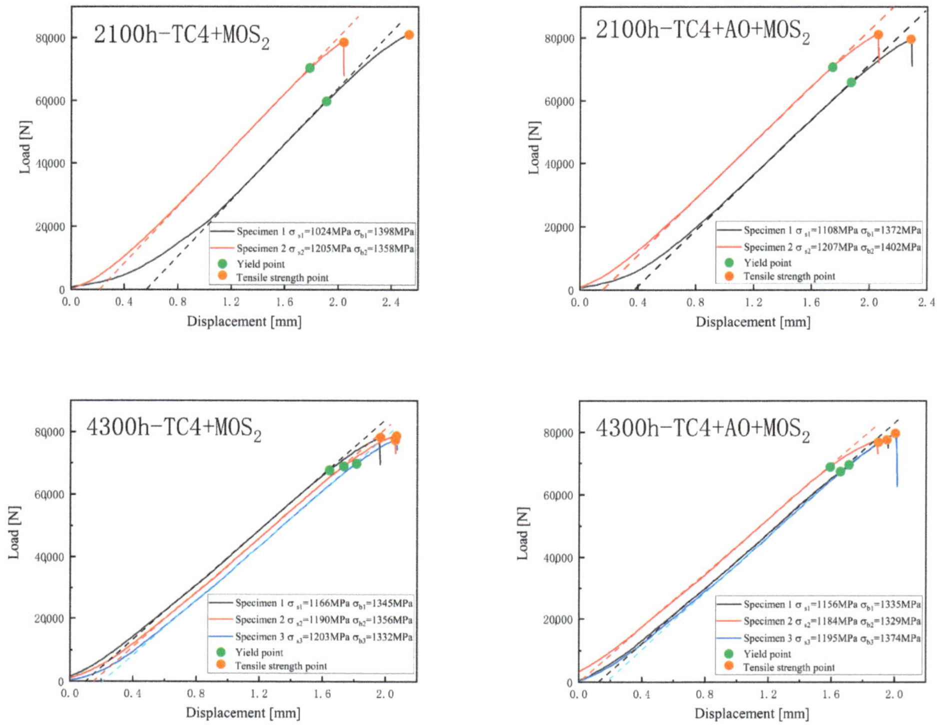

Figure 11. Tensile curves of TC4+MoS_2 and TC4+AO+MoS_2 bolts after different times of coastal hanging corrosion test.

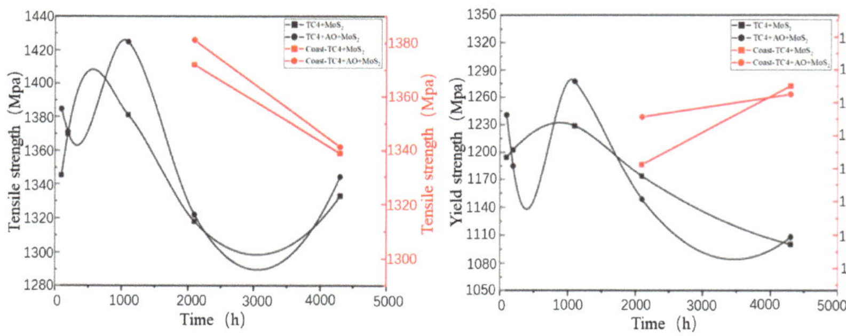

Figure 12. Yield strength and tensile strength of TC4+MoS_2 and TC4+AO+MoS_2 bolts after different times of salt spray corrosion and coastal hanging corrosion.

3.5. Fracture Morphology

The fracture morphology was observed by SEM, revealing the fracture mechanism of the corroded bolts, as shown in Figures 13 and 14. After the corrosion for various durations, the fracture mechanism of TC4 bolts coated with MoS_2 and with the composite treatment of anodic oxidation and MoS_2 coating was still ductile fracturing, featuring prominent propagation zones and shear lips [36]. The cracks originated from the root of the threads, which are marked by red dashed lines and are referred to as the "initiation zone" [37]. The cracks extended linearly from one side of the thread root to the other, typically situated at stress concentration points or material defects such as corrosion sites. High-magnification observations were conducted on the crack propagation zones marked by the red dashed lines and the shear lips marked by yellow dashed lines. At the microscale, distinct dimples were observed on the bolt fracture surfaces. These dimples were tiny voids formed during the plastic deformation process under tension. As the stress increased, these voids gradually expanded and coalesced, ultimately leading to fracturing. The large number of dimples indicated that the bolts' tensile fracture process was ductile fracturing. Furthermore, strips could be discerned around the dimples, representing the trajectories of the crack propagation, which indicated the direction of expansion of the cracks after their initiation, as illustrated by the yellow dashed line icons in the figures [38].

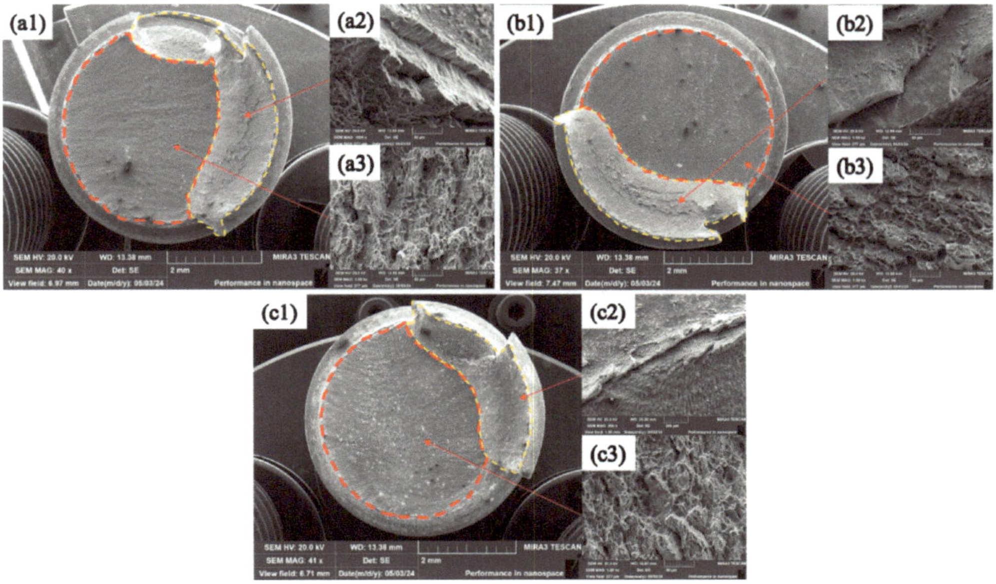

Figure 13. Fracture morphologies of TC4+MoS_2 bolts: (**a1–a3**) after 192 h of salt spray corrosion; (**b1–b3**) after 2100 h of salt spray corrosion; and (**c1–c3**) after 4300 h of salt spray corrosion.

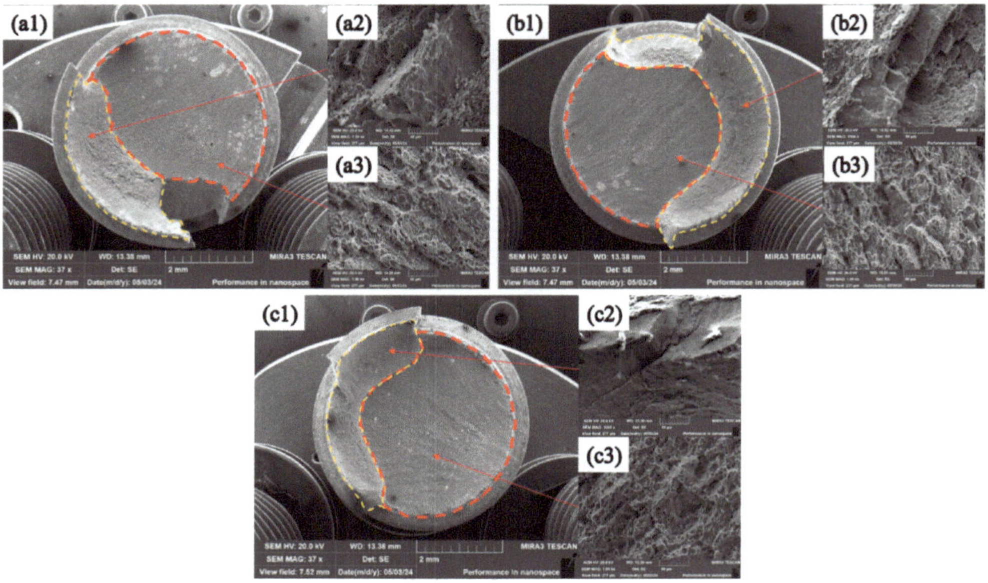

Figure 14. Fracture morphologies of TC4+ac+MoS_2 bolts: (**a1–a3**) after 192 h of salt spray corrosion; (**b1–b3**) after 2100 h of salt spray corrosion; and (**c1–c3**) after 4300 h of salt spray corrosion.

4. Conclusions

A 4300 h salt spray test and a 4300 h coastal hanging corrosion test were carried out on TC4 bolts coated with MoS_2 coating and with a composite treatment of anodic oxidation and MoS_2 coating. By combining the results of reciprocating friction tests and thread friction coefficient measurements, the friction performance of the bolts was investigated. Additionally, the tensile properties and the morphology before and after corrosion were tested. The following conclusions were drawn:

(1) The thickness of the coating on the treated bolt surface was about 42 μm. The reciprocating dry friction coefficients for the MoS_2 coating application and anodic oxidation + MoS_2 coating treatment were 0.166 and 0.192, respectively, and the corresponding friction coefficients of the treated bolt threads were 0.062 and 0.065.

(2) After 4300 h of salt spray corrosion, all the bolts basically did not show any macroscopic rust spots, demonstrating good corrosion resistance resulting from the MoS_2 coating. However, the coating on the bolt surface oxidized, resulting in the peeling of the coating on both the bolt head and the screw rod, which caused the base titanium alloy to oxidize. As the corrosion time increased, the peeling area of the coating of the screw rods of both treated bolts increased significantly. Pulsed anodic oxidation could improve the adhesion of the coating, slowing down the peeling process of the coating during oxidation by revealing a smaller area of peeling off.

(3) When the corrosion time was less than 1100 h, the yield strength and tensile strength fluctuated within a certain range without significant reduction. After 4300 h of corrosion, the decrements in the tensile strength of the TC4+MoS_2 and TC4+AO+MoS_2 bolts were 38 and 26 MPa. The bolts treated with pulsed anodic oxidation had higher tensile strength than those without pulsed anodic oxidation under the same corrosion times. Fracture analysis showed that under the combined effect of corrosion and preloading, the bolt fracture mechanism was still ductile fracturing.

The research results showed that MoS_2 coating is a good means of bolt corrosion protection, and it can effectively reduce the friction coefficient of the bolt surface. Meanwhile, anodic oxidation treatment can improve the long-term effectiveness of MoS_2 coating protection. Therefore, the composite treatment of anodic oxidation + MoS_2 coating is an effective means of preventing the corrosion and seizing of titanium alloy bolts. The results provide scientific data for the maintenance of coated bolts. Longer corrosion tests and friction behavior tests should be further conducted to reveal the effectiveness of MoS_2 coating in protecting bolts from fracturing and wear.

Author Contributions: Conceptualization, D.F., C.L., Q.M., Q.W. and M.X.; methodology, D.F. and M.X.; software, D.F.; validation, D.F., M.X. and W.Y.; formal analysis, R.G. and Y.H.; investigation, D.F.; resources, Y.H.; data curation, D.F., Q.W. and M.X.; writing—original draft preparation, D.F. and C.L.; writing—review and editing, Q.W. and Q.M.; visualization, D.F. and M.X.; supervision, D.F. and Q.W.; project administration, D.F. and Q.W.; funding acquisition, D.F. All authors have read and agreed to the published version of the manuscript.

Funding: This research received no external funding.

Institutional Review Board Statement: Not applicable.

Informed Consent Statement: Informed consent was obtained from all subjects involved in this study.

Data Availability Statement: The original contributions presented in this study are included in the article. Further inquiries can be directed to the corresponding author.

Conflicts of Interest: Author Derong Feng, Maoyang Xie, Weilin Yu, Chao Li, Raolong Guo, Yunpeng Hu are employed by the company Henan Aerospace Precision Manufacturing Co., Ltd. The remaining authors declare that the research was conducted in the absence of any commercial or financial relationships that could be construed as a potential conflict of interest.

Abbreviations

TC4+MoS_2	titanium alloy bolts coated with MoS_2
TC4+AO+MoS_2	titanium alloy bolts treated with anodizing oxidation and MoS_2 coating
TC4	Ti6Al4V
ITER	International Thermonuclear Experimental Reactor
PTFE	polytetrafluoroethylene
SEM	scanning electron microscope
EDS	energy-dispersive spectrometry
COF	coefficient of friction
σ_s	yield strength
μ_{tot}	thread friction coefficient
d_2	pitch diameter
d_b	friction diameter of the nut bearing surface
p	pitch
t_b	torque of end face friction
T_{th}	thread torque
t	tightening torque
f	axial force

References

1. Putyrskii, S.V.; Yakovlev, A.L.; Nochovnaya, N.A. Benefits and Applications of High-Strength Titanium Alloys. *Russ. Eng. Res.* **2018**, *38*, 945–948. [CrossRef]
2. Zhao, Q.; Sun, Q.; Xin, S.; Chen, Y.; Wu, C.; Wang, H.; Xu, J.; Wan, M.; Zeng, W.; Zhao, Y. High-strength titanium alloys for aerospace engineering applications: A review on melting-forging process. *Mater. Sci. Eng. A* **2022**, *845*, 143260. [CrossRef]
3. Liu, X.; Zhang, A.; Zhang, Z.; Liu, K.; Zhu, X.; Chen, W. Cause Analysis of Bolt Seizure Failure and Experimental Study On Influencing Factors. In Proceedings of the 2022 12th International Conference on CYBER Technology in Automation, Control, and Intelligent Systems (CYBER), Baishan, China, 27–31 July 2022; pp. 1142–1145. [CrossRef]

4. Söderman, A.; Törnquist, P.L. Galling is a serious issue in many industries that use stainless steel fasteners- but there are solutions. *Met. Powder Rep.* **2021**, *76*, 194–195. [CrossRef]
5. Croccolo, D.; De Agostinis, M.; Vincenzi, N. Influence of tightening procedures and lubrication conditions on titanium screw joints for lightweight applications. *Tribol. Int.* **2012**, *55*, 68–76. [CrossRef]
6. Podgornik, B.; Hogmark, S.; Sandberg, O. Influence of surface roughness and coating type on the galling properties of coated forming tool steel. *Surf. Coat. Technol.* **2004**, *184*, 338–348. [CrossRef]
7. McDonnell, K.A.; Bley, E.-B.; O'Donoghue, P.; Adeyemi, I.; Pambaguian, L.; Twomey, B. Anti-galling treatemnt for bolts and fasteners using coblast. In Proceedings of the 14th ISMSE & 12th ICPMSE, Biarritz, France, 1–5 October 2018.
8. Wiklund, U.; Hutchings, I. Investigation of surface treatments for galling protection of titanium alloys. *Wear* **2001**, *251*, 1034–1041. [CrossRef]
9. Hua, Y.; Li, F.; Zhang, D.; Liu, T.; Yan, W.; Liu, J.; Zhu, M. Anti-loosening Performance of Three Kind of Coated Bolts under Cyclic Transverse Load. *Lubr. Eng.* **2023**, *48*, 67–73.
10. Li, X.; Zhu, H. Two-dimensional MoS2: Properties, preparation, and applications. *J. Mater.* **2015**, *1*, 33–44. [CrossRef]
11. Lalegani, Z.; Ebrahimi, S.S.; Hamawandi, B.; La Spada, L.; Batili, H.; Toprak, M. Targeted dielectric coating of silver nanoparticles with silica to manipulate optical properties for metasurface applications. *Mater. Chem. Phys.* **2022**, *287*, 126250. [CrossRef]
12. Lincoln, R.L.; Scarpa, F.; Ting, V.P.; Trask, R.S. Multifunctional composites: A metamaterial perspective. *Multifunct. Mater.* **2019**, *2*, 043001. [CrossRef]
13. Oldiges, D.; Hamilton, S. Moly Disulfide in Bolting Applications. In Proceedings of the ASME 2016 Pressure Vessels and Piping Conference, Vancouver, BC, Canada, 17–21 July 2016; Volume 3: Design and Analysis. Paper No. V003T03A007.
14. Xu, X. Study on the Influence of Frictional Coefficient on the Assembly Quality of Bolts. *Heat Treat. Technol. Equip.* **2024**, *45*, 29–32+36. [CrossRef]
15. Zhou, J.; Liu, J.; Ouyang, H.; Cai, Z.; Peng, J.; Zhu, M. Anti-loosening performance of coatings on fasteners subjected to dynamic shear load. *Friction* **2018**, *6*, 32–46. [CrossRef]
16. Gradt, T.; Schneider, T. Tribological Performance of MoS_2 Coatings in Various Environments. *Lubricants* **2016**, *4*, 32. [CrossRef]
17. Cai, S.; Guo, P.; Liu, J.; Zhang, D.; Ke, P.; Wang, A.; Zhu, Y. Friction and Wear Mechanism of MoS2/C Composite Coatings Under Atmospheric Environment. *Tribol. Lett.* **2017**, *65*, 79. [CrossRef]
18. Luo, J.; Cai, Z.B.; Mo, J.L.; Peng, J.F.; Zhu, M.H. Torsional Fretting Wear Behavior of Bonded MoS_2Solid Lubricant Coatings. *Tribol. Trans.* **2015**, *58*, 1124–1130. [CrossRef]
19. Chen, L.; Liu, Z.; Shen, Q. Enhancing tribological performance by anodizing micro-textured surfaces with nano-MoS2 coatings prepared on aluminum-silicon alloys. *Tribol. Int.* **2018**, *122*, 84–95. [CrossRef]
20. Hanjun, H.; Zhen, C.; Xingguang, L.; Xingguo, F.; Yugang, Z.; Kaifeng, Z.; Hui, Z. Effects of substrate roughness on the vacuum tribological properties of duplex PEO/bonded-MoS2 coatings on Ti6Al4V. *Surf. Coat. Technol.* **2018**, *349*, 593–601. [CrossRef]
21. Kim, S.-W.; Jang, J.-S.; Chung, S.-K.; Nam, K.-W.; Yoo, J.-P.; Kim, D.-H. A parametric study of low friction coating by PVD method on the spiralock female thread for ITER application. *Fusion Eng. Des.* **2023**, *192*, 113836. [CrossRef]
22. Liu, J.; Ouyang, H.; Feng, Z.; Cai, Z.; Liu, X.; Zhu, M. Study on self-loosening of bolted joints excited by dynamic axial load. *Tribol. Int.* **2017**, *115*, 432–451. [CrossRef]
23. ArunKumar, S.; Jegathish, V.; Soundharya, R.; JesyAlka, M.; Arul, C.; Sathyanarayanan, S.; Mayavan, S. Evaluating the performance of MoS2 based materials for corrosion protection of mild steel in an aggressive chloride environment. *RSC Adv.* **2017**, *7*, 17332–17335. [CrossRef]
24. Zhang, Y.; Die, J.; Li, F.; Li, H.; Tu, J.; Zhang, K.; Yu, X. Polypyrrole-Modified Molybdenum Disulfide Nanocomposite Epoxy Coating Inhibits Corrosion of Mild Steel. *Coatings* **2023**, *13*, 1046. [CrossRef]
25. Ahn, J.-H.; Lee, J.M.; Cheung, J.-H.; Kim, I.-T. Clamping force loss of high-strength bolts as a result of bolt head corrosion damage: Experimental research A. *Eng. Fail. Anal.* **2016**, *59*, 509–525. [CrossRef]
26. Kim, I.-T.; Lee, J.M.; Huh, J.; Ahn, J.-H. Tensile behaviors of friction bolt connection with bolt head corrosion damage: Experimental research B. *Eng. Fail. Anal.* **2016**, *59*, 526–543. [CrossRef]
27. Wei, S.; Zhao, W.; Xin, H.; Jiang, J.; Yu, Y. Study on the degradation of axial tensile performance of corroded bolts in steel bridges. *Constr. Build. Mater.* **2024**, *450*, 138648. [CrossRef]
28. Wang, H.; Tang, F.; Qin, S.; Tu, K.; Guo, J. Corrosion-Induced Mechanical Degradation of High-Strength Bolted Steel Connection. *J. Mater. Civ. Eng.* **2020**, *32*, 04020203. [CrossRef]
29. Nie, S.; Wang, H.; Yang, B.; Cheng, Z.; Ye, X.; Chen, Z. Corrosion-induced mechanical properties of shear bolted connections in high-strength weathering steel. *Thin-Walled Struct.* **2023**, *190*, 111013. [CrossRef]
30. Siu, J.H.; Li, L.K. An investigation of the effect of surface roughness and coating thickness on the friction and wear behaviour of a commercial MoS2–metal coating on AISI 400C steel. *Wear* **2000**, *237*, 283–287. [CrossRef]
31. Chen, H.; Wang, W.; Le, K.; Liu, Y.; Gao, X.; Luo, Y.; Zhao, X.; Liu, X.; Xu, S.; Liu, W. Effects of substrate roughness on the tribological properties of duplex plasma nitrided and MoS2 coated Ti6Al4V alloy. *Tribol. Int.* **2024**, *191*, 109123. [CrossRef]

32. Bulaqi, H.A.; Mashhadi, M.M.; Geramipanah, F.; Safari, H.; Paknejad, M. Effect of the coefficient of friction and tightening speed on the preload induced at the dental implant complex with the finite element method. *J. Prosthet. Dent.* **2015**, *113*, 405–411. [CrossRef]
33. Joseph, A.; Gautham, V.; Akshay, K.; Sajith, V. 2D MoS2-hBN hybrid coatings for enhanced corrosion resistance of solid lubricant coatings. *Surf. Coat. Technol.* **2022**, *443*, 128612. [CrossRef]
34. Presuel-Moreno, F.; Jakab, M.; Tailleart, N.; Goldman, M.; Scully, J.R. Corrosion-resistant metallic coatings. *Mater. Today* **2008**, *11*, 14–23. [CrossRef]
35. Guo, Q.; Zhao, Z.; Lu, Y.; Xing, Y.; Zhang, F.; Yang, C. Corrosion evolution and axial mechanical performance degradation of corroded M24 high-strength bolts. *J. Constr. Steel Res.* **2024**, *213*, 108411. [CrossRef]
36. Zou, Z.; Zeng, F.; Wu, H.; Liu, J.; Li, Y.; Gu, Y.; Yuan, T.; Zhang, F. The Joint Strength and Fracture Mechanisms of TC4/TC4 and TA0/TA0 Brazed with Ti-25Cu-15Ni Braze Alloy. *J. Mater. Eng. Perform.* **2017**, *26*, 2079–2085. [CrossRef]
37. Feng, D.; Dong, C.; Hu, Y.; Wang, Y.; Ma, J.; Huang, Z.; Wan, Q. Fracture Mode Transition during Assembly of TC4 High-Lock Bolt under Tensile Load: A Combined Experimental Study and Finite Element Analysis. *Materials* **2022**, *15*, 4049. [CrossRef]
38. Cao, R.; Chen, J.H.; Zhang, J. Study on Fracture Mechanisms of TiAl Alloys by In Situ Tensile Tests. *Key Eng. Mater.* **2007**, *353–358*, 34–37. [CrossRef]

materials

MDPI

Article

Numerical Simulation of Galvanic Corrosion and Electrical Insulation for TC4/304 Galvanic Couple

Kaixun Liu, Yuhang Wu, Huicong Liu, Haining Chen and Weiping Li *

School of Materials Science and Engineering, Beihang University, Beijing 100191, China
* Correspondence: liweiping@buaa.edu.cn

Abstract: The galvanic corrosion and electrical insulation between TC4 Ti-alloy and 304 stainless steel coupled in pipe joints were investigated using the finite element method. The results obtained from polarization were applied as boundary conditions. The simulation incorporated secondary current distribution with chemical species transport and laminar flow. COMSOL modeling provided calculated values that showed good agreement with experimental measurements. The model was utilized to examine the influence of geometric and environmental factors, including insulation resistance, insulation distance, pipe diameter, temperature, and electrolyte concentration, on the galvanic corrosion process. The results indicated that electrical insulation performance significantly affects the corrosion process.

Keywords: galvanic corrosion; electrical insulation; finite element; simulation; titanium

Academic Editors: Xiaogang Li, Luigi Calabrese and Arunas Ramanavicius

Received: 8 September 2024
Revised: 16 December 2024
Accepted: 23 December 2024
Published: 25 December 2024

Citation: Liu, K.; Wu, Y.; Liu, H.; Chen, H.; Li, W. Numerical Simulation of Galvanic Corrosion and Electrical Insulation for TC4/304 Galvanic Couple. *Materials* **2025**, *18*, 38. https://doi.org/10.3390/ma18010038

1. Introduction

In maritime engineering, the utilization of a diverse range of pipework systems is important aboard ships. The choice of pipe materials hinges upon several pivotal factors, such as location, operating conditions, and budgetary considerations. It is commonplace in ship pipeline design to incorporate dissimilar metal pipelines. Flanges serve as pivotal components, facilitating the connection of pipes by bolting two flanges together with spacers in between. To mitigate electrochemical corrosion between dissimilar metal pipelines or flanges, insulating spacers are typically incorporated into the initial design to isolate distinct metal components [1]. The adoption of electrical insulation measures can effectively mitigate galvanic corrosion between the steel ship hull and titanium alloy [2]. Polymer insulation applied to the bolted joint between AZ31B and carbon fiber-reinforced composite has been shown to significantly reduce galvanic corrosion [3]. However, galvanic attack may still occur due to the permeation of an aqueous corrosive medium. As service time increases, the insulating properties of these spacers deteriorate, accelerating the corrosion of one of the coupled dissimilar metals.

Numerous experimental and simulation-based studies [4–13] have investigated the effects of various factors on galvanic corrosion, such as potential difference, anode/cathode area ratio [14], distance between cathode and anode, temperature [15], and flow rate [16]. The main conclusions from these studies are summarized as follows. The galvanic current density of the anode increased with the decrease of the anode/cathode area ratio. The galvanic current density and average galvanic current of the anode decreased as the distance between the cathode and anode increased. The corrosion rate became higher with the temperature increased. Zhang et al. [17] studied the galvanic corrosion behavior in a simulated pipeline system. Their study found that the galvanic couple comprising titanium and high-strength steel exhibited an increased current density, which was further amplified with rising temperature and a larger cathode-to-anode area ratio.

However, limited research has been focused on the role of insulating spacers between the anode and cathode in these systems. The increase in the insulating distance results in a longer circuit path, thereby elevating the resistance of the circuit. Song [18] investigated the coupling of magnesium alloy AZ91D with aluminum alloy, stainless steel, and zinc. Their findings showed that the current is non-linearly dependent on spacer width under salt spray conditions, with the current decreasing as the thickness of the insulating gasket increases. Notably, the rapid decrease in current is observed within the first 1 cm, while the decreasing trend slows as the thickness increases further. When the spacer thickness reaches 9 cm, the corrosion current is $22\mu A/cm^2$. The results suggest that while insulating spacers can increase the resistance of the aqueous film on the surface, they may not completely block the electrical path. Simulations have been used effectively to analyze potential and current distributions, providing insights into sensitive areas for galvanic corrosion and estimating its rate [19]. However, most models tend to overlook the effects of resistance and the limitations of mass transfer [20]. Therefore, additional studies are required to incorporate these factors into simulation models to better understand and mitigate galvanic corrosion.

Pan et al. [21] have tested the corrosion potential and current in seawater by connecting resistors with different resistance values between commercially pure titanium TA2 and high-strength steel as well as copper-nickel alloy B10 and high-strength steel and found that the galvanic couples were nearly insulated and the extent of galvanic corrosion was controlled significantly when the galvanic insulation value between the two materials was higher than 10 kΩ and 50 kΩ. Similarly, Zhang et al. [22] have investigated galvanic current density in both static and dynamic seawater environments by connecting different insulated resistances between TA2 and B10 and found that the galvanic corrosion could be avoided when the connecting resistance was higher than 56 kΩ. The performance of electrical insulation between dissimilar metals is directly associated with the resistance of the insulating spacer, but few related models have been reported.

It is crucial to promptly identify the relationship between insulation properties and corrosion behavior and to monitor the electrical insulation between dissimilar metals, especially in real-world environments where corrosion sites extend beyond pipe joints. The corrosion phenomenon is challenging to reproduce, so this model was developed to address the issue and to study the impact of relevant factors. This paper investigates the effect of electrical insulation properties on corrosion behavior, including self-corrosion and galvanic corrosion, by modeling electrical insulation control and corrosion behavior, which can provide a foundation for effective electrical insulation control.

2. Materials and Methods

2.1. Materials

TC4 Ti-ally and 304 stainless steel were used in this study, which is widely used in pipework systems [23]. The TC4 titanium alloy samples were sourced from Xinghua Zengyitai Co., Ltd., Xinghua, China, while the 304 stainless steel samples were obtained from Tianjin Xingang Co., Ltd., Tianjin, China. The chemical compositions from the manufacturer's report are shown in Tables 1 and 2. The specimens were processed into 110 mm × 25 mm × 5 mm and 55 mm × 25 mm × 5 mm plates. The surfaces are shown in Figure 1a,b.

Table 1. Chemical composition of the TC4 Ti-alloy (mass%).

Alloy	Al	V	Fe	Si	C	N	H	O
C4	6.2	4	0.3	0.15	0.1	0.05	0.015	0.2

Table 2. Chemical composition of the 304 stainless steel (mass%).

Alloy	C	Si	Mn	S	P	Cu	Co	B	Ni	Cr
304	0.035	0.66	1.88	0.005	0.023	1	0.06	0.0018	9.27	18.65

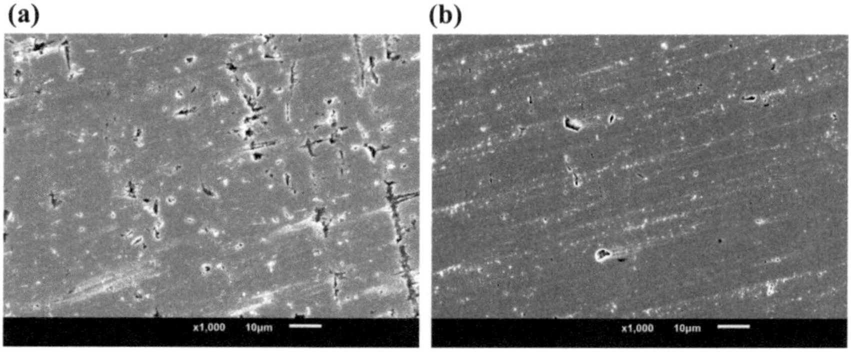

Figure 1. SEM (**a**) TC4 surface; (**b**) 304 surface.

2.2. Polarization and Conductivity Measurement Experiment

Polarization curves of the specimens were measured using a four-channel electrochemical workstation from the Wuhan Corr Test. Before the experiment, the specimens were polished with water-abrasive sandpaper in the order of #240, #600, #800, and #1000 so that no processing traces were basically visible on the surface of the specimens. A three-electrode system was used to measure the polarization curves of stainless steel and titanium alloy specimens, which specimen acted as the working electrode. The platinum and standard calomel electrodes were used as counter electrodes and reference electrodes, respectively. Initially, the open-circuit potentials were measured using an electrochemical workstation, and the stabilized open-circuit potential was recorded. Subsequently, the polarization curves were measured. The conductivity of the electrolyte solution was tested by using a Lei-ci DDS-307A conductivity tester from Shanghai Yidian Co., Ltd., Shanghai, China. An electrolyte solution of 3.5 wt.% NaCl was used in this study. The measurements were conducted at 25 °C.

2.3. Corrosion Test

The stainless steel specimen and titanium alloy specimen were coupled. Two samples are insulated from each other, with the spacing adjustable and ranging from 5 mm to 1 cm, 5 cm, or 10 cm. The surface area of both sides of the test site was 12.5 cm^2. When measuring with the electrochemical workstation, the TC4 titanium alloy specimen was connected as the working electrode, the mercuric electrode was connected to the reference electrode, and the 304 stainless steel was grounded. In the corrosion test, the galvanic corrosion test was chosen. The test interval was set to 50 s, with a total test duration of 10,000 s. Throughout the test, the corrosion current and corrosion potential curves of the galvanic couple were recorded over time. The corrosion test was conducted according to the standard "HB 5374-1987 [24] Method for Measuring Galvanic Current of Different Metals".

2.4. Numerical Model and Boundary Conditions

COMSOL modeling was established for the galvanic corrosion test and electrical insulation control experiment to simulate the corrosion of the two test couples. The model was solved using COMSOL V6.1. In the electrical insulation control experiment, the insulation spacers were placed between two pipes. One terminal with an external short-circuit resistance R was set to connect to the ground, and the potential of the other terminal was set to zero. As depicted in Figure 2b, the blue part on the left represents the titanium alloy pipe, and the green part on the right represents the stainless steel pipe. The equivalent circuit diagram of the galvanic corrosion model is shown in Figure 2c [15]. External resistance includes resistance of the solution between the anode and cathode. As depicted in Figure 2d, the schematic diagram of the simulation concept illustrates that the titanium alloy boundary was grounded after connecting an external short-circuit resistance, while the potential at the stainless steel pipe boundary was set to zero. At this point, the coupling resistance between the two pipes can be adjusted.

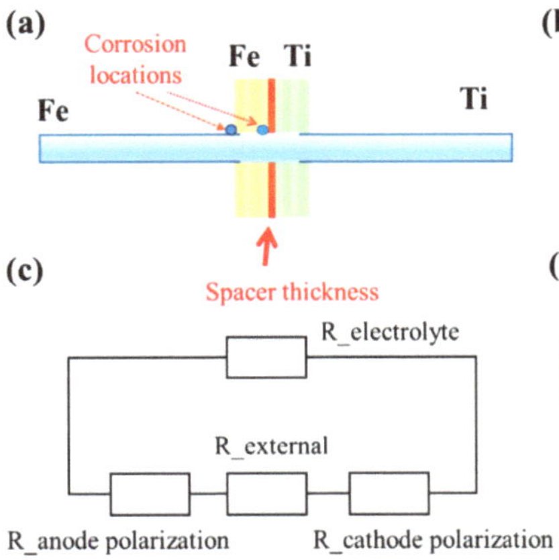

Figure 2. Details of model. (**a**) Diagram of galvanic corrosion; (**b**) diagram of TC4/304 pipe couple; (**c**) equivalent circuit model; (**d**) diagram of the model.

The secondary current distribution with chemical species transport and laminar flow were considered for simulation. The insulation boundary does not allow current to flow through. The conductivity is constant, the electrolyte composition remains near-constant, and the current density is limited by the mass transport of the electroactive species. The current density can be calculated using the sum of the fluxes of all ions. The transport of ionic species can be described by the Nernst–Planck equation [6,25,26]. The charge transfer in the electrolyte is also assumed to obey Ohm's law [27].

$$\vec{I_l} = F\sum_{i=1}^{n} z_i\left(-D_i\nabla c_i - z_i u_i F c_i \nabla\phi_l + c_i\vec{u}\right) = -\sigma_l\nabla\phi_l \tag{1}$$

where $\vec{I_l}$ (A/m^2) denotes the current density vector in the electrolyte, σ_l (S/m) denotes the conductivity, $\nabla\phi_l$ (V/m) denotes the electrolyte potential gradient, F (96,485 C/mol) denotes the Faraday constant, z_i (1) represents charge number, D_i (m^2/s) is the diffusion coefficient, u_i (s·mol/kg) is the mobility, c_i (mol/m^3) is the concentration of the ion and $\vec{u}$ (m/s) is the velocity vector.

The current density due to the electrochemical reactions is also described as a function of the overpotential in the electrode surface. The current density is given by Equation (2). The overpotential η (V) is given by Equation (3) [27,28].

$$-\vec{l_s}\cdot\vec{n} = \vec{l_l}\cdot\vec{n} = \sum(i_0(c_R exp\left(\frac{\alpha_a F\eta}{RT}\right) - c_O exp(\frac{-\alpha_c F\eta}{RT}))) \quad (2)$$

$$\eta = \phi_s - \phi_l - E_{eq} \quad (3)$$

where $\vec{l_s}$ (A/m^2) denotes the current density vector in the electrode, $\vec{n}$ is the normal vector, c_R (1) and c_O (1) are related to the concentration of reducers and oxidizers, and α_a (1) is anodic transfer coefficient. α_c (1) is cathodic transfer coefficient. R (8.314 ($J/(\text{mol}\cdot\text{K})$) denotes the gas constant, T (K) denotes the temperature, ϕ_s (V) is the electrode potential, and E_{eq} (V) is the equilibrium potential of the reaction.

The flowchart of the simulation process is given in Figure 3. When investigating the influence of various factors, the relevant parameters are systematically varied while all other parameters remain constant. Electrochemical parameters are detailed in the results. The parameter setting is listed in Table 3.

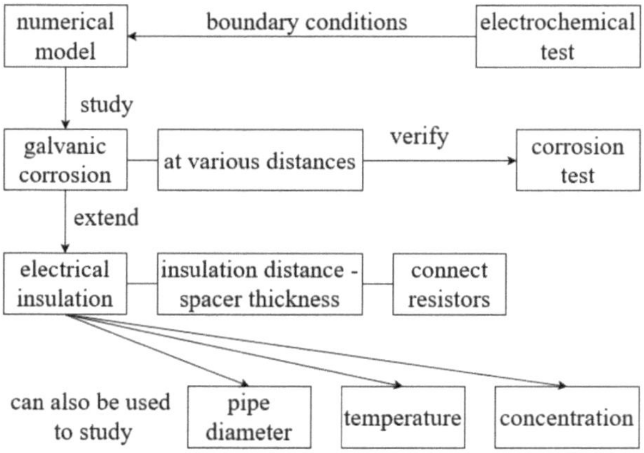

Figure 3. Flowchart.

Table 3. COMSOL model parameters.

Parameters	Value	Unit
d	0.0015	m
T	298	K
D	1×10^{-9}	m^2/s
c_{in}	59.8	mol/m^3
u_{in}	1×10^{-4}	m/s
c_0	59.8	mol/m^3

3. Results

3.1. Electrochemical Test Results

The measured conductivity of 3.5 wt.% NaCl solution is 5.6 S/m at 25 °C. The Tafel polarization curves were plotted against the potential versus logarithmic current density. The polarization curves obtained for TC4 and 304 are shown in Figure 4. The values of equilibrium potential, exchange current density, anodic Tafel slope, and cathodic Tafel slope were obtained from the graph. The results obtained for TC4 and 304 are shown in

Table 4. The results obtained from polarization were applied as boundary conditions. The corrosion potential of TC4 is lower than that of 304.

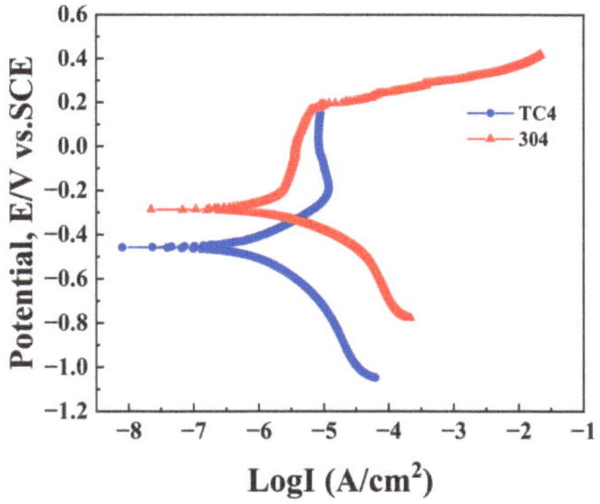

Figure 4. Tafel curve for TC4 and 304 in 3.5 wt.% NaCl solution at 25 °C.

Table 4. Results of Tafel polarization in 3.5 wt.% NaCl solution at 25 °C.

Sample	Equilibrium Potential	Exchange Current Density	Anodic Tafel Slope	Cathodic Tafel Slope
TC4	−0.4842 V	0.0375 A/m^2	0.4667 V/dec	
304	−0.2218 V	0.0393 A/m^2		−0.1718 V/dec

The curves of the corrosion current of different anode and cathode distances obtained through experimental testing over time are shown in Figure 5. To determine the experimental test value of the galvanic current for comparison with the simulated results of galvanic corrosion as the distance between the anode and cathode varies, the average stabilized galvanic current is calculated.

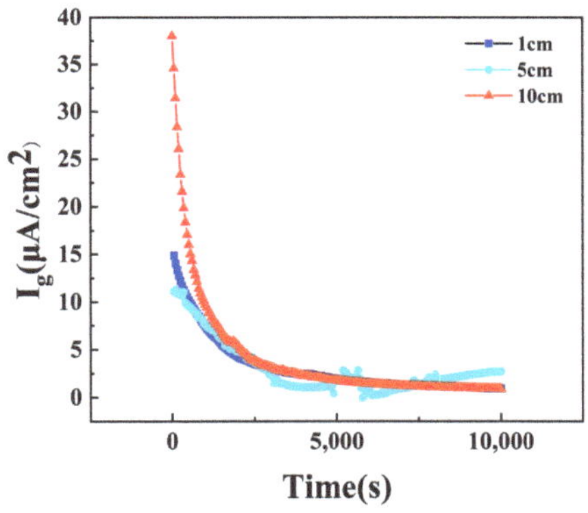

Figure 5. Corrosion current over time in 3.5 wt.% NaCl solution at various anode and cathode distances.

Based on the analysis of mixed potential prediction theory, it is understood that when TC4 forms a galvanic couple with 304, the self-corrosion potential of TC4 is more negative. The dissolution and corrosion rates of TC4 are accelerated under anodic polarization. Conversely, the relatively positive free corrosion potential of 304 causes it to act as the cathode, thereby suppressing the corrosion of 304.

3.2. *Simulation Results*

3.2.1. Simulation Results of Galvanic Corrosion at Various Distances

COMSOL modeling provided calculated values that showed good agreement with experimental measurements. The simulation results and experimental results of corrosion current at various distances are shown in Figure 6. The errors between the simulated results and the experimental results are 0.65%, 0.4%, and 1.32%. The electrochemical field, flow field, and material transfer field are coupled, and the galvanic corrosion model results incorporate concentration polarization and exhibit greater alignment with experimental test findings. Moreover, the model takes into account a more comprehensive range of influencing factors. When the distance is set to 5 cm, the error is minimized. Within the permissible error range, the predicted current density can be used to describe the experimental results.

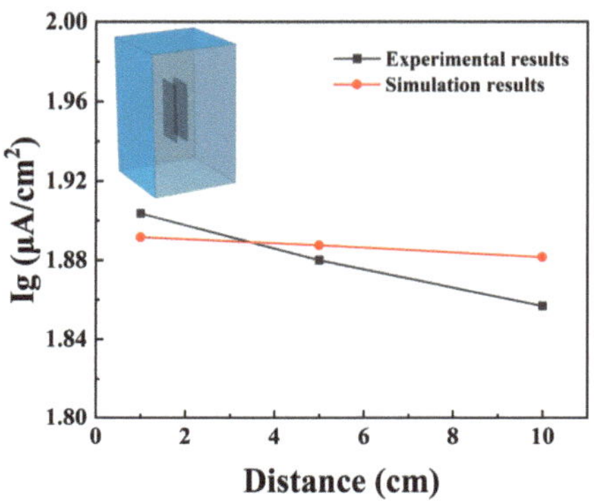

Figure 6. Simulation results and experimental results of corrosion current at various distances.

3.2.2. Effect of Electrical Insulation of Anode/Cathode

The numerical simulations of galvanic corrosion on pipeline models involve direct coupling and are connected at different resistance values. The surface potential distribution of pipe in direct contact and connection with different resistance values are shown in Figure 7. Analysis of the potential distribution reveals that direct contact results in a negative coupling potential, with the anode pipe end exhibiting the most negative potential and the cathode pipe end showing the most positive potential. The maximum difference in coupling potential can reach approximately 30 mV, demonstrating a relatively uniform distribution of coupling potential.

Upon connecting resistors, it is observed that increasing resistance values result in a gradual shift towards positive coupling potentials. For instance, when the simulated resistance value reaches 1000 Ω, the coupling potential approaches −286 mV. Subsequent increases in simulated resistance do not significantly alter the coupling potential, which remains below 1 mV. Furthermore, the difference in coupling potential between pipeline sections diminishes as resistance values rise. Notably, compared to the initial 30 mV coupling potential difference at a connection resistance of 100 Ω, the disparity becomes negligible when the simulated resistance exceeds 1000 Ω. This decreasing trend in galvanic corrosion propensity is evident with increasing resistance.

At a connection resistance of 50 kΩ, a significant shift in the potential distribution of the pipeline occurs. In the anode section on the left, the potential distribution is no longer ordered from low to high, and the most severely corroded segment is no longer the electrical connection point. This transition is attributed to the substantial resistance value of 50 kΩ, which inhibits galvanic corrosion on the titanium alloy pipeline, leading primarily to self-corrosion. Consequently, potential polarization across various pipeline positions becomes more uniform, deviating from the previous axial distribution pattern along the pipeline's length.

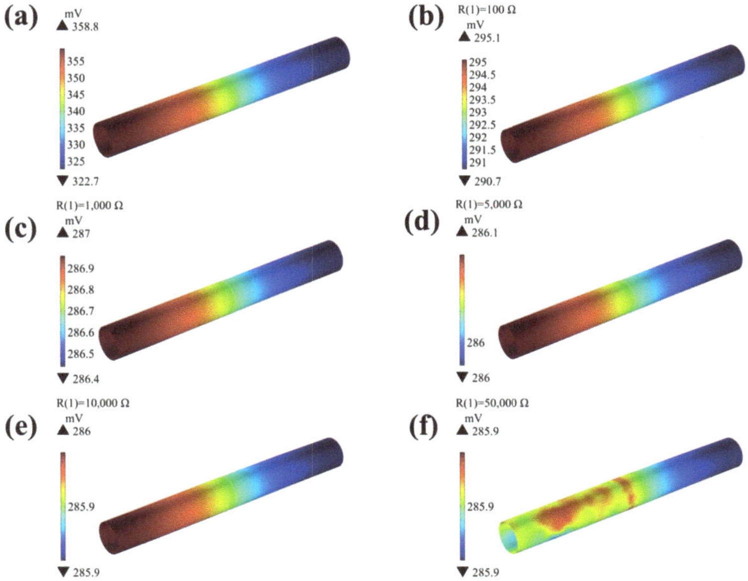

Figure 7. The opposite surface potential distribution of pipe in direct contact and connection with different resistance values: (**a**) 0; (**b**) 100 Ω; (**c**) 1000 Ω; (**d**) 5000 Ω; (**e**) 10,000 Ω; (**f**) 50,000 Ω.

The current density increases along the pipe from both ends toward the center, peaking at the connection point. Within a specific range, an increase in connection resistance results in a corresponding decrease in current density. Upon reaching a connection resistance of 50 kΩ, the current density distribution on the TC4 pipeline becomes uneven, and the maximum current density is no longer observed at the connection point, as shown in Figure 8a. The maximum galvanic corrosion current density varies with the connection resistance, as illustrated in Figure 8b. It is evident that as the resistance value increases, the corrosion current density diminishes. When the resistance surpasses 500 Ω, the corrosion current density falls below 0.3 μA/cm^2, thereby achieving effective insulation control for galvanic corrosion.

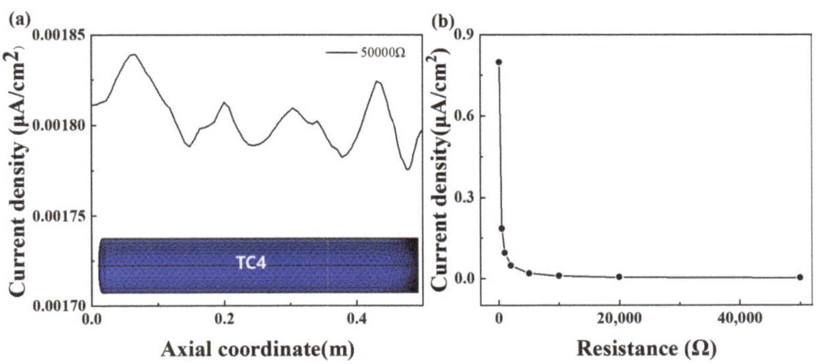

Figure 8. (**a**) The surface current density distribution of TC4 pipe in the electrolyte at a connection resistance of 50,000 Ω; (**b**) maximum current density with different connection resistance values.

3.2.3. Effect of Insulation Distance

The effect of thickness of the insulating spacer thickness between TC4 and 304 on galvanic corrosion is shown in Figure 9. The typical spacer thickness ranges from 1.5 mm to 3 mm. The investigation has been extended to include a range from 0.5 mm to 5 mm. As the spacer thickness increases, the connection resistance correspondingly increases. As the thickness increases from 0.5 mm to 5 mm, the current density decreases, consistent with the current density trend observed in the preceding section. In our design, increasing the spacer thicknesses results in a corresponding increase in connection resistance. It was assumed that the resistance and distance exhibit a linear relationship. The increase in the insulating distance results in a longer circuit path, thereby elevating the resistance of the circuit. Consequently, the impact of increasing the spacer thickness is analogous to that of raising the resistance. The peak anodic current density is non-linearly dependent on spacer width [18].

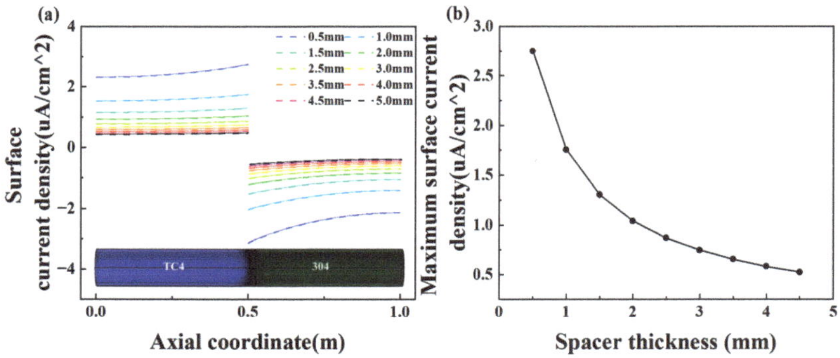

Figure 9. (**a**) The current density distribution along the TC4 and 304 surfaces in the electrolyte with various spacer thicknesses; (**b**) maximum current density with different spacer thicknesses.

3.2.4. Effect of Pipe Diameter

Galvanic corrosion simulations were conducted on pipes with five different diameters, ranging from 10 cm to 30 cm in 5 cm intervals, while other conditions were kept constant. Figure 10 illustrates the axial variation curve of galvanic corrosion current density for these different pipe diameters. It is evident that the current density increases gradually with the enlargement of the pipe diameter, leading to an intensified degree of corrosion. Additionally, as the pipe diameter increases, the galvanic corrosion potential of the TC4 pipeline shifts positively, while the potential of the 304 pipeline shifts negatively due to the hindrance of the corrosion reaction during the cathodic process.

Pipe surface current density increases gradually with the expansion of pipe diameter from 10 cm to 30 cm. This can be attributed to the generation of an electric field in the electrolyte solution resulting from galvanic corrosion within the pipeline. The pipeline subsequently exhibits a shielding effect on the electric field [4]. As the pipe diameter increases, the shielding effect decreases, leading to an increase in the galvanic corrosion current density. Analysis of various calculation points along the TC4 pipeline reveals that the current density at point 3 demonstrates the most substantial variation in response to changes in pipe diameter. In contrast, the variation in current density at point 1, which is located near the connection, is relatively minor when subject to fluctuations in pipe diameter. This observation suggests that changes in pipe diameter significantly impact galvanic corrosion in areas that are distant from the pipe connection.

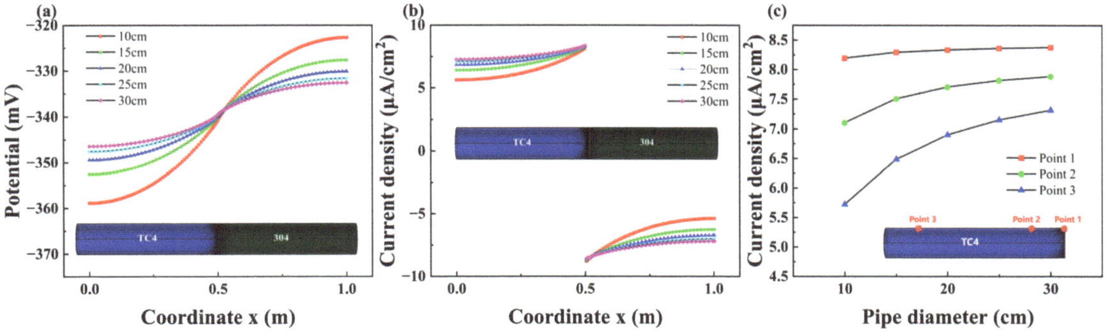

Figure 10. Variations of potential (**a**) and current density (**b**) with pipeline diameter; (**c**) variations of current density with pipe diameter at different points.

3.2.5. Effect of Temperature

The influence of temperature on the galvanic corrosion process is reflected in its impact on electrical conductivity and the electrochemical kinetics equation. Figure 11a,b illustrate the axial distribution of potential and current density during the corrosion process of the pipe in direct contact at different temperatures. It can be observed that as the temperature increases, the polarization value of the anode (TC4) decreases while the polarization value of the cathode (304) increases, leading to an increase in the corrosion current density. In the galvanic corrosion process, elevated temperature enhances cathodic polarization, resulting in an increase in current density.

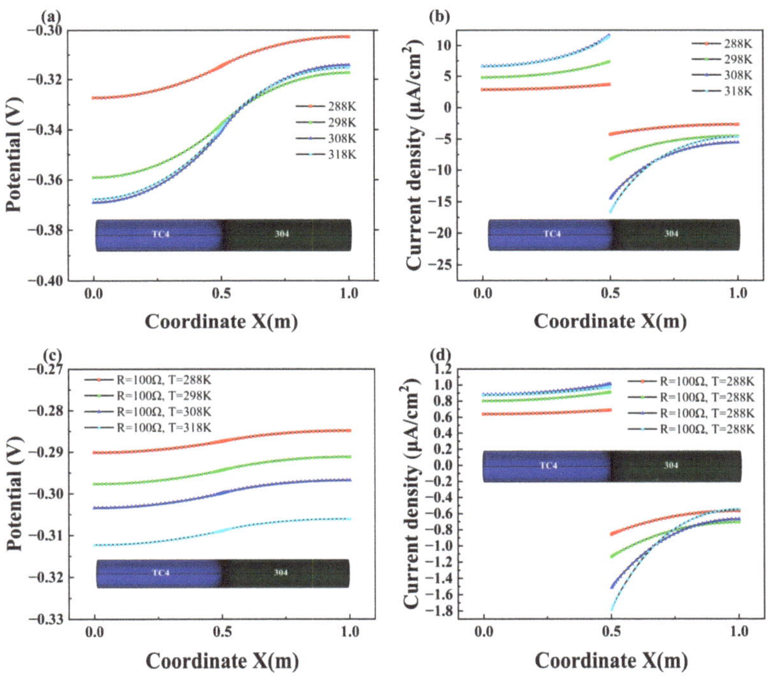

Figure 11. Variations of potential (**a**) and current density (**b**) in direct contact at different temperatures; variations of potential (**c**) and current density (**d**) at a connection resistance of 100 Ω at different temperatures.

When a resistance is connected between the pipelines, the corrosion behavior of the pipelines varies with temperature, as shown in Figure 11c,d. It is evident from the graph that when a 100 Ω resistance is connected, the influence of temperature on the potential variation across different parts of the anode becomes more uniform from Figure 11a,c. Compared to the unconnected resistance scenario, the distribution of potential becomes more uniform. Particularly near the connection, the effect of temperature on the galvanic corrosion current density significantly decreases after connecting the resistance.

As the temperature increases, the anode potential becomes more negative, which results in higher current density for the anode (TC4). The current density increases correspondingly as the temperature increases from 288 K to 308 K. However, the difference between the results at 308 K and 318 K is minimal, as the influence of temperature on the equilibrium potential becomes more pronounced. The impact of temperature on the equilibrium potential should be considered in future research.

3.2.6. Effect of Electrolyte Concentration

The influence of concentration on the galvanic corrosion process is reflected in its effect on conductivity and the behavior of cathodic and anodic corrosion. Figure 12 presents the simulation results of TC4/304 couple corrosion models at varying concentrations. The figure illustrates that as the concentration increases, the anode potential becomes more negative, which leads to a higher current density of the anode (TC4). The current density increases as the concentration rises from 29.4 mol/m^3 to 85.47 mol/m^3. However, the minimal variation between the results at 85.47 mol/m^3 and 119.6 mol/m^3 indicates that the replenished electrolyte is sufficient to compensate for the consumption.

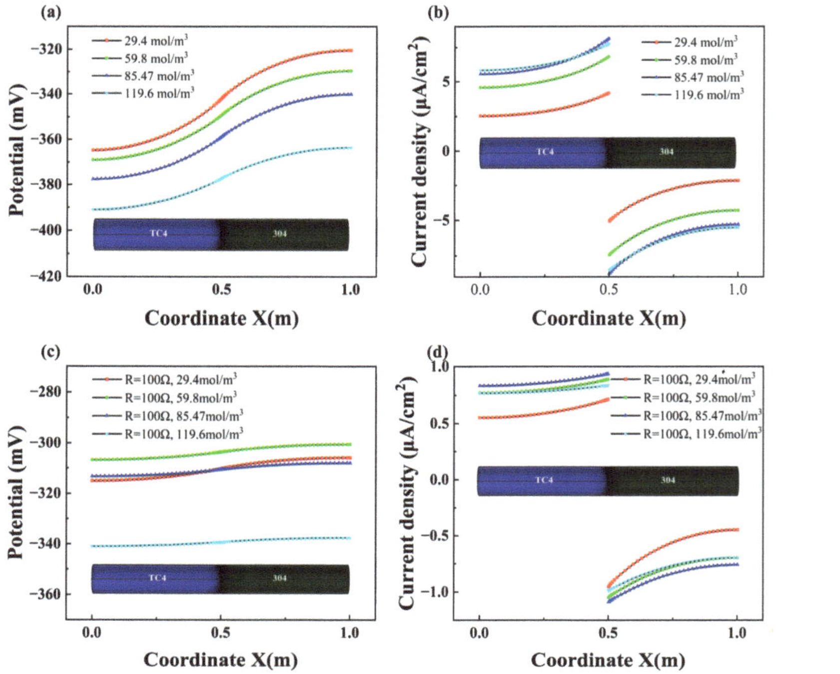

Figure 12. Variations of potential (**a**) and current density (**b**) in direct contact at different electrolyte concentrations; variations of potential (**c**) and current density (**d**) at a connection resistance of 100 Ω at different electrolyte concentrations.

The potential polarization value of the anode pipeline (TC4) decreases as the concentration increases, whereas the polarization value of the cathode pipeline (304) continues to increase. This disparity is attributed to the greater influence of concentration on cathodic polarization behavior compared to anodic polarization. Analysis of the potential and current density distribution curves of pipelines in the TC4/304 couple corrosion model with different concentrations reveals a similar effect of concentration across various sections of the TC4/304 pipeline. With increasing concentration, the magnitude of potential and current changes, both near and far from the connection, remains relatively uniform. This uniformity can be attributed to the fact that the increase in concentration in the galvanic corrosion model of the pipeline primarily governs the polarization of galvanic corrosion as influenced by concentration. Consequently, when the concentration distribution of the pipeline at different concentrations displays minimal variance, the corrosion disparities among different sections are also insignificant. When a 100 Ω resistance is connected, Figure 12c,d depict the current density and potential distribution on the surface of the pipeline at different electrolyte concentrations. The addition of the connecting resistor leads to a reduction in the current density on the pipeline surface. Comparative analysis of the results obtained without adding a resistor shows that after connecting the resistor, the variation in galvanic corrosion current density, especially further from the connection, diminishes with the change in electrolyte concentration.

4. Conclusions

To effectively explain pipeline corrosion, a model was constructed to reproduce the conditions leading to corrosion. The model parameters were derived from polarization and conductivity testing. Once the initial model was established, electrochemical corrosion behavior was verified through experiments, enabling the model to be used for calculating corrosion parameters. The model was extended to consider the influence of various factors on galvanic corrosion in pipelines. The model not only enhances our understanding of the galvanic corrosion mechanism but also predicts the location and current density, providing valuable insights for future structural optimization and electrical insulation control against galvanic corrosion.

1. The numerical simulation model for TC4/304 galvanic corrosion was successfully developed and validated against experimental results. The discrepancy between the simulated galvanic corrosion current and the experimentally measured galvanic current ranges from 0.4% to 1.32% as the distance between the anode and cathode varies. This model was utilized to investigate the influence of geometric and environmental factors on galvanic corrosion.
2. Within a specified range, an increase in insulation resistance resulted in a reduction of the potential difference between the anode and cathode, thereby decreasing the occurrence of galvanic corrosion. Upon reaching a sufficiently high level, the maximum current density at the connection decreased, potentially leading to self-corrosion. The tendency for galvanic corrosion decreases significantly with increasing resistance. When the resistance exceeds 500 Ω, the corrosion current density falls below 0.3 $\mu A/cm^2$, effectively achieving insulation control over galvanic corrosion. As the resistance surpasses 1000 Ω, the difference in coupling potential becomes negligible becomes negligible. Furthermore, when resistance reaches 50,000 Ω, the phenomenon of uneven potential distribution becomes distinctly observable.
3. The surface current density of the pipe increases gradually as the pipe diameter expands from 10 cm to 30 cm. Changes in pipe diameter significantly impact galvanic corrosion in areas distant from the pipe connection. The current density increases correspondingly with rising temperature, from 288 K to 308 K. Similarly, as the

concentration rises from 29.4 mol/m^3 to 85.47 mol/m^3, the current density also increases. However, while the value continues to rise, the variation in current density becomes minimal. Further research will be conducted to investigate the effects of temperature on equilibrium potential, as well as the influence of flow on concentration and conductivity. The introduction of resistance resulted in a deceleration of the changes in potential and current density concerning temperature and concentration.

Author Contributions: Conceptualization, Y.W.; validation, K.L.; formal analysis, K.L. and Y.W.; investigation, Y.W.; resources, W.L.; data curation, Y.W.; writing—original draft preparation, K.L.; writing—review and editing, H.C. and K.L.; supervision, W.L. and H.L.; project administration, H.C. and H.L.; funding acquisition, W.L. All authors have read and agreed to the published version of the manuscript.

Funding: This work was funded by the National Key Research and Development Program of China (2023YFB3408200) and the Defense Industry Technology Development Program (JSHS2022209A001).

Institutional Review Board Statement: Not applicable.

Informed Consent Statement: Not applicable.

Data Availability Statement: The data presented in this study are available upon request from the corresponding author.

Conflicts of Interest: The authors declare no conflicts of interest.

Nomenclature

Symbology	Description	Unit
$\vec{i_l}$	Current density vector in the electrolyte	A/m^2
σ_l	Conductivity	S/m
$\nabla\phi_l$	Electrolyte potential gradient	V/m
F	Faraday constant	C/mol
z_i	Charge number	1
D_i	Diffusion coefficient	m^2/s
u_i	Mobility	s·mol/kg
c_i	Concentration of the ion	mol/m^3
$\vec{u}$	Velocity vector	m/s
η	Overpotential	V
$\vec{i_s}$	Current density vector in the electrode	A/m^2
$\vec{n}$	Normal vector	
c_R	Related to the concentration of reducers	1
c_O	Related to the concentration of oxidizers	1
α_a	Anodic transfer coefficient	1
α_c	Cathodic transfer coefficient	1
R	Gas constant	J/(mol·K)
T	Temperature	K
ϕ_s	Electrode potential	V
E_{eq}	Equilibrium potential of the reaction	V
d	Spacer thickness	m
c_{in}	Inflow molar concentration of electrolyte solution	mol/m^3
u_{in}	Inflow velocity of electrolyte solution	m/s
c_0	Initial molar concentration of electrolyte solution	mol/m^3

References

1. Deshpande, K.B. Effect of aluminium spacer on galvanic corrosion between magnesium and mild steel using numerical model and SVET experiments. *Corros. Sci.* **2012**, *62*, 184–191. [CrossRef]
2. Si, W. An investigation on of Galvanic Corrosion between Exhaust System of Titanium Alloy and Hull Steel. *Dev. Appl. Mater.* **2013**, *28*, 34–38.
3. Jun, J.; Lim, Y.C.; Li, Y.; Warren, C.D.; Feng, Z. Mitigation of Galvanic Corrosion in Bolted Joint of AZ31B and Carbon Fiber-Reinforced Composite Using Polymer Insulation. *Materials* **2021**, *14*, 1670. [CrossRef] [PubMed]
4. Kamble, P.A.; Deshpande, P.P.; Vagge, S.T. Numerical investigation of galvanic corrosion between galvanized steel and mild steel in bolted joint. *Mater. Today Proc.* **2022**, *50*, 1923–1929. [CrossRef]
5. Jia, J.X.; Song, G.; Atrens, A. Influence of geometry on galvanic corrosion of AZ91D coupled to steel. *Corros. Sci.* **2006**, *48*, 2133–2153. [CrossRef]
6. Peng, Y.; Zhao, J.; Liu, Y.; Han, R.; Liu, Z.; Gao, J. Galvanic corrosion between Al-Zn-Mg-Cu alloy and stainless steel in the salt-spray atmosphere. *Mater. Chem. Phys.* **2023**, *294*, 127009. [CrossRef]
7. Wang, Z.; Wang, Y.; Wang, C. Area Ratio of Cathode/Anode Effect on the Galvanic Corrosion of High Potential Difference Coupling in Seawater. In Proceedings of the International Symposium on Application of Materials Science and Energy Materials (SAMSE), Shanghai, China, 17–18 December 2018.
8. Cheng, T.; Huang, H.; Huang, G. Galvanic corrosion behavior between ADC12 aluminum alloy and copper in 3.5 wt% NaCl solution. *J. Electroanal. Chem.* **2022**, *927*, 116984. [CrossRef]
9. Fan, W.; Zhang, Y.; Tian, H.; Wang, W.; Li, W. Corrosion behavior of two low alloy steel in simulative deep-sea environment coupling to titanium alloy. *Colloid Interface Sci. Commun.* **2019**, *29*, 40–46. [CrossRef]
10. Stenta, A.; Basco, S.; Smith, A.; Clemons, C.B.; Golovaty, D.; Kreider, K.L.; Wilder, J.; Young, G.W.; Lillard, R.S. One-dimensional approach to modeling damage evolution in galvanic corrosion. *Corros. Sci.* **2014**, *88*, 36–48. [CrossRef]
11. Khadom, A.A.; Abod, B.M. Mathematical model for galvanic corrosion of steel-copper couple in petroleum waste water in presence of friendly corrosion inhibitor. *J. Appl. Res. Technol.* **2017**, *15*, 14–20. [CrossRef]
12. Yin, L.; Jin, Y.; Leygraf, C.; Birbilis, N.; Pan, J. Numerical Simulation of Micro-Galvanic Corrosion in Al Alloys: Effect of Geometric Factors. *J. Electrochem. Soc.* **2017**, *164*, C75–C84. [CrossRef]
13. Yin, L.; Jin, Y.; Leygraf, C.; Pan, J. Numerical Simulation of Micro-Galvanic Corrosion of Al Alloys: Effect of Chemical Factors. *J. Electrochem. Soc.* **2017**, *164*, C768–C778. [CrossRef]
14. Standish, T.E.; Braithwaite, L.J.; Shoesmith, D.W.; Noel, J.J. Influence of Area Ratio and Chloride Concentration on the Galvanic Coupling of Copper and Carbon Steel. *J. Electrochem. Soc.* **2019**, *166*, C3448–C3455. [CrossRef]
15. Wang Chun, L.; Wu Jiang, H.; Yuan, M. Effect of Temperature on the Galvanic Corrosion of Cu-Ni Alloy/High Strength Steel in Seawater. In Proceedings of the International Symposium on Materials Application and Engineering (SMAE), Chiang Mai, Thailand, 20–21 August 2016.
16. Liu, Q.; Li, H.; Zhou, L. Experimental study of pyrite-galena mixed potential in a flowing system and its applied implications. *Hydrometallurgy* **2009**, *96*, 132–139. [CrossRef]
17. Zhang, H.; Miao, H.; Chen, J.; Mao, F.; Hu, W.; Dong, C.; Xiao, K. Study on galvanic corrosion behavior of researcher pipeline structure. *Int. J. Electrochem. Sci.* **2024**, *19*, 100498.
18. Song, G.L.; Johannesson, B.; Hapugoda, S.; StJohn, D. Galvanic corrosion of magnesium alloy AZ91D in contact with an aluminium alloy, steel and zinc. *Corros. Sci.* **2004**, *46*, 955–977. [CrossRef]
19. Cao, X.; Hu, X. The investigation of micro-galvanic corrosion of SAF 2205 duplex stainless steel based on numerical simulation model and immersion test. *Corros. Sci.* **2022**, *207*, 110601. [CrossRef]
20. Wang, Z.; Bai, Y.; Ma, X.; Xing, S. Numerical simulation of galvanic corrosion for couple of Ti-alloy with Cu-alloy in seawaters. *J. Chin. Soc. Corros. Prot.* **2018**, *38*, 403–408.
21. Pan, D.; Yan, Y.; Gao, X.; Liu, F.; Li, B. Galvanic corrosion and electrical insulation between high strength steel and typical piping materials B10 and TA2. *Corros. Prot.* **2017**, *38*, 589–592+597.
22. Zhang, H.; Li, N.; Xue, J.; Liu, F.; Sun, T. Galvanic corrosion and insulation control between industrial pure titanium and albata. *Corros. Prot.* **2010**, *31*, 615–618.
23. Chen, J.; Zhang, Q. Effect of electrochemical state on corrosion-wear behaviors of TC4 alloy in artificial seawater. *Trans. Nonferrous Met. Soc. China* **2016**, *26*, 1011–1018. [CrossRef]
24. *HB 5374-1987*; Method for Measuring Galvanic Current of Different Metals. Ministry of Aeronautics Industry: Beijing, China, 1987.
25. Rong, H.; Fan, W.; Shan, B.; Yang, J.; Ding, R.; Zhao, X. Galvanic corrosion behavior and numerical simulation of 5083 aluminum alloy and Q235 steel in 3.5% NaCl solution. *Mater. Corros. -Werkst. Und Korros.* **2023**, *74*, 766–776. [CrossRef]
26. Cross, S.R.; Gollapudi, S.; Schuh, C.A. Validated numerical modeling of galvanic corrosion of zinc and aluminum coatings. *Corros. Sci.* **2014**, *88*, 226–233. [CrossRef]

27. Jun, J.; Joshi, V.V.; Crawford, A.; Viswanathan, V.; Leonard, D.N.; Chen, J.; Updadhyay, P.; Lim, Y.C.; Feng, Z. Galvanic corrosion of AZ31B joined to dual-phase steel with and without Zn layer by ultrasonic and friction stir welding. *J. Magnes. Alloy.* **2023**, *11*, 462–479. [CrossRef]
28. Niu, L.; Yang, X.; Si, Q.; Li, Z.; Wang, C. Simulation of corrosion in the riveted zone of aluminium-based skin. *International J. Simul. Model.* **2023**, 2, 279–290. [CrossRef]

materials

Article

Reconstruction of the Passive Layer of AISI 304 and 316 Steel After Scratching

Sylwia Charazińska [1], Andrzej Sikora [2], Beata Malczewska [1] and Paweł Lochyński [1,*]

[1] Institute of Environmental Engineering, Wrocław University of Environmental and Life Sciences, pl. Grunwaldzki 24, 50-363 Wroclaw, Poland; sylwia.charazinska@upwr.edu.pl (S.C.); beata.malczewska@upwr.edu.pl (B.M.)

[2] Department of Nanometrology, Faculty of Electronics, Photonics and Microsystems, Wrocław University of Science and Technology, 50-370 Wrocław, Poland; andrzej.sikora@pwr.edu.pl

* Correspondence: pawel.lochynski@upwr.edu.pl

Abstract: Austenitic stainless steels are used widely in many fields due to their good mechanical properties and high resistance to corrosion. This work focuses on the reconstruction of the passive film after scratching. The purpose of the study was to compare changes in the rate of passive layer reconstruction and to discuss the effect of both the type of material and its electrochemical treatment on the reconstruction of the passive layer for two types of stainless steel: 304 and 316. The XPS tests performed indicate a significantly higher Cr/Fe ratio for the samples after the electropolishing process of 1.41–1.88 compared to the as-received samples of 0.82–0.86. After 2–3 min of sputtering the surface with Ar+ ions, a decrease in chromium content can be observed, with a simultaneous increase in nickel content, visible especially for the electropolished samples. A new approach in the conducted research was to scratch the test samples under controlled conditions, then evaluate the dynamics of the passive layer reconstruction using the AFM method, and then confront the obtained results with XPS measurements for the corresponding samples. For the as-received samples (2B finish) and those after surface treatment, regardless of the level of contamination of the electropolishing process bath, the reconstruction time was similar, which was approximately 2 h, although certain differences in the process dynamics were noticeable.

Keywords: stainless steel; electropolishing; XPS; depth profile; AFM; high-resolution spectra; nano-scratching; passive film

Citation: Charazińska, S.; Sikora, A.; Malczewska, B.; Lochyński, P. Reconstruction of the Passive Layer of AISI 304 and 316 Steel After Scratching. *Materials* **2024**, *17*, 6238. https://doi.org/10.3390/ma17246238

Academic Editor: Xiaogang Li

Received: 3 December 2024
Revised: 14 December 2024
Accepted: 18 December 2024
Published: 20 December 2024

1. Introduction

Stainless steel (SS) is ubiquitous in today's industrial world, as it is generally considered to be corrosion-resistant. This is a result of the relationship between the iron alloy and environmental factors. The corrosion resistance of SS results from the 'passive' chromium-rich oxide layer that forms naturally on the surface of the steel. After its initial formation, its thickness increases over time, which is what its mechanical properties depend on [1–3]. Other properties of the passive layer, such as adherence and stability, determine the SS's level of corrosion resistance [4].

Despite numerous studies, the mechanism of passive layer formation has not been sufficiently understood [2]. In a recent study, Jamebozorgi et al. [5] suggested that the underlying mechanisms of passive layer formation are based on clustering and local detachment of metallic atoms. In contrast, Suresh et al. [6] demonstrated that the formation and breakdown of a passive film layer are mainly controlled by ionic and electronic transport reactions [1,6]. Numerous studies have described the passive film on SS as a bilayer structure [5,7–13]. Passive layers typically exhibit a thickness of a few to several micrometers, are non-porous, show good tightness and chemical resistance, and have good electron conductivity [14]. According to a recent study by Rahimi et al. [15], the cross-section of the passive layer can be separated into an n-type outer layer and a p-type inner layer, based on their chemical composition.

Under typical atmospheric conditions, the passive layer can regenerate when oxygen is present in air. It is produced when oxygen in the air and chromium in the SS undergo a chemical reaction. Chromium oxide is a product of these reactions. Consequently, the inner oxide, which is rich in chromium, is recognized as a corrosion barrier. The fundamental mechanisms in passive layer formation require a more thorough theoretical examination. This is directly related to the limitations of analytical and spectroscopic approaches in investigating nanoscale kinetic processes [5]. Numerous studies have investigated the kinetics of SS oxidation. For example, a recent study by Tcharkhtchi-Gillard et al. [16] concluded that at higher potentials, transpassive behavior was denoted, characterized by a high mass-loss rate and a specific intergranular corrosion morphology. They performed chronoamperometry experiments coupled with mass loss measurements to quantify the oxidation and reduction kinetics. On the other hand, Kim et al. [17] evaluated the growth kinetics of passive films on iron and stainless steel using in situ specular X-ray reflectivity. They reported that in SS, the oxide properties changed as the applied potential varied. Furthermore, Jaffré et al. [18] showed that the electrochemical reactivity of the grown oxide is affected by the chemistry of the environment more than by the mechanical surface treatment. As discussed by Wang et al. [19], oxygen pressure and temperature have the most impact on the oxidation rate.

When a passive layer of 304 or 316 stainless steel is scratched, it can regrow. This process may be hampered if the passive layer is exposed to high temperatures and chloride ions. For instance, the protective capacity of these passive films may be further diminished by a combination of halide ions and surface mechanical damage caused by friction/scratching [1,20]. Cl^- can easily penetrate the passive film and then combine with cations in the passive film to form soluble chloride, breaking the integrity of the passive film, and thus causing pitting corrosion to occur [21]. Additionally, factors influencing the susceptibility of steel to the local failure of the passive layer comprise non-metallic inclusions, the presence of second-phase particles, and structural inhomogeneity [22]. Moreover, Domańska et al. [23] indicated that the observed pitting corrosion of stainless steel samples placed over the surface of municipal wastewater was a consequence of having been in an environment containing hydrogen sulfide. The passive layer can be strengthened by the passivation process, that is, chemical or electrochemical protection of the steel surface [24].

The electropolishing process can be used to remove micro-scratches and clean the metallic surface of stainless steel, followed by the formation of a new protective layer with improved properties. The principles of electrolysis govern the process, which necessitates current density and the use of an electrolyte, usually consisting of sulfuric and orthophosphoric acids. A layer of metal is removed from a workpiece's surface by electropolishing, while an electric current is applied [25]. Previous research has highlighted the importance of electrochemical polishing process parameters [25–29].

As previously reported in the literature, different practices have been applied to prevent the corrosion of metals. Among them, the applications of protective coatings, electroplating, anodizing, control of environmental conditions, and use of anti-corrosion systems or metal alloys are the most common [21,30–35]. Pantoja et al. [36] reported that chemical pretreatments increase the surface energy, primarily because of the removal of most external layers, which causes the passive layer to be strongly Cr-rich. The steel's strength, hardness, hardenability, and ductility are all improved by the addition of nickel, which also lessens the steel's brittleness. Steel that contains molybdenum is more resistant to high temperatures, hydrogen sulfide, and embrittlement. Therefore, the molybdenum content is the primary manner of distinguishing between 304 and 316 grade stainless steel. Other additives often used are silicon, manganese, niobium, nitrogen, sulfur, and phosphorus. Moslehifard et al. [37] found that Co-Cr-Mo alloys exhibit better corrosion resistance than Ni-Cr-based alloys. Wang et al. [38] reported that the addition of nitrogen results in the replacement of nickel which, in turn, causes strong stabilization of the austenitic microstructure. When Xia et al. [39] investigated the effect of adding a small amount of nickel during sintering, they discovered an improvement in corrosion resistance, especially against acids. For the same environment, the addition of molybdenum and tungsten further improved resistance to corrosion. Pardo [40] reported that the addition of molybdenum increased the corrosion resistance of stainless steels in acidic medium, unlike the addition of manganese. On the other hand, Robin et al. [41], when examining the impact of the silicon presence in steel, revealed the presence of silicon in the passive layer, which decreased the chromium content. Additionally, they found that high Si-containing austenitic stainless steels had good corrosion resistance, which may have been caused by a drop in the cathodic reaction rate. In comparison to uncoated specimens, Padhy et al. [9] have reported an increase in the corrosion current density of sputter-deposited TiO_2 on austenitic type 304 L SS in a nitric acid medium.

To assess the chemical composition and thickness of the oxide layer on SS, different methods may be applied, but the Cr/Fe atomic ratio, Cr oxide/Fe oxide ratio, and total oxide thickness are highly relevant indicators. Depending on the emitted signal, the following methods are distinguishable: X-ray photoelectron spectroscopy (XPS—X-ray photoelectron spectroscopy), Auger electron spectroscopy (AES—Auger electron spectroscopy), ultraviolet excited electron spectroscopy (UPS—Ultraviolet Photoelectron Spectroscopy), secondary ion mass spectroscopy (SIMS), ion scattering spectroscopy (ISS), and sputtered neutral particle mass spectroscopy (SNMS) [13,42,43]. The film thickness of the passive layer is a crucial factor. Therefore, the use of appropriate measurement techniques is essential. Yao et al. [44] have proposed a real-time evolution and characterization of passive films, in which they applied magnetic force microscopy, Kelvin probe force microscopy, SEM, EDS, and AES. Studying this phenomenon, they found that changes over time in the passive layer can be divided into three periods, the first nucleation, the second rapid growth, and the third stable growth. In recent years, the most commonly used method for evaluating the passive layer has been XPS [13,38,42]. Information about the depth profile and surface is provided by XPS. It has an analysis depth of roughly 2 to 4 nm and measures the energy spectra of electrons released when the surface is exposed to X-rays. As it is the second most popular technique, AFM offers numerous opportunities to investigate and evaluate the material structure. It is employed to study passive films at the metal/environment interface that are of nanometer order [19,45–48].

Despite the numerous related studies, the issue of the SS reconstruction rate has not received enough attention. The present study's objective was to compare the dynamics of the reconstruction changes with changes in the passive layer reconstruction rate for 304 and 316 steels, including the 2B finish, following electropolishing. With the use of XPS measurements to describe the film's structure, the kind of material and its electrochemical treatment on the passive layer reconstruction have been further examined in this work. Furthermore, samples were scratched under controlled settings in the current study, before the kinetics of the passive layer reconstruction were assessed.

2. Materials and Methods

2.1. Materials

Samples were obtained from cold-rolled 1.5 mm thick AISI 304 (Aperam Stainless France, Isbergues, France) and AISI 316 (Aperam, Genk, Belgium) stainless steel with a 2B surface finish. The chemical material composition of 304 stainless steel was the following (wt%): 18.23% Cr, 8.02% Ni, 1.39% Mn, 0.37% Si, 0.201% Co, 0.066% N, 0.034% P, 0.028% C, 0.002% S, and balance Fe. The composition of 316 steel was, respectively, the following (wt%): 16.57% Cr, 10.04% Ni, 2.03% Mo, 1.32% Mn, 0.032% N, 0.48% Si, 0.030% P, 0.018% C, 0.003% S, and balance Fe.

2.2. Sample Preparation

Tests were conducted on round samples with a diameter of 16.7 mm, with a hole of 1 mm diameter located 1 mm away from the edge of the sample. 304 and 316 stainless steel, as-received samples, were degreased with acetone and then rinsed in distilled water for 20 min in an ultrasonic bath. The acetone rinsing and degreasing procedure was also carried out for the samples prior to the electropolishing process. The process bath for electropolishing contained sulfuric acid and phosphoric acid. The electropolishing process parameters were as follows: temperature 55 °C, process time 15 min, current density 8 A/dm^2. The electropolishing circuit consisted of an electropolished sample placed centrally between two electrodes (cathodes), 2 cm apart on each side. The 304 steel sample was also polished in a bath after long-term operation containing about 6% iron (304-EP6). After the electropolishing process, samples were subjected to multi-stage rinsing in distilled water. The samples, so prepared for further study, were then labeled as containing the type of steel (304 or 316), the surface treatment (AR—as received or EP—electropolished), and the content of Fe contamination in the electropolishing process bath (0 or 6 percent by mass).

2.3. AFM

An atomic force microscope was used for the observation of the passivation layer's reconstruction process in the time domain. Such an approach was possible due to the unique abilities of AFM to utilize the scanning probe as both a tool for surface nanomodification and a measurement device enabling subnanometer-resolution surface 3D quantitative imaging [49,50]. Such unique ability, typical for correlational nanoscopy, enables multifunctional utilization of the setup in a specifically defined submicron area. The nanoscratching procedure, enabling oxidation layer recovery, was performed using a diamond-coated silicon probe DDESP (Bruker, Camarillo, CA, USA). The parameters of the used probe were as follows: material: 0.01 Ω·cm to 0.02 Ω·cm antimony (n) doped silicon; back coating: aluminum; tip coating: conductive diamond, r_{tip} = 150 nm, k = 42 N·m^{-1}, f_{res} = 320 kHz. The tip's diamond coating provided it with durability and wear resistance, which played an essential role in the experiment in terms of surface modification and long-term scanning. The surface imaging data analysis did not reveal any traces of scanning tip wear or contamination during the experiment. The experiment was performed in ambient conditions, while temperature and relative humidity were controlled and maintained at levels 25 °C ± 2 °C and 45% ± 4%.

The parameters of the nanoscratching procedure were selected based on previous experiments, where there had been various probe speeds and heights offset; hence, pressure forces were applied. It should be underlined that the individual processing parameters were related to a certain operation location, while specific grains required some slight approach modifications due to mechanical inhomogeneities of the material. One has to be aware that the force applied to the surface was the subject of a value limitation, while the probe with a certain spring constant must not apply the force, introducing the risk of cantilever damage and hence the destruction of the probe. These particular criteria lead to the different starting depths of the nanoscratches. One example of a 3D image of the sample surface showing the grains' presence is shown in Figure 1a. A large area images were used to determine the spots where the experiment could be performed. The typical speed of the probe during the nanoscratching procedure was set to 0.5 µm/s, while the determined change to the probe's height was set to 200 nm below the previously imaged surface level. The actual force used to modify the surface varied according to the observed area due to different grain orientations in the passivation layer. The above-mentioned procedure incorporated the fabrication of two parallel scratches, oriented vertically, while the fast scanning axis of the microscope was horizontal. An example of a 3D image of the sample surface showing grooves developed using the nanoscratching technique is shown in Figure 1b.

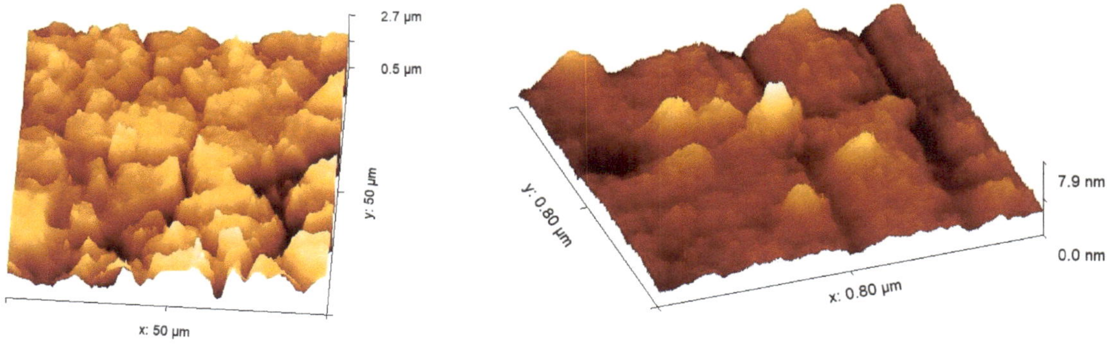

(a) (b)

Figure 1. Examples of the 316-AR sample surface imaged by means of AFM: (**a**) showing the grainy structure of the material, (**b**) after the nanoscratching procedure.

The successfully performed nanoscratching process was followed by the scanning procedure. In order to observe the long-term process of the passivation layer reconstruction, the optimal spot (presence of the surface plateau and lack of near debris) was chosen individually for each set of scratches. Once the location of the best measurement spot was determined, the repetitive profile's acquisition process was launched by repetitive scanning of a single line (slow scanning axis disabled). The measurement was performed continuously, generating the set of files containing time domain surface evolution information. The obtained set of 2D maps provided a Y-axis time-related map of the cut depth changes. By measuring the depth of the fabricated trench at certain time steps, one was able to develop time-related graphs revealing the dynamics of the passivating layer reconstruction. The obtained data show the nanoscratches' depth decrease progress, revealing the passivation layer reconstruction. The plateau level next to the sides of the nanoscratches, used as the reference level, does not change its height.

Each procedure allowed the generation of approx. 70 files, covering a 24 h period (a single file contained 256 lines acquired within a 20 min time period). The data processing involved basic tilt correction and cut depth determination using profile analysis. In order to improve the statistical reliability of acquired data, 10 profiles were averaged. The results of the measurements are shown in Figure 2.

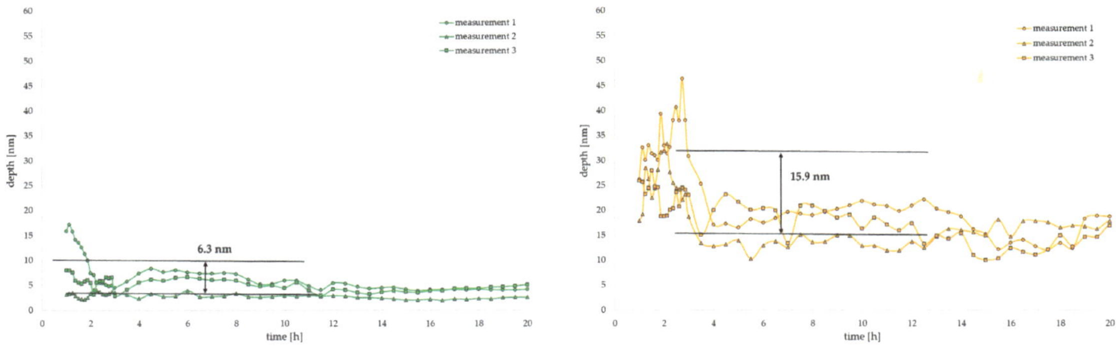

(a) (b)

Figure 2. *Cont.*

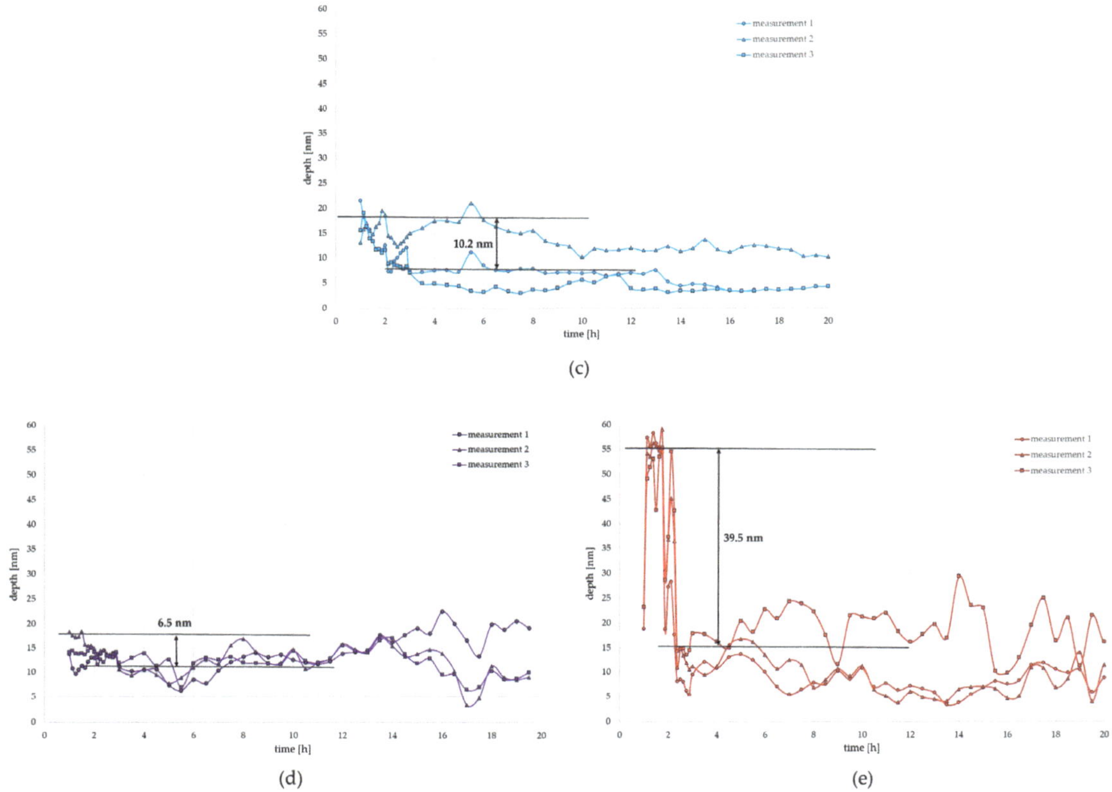

Figure 2. The graphs illustrate the passive layer reconstruction process in the time domain for the following samples: (**a**) 316-AR; (**b**) 316-EP0; (**c**) 304-AR; (**d**) 304-EP0; (**e**) 304-EP6.

The above-mentioned procedure was performed at least three times in different locations, until three data sets were successfully acquired for a certain sample. Such an approach provided a more reliable statistical outcome, taking into account the local (single grain) properties of the sample.

2.4. XPS

XPS analyses were carried out according to the procedure and parameters as described in the authors' previous work in the section Experimental procedures- XPS surface analyses [24]. The XPS analyses were performed using a PHI 5000 VersaProbe spectrometer (ULVAC-PHI, Hagisono, Chigasaki, Kanagawa, Japan) based on the procedures developed at the Institute of Physical Chemistry, Polish Academy of Sciences [51,52].

3. Results and Discussion

The constant advancement in SS technology, which is compelled by the growing demands for operational parameters of different SS alloys, is a key driver among researchers to better comprehend the complexities of the processes taking place on the material's surface. It has been noted that variations in surface modification happen continuously. In this paper, we presented the application of the advanced AFM technique supplemented by the XPS technique. AFM was offered as a technique for monitoring the passivation layer reconstruction process on the surface of SS, before and after electropolishing. To achieve this, a method known as nanoscratching was employed, which involves using a diamond scanning probe to create scratches on a material's surface, breaking the passivation layer's continuity. The AFM application appears to be a very helpful instrument in the analysis of the impact of various conditions on these phenomena, since it makes it possible to observe the dynamics of the rebuilding of the passivation layer in a quantitative manner. AFM observations of the SS passive layer reconstruction rate after nanoscratching enabled the evaluation of changes in the depth of scratches with a subnanometer resolution (Figure 2). The depth provided in Table 1 was calculated as the difference between the initial depth of the nanoscratch and the value where no further change was observed.

Table 1. Summary of AFM results and XPS passive layer composition based on depth profiles.

Sample	AFM Reconstruction	XPS Sputter Time	XPS Etch Depth	Cr/Fe Ratio	C	O	Fe	Ni	Cr	P	S	N	Mo	Si
	[nm]	[min]	[nm]	[-]	[at%]	[at%]	[at%]	[at%]	[at%]	[at%]	[at%]	[at%]	[at%]	[at%]
316-AR	6.3	1.5	2.06	0.82	5.1	50.0	22.3	2.1	18.3	0.5	0.0	0.3	0.4	0.9
		2	2.74	0.80	7.7	44.6	23.7	3.4	19.0	0.4	0.0	0.5	0.7	0.0
316-EP0	15.9	1.5	2.06	1.45	1.4	46.4	19.2	2.4	27.9	0.2	0.1	0.0	0.5	1.9
		2	2.74	1.12	1.4	38.7	24.5	5.3	27.4	0.9	1.1	0.0	0.6	0.2
304-AR	10.2	1.5	2.06	0.79	11.4	52.5	19.2	1.2	15.2	0.0	0.5	0.1	-	-
		2	2.74	0.86	6.9	51.7	20.1	1.6	17.4	0.2	2.0	0.1	-	-
304-EP0	5.6	1.5	2.06	1.88	1.3	43.5	17.2	2.1	32.4	1.1	0.0	2.4	-	-
		2	2.74	1.58	7.4	32.9	20.0	3.5	31.5	0.0	0.5	4.3	-	-
304-EP6	39.5	1.5	2.06	1.41	0.4	45.3	19.2	2.9	27.0	0.2	0.8	4.3	-	-
		2	2.74	1.10	3.8	33.1	25.7	4.6	28.2	0.0	0.0	4.7	-	-

Table 1 summarizes the AFM results and passive layer composition based on XPS depth profiles for the studied samples. Higher carbon percentages of 5.1% and 6.9% were observed for the as-received samples, for steel 316 and 304, respectively. The samples after the electropolishing process had passive layer carbon contents of about 1.4% or less. Similarities between the as-received samples and electropolished samples have also become apparent for the iron and chromium content. The as-received samples obtained higher amounts of iron in the passive layer along with lower chromium values than the electropolished samples, for which less iron and more chromium were noted. It can be seen that the Cr/Fe ratio values were similar, which were 0.82–0.86 for the as-received samples. The electropolishing process increased the chromium/iron ratio. The highest values of close to 1.9 were reached by 304 steel after the electropolishing process in a bath containing 0% contaminants. The other two samples, after the electropolishing process, achieved similar Cr/Fe ratio results of about 1.4. It should be noted that the test samples initially had a lower chromium content in 316 steel than in 304 steel. However, the addition of molybdenum in 316 steel significantly distinguishes it from 304 steel.

In the literature, various techniques have been used in relation to the determination of the dynamics of passive layer changes. Zhang et al. [53] used the TEM (Transmission Electron Microscopy) technique to image passive film growth on 316 stainless steel in 0.1 M NaOH solution. The researchers observed the presence of an ultra-thin film ~4 nm thick on the surface of stainless steel needles after polarization at −0.65 V for 7200 s. Furthermore, they determined that the Fe/Cr ratio increased with increasing applied potential. High-resolution TEM was also used by Tong et al. [54] for studies conducted for AISI 304 L austenitic stainless steel in a borate buffer solution. The authors reported that the thickness of the passive layer was 10–15 nm, 2–4 nm, and 10–18 nm for samples that were in contact with the solution for 1.5 h, 3 h, and 8 h, respectively. They described the results as consistent with those obtained by the transient corrosion current i_{OCP} and photocurrent i_{ph} during passive film growth.

Mujanović et al. [1] have shown that after mechanical depassivation in scratch tests on various stainless steels, repassivation at open circuit potential takes place within the first few tenths of a second after scratching. However, the final healing of the passive layer as well as reaching a steady passive state take much longer. According to our research, the reconstruction of the passive layer of samples 304 and 316 SS took 1–3 h. The passive layer growth rate is higher at the start of the study and decreases over time. Both the width and depth of the cut affect the process's kinetics. Lv et al. [55] used both TEM imaging and EIS (electrochemical impedance spectroscopy) corrosion resistance testing techniques to study the surface of AISI 304 steel. TEM studies conducted for samples cold-rolled at different temperatures made it possible to determine that the sample cold-rolled at -120 °C had the most uniform microstructure. Moreover, EIS studies showed the highest phase degree for this sample, indicating that its passive film is more capacitive, hence a more condensed and intact film. The authors concluded that of the surfaces tested, this one (120 °C) had an equiaxed microstructure and displayed the best corrosion performance. The EIS technique has been employed to evaluate 316 L stainless steel passive film grown in acidic solution by Boissy et al. [56]. The authors used a developed special electrical equivalent circuit for testing at 20 °C, 60 °C, and 80 °C. At the given temperatures, the created circuit made it possible to calculate a diffusion coefficient of about 10^{-14} cm^2/s, which was in line with the values presented in the literature.

Based on literature reports and our own research, it was found that in the case of 304 and 316 stainless steels, the passive layer can usually have a thickness of a few to several nm. However, as stated by [57], the 304 L stainless steel's surface condition influences the semiconductor's properties as well as the passive layer's thickness. Similarly to [53], we observed that the compositions of the passive film were mainly Cr and Fe oxides, based on evaluation of the XPS results.

XPS tests were used to analyze the chemical composition of the samples' surfaces. The tests were carried out for samples made of two types of steel (304 and 316), with both as-received surfaces and after electropolishing in baths with various amounts of contaminants—0% and (for 304 SS) 6% Fe by mass. The chemical composition (at %) of the surface is shown in the depth profiles for the analyzed samples of 316 steel (Figure 3a,b) and 304 steel (Figure 3c–e). The contents of individual elements in the passive layer are shown as percentages. After 2–3 min sputtering the surface with Ar^+ ions, a decrease in chromium content can be observed, with a simultaneous increase in nickel content. This is clearly visible for the samples after the electropolishing process. For all samples tested, the iron content is higher than the chromium content in the bulk. The passive layer composition of both 304 and 316 steels is characterized by a substantially greater Cr/Fe content for the samples after the electropolishing procedure as compared to the value before electropolishing. Furthermore, it was found that the electropolishing process could still be carried out in polluted electrolytes with up to 6% mass iron ions in the electrolyte composition. Despite the significant contamination of the electropolishing bath used, the surface obtained after the process was even characterized by an enhanced composition of the passive layer compared to the surface finish 2B for the as-received samples.

Cr 2p3/2, Fe 2p3/2, and Mo 3d5/2 spectra after sputtering with Ar^+ are presented for samples 316-AR and 316-EP0 (Figure 4). Deconvolutions for samples made of 304 steel and subjected to electropolishing were published by the authors of a previous article [26].

Deconvolutions corresponding to Fe^{2+} oxides are visible for the as-received sample at 707.6 eV and 708.8 eV and for the electropolished sample at 707.0, 708.5, and 709.8 eV (FeO). Peaks corresponding to Fe^{3+} (Fe_2O_3) oxides are present for the as-received sample at 710.6 eV and for the 316-EP0 sample at 711.1 eV. The results obtained in the authors' previous studies for electropolished 304 steel samples indicated the presence of Fe^{2+} oxides at 709.1–709.2 eV and Fe^{3+} oxides at 710.6–711.3 eV [26]. A peak, indicating iron, in metallic form was observed at 706.8 eV and 706.3 eV for the as-received and electropolished samples, respectively. Other researchers have indicated the presence of XPS peaks of metallic iron at

706.6 eV [58]; Fe^{2+} (FeO) between 709.0, 709.6, and 709.7 [59,60]; and Fe^{3+} oxides (Fe_2O_3) between 710.5 and 711.5 eV [60].

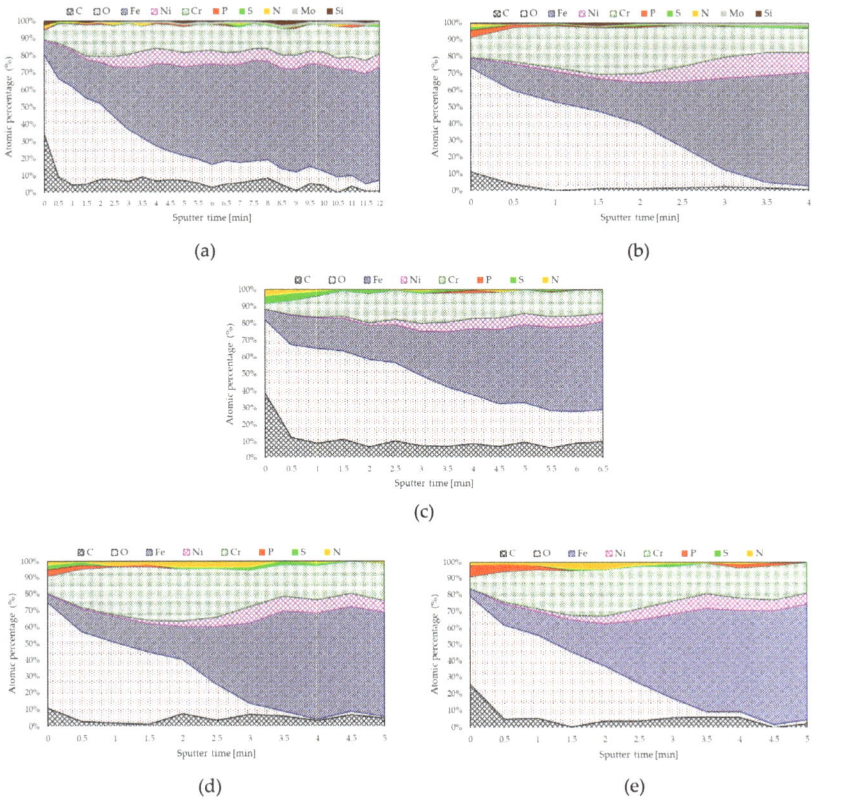

Figure 3. Depth profiles of the surface of samples: (**a**) 316-AR; (**b**) 316-EP0; (**c**) 304-AR; (**d**) 304-EP0; (**e**) 304-EP6.

Chromium is present in metallic form at 573.5 and 573.1 for the as-received and electropolished samples, respectively. In addition, there are peaks corresponding to Cr^{3+} visible for sample 316-AR at 574.3 and 576.5 eV and at 575.5 eV, 576.7 eV for the electropolished sample, as well as for Cr^{6+} (CrO_3) at 578.4 eV, presented for sample 316-EP. Previous studies by the authors have suggested the occurrence of chromium Cr^{3+} oxides on the surface of 304 steel samples in a range of 576.0–576.7 and Cr^{6+} oxides in a range of 578.6–578.9 eV [26]. The values obtained for 316 steel samples are very similar to the previously obtained results. Lai et al. [58] and Yi et al. [61] noted the presence of chromium(III) oxides at 576.6 eV. Other reports suggest the presence of Cr^{3+} oxides (Cr_2O_3) at 576.0–576.8 eV [62,63] and Cr^{6+} (CrO_3) at 578–578.3 eV [64].

Signals associated with molybdenum appear in a range of about 227 to 229 eV. For both the as-received sample and that after the electropolishing process, the signal associated with the metallic form of Mo is visible at 227.5 eV and 227.3 eV, respectively. For both samples, a second peak is observed that is visible at 228.1 eV, and it is not clear which form it may indicate, but it could be associated with the presence of Mo^{3+} or Mo^{4+}. The presence of Mo^{4+} in the form of MoO_2 was noted in the as-received sample at 228.7 eV. For an after-electropolishing sample, two peaks are present at 231,1 eV and 231,3 eV, which can be associated with the Mo^{5+} form. For the electropolished sample, a peak indicating the presence of Mo^{4+} oxide (MoO_2) is seen at 231.5 eV. Other authors conducting tests with 316 steel have noted the presence of XPS peaks associated with the metallic form of molybdenum at 227.0–227.1 eV [65] and 227.8 eV [66]. Molybdenum oxide (MoO_2) peaks were present at 228.6–229.1 eV in a study by Bortolan et al. [65], while Mo^{5+} peaks were located at 231 eV according to data in a study by Wang et al. [66].

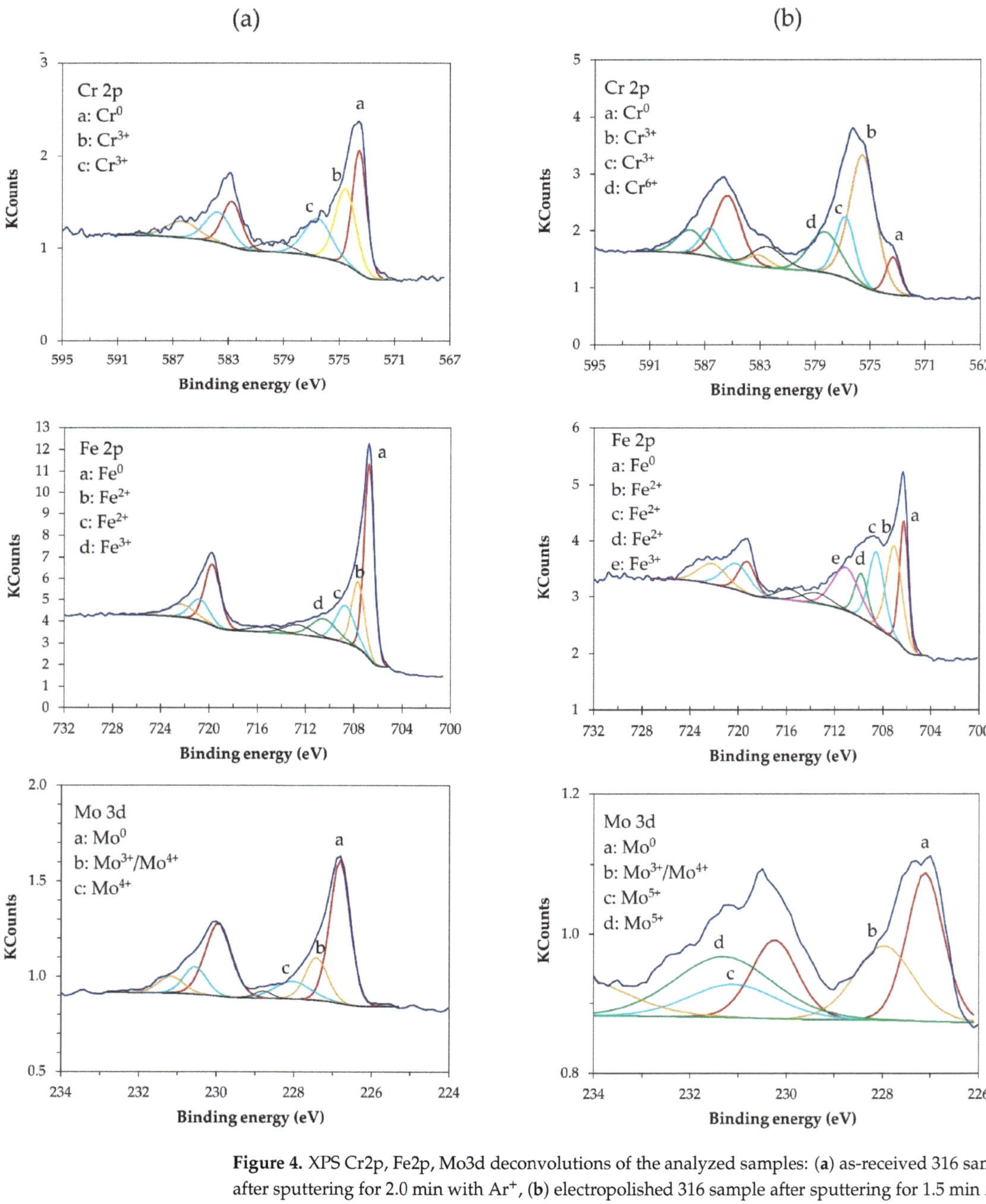

Figure 4. XPS Cr2p, Fe2p, Mo3d deconvolutions of the analyzed samples: (**a**) as-received 316 sample after sputtering for 2.0 min with Ar^+, (**b**) electropolished 316 sample after sputtering for 1.5 min Ar^+.

4. Conclusions

Based on a reconstruction of the passive layer of AISI for as-received samples from 304 and 316 steel and after electropolishing of these steels, samples of 304 and 316 steel after scratching needed between 1 and 3 h to reconstruct the passive layer. During this time, there was a stage of reconstructing the thickness of the passive layer at a level of a few to several nanometers. AFM measurements in the following hours up to the 20th hour did not indicate any further significant changes in the passive layer's thickness. The recovery of the passive layer was markedly different for the 304 steel sample, which had previously been electropolished in an extremely heavily contaminated bath to an iron ion content of 6 wt.%. Between 1 and 2 h after scratching, a thickness of 40 nm was needed. A further several hours confirmed, as with the other samples, that there was no further change in the layer's thickness. The study indicates that irrespective of the level of contamination of the process bath, AISI 304 and 316 steels do not lose their ability to rebuild the passive layer after scratching, even though the level of contamination may affect the rate of rebuilding the passive film. In addition, resistance to the nanoscratching procedure varied for the tested samples, as one could see differences in the depth's starting values. In particular, one might notice that the electropolishing process has increased the submicron homogeneity, with the AFM experiments revealing a much more significant deviation of the initial nanoscratching result, while the processed samples revealed a better repeatability of the outcome. Also, the electropolishing process seems capable of inflicting damage with relative ease on the passivation layer through nanoscale scratching, while the AR samples did not reveal so deep a cut.

Based on our own XPS studies and the relevant literature, the passive film on both AISI 304 and 316 steels is usually a few to several nm thick. The passive layer of austenitic steels owes its resistance and ability to regenerate to the content of chromium oxides and hydroxides; in addition, AISI 316 contains molybdenum oxides in the passive layer formed. XPS studies have shown that the composition of the passive film for both 304 and 316 steels is characterized by a significantly higher Cr/Fe content of 1.41–1.88 for samples after electropolishing compared to values of 0.82–0.86 for as-received samples. Even extremely long electrolyte operation at up to a value of 6 wt.% of iron ions in the electrolyte composition still enables the electropolishing process, resulting in an improved passive layer composition compared to the surface finish 2B for the as-received samples.

Author Contributions: Conceptualization, P.L.; methodology, A.S. and P.L.; investigation, S.C. and A.S.; resources, P.L.; writing—original draft preparation, S.C., B.M., A.S. and P.L.; writing—review and editing, S.C., B.M. and P.L.; visualization, S.C. and A.S.; supervision, P.L. All authors have read and agreed to the published version of the manuscript.

Funding: This research was funded by the National Centre for Research and Development (NCBR) within the project IonsMonit (LIDER/22/0187/L-7/15/NCBR/2016), and the correlational nanoscopy experiment has received funding from the European Union's Horizon 2021 MCSA programme under Grant Agreement No 101086226-ENSIGN. The APC is co-financed by the Wrocław University of Environmental and Life Sciences and Wrocław University of Science and Technology (European Union's Horizon Europe programme under Marie Skłodowska-Curie Actions—Staff Exchanges (SE) grant agreement No 101086226-ENSIGN).

Institutional Review Board Statement: Not applicable.

Informed Consent Statement: Not applicable.

Data Availability Statement: The original contributions presented in this study are included in the article. Further inquiries can be directed to the corresponding author.

Acknowledgments: The authors acknowledge support by Wojciech Lisowski and Marcin Pisarek from the Polish Academy of Sciences who carried out the XPS experiments on a PHI 5000 VersaProbe (ULVAC-PHI) spectrometer.

Conflicts of Interest: The authors declare no conflicts of interest.

References

1. Mujanović, E.; Zajec, B.; Legat, A.; Kosec, T.; Kovač, J.; Mori, G.; Hönig, S.; Zehethofer, G. Depassivation and Repassivation of Stainless Steels by Stepwise PH Change. *Mater. Corros.* **2021**, *72*, 421–433. [CrossRef]
2. Olsson, C.O.A.; Landolt, D. Passive Films on Stainless Steels—Chemistry, Structure and Growth. *Electrochim. Acta* **2003**, *48*, 1093–1104. [CrossRef]
3. Sathyaseelan, V.S.; Rufus, A.L.; Chandramohan, P.; Subramanian, H.; Velmurugan, S. Preparation, Characterization and Dissolution of Passive Oxide Film on the 400 Series Stainless Steel Surfaces. *J. Nucl. Mater.* **2015**, *467*, 89–96. [CrossRef]
4. Liu, M.; Huang, S.; Liu, Z.; Du, C. Investigations on the Passive and Pitting Behaviors of the Multiphase Stainless Steel in Chlorine Atmosphere. *J. Mater. Res. Technol.* **2024**, *28*, 3365–3375. [CrossRef]
5. Jamebozorgi, V.; Rasim, K.; Schröder, C. Computational Chemistry Analysis of Passive Layer Formation and Breakdown Mechanisms in Ferritic Stainless Steels. *Corros. Sci.* **2024**, *235*, 112194. [CrossRef]
6. Suresh, S.; Palogi, C.; Bera, S.; Srinivasan, R. Electrochemical Behavior of Nickel Containing Passive Oxide Films on Carbon Steel in Alkaline Medium. *Thin Solid Films* **2021**, *721*, 138550. [CrossRef]
7. Gui, Y.; Zheng, Z.J.; Gao, Y. The Bi-Layer Structure and the Higher Compactness of a Passive Film on Nanocrystalline 304 Stainless Steel. *Thin Solid Films* **2016**, *599*, 64–71. [CrossRef]
8. Heuer, A.H.; Kahn, H.; Ernst, F.; Michal, G.M.; Hovis, D.B.; Rayne, R.J.; Martin, F.J.; Natishan, P.M. Enhanced Corrosion Resistance of Interstitially Hardened Stainless Steel: Implications of a Critical Passive Layer Thickness for Breakdown. *Acta Mater.* **2012**, *60*, 716–725. [CrossRef]
9. Padhy, N.; Paul, R.; Kamachi Mudali, U.; Raj, B. Morphological and Compositional Analysis of Passive Film on Austenitic Stainless Steel in Nitric Acid Medium. *Appl. Surf. Sci.* **2011**, *257*, 5088–5097. [CrossRef]
10. Maurice, V.; Yang, W.P.; Marcus, P. XPS and STM Study of Passive Films Formed on Fe-22Cr(110) Single-Crystal Surfaces. *J. Electrochem. Soc.* **1996**, *143*, 1182–1200. [CrossRef]
11. Calinski, C.; Strehblow, H.-H. ISS Depth Profiles of the Passive Layer on Fe/Cr Alloys. *J. Electrochem. Soc.* **1989**, *136*, 1328–1331. [CrossRef]
12. Wijesinghe, T.L.S.L.; Blackwood, D.J. Characterisation of Passive Films on 300 Series Stainless Steels. *Appl. Surf. Sci.* **2006**, *253*, 1006–1009. [CrossRef]
13. Jung, R.H.; Tsuchiya, H.; Fujimoto, S. XPS Characterization of Passive Films Formed on Type 304 Stainless Steel in Humid Atmosphere. *Corros. Sci.* **2012**, *58*, 62–68. [CrossRef]
14. Melentiev, R.; Yudhanto, A.; Tao, R.; Vuchkov, T.; Lubineau, G. Metallization of Polymers and Composites: State-of-the-Art Approaches. *Mater. Des.* **2022**, *221*, 110958. [CrossRef]
15. Rahimi, E.; Kosari, A.; Hosseinpour, S.; Davoodi, A.; Zandbergen, H.; Mol, J.M.C. Characterization of the Passive Layer on Ferrite and Austenite Phases of Super Duplex Stainless Steel. *Appl. Surf. Sci.* **2019**, *496*, 143634. [CrossRef]
16. Tcharkhtchi-Gillard, E.; Benoit, M.; Clavier, P.; Gwinner, B.; Miserque, F.; Vivier, V. Kinetics of the Oxidation of Stainless Steel in Hot and Concentrated Nitric Acid in the Passive and Transpassive Domains. *Corros. Sci.* **2016**, *107*, 182–192. [CrossRef]
17. Kim, D.H.; Kim, S.S.; Lee, H.H.; Jang, H.W.; Kim, J.W.; Tang, M.; Liang, K.S.; Sinha, S.K.; Noh, D.Y. Oxidation Kinetics in Iron and Stainless Steel: An in Situ X-Ray Reflectivity Study. *J. Phys. Chem. B* **2004**, *108*, 20213–20218. [CrossRef]
18. Jaffré, K.; Abe, H.; Ter-Ovanessian, B.; Mary, N.; Normand, B.; Watanabe, Y. Influence of Mechanical Surface Treatments on Oxide Properties Formed on 304L Stainless Steel in Simulated BWR and PWR Primary Water. *J. Nucl. Mater.* **2021**, *556*, 153258. [CrossRef]
19. Wang, L.; Seyeux, A.; Marcus, P. Thermal Stability of the Passive Film Formed on 316L Stainless Steel Surface Studied by ToF-SIMS. *Corros. Sci.* **2020**, *165*, 108395. [CrossRef]
20. Peng, S.; Xu, J.; Hu, D.; Xie, Z.H.; Munroe, P. The Impact of Surface Scratches on the Corrosion Behavior of Nanocrystalline High Entropy Alloy Coatings: Electrochemical Experiments and First-Principles Study. *Appl. Mater. Today* **2023**, *31*, 101767. [CrossRef]
21. Liu, M.; Zhu, J.N.; Popovich, V.A.; Borisov, E.; Mol, J.M.C.; Gonzalez-Garcia, Y. Passive Film Formation and Corrosion Resistance of Laser-Powder Bed Fusion Fabricated NiTi Shape Memory Alloys. *J. Mater. Res. Technol.* **2023**, *23*, 2991–3006. [CrossRef]
22. Khatak, H.S. Applications of Fracture Mechanics in Stress Corrosion Cracking and Introduction to Life Prediction Approaches. *Corros. Austenitic Stainl. Steels* **2002**, 190–217. [CrossRef]
23. Lochyński, P.; Domańska, M.; Dziedzic, R.; Hamal, K. Stainless Steel in Municipal Sewage—How to Recognize Favorable Corrosion Conditions. *Materials* **2023**, *16*, 6637. [CrossRef]
24. Chen, M.Z.; Wang, Z.D.; Wu, E.K.; Yang, K.; Zhao, K.; Shi, J.J.; Sun, G.F.; Han, E.H. Electrochemical Passivation Behavior and Surface Chemistry of 316 L Stainless Steel Coatings on NV E690 Steel Fabricated by Underwater Laser Direct Metal Deposition. *Corros. Sci.* **2024**, *229*, 111882. [CrossRef]
25. Łyczkowska-Widłak, E.; Lochyński, P.; Nawrat, G. Electrochemical Polishing of Austenitic Stainless Steels. *Materials* **2020**, *13*, 2557. [CrossRef] [PubMed]
26. Lochyński, P.; Charazińska, S.; Karczewski, M.; Łyczkowska-Widłak, E. A Multi-Factorial Mathematical Model for the Selection of Electropolishing Parameters with a View to Reducing the Environmental Impact. *Sci. Rep.* **2021**, *11*, 9443. [CrossRef]
27. Lochyński, P.; Charazińska, S.; Łyczkowska-Widłak, E.; Sikora, A. Electropolishing of Stainless Steel in Laboratory and Industrial Scale. *Metals* **2019**, *9*, 854. [CrossRef]

28. Yi, R.; Ji, J.; Zhan, Z.; Deng, H. Mechanism Study of Electropolishing from the Perspective of Etching Isotropy. *J. Mater. Process. Technol.* **2022**, *305*, 117599. [CrossRef]
29. Gadalińska, E.; Wronicz, W. Electropolishing Procedure Dedicated to In-Depth Stress Measurements with x-Ray Diffractometry. *Fatigue Aircr. Struct.* **2016**, *2016*, 65–72. [CrossRef]
30. Leimbach, M.; Kurniawan, M.; Zimmermann, C.P.; Höß, C.E.; Fritz, M.; Bund, A. Characterization of Protective Coatings on Stainless Steel Bipolar Plates for PEM Electrolyzers. *ECS Meet. Abstr.* **2022**, *MA2022-02*, 935. [CrossRef]
31. Alparone, G.P.; Penney, D.; Sullivan, J.; Edy, J.; Mills, C. Al_2O_3 Coatings for Protection of Stainless Steel 316L against Corrosion in Zn-Al and Zn-Al-Mg. *Coatings* **2024**, *14*, 606. [CrossRef]
32. Vatani, R.; Zamani-Meymian, M.R.; Ghaffarinejad, A.; Khalilzadeh, S. Corrosion Protection of CrCu Alloy Coating on Stainless Steel. *Surf. Coatings Technol.* **2023**, *474*, 130106. [CrossRef]
33. Krawiec, H.; Vignal, V.; Krystianiak, A.; Gaillard, Y.; Zimowski, S. Mechanical Properties and Corrosion Behaviour after Scratch and Tribological Tests of Electrodeposited Co-Mo/TiO_2 Nano-Composite Coatings. *Appl. Surf. Sci.* **2019**, *475*, 162–174. [CrossRef]
34. Mansfeld, F.; Breslin, C.B.; Pardo, A.; Pérez, F.J. Surface Modification of Stainless Steels: Green Technology for Corrosion Protection. *Surf. Coat. Technol.* **1997**, *90*, 224–228. [CrossRef]
35. Stoychev, S.; Kujawski, D. Methods for Crack Opening Load and Crack Tip Shielding Determination: A Review. *Fatigue Fract. Eng. Mater. Struct.* **2003**, *26*, 1053–1067. [CrossRef]
36. Pantoja, M.; Velasco, F.; Abenojar, J.; Martinez, M.A. Development of Superhydrophobic Coatings on AISI 304 Austenitic Stainless Steel with Different Surface Pretreatments. *Thin Solid Films* **2019**, *671*, 22–30. [CrossRef]
37. Moslehifard, E.; Ghaffari, T.; Mohammadian-Navid, S.; Ghafari-Nia, M.; Farmani, A.; Nasirpouri, F. Effect of Chemical Passivation on Corrosion Behavior and Ion Release of a Commercial Chromium-Cobalt Alloy. *J. Dent. Res. Dent. Clin. Dent. Prospect.* **2020**, *14*, 171–176. [CrossRef] [PubMed]
38. Wang, Y.M.; Voisin, T.; McKeown, J.T.; Ye, J.; Calta, N.P.; Li, Z.; Zeng, Z.; Zhang, Y.; Chen, W.; Roehling, T.T.; et al. Additively Manufactured Hierarchical Stainless Steels with High Strength and Ductility. *Nat. Mater.* **2018**, *17*, 63–70. [CrossRef]
39. Xia, Y.; Schaffer, G.B.; Qian, M. The Effect of a Small Addition of Nickel on the Sintering, Sintered Microstructure, and Mechanical Properties of Ti-45Al-5Nb-0.2C-0.2B Alloy. *J. Alloys Compd.* **2013**, *578*, 195–201. [CrossRef]
40. Pardo, A.; Merino, M.C.; Coy, A.E.; Viejo, F.; Arrabal, R.; Matykina, E. Effect of Mo and Mn Additions on the Corrosion Behaviour of AISI 304 and 316 Stainless Steels in H_2SO_4. *Corros. Sci.* **2008**, *50*, 780–794. [CrossRef]
41. Robin, R.; Miserque, F.; Spagnol, V. Correlation between Composition of Passive Layer and Corrosion Behavior of High Si-Containing Austenitic Stainless Steels in Nitric Acid. *J. Nucl. Mater.* **2008**, *375*, 65–71. [CrossRef]
42. Detriche, S.; Vivegnis, S.; Vanhumbeeck, J.F.; Felten, A.; Louette, P.; Renner, F.U.; Delhalle, J.; Mekhalif, Z. XPS Fast Depth Profile of the Native Oxide Layers on AISI 304, 316 and 430 Commercial Stainless Steels and Their Evolution with Time. *J. Electron. Spectros. Relat. Phenom.* **2020**, *243*, 146970. [CrossRef]
43. Långberg, M.; Örnek, C.; Evertsson, J.; Harlow, G.S.; Linpé, W.; Rullik, L.; Carlà, F.; Felici, R.; Bettini, E.; Kivisäkk, U.; et al. Redefining Passivity Breakdown of Super Duplex Stainless Steel by Electrochemical Operando Synchrotron near Surface X-Ray Analyses. *NPJ Mater. Degrad.* **2019**, *3*, 22. [CrossRef]
44. Yao, J.; Qi, Z.; Dong, C. Real-Time Evolution and Characterization of Passive Films on Individual Ferrite and Austenite Phases of Duplex Stainless Steel. *Electrochem. Commun.* **2022**, *137*, 107265. [CrossRef]
45. Zijlstra, G.; Faber, E.T.; Ocelík, V.; De Hosson, J.T.M. The Growth of a Passive Film on Steel Studied with In-Situ AFM. *WIT Trans. Eng. Sci.* **2017**, *116*, 23–29. [CrossRef]
46. Souier, T.; Martin, F.; Bataillon, C.; Cousty, J. Local Electrical Characteristics of Passive Films Formed on Stainless Steel Surfaces by Current Sensing Atomic Force Microscopy. *Appl. Surf. Sci.* **2010**, *256*, 2434–2439. [CrossRef]
47. Kamachi Mudali, U.; Katada, Y. Electrochemical Atomic Force Microscopic Studies on Passive Films of Nitrogen-Bearing Austenitic Stainless Steels. *Electrochim. Acta* **2001**, *46*, 3735–3742. [CrossRef]
48. Chen, Y.; Tang, P.; Qiao, S.; Liu, S.; Yang, X.; Li, A. Dynamic Breakdown of Passive Films on Stainless Steel during in Situ Thermal Oxidation. *Corros. Sci.* **2022**, *209*, 110799. [CrossRef]
49. Sikora, A. Improvement of the Scanning Area Positioning Repeatability Using Nanomarkers Developed with a Nanoscratching Method. *Meas. Sci. Technol.* **2014**, *25*, 055401. [CrossRef]
50. Sikora, A. High Accuracy and Sensitivity Method of the Observation of the Surface's Morphology Changes by Means of Atomic Force Microscopy with Cyclic, Precise Sample Positioning. *Nanosci. Nanometrology* **2017**, *3*, 6. [CrossRef]
51. Krawczyk, M.; Lisowski, W.; Pisarek, M.; Nikiforow, K. Applied Surface Science Surface Characterization of Low-Temperature Grown Yttrium Oxide. *Appl. Surf. Sci.* **2018**, *437*, 347–356. [CrossRef]
52. Krawczyk, M.; Pisarek, M.; Lisowski, W.; Jablonski, A. Applied Surface Science Surface Studies of Praseodymium by Electron Spectroscopies. *Appl. Surf. Sci.* **2016**, *388*, 691–695. [CrossRef]
53. Zhang, B.; Hao, S.; Wu, J.; Li, X.; Li, C.; Di, X.; Huang, Y. Direct Evidence of Passive Film Growth on 316 Stainless Steel in Alkaline Solution. *Mater. Charact.* **2017**, *131*, 168–174. [CrossRef]
54. Tong, S.; Che, H.L.; Lei, M.K. High-Resolution TEM Characterization of Epitaxial Passivation for a High Nitrogen Face-Centered-Cubic Phase Formed on AISI 304L Austenitic Stainless Steel in Borate Buffer Solution. *Electrochim. Acta* **2021**, *393*, 139075. [CrossRef]

55. Lv, Y.; Luo, H.; Tang, J.; Guo, J.; Pi, J.; Ye, K. Corrosion Properties of Phase Reversion Induced Nano/Ultrafine Grained AISI 304 Metastable Austenite Stainless Steel. *Mater. Res. Bull.* **2018**, *107*, 421–429. [CrossRef]
56. Boissy, C.; Alemany-Dumont, C.; Normand, B. EIS Evaluation of Steady-State Characteristic of 316L Stainless Steel Passive Film Grown in Acidic Solution. *Electrochem. Commun.* **2013**, *26*, 10–12. [CrossRef]
57. Jaffré, K.; Ter-Ovanessian, B.; Abe, H.; Mary, N.; Normand, B.; Watanabe, Y. Effect of Mechanical Surface Treatments on the Surface State and Passive Behavior of 304L Stainless Steel. *Metals* **2021**, *11*, 135. [CrossRef]
58. Lai, W.Y.; Zhao, W.Z.; Yin, Z.F.; Zhang, J. Electrochemical and XPS Studies on Corrosion Behaviours of AISI 304 and AISI 316 Stainless Steels under Plastic Deformation in Sulphuric Acid Solution. *Surf. Interface Anal.* **2011**, *44*, 505–512. [CrossRef]
59. Keller, P.; Strehblow, H. XPS Investigations of Electrochemically Formed Passive Layers on Fe/Cr-Alloys. *Corros. Sci.* **2004**, *46*, 1939–1952. [CrossRef]
60. Freire, L.; Carmezim, M.J.; Ferreira, M.G.S.; Montemor, M.F. The Passive Behaviour of AISI 316 in Alkaline Media and the Effect of PH: A Combined Electrochemical and Analytical Study. *Electrochim. Acta* **2010**, *55*, 6174–6181. [CrossRef]
61. Yi, G.M.; Cheng, Y.H.; Ren, L.; Yang, J.; Wang, Q.; Li, X. Corrosion Behavior of 316 Stainless Steel for Advanced High-Temperature Water-Cooled Nuclear Plant. *Surf. Interface Anal.* **2022**, *54*, 767–774. [CrossRef]
62. Grosvenor, A.P.; Kobe, B.A.; Biesinger, M.C.; McIntyre, N.S. Investigation of Multiplet Splitting of Fe 2p XPS Spectra and Bonding in Iron Compounds. *Surf. Interface Anal.* **2004**, *36*, 1564–1574. [CrossRef]
63. Galtayries, A.; Warocquier-Clerout, R.; Nagel, M.-D.; Marcus, P. Fibronectin Adsorption on Fe–Cr Alloy Studied by XPS. *Surf. Interface Anal.* **2006**, *4*, 186–190. [CrossRef]
64. Cano, E.; Martı, L.; Simancas, J.; Pe, F.J. Influence of N, Ar and Si Ion Implantation on the Passive Layer and Corrosion Behaviour of AISI 304 and 430 Stainless Steels. *Surf. Coat. Technol.* **2006**, *200*, 5123–5131. [CrossRef]
65. Bortolan, C.C.; Copes, F.; Shekargoftar, M.; de Oliveira Fidelis Sales, V.; Paternoster, C.; Contri Campanelli, L.; Giguère, N.; Mantovani, D. Electrochemical and in Vitro Biological Behaviors of a Ti-Mo-Fe Alloy Specifically Designed for Stent Applications. *Biomater. Biosyst.* **2023**, *10*, 100076. [CrossRef] [PubMed]
66. Wang, Q.Y.; Wang, X.Z.; Luo, H.; Luo, J.L. A Study on Corrosion Behaviors of Ni-Cr-Mo Laser Coating, 316 Stainless Steel and X70 Steel in Simulated Solutions with H_2S and CO_2. *Surf. Coat. Technol.* **2016**, *291*, 250–257. [CrossRef]

materials

MDPI

Article

On the Possibility of the Deformation of Mg and Alloys Without Preheating of Initial Billets: Understanding Their Corrosion Performance via Electrochemical Tests

Anna Dobkowska [1,*] and Jiri Kubasek [2,*]

[1] Faculty of Materials Science and Engineering, Warsaw University of Technology, Woloska, 02-507 Warsaw, Poland

[2] Department of Metals and Corrosion Engineering, University of Chemistry and Technology, Technicka 5, 168 28 Prague 6, Czech Republic

* Correspondence: anna.dobkowska@pw.edu.pl (A.D.); jiri.kubasek@vscht.cz (J.K.)

Abstract: Due to limited slip systems activated at room temperature, the plastic deformation of Mg and its alloys without any preheating of initial billets is significantly limited. To overcome those issues, new methods of severe plastic deformation are discovered and developed. One such example is extrusion with an oscillating die, called KoBo. This method, due to the oscillations of reversible die located at the end of extruded, introduces material into the plastic flow, and thus, enables deformation without preheating of the initial billets of metals that are hard to deform. Such solution is important from an industrial point of view and may lead to serious savings and reduction in carbon dioxide emission to the atmosphere. Therefore, this paper focuses on the possibility of KoBo extrusion of hcp-structured Mg alloys with different chemical compositions and includes comparison of their corrosion resistance using short-term electrochemical tests. In order to have a broad view of the problem presented, we compared the electrochemical behavior of the following groups of Mg materials: pure Mg, Mg-Al-Zn, Mg-Li, and Mg-Y-RE. It was stated that the KoBo method performed at room temperature improves the corrosion resistance of pure Mg when compared to the initial billet and the alloys which belong to the Mg-Al-Zn, Mg-Li, and Mg-Y-RE series. The presented study shows that different corrosion trends are observed for traditionally deformed alloys, and they significantly vary from nascent developments, such as KoBo extrusion. Therefore, it is crucial to widely study those methods because it may be a path leading to long-lasting solution to the formability limitations of Mg-based metallic systems.

Keywords: magnesium alloys; plastic deformation; microstructure; corrosion

Citation: Dobkowska, A.; Kubasek, J. On the Possibility of the Deformation of Mg and Alloys Without Preheating of Initial Billets: Understanding Their Corrosion Performance via Electrochemical Tests. *Materials* **2024**, *17*, 6182. https://doi.org/10.3390/ma17246182

Academic Editor: Xiaogang Li

Received: 12 November 2024
Revised: 3 December 2024
Accepted: 16 December 2024
Published: 18 December 2024

1. Introduction

The most appropriate way to describe the role of grain size in the corrosion behavior of materials after deformation is to analyze pure metals to avoid the effect of second phases and their anodic or cathodic contribution to corrosion mechanisms. Pure Mg has a standard potential of −2.373 V vs. SHE, while its free corrosion potential is more positive (toward −1.6 V), indicating that a surface layer is present on the pure Mg, limiting its contact with the solution [1,2]. This corrosion layer is not fully protective, but limits corrosion reactions on the substrate of Mg metal to some extent. To date, many researchers focused on finding clear relationship between grain size and corrosion behavior for Mg alloys [3–6]. As is commonly known, grain size is a predominant factor affecting corrosion resistance, and it might be expected that a universal law describing it would exist. Unfortunately, this is impossible for plastically deformed polycrystalline materials because grain refinement itself is also related to the dislocation density and distribution, as well as the internal strain induced during deformation [7]. Another important factor is the tremendous role of an intermetallic phase, β-$Mg_{17}Al_{12}$, in mechanical and corrosion properties, which has already been described [8–10]. When AZ61 (6 wt.% Al, 1 wt.% Zn) was subjected to ECAP, precipitation strengthening was observed [11]. The large proportion of β-$Mg_{17}Al_{12}$ hindered dislocation movement, leading to an increased dislocation density, thus causing work hardening. Nevertheless, such a large amount of this type of intermetallic phase may promote corrosion reactions by forming microgalvanic reactions with the Mg matrix [12]. From the opposite point of view, β-$Mg_{17}Al_{12}$ may form a barrier against spreading corrosion reactions [13–15]. Therefore, the dual role of β-$Mg_{17}Al_{12}$ in the corrosion resistance of AZ-type alloys needs to be further examined. When it comes to dual-phase structured Mg-Li alloys, two crucial factors dominating their corrosion performance have been recently described. Zeng et al. [16] suggested that Li leads to the formation of a thin film, which may play a protective role on the surface. It was also proposed that corrosion of the dual-phase structured AZ31+7.5Li traditionally extruded at 350 °C is a result of microgalvanic interaction between the α(Mg) and β(Li) phases and their relative distribution [17]. The more equal the fractions of α(Mg) and β(Li), the less intense were the resulting corrosion reactions. Besides the mentioned factors, texture intensity also affects the corrosion behavior especially in alloys containing Y and RE as alloying elements.

All of the above mentioned factors are typical for Mg alloys plastically deformed at high temperature. Lastly, our research groups proved that Mg and its alloys may be formed without preheating of the initial billet which is very unusual for hcp-structured metals [18]. The method used for the plastic deformation of Mg alloys at room temperature is a modified extrusion with an oscillating die located at the end of extruded (as described in [19], called KoBo. Since Mg and its alloys are characterized by poor corrosion characteristics, which limit their widespread use in industrial applications, it is of crucial importance to carry out a systematic study about the factors affecting their corrosion after extrusion using nascent developments, such as KoBo, which may lead to many economic benefits in industrial applications (elimination of furnaces from production routes, reduction in gas emission, faster deformation processes, etc.). Therefore, the main aim of this paper is to select and describe the predominant factors influencing various groups of KoBo-deformed Mg alloys and to specify the most corrosion-resistant group of Mg alloys after KoBo extrusion based on fast electrochemical corrosion tests.

2. Materials and Methods

The electrochemical behavior indicating corrosion resistance of various Mg alloys processed using extrusion with an oscillating die (KoBo) without preheating of the initial billet is investigated and compared. The materials for this research were selected based on our previous studies, where the corrosion resistance of pure Mg [20], AZ31 [21], AZ61 [22], AZ31+4Li [23], AZ31+7.5Li [24], and WE43 [25] extruded using KoBo at various extrusion ratios has been described. For this research, the alloys with the highest corrosion resistance analyzed in the previous studies have been chosen, and they are specified in Table 1.

Table 1. Selection of Mg materials in order to make a comparative analysis of their corrosion resistance.

Materials	Pure Mg	AZ31	AZ61	AZ31+4Li	AZ31+7.5Li	WE43
Extrusion ratio	7:1	10:1	7:1	10:1	10:1	5:1

The electrochemical measurements were performed in naturally aerated 0.01 M NaCl solution using a Gamry FAS1 600+ potentiostat equipped with three electrodes: platinum as the counter electrode, Ag/AgCl as the reference electrode, and the measured sample as the working electrode. The electrolyte was made up using analytical-grade reagents and distilled water. The samples were immersed for 1 h, and electrochemical impedance spectroscopy was performed over a frequency range of 0.01 Hz–10,000 Hz. Immediately afterward, potentiodynamic polarization tests were carried out over a range from 0.2 V below E_{OCP} to 1.0 V vs. Ref (a scan rate of 5 mV/s was used). At least three tests were performed for each alloy. The EIS and polarization curves were fitted using the Gamry Echem software version 5.58. Corrosion damage was characterized after 1 h of immersion in 0.1 M NaCl. Before electrochemical tests, samples were polished using water-free diamond suspensions (3 and 1 μm). The morphology of corrosion damage was observed, after chemical removal of the corrosion products in aqueous CrO_3 solution (4 min, 180 g/L CrO_3), using a scanning electron microscope (SEM, Hitachi SU8000, Japan). Additionally, the corrosion rate of the materials was calculated using the hydrogen release method [26–29]. To ensure the reproducibility of the results, at least three specimens from each composition were tested.

3. Results

The evaluation of the open circuit potential (E_{OCP}) of the investigated materials during 1 h of immersion in 0.1 M NaCl is presented in Figure 1. The highest values of the E_{OCP} were observed for AZ31 and AZ61, and their average values were −1.45 V/Ref and −1.46 V/Ref, respectively. The 4 and 7.5 wt.% of Li addition shifted the E_{OCP} toward negative values: −1.53 V/Ref and −1.54 V/Ref, respectively. The lowest values of E_{OCP} were observed for pure Mg and WE43: −1.56 V/Ref and −1.65 V/Ref, respectively. Inspection of the anodic branch of the polarization curves in Figure 2 determined the active dissolution of the AZ61, while the rest of the alloys underwent localized corrosion. A characteristic plateau was observed on the anodic branch of AZ31, WE43, pure Mg, AZ31+4Li, and AZ31+7.5Li, and a typical inflection point indicating the breakdown potential, E_b, was also noted (labeled by the red arrows in Figure 2). The difference between E_b and E_{corr} describes the resistivity to pitting corrosion, ΔE [30,31]; the broader the difference, the higher the resistance to pitting. The numerical values of the resistivity to pitting are shown in Table 2. Undoubtedly, the highest pitting susceptibilities were in the cases of pure Mg and the alloys with Li addition. The most stable corrosion layers were formed on the AZ31 and WE43.

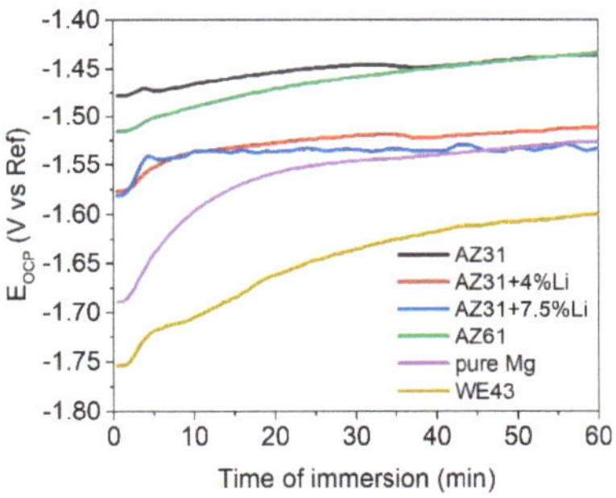

Figure 1. Open circuit potential (E_{OCP}) evolution during 1 h of immersion in naturally aerated 0.1 M NaCl.

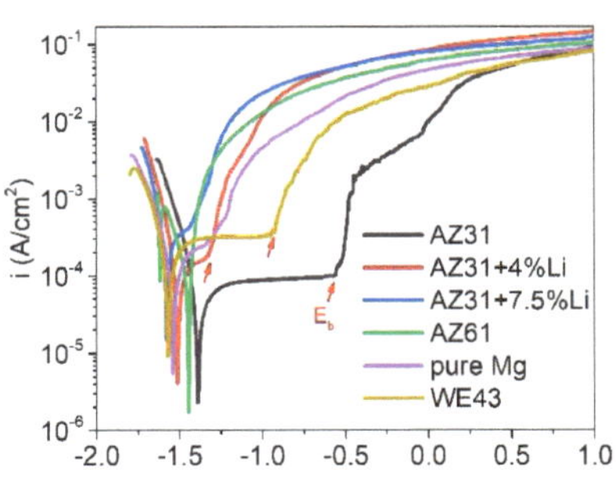

Figure 2. Potentiodynamic polarization curves recorded after 1 h of immersion in naturally aerated 0.1 M NaCl.

Table 2. The characteristic parameters calculated based on the extrapolation of linear polarization (E_{corr}—corrosion potential, E_b—breakdown potential).

Parameters	Pure Mg	AZ31	AZ61	AZ31+4Li	AZ31+7.5Li	WE43
E_{corr} (V vs. Ref)	−1.54	−1.39	−1.44	−1.51	−1.58	−1.57
E_b (V vs. Ref)	−1.34	−0.56	-	−1.33	−1.45	−0.94
ΔE	0.20	0.83	-	0.18	0.13	0.63

The Nyquist plots recorded for the investigated materials after 1 h of immersion in naturally aerated 0.1 M NaCl are shown in Figure 3a, while corresponding Bode plots are depicted in Figure 3b. All samples exhibited two semicircles, corresponding to two time constants, with the exception of AZ31+7.5Li, where the small radius of the Nyquist plot and intense inductive loop may indicate strong localized corrosion and/or precipitation of corrosion products on the surface. The Nyquist plots for pure Mg, AZ31, AZ61, AZ31+4Li, and WE43 were composed of two semicircles registered at a high and medium frequency range. The first one was related to the charge transfer resistance at the double layer interface, while the second one corresponded to the corrosion products' layer [32,33]. The equivalent electrical circuits given in Figure 3c,d were used to fit data for these samples. Here, R_S is the solution resistance; R_{ct} is the charge transfer resistance, and the corresponding capacitance is defined by the constant phase element (CPE_{ct}). The R_f and CPE_f variables stand for the resistance of the corrosion products' layer. In the given equivalent electrical circuits, CPE is used to compensate for the nonuniformity of the corrosion system [34]. The results from the fitting are depicted in Table 2. The equivalent electrical circuits given in Figure 3d were used to fit the data recorded for AZ31+7.5Li. They were composed of a capacitive loop described by R_{ct} and CPE_{ct} and an inductive loop defined by R_L and L, with the latter loop being most probably related to the initial stage of localized corrosion [35,36], although it can also be related to the precipitation of corrosion products. The behavior of the analyzed samples, with the exception of Mg-7.5Li, was very similar to each other in the analyzed medium and exhibited the presence of a protective film. The complex nature of the fitted data was supported by the corrosion rate calculated based on hydrogen release tests, Figure 4. Both the EIS results and estimated corrosion rates showed a similar tendency with respect to the corrosion behavior of the investigated materials. The least resistant was dual-structured AZ31+7.5Li, where strong dissolution of the alloy was observed. Its corrosion rate was found to be 23.2 ± 1.2 mm/year. The rest of the materials were less active in the analyzed solution. The most resistant were AZ31 and AZ31+4Li, with corrosion rates of 11.0 ± 0.9 mm/year and 11.0 ± 2.1 mm/year, respectively. The higher standard deviation out of these two samples was observed for the alloy containing Li. These alloys exhibited the highest values of R_{CT} and R_f, suggesting that the most stable corrosion product layers are formed on their surfaces (Table 3). Less resistant than AZ31 and AZ31+4Li were AZ61, WE43, and pure Mg, and their corrosion rates were comparable to each other: 16.0 ± 2.2, 14.4 ± 1.2, and 16.0 ± 1.1 mm/year, respectively.

Table 3. Parameters fitted from EIS results using data shown in Figure 3.

Materials	R_s (Ω·cm²)	R_{ct} (Ω·cm²)	CPE_{ct} (μSsᵃ/cm²)	a_1	R_L (Ω·cm²)	L (H·cm²)	R_f (Ω·cm²)	CPE_f (μSsᵃ/cm²)	a_2
Pure Mg	61	1242	0.000018	0.94	N/A	N/A	634	0.000840	0.84
AZ31	22	2266	0.000008	0.85	N/A	N/A	1420	0.000120	0.86
AZ61	10	1454	0.000019	0.94	N/A	N/A	1016	0.001688	072
AZ31+4Li	17	2357	0.000009	0.93	N/A	N/A	1603	0.000155	0.71
AZ31+7.5Li	17	183	0.000025	0.88	691	182	N/A	N/A	N/A
WE43	14	1201	0.000027	0.92	N/A	N/A	1294	0.001756	0.76

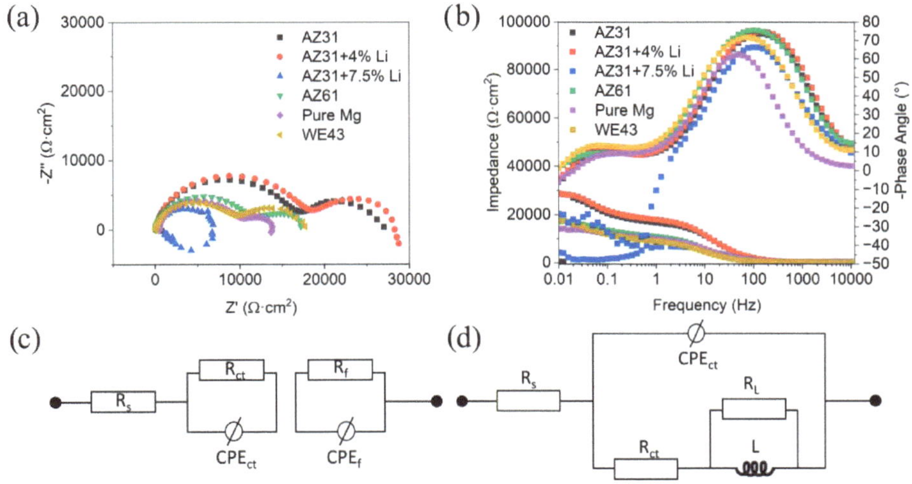

Figure 3. Electrochemical impedance spectroscopy (EIS) results recorded after 1 h of immersion in naturally aerated 0.1 M NaCl and presented in the form of (**a**) Nyquist plots, (**b**) Bode plots, (**c**) equivalent electrical circuit used for pure Mg, AZ31, AZ61, AZ31+4Li, and WE43 fitting, and (**d**) equivalent electrical circuit used for AZ31+7.5Li fitting.

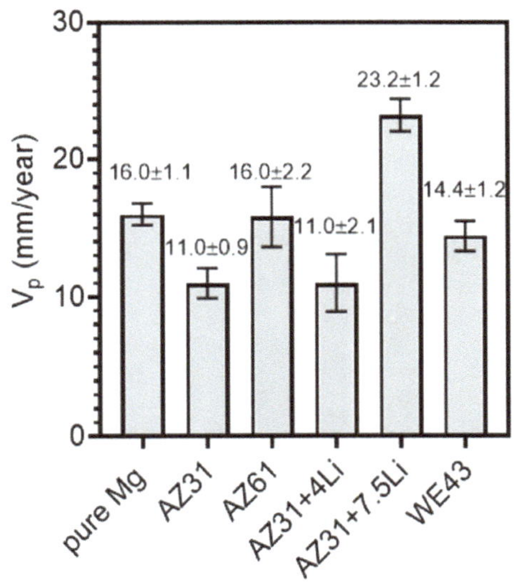

Figure 4. Corrosion rate calculated based on the hydrogen release method.

Figure 5a–f show the surface morphologies of the corroded samples after removing the corrosion products. Many small and round pits were observed on the surface of pure Mg (Figure 5a); in random locations, they propagated into the depth of the material (inset in Figure 5a). In the case of AZ31 (Figure 5b) and AZ61 (Figure 5c), besides small round pits, larger and deeper pitting holes interlinked with one another on the surface were observed. The pits propagated on the surface of AZ31 are as deep as those formed on pure Mg (inset in Figure 5b). As shown in the inset in Figure 5c, corrosion damage in triple points was formed on the AZ61. The surface of the AZ31 with Li was more damaged; however, the corrosion damage differed depending on the amount of Li present. On the surface of the single-α(Mg)-phase AZ31+4Li, corrosion in the form of threads proceeding along the grain boundaries was observed (Figure 5d). The dual-phase structured alloy exhibited a different mode of degradation related to microgalvanic corrosion between the α(Mg) and β(Li) phases (Figure 5e). This phenomenon is explained in detail in our previous study [24]. In the case of WE43, the corrosion mechanism propagated along the grain boundaries (Figure 5f). The corroded grain boundaries are depicted at higher magnification in the inset in Figure 5f.

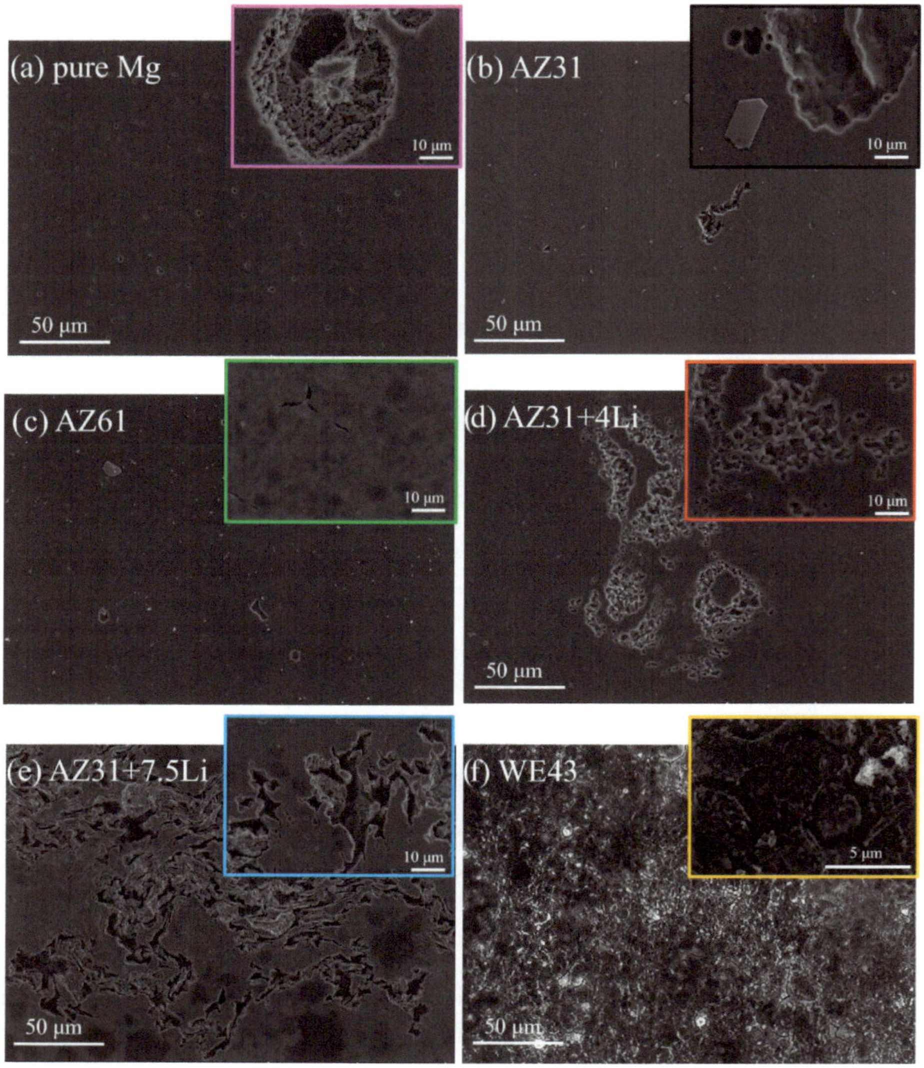

Figure 5. Corrosion damage characterization after chemical removal of corrosion products after 1 h of immersion in 0.1 M NaCl: (**a**) pure Mg, (**b**) AZ31, (**c**) AZ61, (**d**) AZ31+4Li, (**e**) AZ31+7.5Li, and (**f**) WE43.

4. Discussion

The extrusion with an oscillating die (KoBo) method exhibits great potential for Mg alloy deformation without preheating of the initial billet. It may significantly reduce the grain size and redistribute secondary phases in the microstructure. It is noteworthy that the grain refinement does not depend on the extrusion ratio, but on the intensity of dynamic recrystallization (DRX) triggered by the oscillations of the reversible die. As a result of this phenomenon, the corrosion resistance of Mg alloys changes in a nonlinear manner. Of course, the corrosion resistance of KoBo-extruded alloys is dependent on the alloy chemistry, but based on the data published so far, the predominant role seems to be played by grain size and dislocation distribution.

Based on the presented results of electrochemical behavior and hydrogen release, the corrosion resistance of the analyzed materials may be ordered as follows: AZ31 > AZ31+4Li >> WE43 > pure Mg > AZ61 >>> AZ31+7.5Li. The order of the corrosion resistance of each material were analyzed based on the detailed description of the microstructure presented in our previous works, and the results are shortly summarized in Table 4.

Table 4. Summary of the basic microstructural features characteristic for the investigated materials.

Material	Phase Structure	Grain Size (μm)	Second Phases	Grain Orientation	Ref.
Pure Mg	α(Mg) phase	3.88 μm; uniformly distributed	N/A	$(10\bar{1}0)$ and (2110)	[21]
AZ31	α(Mg) phase	4.38 μm; uniformly distributed	Coarse Al_5Mn_8	$(10\bar{1}0)$ and (2110) with some randomly present (0001)	[21]
AZ61	α(Mg) phase	6.6 μm uniformly distributed	Coarse Al_5Mn_8 and nano β-$Mg_{17}Al_{12}$ located at grain boundaries	$(10\bar{1}0)$ and (2110)	[22]
AZ31+4Li	α(Mg) phase	3.24 μm; uniformly distributed	Coarse AlLi and β-$Mg_{17}Al_{12}$ (nano) distributed in the grain interiors	$(10\bar{1}0)$ and (2110)	[23]
AZ31+7.5Li	α(Mg)+β(Li)	No EBSD data, based on SEM a few μm	Coarse AlLi and $MgLi_2Al$ with size around 1 μm, β-$Mg_{17}Al_{12}$ (nano) distributed in the grain interiors	No EBSD data	[24]
WE43	α(Mg) phase	3.3 μm; uniformly distributed	$Mg_{41}Nd_5$, $Mg_{24}Y_5$, $Mg_{12}Nd$, and nano ternary Mg-Nd-Y	Random	[25]

As per results of this work, lower corrosion rate estimated via electrochemical measurements was observed for α(Mg) alloys than for dual-structured α(Mg)+β(Li) alloys. The most resistant in the analyzed medium was AZ31. As shown in [21], it has a grain size of 4.38 μm and a preferential orientation of the grain toward $(10\bar{1}0)$ and (2110). The Al fully dissolved in the solution, and no β-$Mg_{17}Al_{12}$ was formed. Similar level of corrosion resistance was observed for the AZ31 alloyed with 4 wt.% of Li. In this case, nanosized β-$Mg_{17}Al_{12}$ distributed in the grain interiors was observed, creating a barrier against spreading corrosion reactions caused by the Li in the alloy [23]. Lower corrosion resistance and also corrosion products being formed on the surface in some kind of "protective" layer, was characteristic for WE43, pure Mg, and AZ61. The presence of nanosized β-$Mg_{17}Al_{12}$ at the grain boundaries promotes microgalvanic corrosion in AZ61, thus lowering its corrosion resistance when compared to AZ31 [22]. Surprisingly, pure Mg had a higher corrosion rate than AZ31. Since the grain sizes for pure Mg and AZ31 were similar, the factor favoring higher corrosion resistance lay in the crystallographic orientation of the grains formed in AZ31. In this case, besides grains oriented to $(10\bar{1}0)$ and (2110), grains with a (0001) orientation were present. In this relation, the grains with an (0001) orientation parallel to the corroded surface are more corrosion resistant than grains with orientations $(10\bar{1}0)$ and (2110) [37].

During KoBo, materials enters to the state of plastic flow, therefore the reversibly oscillating of the die at the end of the extruded enables deformation of Mg alloys. Simultaneously, the presence of the complex strain conditions formed as a result of the die oscillations resulted in the accumulation of high density of defects in the microstructure of the extruded material. This in turn led to the pile up of dislocations in the refined grains. As stated in [21], in the KoBo extruded AZ31 alloy, the main factors affecting corrosion resistance were dislocation density and its distribution. Consequently, despite the KoBo extrusion reducing grain size from a coarse to a micrometer scale, it did not improve the corrosion resistance of the alloy. A similar situation was observed for the AZ61 [22,38]; however, strong microgalvanic reactions between β-$Mg_{17}Al_{12}$ and the Mg matrix occurred, overwhelming the role of other microstructural components in its corrosion resistance [39]. WE43 has a completely different chemical composition than that of the other Mg alloys, and here, the predominant role in the corrosion mechanism may be caused by the presence of tiny precipitates located in this alloy [25], which has a corrosion resistance similar to AZ61 and pure Mg. Nevertheless, the role of small second phases must be further analyzed.

5. Conclusions

Based on the performed tests and reviewed data, the following may be concluded:

- KoBo extrusion enables the deformation of Mg and its alloys without preheating of the initial billet. The obtained materials are without cracking and reduce the diameter of the ingots significantly, resulting in the production of long elements, such as rods or wires. The next step of the KoBo improvement should be research on the possibility of thin-walled structure fabrication;
- KoBo extrusion improves the corrosion resistance of the initial ingots of Mg and its alloys; however, there is no clear relationship between grain refinement and corrosion resistance improvement, and smaller grain size does not always lead to higher corrosion resistance. The properties of the specific alloy must be treated as a separate case, since the intensity of the dynamic recrystallization is related to the extrusion ratio and strain induced by the reversible die;
- The corrosion resistance of the KoBo-extruded alloys depends mainly on the alloy chemistry and its microstructure; however, it is also dominated by the formation of stresses during extrusion with an oscillating die. Importantly, we do not have any clear information about the specific temperature inside the extruder. The effect of the internal heating resulting from the friction of the material and the extruder during processing should be further investigated;
- In the current research on magnesium alloys, the imperative issue of balancing the structural characteristics and mechanical properties of the alloys remains to be resolved. A reduction in the deformation temperature of Mg alloys will generate savings and reduce CO_2 gas emissions, which is crucial to ensure the sustainable growth of industry.

Author Contributions: Conceptualization, methodology, supervision, formal analysis, investigation, and writing—original draft preparation, A.D.; investigations, editing of the original draft, resources, and funding acquisition, J.K. All authors have read and agreed to the published version of the manuscript.

Funding: This research was partially funded by the Czech Science Foundation (GACR), project number 22-22248S.

Institutional Review Board Statement: Not applicable.

Informed Consent Statement: Not applicable.

Data Availability Statement: The original contributions presented in this study are included in the article. Further inquiries can be directed to the corresponding authors.

Conflicts of Interest: The authors declare no conflicts of interest.

References

1. Atrens, A.; Song, G.L.; Cao, F.; Shi, Z.; Bowen, P.K. Advances in Mg Corrosion and Research Suggestions. *J. Magnes. Alloys* **2013**, *1*, 177–200. [CrossRef]
2. Atrens, A.; Song, G.L.; Liu, M.; Shi, Z.; Cao, F.; Dargusch, M.S. Review of Recent Developments in the Field of Magnesium Corrosion. *Adv. Eng. Mater.* **2015**, *17*, 400–453. [CrossRef]
3. Ralston, K.D.; Birbilis, N.; Davies, C.H.J. Revealing the Relationship between Grain Size and Corrosion Rate of Metals. *Scr. Mater.* **2010**, *63*, 1201–1204. [CrossRef]

4. Bahmani, A.; Arthanari, S.; Shin, K.S. Formulation of Corrosion Rate of Magnesium Alloys Using Microstructural Parameters. *J. Magnes. Alloys* **2020**, *8*, 134–149. [CrossRef]
5. Ben-Hamu, G.; Eliezer, D.; Shin, K.S.; Cohen, S. The Relation between Microstructure and Corrosion Behavior of Mg-Y-RE-Zr Alloys. *J. Alloys Compd.* **2007**, *431*, 269–276. [CrossRef]
6. Wang, L.; Shinohara, T.; Zhang, B.P. Influence of Chloride, Sulfate and Bicarbonate Anions on the Corrosion Behavior of AZ31 Magnesium Alloy. *J. Alloys Compd.* **2010**, *496*, 500–507. [CrossRef]
7. Miyamoto, H. Corrosion of Ultrafine Grained Materials by Severe Plastic Deformation, an Overview. *Mater. Trans.* **2016**, *57*, 559–572. [CrossRef]
8. Hoog, C.O.; Birbilis, N.; Estrin, Y. Corrosion of Pure Mg as a Function of Grain Size and Processing Route. *Adv. Eng. Mater.* **2008**, *10*, 579–582. [CrossRef]
9. Jiang, B.; Xiang, Q.; Atrens, A.; Song, J.; Pan, F. Influence of Crystallographic Texture and Grain Size on the Corrosion Behaviour of As-Extruded Mg Alloy AZ31 Sheets. *Corros. Sci.* **2017**, *126*, 374–380. [CrossRef]
10. Gollapudi, S. Grain Size Distribution Effects on the Corrosion Behaviour of Materials. *Corros. Sci.* **2012**, *62*, 90–94. [CrossRef]
11. Shan, Z.; Yang, J.; Fan, J.; Zhang, H.; Zhang, Q.; Wu, Y.; Dong, H. Microstructure Evolution and Mechanical Properties of an AZ61 Alloy Processed with TS-ECAP and EPT. *Mater. Sci. Eng. A* **2020**, *780*, 139195. [CrossRef]
12. Atrens, A.; Liu, M.; Zainal Abidin, N.I. Corrosion Mechanism Applicable to Biodegradable Magnesium Implants. *Mater. Sci. Eng. B Solid-State Mater. Adv. Technol.* **2011**, *176*, 1609–1636. [CrossRef]
13. Lunder, O.; Lein, J.E.; Aune, T.K.; Nisancioglu, K. Role of Mg17Al12phase in the Corrosion of Mg Alloy AZ91. *Corrosion* **1989**, *45*, 741–748. [CrossRef]
14. Pardo, A.; Merino, M.C.; Coy, A.E.; Viejo, F.; Arrabal, R.; Feliú, S. Influence of Microstructure and Composition on the Corrosion Behaviour of Mg/Al Alloys in Chloride Media. *Electrochim. Acta* **2008**, *53*, 7890–7902. [CrossRef]
15. Ubeda, C.; Garces, G.; Adeva, P.; Llorente, I.; Frankel, G.S.; Fajardo, S. The Role of the Beta-Mg17Al12 Phase on the Anomalous Hydrogen Evolution and Anodic Dissolution of AZ Magnesium Alloys. *Corros. Sci.* **2020**, *165*, 108384. [CrossRef]
16. Zeng, R.C.; Sun, L.; Zheng, Y.F.; Cui, H.Z.; Han, E.H. Corrosion and Characterisation of Dual Phase Mg-Li-Ca Alloy in Hank's Solution: The Influence of Microstructural Features. *Corros. Sci.* **2014**, *79*, 69–82. [CrossRef]
17. Dobkowska, A.; Adamczyk-Cieślak, B.; Kubásek, J.; Vojtěch, D.; Kuc, D.; Hadasik, E.; Mizera, J. Microstructure and Corrosion Resistance of a Duplex Structured Mg–7.5Li–3Al–1Zn. *J. Magnes. Alloys* **2021**, *9*, 467–477. [CrossRef]
18. Dutkiewicz, J.; Kalita, D.; Maziarz, W.; Tański, T.; Borek, W.; Ostachowski, P.; Faryna, M. Effect of KOBO Extrusion and Following Cyclic Forging on Grain Refinement of Mg–9Li–2Al–0.5Sc Alloy. *Met. Mater. Int. Korean Inst. Met. Mater.* **2020**, *26*, 1004–1014. [CrossRef]
19. Bochniak, W.; Korbel, A. KOBO Type Forming: Forging of Metals under Complex Conditions of the Process. *J. Mater. Process. Technol.* **2003**, *134*, 120–134. [CrossRef]
20. Dobkowska, A.; Adamczyk-cie, B.; Aydee, M.; Garcia, G. Effect of High Deformation without Preheating on Microstructure and Corrosion of Pure Mg. *Metals* **2024**, *14*, 949. [CrossRef]
21. Dobkowska, A.; Adamczyk-Cieślak, B.; Chlewicka, M.; Towarek, A.; Zielińska, A.; Koralnik, M.; Kuc, D.; Mizera, J. Evolution of Microstructure Dependent Corrosion Properties of Ultrafine AZ31 under Conditions of Extrusion with a Forward Backward Oscillating Die. *J. Mater. Res. Technol.* **2022**, *18*, 4486–4496. [CrossRef]
22. Dobkowska, A.; Zielińska, A.; Donik, Č.; Łojkowski, M.; Adamczyk-Cieślak, B. Microstructure and Properties of an AZ61 Alloy after Extrusion with a Forward-Backward Oscillating Die without Preheating of the Initial Billet. *J. Alloys Compd.* **2023**, *952*, 169843. [CrossRef]
23. Dobkowska, A.; Koralnik, M.; Adamczyk-Cieślak, B.; Kuc, D.; Chromiński, W.; Kubasek, J.; Mizera, J. The Effect of Extrusion Ratio on the Corrosion Resistance of Ultrafine-Grained Mg-4Li-3Al-Zn Alloy Deformed Using Extrusion with a Forward-Backward Oscillating Die. *J. Mater. Eng. Perform.* **2022**, *31*, 8932–8939. [CrossRef]
24. Dobkowska, A.; Adamczyk-Cieślak, B.; Koralnik, M.; Chromiński, W.; Kubasek, J.; Ciftci, J.; Kuc, D.; Mizera, J. Corrosion Behavior of Fine-Grained Mg-7.5Li-3Al-1Zn Fabricated by Extrusion with a Forward-Backward Rotating Die (KoBo). *J. Magnes. Alloys* **2021**, *10*, 811–820. [CrossRef]
25. Dobkowska, A.; Zielińska, A.; Paulin, I.; Donik, C.; Koralnik, M.; Adamczyk-Cieślak, B.; Wieczorek-Czarnocka, M.; Kuc, D.; Kubasek, J.; Mikuszewski, T.; et al. Microstructural, Corrosion and Mechanical Properties of a WE43 Alloy: Conventional Extrusion versus SPD. *J. Alloys Compd.* **2024**, *976*, 173090. [CrossRef]
26. Kirkland, N.; Birbilis, N.; Staiger, M. Assessing the Corrosion of Biodegradable Magnesium Implants: A Critical Review of Current Methodologies and Their Limitations. *Acta Biomater.* **2012**, *61*, 65–69. [CrossRef]
27. Birbilis, N.; King, A.D.; Thomas, S.; Frankel, G.S.; Scully, J.R. Evidence for Enhanced Catalytic Activity of Magnesium Arising from Anodic Dissolution. *Electrochim. Acta* **2014**, *132*, 277–283. [CrossRef]
28. Ghali, E.; Dietzel, W.; Kainer, K.-U. Testing of General and Localized Corrosion of Magnesium Alloys: A Critical Review. *J. Mater. Eng. Perform.* **2004**, *13*, 517–529. [CrossRef]
29. Jeong, Y.S.; Kim, W.J. Enhancement of Mechanical Properties and Corrosion Resistance of Mg-Ca Alloys through Microstructural Refinement by Indirect Extrusion. *Corros. Sci.* **2014**, *82*, 392–403. [CrossRef]
30. Esmailzadeh, S.; Aliofkhazraei, M.; Sarlak, H. Interpretation of Cyclic Potentiodynamic Polarization Test Results for Study of Corrosion Behavior of Metals: A Review. *Prot. Met. Phys. Chem. Surfaces* **2018**, *54*, 976–989. [CrossRef]

31. Benedetti, A.; Cirisano, F.; Delucchi, M.; Faimali, M.; Ferrari, M. Potentiodynamic Study of Al-Mg Alloy with Superhydrophobic Coating in Photobiologically Active/Not Active Natural Seawater. *Colloids Surfaces B Biointerfaces* **2016**, *137*, 167–175. [CrossRef] [PubMed]
32. King, A.D.; Birbilis, N.; Scully, J.R. Accurate Electrochemical Measurement of Magnesium Corrosion Rates; A Combined Impedance, Mass-Loss and Hydrogen Collection Study. *Electrochim. Acta* **2014**, *121*, 394–406. [CrossRef]
33. Li, C.Q.; Xu, D.K.; Chen, X.B.; Wang, B.J.; Wu, R.Z.; Han, E.H.; Birbilis, N. Composition and Microstructure Dependent Corrosion Behaviour of Mg-Li Alloys. *Electrochim. Acta* **2018**, *260*, 55–64. [CrossRef]
34. Jamesh, M.I.; Wu, G.; Zhao, Y.; McKenzie, D.R.; Bilek, M.M.M.; Chu, P.K. Electrochemical Corrosion Behavior of Biodegradable Mg-Y-RE and Mg-Zn-Zr Alloys in Ringer's Solution and Simulated Body Fluid. *Corros. Sci.* **2015**, *91*, 160–184. [CrossRef]
35. Xu, D.K.; Han, E.H. Effect of Quasicrystalline Phase on Improving the Corrosion Resistance of a Duplex Structured Mg-Li Alloy. *Scr. Mater.* **2014**, *71*, 21–24.
36. Tian, G.; Wang, J.; Xue, C.; Wang, S.; Yang, X.; Su, H.; Li, Q.; Li, X.; Yan, C.; Yang, Z. Improving Corrosion Resistance of Mg–Li Alloys by Sn Microalloying. *J. Mater. Res. Technol.* **2023**, *26*, 199–217. [CrossRef]
37. Ma, H.; Wu, L.; Liu, C.; Liu, M.; Wang, C.; Li, D.; Chen, X.Q.; Dong, J.; Ke, W. First-Principles Modeling of the Hydrogen Evolution Reaction and Its Application in Electrochemical Corrosion of Mg. *Acta Mater.* **2020**, *183*, 377–389. [CrossRef]
38. Bahmani, A.; Lotfpour, M.; Taghizadeh, M.; Kim, W.J. Corrosion Behavior of Severely Plastically Deformed Mg and Mg Alloys. *J. Magnes. Alloys* **2022**, *10*, 2607–2648. [CrossRef]
39. Atrens, A.; Liu, M. Corrosion of Magnesium (Mg) Alloys and Metallurgical Influence. *Corros. Magnes. Alloys* **2011**, 117–165. [CrossRef]

materials

MDPI

Article

Effect of ZrB_2 Content on the Properties of Copper Matrix Composite

Iwona Sulima [1], Remigiusz Kowalik [2,*], Michał Stępień [2] and Paweł Hyjek [1]

1 Institute of Technology, University of the National Education Commission, Krakow, Podchorazych 2 Str., 30-084 Krakow, Poland; iwona.sulima@uken.krakow.pl (I.S.); pawel.hyjek@uken.krakow.pl (P.H.)

2 Faculty of Non-Ferrous Metals, AGH University of Krakow, Mickiewicz 30 Av., 30-059 Krakow, Poland; mstepien@agh.edu.pl

* Correspondence: rkowalik@agh.edu.pl

Abstract: This study examined the physical, mechanical, tribological, and corrosion properties of copper metal matrix composites reinforced with zirconium diboride (ZrB_2). Cu-xZrB_2 composites (x = 0.5, 10, 15, 20 wt.%) were produced by the ball-milling process and spark plasma sintering (SPS). Introducing ZrB_2 particles into copper matrix composites significantly improves their mechanical and tribological properties while deteriorating their density, porosity, and corrosion properties. It was shown that the relative densities of the composites gradually decreased from 96% to 90%, with an increase in the ZrB_2 content to 20 wt.%. Likewise, hardness, compressive strength, and wear resistance improved with increasing ZrB_2 content in the copper matrix. Corrosion resistance tests in a 0.05 M sulfuric acid environment showed a disproportionate decrease in the resistance of this composite with an increase in the concentration of the ceramic phase compared to other environments.

Keywords: copper matrix composites; ZrB_2; powder metallurgy; mechanical properties; wear resistance; corrosion resistance

Citation: Sulima, I.; Kowalik, R.; Stępień, M.; Hyjek, P. Effect of ZrB_2 Content on the Properties of Copper Matrix Composite. *Materials* **2024**, *17*, 6105. https://doi.org/10.3390/ma17246105

Academic Editor: Xiaogang Li

Received: 19 November 2024
Revised: 2 December 2024
Accepted: 9 December 2024
Published: 13 December 2024

1. Introduction

For modern technology and industry needs, continuous development of new materials with unique properties is required. Their production by conventional methods is often very difficult or impossible. Such materials include, among others, metal matrix composites, which are a very dynamically developing group of materials [1]. The most commonly used matrix materials of composites are magnesium [2], aluminum [3], titanium [4], iron [5], nickel [6], and copper [7]. Typical reinforcing phases are carbides (TiC, SiC) [8,9], oxides (ZrO_2, Al_2O_3, TiO_2, SiO_2) [10–12], nitrides (Si_3N_4, BN) [13,14] and borides (TiB_2, ZrB_2) [15,16]. In recent years, there has been a new trend of using attractive ceramic materials, such as borides, as reinforcement for metal composites [15,17–19]. Among these materials, zirconium diboride is often chosen due to its favorable combination of properties, which include a high melting point, low density, high modulus of elasticity, excellent hardness, electrical and thermal stability, as well as excellent wear resistance and good corrosion resistance [20–22]. Copper matrix composites are materials whose use requires excellent mechanical properties as well as high electrical or thermal conductivity. Mechanical properties can be effectively improved by adding a ceramic reinforcing phase to the copper matrix. However, this often affects the deterioration of the conductivity of the composites. Therefore, a crucial aspect in the design of copper matrix composites is to produce a material with good mechanical properties and good conductivity [23,24].

Copper matrix composites can be effectively manufactured by various powder metallurgy techniques, including conventional sintering [25], hot pressing (HP) [26], hot isostatic pressing (HIP) [27], and Spark Plasma Sintering (SPS) [28,29]. Table 1 presents selected data from the literature [23,30–33] regarding the influence of the type and amount of ceramic phases on the density and hardness of copper matrix composites.

SPS is an advanced manufacturing technology that enables the production of very dense composite materials. The use of this modern sintering technology allows for the reduction of the sintering temperature, shortens the sintering time, and allows the sintering process to be conducted without the use of activators. This positively impacts the economic aspect of manufactured composite materials [30,34]. In the literature [31–33,35,36], several studies have focused on producing copper matrix composites using the SPS method. Long et al. [32] sintered Cu–NbC composites using spark plasma sintering temperatures between 900 °C and 1000 °C for 15 min. The results showed that the composites' density, hardness, and electrical conductivity proved to be a function of temperature. Simultaneously, it was observed that the hardness increased with the increase in the NbC phase while the electrical conductivity deteriorated. In turn, Wang et al. [31] investigated the effect of various SPS parameters, temperature (600–900 °C), and pressure (10–80 MPa) on the electrical and mechanical properties of Cu-TiC composites. It was shown that the increase in sintering temperature significantly improves the density, microhardness, and electrical conductivity. Analyzing the results, it was noticed that density, microhardness, and electrical conductivity showed a positive correlation with the sintering pressure. Soloviova et al. produced Cu-(LaB_6-TiB_2) composites using the SPS method. They obtained a twofold increase in the density of the material as a function of the sintering temperature. A positive effect of the temperature change from 850 °C to 1050 °C of sintering on the hardness and electrical conductivity of the composites was demonstrated.

Table 1. The influence of ceramic particles on density and hardness of the Cu-based composites.

Cu-Based Composites	Sintering Condition	Density	Hardness	Ref.
Cu-5 vol.% TiC	SPS; 800 °C; 10–80 MPa; 5 min	7.0–8.6 g/cm^3	125–268 HV1	[31]
Cu-1 vol.% NbC	SPS; 1000 °C; 100 MPa; 10 min	98.1%	162 HV	[32]
Cu-5 vol.% NbC		97.0%	268 HV	
Cu-15 vol.% NbC		96.8%	395 HV	
Cu-25 vol.% NbC		97.0%	452 HV	
Cu-1 vol.% Al_2O_3	SPS; 700 °C; 10–50 MPa; 5 min	93.2%	77 HV0.3	[33]
Cu-5 vol.% Al_2O_3		92.8%	125 HV0.3	
Cu-7 vol.% Al_2O_3		86.1%	75 HV0.3	
Cu-1 wt.% ZrB_2	Hot-pressed sintering; 840 °C; 25 MPa; 2 h	96%	69 HV0.2	[37]
Cu-3 wt.% ZrB_2		95%	84 HV0.2	
Cu-5 wt.% ZrB_2		92.5%	93 HV0.2	
Cu-7 wt.% ZrB_2		91.8%	100.8 HV0.2	
Cu-9 wt.% ZrB_2		91.3%	82 HV0.2	
Cu-2 wt.% Mo_2C	Hot-pressed sintering; 880 °C; 20 MPa; 10 min	91.5%	58.9 HV	[38]
Cu-5 wt.% Mo_2C		91.2%	65.8 HV	
Cu-7 wt.% Mo_2C		91.3%	69.6 HV	
Cu-1wt.% Y_2O_3	SPS; 900 °C; 50 MPa; 5 min	93.9%	101.3 HV0.1	[39]
Cu-3 wt.% Y_2O_3		100%	125.7 HV0.1	
Cu-5 wt.% Y_2O_3		100%	140.5 HV0.1	

In this paper, Cu-ZrB_2 composites were prepared using the Spark Plasma Sintering (SPS) method. The influence of different zirconium diboride content (5, 10, 15, and 20 wt.%) on the properties of sintered materials, such as density, portability, Young's modulus, hardness, compressive strength, corrosion resistance, and abrasive wear resistance, was analyzed. The results will help to understand the importance and influence of the amount of zirconium diboride in optimizing the properties of such composite materials.

2. Materials and Methods

This study used commercially available Cu powder (10 µm, 99.9 wt.% purity, Kamb Import Export, Warsaw, Poland) and ZrB2 powder (2.5–5.5 µm, 99.9 wt.%, H.C. Starck Tungsten GmbH, Goslar, Germany) to fabricate composites. The microstructures of raw powders are shown in Figures 1 and 2, which correspond to the results of the particle size tests. The morphology of the starting powders was evaluated using scanning electron microscopy (JEOL JSM 6610LV, Tokyo, Japan). Measurements of the particle size distribution of the starting powders and composite mixtures were performed in polypropylene alcohol using the SALD-7500 nano analyzer (Shimadzu Corporation, Kyoto, Japan) with Wing SALD II software (Version 3.4.9, Shimadzu Corporation, Kyoto, Japan). Measurements were carried out with a measurement step of 1 s.

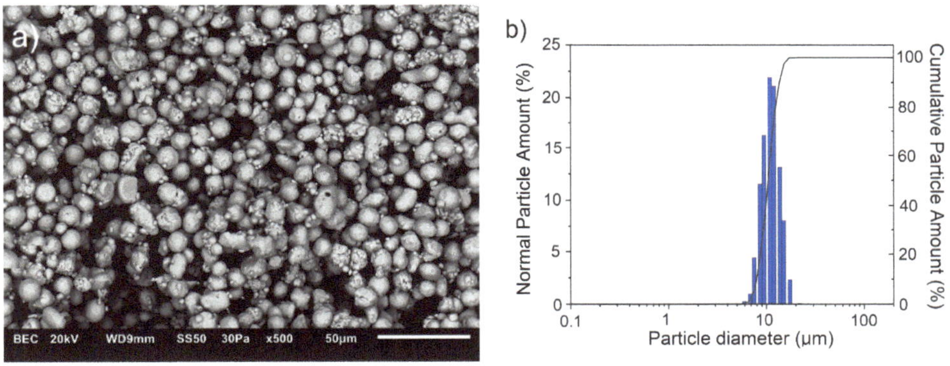

Figure 1. (**a**) SEM images of copper powder with (**b**) corresponding particle diameter distribution.

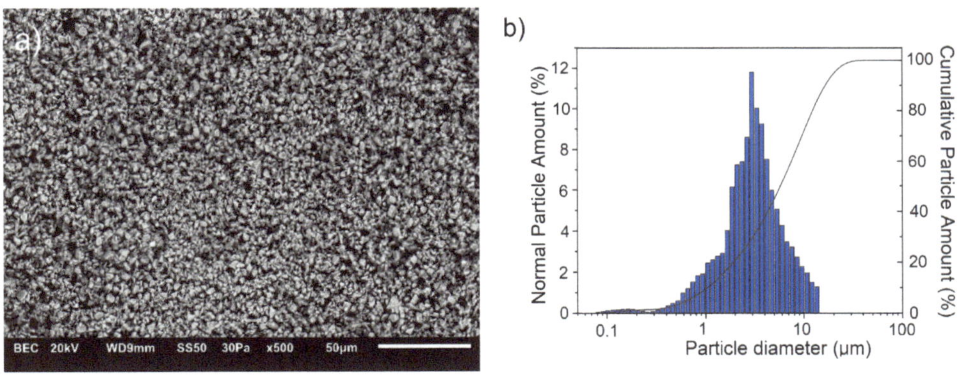

Figure 2. (**a**) SEM images of diboride zirconium powder with (**b**) corresponding particle diameter distribution.

Next, the mixture powders were milled using a Fritsch Pulveristte 5 planetary ball mill (Fritsch GmbH, Idar-Oberstein, Germany) with a ball-to-powder weight ratio of 5:1. The container and balls were made of tungsten carbide. Milling was carried out at a speed of 200 rpm for 10 h with a repeating cycle: 20 min of milling—10 min break. Four powder mixtures were prepared, which contained the following amounts of the ZrB_2: 5 wt.%, 10 wt.%, 15% wt.%, and 20 wt.%. Figure 3 shows the powder mixture's morphology and particle size distribution containing 20 wt.% ZrB_2 after 10 h of milling.

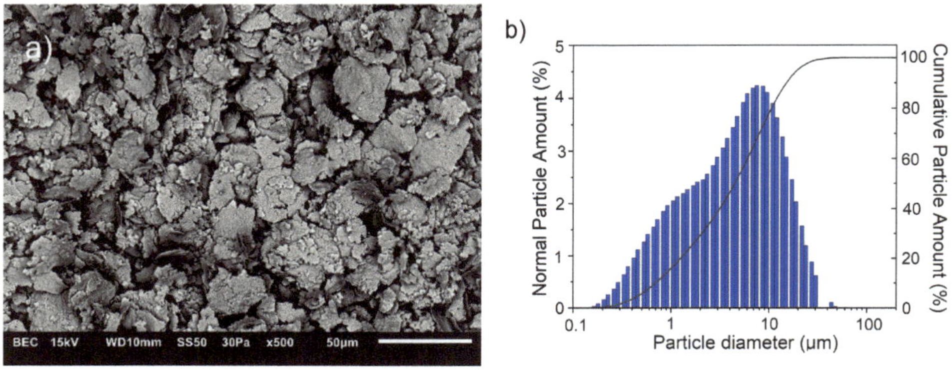

Figure 3. (**a**) SEM images with (**b**) corresponding particle diameter distribution of the powders with 20% ZrB_2 after 10 h milling.

The LSP 100 laboratory sinter press (Dr. Frisch, Fellbach, Germany) was used for sintering the copper and Cu-ZrB_2 composites. The powders were sintered at 35 MPa in an argon atmosphere at 900 °C for 3 min. The furnace heating rate was kept at 100 °C/min. Figure 4 shows the heating curve registered during the SPS process.

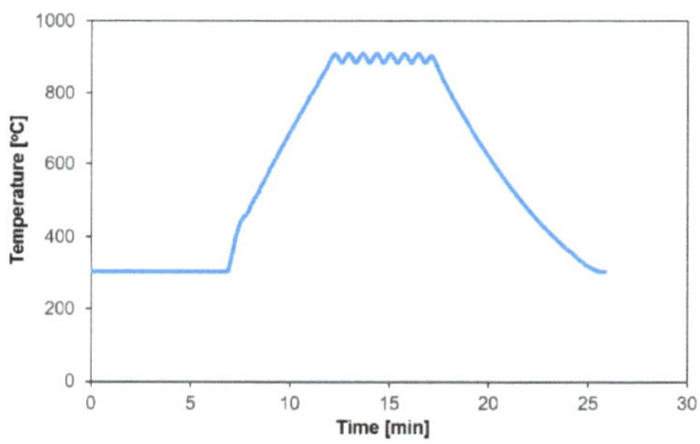

Figure 4. Heating curve of sintering Cu-5%ZrB_2 composite.

The sintered samples were characterized by X-ray diffraction (XRD) to identify the phases. The studies were performed with a Rigaku Miniflex II X-ray diffractometer using Cu-K radiation (Tokyo, Japan). Microstructure characterization was carried out using the OLYMPUS LEXT OLS5100 confocal microscope (Olympus, Tokyo, Japan) and scanning electron microscopy (SEM, JEOL JSM 6610LV, Tokyo, Japan) equipped with energy-dispersive spectroscopy (EDS, Aztec, Oxford Instruments, Oxford, UK).

Hardness was measured using NEXUS 4000 Vicker's microhardness tester (Innovatest, EuropeBV, Maastricht, The Netherlands) with a load of 2.942 N. A minimum of ten measurements on each sample were implemented for hardness evaluations. Compression tests were conducted on an INSTRON TT-DM testing machine (Norwood, MA, USA) with a crosshead speed of 1×10^{-4} mm/s. The tests were carried out at room temperature in accordance with the requirements of the ISO standard [40].

The corrosion test was carried out using a 0.05 M H_2SO_4 solution. All measurements were conducted at 25 $\pm$ 1 °C in a deaerated solution. The solution was deaerated by bubbling argon (175 mL/min) 30 min before sample mounting, and argon flow continued throughout the experiment. The exanimated surface was polished on a P800 grid of abrasive paper, cleaned in an ultrasonic bath in ethanol, and dried in a hot air stream (60 °C). A conventional three-electrode system was applied with Ag/AgCl (3M KCl) as the reference electrode and the platinum wire as a counter electrode. The experiments were performed on the potentiostat of AUTOLAB PGSTAT128 (Metrohm AG, Herisau, Switzerland). Before each polarization test, a sample was conditioned for 30 min in the solution to obtain an open current potential (OCP) value. The potentiodynamic polarization scan was performed from −0.20 V vs. OCP in an anodic direction at a scan rate of 1 mV/s. The direction was changed when the current density reached 5 mA/cm^{2,} and the scan was ended at the initial potential value. The measurements were repeated three times to obtain more reliable results, and the measurement uncertainty was given as standard deviation.

The wear behavior of the sintered materials was assessed using a ball-on-disc tribometer (Elbit Company, Koszyce Małe, Poland) as per the ISO standard [41]. Wear tests were conducted in a dry environment. Wear experiments were carried out under the following conditions: time of 10,000 s, load of 5 N, distance of 1000 m, and rotational speed of 200 rpm. A counter-sample made of Al_2O_3 (diameter of 3.175 mm) was used. Before each test, the samples and the steel balls were cleaned with acetone in an ultrasonic washer. The friction coefficient was measured continuously during the test following the established wear conditions. Mass loss (Δm) was determined using the weight loss method. The samples were weighed before and after each wear test using an analytical balance RADWAG AS 220/C/2 (Radwag, Radom, Poland) with a precision of 0.1 mg. Next, the mass loss was calculated.

In the next step, the wear track was observed using an OLYMPUS LEXT OLS5100 confocal microscope and scanning electron microscopy (SEM) using Jeol JSM 6610 LV. The average width of the wear path was determined based on 50 measurements taken for each analyzed wear. Figure 5 shows a part of the wear track marked with sample width measurements. The wear volume of the disc was determined, and then the wear coefficient was calculated according to the formula [41]:

$$W_{V(disc)} = \frac{V_{disc}}{F_n{*}L} \tag{1}$$

where:

F_n—applied load [N]

V_{disc}—wear volume of disc specimen [mm^3]

L—sliding distance [m].

Figure 5. Example measurements of the average width of wear track for Cu-5%ZrB_2 composite.

3. Results and Discussion

3.1. Physical and Mechanical Properties and Microstructure of the Sintered Materials

Table 2 shows the changes in measured density, porosity, and Young's modulus of sintered copper and composites. It was shown that the relative density of Cu-xZrB_2 composites (x = 0, 5, 10, 15, 20 wt.%) decreased with the increasing amount of ZrB_2 particles, which were 98%, 96%, 94%, 91%, and 90%, respectively. Simultaneously, porosity increased with the increase in the reinforcing phase. Similar results were reported for other ceramic phase-reinforced copper matrix composites. Shaik and Golla [42] found that the relative density of copper matrix composites decreased from 99.76% for pure copper to 96.02% for the Cu-10 wt.% ZB_2 composite. Zhang et al. [43] showed that in Cu-YB_2C_2 composites, the relative density decreased from 99.86% for pure copper to 94.15% for the composite containing 20 wt.% YB_2C_2.

Table 2. Influence of the ZrB_2 content on the properties of sintered materials.

Sintered Materials	Apparent Density [g/cm^3]	Relative Density [%]	Open Porosity [%]	Young's Modulus [GPa]	Hardness HV0.3 [-]
Cu-0% ZrB_2	8.72	98	0.12	107	61.2
Cu-5% ZrB_2	8.41	96	1.91	120	105
Cu-10% ZrB_2	8.06	94	3.45	131	133
Cu-15% ZrB_2	7.59	91	6.86	141	140
Cu-20% ZrB_2	7.34	90	8.85	148	161

It was observed that the values of Young's modulus and hardness depended on the amount of ZrB_2 in the matrix. Young's modulus gradually increased with the amount of ZrB_2 in the copper matrix, from 107 GPa for pure copper to 148 GPa for Cu-20% ZrB_2 composites. The hardness measurement results indicated that adding 5 wt.% of zirconium diboride nearly doubled the hardness compared to copper (Cu-0% ZrB_2). Fine hard ceramic particles ZrB_2 in the copper matrix can act as obstacles during the pressing of the indenter, contributing to the increase in the hardness of the prepared composite. Consequently, increasing the content of the reinforcing phase provides more excellent resistance to plastic deformation, thus increasing the hardness of the composite. Therefore, the composites containing 20 wt.% ZrB_2 showed three times higher hardness than sintered copper (Cu-0% ZrB_2). The hardness of copper and the Cu-20% ZrB_2 composite was 61.2 HV0.3 and 161 HV0.3, respectively.

Figure 6 shows the summary results of compressive strength tests of sintered materials. Similar characteristics of the true stress–strain curves were obtained for all composites with different plastic strain ranges. It was observed that, after reaching the maximum true stress, the stress value gradually decreased for each tested composite material. For comparison, sintered copper samples were deformed plastically during the tests and did not crack or break. Therefore, copper samples were deformed to the appropriate level of deformation. It was observed that the presence of hard ZrB_2 particles in the copper matrix affects the mechanical properties of the composites. Based on the results of experimental studies obtained at room temperature, it was found that with the increase in the amount of ZrB_2 phase in the matrix, the compressive strength of the tested materials improves, while the plastic properties deteriorate. The maximum true stress value was the highest for composites containing 20% ZrB_2 and 364 MPa. Another effect was observed in the studies of Zhai et al. [44], where the reinforcing phase of the copper matrix was carbides (TiC). It was shown that the ductility and strength of the composite were the best when the mass fraction of TiC was 10%. For this composite, the maximum compressive strength was 650 MPa. Then, the compressive strain decreased with the increase of TiC content. When the mass fraction of TiC reached 25 wt.%, the composite's yield strength and compressive strain decreased rapidly.

Figures 7–9 show the comparative microstructures of sintered composites. EDS and X-ray techniques were used to verify the phase composition of the sintered composites. The reinforcing phase ZrB_2 is distributed homogeneously throughout the volume of the copper matrix. The results of phase analyses confirmed the presence of the ZrB_2 phase for all sintered composites (Figure 10). The identified WC phase (Figure 10) is a contamination resulting from the milling of composite powders in a tungsten carbide container.

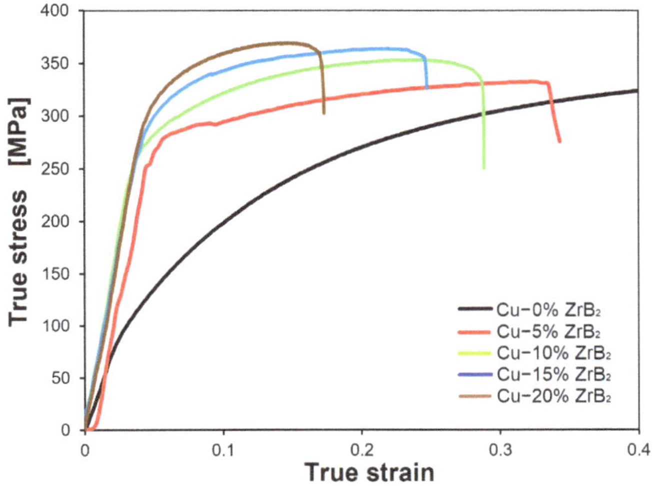

Figure 6. Compressive stress–strain curves of the sintered materials.

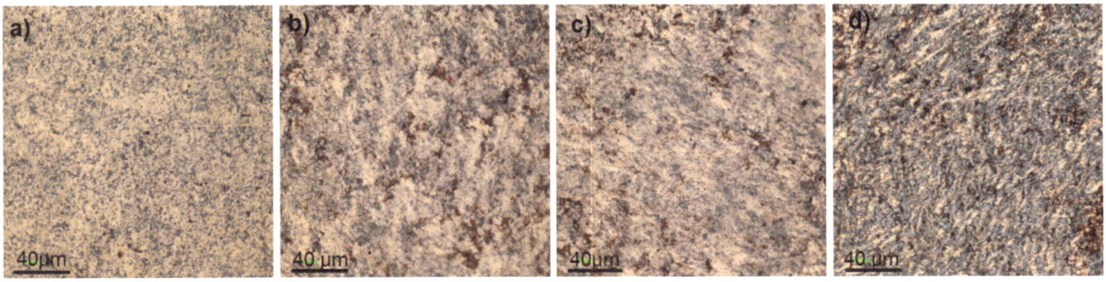

Figure 7. Microstructure of copper matrix composites with: (**a**) 5% ZrB_2, (**b**) 10% ZrB_2, (**c**) 15% ZrB_2, and (**d**) 20% ZrB_2 obtained from laser microscope.

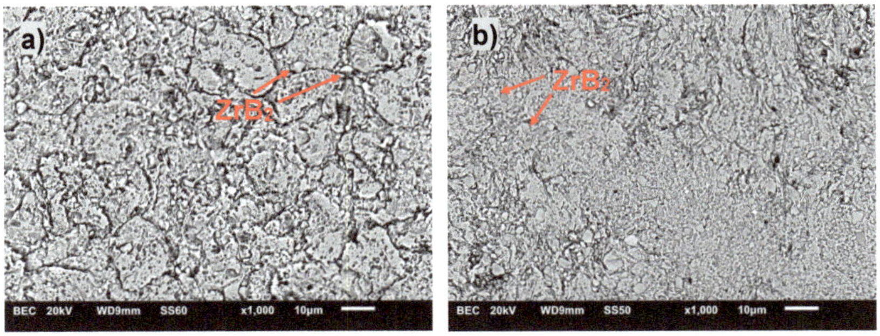

Figure 8. *Cont.*

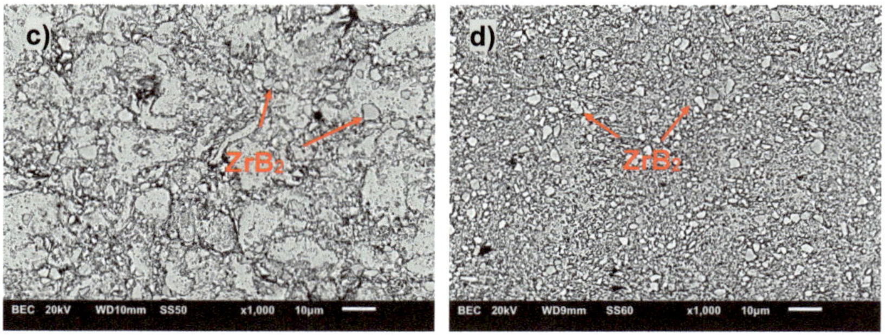

Figure 8. Microstructure (SEM) of copper matrix composites with: (**a**) 5% ZrB_2, (**b**) 10% ZrB_2, (**c**) 15% ZrB_2, and (**d**) 20% ZrB_2.

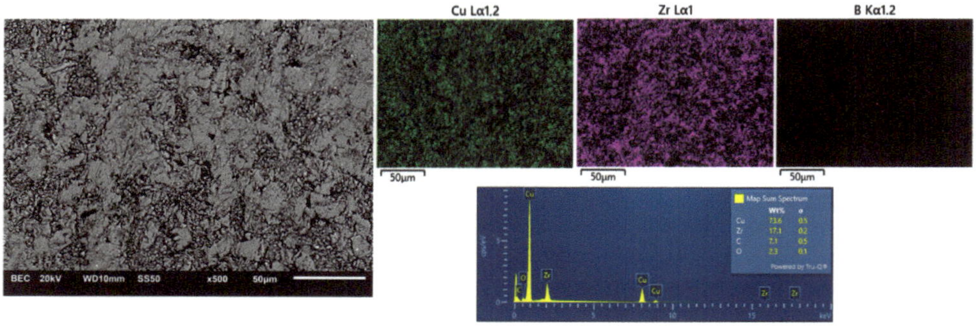

Figure 9. Selected microstructure (SEM) and EDS composition analysis of copper-20% ZrB_2 composite.

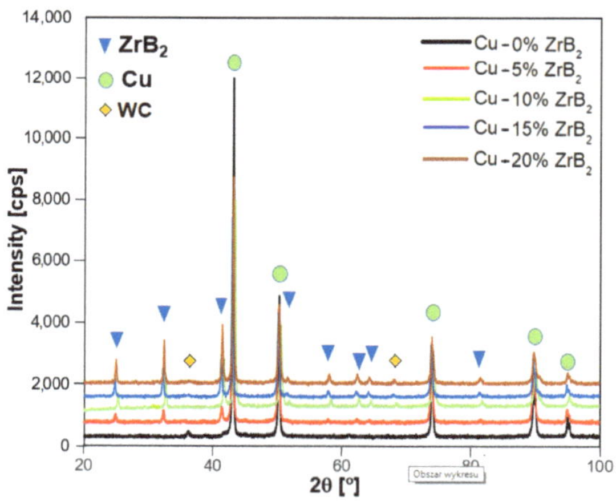

Figure 10. XRD patterns of copper matrix composites with different ZrB_2 content.

3.2. Corrosion Behavior

In general, copper cannot oxidize in an acid solution with hydrogen displacement without dissolved oxygen. The reactions occurring during this process are presented in Equations (2)–(4), with the standard potentials vs. SHE (relative to Ag/AgCl 3 M KCl). The corrosion potential in a deaerated acid solution is determined by the two reactions: anodic copper dissolution (2) and cathodic hydrogen evolution (3).

$$Cu - 2e \rightarrow Cu^{2+} E^0 = 0.337(+0.127)[V] \tag{2}$$

$$2H^{+} + 2e^{-} \rightarrow H_2E^0 = 0.000(-0.210)[V] \quad (3)$$

$$O_2 + 4H^{-} + 4e^{-} \rightarrow 2H_2OE^0 = 1.230(+1.030)[V] \quad (4)$$

The slightly positive shift of open circuit potential (Figure 11) in a sintered Cu sample without ZrB_2 compared to the solid copper sample may elucidate increased oxygen content on the surface of powder particles and trapped in the pores. According to potential mixed potential theory, without Reaction (4), the potential is shifted in the direction of Reactions (2) and (3). As time passes and the oxygen bound on the surface of the powder sample particles is depleted, the equilibrium potential tends to the potential of a solid pure copper sample. However, this mechanism cannot explain the significant shift in the open circuit potential towards the negative side of samples by adding the ZrB_2 ceramic phase. In a similar acid sulfate solution (0.25 M Na_2SO_4 + 0.05 M H_2SO_4, pH = 1), pure ZrB_2 ceramics show corrosion potential of about −0.4 V vs. Ag/AgCl electrode and expected corrosion current on 3 $\mu A/cm^2$ [45]. So, it is possible, based on the mixed potential theory, that OCP (also corrosion potential obtained from Tafel extrapolation) is shifted by the presence of the ceramic phase, not by the effect caused by the porosity of the samples.

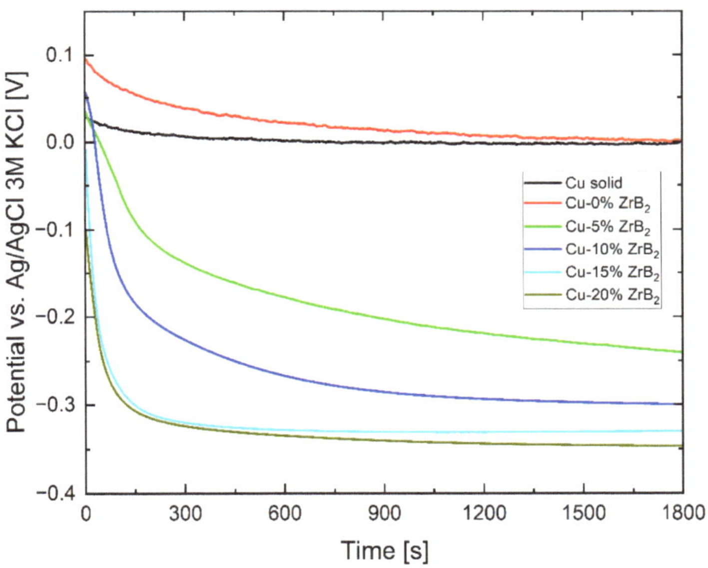

Figure 11. Open circuit potential recorded in 0.05 M H_2SO_4.

Polarization curves collected during the experiment are presented in Figure 12. On their basis, the corrosion potential and current, as well as the cathodic and anodic Tafel constants, were calculated using Tafel extrapolation, and the results are gathered in Table 3. The solid Cu sample and the sintered Cu sample without adding ZrB_2 do not differ significantly, apart from the difference in the slope of the cathode part, which is probably related to diffusion. The polarization curves show a significant difference in the corrosion behavior of samples with and without adding the ZrB_2 ceramic phase. The corrosion potential of samples with ZrB_2 shifted about 300 mV toward the less noble direction (according to the OCP value). The corrosion current also increases strongly with the increase in the content of the ZrB_2 phase. This increase is disproportionately large compared to similar samples tested in chloride solutions (3.5 wt.% NaCl) with a pH close to natural [46]. In a chloride environment, the corrosion current increased approximately 40 times in the worst possible case, while in a sulfate environment, it increased approximately 300 times.

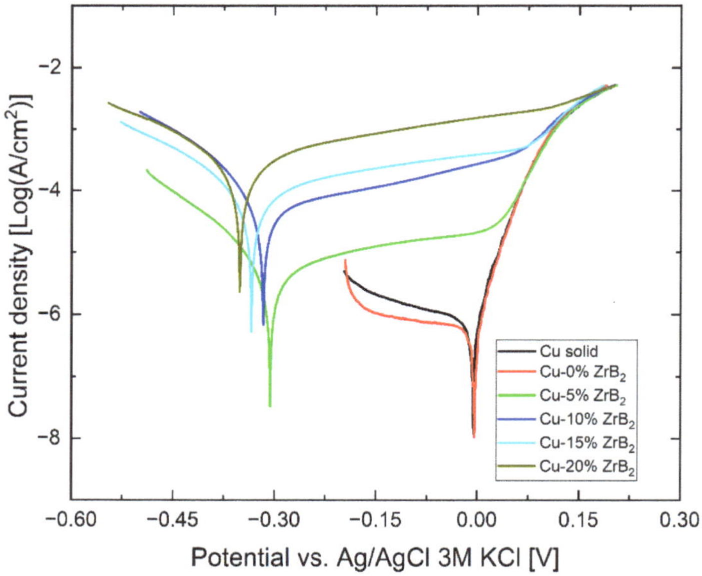

Figure 12. Polarization curves recorded on Cu and sintered Cu-ZrB_2 composites obtained in deaerated 0.05 H_2SO_4 solution.

Table 3. Potentiodynamic polarization parameters for copper and Cu-ZrB_2 composites in 0.05 M H_2SO_4 solution.

Sample	OCP [mV]	E_{corr} [mV]	I_{corr} $\mu A/cm^2$	β-Anodic Tafel Const.	β Cathodic Tafel Const.
Cu solid	4 ± 4	−3 ± 5	0.7 ± 0.1	33 ± 1	−361 ± 20
Cu-0% ZrB_2	1 ± 3	−8 ± 3	0.6 ± 0.05	31 ± 1	−610 ± 27
Cu-5% ZrB_2	285 ± 20	302 ± 9	7.3 ± 2	595 ± 20	−120 ± 15
Cu-10% ZrB_2	290 ± 7	315 ± 8	58 ± 7	509 ± 48	−107 ± 9
Cu-15% ZrB_2	340 ± 10	345 ± 10	133 ± 20	650 ± 70	−178 ± 5
Cu-20% ZrB_2	340 ± 15	350 ± 12	289 ± 27	467 ± 20	−193 ± 8

Moreover, the increase in the corrosion current for ZrB_2 in a sulfate environment (0.1 M H_2SO_4) in a 316L stainless steel matrix is also approximately two times lower than in this case [5]. The corrosion process of ZrB_2 in acid solutions is most likely responsible for such a sharp increase in corrosion current, where ZrB_2 is decomposed according to Reactions (4) proposed by Monticelli [42]

$$ZrB_2 + 8H_2O \rightarrow ZrO_2 \downarrow + 2H_3BO_4 + 10H^+ + 10e^- \quad (5)$$

This process produces boric acid, which further acidifies the corrosion environment, enhancing the effect of crevice corrosion in the pores. In the case of a stainless steel matrix, under favorable conditions, matrix corrosion can be slowed by passivation. However, the copper matrix needs to provide such a chance. Moreover, the stability of the oxide layer formed on ZrB_2 particles, which slows down the corrosion process, depends on the type of soluble complex and the pH value. In the case of acid sulfate solution (reaction 6), the equilibrium is shifted towards the right side due to higher complex stability [45,47].

$$ZrO_2 \downarrow + 2SO_4^{2-} + 2H^+ \leftrightarrow [ZrO(SO_4)_2]^{2-} + H_2O \quad (6)$$

This accelerates corrosion in sulfate solutions, but this phenomenon is inhibited in chloride solutions due to the low stability of the $[ZrCl_6]^{2-}$ complex. Moreover, the lack of oxygen dissolved in water additionally inhibits the formation of ZrO_2, which also accelerates the corrosion process of ZrB_2.

On the other hand, the presence of oxygen would accelerate the corrosion of the copper matrix. Therefore, the selected test conditions are extremely unfavorable from the point of view of the tested material. However, due to its good mechanical properties, it is worth selecting appropriate corrosion inhibitors in subsequent work.

3.3. Wear Behavior

The results of the friction coefficient as a function of time are shown in Figure 13. It was found that the friction coefficient of copper matrix composites decreased significantly when a higher ZrB_2 content was used. Comparatively, for pure copper (Cu-0% ZrB_2) and the Cu-20% ZrB_2 composite, the friction coefficient was 0.76 and 0.44, respectively. It was observed that the steady-state friction coefficient was achieved after about 4000 s for pure Cu, while for Cu-ZrB_2 composites, the friction coefficient stabilized after about 2000 s. Similar relationships were observed in other studies [42,44]. Shaik et al. [42] studied Cu-ZrB_2 composites produced using a hot pressing method using the pin-on-disc method. They showed that the friction coefficient of Cu decreased from 0.56 to 0.16 after adding 10% ZrB_2. In their studies, Zhou et al. [48] introduced a reinforcing phase WS_2 (20 wt.%) with particle sizes of 0.6 and 5.0 μm into the copper matrix, which allowed them to achieve the friction coefficient of 0.158 and 0.172, respectively.

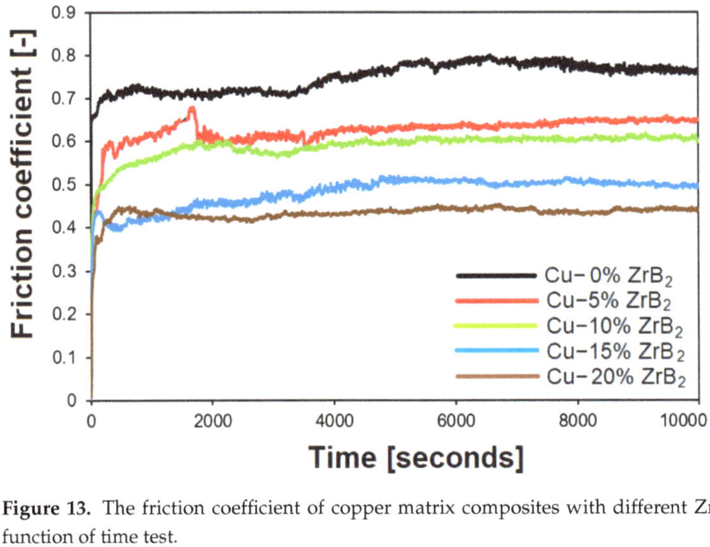

Figure 13. The friction coefficient of copper matrix composites with different ZrB_2 content as a function of time test.

The conducted tests (Figure 14) showed that copper (Cu-0% ZrB_2) shows the largest mass loss and, as expected, is characterized by the highest specific wear rate (3.48×10^{-6} mm^3/Nm). It can be seen that adding only 5 wt.% ZB_2 causes a more than two-fold decrease in the specific wear rate value (1.28×10^{-6} mm^3/Nm). Adding more of the reinforcing phase, ZrB_2 causes a gradual reduction in the specific wear rate. For composites containing 20 wt.% ZrB_2, the specific wear rate is 0.42×10^{-6} mm^3/Nm. Based on the results, it was found that the wear resistance of copper-based composites increases with the increase of ZrB_2 content. This is because the Al_2O_3 ball can easily penetrate the soft copper matrix during the ball-on-disc test, resulting in excessive material removal from the worn surface.

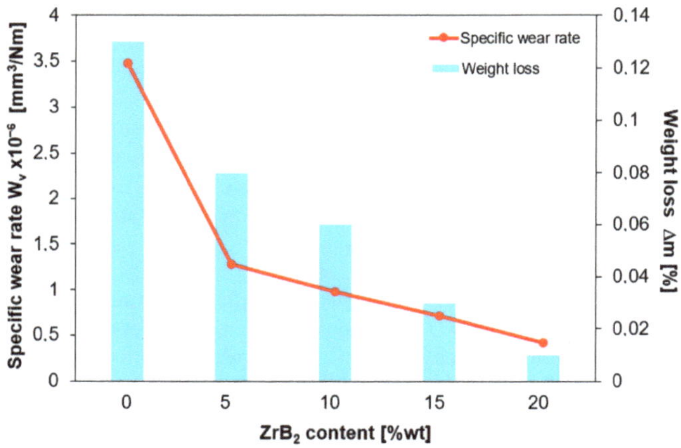

Figure 14. Variation of specific wear rate and weight loss as a function of ZrB_2 content in the copper matrix.

In contrast, material removal is more limited in sintered composites because the ZrB_2 reinforcing phase acts as a barrier and protects the copper matrix from the Al_2O_3 ball. By analyzing the results of the mass loss measurement results, it can be concluded that its value decreases with the increase in the reinforcing phase ZrB_2. This is because the addition of ceramic particles ZrB_2 significantly increases the hardness of copper, as shown in Table 1. Similar relationships were demonstrated in other studies of copper matrix composites [42–44]. Fathy et al. [49] showed that pure copper exhibits an exceptionally high wear rate compared to Cu-ZrB_2 nanocomposites. Their studies using the pin-on-disc method used different normal loads and sliding speeds. Generally, it was shown that the wear rate of nanocomposites decreases with the increase of ZrO_2 content (0, 3, 6, and 9 wt.%). In turn, Tjong and Lau [50,51] showed that adding 5% SiC effectively affects the wear resistance of Cu-SiC composites. A further increase in SiC content resulted in a significant reduction in volume loss, especially in the case of the composite containing 20 vol.% SiC.

Figure 15 shows examples of wear path microstructures for sintered composites. Generally, for all Cu-xZrB_2 composites (x = 5, 10, 15, 20 wt.%), a very similar material removal mechanism by the Al_2O_3 ball from the wear track was observed. In the area of abrasion, scratches and abrasive grooves are visible. Additionally, the worn surfaces of the composites were subject to plastic deformation, which resulted in delamination of the material. During the tests, the material was permanently deformed and gradually abraded to the edge of the wear track. This suggests a mixed nature of wear of the tested materials. Microstructural observations indicate abrasive and adhesive wear of the composite materials.

Figure 16 shows examples of wear tracks and their profiles measured for copper and composites with different ZrB_2 contents. 3D analyses of the profiles showed differences in the dimensions of the wear trace. A decrease in the width of the wear track was observed with the increase in the amount of the reinforcing phase (Figure 17). These observations confirmed the results of wear measurements of the sintered composite samples. As mentioned earlier, adding more ZrB_2 to the copper matrix causes a gradual decrease in the specific wear rate. The hard ZrB_2 reinforcing phase protects the copper matrix during friction. The ZrB_2 particles resist the ceramic surface of the ball (Al_2O_3). Consequently, the presence of hard phases contributes to the composite's increased hardness and wear resistance, which leads to a smaller loss of mass of the composite material.

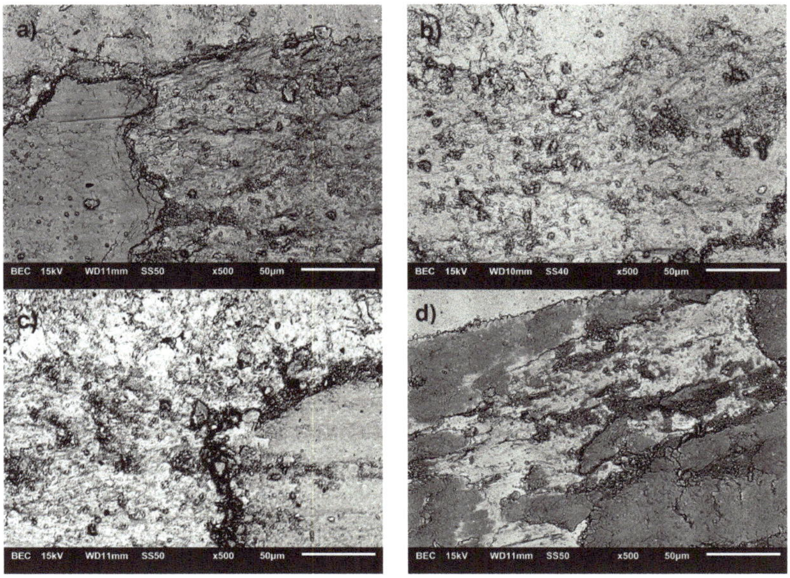

Figure 15. SEM micrograph of worn surface of copper matrix composites with: (**a**) 5% ZrB_2, (**b**) 10% ZrB_2, (**c**) 15% ZrB_2, and (**d**) 20% ZrB_2.

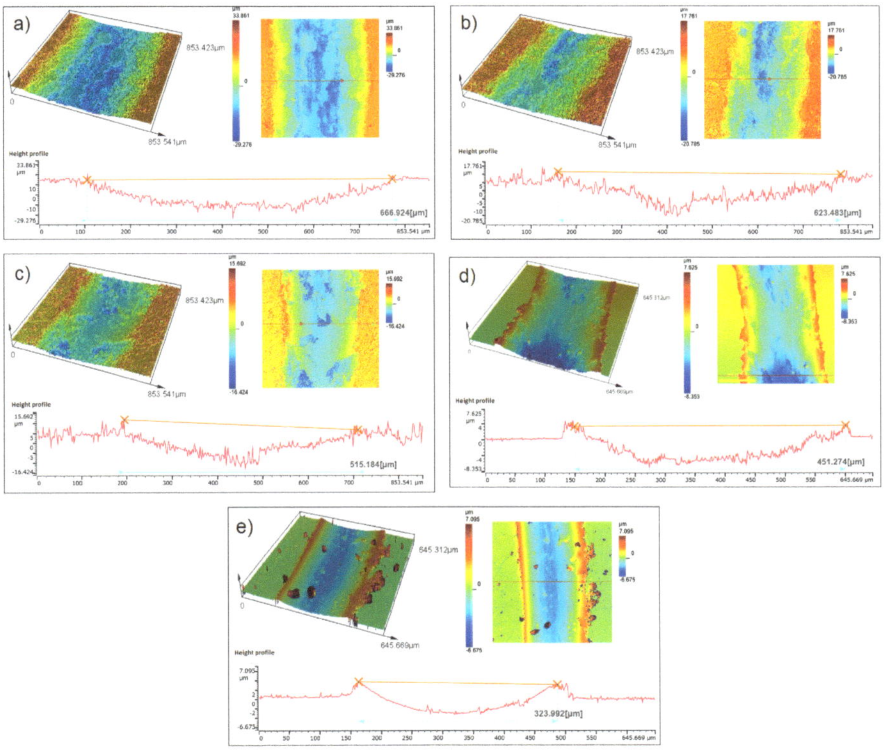

Figure 16. 3D laser morphology of worn surface: (**a**) Cu-0% ZrB_2, (**b**) Cu-5% ZrB_2 composite, (**c**) Cu-10% ZrB_2 composite, (**d**) Cu-15% ZrB_2 composite and (**e**) Cu-20% ZrB_2 composite.

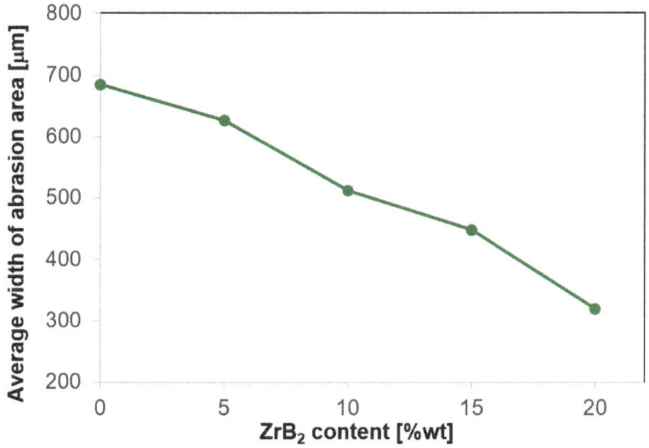

Figure 17. Variation of the average width of wear track as a function of ZrB_2 content in the copper matrix.

4. Conclusions

1. Cu matrix composites reinforced with ZrB_2 particles were successfully produced by Spark Plasma Sintering.
2. Using ZrB_2 as a reinforcing phase of the copper matrix improves its mechanical and tribological properties while decreasing its density, porosity, and corrosion resistance.
3. Sintered copper without a reinforcing phase was characterized by the highest relative density (98%) and the lowest porosity (0.12). The relative density of the composites decreased from 96.0 to 90%, respectively, for 5–20% ZrB_2 content.
4. Composites with a higher ZrB_2 content exhibit higher hardness, lower friction coefficient values, and specific wear rate values. The Cu-ZrB_2 composites show very poor corrosion resistance in a sulphuric acid environment, probably due to the corrosion of the ceramic phase.

Author Contributions: Conceptualization, I.S. and R.K.; Methodology, M.S. and P.H.; Validation, M.S. and P.H.; Formal analysis, I.S. and M.S.; Investigation, I.S., M.S. and P.H.; Resources, I.S. and R.K.; Writing—original draft, I.S. and M.S.; Writing—review & editing, R.K.; Visualization, I.S.; Supervision, R.K.; Project administration, I.S.; Funding acquisition, I.S. All authors have read and agreed to the published version of the manuscript.

Funding: This research was funded by funds from the Institute of Technology, University of the National Education Commission (WPBU/2024/03/00009). The project for purchasing scientific and research equipment: "Innovative research and scientific platform for a new class of nanocomposites", financed by the Ministry of Education and Science, contract number 7216/IA/SP/2021.

Institutional Review Board Statement: Not applicable.

Informed Consent Statement: Not applicable.

Data Availability Statement: The original contributions presented in this study are included in the article. Further inquiries can be directed to the corresponding author.

Conflicts of Interest: The authors declare no conflict of interest.

References

1. Ibrahim, I.; Mohamed, F.; Lavernia, E. Particulate reinforced metal matrix composites—A review. *J. Mater. Sci.* **1991**, *26*, 1137–1156. [CrossRef]
2. Fayomi, J.; Popoola, A.P.I.; Popoola, O.M.; Aigbodon, V.S. Magnesium-based composite by nano-nucleation of β-$Mg_{17}Al_{12}$ using spark plasma sintering route for advanced structural application. *J. Mater. Res. Technol.* **2023**, *24*, 1547–1561. [CrossRef]

3. Kumar, S.D.; Ravichandran, M.; Jeevika, A.; Stalin, B.; Kailasanathan, C.; Karthick, A. Effect of ZrB_2 on microstructural, mechanical and corrosion behaviour of aluminium (AA7178) alloy matrix composite prepared by the stir casting route. *Ceram. Int.* **2021**, *47*, 12951–12962. [CrossRef]
4. Wei, S.; Zhang, Z.; Wang, F.; Shen, X.; Cai, H.; Lee, S.; Wang, L. Effect of Ti content and sintering temperature on the microstructures and mechanical properties of TiB_2 reinforced titanium composites synthesized by SPS process. *Mater. Sci. Eng. A* **2013**, *560*, 249–255. [CrossRef]
5. Stępień, M.; Sulima, I.; Hyjek, P.; Kowalik, R. Evaluation of the corrosion resistance of spark plasma sintered stainless steel 316L matrix composites with zirconium diboride in sulfuric acid. *Arch. Civ. Mech. Eng.* **2022**, *22*, 127. [CrossRef]
6. Hyjek, P.; Jaworska, L. Application of SHS in the Manufacture of $(NiAl/Ni_3Al)/TiB_2$ Composite. *Metall. Mater. Trans. A* **2019**, *50*, 3724–3735. [CrossRef]
7. Zhimeng, T.; Zemin, W.; Lei, X.; Libo, Z.; Zhaohui, H.; Jianhua, L. Thermal and tribological properties of MoS_2 doped graphite/copper composites by microwave sintering. *J. Mater. Res. Technol.* **2021**, *15*, 6001–6010. [CrossRef]
8. Akhtar, F.; Askari, S.J.; Shah, K.A.; Du, X.; Guo, S. Microstructure, mechanical properties, electrical conductivity and wear behavior of high volume TiC reinforced Cu–matrix composites. *Mater. Charact.* **2009**, *60*, 327–336. [CrossRef]
9. Akbarpour, M.R.; Gazani, F.; Mirabad, H.M.; Khezri, I.; Moeini, A.; Sohrabi, N.; Kim, H.S. Recent advances in processing, and mechanical, thermal and electrical properties of Cu-SiC metal matrix composites prepared by powder metallurgy. *Prog. Mater. Sci.* **2023**, *140*, 101191. [CrossRef]
10. Yadav, M.; Kumaraswamidhas, L.A.; Singh, S.K. Investigation of solid particle erosion behavior of Al-Al_2O_3 and Al-ZrO_2 metal matrix composites fabricated through powder metallurgy technique. *Tribol. Int.* **2022**, *172*, 107636. [CrossRef]
11. Wąsik, A.; Leszczyńska-Madej, B.; Madej, M. Effects of TiO_2 Nanoparticle Addition on Microstructure and Selected Properties of Ag–xTiO_2 Composites. *Materials* **2021**, *14*, 4846. [CrossRef] [PubMed]
12. Gao, S.; Wang, Q.; Guo, H.; Liu, X.; Li, G.; Dong, L.; Song, K. High strength and high electrical conductivity Cu composites reinforced by SiO_2 aerogel particles. *J. Mater. Res. Technol.* **2023**, *24*, 2346–2356. [CrossRef]
13. Kirubagharan, R.; Ramesh, C.; Ranjith, M.; Kumar, V.S.; Soundhirarajan, N. Mechanical behaviour of aluminium based metal matrix composite reinforced with Si_3N_4 and B_4C. *Mater. Today Proc.* **2020**, *33*, 3347–3352. [CrossRef]
14. Kushwah, R.; Jayan, M.; Kanagasabapathy, H.; Kumar, T.C.A.; Halder, S.; Begum, N.; Rajkumar, S. The impact of boron nitride (BN) on tribological behaviour of AZ84 magnesium matrix composites. *Mater. Today Proc.* **2022**, *59*, 1271–1276. [CrossRef]
15. Sulima, I.; Hyjek, P.; Podsiadło, M.; Boczkal, S. Effect of Zirconium Diboride and Titanium Diboride on the Structure and Properties of 316L Steel-Based Composites. *Materials* **2023**, *16*, 439. [CrossRef] [PubMed]
16. Bacon, D.H.; Edwards, L. Fatigue and fracture of a 316 stainless steel metal matrix composite reinforced with 25% titanium diboride. *Int. J. Fatigue* **2012**, *48*, 39–47. [CrossRef]
17. Guo, X.; Jia, L.; Lu, Z.; Xie, H.; Kondoh, K. Enhanced combination of strength and electrical conductivity properties with CrB_2 reinforced Cu matrix composites fabricated by powder metallurgy. *Mater. Today Commun.* **2024**, *38*, 107980. [CrossRef]
18. Li, Y.; Fang, W.; Guo, S.; Duan, M.; Luo, B.; Li, Y.; Liu, J.; Gu, J.; Peng, Y.; Zhou, Q.; et al. In-situ reinforced Ti6Al4V-TiB_2 titanium matrix composite fabricated by selective laser melting: Microstructure, mechanical properties and heat treatment. *Mater. Charact.* **2024**, *212*, 113952. [CrossRef]
19. Demirtaş, H.; Karakulak, E.; Nadendla, H.B. Production and characterization of in-situ Al-NbB_2 composites. *Mater. Lett.* **2023**, *330*, 133377. [CrossRef]
20. Fahrenholtz, W.G.; Hilmas, G.E.; Talmy, I.G.; Zaykoski, J.A. Refractory diborides of zirconium and hafnium. *J. Am. Ceram. Soc.* **2007**, *90*, 1347–1364. [CrossRef]
21. Mroz, C. Zirconium Diboride. *Am. Ceram. Soc. Bull.* **1994**, *73*, 141.
22. Sivakumar, S.; Golla, B.R.; Rajulapati, K.V. Influence of ZrB2 hard ceramic reinforcement on mechanical and wear properties of aluminum. *Ceram. Int.* **2019**, *45*, 7055–7070. [CrossRef]
23. Jiang, Y.; Wang, C.; Liang, S.; Ren, J.; Du, X.; Liu, F. TiB_2(-TiB)/Cu in-situ composites prepared by hot-press with the sintering temperature just beneath the melting point of copper. *Mater. Charact.* **2016**, *121*, 76–81. [CrossRef]
24. Ružić, J.; Stašić, J.; Rajković, V.; Božić, D. Synthesis, microstructure and mechanical properties of ZrB_2 nano and microparticle reinforced copper matrix composite by in situ processings. *Mater. Des.* **2014**, *62*, 409–415. [CrossRef]
25. Prajapati, P.K.; Chaira, D. Fabrication and characterization of Cu–B4C metal matrix composite by powder metallurgy: Effect of B4C on microstructure, mechanical properties and electrical conductivity. *Trans. Indian Inst. Met.* **2018**, *72*, 673–684. [CrossRef]
26. Roosta, M.; Baharvandi, H. The comparison of W/Cu and W/ZrC composites fabricated through hot-press. *Int. J. Refract. Met. Hard Mater.* **2010**, *28*, 587–592. [CrossRef]
27. Zhang, G.; Jiang, X.; Qiao, C.; Shao, Z.; Zhu, D.; Zhu, M. Investigation of the microstructure and mechanical properties of copper-graphite composites reinforced with single-crystal α-Al2O3 fibres by hot isostatic pressing. *Materials* **2018**, *11*, 982. [CrossRef]
28. Sule, R.; Olubambi, P.A.; Sigalas, I.; Asante, J.K.O.; Garrett, J.C. Effect of SPS consolidation parameters on submicron Cu and Cu-CNT composites for thermal management. *Powder Technol.* **2014**, *258*, 198–205. [CrossRef]
29. Sulima, I.; Boczkal, G. Processing and Properties of ZrB_2-Copper Matrix Composites Produced by Ball Milling and Spark Plasma Sintering. *Materials* **2023**, *16*, 7455. [CrossRef] [PubMed]

30. Tokita, M. Progress of Spark Plasma Sintering (SPS) Method, Systems, Ceramics Applications and Industrialization. *Ceramics* **2021**, *4*, 160–198. [CrossRef]
31. Wang, F.; Li, Y.; Wang, X.; Koizumi, Y.; Kenta, Y.; Chib, A. In-situ fabrication and characterization of ultrafine structured Cu-TiC composites with high strength and high conductivity by mechanical milling. *J. Alloys Compd.* **2016**, *657*, 122–132. [CrossRef]
32. Long, B.D.; Othman, R.; Umemoto, M.; Zuhailawati, H. Spark plasma sintering of mechanically alloyed in situ copper–niobium carbide composite. *J. Alloys Compd.* **2010**, *505*, 510–515. [CrossRef]
33. Dash, K.; Ray, B.C.; Chaira, D. Synthesis and characterization of copper–alumina metal matrix composite by conventional and spark plasma sintering. *J. Alloys Compd.* **2012**, *512*, 78–84. [CrossRef]
34. Tokita, M. Industrial applications for functionally graded materials fabricated by spark plasma sintering (SPS) systems. *Sci. Technol.* **2000**, *3*, 251–256.
35. Soloviova, T.; Solodkyi, I.; Loboda, P.I. Spark plasma sintering of Cu-(LaB6-TiB_2) metal-ceramic composite and its physical-mechanical properties. *J. Superhard Mater.* **2019**, *41*, 213–220. [CrossRef]
36. Cipolloni, G.; Pellizzari, M.; Molinari, A.; Hebda, M.; Zadra, M. Contamination during the high-energy milling of atomized copper powder and its effects on spark plasma sinteing. *Powder Technol.* **2015**, *275*, 51–59. [CrossRef]
37. Wang, C.; Lin, H.; Zhang, Z.; Li, W. Fabrication, Interfacial characteristics and strengthening mechanisms of ZrB_2 microparticles reinforced Cu composites prepared by hot-pressed sintering. *J. Alloys Compd.* **2018**, *748*, 546–552. [CrossRef]
38. Zhang, X.; Qian, L.; Bin, L.; Xi, C.; You, L.; Zhen, H.; Huan, L. Enhanced softening resistance and mechanical properties of Mo_2C particle-reinforced Cu-matrix composites. *Compos. Struct.* **2023**, *305*, 11650. [CrossRef]
39. Qin, Y.Q.; Tian, Y.; Zhuang, Y.; Luo, L.M.; Zan, X.; Wu, Y.C. Effects of solid–liquid doping and spark plasma sintering on the microstructure and mechanical properties of Y_2O_3-doped copper matrix composites. *Vacuum* **2021**, *192*, 110436. [CrossRef]
40. *ISO 6892-1:2019;* Metallic Materials—Tensile Test, Part 1: Method of Test at Room Temperature. ISO: Geneva, Switzerland, 2019.
41. *ISO 20808:2004(E);* Part 1: Test Method International Standard, Fine Ceramics (Advanced Ceramics, Advanced Technical Ceramics)—Determination of Friction and Wear Characteristics of Monolithic Ceramics by Ball-on-Disc Method. ISO: Geneva, Switzerland, 2004.
42. Shaik, M.; Golla, B. Two body abrasion wear behaviour of Cu–ZrB_2 composites against SiC emery paper. *Wear* **2020**, *450–451*, 203260. [CrossRef]
43. Zhang, H.; Li, Y.; Xu, Z.; Fu, Y.; Li, M.; Zhao, G.; Chen, J. The microstructure, mechanical properties and tribological behaviors of YB_2C_2 reinforced Cu matrix composites prepared by spark plasma sintering. *Compos. Part A* **2024**, *187*, 108500. [CrossRef]
44. Zhai, Z.; Dong, H.; Li, D.; Wang, Z.; Sun, C.; Chen, C. Effect of TiC Particles on the Properties of Copper Matrix Composites. *Inorganics* **2024**, *12*, 120. [CrossRef]
45. Monticelli, C.; Bellosi, A.; Collec, M.D. Electrochemical Behavior of ZrB_2 in Aqueous Solutions. *J. Electrochem. Soc.* **2004**, *151*, B331–B339. [CrossRef]
46. Sulima, I.; Stępień, M.; Hyjek, P.; Boczkal, S.; Kowalik, R. Mechanical, Corrosion and Wear Characteristics of Cu-Based Composites Reinforced with Zirconium Diboride Consolidated by SPS. *Metals* **2024**, *14*, 974. [CrossRef]
47. Monticelli, C.; Zucchi, F.; Pagnoni, A.; Colle, M.D. Corrosion of a zirconium diboride/silicon carbide composite in aqueous solutions. *Electrochim. Acta* **2005**, *50*, 3461–3469. [CrossRef]
48. Zhou, J.; Ma, C.; Kang, X.; Zhang, L.; Liu, X. Effect of WS_2 particle size on mechanical properties and tribological behaviors of Cu−WS_2 composites sintered by SPS. *Trans. Nonferrous Met. Soc. China* **2018**, *28*, 1176–1185. [CrossRef]
49. Fathy, A.; Elkady, O.; Abu-Oqail, M. Synthesis and characterization of Cu–ZrO_2 nanocomposite produced by thermochemical process. *J. Alloys Compd.* **2017**, *719*, 411–419. [CrossRef]
50. Tjong, S.C.; Lau, K.C. Tribological behaviour of SiC particle-reinforced copper matrix composites. *Mater. Lett.* **2000**, *43*, 274–280. [CrossRef]
51. Tjong, S.C.; Lau, K.C. Abrasive wear behavior of TiB_2 particle-reinforced copper matrix composites. *Mater. Sci. Eng.* **2000**, *282A*, 183–186. [CrossRef]

materials

MDPI

Article

Hydrogen Diffusion in Deformed Austenitic TRIP Steel—A Study of Mathematical Prediction and Experimental Validation

Christian Hempel [1], Marcel Mandel [2,*], Caroline Quitzke [3], Marco Wendler [3], Thilo Kreschel [3], Olena Volkova [3] and Lutz Krüger [2]

[1] Sunfire GmbH, Gasanstaltstr. 2, 01237 Dresden, Germany; christian.hempel@sunfire.de
[2] Institute of Materials Engineering, Technische Universität Bergakademie Freiberg, Gustav-Zeuner Str. 5, 09599 Freiberg, Germany; krueger@ww.tu-freiberg.de
[3] Institute of Iron and Steel Technology, Technische Universität Bergakademie Freiberg, Leipziger Str. 34, 09599 Freiberg, Germany; caroline.quitzke@iest.tu-freiberg.de (C.Q.); marco.wendler@iest.tu-freiberg.de (M.W.); thilo.kreschel@iest.tu-freiberg.de (T.K.); olena.volkova@iest.tu-freiberg.de (O.V.)
* Correspondence: mandel@iwt.tu-freiberg.de

Abstract: This study focuses on the effect of pre-deformation on hydrogen diffusion and hydrogen embrittlement of the high alloy austenitic TRIP steel X3CrMnNiMo17-8-4. Different cold-rolled steel sheets with thicknesses of ≤400 μm were electrochemically charged on both sides in 0.1 M sodium hydroxide with hydrogen for two weeks. Comparative measurements on uncharged and immersed samples prove that hydrogen causes embrittlement in this steel for all investigated states. The embrittlement increases with increasing pre-deformation and is accompanied by deformation-induced martensite formation. The corresponding fractured surfaces were examined using electron microscopy and compared to modelled hydrogen distributions with previously determined diffusion coefficients. For this purpose, various diffusion coefficients are determined using the Devanathan–Stachurski permeation test and hot extraction in order to describe the diffusion process. The hydrogen concentration profiles and the fractographic analyses show a good agreement, so this study provides a basis for estimating the embrittlement behaviour for later application.

Keywords: cold rolling; high alloy austenitic steel; hydrogen diffusion modelling; hydrogen embrittlement; TRIP

Citation: Hempel, C.; Mandel, M.; Quitzke, C.; Wendler, M.; Kreschel, T.; Volkova, O.; Krüger, L. Hydrogen Diffusion in Deformed Austenitic TRIP Steel—A Study of Mathematical Prediction and Experimental Validation. *Materials* **2024**, *17*, 6114. https://doi.org/10.3390/ma17246114

Academic Editor: Xiaogang Li

Received: 29 October 2024
Revised: 22 November 2024
Accepted: 11 December 2024
Published: 13 December 2024

1. Introduction

Hydrogen is said to be one of the key options for decarbonisation [1,2] to achieve the ambitious goals of the energy transition. In this context, new components and devices are being developed for hydrogen applications [3,4], but it is also conceivable to use existing systems [5]. Due to the fact that hydrogen is the smallest atom, it can penetrate into several materials, e.g., polymers [6], ceramics [7] or metals [8], and can also stimulate chemical reactions and subsequently generate a wide variety of corrosion products [9]. One of the pivotal questions is if and how dissolved hydrogen affects the used materials [10,11]. It is well known that hydrogen interacts with materials, especially metals, and can change the mechanical properties [12,13]. This effect, also known as hydrogen embrittlement (HE), is based on the hydrogen uptake and its diffusion behaviour, which depends on a large number of influencing factors [14–16]. As known for steels, susceptibility to HE is related to the apparent hydrogen diffusion coefficient. Ferritic steels that exhibit a relatively high diffusion coefficient (D_{app}~2×10^{-7} cm^2 s^{-1} [17]) are more susceptible to HE than austenitic steels (D_{app}~3×10^{-12} cm^2 s^{-1} [18]), which exhibit a significant lower diffusion coefficient [19]. For two- or multiphase systems, the diffusion coefficients are usually between those of the single components and depend on the phase composition [20,21].

With regard to modern steel development, weight and cost savings are often achieved through the use of advanced high-strength steels (AHSSs). This class of steels includes transformation-induced plasticity (TRIP) steels that show a deformation-induced martensite formation from metastable austenite during plastic deformation [22]. This phase transformation leads to enhanced mechanical properties, e.g., higher strength [23,24]. According to this phase transformation, TRIP steels appear as two- or multiphase systems. The use of such steels in hydrogen applications is attractive because of their good mechanical properties, such as high strain hardening, high yield and tensile strength and fracture toughness [25]. Investigations on the HE of pre-strained TRIP steels have already been part of the literature [26–28]. In general, all studies agree that a pre-deformation leads to a higher susceptibility to hydrogen embrittlement.

In addition to classic material testing, theoretical calculations for describing HE have also been established [29–32]. With such calculations, the fracture behaviour can be characterised, and an estimation of the durability of materials under real conditions is possible, which is a main task when making material decisions.

An essential part of this study is to investigate the effect of cold working/phase transformation on susceptibility to HE of the newly developed TRIP steel X3CrMnNiMoN17-8-4. The steel is characterised by excellent strength, formability and corrosion resistance, which makes it very interesting as a pipeline material for applications in the oil and gas industry. For this purpose, different pre-strained material states were electrochemically charged with hydrogen and subsequently analysed by tensile tests. In addition, extensive experiments were carried out using hot extraction in order to describe the hydrogen distribution after charging and, finally, at the stage of mechanical testing. The results of hot extraction were compared to calculations of the hydrogen distribution along the steel cross section and finally correlated to fractographic analyses of the fracture surfaces.

2. Materials and Methods

2.1. Materials

In this study, the high alloy austenitic steel X3CrMnNiMoN17-8-4 was investigated, and its chemical composition is given in Table 1. A precise description of the manufacturing process is depicted in detail by Quitzke et al. [33].

Table 1. Chemical composition of the investigated high alloy austenitic stainless steel in wt.%.

C	Cr	Mn	Ni	Mo	Si	N	Fe	Traces
0.03	17.4	7.48	3.63	0.56	0.15	0.19	bal.	S, P, Al, Nb, Ti, Sn < 0.01 Co, Cu, W < 0.03 V < 0.06

In order to achieve thin steel sheets, a hot-rolled strip of thickness h = 4 mm was cold-rolled in a quarto stand. Due to work hardening, intermediate stress relief annealing was necessary after several rolling passes. However, the actual starting points after the last solution annealing are summarised in Table 2. The given true strains ε were calculated by Equation (1) as follows:

$$\varepsilon = \left| \ln \frac{h_1}{h_0} \right| \quad (1)$$

where h_0—thickness before rolling; and h_1—thickness after rolling.

Table 2. Strip thickness before and after cold rolling with a corresponding number of roll passes and calculated true strain.

Strip Thickness h_0 Before Cold Rolling [µm]	Strip Thickness h_1 After Cold Rolling [µm]	Number of Roll Passes	True Strain
400	400	-	0
400	292	1	0.32
450	305	2	0.39
570	350	3	0.49

2.2. Hydrogen Charging

Samples with a dimension of 55 × 55 mm were electrochemically charged with hydrogen for two weeks in a double-cell test set-up, according to Devanathan and Stachurski (DS cell) [34]. Before charging, all samples were coated with palladium to prevent oxide layers and anodic metal dissolution [35,36]. The permeation experiments were carried out in a 0.1 M sodium hydroxide solution within both cells. In order to reduce hydrogen recombination, an amount of 25 mg L^{-1} of arsenic oxide was added as recombination poison [37]. The charged area was A = 13.46 cm^2 (d = 4.14 cm). A current density of $i_c = -10$ mA cm^{-2} was applied on both sides to achieve a uniform hydrogen distribution within the steel samples.

Furthermore, various samples with a pre-deformation of $\varepsilon = 0.49$ were charged for 4, 7, 14, 21 and 36 days. The total amount of diffusible hydrogen was determined using hot extraction (see Section 2.5) and compared to the mathematical calculations. In addition to the determination of hydrogen concentration during charging, the hydrogen amount for the 36-day charged sample was also determined during discharging after 3, 7, 14, 28 and 56 days. Therefore, the charged sample was cut into pieces, and the extraction was performed as triple repeat determination. Similar tests were performed for the other material states.

2.3. Tensile Tests

Three samples were cut from the charged area using electrical discharge machining, as schematically shown in Figure 1a. The sample geometry is given in Figure 1b. The tensile tests were performed using a miniature load frame (Kammrath & Weiss, Schwerte, Germany) at a linear velocity of 10 µm s^{-1}. The corresponding strain measurements were evaluated by an optical extensometer based on digital image correlation using the microDAC system (Chemnitzer Werkstoffmechanik, Chemnitz, Germany). Three different experimental conditions were chosen to examine the influence of hydrogen charging as well as the influence of corrosion caused by the electrolyte. The first one was the uncharged initial state, the second one was the two-week hydrogen-charged state and the third one was tested after two weeks of immersion in the 0.1 M sodium hydroxide solution. Due to the necessary manufacturing process, the mechanical tests were carried out two weeks after the charging was stopped so that the effusion of hydrogen was considered as well. Based on the high hydrogen charging test times at room temperature, the condition of a two-week hydrogen charging was chosen as an average testing time, which should show a dependence of the embrittlement depth on the degree of deformation. The fractographic analyses carried out after the tensile test reveal this dependence very well. It is assumed that if the charging time is extended, the embrittlement front penetrates deeper into the material, the entire sample cross section becomes homogeneously brittle and it becomes more difficult to demonstrate the dependence on the degree of deformation. Nevertheless, varying environmental conditions and the extension of the charging time are recommended to gain a deeper understanding of the influence of hydrogen on TRIP steels. However, these far-reaching analyses are not the subject of this study.

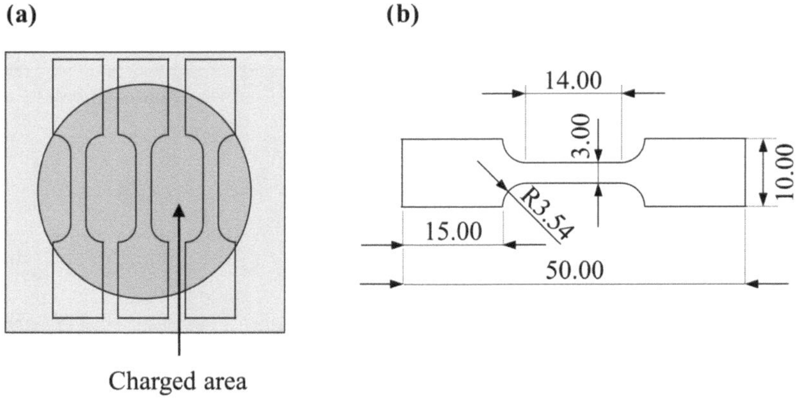

Figure 1. (**a**) Schematic drawing of tensile test samples extracted from the hydrogen-charged area; (**b**) geometry of the used tensile samples according to [38] (Unit: mm).

2.4. Fractography

For the fractographic SEM analyses, a MIRA 3 XMU (TESCAN, Brno, Czech Republic) was used to analyse the HE and to compare the size of the embrittled areas using mathematical calculations.

2.5. Hydrogen Determination

The amount of diffusible hydrogen was determined via hot extraction using nitrogen as carrier gas at 1373 K in a Rosemount H2A (Fisher-Rosemount, Langenfeld, Germany) gas analyser. For quantification, a thermal conductivity detector was used.

3. Mathematics

Based on the experimental investigations and results, a calculation of the hydrogen concentration profiles in dependence on the degree of deformation was performed. The nomenclature of the used parameters is given in Table 3.

Table 3. Applied nomenclature in this study.

Symbol	Meaning
A	Charged area, 13.46 cm^2
c	Concentration of hydrogen, wt.%
c_0	Hydrogen concentration before charging, wt.%
c_1	Hydrogen concentration at equilibrium, wt.%
c_H	Overall hydrogen concentration in a sample, wt.%
D	Diffusion coefficient
D_{app}	Apparent diffusion coefficient
E_a	Activation energy
k_0	Pre-exponential factor
L	Length of diffusion/thickness of measured sheets
t	Time
x	Position vector

The basis for the calculations of the diffusion processes are Fick's laws I and II. The relation between sheet thickness ($L \leq 400$ μm) and the hydrogen-charged area with a diameter of d = 4.2 cm allows for an assumption of a unidirectional diffusion along the sheet cross section. The location-dependent Fick's law I in one dimension is given in Equation (2), and the corresponding Fick's law II, which describes the change in concentration with respect to the time, is given in Equation (3).

$$j(x,t) = -D\,\frac{\partial c(x,t)}{\partial x} \tag{2}$$

$$\frac{\partial c}{\partial t} = -D\,\frac{\partial^2 c(x,t)}{\partial x^2} \tag{3}$$

For a both-sided charging of plane sheets, an analytical solution based on a Laplace transformation that considers particular boundary conditions was stated by Crank [39] and is given in Equation (4). The boundary conditions are defined by the following:

1. A uniform hydrogen distribution in the material at the beginning of charging.
2. A symmetric and steady concentration of hydrogen at the entry sides.

$$\frac{c(x)-c_0}{c_1-c_0} = 1-\frac{4}{\pi}\sum_{n=0}^{\infty}\frac{(-1)^n}{2n+1}exp\left(-\frac{D\,(2n+1)^2\pi^2 t}{4L^2}\right)cos\left(\frac{(2n+1)\pi x}{2L}\right) \tag{4}$$

where $-L < x < L$, and for a given diffusion coefficient and sample geometry, a concentration profile can be calculated for a specific charging time.

As mentioned above, the tensile tests could not have been performed during or immediately after charging. Therefore, it is also necessary to evaluate the discharging process of hydrogen. For this purpose, Equation (5), which is also given by Crank [39], was applied to describe the changes in concentration for a given initial hydrogen distribution. The equation is valid if the hydrogen concentration at the entry/exit side is zero, which is assumed when the charging process is stopped.

$$c'(x) = \frac{2}{L}\sum_{n=1}^{\infty} sin\left(\frac{n\pi x}{\pi}\right)exp\left(-\frac{D\pi^2 n^2 t}{L^2}\right)\int_0^L c(x)sin\left(\frac{n\pi x}{\pi}\right)dx \tag{5}$$

Finally, when using the calculated concentration profiles, the total amount of dissolved hydrogen is evaluable by integration (Equation (6)). In this study, the integrals were approximated via the trapezoidal method. Figure 2 shows an example of the calculated hydrogen distribution in a plane sheet after hydrogen charging and subsequent discharging. In this study, the step size of the position vector was chosen to be 1 μm for each calculation.

$$c_H = \int_0^L c(x)\,dx \tag{6}$$

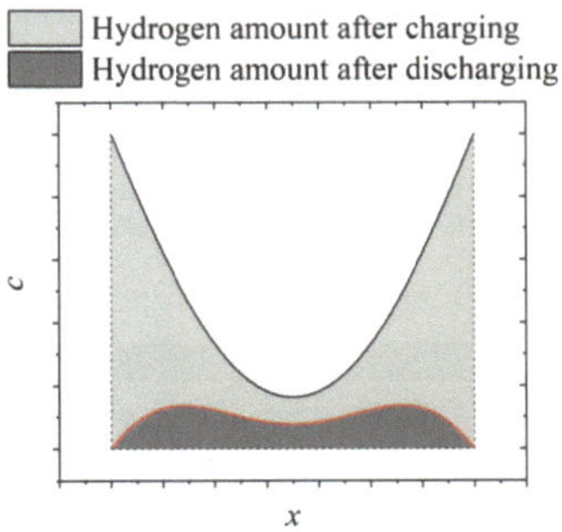

Figure 2. Schematic view of calculated hydrogen distributions in a plane sheet after hydrogen charging and subsequent discharging, grey areas indicate the integral of curves, representing the total amount of hydrogen. Black/red line—concentration profile *c*(x) after charging/discharging.

All calculations were determined using the MATLAB software R2018b by MathWorks (Natick, MA, USA). In conjunction with sum expressions, the Symbolic Math Toolbox was used as an additional tool. Our own designed MatLab scripts were used for the optimisation calculations based on the algorithm of Lagarias et al. [40]. Moreover, the calculations were carried out at room temperature and required the apparent diffusion coefficient D_{app} and the hydrogen equilibrium concentration c_1 as input parameters. To evaluate D_{app}, hydrogen permeation tests at increased temperatures of 323 K, 338 K and 353 K were performed using a Devanathan–Stachursky test set-up, and the apparent diffusion coefficient at room temperature was extrapolated using the Arrhenius relationship. Details of the method are described elsewhere, and the interested reader is referred to ref. [41]. In a similar manner, c_1 was extrapolated, applying Equations (4) and (6) to experimental results for the non-equilibrium state.

4. Results and Discussion

4.1. Modelling of Hydrogen Distribution and Determination of Hydrogen Concentration

The parameters necessary for modelling the hydrogen distribution in the samples are partially published elsewhere [41]. The pre-exponential factor and the activation energies were determined in a DS cell experiment for a temperature range from 323.15 K to 353.15 K in order to shorten the test duration. The Arrhenius equation (Equation (7)) is afterwards used to extrapolate the diffusion coefficient at room temperature (RT = 293.15 K). The calculated values are summarised in Table 4.

$$D_{app} = k_0 \cdot exp\left(-\frac{E_a}{R \cdot 293.15\ K}\right) \tag{7}$$

Table 4. Pre-exponential factors and activation energies for hydrogen diffusion in pre-deformed X3CrMnNiMoN17-8-4 determined by DS cell and calculated apparent diffusion coefficients at RT according to [40].

True Strain	First Charging			Discharging		
	k_0/cm^2 s^{-1}	E_a/kJ mol^{-1}	D_{app} at RT/cm^2 s^{-1}	k_0/cm^2 s^{-1}	E_a/kJ mol^{-1}	D_{app} at RT/cm^2 s^{-1}
0	2.6×10^{-5}	42.8	6.2×10^{-13}	1.3×10^{-7}	23.4	8.9×10^{-12}
0.32	5.1×10^{-5}	39.7	4.4×10^{-12}	2.1×10^{-7}	21.4	3.2×10^{-11}
0.39	1.3×10^{-5}	33.7	1.3×10^{-11}	4.3×10^{-7}	20.8	8.4×10^{-11}
0.49	3.8×10^{-7}	20.6	4.2×10^{-11}	5.8×10^{-7}	21.1	1.0×10^{-10}

The initial hydrogen concentration was determined to be c_0 = 0.93 ± 0.27 ppm and assumed to be uniform. The value was subtracted from the data, which were determined by hot extraction measurements.

In order to verify the extrapolated diffusion coefficients, various samples of the pre-deformed material state ε = 0.49 were charged with hydrogen for different periods, and the total amount of dissolved hydrogen was measured. Figure 3a shows the experimental data of these measurements. Due to the fact that the equilibrium was not reached, the value of the equilibrium hydrogen concentration c_1 was fitted to the experimental data by substituting Equation (4) in Equation (6) and by using the extrapolated apparent diffusion coefficient at RT. The equilibrium hydrogen concentration was determined to be about c_1 = 780 ppm. With respect to the fact that hydrogen concentrations higher than 1000 ppm have been reported for pure austenitic steels under similar charging conditions [42], the given value is reasonable. Figure 3b represents the corresponding hydrogen distributions that were calculated by Equation (4) for the different charging times in consideration of D_{app} and c_1.

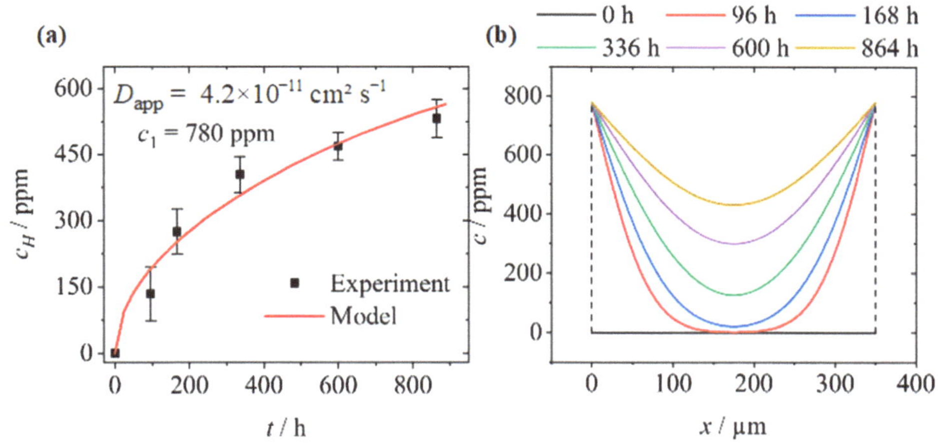

Figure 3. (**a**) Hydrogen concentrations in ε = 0.49 pre-deformed X3CrMnNiMoN17-8-4 and corresponding mathematical fit using Equations (4) and (6); (**b**) calculated hydrogen distribution profiles in pre-deformed ε = 0.49 samples after different durations of charging at RT using the extrapolated equilibrium concentration c_1 = 780 ppm and the extrapolated apparent diffusion coefficient D_{app} = 4.2 × 10^{-11} cm^2 s^{-1}.

Figure 4 shows the measured hydrogen amount during discharging in samples that were charged for 36 days and stored in a desiccator. As described in Section 3, Equations (5) and (6) were used to calculate the theoretical hydrogen distributions and overall hydrogen concentrations during the effusion of hydrogen. In the same way, as for the charging process, the Arrhenius equation (Equation (7)) was used to extrapolate the apparent diffusion coefficient at RT for the effusion of hydrogen. The calculated values from the DS cell experiments are summarised in Table 4. When comparing the values of hydrogen, determined via hot extraction (Figure 4, black squares), to the calculated outward diffusion using D_{app} = 1.0 × 10^{-10} cm^2 s^{-1} of DS cell experiments (Figure 4, red line), it is noticeable that there is a clear deviation. When using the DS cell, the measured hydrogen effusion was much faster than during storage at low humidity atmosphere. There are two possible explanations for this discrepancy. On one hand, mainly weakly bound hydrogen is detected in the DS cell measurements, e.g., hydrogen that is interstitially dissolved in the lattice. Whether hydrogen diffuses out of traps during discharging in the DS cell, and if so, how much hydrogen, can hardly be said without further methods. Due to the higher binding energy, trapped hydrogen diffuses more slowly out of the samples. On the other hand, the both-sided charged samples that were used for hot extraction were cleaned after their removal and stored in low-humidity air. Larger parts of the palladium layer were detached during cleaning, whereby it can be expected that a passive oxide has subsequently formed on the underlying metal surface. The literature provides diffusion coefficients for oxides that are some orders of magnitude lower even than that for austenite, e.g., $D_{Cr_2O_3,RT} \approx 8 \times 10^{-18}$ cm^2 s^{-1} [43]. Fe_2O_3 is described as a nearly entire diffusion barrier for hydrogen [44]. In order to obtain an impression of the hydrogen effusion actually taking place, Equations (5) and (6) were fitted to the experimental data by variation of D_{app}, resulting in a diffusion coefficient of $D_{app,\ Hot\ extraction}$ = 1.6 × 10^{-11} cm^2 s^{-1} that describes the experimental findings much better (Figure 4, blue line).

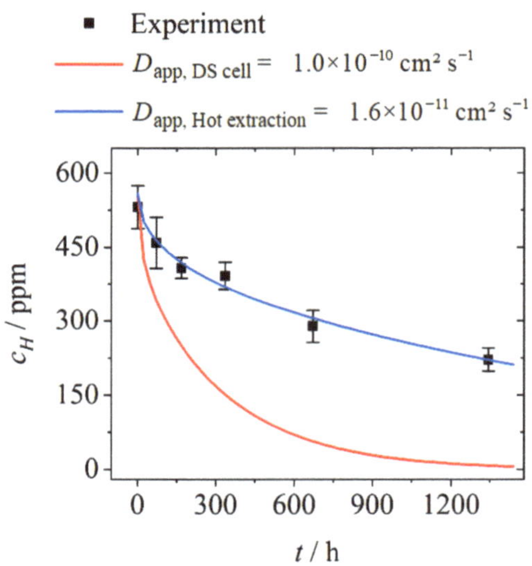

Figure 4. Experimentally determined hydrogen concentrations during discharging at atmosphere in $\varepsilon = 0.49$ pre-deformed steel sample after 36 days of hydrogen charging and corresponding mathematical fit using the extrapolated apparent diffusion coefficients D_{app} determined by DS cell (red line) and by hot extraction (blue line).

Using the results of the hot extraction experiments, the apparent diffusion coefficients for the states ε = 0, 0.32 and 0.39 were determined in the same way. The values are summarised in Table 5.

Table 5. Hot extraction-determined apparent diffusion coefficients for the effusion of hydrogen out of pre-deformed X3CrMnNiMoN17-8-4.

True Strain	$D_{app,\ Hot\ extraction}$ at RT/cm^2 s^{-1}
0	2.7×10^{-13}
0.32	3.0×10^{-12}
0.39	3.8×10^{-12}
0.49	1.6×10^{-11}

For the calculation of the hydrogen distribution profiles after two weeks of charging, the diffusion coefficients provided in Table 4 and Equation (4) were used. The values provided in Table 4 were determined in the preliminary study by Ref. [41] and exhibit a significant drop for the pre-exponential factor k_0 for $\varepsilon = 0.49$. It is assumed that this effect is due to the increased martensite content and, consequently, higher electrochemical corrosion reaction and passive oxide formation, which ultimately hinders the hydrogen diffusion process. The diffusion coefficients given in Table 5 and Equation (5) were used to calculate the concentration profiles after two weeks of discharging, which corresponds to the time of the tensile tests. The results are shown in Figure 5. The hydrogen distribution profiles prove that the penetration depth increases when the degree of deformation increases. This effect is related to the increased diffusion coefficient of hydrogen in martensite [45] when compared to austenite [42]. The profiles also confirm that the hydrogen remained in the samples even after the two weeks of effusion.

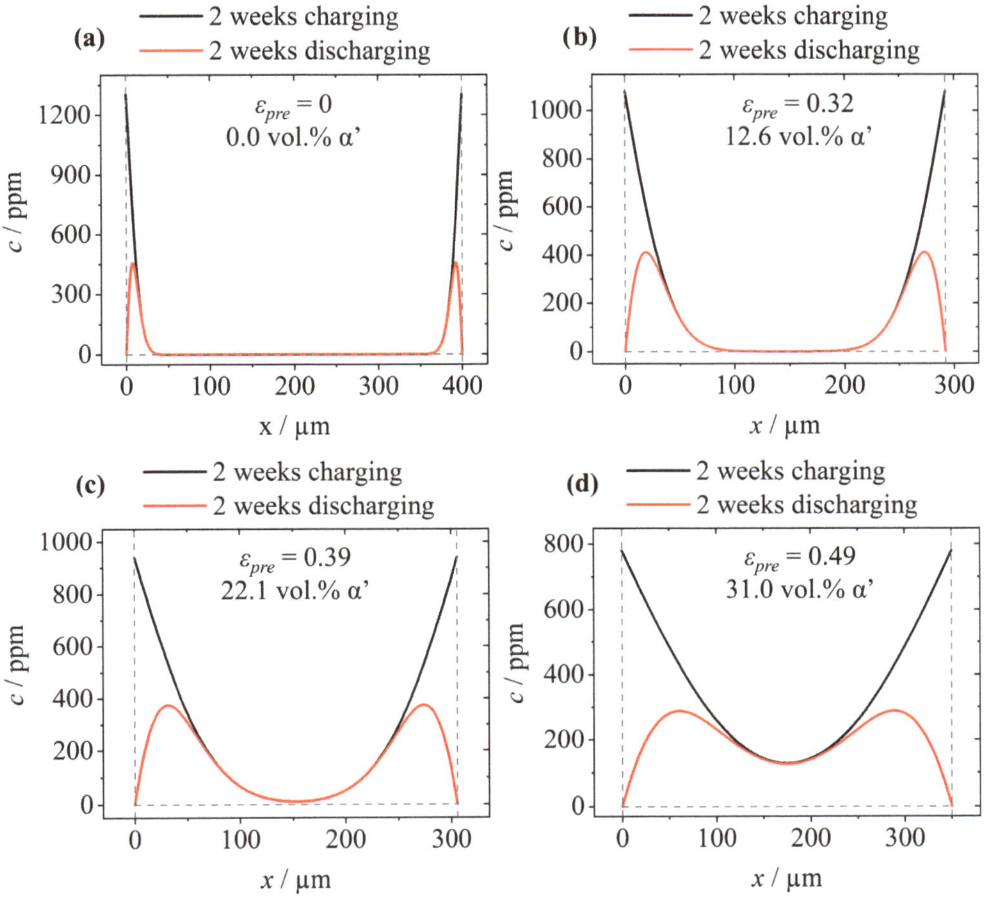

Figure 5. Hydrogen distributions in different pre-deformed states of X3CrMnNiMoN17-8-4 after two weeks of electrochemical charging (black line) and after two weeks of discharging (red line); (**a**) ε = 0, (**b**) ε = 0.32, (**c**) ε = 0.39, and (**d**) ε = 0.49. Dashed lines indicate the sample width used for calculation.

Based on the distribution profiles, the calculated and experimental total amounts of hydrogen $c_{H,calc}$ and $c_{H,exp}$ after two weeks of charging and two weeks of discharging were determined by Equation (6). The results are summarised in Table 6. The comparison shows good agreement between calculation and experiment and confirms the appropriateness of the chosen mathematical approach.

Table 6. Comparison of calculated and experimentally determined hydrogen amounts in different pre-strained X3CrMnNiMoN17-8-4.

True Strain	2 Weeks of Charging		2 Weeks of Discharging (Tensile Tests)	
	$c_{H,exp}$/ppm	$c_{H,calc}$/ppm	$c_{H,exp}$/ppm	$c_{H,calc}$/ppm
0	63.8 ± 17.3	64.3	35.5 ± 9.5	39.5
0.32	195.4 ± 25.0	191.7	100.6 ± 16.2	109.9
0.39	262.6 ± 35.3	271.7	190.1 ± 18.1	186.0
0.49	403.8 ± 41.1	357.5	177.2 ± 14.3	198.7

4.2. Mechanical Testing

The engineering stress–strain plots of the uncharged initial state and pre-deformed material states are presented in Figure 6, indicating a significant change in mechanical properties due to the pre-deformation. The pre-strain effectively enhances both yield strength (YS) and ultimate tensile strength (UTS), whereas the ductility decreases, reflected in a decrease in total elongation (TE).

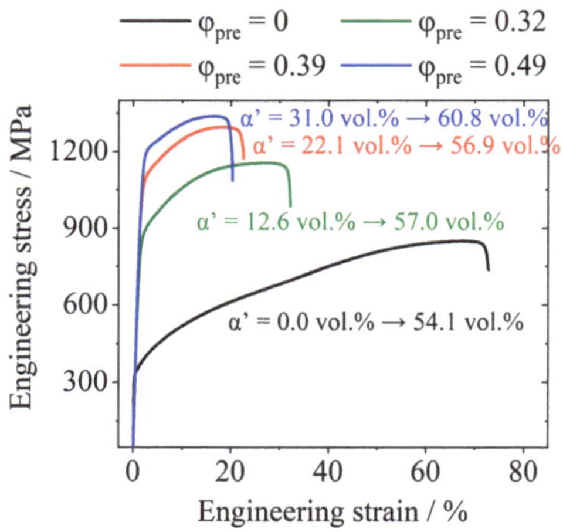

Figure 6. Results of tensile tests for X3CrMnNiMoN17-8-4 at different states of pre-deformation. The values of α' provided for each curve give the initial α' content received after cold rolling and the final α' content determined after tensile testing.

During pre-strain, the dislocation density increases, and parts of the metastable austenite transform to α'-martensite [33,41,46–49]. Both dislocations and the new interfaces interfere with dislocation movement and cause an increase in YS and UTS. The strain-induced martensite has much higher strength than austenite and less ductility. The decrease in ductility and, thus, in TE also results from impeded dislocation movement possibilities. In addition, it is noticeable that the total ratio of α'-martensite after pre-deformation and tensile tests increases with higher degrees of previous cold rolling. The observation is explained by the fact that cold rolling does not represent a uniaxial load. Thus, the values of true strain until the samples fracture and the amount of α'-martensite slightly increase.

Figure 7 presents the engineering stress–strain curves of the samples that were uncharged, hydrogen charged in 0.1 M sodium hydroxide for two weeks and immersed in 0.1 M sodium hydroxide for two weeks. For the undeformed state $\varepsilon = 0$ (see Figure 7a), it is clear that immersion in 0.1 M sodium hydroxide solution does not have a significant effect on the mechanical properties. The result confirms the good corrosion resistance of CrNi steels, which is also proved by electrochemical tests in the literature [50]. When comparing the annealed state to the cold-worked states, an opposite effect is noticed. In particular, TE decreases as a result of immersion of the pre-deformed samples in the sodium hydroxide solution (see Figure 7b–d). It is known that the corrosion resistance decreases as a result of plastic deformation [51,52]. A similar observation is made in this study, as TE decreases stronger in dependence on pre-deformation in comparison to the not immersed state and also increased pre-deformation. The α'-martensite that is induced during cold rolling promotes a pile-up of dislocations. This causes higher local potential differences that are associated with higher corrosion susceptibility. Kolodii [53] also confirms this observation by thermomechanical calculations. The corrosive attack leads to microcracks, which in turn leads to localised stresses that finally promote a failure during the tensile test. In addition, the corrosive attack also promotes the formation of hydrogen, which can embrittle the material as well. The corrosive attack also reveals that the applied palladium coating cannot be considered as perfectly dense. Nevertheless, it should also be noted that the effect of corrosion is noticed in the range of higher elongations. YS and UTS do not deteriorate significantly for the pre-deformed samples during two weeks of immersion.

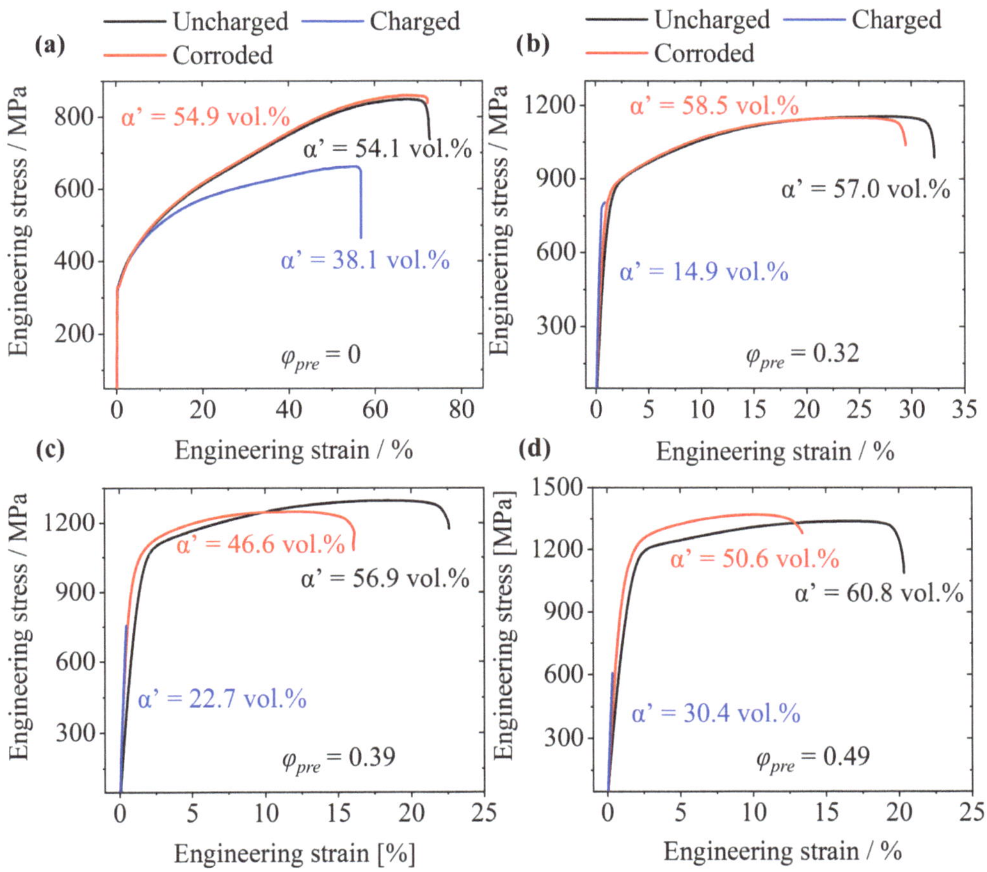

Figure 7. Tensile test results for X3CrMnNiMoN17-8-4 with and without hydrogen charging and after 2 weeks of immersion in 0.1 M sodium hydroxide solution with different pre-deformation; (**a**) $\varepsilon_{pre} = 0$, (**b**) $\varepsilon_{pre} = 0.32$, (**c**) $\varepsilon_{pre} = 0.39$, and (**d**) $\varepsilon_{pre} = 0.49$.

In contrast, the hydrogen has a significant embrittlement effect on all material states (see Figure 7a–d). For the undeformed state, this is primarily reflected in a reduction in UTS and TE (Figure 7a). As shown in Figure 5a, the theoretically calculated hydrogen penetration reaches a depth of approximately 40 µm. Thus, hydrogen does not affect the entire cross section, and a ductile deformation occurs. The distinct increase in α'-martensite during tensile testing also shows that the samples are ductile even after being charged with hydrogen. When the pre-deformation further increases, the ductility decreases rapidly. For the material state $\varepsilon = 0.32$, the yield point is just reached, and there is even a slight increase in the martensite fraction. The two material states with the highest degree of deformation already break in the range of elastic deformation. The rupture stress even decreases for the pre-deformed states, though the tensile strength originally increased. It is stated that the embrittlement caused by hydrogen increases when the degree of deformation increases. Similar results were also obtained by Hojo et al. [26], who examined a pre-strained TRIP-assisted bainitic ferrite steel. Two main reasons are given for the obtained observations.

The first reason is that the phase transformation from austenite (fcc lattice) to α'-martensite (bcc lattice) leads to a significantly higher apparent diffusion coefficient. Considering that hydrogen occupies tetrahedral voids in bcc [54] and octahedral in fcc [55], the faster hydrogen diffusion is a consequence. The occupied voids in the bcc lattice are smaller (lower binding energy [56]) and have a shorter distance to each other (lower migration energy [57]). As a result of a higher diffusion coefficient, hydrogen can penetrate the material more quickly and is able to reach more traps, which are primarily responsible for HE [58,59]. This, in turn, leads to a higher hydrogen-affected cross section affected and consequently to a higher susceptibility for HE. The second reason is that the pre-strain causes a significant increase in dislocation density, which acts as a hydrogen trap [60]. As Novak et al. [61] showed in a micro-mechanical model, dislocations can act as a main initiation point for an intergranular fracture during the mechanical loading of hydrogen-charged samples. It is concluded that after charging a material of high dislocation density, the number of hydrogen-filled dislocations increases, which finally increases the susceptibility to HE.

4.3. Fractographic Analysis and Comparison to Calculated Hydrogen Distribution Profiles

Figure 8a shows the fracture surface of the uncharged initial state. A ductile fracture is observed for all uncharged samples, which is expressed by the formation of dimples [62]. In contrast, Figure 8b shows the fracture surface of the undeformed state ($\varepsilon = 0$) after two weeks of hydrogen charging, revealing a ductile fracture in the centre of the sample and a brittle fracture at the top and the bottom where the hydrogen penetrated.

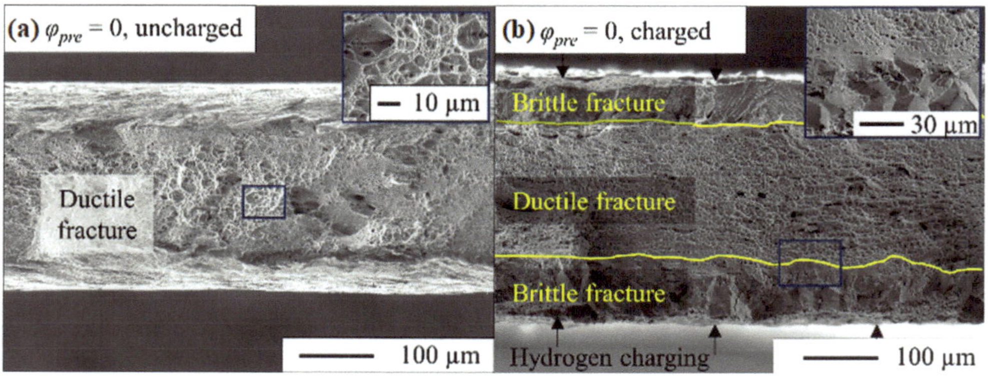

Figure 8. Comparison of fractured surfaces after tensile testing of the X3CrMnNiMoN17-8-4 in the (**a**) uncharged state and (**b**) after two weeks of hydrogen charging.

The areas of brittle fracture show typical indications of hydrogen-assisted cracking, and in Figure 9, an overview of the main differences in ductile fracture is shown in detail. A significant characteristic is the gaping grain boundaries (Figure 9a), which are associated with intergranular fracture along prior austenite grain boundaries. This phenomenon is linked to the theory of hydrogen-enhanced decohesion (hereinafter HEDE) [63,64]. The HEDE assumes that hydrogen reduces the interatomic bonding force in metals between adjacent crystallographic planes. During the fracture propagation, hydrogen diffuses preferentially to energetically favourable areas (e.g., grain or phase boundaries), lowering the cohesive strength and surface energy. A further characteristic, also explained by the HEDE theory, is the microcrack formed along a non-metallic inclusion, shown in Figure 9b. Furthermore, residual ductility is observed on the grain surfaces and is known as "crow's feet", see Figure 9c, which is also typical for HE [65,66]. These fine plastic parts are explained by a local drop in yield stress due to hydrogen and an increased mobility of dislocation movement. The mechanism is known as hydrogen-enhanced local plasticity (HELP) [67,68].

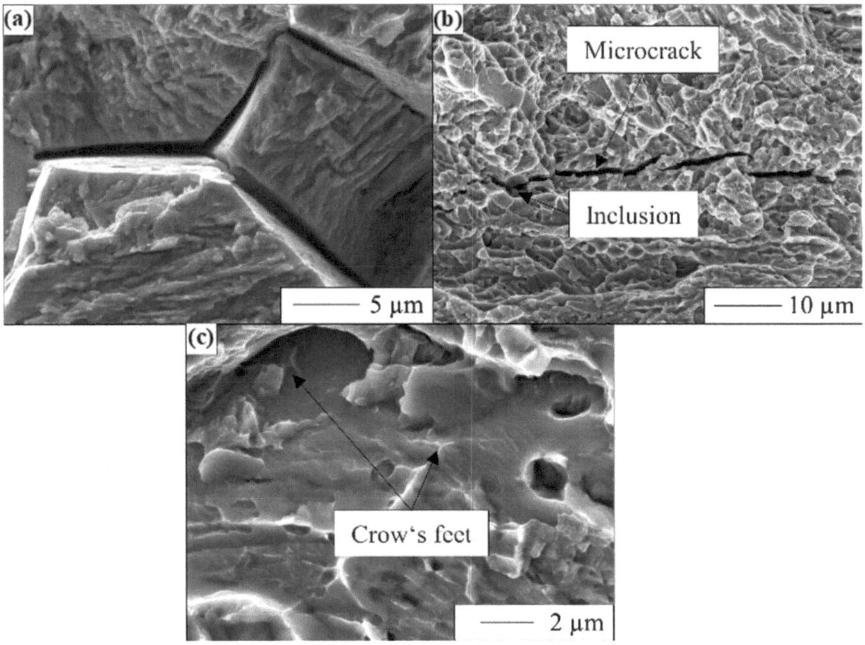

Figure 9. Characteristics of HE on the fracture surface of hydrogen charged X3CrMnNiMoN17-8-4 after tensile testing. (**a**) Gaping grain boundaries, (**b**) microcracks around inclusions, and (**c**) fine ductile hairlines known as "crow's feet".

Figure 10 shows the cross sections of the fractured surfaces of undeformed and pre-deformation states after two weeks of hydrogen charging. It is clear that the ratio of the brittle fracture surface compared to the ductile fracture surface increases when the degree of deformation increases. While many dimples are found in the undeformed state (Figure 10a), brittle fracture occurs almost over the entire cross section at the highest pre-deformed state with $\varepsilon = 0.49$ (Figure 10d). In principle, it is assumed that hydrogen propagates faster in the deformation-induced martensite due to its higher diffusion rate.

These observations are in good agreement with the calculated concentration profiles presented in Figure 5. Overlaying the profiles with the fractographic analyses exhibits the best correlations for the pre-deformed states (Figure 10b–d), as these show little or no plastic deformation. The discrepancy of the undeformed state (Figure 10 a) is related to the determination of the diffusion coefficient on a significantly thinner metal sheet (see Ref. [41] with $h_0 = 80$ µm) that was used for the concentration profile calculation. Although the chemical composition is almost identical, differences in the grain size distribution of the sample used in Ref. [41] and the sample used in this study influence the hydrogen diffusion behaviour significantly. In addition, the charged samples of the initial state also show plastic deformation and as a result, the thickness of the cross section changed. The penetration depth of the hydrogen was simply assumed to be constant in Figure 10a, but it does not correspond to reality. For this reason, the hydrogen concentration profiles of the undeformed state are considered a rough approximation.

It is important to mention that the fractographic images exhibit a very homogeneous hydrogen embrittlement front and penetration depth on both sides for all degrees of deformation. This characteristic is clearly attributed to the experimental procedure used for the hydrogen charging in the DS cell. The homogeneity of the electric field, as well as the high current applied during the charging, promote a strong vertical hydrogen diffusion and show homogeneity in the material degradation behaviour, which afterwards can be theoretically calculated very well with the suggested approach.

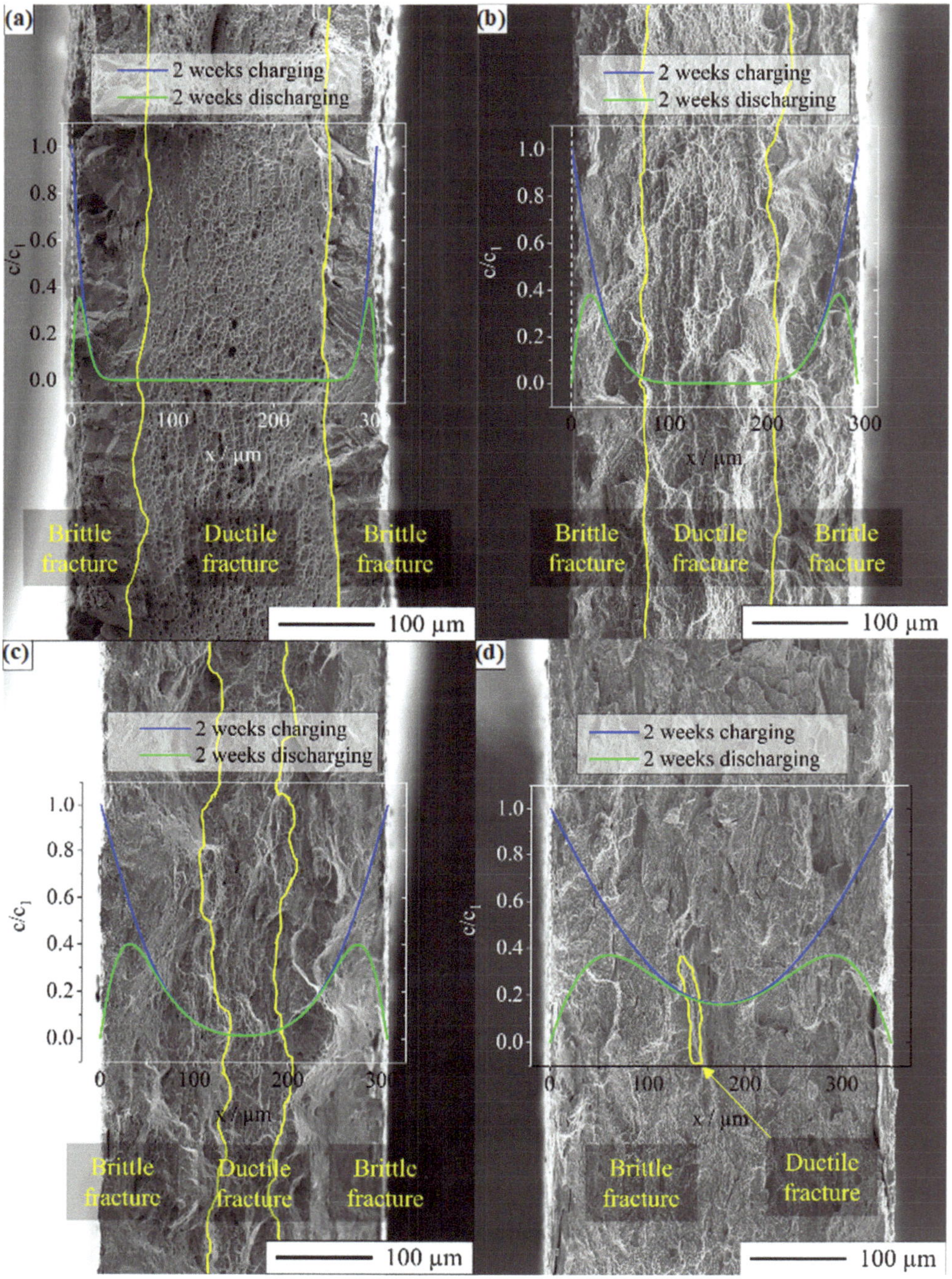

Figure 10. Comparison of fracture surfaces and calculated hydrogen distributions after two weeks of charging and two weeks of discharging in the samples of X3CrMnNiMoN17-8-4 at different degrees of deformation with (**a**) $\varepsilon = 0$, (**b**) $\varepsilon = 0.32$. (**c**) $\varepsilon = 0.39$, and (**d**) $\varepsilon = 0.49$.

Nevertheless, when transferring the results to the application, it is recommended to consider an inhomogeneity of the hydrogen distribution on the material surface and thus a local risk of embrittlement. Furthermore, homogeneous material deformation cannot be assumed under application conditions, so structure-related different penetration depths and thus local embrittlement effects can be more pronounced and must also be considered. As a consequence, the calculation of the concentration profile becomes significantly more demanding.

The consideration of inhomogeneous hydrogen distribution and a varying degree of deformation within the material are the subject of future research initiatives, whereby the finite element method (FEM) is also used as a mathematical tool for theoretical calculations.

5. Conclusions

In this study, the effect of both-sided electrochemical hydrogen charging on the mechanical properties under the influence of cold working on TRIP steel sheets of thicknesses ≤ 400 μm was investigated. In addition, calculations based on DS cell and hot extraction experiments were carried out and compared to the results of fractography. The results of the study are summarised as follows:

- The susceptibility to HE increases with increasing pre-strain for the investigated steel sheets. The effect is attributed to the increased diffusion coefficient in α'-martensite and the increased dislocation density.
- DS cell experiments are useful for the description of hydrogen charging but not useful when calculating the real hydrogen effusion behaviour.
- Hot extraction experiments are adequate to calculate diffusion coefficients for hydrogen diffusion during charging and for the period between charging und mechanical testing.
- The determined diffusion coefficients are used to calculate the hydrogen distribution profiles at the moment of tensile loading. The depth of penetration in the concentration profiles corresponds well to the brittle fracture in the fractographic sample analysis.
- The lower the pre-deformation of the steel and thus the deformation-induced martensite, the lower the hydrogen penetration depth.

Author Contributions: Conceptualisation, C.H.; Data curation, C.H., C.Q. and M.W.; Formal analysis, C.H. and M.W.; Funding acquisition, M.M., O.V. and L.K.; Investigation, C.H., C.Q. and T.K.; Methodology, C.H.; Project administration, O.V. and L.K.; Software, C.H.; Supervision, M.M., M.W., O.V. and L.K.; Validation, C.H.; Visualisation, C.H.; Writing—original draft, C.H.; Writing—review and editing, C.H. and M.M. All authors have read and agreed to the published version of the manuscript.

Funding: We are grateful for the financial support of the Deutsche Forschungsgemeinschaft (DFG) in the context of the Collaborative Research Centre 799 (transfer-project T5, funding no. 54473466).

Institutional Review Board Statement: Not applicable.

Informed Consent Statement: Not applicable.

Data Availability Statement: The datasets presented in this article are not readily available because the data are part of an ongoing study. Requests to access the datasets should be directed to mandel@iwt.tu-freiberg.de.

Acknowledgments: We express our gratitude to D. Seifert and J. Freudenberger (Leibniz IFW Dresden) for experimental support/cold rolling. Last but not least, the authors are very grateful to Robert Lehnert and PD Anja Weidner for their experimental support during the execution of tensile tests.

Conflicts of Interest: Author Christian Hempel was employed by the Sunfire GmbH. The remaining authors declare that the research was conducted in the absence of any commercial or financial relationships that could be construed as a potential conflict of interest.

References

1. Parra, D.; Valverde, L.; Javier Pino, F.; Patel, M.K. A review on the role, cost and value of hydrogen energy systems for deep decarbonisation. *Renew. Sustain. Energy Rev.* **2019**, *101*, 279–294.
2. Sgobbi, A.; Nijs, W.; De Miglio, R.; Chiodi, A.; Gargiulo, M.; Thiel, C. How far away is hydrogen? Its role in the medium and long-term decarbonisation of the European energy system. *Int. J. Hydrogen Energy* **2016**, *41*, 19–35.
3. Posdziech, O.; Schwarze, K.; Brabandt, J. Efficient hydrogen production for industry and electricity storage via high-temperature electrolysis. *Int. J. Hydrogen Energy* **2019**, *44*, 19089–19101.
4. Cipriani, G.; Di Dio, V.; Genduso, F.; La Cascia, D.; Liga, R.; Miceli, R.; Galluzzo, G.R. Perspective on hydrogen energy carrier and its automotive applications. *Int. J. Hydrogen Energy* **2014**, *39*, 8482–8494.
5. Wikramanayake, E.; Acharya, P.V.; Kapner, M.; Bahadur, V. Green hydrogen-based energy storage in Texas for decarbonization of the electric grid. In Proceedings of the IEEE GreenTech Conference, Virtual, 7–9 April 2021; pp. 409–415.
6. Jung, J.K.; Kim, I.G.; Kim, K.T. Evaluation of hydrogen permeation characteristics in rubbery polymers. *Curr. Appl. Phys.* **2021**, *21*, 43–49.
7. Guo, X.-H.; Shi, S.-Q.; Qiao, L.-J. Simulation of hydrogen diffusion and initiation of hydrogen-induced cracking in PZT ferroelectric ceramics using a phase field model. *J. Am. Ceram. Soc.* **2007**, *90*, 2868–2872.
8. Louthan, M.R.; Caskey, G.R. Hydrogen transport and embrittlement in structural metals. *Int. J. Hydrogen Energy* **1976**, *1*, 291–305.
9. Gao, Y.; Pastrana, A.P.C.; Manogharan, G.; van Duin, A.C.T. Molecular dynamics study of melting, diffusion, and sintering of cementite chromia core–shell particles. *Comput. Mater. Sci.* **2021**, *199*, 110721.
10. Griffiths, S.; Sovacool, B.K.; Kim, J.; Bazilian, M.; Uratani, J.M. Industrial decarbonization via hydrogen: A critical and systematic review of developments, socio-technical systems and policy options. *Energy Res. Soc. Sci.* **2021**, *80*, 102208.
11. Laadel, N.-E.; El Mansori, M.; Kang, N.; Marlin, S.; Boussant-Roux, Y. Permeation barriers for hydrogen embrittlement prevention in metals—A review on mechanisms, materials suitability and efficiency. *Int. J. Hydrogen Energy* **2022**, *47*, 32707–32731.
12. Lynch, S. Hydrogen embrittlement phenomena and mechanisms. *Corros. Rev.* **2012**, *30*, 105–123.
13. Robertson, I.M.; Sofronis, P.; Nagao, A.; Martin, M.L.; Wang, S.; Gross, D.W.; Nygren, K.E. Hydrogen embrittlement understood. *Metall. Mater. Trans. A* **2015**, *46*, 2323–2341.
14. Zan, N.; Ding, H.; Guo, X.; Tang, Z.; Bleck, W. Effects of grain size on hydrogen embrittlement in a Fe-22Mn-0.6 C TWIP steel. *Int. J. Hydrogen Energy* **2015**, *40*, 10687–10696. [CrossRef]
15. Takasawa, K.; Ikeda, R.; Ishikawa, N.; Ishigaki, R. Effects of grain size and dislocation density on the susceptibility to high-pressure hydrogen environment embrittlement of high-strength low-alloy steels. *Int. J. Hydrogen Energy* **2012**, *37*, 2669–2675. [CrossRef]
16. Koyama, M.; Akiyama, E.; Lee, Y.-K.; Raabe, D.; Tsuzaki, K. Overview of hydrogen embrittlement in high-Mn steels. *Int. J. Hydrogen Energy* **2017**, *42*, 12706–12723.
17. Hadam, U.; Zakroczymski, T. Absorption of hydrogen in tensile strained iron and high-carbon steel studied by electrochemical permeation and desorption techniques. *Int. J. Hydrogen Energy* **2009**, *34*, 2449–2459.
18. Perng, T.-P.; Altstetter, C.J. Effects of deformation on hydrogen permeation in austenitic stainless steels. *Acta Metall.* **1986**, *34*, 1771–1781.
19. Perng, T.P.; Altstetter, C.J. Comparison of hydrogen gas embrittlement of austenitic and ferritic stainless steels. *Metall. Mater. Trans. A* **1987**, *18*, 123–134.
20. Owczarek, E.; Zakroczymski, T. Hydrogen transport in a duplex stainless steel. *Acta Mater.* **2000**, *48*, 3059–3070.
21. Claeys, L.; Depover, T.; De Graeve, I.; Verbeken, K. Electrochemical hydrogen charging of duplex stainless steel. *Corrosion* **2019**, *75*, 880–887.
22. Olson, G.B.; Azrin, M. Transformation behavior of TRIP steels. *Metall. Mater. Trans. A* **1978**, *9*, 713–721. [CrossRef]
23. Wendler, M.; Weiß, A.; Krüger, L.; Mola, J.; Franke, A.; Kovalev, A.; Wolf, S. Effect of manganese on microstructure and mechanical properties of cast high alloyed Cr M n Ni-N steels. *Adv. Eng. Mater.* **2013**, *15*, 558–565. [CrossRef]
24. Schröder, C.; Weiß, A.; Volkova, O. Development of a new stainless austenitic CrMnNi steel using TRIP/TWIP effect for a high cold formability: Entwicklung eines neuartigen nichtrostenden, austenitischen CrMnNi-Stahls mit hoher Kaltumformbarkeit unter Ausnutzung von TRIP/TWIP-Effekten. *Mater. Werkst.* **2018**, *49*, 577–590. [CrossRef]
25. Dwivedi, S.K.; Vishwakarma, M. Effect of hydrogen in advanced high strength steel materials. *Int. J. Hydrogen Energy* **2019**, *44*, 28007–28030. [CrossRef]
26. Hojo, T.; Kumai, B.; Koyama, M.; Akiyama, E.; Waki, H.; Saitoh, H.; Shiro, A.; Yasuda, R.; Shobu, T.; Nagasaka, A. Hydrogen embrittlement resistance of pre-strained ultra-high-strength low alloy TRIP-aided steel. *Int. J. Fract.* **2020**, *224*, 253–260. [CrossRef]
27. Ronevich, J.A.; De Cooman, B.C.; Speer, J.G.; De Moor, E.; Matlock, D.K. Hydrogen effects in prestrained transformation induced plasticity steel. *Metall. Mater. Trans. A* **2012**, *43*, 2293–2301. [CrossRef]
28. Wang, Y.; Wang, X.; Gong, J.; Shen, L.; Dong, W. Hydrogen embrittlement of catholically hydrogen-precharged 304L austenitic stainless steel: Effect of plastic pre-strain. *Int. J. Hydrogen Energy* **2014**, *39*, 13909–13918. [CrossRef]
29. Dadfarnia, M.; Sofronis, P.; Neeraj, T. Hydrogen interaction with multiple traps: Can it be used to mitigate embrittlement? *Int. J. Hydrogen Energy* **2011**, *36*, 10141–10148. [CrossRef]
30. Olden, V.; Thaulow, C.; Johnsen, R. Modelling of hydrogen diffusion and hydrogen induced cracking in supermartensitic and duplex stainless steels. *Mater. Des.* **2008**, *29*, 1934–1948. [CrossRef]

31. Wang, Z.; Liu, J.; Huang, F.; Bi, Y.; Zhang, S. Hydrogen diffusion and its effect on hydrogen embrittlement in DP steels with different martensite content. *Front. Mater.* **2020**, *7*, 620000. [CrossRef]
32. Mente, T.; Bollinghaus, T. Modeling of hydrogen distributionin a duplex stainless steel. *Weld. World* **2012**, *56*, 66–78. [CrossRef]
33. Quitzke, C.; Schröder, C.; Ullrich, C.; Mandel, M.; Krüger, L.; Volkova, O.; Wendler, M. Evaluation of strain-induced martensite formation and mechanical properties in N-alloyed austenitic stainless steels by in situ tensile tests. *Mater. Sci. Eng. A* **2021**, *808*, 140930. [CrossRef]
34. Devanathan, M.A.V.; Stachurski, Z. The adsorption and diffusion of electrolytic hydrogen in palladium. *Proc. R. Soc. Lond. Ser. A Math. Phys. Sci.* **1962**, *270*, 90–102.
35. Manolatos, P.; Jerome, M.; Galland, J. Necessity of a palladium coating to ensure hydrogen oxidation during electrochemical permeation measurements on iron. *Electrochim. Acta* **1995**, *40*, 867–871. [CrossRef]
36. Pyun, S.-I.; Oriani, R.A. The permeation of hydrogen through the passivating films on iron and nickel. *Corros. Sci.* **1989**, *29*, 485–496. [CrossRef]
37. Aten, A.H.W.; Zieren, M. The Poisoning of Hydrogen Electrodes. *Trans. Am. Electrochem. Soc.* **1930**, *58*, 153. [CrossRef]
38. *DIN 50125:2016-12*; Testing of Metallic Materials—Tensile Test Pieces. Beuth Verlag GmbH: Berlin, Germany, 2016.
39. Crank, J. *The Mathematics of Diffusion*; Clarendon Press: Oxford, UK, 1975.
40. Lagarias, J.C.; Reeds, J.A.; Wright, M.H.; Wright, P.E. Convergence properties of the Nelder--Mead simplex method in low dimensions. *SIAM J. Optim.* **1998**, *9*, 112–147. [CrossRef]
41. Hempel, C.; Mandel, M.; Schröder, C.; Quitzke, C.; Schimpf, C.; Wendler, M.; Volkova, O.; Krüger, L. Influence of microstructure on hydrogen trapping and diffusion in a pre-deformed TRIP steel. *Int. J. Hydrogen Energy* **2023**, *48*, 4906–4920. [CrossRef]
42. Ningshen, S.; Uhlemann, M.; Schneider, F.; Khatak, H.S. Diffusion behaviour of hydrogen in nitrogen containing austenitic alloys. *Corros. Sci.* **2001**, *43*, 2255–2264. [CrossRef]
43. Chen, C.; Yu, H.; Zheng, S. First-principles study of hydrogen diffusion mechanism in Cr_2O_3. *Sci. China Technol. Sci.* **2011**, *54*, 88–94. [CrossRef]
44. Xing, B.; Wu, J.; Cheng, J.; Zhang, L.; Wu, M. Hydrogen diffusion in α-Fe_2O_3: Implication for an effective hydrogen diffusion barrier. *Int. J. Hydrogen Energy* **2020**, *45*, 32648–32653. [CrossRef]
45. Hinds, G.; Zhao, J.; Turnbull, A. Hydrogen diffusion in super 13% chromium martensitic stainless steel. *Corrosion* **2005**, *61*, 348–354. [CrossRef]
46. Martin, S.; Wolf, S.; Decker, S.; Krüger, L.; Martin, U. Deformation Bands in High-Alloy Austenitic 16Cr6Mn6Ni TRIP Steel: Phase Transformation and Its Consequences on Strain Hardening at Room Temperature. *Steel Res. Int.* **2015**, *86*, 1187–1196. [CrossRef]
47. Wendler, M.; Hauser, M.; Fabrichnaya, O.; Krüger, L.; Weiß, A.; Mola, J. Thermal and deformation-induced phase transformation behavior of Fe–15Cr–3Mn–3Ni–0.1 N–(0.05–0.25) C austenitic and austenitic–martensitic cast stainless steels. *Mater. Sci. Eng. A* **2015**, *645*, 28–39. [CrossRef]
48. Alsultan, S.; Quitzke, C.; Cheng, Z.; Krüger, L.; Volkova, O.; Wendler, M. Strain-Induced Martensite Formation and Mechanical Properties of Fe–19Cr–4Ni–3Mn–0.15N–0.15C Austenitic Stainless Steel at Cryogenic Temperature. *Steel Res. Int.* **2021**, *92*, 2000611. [CrossRef]
49. Quitzke, C.; Schröder, C.; Hempel, C.; Mandel, M.; Krüger, L.; Volkova, O.; Wendler, M. Determination of Martensite Formation Rate by Magnetic In Situ Measurement in CrMnNi–N Steel. *Steel Res. Int.* **2023**, *94*, 2200859. [CrossRef]
50. Elsener, B.; Addari, D.; Coray, S.; Rossi, A. Stainless steel reinforcing bars–reason for their high pitting corrosion resistance. *Mater. Corros.* **2011**, *62*, 111–119. [CrossRef]
51. Peguet, L.; Malki, B.; Baroux, B. Influence of cold working on the pitting corrosion resistance of stainless steels. *Corros. Sci.* **2007**, *49*, 1933–1948. [CrossRef]
52. Zhao, Y.; Wang, Y.; Li, X.; Zhang, W.; Tang, S.; Liu, Z.J. Effects of plastic straining on the corrosion resistance of TRIP-aided lean duplex stainless steels. *Mater. Sci.* **2018**, *53*, 9258–9272. [CrossRef]
53. Kolodii, B.I. Theoretical investigation of the interaction of a deformed metal with a corrosion medium. *Mater. Sci.* **2000**, *36*, 884–891. [CrossRef]
54. Hirth, J.P. Effects of hydrogen on the properties of iron and steel. *Metall. Mater. Trans. A* **1980**, *11*, 861–890. [CrossRef]
55. Ulmer, D.G.; Altstetter, C.J. Phase relations in the hydrogen-austenite system. *Acta Metall.* **1993**, *41*, 2235–2241. [CrossRef]
56. He, Y.; Li, Y.; Chen, C.; Yu, H. Diffusion coefficient of hydrogen interstitial atom in α-Fe, γ-Fe and ε-Fe crystals by first-principle calculations. *Int. J. Hydrogen Energy* **2017**, *42*, 27438–27445. [CrossRef]
57. Hirata, K.; Iikubo, S.; Koyama, M.; Tsuzaki, K.; Ohtani, H. First-principles study on hydrogen diffusivity in BCC, FCC, and HCP iron. *Metall. Mater. Trans. A* **2018**, *49*, 5015–5022. [CrossRef]
58. Bernstein, I.M.; Pressouyre, G.M. The role of traps in the microstructural control of hydrogen embrittlement of steels. In *Hydrogen Degradation of Ferrous Alloys*; Oriani, R.A., Hirth, J.P., Smialowski, M., Eds.; Noyes Publication: Park Ridge, NJ, USA, 1985.
59. Sofronis, P.; Birnbaum, H.K. Mechanics of the hydrogendashdislocationdashimpurity interactions—I. Increasing shear modulus. *J. Mech. Phys. Solids* **1995**, *43*, 49–90. [CrossRef]
60. Tien, J.K.; Thompson, A.W.; Bernstein, I.M.; Richards, R.J. Hydrogen transport by dislocations. *Metall. Mater. Trans A* **1976**, *7*, 821–829. [CrossRef]
61. Novak, P.; Yuan, R.; Somerday, B.P.; Sofronis, P.; Ritchie, R.O. A statistical, physical-based, micro-mechanical model of hydrogen-induced intergranular fracture in steel. *J. Mech. Phys. Solids* **2010**, *58*, 206–226. [CrossRef]

62. Garrison, W.M., Jr.; Moody, N.R. Ductile fracture. *J. Phys. Chem. Solids* **1987**, *48*, 1035–1074. [CrossRef]
63. Troiano, A.R. The role of hydrogen and other interstitials in the mechanical behavior of metals. *Metallogr. Microstruct. Anal.* **2016**, *5*, 557–569. [CrossRef]
64. Oriani, R.A.; Josephic, P.H. Equilibrium aspects of hydrogen-induced cracking of steels. *Acta Metall.* **1974**, *22*, 1065–1074. [CrossRef]
65. Eliaz, N.; Shachar, A.; Tal, B.; Eliezer, D. Characteristics of hydrogen embrittlement, stress corrosion cracking and tempered martensite embrittlement in high-strength steels. *Eng. Fail. Anal.* **2002**, *9*, 167–184.
66. Pohl, M. Failure Analysis to Clarify Unambiguous Hydrogen Fracturing in Steels. *Metallogr. Microstruct. Anal.* **2017**, *6*, 282–288.
67. Beachem, C.D. A new model for hydrogen-assisted cracking (hydrogen "embrittlement"). *Metall. Trans.* **1972**, *3*, 441–455.
68. Tabata, T.; Birnbaum, H.K. Direct observations of the effect of hydrogen on the behavior of dislocations in iron. *Scr. Metall.* **1983**, *17*, 947–950.

materials

MDPI

Article

Corrosion Behavior of Copper Foil on PCB Substrates Under Atmospheric Environment in Sichuan-Tibet Region of China

Xueqing Yang [1], Haifeng Jin [2], Dongfang Jia [1], Jinke Wang [1], Zhibin Chen [1], Rui Tian [1], Shiyao Du [1] and Lingwei Ma [1,*]

[1] National Materials Corrosion and Protection Data Center, University of Science and Technology Beijing, Beijing 100083, China; yang_xueqing1024@163.com (X.Y.); df_jia177@163.com (D.J.); d202110710@xs.ustb.edu.cn (J.W.); chenzhibin0131@163.com (Z.C.); tr19990913@163.com (R.T.); b24054642@ustb.edu.cn (S.D.)

[2] Marine Design and Research Institute of China, Shanghai 200011, China; 13120172565@163.com

* Correspondence: mlw1215@ustb.edu.cn

Abstract: Copper foil is widely used in electronic components and devices. This study investigates the corrosion behavior of copper foil on printed circuit boards exposed for one year in a closed atmospheric environment across 22 different sites in the Sichuan-Tibet region. Through electrochemical, SEM/EDS, and XRD analyses, the corrosion behavior of copper foil material across the five selected sites (Meishan, Mangkang, Luding, Batang, and Panzhihua) and the influence of environmental factors were discussed. The results show that copper foils in areas with a large temperature change, higher humidity, and more rainfall exhibit more severe corrosion. The corrosion products primarily form a double-layer structure; the bottom layer consists of relatively stable Cu_2O, while the type of corrosion product in the upper layer is strongly influenced by the local climatic environment, predominantly containing $CuSO_4$, $CuSO_4(OH)$, and $Cu_2Cl(OH)_3$. In dry areas, copper oxides tend to form, whereas in humid areas, copper sulfides are more likely to grow.

Keywords: printed circuit boards; copper foil; atmospheric corrosion; electrochemistry; Sichuan-Tibet region

Citation: Yang, X.; Jin, H.; Jia, D.; Wang, J.; Chen, Z.; Tian, R.; Du, S.; Ma, L. Corrosion Behavior of Copper Foil on PCB Substrates Under Atmospheric Environment in Sichuan-Tibet Region of China. *Materials* **2024**, *17*, 6039. https://doi.org/10.3390/ma17246039

Academic Editor: Hendra Hermawan

Received: 4 November 2024
Revised: 4 December 2024
Accepted: 6 December 2024
Published: 10 December 2024

1. Introduction

Due to its excellent electrical conductivity, thermal conductivity, ease of processing, and reliable soldering, copper is widely used as an electrical connection carrier for various components and circuits [1,2]. In the electronics field, copper foil is commonly used as a conductive material for the substrate of printed circuit boards (PCBs), serving as the conductive and grounding layers of the circuit board [3]. Copper foil layers, known as copper-clad laminates, are prone to oxidation or electrochemical corrosion in atmospheric environments [4,5]. As electronic circuits and components evolve towards higher density, miniaturization, and higher integration, copper foil on PCBs faces increasing challenges [6]. In complex atmospheric environments, the corrosion of copper foil can affect the performance of electronic devices to varying degrees, such as reducing the conductivity of the copper foil, affecting its electrical performance and reliability, and even causing short or open circuit behavior [7].

Atmospheric corrosion refers to electrochemical corrosion that occurs within a thin condensed water film in an atmospheric environment [8]. The influence of humidity on the condensation phenomenon in the atmospheric environment is evident [9,10]. Starting at approximately 60% relative humidity (RH), a very thin film of moisture forms on the metal surface, leading to the initiation of corrosion, which is known as critical humidity. As the relative humidity increases, the thickness of the water film on the metal surface also increases. Hygroscopic pollutants, such as dust, absorb moisture from the environment, increasing the humidity on the metal surface and accelerating the corrosion process [11–15].

This form of corrosion is ubiquitous. Although corrosion in a closed atmospheric environment is generally less severe than that occurring outdoors, it remains an important research area, particularly in the electronics field, where even minor corrosion damage can cause electronic equipment failure.

Currently, research on the corrosion behavior of copper primarily focuses on outdoor atmospheric exposure environments (such as humid marine and dry areas) and indoor accelerated corrosion tests [16–20]. For example, Mendoza et al. [21] investigated the structure of copper corrosion products in different atmospheric environments in the tropical climate of Cuba, showing that the products were primarily Cu_2O, $Cu_2Cl(OH)_3$, $Cu_4SO_4(OH)_6{\cdot}2H_2O$, and $Cu_4SO_4(OH)_6$. Kong et al. [22] conducted a three-year atmospheric corrosion study on pure copper in Turpan, Xinjiang, China, and found that the corrosion product layer formed on the surface of copper could enhance its corrosion resistance, but the protective effect decreased with increasing temperature. Feng et al. [23] conducted outdoor exposure experiments on PCB-Cu in a tropical marine atmosphere and found that long-term exposure led to the formation of a double-layer structure, with a dense oxide layer forming internally and loose corrosion products forming externally. Wu et al. [1] analyzed the corrosion failure of copper wires in outdoor junction boxes of substations, identified $Cu_2Cl(OH)_3$ products in field investigations, and determined that chloride ions and relative humidity are the key factors affecting the corrosion behavior of copper wires. However, there is limited research on the corrosion behavior of copper foil on PCBs in a closed atmospheric environment.

With the accelerated deployment of 5G networks, the construction of communication infrastructure, and the technological advancements in the Sichuan-Tibet region, the performance of copper foil on PCBs is increasingly critical for the longevity of electronic network equipment [24]. The environmental characteristics of the Sichuan-Tibet region, with its high altitude, plateau climate, strong ultraviolet radiation, extreme dryness, and large temperature change, endow higher demands on the corrosion protection of circuit board copper foil. In Tibet, the environment is characterized by long sunshine hours, strong radiation, low temperatures, significant temperature fluctuations, distinct wet and dry seasons, significant nighttime rainfall, relatively dry winters and springs, frequent strong winds, high altitude, and low atmospheric pressure [24]. In contrast, the environment in the Sichuan region is characterized by high precipitation but with dry winter, spring droughts, summer floods, and autumn rains. The annual precipitation distribution is uneven, with about 70% of the rainfall from early June to October each year. Moreover, the pH values of rainfall across various cities in Sichuan fluctuate between 5.09 and 7.54 throughout the year, with acid rain having a pH of 4.82 and a frequency of up to 11.3% [25,26]. To better understand the service differences of copper foil in the Sichuan-Tibet region, accumulating the corrosion data and studying the corrosion behavior of copper foil in typical atmospheric environments across the region is of great significance. Therefore, in this study, we investigate the service differences of copper foil on circuit boards across 22 outdoor exposure sites in the Sichuan-Tibet region, aiming to study the impact of atmospheric corrosion and the environmental characteristics of different sites on the corrosion behavior of copper foil.

2. Materials and Methods

2.1. Experimental Materials

To investigate the service differences of copper foil in the Sichuan-Tibet region, 22 sites were selected (as shown in Figure 1), including Ganzi, Meishan, Chengdu, Luzhou, Xichang, Dazhou, Jinsha River, Panzhihua, Jingzhou, Yajiang, Batang, Guang'an, Danba, Yibin, Ya'an, Luding, Jilong, Mozhu, Lantsang, Xvmu, Mangkang, and Chawu. The latitude and longitude of these sites are provided in Table 1. The copper foil was exposed to a closed atmospheric environment at each site for one year (see Figure 2a) to study the patterns of atmospheric corrosion and the influence of site-specific environmental characteristics on corrosion behavior.

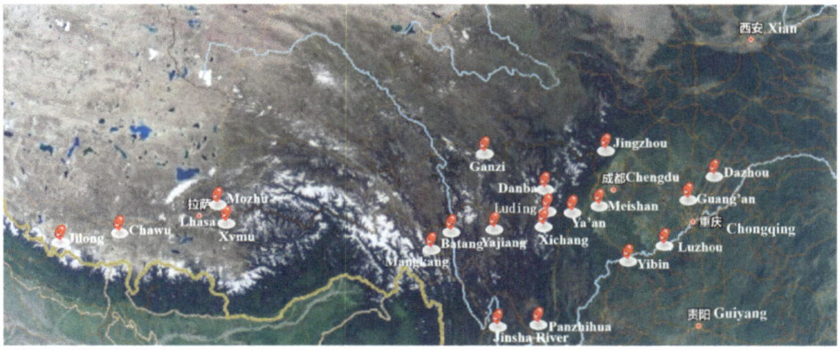

Figure 1. Location maps of different sites.

Table 1. Latitude and longitude of different sites.

Site	Longitude and Latitude	Site	Longitude and Latitude
Meishan	N 29°24′ E 102°49′	Yajiang	N 29°03′ E 100°19′
Chengdu	N 30°05′ E 102°54′	Jilong	N 28°03′ E 84°35′
Mangkang	N 28°37′ E 98°00′	Chawu	N 28°47′ E 87°24′
Dazhou	N 30°19′ E 106°39′	Danba	N 30°24′ E 101°17′
Mozhu	N 29°50′ E 91°44′	Luding	N 29°54′ E 101°46′
Yibin	N 27°50′ E 103°36′	Jinsha River	N 26°32′ E 102°51′
Jingzhou	N 31°25′ E 102°49′	Lantsang	N 22°45′ E 100°57′
Xichang	N 27°32′ E 101°46′	Batang	N 28°46′ E 98°58′
Ya'an	N 28°51′ E 101°55′	Xvmu	N 27°98′ E 85°98′
Guang'an	N 30°01′ E 105°56′	Ganzi	N 27°58′ E 97°22′
Luzhou	N 27°39′ E 105°08′	Panzhihua	N 26°05′ E 101°08′

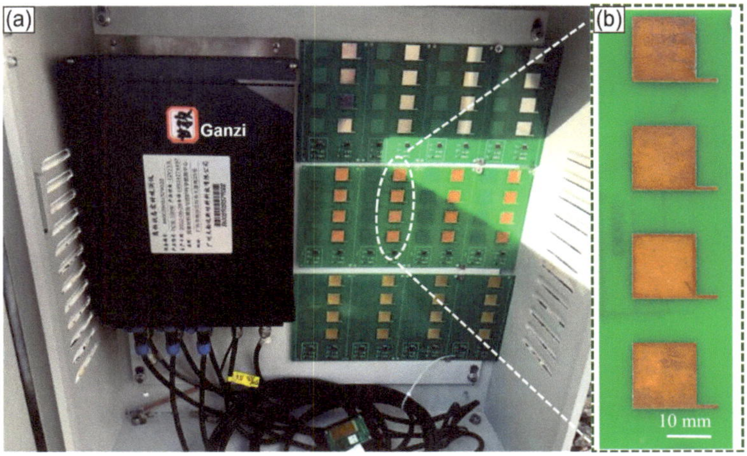

Figure 2. (**a**) Copper foil placement site and (**b**) photos of the circuit board under a closed atmosphere in the Sichuan-Tibet area.

The experimental material, electrolytic copper foil for circuit boards, has a purity exceeding 99.95%, with dimensions of 15 × 15 mm^2 and a thickness of 35 μm. The macroscopic images of the copper foil are shown in Figure 2b. The base layer is an FR-4 board, composed of epoxy resin and glass fiber cloth laminated together. The exposure period for the samples was from January 2023 to January 2024.

The detailed experimental design is described as follows:

Experimental objectives: To evaluate the atmospheric corrosion resistance of copper foil materials, explore corrosion patterns, and investigate the effects of environmental factors on the performance of copper foil materials.

Experimental materials: Integrated circuit board copper foil.

Exposure environment: Test chambers simulating atmospheric conditions in the Sichuan-Tibet region of China.

Exposure duration: One year.

Environmental data: Temperature, rainfall, and atmospheric pollutants.

Analytical techniques: Scanning electron microscopy (SEM), energy-dispersive X-ray spectroscopy (EDS), and X-ray diffraction (XRD).

Performance analysis: Electrochemical impedance spectroscopy (EIS). For each test, three parallel samples were measured to ensure data reliability.

2.2. Experimental Methods

The copper foils were exposed to the closed atmospheric environment of each site in the Sichuan-Tibet region for one year. After the experiment, macroscopic corrosion morphologies were taken using a digital camera. A laser confocal microscope was employed to observe and analyze the microscopic corrosion morphology, and an SEM (ZEISS Sigma 300, Oberkochen, Germany) was used to examine the corrosion products on the sample surface. Additionally, EDS was conducted to analyze the elemental composition of the corrosion products, and XRD (Rigaku SmartLab SE, Tokyo, Japan) was used to analyze the phases of the corrosion products. The XRD scanning rate was set to 2°/min, and the scanning range was 10° to 80°.

EIS was performed according to the ASTM G106 standard at room temperature using a CHI660E electrochemical workstation. The copper foil served as the working electrode, a platinum sheet was used as the counter electrode, and a saturated calomel electrode (SCE) was employed as the reference electrode. The working area was 15 × 15 mm^2. Before testing, the open circuit potential (OCP) was stabilized for 30 min. All the EIS measurements were conducted at a frequency range of 10^5 to 10^{-2} Hz, with a sinusoidal voltage perturbation amplitude of 10 mV relative to the OCP. The electrolyte solution was 3.5 wt.% NaCl solution. For each test, three parallel samples were measured to ensure the reliability of the data. The working process of the electrochemical workstation is shown in Figure 3. The EIS data were fitted using the ZSimpWin 3.6 software.

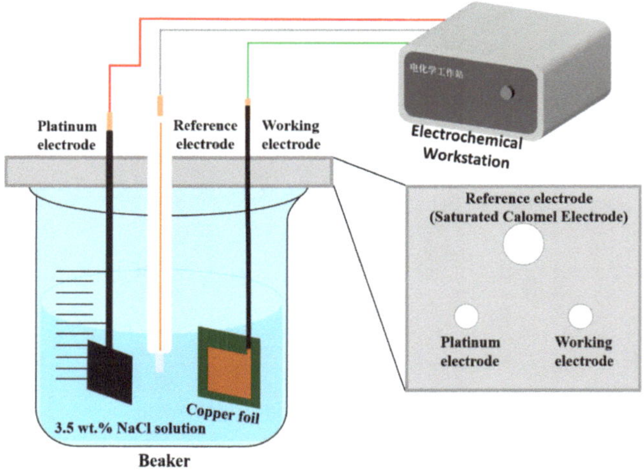

Figure 3. Electrochemical workstation working diagram.

3. Results and Discussion

3.1. Macroscopic Corrosion Morphology

Figure 4 shows the macroscopic corrosion morphology of integrated circuit board copper foils after one year of exposure in a closed atmospheric environment across 22 sites. Due to the different climatic conditions, the samples exhibited varying degrees of corrosion across different sites, with noticeable changes in surface color and a loss of metallic luster. The corrosion behavior in Meishan, Mozhu, Mangkang, and Chengdu was relatively mild, followed by Yibin, Luzhou, Xichang, and Jingzhou. The most severe corrosion occurred in Lantsang, Batang, and Panzhihua. For example, the surface of the copper foil in Meishan exhibited slight corrosion scratches, with the surface color remaining close to the original copper color, and the oxide layer was relatively uniform. As the corrosion intensified, such as in Batang, localized areas of the copper foil surface showed an accumulation of white corrosion products, with the sample color primarily reddish-brown. The samples from Panzhihua showed the most severe corrosion, with the brownish-black corrosion products adhering to the base material.

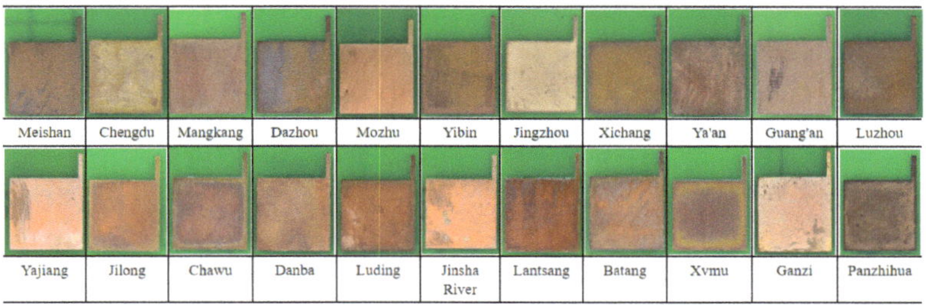

Figure 4. Macroscopic morphology of corrosion of samples after 1-year exposure to a closed atmospheric environment across 22 sites.

3.2. Electrochemical Analysis

The protective effect of the corrosion products formed on the metal substrate is closely related to the nature of these products. Electrochemical testing can effectively assess the corrosion resistance of metals. EIS was used to characterize the corrosion resistance of the copper foil on circuit boards, with the low-frequency impedance at 0.01 Hz ($|Z|_{0.01Hz}$) serving as an important indicator of the sample's anti-corrosion performance. The low-frequency impedance values were measured after immersing the samples in a 3.5 wt.% NaCl solution for 3 days, as shown in Figure 5. The results indicate that the $|Z|_{0.01Hz}$ values of the samples from Meishan, Chengdu, Mozhu, and Dazhou were relatively high, reaching up to 12,000 $\Omega \cdot cm^2$. The $|Z|_{0.01Hz}$ values from the Luding, Danba, Luzhou, and Yajiang samples were moderate, while those from Batang and Panzhihua were the lowest, falling below 1000 $\Omega \cdot cm^2$. Based on the macroscopic morphology and low-frequency impedance data, samples from Meishan, Mangkang, Luding, Batang, and Panzhihua were selected for further investigation.

Figure 6 shows the EIS data of the copper foil exposed for one year at five stations—Meishan, Mangkang, Luding, Batang, and Panzhihua. In the Nyquist plot (Figure 6a), all samples exhibit a single capacitive arc. However, the impedance of the samples from Panzhihua displays a diffusion tail in the low-frequency region, corresponding to Warburg impedance. This likely includes the diffusion of soluble metal ions from the sample surface to the anode in the solution, as well as the diffusion of dissolved oxygen toward the cathode in the opposite direction. Among the five sites, the Meishan samples exhibited the largest capacitive arc radius, indicating the best corrosion resistance, while the Panzhihua samples showed the smallest capacitive arc radius with the most severe corrosion behavior. In the Bode-impedance modulus plot (Figure 6b), the $|Z|_{0.01Hz}$ values for the Meishan and Mangkang samples were similar, at 12,000 $\Omega \cdot cm^2$ and 10,400 $\Omega \cdot cm^2$, respectively.

The copper foil exposed at Luding and Batang followed, with values of 4500 Ω·cm^2 and 2840 Ω·cm^2, respectively. The sample from Panzhihua had the smallest $|Z|_{0.01Hz}$ value, at only 802 Ω·cm^2, indicating the worst corrosion resistance. In the phase angle plot (Figure 6c), two time constants were observed on the sample surfaces, corresponding to the corrosion product film and the interface corrosion reaction. The width of the phase angle change curve reflects the tendency of the sample to corrode; a wider phase angle frequency range indicates better corrosion resistance. Overall, the $|Z|_{0.01Hz}$ values decreased in the following order: Meishan, Mangkang, Luding, Batang, and Panzhihua, while the phase angle range narrowed, indicating a gradual decline in the corrosion resistance of the sample surfaces across these five sites.

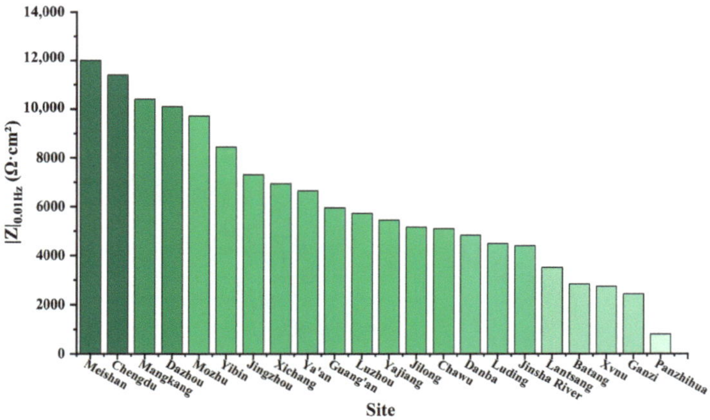

Figure 5. Low-frequency impedance values of copper foil samples placed in a closed atmospheric environment across different stations for 1 year.

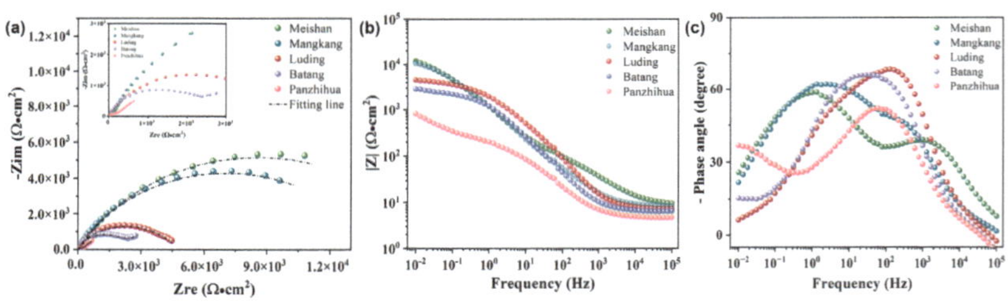

Figure 6. (**a**) Nyquist diagram, (**b**) Bode-impedance modulus plot, and (**c**) Bode-phase angle plot of the samples after the closed atmospheric environment exposure experiment across 5 sites.

As shown in Figure 7, two equivalent circuits were used to fit the EIS results. The equivalent circuit shown in Figure 7a was employed to fit the EIS results of samples from the Meishan, Markam, and Luding sites. Meanwhile, the equivalent circuit in Figure 7b was used for fitting the EIS data from the Batang and Panzhihua sites due to the apparent diffusion tail observed in the low-frequency region, indicating the occurrence of surface diffusion processes. In the equivalent circuits, R_s represents the electrolyte resistance, R_{ct} represents the charge transfer resistance on the metal substrate, R_f represents the resistance of the surface corrosion product layer, Q_{dl} and Q_f are, respectively, the double-layer capacitance and the capacitance of the corrosion product layer, and W is the Warburg impedance in the low-frequency region. When using the equivalent circuits for fitting, the chi-square values all reach the order of magnitude of 10^{-3}, indicating the applicability of the equivalent circuits. The fitting results of the EIS data are shown in Table 2. The total resistance R_t is defined as the sum of the R_{ct} and R_f, which can be used to evaluate the overall corrosion resistance of the samples. The fitting results of R_f, R_{ct}, and R_f show a consistent variation trend. Specifically, the resistance values of the samples across the five sites of Meishan, Markam, Luding, Batang, and Panzhihua gradually decrease.

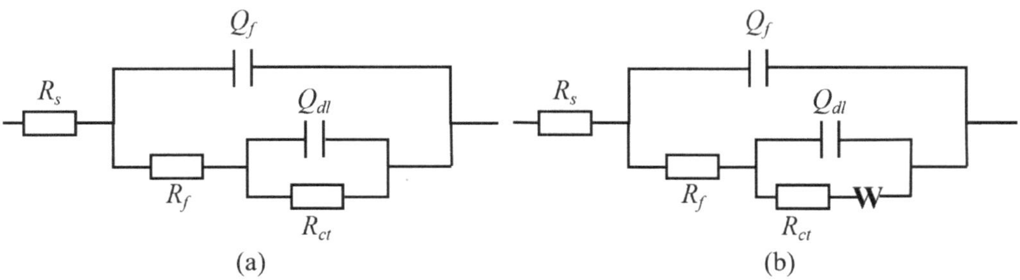

Figure 7. Equivalent circuit used to fit EIS data for copper in the following: (**a**) Meishan, Mangkang, Luding; and (**b**) Batang, Panzhihua.

Table 2. EIS fitting results of sample surfaces across different sites.

Sites	R_s ($\Omega \cdot cm^2$)	Y_f ($S \cdot s^n \cdot cm^{-2}$)	n_f	R_f ($\Omega \cdot cm^2$)	Y_{dl} ($S \cdot s^n \cdot cm^{-2}$)	n_{dl}	R_{ct} ($\Omega \cdot cm^2$)	Y_W ($S \cdot s^{0.5} \cdot cm^{-2}$)	R_t ($\Omega \cdot cm^2$)
Meishan	8.43	1.05×10^{-4}	0.64	1.62×10^3	1.45×10^{-4}	0.73	1.70×10^4	-	18,620
Mangkang	7.90	2.23×10^{-4}	0.68	4.75×10^2	1.42×10^{-5}	1.00	1.38×10^4	-	14,275
Luding	7.30	3.90×10^{-5}	0.80	5.34×10^2	1.23×10^{-4}	0.53	4.31×10^3	-	4844
Batang	6.46	1.52×10^{-5}	1.00	4.57×10^0	1.41×10^{-4}	0.73	2.29×10^3	3.95×10^{-3}	2294.57
Panzhihua	6.04	5.89×10^{-4}	0.69	5.13×10^2	3.79×10^{-3}	0.62	1.71×10^3	3.22×10^{-3}	2223

3.3. Microscopic Corrosion Morphology and Composition Analysis

To assess the degree of corrosion on the sample surface, Figure 8 presents the surface morphology and 3D confocal images of the samples after one year of exposure at Meishan, Mangkang, Luding, Batang, and Panzhihua. Among these sites, the copper foil surface in Meishan exhibited no significant scratches, with only a few irregular spot-like protrusions of corrosion products and the lowest surface roughness. This is likely due to the relatively mild environmental conditions, which helped maintain its high corrosion resistance and significantly slowed the progression of localized corrosion. In Mangkang, noticeable scratches appeared on the sample surface, with reddish-brown corrosion products accumulating in the scratches, identified as copper oxides [16,27]. In Luding, the sample surface showed prominent scratches, accompanied by large patches of reddish-brown and black, irregularly shaped corrosion products. The roughness was significantly higher in areas containing black corrosion products. The corrosion became more severe in Batang, where the corrosion product layer exhibited a distinct double-layer structure: the bottom layer comprised reddish-brown corrosion products, while the upper layer featured white stripe-like corrosion products [28]. The most severe corrosion was observed in Panzhihua, which has the highest average annual temperature and the lowest rainfall. The sample surface displayed complex and unevenly distributed corrosion products, including irregular black corrosion product protrusions, primarily attributed to the further oxidation of copper, forming CuO [29].

Figure 9 shows the SEM images of corrosion products and the corresponding EDS spectra of samples. In the EDS spectra, region A corresponds to the thicker corrosion product protrusions, while region B represents corrosion products closer to the copper substrate. In Figure 9a,b, the morphology of the corrosion product of the Meishan sample appeared as layered, dot-like protrusions. The similar elemental composition in regions A and B indicates relatively uniform corrosion with low sulfur content. The corrosion products were primarily composed of copper oxides [30]. In Figure 9c,d, the SEM images of the Mangkang sample revealed significant compositional differences between regions A and B, indicating a double-layer structure in the corrosion products. The first layer of corrosion products in region B was relatively thin, mainly consisting of copper oxides, while the second layer in region A was thicker and loosely distributed, primarily consisting of copper oxides and sulfides. In Figure 9e,f, the sulfur and oxygen contents in region A of the sample from Luding were significantly higher than in region B, indicating that the corrosion products in region A were likely composed of both copper oxides and sulfides. In Figure 9g,h, the oxygen content in the corrosion products of the Batang sample was similar to that of the samples from Luding and Mangkang, but the sulfur content was relatively high, indicating that the corrosion products formed among these three sites had similar structures. However, the Batang sample had more sulfide content in the second layer of corrosion products. In Figure 9i,j, the sample from Panzhihua showed several surface cracks, which reduced the adhesion of the oxide layer to the substrate, accelerating the corrosion process. The EDS analysis revealed that the corrosion products primarily consisted of oxides with a small amount of sulfides. Overall, the formation of the loosely distributed corrosion products on the sample surfaces involved the interweaving of lamellar products, which eventually formed tightly connected, cluster-like corrosion products.

Figure 10 shows the elemental distribution on the surface of corrosion products of copper foil under a closed atmospheric environment across five typical sites. It can be seen that sulfur elements mainly exist in the raised area of corrosion products, while oxygen elements are primarily present in the underlying corrosion products. Among the sites, the Meishan sample exhibits the lowest sulfur content with a thin layer of corrosion products. Moreover, the corrosion products at Meishan and Panzhihua exhibit a more uniform distribution of oxygen elements. The overall elemental distribution results are consistent with the EDS results discussed earlier.

Figure 11 shows the XRD test results of corrosion products on copper foils. Combining the results mentioned above, the main elements of the rust layer on the sample surface were Cu, O, and S, and the corrosion products mainly consisted of copper oxides and sulfides. The XRD results revealed peaks corresponding to copper oxides (Cu_2O and CuO), copper sulfides ($CuSO_4(OH)$ and $CuSO_4$), and $CuCl(OH)_3$, respectively. The corrosion products on the Meishan samples consisted exclusively of Cu_2O. For the samples from Luding and Batang, distinct peaks of sulfides and chlorides, including $CuSO_4$, $CuSO_4(OH)$, and $Cu_2Cl(OH)_3$, were observed, corresponding to the white corrosion products identified in the macroscopic morphology. In contrast, the corrosion products on the Panzhihua samples primarily consisted of copper oxides, with no significant sulfides observed.

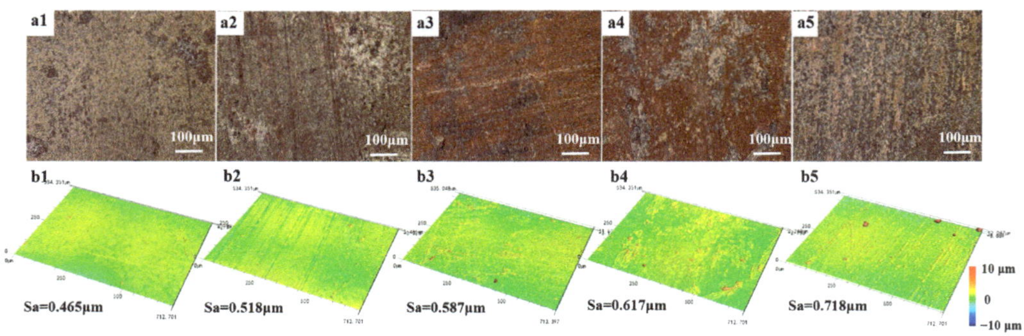

Figure 8. The micro-topography and surface roughness of samples placed in a closed atmospheric environment across different stations for 1 year: (**a1**,**b1**) Meishan, (**a2**,**b2**) Mangkang, (**a3**,**b3**) Luding, (**a4**,**b4**) Batang, and (**a5**,**b5**) Panzhihua.

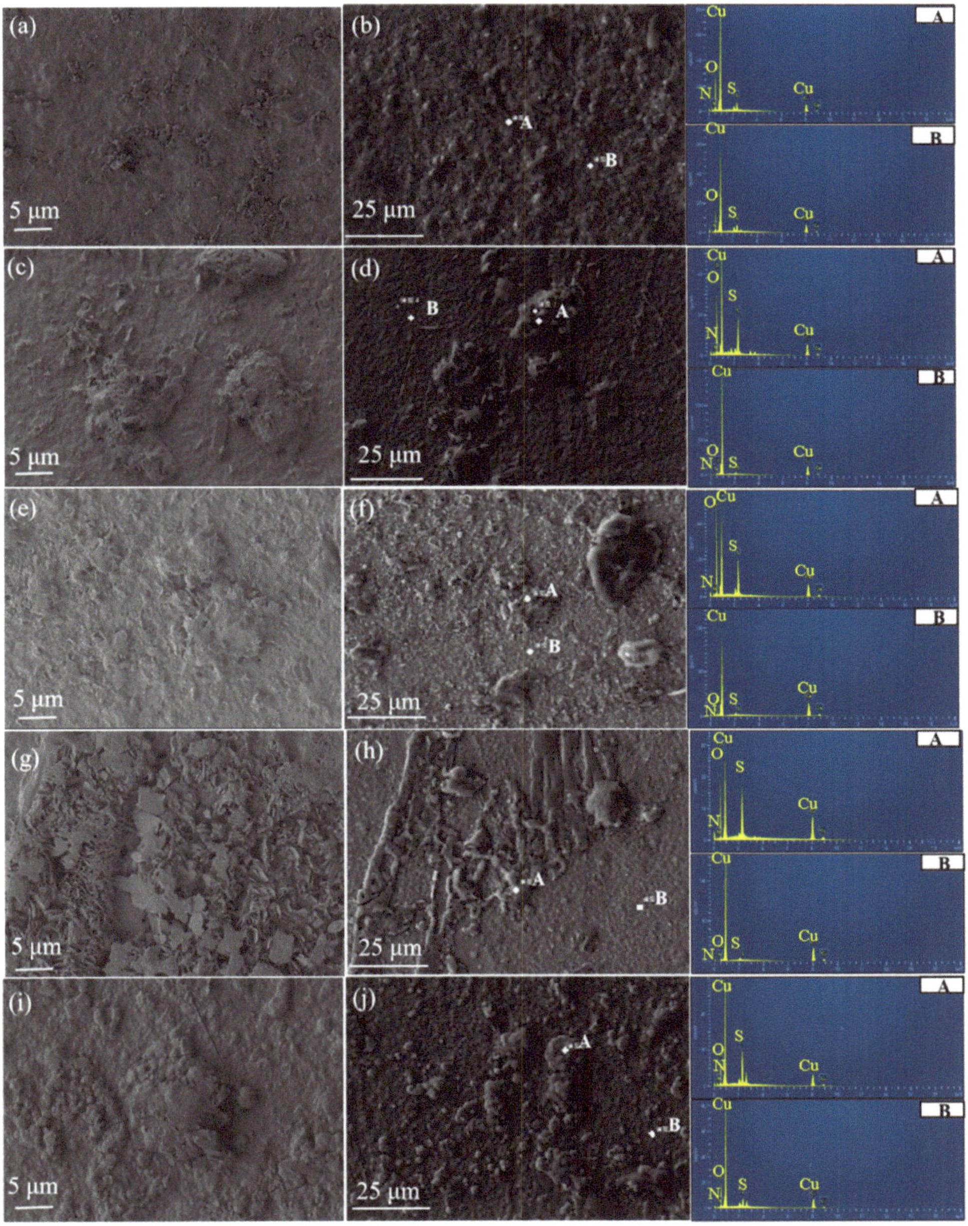

Figure 9. SEM images of corrosion products and EDS images of corresponding corrosion products at point A and point B after samples were placed in a closed atmospheric environment across different stations for 1 year: (**a**,**b**) Meishan, (**c**,**d**) Mangkang, (**e**,**f**) Luding, (**g**,**h**) Batang, and (**i**,**j**) Panzhihua.

Figure 10. Elemental distribution of corrosion products on the surface of copper foil samples: (**a**) Meishan, (**b**) Mangkang, (**c**) Luding, (**d**) Batang, and (**e**) Panzhihua.

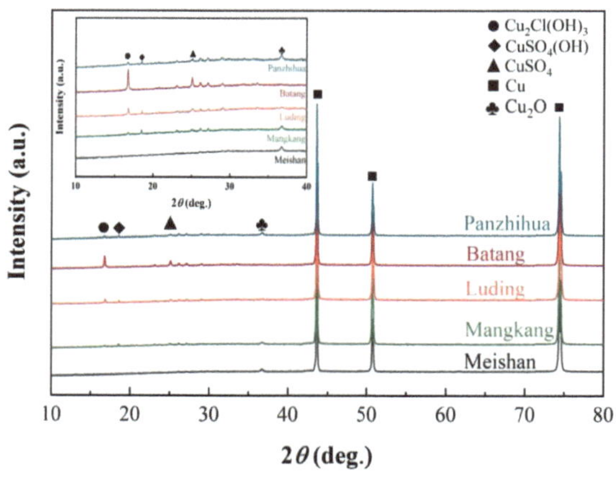

Figure 11. XRD patterns of corrosion products on copper foil surface across different sites.

3.4. Discussion of Environmental Factors

The degree of corrosion of copper foil samples is strongly related to the environmental factors of the region where they were exposed to the atmosphere, such as temperature, salt dust, relative humidity, and pH values [5,31]. Figures 12 and 13 focus on the meteorological data for the year 2023 in the five regions of Meishan, Mangkang, Luding, Batang, and Panzhihua. Figure 12 shows the monthly average temperature and total precipitation at the five stations, while Figure 13 presents the annual average temperature and total precipitation among the five sites. Meishan is located in the western part of the Sichuan Basin and features a subtropical humid monsoon climate with four distinct seasons. The monthly temperatures exhibit an unimodal variation, ranging from an average of 6.3–7.5 °C in January to 25.7–26.7 °C in July. The annual precipitation exceeds 1000 mm. Mangkang, located in the southeastern part of the Qinghai-Tibet Plateau, experiences the lowest temperatures, with an average temperature of −1 °C in January and 15 °C in July, and an annual average temperature of 7.2 °C. Precipitation is unevenly distributed throughout the year, resulting in mild, humid summers and cold, dry winters, with a frost period lasting up to 270 days. Luding is a typical alpine gorge region, with an annual average temperature of 15 °C, annual precipitation of approximately 1708 mm, and 332 frost-free days annually. In Batang, precipitation increases significantly in July and August, with an annual average temperature of 13 °C. Panzhihua, characterized by the highest temperature and lowest rainfall, has a typical dry–hot valley climate featuring a long summer, no winter, extended sunshine hours, and intense solar radiation.

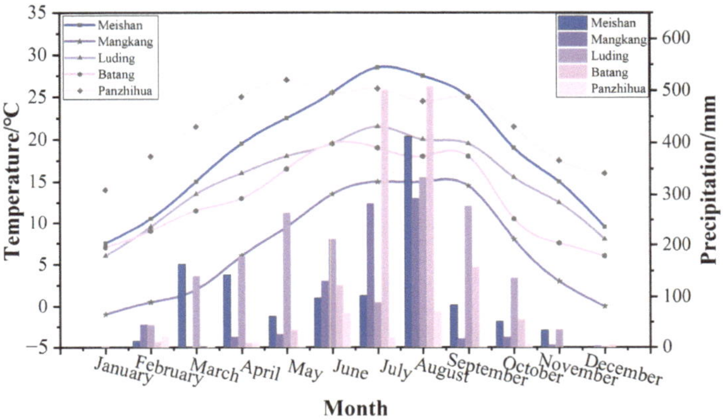

Figure 12. Monthly mean temperature and monthly precipitation across five stations in 2023.

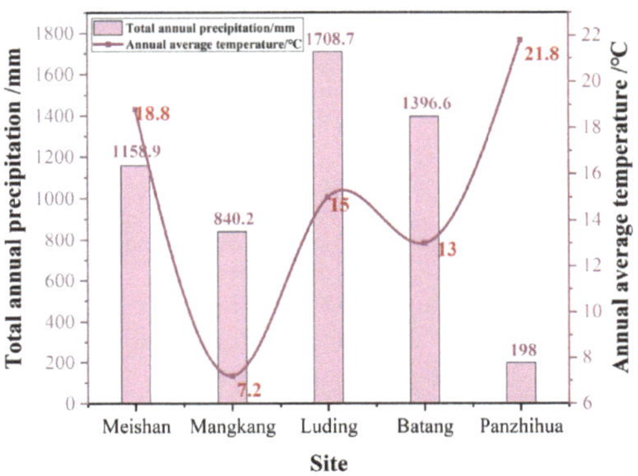

Figure 13. Annual mean temperature and annual precipitation across five stations in 2023.

Based on the analysis of the composition and morphology of corrosion products at the five typical stations in the Sichuan-Tibet region, temperature and relative humidity are crucial factors affecting the corrosion of copper foil in a closed atmospheric environment. The samples in areas with higher rainfall and larger variations in temperature and humidity exhibited poorer stability. Additionally, SO_2 is one of the most severe atmospheric corrosion gases, especially when SO_2 and water vapor co-exist, leading to severe corrosion. This is because when dust and atmospheric pollutants deposit on the copper foil surface, the critical relative humidity of the copper foil surface is significantly reduced, making it easier for an electrolyte film to form and cause corrosion on the copper foil. In terms of corrosion product composition, copper oxides tend to form in dry–hot regions, while copper sulfides are more likely to form in humid–hot areas. For example, in Batang and Luding, where precipitation and relative humidity are high, the adsorption of SO_2 gas from the air promotes the formation of sulfides, thereby exacerbating corrosion. In Panzhihua, despite the highest temperatures, the corrosion products mainly consisted of oxides, with very few sulfides, primarily due to the dry–hot climate and low relative humidity. Furthermore, in regions with adsorptive sea salt ions in the atmosphere, the corrosion products on copper foil may include basic copper chlorides [32].

3.5. Discussion of Corrosion Mechanism

The corrosion of copper foil in an atmospheric environment primarily follows an electrochemical corrosion mechanism consisting of a cathode, anode, electronic conductor, and ion conductor (electrolyte on the metal surface). This process can be considered a short-circuited galvanic cell, primarily comprising the anodic reaction of copper dissolution with electron loss and the cathodic reaction of oxygen reduction or water decomposition within the liquid film. The atmospheric factors affecting corrosion in these five sites are mainly oxygen, water, and sulfur dioxide (as shown in Figure 14a). During the corrosion process, the dissolved copper ions react with the atmospheric components to form multiple phases, including Cu^+ and Cu^{2+} valence states [33]. The main reactions are as follows:

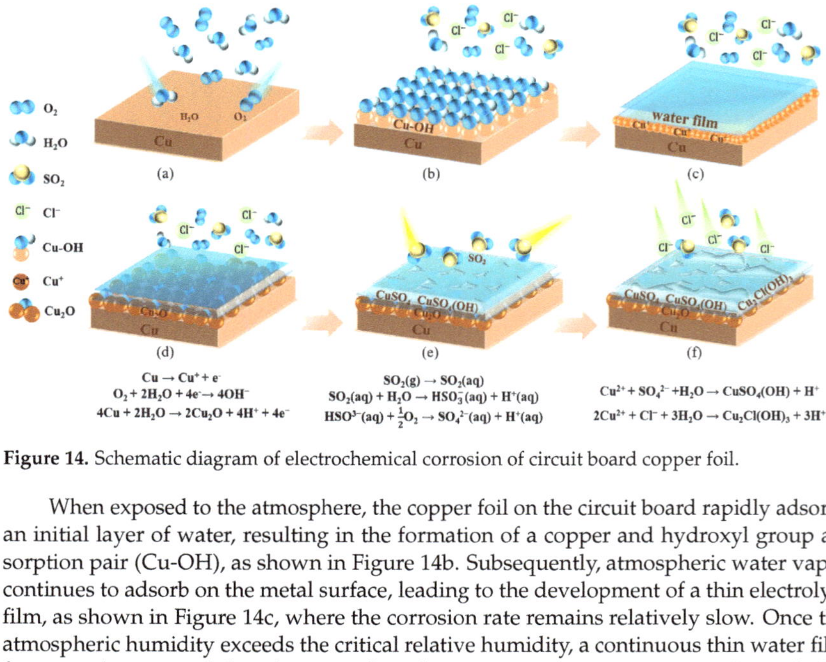

Figure 14. Schematic diagram of electrochemical corrosion of circuit board copper foil.

When exposed to the atmosphere, the copper foil on the circuit board rapidly adsorbs an initial layer of water, resulting in the formation of a copper and hydroxyl group adsorption pair (Cu-OH), as shown in Figure 14b. Subsequently, atmospheric water vapor continues to adsorb on the metal surface, leading to the development of a thin electrolyte film, as shown in Figure 14c, where the corrosion rate remains relatively slow. Once the atmospheric humidity exceeds the critical relative humidity, a continuous thin water film forms on the copper foil surface, significantly increasing the corrosion rate. Beneath this thin electrolyte film, the copper anodic dissolution reaction occurs first, forming Cu^+/Cu^{2-}, while the cathodic reaction mainly involves oxygen reduction or water decomposition, precipitating Cu_2O on the copper surface, as shown in Figure 14d.

$$Cu \rightarrow Cu^+ + e^- \tag{1}$$

$$O_2 + 2H_2O + 4e^- \rightarrow 4OH^- \tag{2}$$

$$4Cu + 2H_2O \rightarrow 2Cu_2O + 4H^+ + 4e^- \tag{3}$$

Subsequently, weak anodic reactions occur simultaneously on the cathode plate, forming a CuO product layer. In humid air, CuO readily reacts with SO_2 to produce sulfides.

$$2Cu_2O + O_2 \rightarrow 4CuO \tag{4}$$

When the adsorptive salt ions, dust, atmospheric pollutants, and corrosion products deposit on the copper foil surface, the critical relative humidity of the copper surface decreases significantly. This promotes the formation of an electrolyte film, further accelerating copper foil corrosion. In regions with higher humidity, the electrolyte film on the copper foil surface is relatively thicker, facilitating the dissolution of SO_2 from the atmosphere. The dissolution process of SO_2 is shown in Figure 14e, with the reactions as follows:

$$SO_2(g) \rightarrow SO_2(aq) \tag{5}$$

$$SO_2(aq) + H_2O \rightarrow HSO_3^-(aq) + H^+(aq) \tag{6}$$

$$HSO_3^-(aq) + \frac{1}{2}O_2 \rightarrow SO_4^{2-}(aq) + H^+(aq) \tag{7}$$

HSO_3^-, SO_4^{2-}, H^+, and $SO_2{\cdot}nH_2O$ co-exist in the water film on the copper foil surface, making the water film acidic and accelerating copper corrosion. Therefore, in acidic electrolytes, Cu_2O in areas with poorer local stability is damaged [34], and the surface oxide film is destroyed, becoming the anodic region of the corrosion microcell and further corroding and dissolving to form $CuSO_4$ and $CuSO_4(OH)$:

$$Cu_2O + 2H^+ \rightarrow Cu + Cu^{2+} + H_2O \tag{8}$$

$$Cu^{2+} + SO_4^{2-} + H_2O \rightarrow CuSO_4(OH) + H^+ \tag{9}$$

When Cl^- is present in the environment, the copper ions generated by the dissolution of Cu_2O react with it to form $Cu_2Cl(OH)_3$:

$$2Cu^{2+} + Cl^- + 3H_2O \rightarrow Cu_2Cl(OH)_3 + 3H^+ \tag{10}$$

When both SO_2 and Cl^- exist on the copper foil surface, the atmospheric corrosion of copper occurs, as shown in Figure 14f. On the one hand, the presence of sea salt ions can increase the conductivity of the film; on the other hand, SO_2 dissolved in the film reduces its pH, and these two effects function together to enhance the corrosivity of the film, thereby accelerating the corrosion process of the copper foil.

Based on the above discussion and analysis, the corrosion process of copper foil can be summarized as follows: Initially, Cu_2O forms preferentially, making the copper foil relatively stable with a low corrosion rate, primarily manifesting as localized corrosion. Over time, Cu_2O oxidizes to CuO, but due to the deposition of atmospheric pollutants and particles, CuO can quickly react with SO_2 in humid air to form sulfides, while the adhesion of Cu_2O to the substrate weakens, reducing its protective effect. When both the atmospheric pollutants, SO_2 and sea salt ions, deposit on the sample surface, products such as $Cu_2Cl(OH)_3$, $CuSO_4$, and $CuSO_4(OH)$ form on the copper foil surface. These corrosion products are often loose and unevenly distributed, and the Cu_2O corrosion product layer adjacent to the substrate cracks, causing instability in the copper foil substrate and accelerating corrosion.

4. Conclusions

This study analyzed the results of one-year exposure experiments conducted in a closed atmospheric environment across five typical sites in the Sichuan-Tibet region on copper foil.

(1) Among the five regions, the corrosion product layer on the copper foil surface in Meishan was denser and more uniform, and the type of corrosion product was relatively

single. While in Mangkang, Luding, Batang, and Panzhihua, the corrosion products were loosely distributed and the surface roughness of the samples increased successively.

(2) The composition analysis of the corrosion products revealed that Cu_2O was the main corrosion product. A dense and uniform Cu_2O product layer provided potential protection to the substrate. Subsequently, the deposition of pollutants and particles in the humid atmosphere caused the destruction of Cu_2O in less stable areas, leading to the breakdown of the surface oxide film and the formation of anodic regions in corrosion microcells. This resulted in further corrosion and dissolution, forming sulfides such as $CuSO_4$ and $CuSO_4(OH)$. Additionally, the presence of $Cu_2Cl(OH)_3$ in the corrosion products from Luding and Batang suggests the existence of Cl^- in the atmosphere at these sites. In Panzhihua, the corrosion products primarily consisted of Cu_2O, with a very small amount of CuO, which quickly reacted with SO_2 in humid air to generate sulfides.

(3) The presence of both SO_2 and water vapor resulted in severe corrosion. The samples located in areas with higher rainfall and greater temperature and humidity fluctuations demonstrated reduced stability. The dry–hot regions tend to form copper oxides, whereas the humid-hot regions facilitate the formation of copper sulfides.

In summary, the corrosion of copper foil is susceptible to environmental factors such as air, moisture, acids, and bases, leading to loss of luster and spot formation and significantly affecting its service life and aesthetics. Therefore, this study offers a theoretical basis for enhancing the service life and performance of copper foil in the Sichuan-Tibet region.

Author Contributions: Writing—original draft, visualization, methodology. H.J.: supervision, resources, methodology. D.J.: formal analysis, data curation. J.W.: writing—review and editing, resources. Z.C.: investigation, formal analysis, data curation. R.T.: methodology, conceptualization. S.D.: data curation, conceptualization. L.M.: writing—review and editing, supervision, resources, project administration; methodology, validation, X.Y. All authors have read and agreed to the published version of the manuscript.

Funding: This work is financially supported by the National Science and Technology Resources Investigation Program of China (Grant No. 2021FY100603) and the National Natural Science Foundation of China (No. 52371049).

Informed Consent Statement: Not applicable.

Data Availability Statement: The original contributions presented in this study are included in the article. Further inquiries can be directed to the corresponding authors.

Conflicts of Interest: The authors declare that they have no known competing financial interests or personal relationships that could have appeared to influence the work reported in this paper.

References

1. Wu, T.; Zhou, Z.; Xu, S.; Xie, Y.; Huang, L.; Yin, F. A Corrosion Failure Analysis of Copper Wires Used in Outdoor Terminal Boxes in Substation. *Eng. Fail. Anal.* **2019**, *98*, 83–94. [CrossRef]
2. Qi, D.; Cheng, R.; Du, X.; Chen, Y.; Zhang, Z.; Zhang, J. Review on Atmospheric Corrosion of Copper and Copper Alloys. *J. Chin. Soc. Corros. Prot.* **2014**, *34*, 389–398. [CrossRef]
3. Gen, W.; Chen, X.; Hu, A.; Li, M. Effect of Ag on Oxidation of Cu-Base Leadframe. *Microelectron. Reliab.* **2011**, *51*, 866–870. [CrossRef]
4. Vera, R.; Delgado, D.; Rosales, B.M. Effect of Unusually Elevated SO_2 Atmospheric Content on the Corrosion of High Power Electrical Conductors Part 3.: Pure Copper. *Corros. Sci.* **2008**, *50*, 1080–1098. [CrossRef]
5. Vera, R.; Delgado, D.; Rosales, B.M. Effect of Atmospheric Pollutants on the Corrosion of High Power Electrical Conductors—Part 2. Pure Copper. *Corros. Sci.* **2007**, *49*, 2329–2350. [CrossRef]
6. Ghosh, S. Three-Dimensional Microplate Formation with Evaporating Nanoparticle Suspensions on Superhydrophobic Surfaces. *Colloids Surf.-Physicochem. Eng. Asp.* **2017**, *529*, 901–906. [CrossRef]
7. Vimala, J.S.; Natesan, M.; Rajendran, S. Corrosion and Protection of Electronic Components in Different Environmental Conditions—An Overview. *Open Corros. J.* **2009**, *2*, 105–113. [CrossRef]
8. Hu, Y.; Xiao, K.; Yan, L.; Hao, X.; Huang, L.; Lou, Y. Effect of Fungus, *Aspergillus* Sp. F1-1, on the Corrosion Behavior of PCB-HASL in Humid Atmospheric Environment. *Surf. Topogr.-Metrol. Prop.* **2022**, *10*, 015022. [CrossRef]
9. Li, C.; Ma, Y.; Li, Y.; Wang, F. EIS Monitoring Study of Atmospheric Corrosion under Variable Relative Humidity. *Corros. Sci.* **2010**, *52*, 3677–3686. [CrossRef]

10. Tsuru, T.; Nishikata, A.; Wang, J. Electrochemical Studies on Corrosion under a Water Film. *Mater. Sci. Eng. A* **1995**, *198*, 161–168. [CrossRef]
11. Wang, J.; Bai, Z.; Xiao, K.; Gao, X.; Yi, P.; Dong, C.; Wu, J.; Wei, D. Influence of Atmospheric Particulates on Initial Corrosion Behavior of Printed Circuit Board in Pollution Environments. *Appl. Surf. Sci.* **2019**, *467*, 889–901. [CrossRef]
12. Jonsson, M.; Persson, D.; Thierry, D. Corrosion Product Formation during NaCl Induced Atmospheric Corrosion of Magnesium Alloy AZ91D. *Corros. Sci.* **2007**, *49*, 1540–1558. [CrossRef]
13. Wiesinger, R.; Martina, I.; Kleber, C.; Schreiner, M. Influence of Relative Humidity and Ozone on Atmospheric Silver Corrosion. *Corros. Sci.* **2013**, *77*, 69–76. [CrossRef]
14. Fan, Z.; Zhou, H.; Wang, X.; Song, J.; Shi, J.; Liu, C.; Chen, J. Study on eutectoid phase corrosion behavior and rust layer protection mechanism of bronze coins from the Western Han Dynasty (first century BC) in China. *J. Mater. Sci.* **2024**, *59*, 14676–14691. [CrossRef]
15. Liao, J.; Miyawaki, Y. Atmospheric Corrosion Resistance of a Diecast Mg-Al-Ca-Si Magnesium Alloy with Continuous Intermetallic Phases. *J. Mater. Eng. Perform.* **2023**, *32*, 2813–2823. [CrossRef]
16. An, B.; Zhang, X.; Han, E. Corrosion behavior of pure copper during initial exposure stage in atmosphere of Shenyang City. *ACTA Metall. Sin.* **2007**, *43*, 77–81.
17. Wu, J.; Li, X.; Dong, C.; Zhang, S.; Zhou, J.; Jun, W. Initial corrosion behavior of copper and brass in tropical maritime atmospheric environment. *J. Chin. Soc. Corros. Prot.* **2012**, *32*, 70–74.
18. Nairn, J.; Skennerton, S.; Atrens, A. Comparative atmospheric corrosion of primary and cold rolled copper in Australia. *J. Mater. Sci.* **2003**, *38*, 995–1005. [CrossRef]
19. Wang, J.; Zhou, X.; Xie, Y.; Xu, S.; Ouyang, Z.; Wang, X.; Li, D.; Li, C.; Wu, T. Atmospheric corrosion of tin coating on T2 copper in Xiangtan, China. *Corros. Commun.* **2024**, *16*, 52–60. [CrossRef]
20. Xiao, K.; Gao, X.; Yan, L.; Yi, P.; Zhang, D.; Dong, C.; Wu, J.; Li, X. Atmospheric Corrosion Factors of Printed Circuit Boards in a Dry-Heat Desert Environment: Salty Dust and Diurnal Temperature Difference. *Chem. Eng. J.* **2018**, *336*, 92–101. [CrossRef]
21. Mendoza, A.R.; Corvo, F.; Gómez, A.; Gómez, J. Influence of the corrosion products of copper on its atmospheric corrosion kinetics in tropical climate. *Corros. Sci.* **2004**, *46*, 1189–1200. [CrossRef]
22. Kong, D.; Dong, C.; Fang, Y.; Xiao, K.; Guo, C.; He, G.; Li, X. Copper corrosion in hot and dry atmosphere environment in Turpan, China. *Trans. Nonferr. Met. Soc. China* **2016**, *26*, 1721–1728. [CrossRef]
23. Feng, Y.; Bai, Z.; Yao, Q.; Zhang, D.; Song, J.; Dong, C.; Wu, J.; Xiao, K. Corrosion Behavior of Printed Circuit Boards in Tropical Marine Atmosphere. *Int. J. Electrochem. Sci.* **2019**, *14*, 11300–11311. [CrossRef]
24. Zhao, Q.; Liu, X.; Wang, H.; Zhu, Y.; An, Y.; Yu, D.; Qi, J. Research Progress in Corrosion Protection Technology for Electronic Components. *Metals* **2023**, *13*, 1508. [CrossRef]
25. Wang, C.X.; Wang, M.T.; Wang, S.J. Spatial and Temporal Variation Characteristics of Different Grades of Precipitation in Qinghai-Tibet Plateau. *Chin. J. Appl. Environ. Biol.* **2022**, *28*, 829–835.
26. Deng, M.; Lu, R.; Li, C. Contrasts between the Interannual Variations of Extreme Rainfall over Western and Eastern Sichuan in Mid-Summer. *Adv. Atmos. Sci.* **2022**, *39*, 999–1011. [CrossRef]
27. Yi, C.; Du, X.; Yang, Y.; Chen, Y.; Wei, G.; Yang, Z.; Zhang, Z. Study on the Initial Atmospheric Corrosion Behavior of Copper in Chloride-Containing Environments. *Int. J. Electrochem. Sci.* **2017**, *12*, 3597–3613. [CrossRef]
28. de la Fuente, D.; Simancas, J.; Morcillo, M. Morphological Study of 16-Year Patinas Formed on Copper in a Wide Range of Atmospheric Exposures. *Corros. Sci.* **2008**, *50*, 268–285. [CrossRef]
29. Rice, D.W.; Peterson, P.; Rigby, E.B.; Phipps, P.B.P.; Cappell, R.J.; Tremoureux, R. Atmospheric Corrosion of Copper and Silver. *J. Electrochem. Soc.* **1981**, *128*, 275. [CrossRef]
30. Pan, C.; Lv, W.; Wang, Z.; Su, W.; Wang, C.; Liu, S. Atmospheric Corrosion of Copper Exposed in a Simulated Coastal-Industrial Atmosphere. *J. Mater. Sci. Technol.* **2017**, *33*, 587–595. [CrossRef]
31. Fang, W.; Chen, H.; Liu, X. Research on Corrosion Causes of Common Metal Material in Power Grid. *IOP Conf. Ser. Earth Environ. Sci.* **2020**, *512*, 012107. [CrossRef]
32. Yi, P.; Dong, C.; Xiao, K.; Li, X. Surface Failure Analysis of a Field-Exposed Copper-Clad Plate in a Marine Environment with Industrial Pollution. *Appl. Surf. Sci.* **2017**, *399*, 608–616. [CrossRef]
33. Wang, X.; Su, H.; Xie, Y.; Wang, J.; Feng, C.; Li, D.; Wu, T. Atmospheric Corrosion of T2 Copper and H62 Brass Exposed in an Urban Environment. *Mater. Chem. Phys.* **2023**, *299*, 127487. [CrossRef]
34. Yi, P.; Xiao, K.; Ding, K.; Yan, L.; Dong, C.; Li, X. Initial Corrosion Behavior of a Copper-Clad Plate in Typical Outdoor Atmospheric Environments. *Electron. Mater. Lett.* **2016**, *12*, 163–170. [CrossRef]

materials

Article

Steel Catenary Riser Fatigue Assessment: Fracture Mechanics Approach Versus *S–N* Curve Method

Niantao Zhang, Caiyan Deng, Wenqiang Zhang *, Siyuan Li and Baoming Gong

Department of Materials Science and Engineering, Tianjin University, Tianjin 300072, China; 1016208055@tju.edu.cn (N.Z.); dengcy@tju.edu.cn (C.D.); 1020208066@tju.edu.cn (S.L.); gong_bm@tju.edu.cn (B.G.)
* Correspondence: wqzhang2022@tju.edu.cn; Tel.: +86-22-85356663

Abstract: In this paper, the fatigue resistance of a full-scale Steel Catenary Riser (SCR) girth weld is investigated using the Strength–Number of cycles (*S–N*) curve method based on weld formation quality and fracture mechanics approaches. The test results, presented in the form of *S–N* curves, are superior to the design curve E in BS 7608. Compared with the *S–N* curve determined by a resonant bending rig, the analytical fracture mechanics, i.e., engineering critical assessment (ECA) based on BS 7910, can provide a rational estimation of full-scale girth welds. For the numerical methods, the short crack growth phase is crucial to improving the accuracy and reliability of the assessment. For the girth weld with a concave root, the geometries of the weld cap are the predominant factors for fatigue life. Although the crack initiation site is always located at the outer surface regardless of the flushed or welded caps, the weld grinding treatment is still effective in promoting fatigue life.

Keywords: steel catenary risers (SCRs); fatigue assessment; *S–N* curve; finite element analysis (FEA); failure assessment diagram (FAD)

Citation: Zhang, N.; Deng, C.; Zhang, W.; Li, S.; Gong, B. Steel Catenary Riser Fatigue Assessment: Fracture Mechanics Approach Versus *S–N* Curve Method. *Materials* **2024**, *17*, 5677. https://doi.org/10.3390/ma17225677

Academic Editor: Carmine Maletta

Received: 17 October 2024
Revised: 9 November 2024
Accepted: 19 November 2024
Published: 20 November 2024

1. Introduction

Steel Catenary Risers (SCRs) are free-hanging riser suspended from floating facilities to the seabed in the form of a catenary without buoys devices, as shown in Figure 1 [1]. They have been widely used in offshore oil and gas development in deep-water fields compared to other common riser forms such as flexible pipes, riser towers, and top tensioned risers, due to their reliable performance, easy manufacturing, and convenient installation [2–4]. SCRs usually consist of numerous individual segments, i.e., standard-size pipes, which are conveniently and effectively joined by girth welds between adjacent segments. However, the girth weld is usually still the 'weakest' link for fatigue resistance due to the inevitable stress concentration, weld defects, and residual stresses. As a result of the continuously dynamic environmental and operational loads, SCRs inevitably suffer severe cyclic vibration along the pipeline axis direction over time during service, making them vulnerable to fatigue failure, even fracture at the point where the 'weakest' link is located. Once a part of an SCR is cracked or broken at the location of the girth weld, the oil or natural gas it transports will leak, resulting in serious environmental pollution, ecological disasters, and huge economic losses [5,6].

It is essential to assess the fitness-for-service of SCR structures to ascertain if a known flaw and geometrical discontinuity in the girth weld are likely to fail the structure under cyclic load before the designed in-service life [7], which is normally designed for a life ranging from 10 years to 40 years [8]. There are two widely accepted approaches to fatigue assessment and design of welded joints and components: the *S–N* curve method and the crack propagation method [9,10]. The most common analysis approach for fatigue life prediction is based on the *S–N* curve method, where various fatigue driving parameters may be used, including nominal stress, structural hot-spot stress, and effective notch stress (strain) [11]. The *S–N*-based approach is determined by fatigue testing of the considered welded details and the linear damage hypothesis. The targeted fatigue life at each stress level is determined using the equations recommended in IIW-2259-15:2016 [9], BS 7608:2014 [12], or DNV-RP-C203:2011 [10] with the adjustment at a 97.7% confidence interval. The fatigue design rules of the *S–N* curve are mainly based on data generated from tests on either beams or small-scale plate specimens incorporating the designated weld details, where the verification is performed so that the predicted fatigue life is superior to the target structures with a required safety margin. However, the *S–N* curve method is not able to analyze the fatigue process itself, and the geometrical details and loading histogram are usually omitted or approximated. Therefore, the redundant design based on the *S–N* curve is inevitable. In contrast, the fracture mechanics-based assessment approach assumes the fatigue life of welded joints and components is mainly comprised of the crack propagation from the pre-existed welding defects as described by Paris law and its modifications, whilst the crack initiation is negligible. Fracture mechanics concepts are valuable for assessing the impact of pre-existing welding defects, whereas the S–N curve approach does not consider the actual morphology of these defects. Maljaars et al. evaluated the failure probability of structures under fatigue loading using Linear Elastic Fracture Mechanics (LEFM) [13]. Applying fracture mechanics to the fatigue assessment of offshore pipelines, such as SCRs, is particularly useful for determining whether a sufficient time interval exists between crack detection during in-service inspections and the onset of unstable fracture [14].

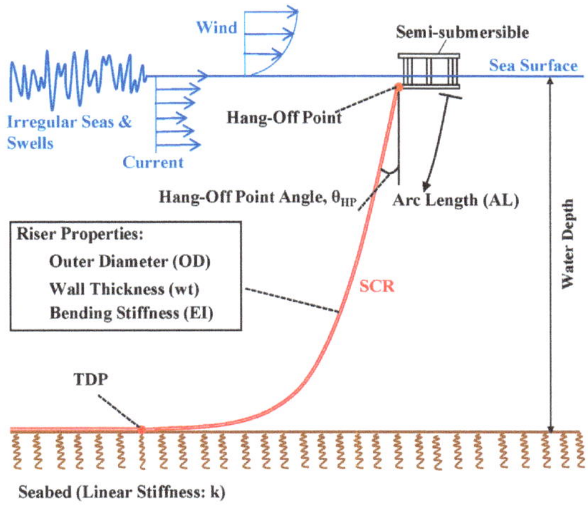

Figure 1. Overall view of the floating system with SCR [1].

In this paper, the full-scale fatigue behaviors of SCR girth welds are tested by a resonant bending rig and plotted in the form of stress versus fatigue life. The welded qualification-related *S–N* curve method according to BS 7608 is compared to the fracture mechanics-based assessment method, i.e., Engineering Critical Assessment (ECA) by BS 7910 and crack growth simulation by FRAC3D incorporated with NASGRO equation, and the accuracy is evaluated in terms of experimental fatigue lives. The comparative investigation of these methods is performed and discussed at the end of this paper.

2. Materials and Methods

2.1. Material Properties and Welding Process

The pipeline of the SCR in this study is made of API 5L X65 material (Hengyang Valin Steel Tube Co., LTD., Hengyang, China) with an outside diameter of 168.3 mm and a thickness of 18.3 mm. An ER70S-6 material (Jinqiao Welding Materials. Tianjin, China) solid wire and E81T1-Ni1M JH4 material (Jinqiao Welding Materials, Tianjin, China) flux-cored wire with a diameter of 1.2 mm were employed as the Gas Metal Arc Welding (GMAW) root and filler wire, respectively [15]. The chemical compositions of the API 5L X65 and the wires are provided in Table 1 for reference [15]. The material properties for the pipeline steel are as follows: the minimum yield strength, σ_s, is 506 MPa; the tensile strength, σ_u, is 580 MPa; the modulus of elasticity, *E*, is 210 GPa. The fracture toughness of API X65 steel is taken as 109.9 MPa·m$^{1/2}$ according to the software of FRANC3D V8.4.5. During the welding process, the GMAW/Pulse Gas Tungsten Arc Welding (GTAW-P) double-sided root welding process was employed for welding the root welds [15]. Filler metal was utilized in the GTAW root welding and no filler in the GTAW-P root welding process. Subsequently, filling and cover welding were carried out. The detailed parameters of welding processes are provided in Table 1. For the comparative study, weld reinforcements of the SCR girth welds were ground flush by the post-weld treatment in contrast with the as-welded ones at each load level.

Table 1. The welding parameters of SCR girth weld.

	Current/A	Voltage/V	Feeding Speed /cm/min	Welding Speed /mm/min	Interlayer Temperature/°C
Double-sided Root welding	MIG:200 TIG:280	MIG:18 TIG:10	/	270	70
Fill	300	14	75	110–120	200–250
Weld cap1	230	13	45	120	220
Weld cap2	220	13	45	110	225

2.2. *Resonance Bending Test for Full Scale Girth Weld Joint*

In this paper, a resonant bending rig developed by the Welded Structure Laboratory at Tianjin University was used to study the fatigue life of full-scale girth welds in SCRs, as shown in Figure 2. The pipe, 5530 mm in length, is subjected to excitation with a resonance frequency between 25 Hz and 35 Hz so that each girth weld can undergo alternative loading with stress ratio *R* = 0. Since a standing wave produced by the eccentric mass excites the bending of the pipe, the supporting framework is located at the wave node of the first eigenmode of the pipe, and the weight is carried vertically by air springs. Moreover, the internal water pressure is introduced to achieve higher axial mean stress than zero. When water leakage appears or the pressure drops immediately, a through-thickness crack may exist, and the test is stopped (see Figure 3 for the case).

Figure 2. Resonance bending fatigue testing platform, Tianjin University (equipment no.: GWT0614; dimensions: 15 m × 1.3 m × 2.0 m; range of the test pipeline diameter: 168.3–355.6 mm; range of the test pipeline length: 6–12 m; frequency: 0–25 Hz).

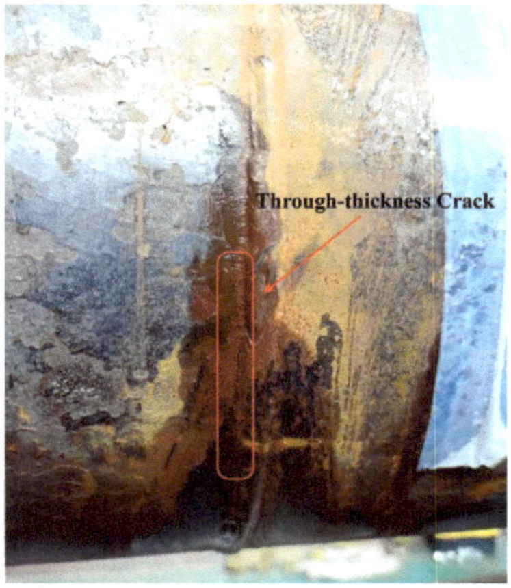

Figure 3. The occurrence of a through-thickness crack after internal water pressure leak.

The tested pipe is joined by two girth welds with a distance of 800 mm plus the width of weldment, as shown in Figure 4. The cyclic stress difference between two girth welds is negligible, as validated by strain gauge measurements. During the testing, eight strain gauges are arranged circumferentially to monitor the axial strain variations positioned 100 mm far from each side of the girth welds, and the illustration of the arrangement is shown in Figure 5. An integrated monitor system is used to acquire the online experimental data, including internal pressure monitored by pressure transducer EPU-2500/YPR-8 (Shanghai Zhendan Sensor Instrument Factory, Shanghai, China) and strain by resistance strain gauge DH5922D (Donghua Test, Jingjiang, China).

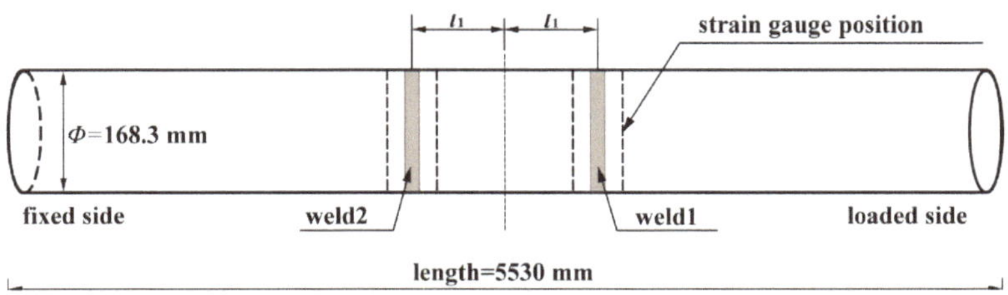

Figure 4. Schematic diagram of resonance bending fatigue test of full-size SCR (l_1 = 400 mm).

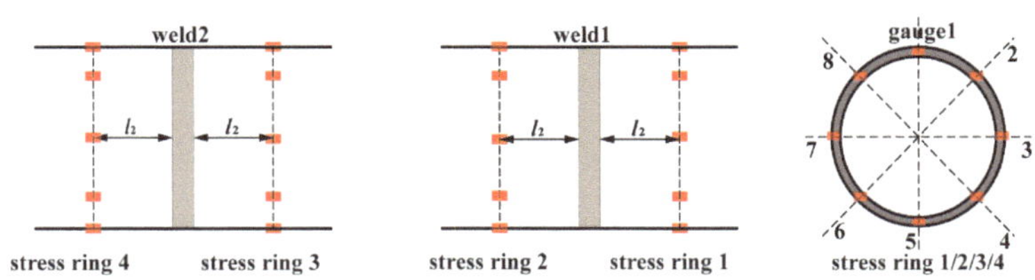

Figure 5. Schematic diagram of strain gauge arrangement scheme (l_2 = 100 mm).

The average applied bending stress range was calculated by averaging the strain gauge readings from the adjacent rings and linearly interpolating them to the weld position. Three levels of fatigue stress (high, medium, and low) were used for the tests. Also, the targeted fatigue life at each stress level was determined using the recommendation in BS 7608: 2014, and the adjustment was performed to obtain the failure cycle at a 97.7% confidence interval as described in BS 7608:2014. Post-mortems were performed for both the weld cap and root for crack examination after each test. Moreover, both the measurement of misalignment and macro fractography analysis for the failed strings were conducted to identify the crack initiation and propagation characteristics.

3. Fatigue Assessment Based on Fracture Mechanics

The fracture mechanics method describes the entire crack growth history from a small initial crack toward the final failure, which is an indispensable tool once an unavoidable crack or defect is detected and sized. Recently, several standards related to the fatigue assessment based on the fracture mechanics principle have been widely used in ocean engineering, including BS 7910:2013 [16], DNV C203:2011 [10], and IIW-2259-15:2016. In this section, the study of fatigue resistance assessment of the SCR in terms of a numerical model by NASGRO and the analytical failure assessment diagram method is performed.

3.1. *The NASGRO Model by FRANC3D*

To achieve a reliable prediction for the growth of a fatigue crack in the high cycle regime, crack propagation behaviors with both scales comparable to the microstructural grain size and the local plastic zone size are essential. The well-known NASGRO equation for fatigue life prediction based on LEFM is an analytical crack growth rate equation that incorporates several features observed in real materials, such as sensitivity to near-threshold and near-critical growth, sensitivity to the stress ratio, and small crack sensitivity [17–19]. The basic formula of the NASGRO crack growth rate equation is expressed as follows:

$$\frac{\mathrm{d}a}{\mathrm{d}N} = C\left[\left(\frac{1-f}{1-R}\right)\Delta K\right]^{\mathrm{n}} \frac{\left(1-\frac{\Delta K_{\mathrm{th}}}{\Delta K}\right)^{\mathrm{p}}}{\left(1-\frac{\Delta K_{\mathrm{max}}}{K_{\mathrm{c}}}\right)^{\mathrm{q}}} \tag{1}$$

where C, n, p, and q all are empirical constants; f is the crack opening function associated with plasticity-induced crack closure, which has been defined by Newman [20]; ΔK_{th} is the threshold stress intensity factor range below which there is no crack growth. It is approximated by the following empirical formulae:

$$\Delta K_{\mathrm{th}} = \begin{cases} \Delta K_1^* \left[\frac{1-R}{1-f}\right]^{(1+RC_{\mathrm{th}}^p)} / (1-A_0)^{(1-\mathrm{R})C_{\mathrm{th}}^p} & R \geq 0 \\ \Delta K_1^* \left[\frac{1-R}{1-f}\right]^{(1+RC_{\mathrm{th}}^m)} / (1-A_0)^{(C_{\mathrm{th}}^p - \mathrm{R}C_{\mathrm{th}}^m)} & R < 0 \end{cases} \tag{2}$$

and ΔK_1^* can be calculated as follows:

$$\Delta K_1^* = \Delta K_1 \left[\frac{a}{a+a_0}\right]^{1/2} \tag{3}$$

where C_{th} is an empirical constant, ΔK_1 is the observed threshold for a high-stress ratio, and a_0 is a small crack parameter, also known as the intrinsic crack length (usually 0.0381 mm). K_{c} is the fracture toughness that should be chosen via the plane-strain fracture toughness, K_{1c}, the plane-stress toughness, K_{1e}, or calculated as follows:

$$K_{\mathrm{c}} = K_{\mathrm{Ic}}\left(1 + B_{\mathrm{k}} e^{-(A_{\mathrm{k}} t / t_0)^2}\right) \tag{4}$$

where $t_0 = 2.5(K_{\mathrm{Ic}}/\sigma_{\mathrm{ys}})^2$, A_{k} and B_{k} are empirical constants. The thickness, t should be specified on the basis of engineering judgment. In this work, the plane-strain fracture toughness, K_{Ic}, is chosen. The parameters required in the NASGRO equation are provided in Table 2.

Table 2. The parameters of X65 steel required in the NASGRO equation.

Parameters	Value	Parameters	Value
σ_u/MPa	580	a_0/mm	0.0381
σ_s/MPa	506	n	2.8
K_{Ic}/MPa·m$^{1/2}$	109.9	c	1.23×10^{-12}
ΔK_0/MPa·m$^{1/2}$	6.59	P	0.5
A_k	0.75	q	0.5
B_k	0.5	Cth	2.2

3.2. FAD Method in BS 7910

Failure Assessment Diagram (FAD) and Crack Driving Force (CDF) approaches are two basic engineering assessment methods that are aimed at providing conservative assessments of the severity of crack-like flaws. This technique originates from the work of the British Central Electricity Generating Board (CEGB) and its R6 assessment procedure. The FAD approach is recognized as an efficient methodology to assess the potential interaction between brittle fracture and plastic collapse of loaded structural components that contain cracks, where a rough geometry-independent failure line is constructed by normalizing the CDF by the material fracture resistance [21]. More specifically, the vertical axis of the FAD compares the applied loading, in terms of the corresponding CDFs normalized by the fracture toughness of the material; the horizontal axis is the ratio of the applied primary load to that required for plastic collapse. The assessment of the component is then based on the relative location of a geometry-dependent assessment point with respect to this Failure Assessment Line (FAL), which is considered safe as long as the assessment point lies within the area below the FAL, and potentially unsafe if it is located on the line or outside the shaded area, as illustrated in Figure 6 [22]. Many studies have been done to verify the efficiency of the FAD concept in evaluating the mechanical integrity of components with flaws. The BS 7910 is the most widely used engineering code to evaluate the acceptability of flaws in metallic structures [23]. In Figure 6, Option 1, presented as the function of L_r, is recommended for general use in cases where knowledge regarding the material properties is limited. It is a conservative procedure that is relatively simple to employ and does not require detailed stress–strain data for the materials being analyzed; Option 2 is based on the use of a material-specific stress–strain curve; Option 3 uses numerical analysis to generate a FAD and is not confined to use with materials showing ductile tearing.

(1) Option 1 of failure assessment procedures: basic level

$$f(L_r) = \left(1 - 0.14L_r^2\right)\left[0.3 + 0.7\exp\left(-0.65L_r^6\right)\right] \quad L_r \le L_{r,max} \tag{5}$$

(2) Option 2 of failure assessment procedures: standard level

$$f(L_r) = \left(\frac{E\varepsilon_{ref}}{L_r\sigma_s} + \frac{L_r^3\sigma_s}{2E\varepsilon_{ref}}\right)^{\frac{1}{2}} \quad L_r \le L_{r,max} \tag{6}$$

where ε_{ref} is the reference strain.

(3) Option 3 of failure assessment procedures: advanced level

$$f(L_r) = \sqrt{\frac{J_{el}}{J_{el} + J_{pl}}} \quad L_r \le L_{r,max} \tag{7}$$

where J_{el} and J_{ep} are the elastic and plastic components of J-integral, respectively.

The cut-off line of the FAC for the above procedures is defined as:

$$L_{r,max} = \frac{\sigma_s + \sigma_u}{2\sigma_s} \tag{8}$$

When the value of L_r exceeds the $L_{r,max}$, the value of K_r is taken as zero. In this paper, Option 1 obtained from the bottom-bound fitting of the FAD curve family generated using Equation (5) for various material constitutive relations was employed.

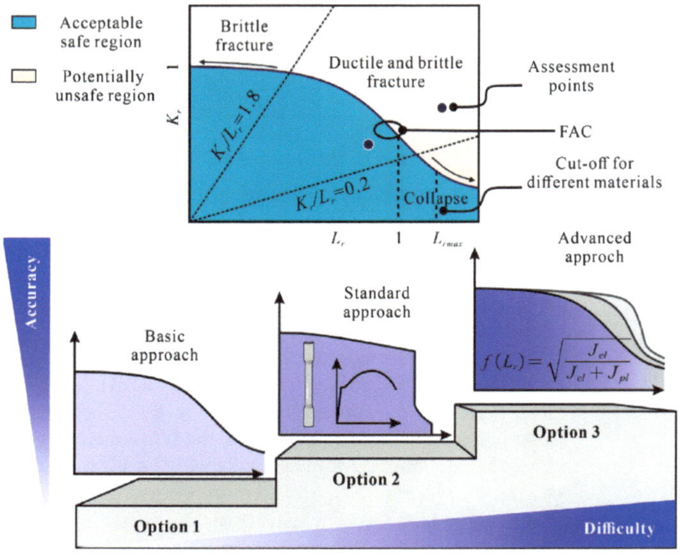

Figure 6. Schematic of Failure Assessment Diagram [19].

4. Result

4.1. S–N Curve and Comparison with BS 7608

The test results obtained from the full-scale pipes are presented in Table 3 and Figure 7, respectively. In this study, the fatigue life was defined as the number of cycles required to produce a through-wall fatigue crack or greater than 10^7 (referred to as run-out). The design curve D in BS 7608:2014 is designated as the targeted qualification of outer diameter for the tests. Meanwhile, the design curve E in BS 7608:2014 with the adjusted value at a 97.7% confidence interval was also plotted in Figure 7 for comparison. The test results are generally better than the BS 7608: 2014 Class E design *S–N* curve. Both the low ($\Delta\sigma$ = 68.9 MPa) and medium ($\Delta\sigma$ = 103.4 MPa) stresses cannot produce final failure until an endurance larger than 3.2×10^7 cycles and 1.2×10^7 cycles, respectively. Therefore, only the tests at high loading levels are studied further for comparison with other methods. More specifically, it is found in Figure 8 that fatigue crack initiated at the toe of the convex weld root and grew through the wall thickness for the external polished pipe (Specimen H-1), while the cap weld toe was the crack initiation site and then propagated perpendicular to the wall thickness for the as-weld pipe (Specimen H-2). In contrast, it is found in Figure 9 that all the cracks initiate at the external surface for the flushed and as-welded girth due to the concave weld roots, and many more failure cycles are achieved. Therefore, the geometrical concentration introduced by the girth weld root is predominant, and the crack initiation site depends on weld root convexity. Moreover, although the reinforcement flush treatment can extend the fatigue life for both convex and concave weld roots, Specimen H-1 still exhibits an unexpectedly shorter fatigue life compared to specimen H-2. This is because crack initiation at the internal surface of the pipe shields the life extension effect of the cap flush. The crack initiation transitions from the external surface to the internal surface of the pipe, which is more difficult to detect during routine inspection and may result in catastrophic failure of the structures.

Table 3. The results of stress versus cycle life by resonance bending fatigue test.

Specimen No.	Stress Range $\Delta\sigma$/MPa	Internal Pressure σ_{mean}/MPa	Loading Cycle	Post-Weld Treatment
L-1	68.9	72	32,003,250 (run-out)	Cap flush
L-2	68.9	65	32,900,204 (run-out)	As-weld
M-1	103.4	68	12,965,535 (run-out)	Cap flush
M-2	103.4	71	13,555,804 (run-out)	As-weld
H-1	172.4	67	559,600	Cap flush
H-2	173.0	67	755,400	As-weld
H-3	193.0	70	1,150,000	As-weld
H-4	193.0	70	3,050,000	Cap flush

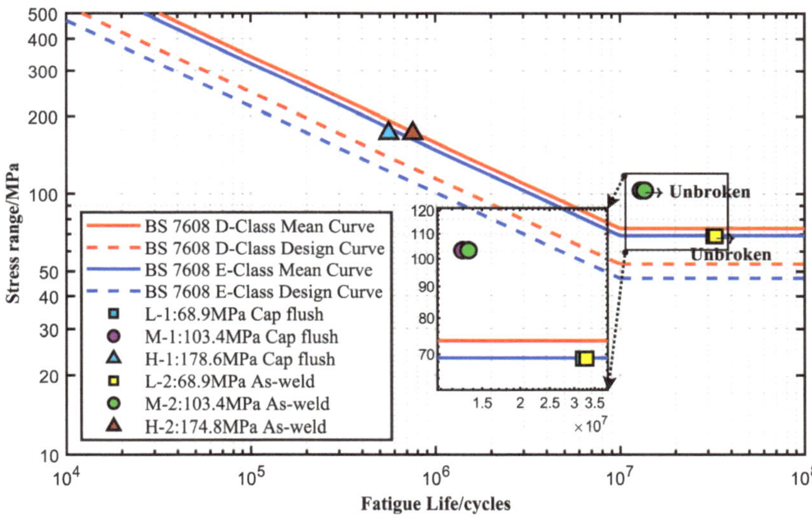

Figure 7. The plots of stress level versus fatigue life for six girth welds in relation to BS 7608: 2014 Class D and E curves.

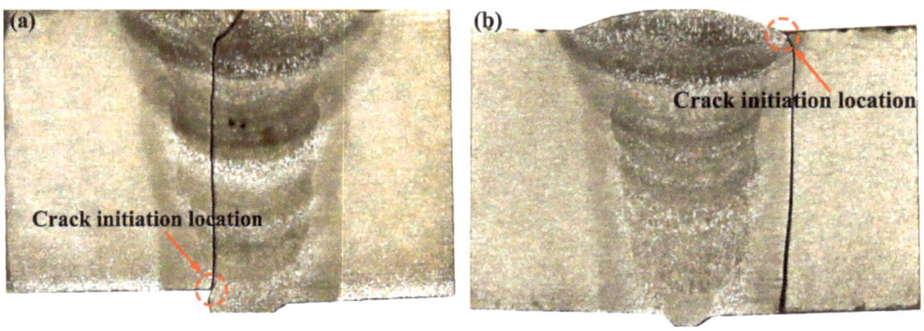

Figure 8. Crack growth direction along the wall thickness for girth welds with convex root: (**a**) H-1: cap flush; (**b**) H-2: as-weld.

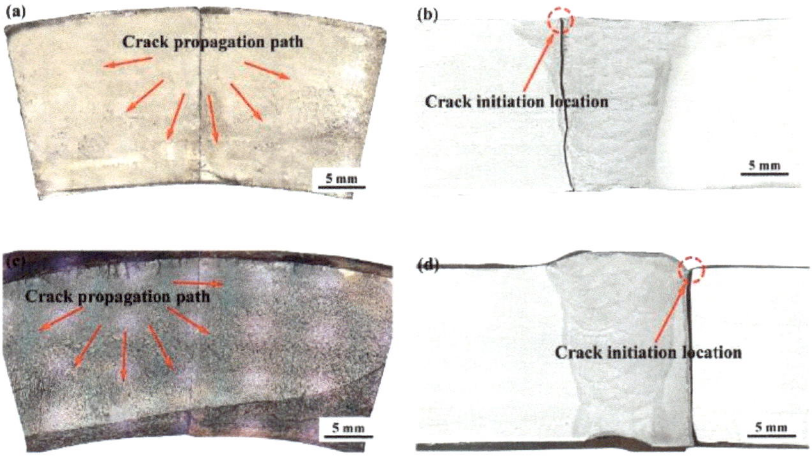

Figure 9. Crack growth path along the wall thickness and crack initiation location for girth welds with concave root: (**a**,**b**) H-3: cap flush; (**c**,**d**) H-4: as-weld.

4.2. *Numerical Prediction Based on the Crack Growth Model*

In order to simulate the crack growth using linear elastic fracture mechanics (LEFM) and evaluate the fatigue life of the SCRs, Finite Element Analysis (FEA) was performed on the three-dimensional model of the pipeline with the combination of the commercial software ABAQUS 2022 and FRANC3D V8.4.5 [24,25]. FRANC3D, a fracture mechanics tool, is designed to analyze crack growth in Finite Element (FE) models generated by ABAQUS. Its primary function involves inserting and extending crack tip elements, leveraging quadratic order elements, such as 20-node solid quadratic elements, in three-dimensional models to accurately represent the stress fields near the crack tip. ABAQUS serves as the solver for the models modified by FRANC3D, facilitating the calculation to generate the output database corresponding to various load conditions and crack geometrics. Subsequently, the Stress Intensity Factors (SIFs) along the crack front are calculated based on the displacement field extracted from the output database using the M-Integral in FRANC3D. This process also enables the determination of the next crack growth increment, allowing for continuous extension of the crack.

The model is partitioned for ease of meshing and locating of the crack. Detailed focused meshes are required around the crack tip for accurate contour integral evaluation. Abaqus integrates around the ring of element (contour) enclosing the crack tip to determine the Stress Intensity Factors (SIFs). Contour integrals are evaluated for multiple rings of elements surrounding the crack tip node. The first contour is formed from elements directly connected to the crack tip node. Each subsequent contour is created by offsetting one element away from the previous contour. The crack aspect ratio (crack depth/crack half-length, $a/2c$) is assumed as 0.1 based on recommendations in [26]. Cracks in SCRs typically start from a long shallow surface flaw situated at the interface of weld and parent metal, and then grow gradually as a semi-elliptical flaw until it penetrates the wall thickness (see Figure 8).

4.2.1. Equivalence of Load Condition

Online experimental data of the stress state was acquired from the integrated monitor system via the strain gauges arranged circumferentially as shown in Figure 5. The average of the maximum stress range recorded by each strain gauge throughout the test is shown as a stress circle in Figure 10. Compared with the numerical magnitude of the stress range, the difference in the stress range at different strain gauge positions is negligible. Therefore, for specimen H-1 and specimen H-2, the load conditions at both ends of the girth weld in the resonance bending test condition are equivalent to the stress ranges of 178.6 MPa and 174.8 MPa, respectively. The equivalent load conditions are later employed in the simulation analysis of crack propagation.

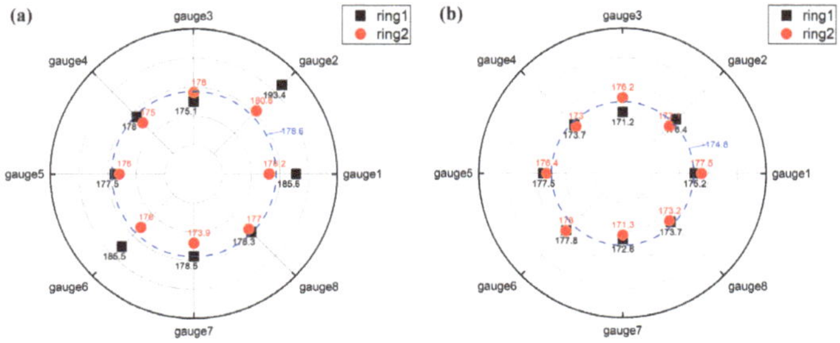

Figure 10. Average stress range distribution at both ends of the girth weld at the internal water pressure leak: (**a**) specimen H-1; (**b**) specimen H-2 (unit: MPa).

4.2.2. Geometry Simplification of the Pipeline

The geometric model should be determined according to the actual geometry of the pipeline for accurate fatigue life prediction. Since the weld profile in terms of weld reinforcement height, toe radius, and toe angle varies along the location of the grith weld, it is thus impractical to model the actual geometrical discontinuity in the numerical analyses. However, the effects of weld reinforcement cannot be omitted when a shallow surface flaw ($z/B < 0.2$) is located around the stress concentration of the weld toe or weld root. Therefore, a plain pipeline model instead of the actual geometry of the SCR is employed, while a stress correction factor, M_k, multiplying the actual axial stresses monitored by strain gauges is applied to compensate for the stress concentration effect caused by the geometric discontinuity at both weld cap and weld root toes, which is regarded as geometry simplification as shown in Figure 11. More specifically, when the real structure of the weld profile is neglected in the process of model simplification, the corresponding axial stress P is adjusted by multiplying it with M_k calculated via empirical formula. Finally, the new geometric model, which is equivalent to the original, has been constructed, and it provides considerable convenience for both modeling and insertion of initiation in the subsequent subsection. M_k is defined according to BS 7910: 2013 as follows:

$$M_k = v(z/B)^w \tag{9}$$

where v and w have different values corresponding to flaws at the toes of full penetration or attachment welds, respectively. M_k and the detailed size of the macrographs of the weld are listed in Table 4.

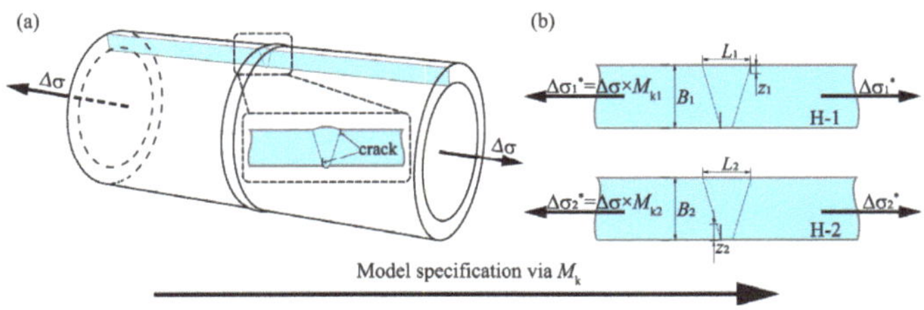

Figure 11. The model simplification approach: (**a**) Cross-section of circumferential weld in the actual pipeline; (**b**) Considering the stress concentration effect due to weld reinforcement by a correction factor, M_k.

Table 4. The calculation results of M_k at the axial loading mode.

Specimen	Crack Location	*L*/mm	*B*/mm	*z*/mm	*L*/*B*	*z*/*B*	v	w	M_k
H-1	Weld root toe	3.2	18.3	0.1	≤2	$\leq 0.05(L/B)^{0.55}$	0.3185	−0.31	1.6
H-2	Weld cap toe	14.1	18.3	0.1	≤2	$\leq 0.05(L/B)^{0.55}$	0.4753	−0.31	2.4
H-3	Weld cap toe	/	/	0.1	/	/	/	/	1.0
H-4	Weld cap toe	15.5	18.3	0.1	≤2	$\leq 0.05(L/B)^{0.55}$	0.4876	−0.31	2.5

4.2.3. Modeling and Insertion of the Initial Crack

To simulate the crack growth and evaluate the fatigue life using the NASGRO equation, an initial surface crack was introduced into stress concentration zones, such as the toe of the weld root (H-1) or weld cap (H-2), in accordance with the macrograph of the weld cross-section. The numerical analyses were conducted for six cases as tabulated in Table 5 and illustrated in Figure 12. Only half of the FE model was modeled using eight-node linear brick elements (C3D8) due to the symmetry. The process of geometric simplification has been detailed in Figure 11, and the transition mesh was applied to the welded joint section of the global model as shown in Figure 12b. As shown in Figure 12c, a sub-model was separated from the global model, and the corresponding boundary conditions were introduced into FRANC3D, where half of the long shallow crack was inserted with the sizes of d = 1.00 mm, e = 0.11 mm, and r_c = 0.10 mm, respectively.

Table 5. FEA model specifications.

Specimen	Crack Location	M_k
H-1	Weld root	1.0
H-1	Weld root	1.6
H-1	Weld root	1.0
H-2	Weld toe	1.0
H-2	Weld toe	2.4
H-2	Weld toe	1.0
H-3	Weld toe	1.0
H-4	Weld toe	2.5

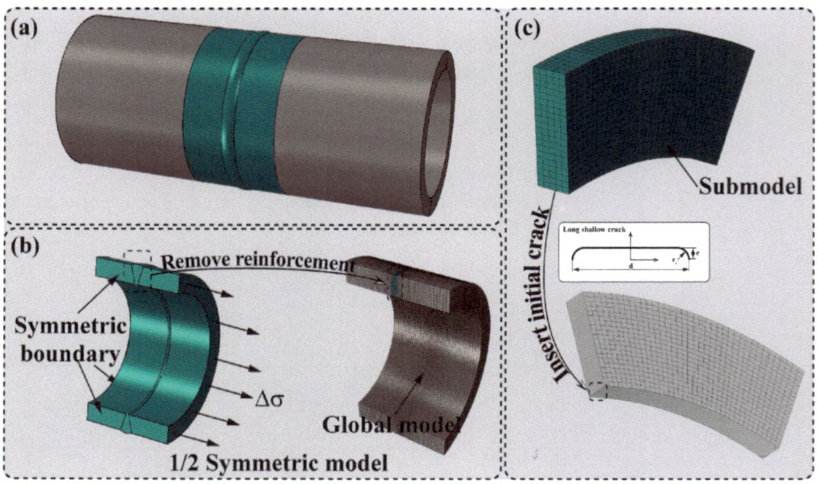

Figure 12. (**a**) An actual pipeline; (**b**) Details of the 1/2 symmetric model; (**c**) Insertion of initial crack.

4.2.4. Simulation of Fatigue Crack Growth

There are several different criteria to calculate the direction of crack growth, and one of the most widely used criteria is the maximum tangential stress (MTS) criterion proposed by Erdogan and Sih [27]. The results indicate that the value of K_I is larger than K_{II} and K_{III} as shown Figure 13, suggesting that mode-I (opening mode) predominantly governs the crack growth process [28]. In FRANC3D, the SIFs evolution, as shown in Figure 13b, was constructed using the increment of crack depth and the calculated SIF values at the crack front nodes. The crack distance is normalized along the crack front as shown in the upper inset of Figure 13a, where point G represents the intersection point of the symmetric boundary and the crack front, i.e., the intermediate point of the complete crack, and point H represents intersection point of the inner surface of the pipeline in specimen H-1 (the outer surface of pipeline in specimen H-2) and the crack front as shown in the lower inset of Figure 13a. The SIFs result shows the crack type is the opening mode under the uniaxial load condition, and the value of K_I is above the threshold values ΔK_{th}. The propagation path and SIFs distribution along the crack front can be seen from Figure 13b. Also, as the crack grows, K_I of point G becomes larger. As a result, at the final step of crack propagation through the thickness, the SIFs reach their maximum value.

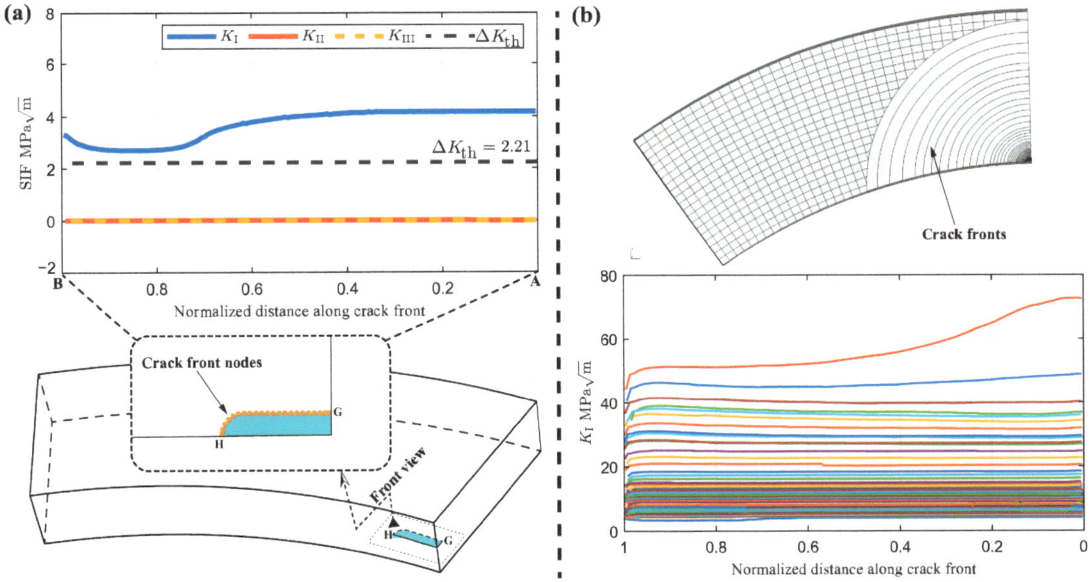

Figure 13. (**a**) The SIFs along initial crack front; (**b**) along crack front for the propagation of crack.

Figure 14 illustrates the trend in fatigue life and SIFs at point G as crack depth increases. During the crack growth process, when the crack depth exceeds 8 mm, the SIFs of specimen H-1 are distinctly higher than that of specimen H-2, and the gap between them progressively widens. The higher SIFs lead to a reduction in fatigue life, as illustrated in Figure 14, where specimen H-2 shows a longer fatigue life compared to specimen H-1. This finding is consistent with the results of the resonance bending test for the full-scale girth weld joint, where the fatigue life extension effect of the reinforcement flush treatment is masked.

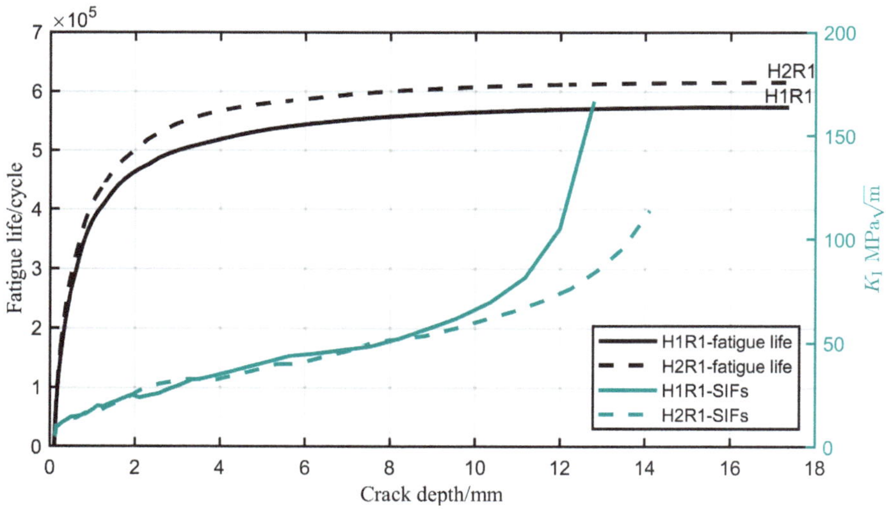

Figure 14. The variation trend of fatigue life and SIFs with the increasing crack depth.

4.2.5. Fatigue Life Prediction

As discussed before, it is found that the compensation for the stress concentration effect by M_k has a significant influence on the axial load, as shown in Figure 15. The fatigue life of the pipeline under different conditions is summarized in Table 6. Comparing the calculated total number of cycles to failure against that without the stress correction factor, a noticeable variation is found in the predicted fatigue life for specimens H-1 and H-2, which decrease by approximately 80.4% and 98.4%, respectively, at R = 0.1.

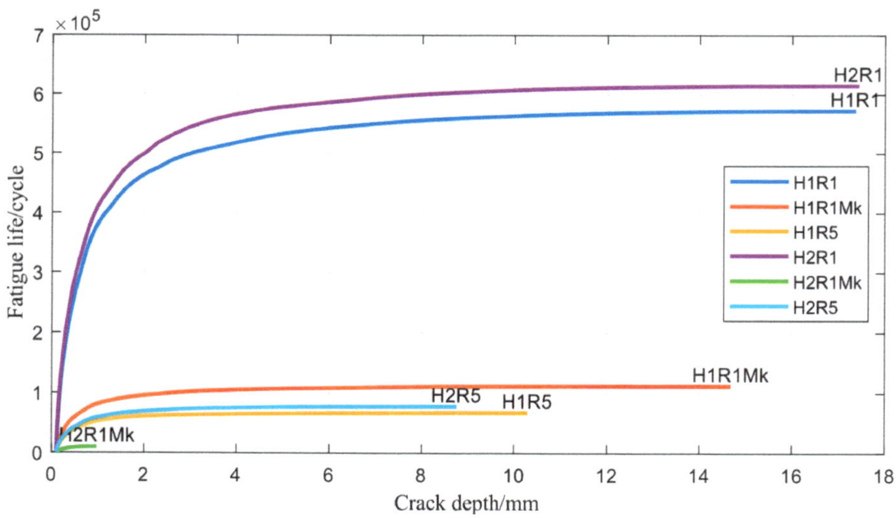

Figure 15. Number of cycles as a function of crack depth on the path for 50% normalized distance in the crack front (H1 and H2 represent specimens H-1 and H-2; R1 and R5 represent stress ratio of 0.1 and 0.5, respectively; Mk represents stress correction factor).

Table 6. Predicted fatigue life at R = 0.1 and 0.5 by numerical method.

Specimen	Stress Ratio	M_k	Fatigue Life/Cycle
H-1	0.1	1	573,012
	0.5	1	67,642
	0.1	1.6	112,353
H-2	0.1	1	614,744
	0.5	1	77,834
	0.1	2.4	10,011
H-3	0.1	1	551,000
	0.5	1	69,000
H-4	0.1	1	556,452
	0.5	1	71,872
	0.1	2.5	9144

4.3. Assessment by FAD

For the structural assessment based on the FAD method, a pipe riser with a longitudinal semi-elliptical crack oriented in the axial direction of the pipe on its outer or internal surface was idealized. The length and depth of the semi-elliptical crack were characterized by $2c$ and a with the assumed aspect ratio $a/c = 0.1$, which is consistent with the initial crack in the numerical simulations. Herein, the welding misalignment was also taken into account.

4.3.1. Stress Intensity Factor Solution

The fracture mechanics-based Engineering Critical Assessment (ECA) approach assumes that a flaw can be idealized as a sharp-tipped crack that propagates in accordance with the law relating the crack growth rate, da/dN, and the range of stress intensity factor, ΔK, for the material containing the flaw, where the correlation law follows the Paris formula as follow:

$$\frac{da}{dN} = C(\Delta K)^m \tag{10}$$

where C and m are a constant that depends on the material and the applied conditions, including the environment and cyclic frequency. Herein, the two-stage fatigue crack growth relationship is employed, with stage one near the crack growth threshold and stage two beyond it. The values of the constant C_1 and m_1 in stage one are 2.1×10^{-17} and 5.1, respectively, and the values of the constant C_2 and m_2 in stage two are 1.29×10^{-12} and 2.88, respectively. The SIF solution for fatigue assessments is given by:

$$\Delta K_{\mathrm{I}} = Y(\Delta\sigma)\sqrt{\pi a} \tag{11}$$

$$(Y\Delta\sigma)_{\mathrm{p}} = Mf_{\mathrm{w}}\{k_{\mathrm{tm}}M_{\mathrm{km}}M_{\mathrm{m}}\Delta\sigma_{\mathrm{m}} + k_{\mathrm{tb}}M_{\mathrm{kb}}M_{\mathrm{b}}[\Delta\sigma_{\mathrm{b}} + (k_{\mathrm{m}} - 1)\Delta\sigma_{\mathrm{m}} \tag{12}$$

where the subscript p indicates the primary stress and the subscript m, b indicates the membrane stress and bending stress, respectively. Only the effect of membrane stress on fatigue life is considered. The influence of misalignment detailed in Table 7 is taken into account by k_{m} in Equation (14) and expressed by:

$$k_{\mathrm{m}} = \frac{\Delta P_{\mathrm{m}} + \Delta\sigma_{\mathrm{s}}}{\Delta P_{\mathrm{m}}} = 1 + \frac{\Delta\sigma_{\mathrm{s}}}{\Delta P_{\mathrm{m}}} \tag{13}$$

$$\frac{\sigma_{\mathrm{s}}}{P_{\mathrm{m}}} = \frac{6e}{B_1(1-\nu^2)}\left[\frac{1}{1+(B_2/B_1)^{1.5}}\right]\sigma_{\mathrm{s}}/P_m < 1 \tag{14}$$

Table 7. Misalignment parameters.

Specimen	Fracture Position	Welding Misalignment/mm
H-1	Inner	1.008
H-2	Outer	0.272
H-3	Outer	0.0
H-4	Outer	0.0

4.3.2. Fatigue–Fracture Assessment

The assessment procedures were implemented using the commercial software Crack-WISE 5.0, which allows both the fatigue and fracture assessments to be performed. The assessment in terms of fatigue life versus crack depth is shown in Figure 16, and the corresponding parameters and results are listed in Table 8. It was found that the fatigue life tends to flatten out as the crack grows, and the real-time crack size after each crack depth increment, Δa, is evaluated for fracture instability or plastic collapse, as shown in Figure 17, to ensure its integrity. The structure is considered to have failed once the assessment point (L_r, K_r) exceeds the area below the FAC, as designed using Option 1 of BS 7910:2013, then the number of load cycles is regarded as the failure life. Meanwhile, the loci of the assessment point under different load conditions can be found in Figure 17, and the final failure size of the crack and fatigue life are summarized in Table 8.

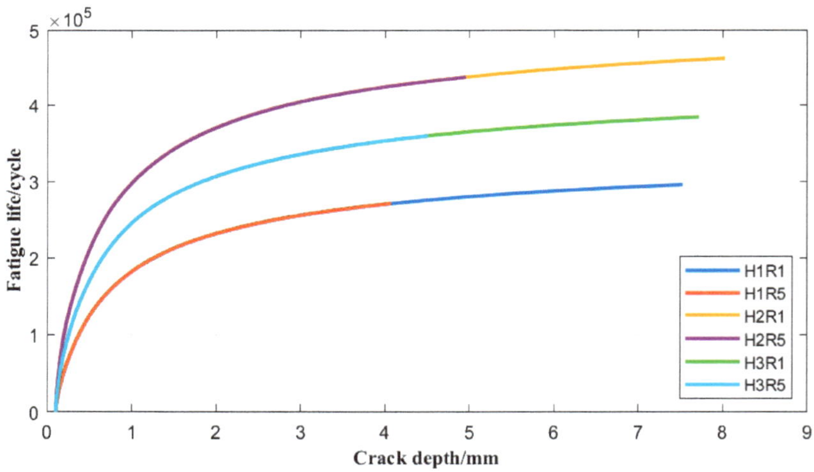

Figure 16. Fatigue life versus crack depth in the BS 7910 assessment considering misalignment.

Table 8. The results of the final failure size of crack and fatigue life predicted by BS 7910.

Specimen No.		Stress Ratio/*R*	Failure Crack Size/mm		Fatigue Life *N*/Cycle
			a	2*c*	
H-1	H1R1	0.1	7.31	17.74	295,000
	H1R5	0.5	3.99	9.25	270,000
H-2	H2R1	0.1	7.85	18.52	462,000
	H2R5	0.5	4.86	11.14	437,000
H-3	H3R1	0.1	7.52	17.41	382,000
	H3R5	0.5	4.42	10.03	358,000

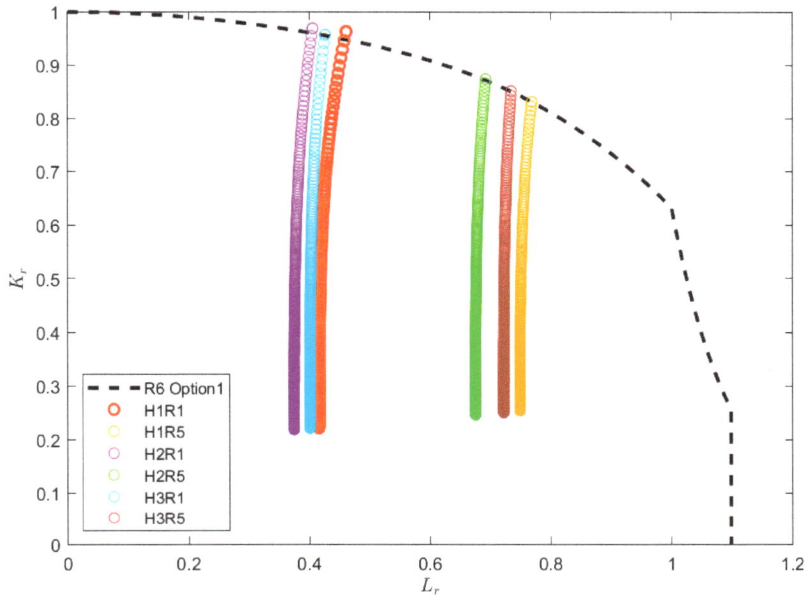

Figure 17. The iteration of fracture assessment points with welding misalignment.

4.4. Comparative Study of Assessment Methods

Based on the results of the different assessment methods described above, the percentage deviation from the experiment is obtained and tabulated in Table 9, where a negative sign indicates lower than the experimental results and a positive sign indicates higher than the experimental results. It is found that when the stress concentration effect is neglected, the numerical prediction results are closer to the experimental results, with a better percentage deviation of even +2.38%. Therefore,M_k has a significant effect on the fatigue life of specimen H-1 and specimen H-2, leading to a decrease of approximately 79.93% and 98.67%, respectively, at R = 0.1 compared to the experimental results. There is a relatively high deviation for specimens H-3 and H-4, and the percentage deviations with the stress ratio of 0.1 are −52.09% and −81.76%, respectively. When the M_k is considered, the deviation is even greater, almost 100 percent. As for the FAD method, the percentage deviation remains at −47.29% and −38.85% for specimen H-1 and specimen H-2, respectively, and is −66.78 for both specimen H-3 and specimen H-4.

Table 9. Percentage deviation from experiment results across different assessment methods.

Specimen No.	FEA			FAD
	Stress Ratio/*R*	M_k	Percentage Deviation/%	Percentage Deviation/%
H-1	0.1	1.0 1.6	+2.38 −79.93	−47.29
H-2	0.1	1.0 2.4	−18.63 −98.67	−38.85
H-3	0.1	1.0	−52.09	−66.78
H-4	0.1 0.1	1.0 2.5	−81.76 −99.70	−66.78

5. Summary

In this work, the full-scale fatigue behaviors of SCR girth welds are tested by a resonant bending rig. The test results are presented in the form of *S–N* curves and are superior to the design curves E in BS 7608. The cap flush is still an effective post-weld treatment to improve the fatigue behavior of full-scale girth weld, and its benefit is dependent on the original weldment profiles. Moreover, the fatigue assessment methods based on fracture mechanics were performed, and the comparison with the welded qualification-related *S–N* curve was studied. The summary is concluded as follows:

(a) The geometrical concentration introduced by full-scale girth weld is predominant for fatigue behaviors, and the competition for crack initiation exists between the toes of the weld root and cap, which depends on the weld root convexity;
(b) The analytical results based on fracture mechanics are able to provide a rational estimation of full-scale girth weld, although the conservative prediction is still not trivial. The M_k, which refers to the stress correction factor recommended, may overestimate the three-dimensional stress intensity factor for internal surface cracks located at the girth pipe root toe. To improve the accuracy, the new formula of M_k for both weld cap and root toes shall be determined by three-dimensional finite element simulation;
(c) For the numerical assessment (NASGRO equation) methods, the assessed fatigue life is relatively lower than the experimental results due to the neglected short crack growth. However, for the well-manufactured welded joint, it is widely recognized that the crack initiation or short crack growth phase amounts to most of the fatigue life. However, the present fracture mechanics approach is still under development to incorporate the proper short crack propagation theory into the fatigue assessment to improve their accuracy and ratability;
(d) For internal surface cracks located at the girth pipe root toe, to assess the impact of various parameters on the M_k, three-dimensional finite element models are developed. These models take different types of welding misalignment, a range of variable-wall-thickness ratios, and initial crack depths into account. The findings reveal that the stress intensity correction factor Mk tends to increase with the rise in misalignment and variable-wall-thickness ratio.

Author Contributions: Conceptualization, supervision, B.G. and C.D.; methodology, formal analysis, B.G. and S.L.; validation, data curation, investigation, visualization, W.Z. and N.Z.; writing—original draft preparation, W.Z. and N.Z.; writing—review and editing, C.D., B.G. and S.L. All authors have read and agreed to the published version of the manuscript.

Funding: This research received no external funding.

Institutional Review Board Statement: Not applicable.

Informed Consent Statement: Not applicable.

Data Availability Statement: The original contributions presented in this study are included in the article. Further inquiries can be directed to the corresponding author.

Conflicts of Interest: The authors declare that they have no known competing financial interests or personal relationships that could have influenced the work reported in this study.

References

1. Hejazi, R.; Grime, A.; Randolph, M.; Efthymiou, M. A Bayesian machine learning approach to rapidly quantifying the fatigue probability of failure for steel catenary risers. *Ocean Eng.* **2021**, *235*, 109353. [CrossRef]
2. Zou, J. Semisubmersible platforms with Steel Catenary Risers for Western Australia and Gulf of Mexico. *Ocean Syst. Eng.* **2012**, *2*, 99–113. [CrossRef]
3. Xie, W.; Liang, Z.; Jiang, Z.; Pan, J.; Hu, Z.; Cabrera, J. A study on the dynamic responses of a steel catenary riser transporting varying-density flow and subjected to regular waves. *Ocean Eng.* **2022**, *265*, 112670. [CrossRef]
4. Wang, W.; Wu, J.; Yan, F.; Zhang, H.; Zhang, D. Defect assessment of steel catenary risers considering welding residual stress. *Ocean Eng.* **2024**, *295*, 116765. [CrossRef]

5. Ibrion, M. Learning from failures: Accidents of marine structures on Norwegian continental shelf over 40 years time period. *Eng. Fail. Anal.* **2020**, *111*, 104487. [CrossRef]
6. Figueredo, A.K.M. Subsea pipelines incidents prevention: A case study in Brazil. *J. Loss Prev. Process Ind.* **2023**, *83*, 105007. [CrossRef]
7. Milne, I.; Ainsworth, R.A.; Dowling, A.R.; Stewart, A.T. R6-Assessment of the integrity of structures containing defects. *Int. J. Press. Vessel. Pip.* **1988**, *32*, 3–104. [CrossRef]
8. Guo, B. (Ed.) *Offshore Pipelines*; GPP: Amsterdam, The Netherlands; Boston, MA, USA, 2005.
9. *IIW-2259-15*; Recommendations for Fatigue Design of Welded Joints and Components. International Institute of Welding; Springer International Publication: Cham, Switzerland, 2016.
10. *DNV-RP-C203*; Fatigue Design of Offshore Steel Structures. Det Norske Veritas Germanischer Lloyd: Høvik, Norway, 2011.
11. Dong, Y. Improved effective notch strain approach for fatigue reliability assessment of load-carrying fillet welded cruciform joints in low and high cycle fatigue. *Mar. Struct.* **2021**, *75*, 102849. [CrossRef]
12. *BS 7608*; Guide to Fatigue Design and Assessment of Steel Products. British Standards Institution, BSI Standards Publication: London, UK, 2014.
13. Maljaars, J. Probabilistic model for fatigue crack growth and fracture of welded joints in civil engineering structures. *Int. J. Fatigue* **2012**, *38*, 108–117. [CrossRef]
14. Kyriakides, S.; Corona, E. *Mechanics of Offshore Pipelines*; Elsevier: Amsterdam, The Netherlands; Boston, MA, USA, 2007.
15. Wu, S.; Feng, J.; Cheng, F.; Wang, Z.; Wang, D. Research on root appearance and fatigue life of steel catenary riser (SCR) using GMAW/GTAW-P double-sided root welding process. *Int. J. Press. Vessel. Pip.* **2023**, *201*, 104866. [CrossRef]
16. *BS 7910*; Guide to Methods for Assessing the Acceptability of Flaws in Metallic Structures. British Standards Institution, BSI Standards Publication: London, UK, 2013.
17. Maierhofer, J.; Pippan, R.; Ganser, H.-P. Modified NASGRO equation for short cracks and application to the fitness-for-purpose assessment of surface-treated components. *Procedia Mater. Sci.* **2014**, *3*, 930–935. [CrossRef]
18. Forman, G.; Mettu, R. Behavior of Surface and Corner Cracks Subjected to Tensile and Bending Loads in Ti-6A1-4V Alloy. 1990. Available online: https://ntrs.nasa.gov/citations/19910009960 (accessed on 15 September 2024).
19. Mettu, S.R.; Shivakumar, V.; Forman, R.G.; McMahon, J.J.; Johnson, N.; Newman, J.C.; Langley, N. NASGRO 3.0—A Software for Analyzing Aging Aircraft. 1995. Available online: https://ntrs.nasa.gov/citations/19990028759 (accessed on 10 July 2024).
20. Newman, J.C. A crack opening stress equation for fatigue crack growth. *Int. J. Fract.* **1984**, *24*, R131–R135. [CrossRef]
21. Zerbst, U.; Schwalbe, K.-H.; Ainsworth, R.A. 7.01—An Overview of Failure Assessment Methods in Codes and Standards. In *Comprehensive Structural Integrity*; Milne, I., Ritchie, R.O., Karihaloo, B., Eds.; Pergamon: Oxford, UK, 2003; pp. 1–48. [CrossRef]
22. Li, S.; Gong, B.; Dai, L.; Deng, C.; Di, X. Data-driven probabilistic failure assessment curve based on similitude principle. *Int. J. Solids Struct.* **2024**, *295*, 112819. [CrossRef]
23. Fajuyigbe, A.; Brennan, F. Fitness-for-purpose assessment of cracked offshore wind turbine monopile. *Mar. Struct.* **2021**, *77*, 102965. [CrossRef]
24. Canale, G.; Kinawy, M.; Maligno, A.; Sathujoda, P.; Citarella, R. Study of Mixed-Mode Cracking of Dovetail Root of an Aero-Engine Blade Like Structure. *Appl. Sci.* **2019**, *9*, 3825. [CrossRef]
25. Mangardich, D.; Abrari, F.; Fawaz, Z. Modeling crack growth of an aircraft engine high pressure compressor blade under combined HCF and LCF loading. *Eng. Fract. Mech.* **2019**, *214*, 474–486. [CrossRef]
26. Dong, P.; Hong, J.K.; Osage, D.A.; Dewees, D.; Prager, M. The master *S-N* curve method: An implementation for fatigue evaluation of welded components in the ASME B&PV code, section VIII, division 2 and API 579-1/ASME FFS-1. In *Welding Research Council Bulletin (TN.523)*; Welding Research Council: New York, NY, USA, 2010.
27. Erdogan, F.; Sih, G.C. On the crack extension in plates under plane loading and transverse shear. *J. Basic Eng.* **1963**, *85*, 519–525. [CrossRef]
28. Delkhosh, E.; Khurshid, M.; Barsoum, I.; Barsoum, Z. Fracture mechanics and fatigue life assessment of box-shaped welded structures: FEM analysis and parametric design. *Weld. World* **2020**, *64*, 1535–1551. [CrossRef]

materials

Article

Phase, Microstructure and Corrosion Behaviour of $Al_{0.3}FeCoNiCr_x$ High-Entropy Alloys via Cr Addition

Mengyao Chen [1,†], Haicheng Shen [2,†], Cheng Wen [2,3,*], Nanchuan Wang [2], Yuwan Tian [2,3], Weihua Zhong [1] and Haiyou Huang [4]

1 China Institute of Atomic Energy, Beijing 102413, China
2 College of Mechanical Engineering, Guangdong Ocean University, Zhanjiang 524000, China
3 Zhanjiang Key Laboratory of Corrosion and Protection for Ocean Engineering and Equipment, Guangdong Ocean University, Zhanjiang 524000, China
4 Institute for Advanced Materials and Technology, University of Science and Technology Beijing, Beijing 100083, China
* Correspondence: wcheng.3jia@163.com
† These authors contributed equally to this work.

Abstract: The microstructure evolution of $Al_{0.3}FeCoNiCr_x$ (x = 0, 0.3, 0.5, 1, 1.5) high-entropy alloys (HEAs) were studied using X-Ray diffraction technique and scanning electron microscope equipped with energy dispersive spectrometer. The short-term and long-term corrosion behaviours of these alloys in 3.5 wt.% NaCl solution were studied by electrochemical impedance spectroscopy, potentiodynamic polarisation measurement, immersion test and corrosion morphology analysis. The results show that all the designed HEAs present single-phase FCC structure, and the increase in Cr content changes the microstructural morphology from cellular to a typical dendritic–interdendritic state. Without the influence of phase transformation, the corrosion resistance of $Al_{0.3}FeCoNiCr_x$ HEAs gradually increases with the increase in Cr content. Our designed alloys exhibit excellent corrosion resistance compared to the existing HEAs in the AlFeCoNiCr composition system.

Keywords: AlFeCoNiCr high-entropy alloy; corrosion; Cr effect; microstructure

Citation: Chen, M.; Shen, H.; Wen, C.; Wang, N.; Tian, Y.; Zhong, W.; Huang, H. Phase, Microstructure and Corrosion Behaviour of $Al_{0.3}FeCoNiCr_x$ High-Entropy Alloys via Cr Addition. *Materials* **2024**, *17*, 5259. https://doi.org/10.3390/ma17215259

Academic Editor: Daniela Kovacheva

Received: 1 October 2024
Revised: 16 October 2024
Accepted: 23 October 2024
Published: 29 October 2024

1. Introduction

The design of traditional alloys typically relies on one or two primary elements, with the minor addition of other elements to regulate properties such as strength, ductility, corrosion resistance and so on, such as steel, aluminium alloys and titanium alloys. This single-primary-element design approach inevitably imposes limitations on alloy composition design. In view of this, Yeh proposed the concept of high-entropy alloys (HEAs) in 2004 [1], which were denoted as multi-principal elements alloys composed of five or more elements, with each element's content ranging from 5 at.% to 35 at.%. HEAs concept significantly expands the compositional space for alloy design and provides new ideas for the development of novel alloy materials. Compared to traditional alloys, HEAs exhibit a range of exceptional properties, such as outstanding corrosion resistance, remarkable high-temperature strength, and good low-temperature plasticity [2–6]. These characteristics suggest significant potential for application in fields such as aerospace and marine structures. Developing new corrosion-resistant alloy materials is crucial for enhancing the durability of marine engineering structures and advancing the marine industry. HEAs, noted for their excellent corrosion resistance, have garnered considerable attention in recent years, with extensive studies on the oceanic corrosion behaviours of transition metal HEAs, typically as the AlFeCoNiCr composition system.

By incorporating elements such as Al, Mo, Ti and Cu, researchers have investigated the corrosion behaviour evolution of FeCoNiCr-based HEAs in various solutions containing Cl^-, acids and bases [7–9]. For instance, Shi et al. [10] examined the effects of increased Al content on the microstructural evolution and corrosion performance of $Al_xCoCrFeNi$ HEAs, revealing that a higher Al content increases the proportion of the (Al, Ni)-rich and Cr-poor BCC phase, which reduces the alloy's resistance to localised corrosion. The research of Zhou et al. [11] has shown that the addition of Ti promotes the formation of an oxide film, thereby reducing the corrosion current density of FeCoNiCrTi alloys in a 3.5% NaCl solution. Chou et al. [12] investigated the pitting behaviour of $Co_{1.5}CrFeNi_{1.5}Ti_{0.5}Mo_{0.1}$ alloys in Cl^- containing solutions, elucidating the mechanism by which the formation of Mo oxides enhances the alloy's pitting resistance. Additionally, Liu et al. [13] found that a moderate increase in Mo content in $Al_xCrFeNi_{2.5}Mo_{1-x}$ HEAs inhibits the formation of the (Al, Ni)-rich and Cr-poor BCC phase, thereby improving the alloy's corrosion resistance. However, Linder et al. [8] revealed that increasing the Mo content in $(FeCoNiCr)_{1-x}Mo_x$ alloys resulted in a detrimental effect on the alloy's corrosion performance due to an increase in the proportion of the σ phase. Most research shows that the incorporation of different elements impacts the phase structure and microstructure evolution of HEAs with various composition systems, leading to differences in corrosion behaviour. In fact, the effect of the same element on the corrosion resistance of different HEA systems often varies. The positive effect of some elements on corrosion resistance in a certain composition system may be negative in another system, typically, such as Mo for HEAs in FeNiCrAl-base [13], FeNiCrCo-base [12] and FeNiCrCoTi-base [8] composition system.

As a typical passivation-promoting element for traditional corrosion-resistant steel design, Cr addition is usually considered to be favourable to corrosion resistance improvement of HEAs. However, Cr is also prone to inducing compositional segregation and the precipitation of Cr-rich phases, which can adversely affect the localised corrosion resistance of HEAs. From the investigation of Chai et al. [14], the addition of excessive Cr led to element segregation and an uneven distribution of Cr-rich and Cr-poor regions, ultimately resulting in a decrease in the corrosion resistance of $FeCoNiCr_x$ HEAs. Tsau et al. [15] reported that the corrosion resistance of Cr-modified FeCoNi alloys, specifically FeCoNiCr with high Cr content, was inferior to that of FeCoNi without Cr in both 1 M H_2SO_4 and NaCl solutions. The addition of Cr led to the formation of hexagonal close-packed (HCP) precipitates within the alloy matrix, contributing to the deterioration of corrosion performance. The study by Jiang et al. [16] indicated that the effect of Cr addition on the corrosion performance of $AlFeCoNiCr_x$ HEAs coatings exhibited a nonlinear relationship. According to the results of electrochemical tests and immersion tests, the corrosion resistance of the alloy coatings in 3.5 wt.% NaCl solutions initially increase with the increasing Cr content, followed by a subsequent decline. The authors asserted that the enhanced corrosion resistance of $AlFeCoNiCr_x$ coatings was a result of the combined effects of Cr and Al elements. The increase in Cr content within the passivation film cannot compensate for the decrease in corrosion resistance caused by the reduction of Al content. Consequently, the coating with the highest Cr content demonstrated poorer corrosion resistance. Qiu et al. [17] also pointed out that Cr segregation can lead to the precipitation of the Cr-rich σ phase, which makes the Cr-depleted HEAs matrix more prone to corrosion. Additionally, due to the interaction between multiple elements in HEAs for different composition systems, the effect of the Cr element on improving corrosion resistance is still vague and needs to be investigated. Actually, phase transformation in HEAs caused by Cr addition can influence the corrosion performance of HEAs and may lead to galvanic corrosion between the solid solution matrix and the precipitation phase [18]. The corrosion behaviour of Cr-containing HEAs exhibits unique characteristics.

The development of homogeneous solid solution alloys that minimise structural heterogeneities represents the most effective strategy for designing corrosion-resistant alloys (CRAs) with the highest potential for superior corrosion resistance [19]. One significant approach in the design of CRAs has been the development of homogeneous solid solution alloys that emphasise phase and compositional stability while minimising structural non-uniformities.

Previous studies on Cr-containing HEAs did not pay attention to this point, resulting in excessive Cr content addition that actually reduces the corrosion resistance. The goal of this study is to gain a deeper understanding of the corrosion behaviour of HEAs as a function of chemical composition (specifically Cr itself) while maintaining a fixed single-phase structure. We intend to eliminate the influence of phase transformation caused by increased Cr content on corrosion behaviour and explore the enhancing effect of the Cr element itself on corrosion performance; thus, a series of single-phase HEAs, denoted as $Al_{0.3}FeCoNiCr_x$, with varying Cr contents were designed based on empirical rules. The phase structure and microstructural characteristics of the alloys were analysed using techniques such as X-Ray diffraction (XRD) and scanning electron microscopy (SEM). Furthermore, the effects of increased Cr content on the short-term and long-term corrosion behaviour of the alloys were investigated through potentiodynamic polarization measurements, impedance analysis and immersion tests. The positive effect of Cr addition on the corrosion performance of HEAs with single-phase solid solution is demonstrated. In comparison to the reported HEAs within the AlFeCoNiCr compositional system, our designed single-phase alloys show outstanding corrosion resistance in marine environments.

2. Materials and Methods

2.1. HEAs Preparation

The $Al_{0.3}FeCoNiCr_x$ (x = 0, 0.3, 0.5, 1 and 1.5) HEAs were prepared by arc melting in a Ti-gettered argon atmosphere, utilizing high-purity Al, Fe, Co, Ni and Cr elements (purity $\geq$ 99.9 wt.%). The melting current is around 200A, with a slight difference for the five alloys. During each melting process, the alloys were maintained in a liquid state for 3 min, simultaneously conducting electromagnetic stirring to accelerate liquid flow. The ingot was then rapidly solidified in a copper mould. The alloys were melted five times, and the ingot was flipped after each melting process to ensure chemical homogeneity. Square pieces measuring 10 mm $\times$ 10 mm $\times$ 5 mm were cut from the bulk ingot for phase structure and microstructural analysis. For simplification, the five alloys were designated as Cr0 (x = 0), Cr3 (x = 0.3), Cr5 (x = 0.5), Cr10 (x = 1) and Cr15 (x = 1.5), respectively.

2.2. Characterization of Phase and Microstructure

The phase structures of the $Al_{0.3}FeCoNiCr_x$ HEAs were detected with an X-Ray diffraction technique (XRD) using Bruker D8 Advancey equipment (Karlsruhe, Germany), and the detailed parameters were as follows: Cu Kα radiation, scanning angle 20~100° and scanning rate 6°/min. A scanning electron microscope (SEM) equipped with EDS was employed to characterise the samples' microstructure, pitting corrosion morphology, and element distribution. The specimens for microstructure observation were ground using silicon carbide abrasive papers from grit 240 to 2000 and polished with 1.5 μm diamond polishing agents, then etched in a mixture of nitric acid and hydrofluoric acid aqueous solution with a volume ratio of 1:1:2.

2.3. Electrochemical Measurement and Immersion Test

The specimens for the electrochemical measurements were cut into 10 mm $\times$ 10 mm $\times$ 5 mm bulks and cold-mounted in epoxy resin with copper wires spot-welded to the back for electrical contact. Prior to each electrochemical experiment, the exposed surface of the specimens was wet ground with a series of SiC papers ranging from 240 to 2000 grit, thoroughly cleaned with alcohol, acetone and deionised water in sequence. Both electrochemical polarisation and electrochemical impedance spectroscopy (EIS) were conducted using an Autolab 302N electrochemical workstation in a three-electrode system (Herisau, Switzerland). The specimen served as the working electrode, a saturated calomel reference electrode (SCE) was used as the reference electrode, and a platinum counter electrode was used. During the experiment, the working electrode was initially cathodically polarised at −800 mV vs. SCE for 5 min to reduce any possible surface oxides. The sample was then allowed to corrode freely for 30 min to reach a quasi-stationary open circuit potential

(OCP). The potentiodynamic-polarisation tests were performed at scanning rates varying from 10 mV/min, starting from an initial potential of −250 mV vs. OCP until the current density reached a maximum of 1 mA/cm^2. The EIS tests were carried out at the OCP with a sinusoidal potential amplitude of 10 mV, operating from 100 kHz to 10 mHz.

Additionally, to further investigate the long-term corrosion behaviour of the designed HEAs, the electrochemical samples were immersed in a 3.5 wt.% NaCl solution at room temperature for 0, 1, 3, 5 and 7 days. Subsequently, the impedance was tested at each interval. The surface morphologies of the alloys after the immersion tests were also analysed using SEM.

3. Results and Discussion

3.1. Phase Structure

The XRD patterns presented in Figure 1 indicate that all the alloys exhibit a simple FCC solid-solution structure, characterised by three distinctive peaks corresponding to the (1, 1, 1), (2, 0, 0) and (3, 1, 1) crystal planes, as anticipated. Exploring the phase formation rules and predicting the phase structure has long been a pivotal topic in the realm of HEAs. To address this, researchers have proposed numerous empirical parameters for phase prediction in HEAs, primarily derived from Hume-Rothery rules and thermodynamic criteria. These criteria are determined by the physicochemical and thermodynamic properties of the constituent alloying elements in HEAs, with the parameter ranges for phase selection often being empirical values based on diverse datasets.

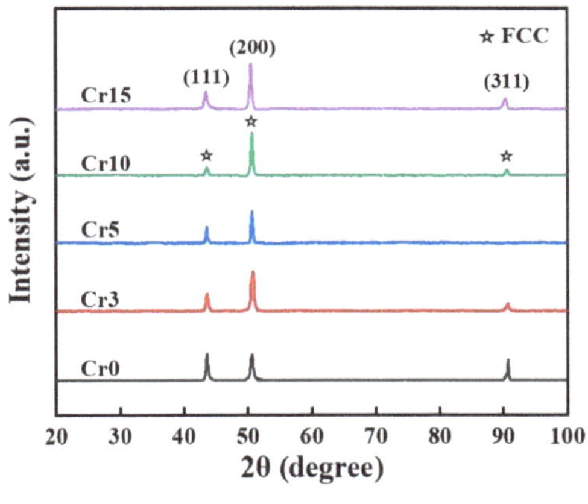

Figure 1. XRD patterns of the studied HEAs.

The calculated parameters for the studied HEAs are detailed in Table 1, from which it is evident that all five alloys meet the thermodynamic parameter criteria for solid solution formation in HEAs, specifically $-22 < \Delta H_{mix} < 7$ and $11 < \Delta S_{mix} < 19.5$, as referenced in [20]. It is widely acknowledged that δr (atomic size mismatch) serves as a valuable empirical parameter for phase prediction in HEAs. However, varying research groups have occasionally proposed different threshold values. Ren et al. [21] asserted that solid solutions were formed when $\delta r < 0.028$ and $\Delta H_{mix} > -8.8$, significantly narrowing the δr-ΔH_{mix} space compared to Yang's research ($\delta r < 0.066$) [22]. This divergence arises because Yang's work classifies ordered solid solutions as solid solutions rather than intermetallic compounds. Kube's research [23] highlights that HEAs with increasing δr values tend to favour a BCC structure over FCC, attributed to the BCC structure's capacity to accommodate larger atomic size differences with lower strain energy. As seen in Table 1, the studied alloys exhibit low δr values, signifying a high tendency for the formation of stable FCC solid solutions, which aligns with the XRD results.

Table 1. Empirical parameters calculated for the studied HEAs: ΔH_{mix}—mixing enthalpy, ΔS_{mix}—mixing entropy, δr—atomic size mismatch, VEC—valence electron concentration, and $\Delta\chi_{Allen}$—Allen electronegativity difference.

Alloys	ΔH_{mix}	ΔS_{mix}	δr	VEC	$\Delta\chi_{Allen}$
$Al_{0.3}FeCoNi$ (Cr0)	−6.83	10.84	0.0425	8.45	3.97%
$Al_{0.3}FeCoNiCr_{0.3}$ (Cr3)	−7.13	12.32	0.0408	8.25	4.63%
$Al_{0.3}FeCoNiCr_{0.5}$ (Cr5)	−7.23	12.65	0.0398	8.13	4.92%
$Al_{0.3}FeCoNiCr$ (Cr10)	−7.27	12.83	0.0376	7.88	5.37%
$Al_{0.3}FeCoNiCr_{1.5}$ (Cr15)	−7.14	12.61	0.0357	7.69	5.95%

Valence electron concentration (VEC) has been proposed as a crucial parameter to control the phase stability of HEAs, distinguishing between FCC and BCC solid solutions [24]. The stability of the FCC phase is observed at higher VEC (>8), whereas the BCC phase is stable at lower VEC (<6.87). By examining the relationship between electronic concentration and solid solution phase formation in HEAs, Battezzati et al. [25] provided a new guideline for selecting the set of elements to obtain different solid solution phases of HEAs based on data analysis using a broader range of alloy samples. In this research, the VEC value requirement for HEAs forming a single-phase FCC structure is greater than 7.5, and all the studied alloys meet this threshold. However, atoms with different electron affinities tend to form ordered intermetallics according to the Hume-Rothery rules. Guo's research [20] points out that $\Delta\chi$ (Pauling electronegativity difference) seems to have no significant influence on solid solution phase formation in HEAs. However, the newly proposed $\Delta\chi_{Allen}$ (Allen electronegativity difference) indicates that for HEAs with $0.001 < \delta r < 0.006$ and $3\% < \Delta\chi_{Allen} < 6\%$, only solid solutions are formed [25]. The XRD results are consistent with this parameter limit, reinforcing the importance of $\Delta\chi_{Allen}$ in predicting the phase stability of HEAs.

Similar parameters like Ω [22], Φ [26], Λ [27], s_m and K_m [28] were also proposed from the aspects of thermodynamics, lattice distortion and electronic behaviour of components to try to explain the phase formation of HEAs. For the studied HEAs, Cr is an element that stabilises the BCC solid solution phase [29]. As the content of Cr increases, δr continuously decreases, which enhances the tendency of the alloy to form an FCC solid solution. Conversely, the continual decrease in VEC reduces the inclination of the alloy to form an FCC solid solution. Additionally, the increasing $\Delta\chi_{Allen}$ raises the propensity for the formation of intermetallic compounds in the alloy. In sum, phase prediction and design of HEAs is a complex issue that has yet to be solved. Even though there are no parameters that can accurately determine the HEAs phases up to now, the parametric approach is still considered a good reference for rapid composition design in HEAs.

3.2. Microstructure

The cross-sections perpendicular to the direction of heat flow of the as-cast samples were used for the microstructure observation. Figure 2 displays the SEM micrographs of $Al_{0.3}FeCoNiCr_x$ alloys in secondary electronic mode. It can be seen that Cr0, Cr3 and Cr5 present a uniform and fine cellular structure. With the increase in Cr content, the alloys Cr10 and Cr15 exhibit a typical dendritic–interdendritic state. The increase in Cr content changes the microstructural morphology from cellular to dendrite, similar to the previous research on microstructure evolution for Al-free $FeCoNiCr_x$ HEAs [14]. From the viewpoint of solidification, the cellular-to-dendrite transition is mainly attributed to the instability of the solid/liquid interface caused by element microsegregation, which may be detrimental to the localised corrosion resistance of alloys. We will discuss it in the following part. As shown in Figure 3, we utilised the EDS technique for the analysis of the alloy composition, with a comparison of the detected elemental concentrations and the nominal compositions presented in Table 2. It can be observed that the average composition of the prepared alloy is close to the designed composition, with deviations in elemental concentrations generally within 5%. Furthermore, the elemental analysis indicates that with the increase in Cr content, there is a tendency for greater compositional segregation in the alloy, although the degree of segregation remains minimal. It can be inferred that, for the alloys in this study, the effects of compositional segregation and phase structure changes due to the increase in Cr on the corrosion performance are relatively small. Thus, we can explore the influence of Cr itself on the alloy's corrosion behaviour and corrosion resistance.

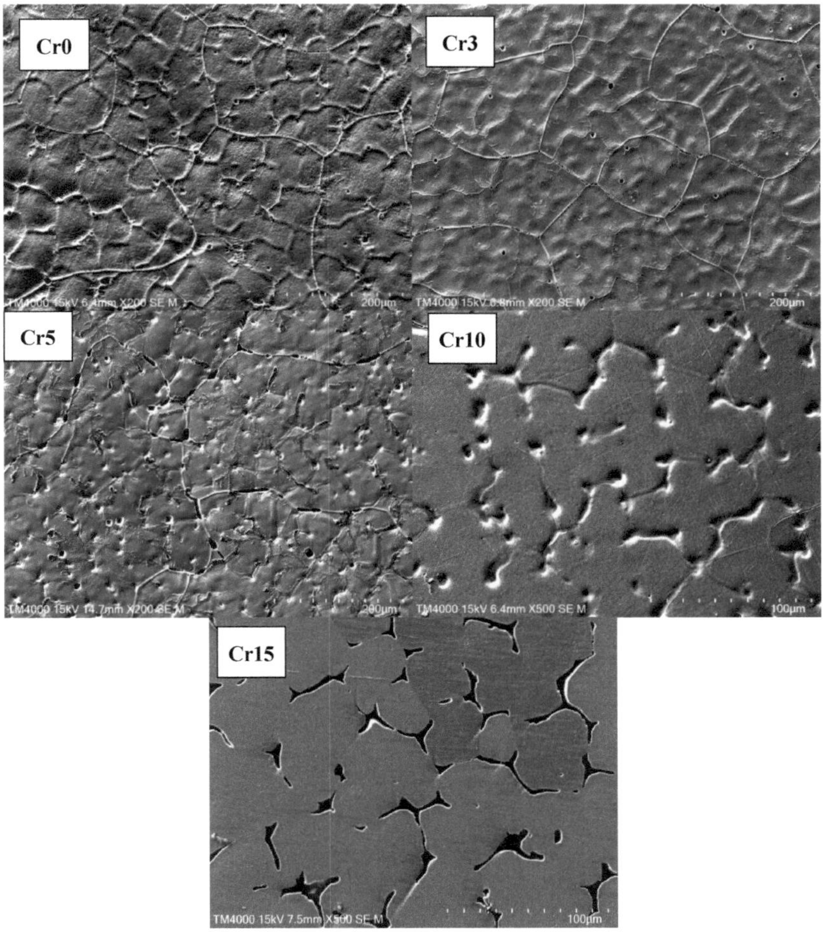

Figure 2. SEM micrographs of the studied HEAs.

3.3. *Short-Term Corrosion Behaviour*

Figure 4a displays the typical potentiodynamic polarisation curves of the $Al_{0.3}FeCoNiCr_x$ HEAs in 3.5 wt.% NaCl. The corrosion current densities of the alloys, determined using the Tafel extrapolation method, are presented in Table 3. From Figure 4a, it is evident that the designed alloy Cr0 without Cr exhibits pronounced active corrosion behaviour. The anodic current density gradually increases with the potential, showing no signs of depassivation or pitting corrosion. In contrast, all the Cr-containing alloys exhibited passivity and breakdown behaviour, as indicated by a slowly increasing anodic current density followed by a sudden surge in current when the potential reached a certain value. It can be seen that all four alloys have obvious passivation zones, and the polarisation curves directly transition from the Tafel zone to the passivation zone without any activation passivation transition process, indicating the natural formation of a protective passivation film under open circuit potential. In the passivation zone, the passivation current density of the four alloys is relatively low, with an order of magnitude of 10^{-6} A/cm^2.

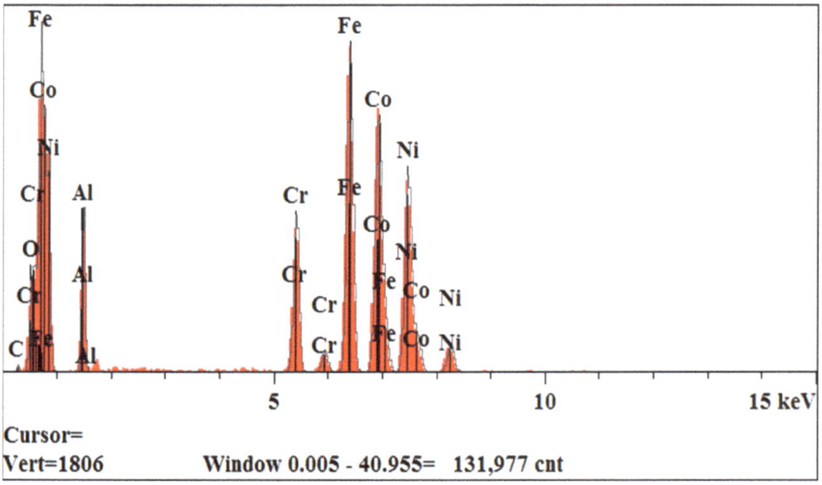

Figure 3. Element detection of the studied HEAs using the EDS technique.

Table 2. EDS analysis result of the studied alloys.

Alloys \ Element	Content, at.%	Al	Fe	Co	Ni	Cr
Cr0	Nominal	9.1	30.3	30.3	30.3	0
	EDS	8.96 ± 0.11	29.65 ± 0.27	30.12 ± 0.21	31.27 ± 0.19	0
Cr3	Nominal	8.33	27.78	27.78	27.78	8.33
	EDS	7.57 ± 0.12	27.15 ± 0.28	27.90 ± 0.18	29.18 ± 0.20	8.20 ± 0.06
Cr5	Nominal	7.88	26.32	26.32	26.32	13.16
	EDS	7.69 ± 0.13	26.95 ± 0.27	25.87 ± 0.16	25.62 ± 0.17	13.87 ± 0.08
Cr10	Nominal	6.96	23.26	23.26	23.26	23.26
	EDS	6.38 ± 0.12	23.34 ± 0.25	24.18 ± 0.17	22.99 ± 0.15	23.11 ± 0.15
Cr15	Nominal	6.26	20.83	20.83	20.83	31.25
	EDS	5.87 ± 0.09	21.26 ± 0.24	20.08 ± 0.14	20.44 ± 0.13	32.35 ± 0.18

Table 3. Polarisation test results of the studied alloys in 3.5 wt.% NaCl solution.

Alloys	i_{corr}, μA/cm^2	E_{pit}, mV	E_{corr}, mV	E_{pit}-E_{corr}, mV
Cr0	0.798 (±0.157)	/	−294 (±12)	/
Cr3	0.145 (±0.037)	84 (±5)	−242 (±11)	326 (±16)
Cr5	0.141 (±0.042)	200 (±11)	−245 (±12)	445 (±23)
Cr10	0.129 (±0.036)	308 (±13)	−169 (±9)	477 (±22)
Cr15	0.071 (±0.014)	412 (±20)	−166 (±9)	578 (±29)

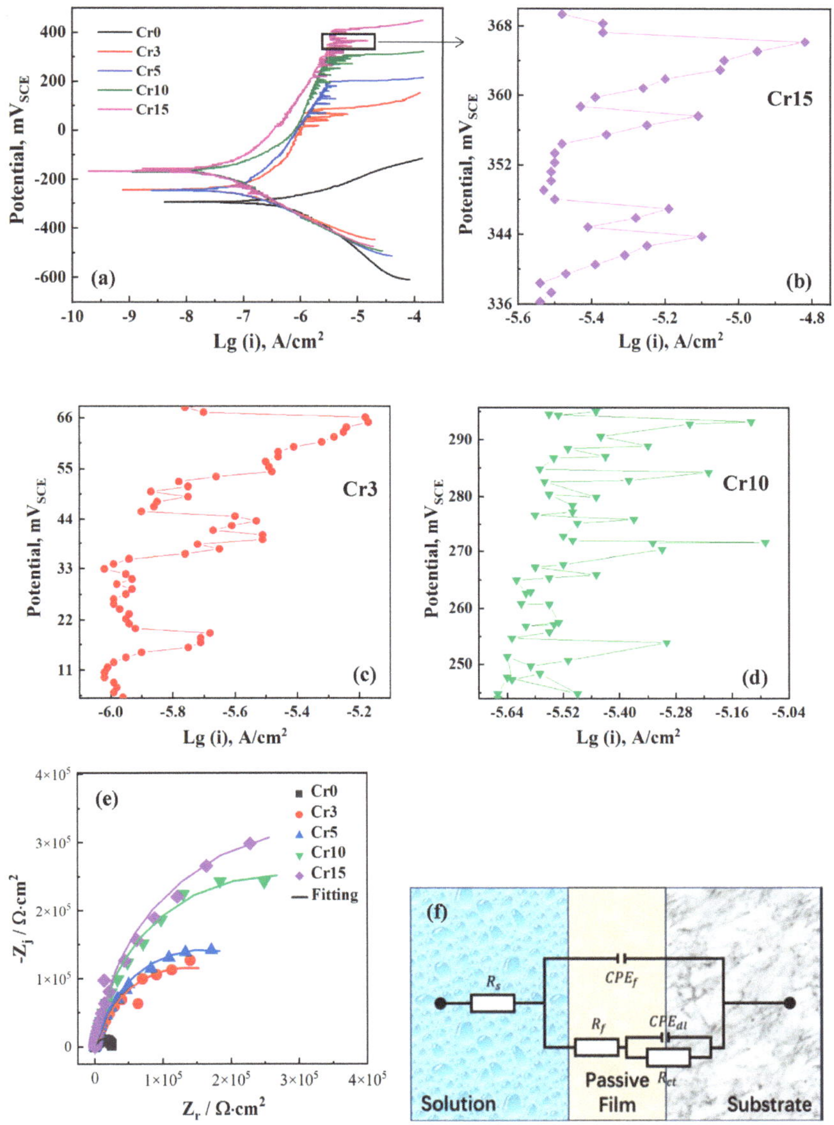

Figure 4. Electrochemical test result: (**a**) Potentiodynamic polarisation curves; (**b**) Localised magnification of polarisation test within the black box for Cr15 (**b**), Cr3 (**c**) and Cr10 (**d**); (**e**) Nyquist plots; (**f**) equivalent electron circuit model of the EIS data fitting.

The Cr element combines a relatively high heat of adsorption for oxygen (or hydroxide) with a comparatively low Cr-Cr bond strength [30], thus the enhancement of passivation in alloys can be achieved through Cr addition. Although Al belongs to this type of element, a passive state is absent in the curves obtained for the Cr0 alloy with 9 at.% Al content due to the lack of Cr. Even with higher Al content, there is no evidence of passivation behaviour in the $AlCoCr_{0.5}FeNi$ alloy coating with low Cr content [16]. Actually, compared to Al, Cr plays a decisive role in improving corrosion resistance in such HEA composition systems through the formation of a steady-state passivation film. For the studied HEAs, higher Cr content indicates higher breakdown potential. According to Table 3, the larger passivation interval further demonstrates the positive effect of Cr addition on corrosion protection for the single-phase HEAs. It is generally considered that the Cr content should be more than 12 at.% to ensure passivation for traditional Fe-Cr-Ni alloys. From the previous research of Lu et al. [31], HEAs containing 6 at.% Cr showed signs of passivity, which was significantly lower than the Cr threshold passivation level. Our work also demonstrates that from Cr0 to Cr3, an increase of 8.3 at.% in Cr content can achieve the passivation protection of alloys.

From Figure 4a, for all the Cr-containing HEAs, at potentials of 100 to 200 mV below their pitting potentials, numerous current transients are evident in the passive region, indicative of metastable pitting [31–33]. The occurrence of metastable pitting can be attributed to the fact that, along with the increase in applied potential, the insoluble oxides in the passive film may be transformed into soluble chlorides due to O in the oxides being replaced by Cl^- in the NaCl solution. This transformation leads to the initiation of pitting and an increase in anodic current density. However, under the strong passivation effect of the Cr element, the pitting is repaired, resulting in a subsequent decline in anodic current density. A further increase in the anodic overpotential progressively raises both the magnitude and number of the metastable pits until the current suddenly increases. Typically, Figure 4b–d depict the localised magnification of the polarisation curve for the Cr15, Cr3 and Cr10 alloys, which belongs to a typical class of current density fluctuation behaviour observed in metastable pitting corrosion for traditional CRAs [32]. The current density slowly increases due to the anodic dissolution of the metal inside the pitting corrosion during the development of metastable pitting corrosion. Once re-passivation occurs after a period of time due to metal dissolution, the current density sharply decreases. In fact, the higher the Cr content, the faster the response to re-passivation repair is. As the Cr content increases, the current fluctuation from metal dissolution to re-passivation decreases, indicating the improved stability of the protective passivation film. The stability of re-passivation repair brought by the increase in Cr content is very important for the long-term corrosion of the investigated HEAs. Li et al. [34] also observed similar electrochemical behaviours in the polarisation measurements. The frequent breakdown and re-passivation events in the passive region suggest that the passive films formed on these alloys are susceptible to Cl^--induced instabilities. This susceptibility is likely associated with structural defects in the as-cast condition, such as shrinkage porosity and shrinkage cavities, rather than element segregation. A similar phenomenon also occurs in HEAs like CoCrFeNi and CoCrFeNiMn, which possess uniform compositions after deformation and heat treatment [35].

In terms of corrosion current density, it is evident that Cr significantly enhances the corrosion resistance of HEAs, and this improvement increases with rising Cr content, which is similar to the behaviour observed in Cr-containing steel. The corrosion current densities for Cr3, Cr5, and Cr10 decrease by approximately 5.5 times compared to that of Cr0, while the corrosion current density for Cr15 decreases by about 11 times relative to Cr0. It can be seen that the addition of merely 8.3 at.% Cr (Cr3) to the $Al_{0.3}FeCoNiCr_x$ HEAs can achieve a notable enhancement in corrosion resistance. However, the Cr content was further increased to 23.3 at.% (Cr10) does not yield significant changes in corrosion resistance. When the Cr content is raised to 31.3.% (Cr15), the corrosion resistance of the alloy improves once more. From the above analysis, although the corrosion resistance of the designed alloys continues to improve with the increased Cr, the degree of improvement weakens. The corrosion performance indicators corresponding to the unit Cr addition are decreasing, especially when the cellular-dendritic transition occurs (Cr5 to Cr10). The most obvious change in pitting potential (E_{pit}) and passivation interval (E_{pit}-E_{corr}) is observed, indicating that the excess Cr addition may lead to localised corrosion due to the formation of dendritic microstructure. From Figure 4c,d, compared to the Cr3 alloy with a cellular microstructure, the Cr10 alloy exhibits more severe current fluctuations during the metastable pitting stage. The cellular-dendritic transformation caused by excess Cr addition is unfavourable to localised corrosion, which may not be compensated by the passivation-promoting effect of the Cr element itself.

Figure 4e depicts the EIS results of the studied alloys. From the Nyquist plot, all impedances exhibit similar capacitive arcs, indicating that the alloys possess comparable interfacial electrical properties and no mass transportation processes are observed [36–38]. It is generally accepted that a larger radius of the capacitive arc corresponds to better corrosion resistance of the alloys [36,39,40]. Therefore, the increase in the radius of the capacitive arc with rising Cr content also suggests that Cr addition effectively enhances the corrosion resistance of the present HEAs. Furthermore, the EIS data were fitted using the circuit shown in Figure 4f. Due to the passivation behaviour observed in the polarisation curves and the broad phase angle peak evident in the Bode plot, a fitting circuit with two time constants was employed. The peak in the phase angle-frequency plot in the 1~10 Hz range corresponds to the charge transfer mechanism and the double layer capacitance effect at the metal surface, while the notable dispersion phenomenon or recognisable time constant at around 100 Hz correlates with the properties of the surface film on the metal. Consequently, in Figure 4f, Rs represents the solution resistance, while CPEf and Rf denote the film capacitance and film resistance of the metal surface, respectively. CPEdl and Rct represent the double-layer capacitance and charge transfer resistance at the metal surface.

The EIS data-fitting results are shown in Table 4; it also presents the polarisation resistance (Rp = Rf + Rct) to characterise the corrosion resistance of the passivated samples [41–43]. It is evident that the values of Rct are 1~3 orders of magnitude higher than Rf, indicating that the charge transfer process is a significant factor controlling the overall corrosion kinetics in this system. The Rct values for Cr3, Cr5, Cr10 and Cr15 are approximately 14, 16, 28 and 36 times higher than that of Cr0, respectively, and the Rp values are about 10, 12, 21 and 26 times higher, respectively. The impedance analysis also indicates that there is an addition of merely 8.3 at.% Cr (Cr3) can significantly enhance the corrosion resistance of the present HEAs. Further increasing the Cr content to 13.2 at.% (Cr5) results in a marginal change in corrosion resistance while increasing the Cr content to 23.3 at.% (Cr10) improves the corrosion resistance again. This conclusion is consistent with the results from the polarisation curves. Overall, the results from polarisation curves and EIS corroborate each other, fully validating the effectiveness of Cr in reducing the corrosion rate of HEAs. Moreover, this enhancing effect persists and even strengthens with increasing Cr content for the present single-phase alloys.

Table 4. The fitting parameter of EIS results for the studied alloys.

Alloys	Rs, $\Omega \cdot cm^{-2}$	CPE_f, $\mu\Omega^{-1} \cdot cm^2 \cdot S^n$	n1	Rf, $k\Omega \cdot cm^{-2}$	CPE_{dl}, $\mu\Omega^{-1} \cdot cm^2 \cdot S^n$	n2	Rct, $k\Omega \cdot cm^{-2}$	Rp, $k\Omega \cdot cm^{-2}$
Cr0	7.1 ± 0.09	36.4 ± 0.08	0.95 ± 0.002	7.6 ± 0.01	121.1 ± 0.038	1.00 ± 0.002	17.5 ± 0.91	25.1 ± 0.92
Cr3	7.9 ± 0.08	18.7 ± 0.05	1.00 ± 0.002	22.9 ± 0.03	29.3 ± 0.012	0.90 ± 0.002	240.5 ± 12.07	263.4 ± 12.10
Cr5	7.3 ± 0.12	18.4 ± 0.04	1.00 ± 0.003	27.4 ± 0.02	22.6 ± 0.009	0.91 ± 0.002	285.7 ± 14.91	313.1 ± 14.93
Cr10	7.5 ± 0.13	15.3 ± 0.03	1.00 ± 0.002	28.3 ± 0.03	14.3 ± 0.004	0.96 ± 0.003	496.5 ± 23.65	524.8 ± 23.68
Cr15	8.1 ± 0.15	16.1 ± 0.04	1.00 ± 0.002	36.5 ± 0.03	14.2 ± 0.005	0.95 ± 0.002	628.5 ± 25.32	665.0 ± 25.35

3.4. Long-Term Corrosion Behaviour

To further investigate the long-term corrosion behaviour of the designed HEAs, the electrochemical samples were immersed in a 3.5 wt.% NaCl solution at room temperature for seven days. Non-destructive EIS tests were conducted for the five alloy samples on 0, 1, 3, 5, and 7 d, with results displayed in Figure 5. It can be observed that the impedance value and the radius of the capacitive arc for the Cr0 alloy decrease on the first day, indicating that corrosion occurs on its surface. After day 3, the impedance value and the radius of the capacitive arc increase, which may be associated with a certain amount of corrosion products covering the sample surface. From day 3 onwards, the changes in impedance value and radius of the capacitive arc for Cr0 are minimal, and the low-frequency impedance stabilise at 25 $k\Omega \cdot cm^2$, suggesting that it experiences continuous corrosion at a relatively stable rate.

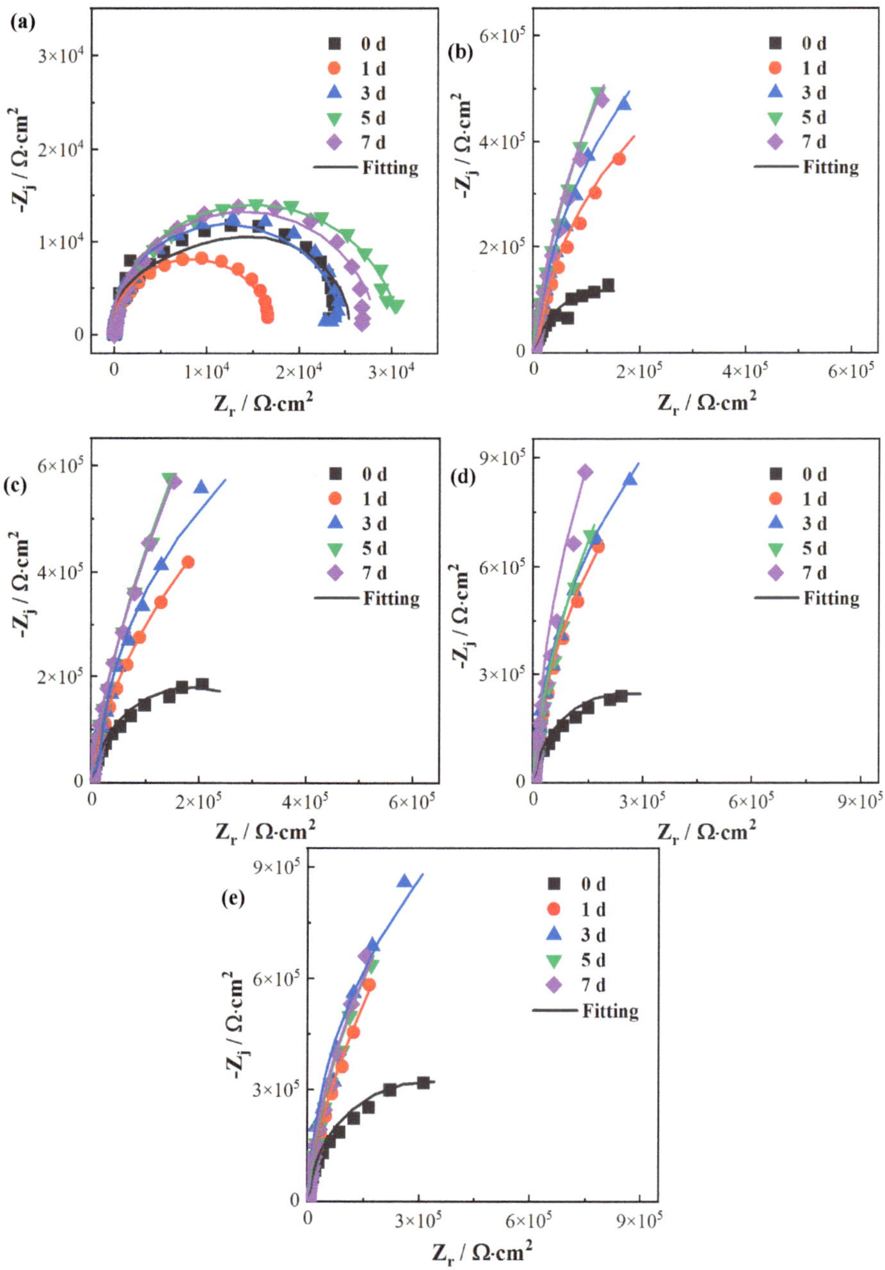

Figure 5. Nyquist plot of the five alloys immersed in 3.5 wt.% NaCl solution for 0, 1, 3, 5 and 7 d: (**a**) Cr0, (**b**) Cr3, (**c**) Cr5, (**d**) Cr10, (**e**) Cr15.

In comparison, Cr3 and Cr5 exhibit significant differences from Cr0. Firstly, their impedance values and capacitive arc radii markedly increase on day 1, indicating that the addition of a small amount of Cr effectively promotes passivation of the metal surface in a Cl^--containing solution environment. During days 3 to 7, the impedance values and capacitive arc radii continue to increase slowly, suggesting that the passivation film on the metal surface thickens, becomes denser and increases in protective quality. Towards the later stages of immersion, the low-frequency impedance values for Cr3 and Cr5 notably increase to 200 kΩ·cm^2, which represents an order of magnitude improvement compared to the 25 kΩ·cm^2 of Cr0, thus confirming that a small amount of Cr significantly enhances the corrosion resistance of the studied HEAs. For Cr10 and Cr15, the evolution of their impedance spectra was similar to that of Cr3 and Cr5, with slight differences: the impedance spectra for Cr10 and Cr15 reached elevated values on the first day of immersion and remained relatively stable during the immersion period from days 3 to 7. This indicates that an increase in Cr content can accelerate the rate of passivation on the metal surface.

Similarly, the EIS data of the alloy samples were fitted using the equivalent circuit with two time constants shown in Figure 4c, and the results are presented in Table 5. It can be seen that the variation pattern of the polarisation resistance for Cr0, Cr3, Cr5, Cr10 and Cr15 is consistent with that of the low-frequency impedance values: all of them show a slowly increasing trend over time, indicating that their surface corrosion rates are gradually decreasing. The polarisation resistance of Cr0 remains in the order of 10 kΩ·cm^2, while the polarisation resistance of the HEAs with Cr addition is in the order of 1 MΩ·cm^2, and may even reach 10 MΩ·cm^2 in the later stages, further confirming the continuous role of the alloying element Cr in enhancing the corrosion resistance of the studied HEAs.

Table 5. The fitting parameter of EIS results for the studied alloys after immersion test.

Alloys	Time	Rs, Ω·cm^{-2}	CPE$_f$, μΩ^{-1}·cm^2·S^n	n1	Rf, kΩ·cm^{-2}	CPE$_{dl}$, μΩ^{-1}·cm^2·S^n	n2	Rct, kΩ·cm^{-2}	Rp, kΩ·cm^{-2}
Cr0	0 d	8.6 ± 0.15	24.5 ± 0.06	1.00 ± 0.003	15.6 ± 0.02	104.9 ± 0.035	1.00 ± 0.003	10.2 ± 0.51	25.8 ± 0.53
	1 d	8.2 ± 0.14	26.3 ± 0.07	0.96 ± 0.002	0.01 ± 0.00	16.8 ± 0.005	0.94 ± 0.003	17.5 ± 0.92	17.51 ± 0.92
	3 d	8.2 ± 0.16	16.3 ± 0.04	1.00 ± 0.003	0.01 ± 0.00	17.7 ± 0.006	0.94 ± 0.002	24.9 ± 1.37	24.91 ± 1.37
	5 d	8.0 ± 0.15	18.0 ± 0.05	0.98 ± 0.002	0.01 ± 0.00	25.8 ± 0.012	0.90 ± 0.002	31.0 ± 1.55	31.01 ± 1.55
	7 d	7.5 ± 0.13	24.6 ± 0.07	0.96 ± 0.002	0.001 ± 0.00	15.5 ± 0.006	0.94 ± 0.002	28.6 ± 1.33	28.60 ± 1.33
Cr3	0 d	7.9 ± 0.14	18.6 ± 0.04	1.00 ± 0.002	22.6 ± 0.03	29.3 ± 0.013	0.90 ± 0.002	240.8 ± 12.07	263.4 ± 12.10
	1 d	8.0 ± 0.15	14.6 ± 0.03	1.00 ± 0.003	49.5 ± 0.05	16.3 ± 0.005	0.94 ± 0.003	1252.0 ± 67.56	1301.5 ± 67.61
	3 d	7.7 ± 0.08	15.7 ± 0.03	1.00 ± 0.002	44.2 ± 0.04	10.8 ± 0.004	0.92 ± 0.002	1879.0 ± 98.56	1923.2 ± 98.60
	5 d	6.2 ± 0.06	24.1 ± 0.08	0.95 ± 0.003	738.2 ± 0.81	2.1 ± 0.000	0.60 ± 0.001	6622.0 ± 313.6	7360.2 ± 314.41
	7 d	6.2 ± 0.06	16.1 ± 0.04	1.00 ± 0.003	49.5 ± 0.05	11.5 ± 0.004	0.93 ± 0.003	2758.0 ± 147.1	2807.5 ± 147.15
Cr5	0 d	7.5 ± 0.08	15.7 ± 0.04	1.00 ± 0.002	35.7 ± 0.03	18.9 ± 0.008	1.00 ± 0.003	331.4 ± 17.56	367.1 ± 17.59
	1 d	11.8 ± 0.14	24.0 ± 0.07	0.95 ± 0.002	385.1 ± 0.42	5.1 ± 0.002	0.76 ± 0.001	1400.0 ± 81.22	1785.1 ± 81.64
	3 d	6.2 ± 0.08	11.2 ± 0.03	1.00 ± 0.003	59.0 ± 0.06	13.1 ± 0.005	1.00 ± 0.003	1594.0 ± 87.35	1653.0 ± 87.41
	5 d	5.4 ± 0.07	19.6 ± 0.06	0.96 ± 0.002	34.7 ± 0.04	2.4 ± 0.000	0.68 ± 0.001	5662.0 ± 280.2	5696.7 ± 280.24
	7 d	5.2 ± 0.07	20.3 ± 0.05	0.96 ± 0.002	383.1 ± 0.36	1.7 ± 0.000	0.49 ± 0.001	17,000 ± 755.6	17,383.1 ± 755.96
Cr10	0 d	7.9 ± 0.09	13.4 ± 0.04	1.00 ± 0.003	39.1 ± 0.05	15.1 ± 0.005	1.00 ± 0.003	461.5 ± 26.97	500.6 ± 27.02
	1 d	25.8 ± 0.28	14.5 ± 0.04	1.00 ± 0.002	62.6 ± 0.06	6.4 ± 0.003	0.92 ± 0.002	3061.0 ± 161.6	3123.6 ± 162.2
	3 d	7.2 ± 0.07	9.9 ± 0.02	1.00 ± 0.003	70.2 ± 0.08	6.6 ± 0.003	1.00 ± 0.003	3015.0 ± 170.4	3085.2 ± 171.2
	5 d	7.2 ± 0.07	15.8 ± 0.04	0.97 ± 0.003	138.9 ± 0.15	3.2 ± 0.000	0.87 ± 0.002	4745.0 ± 245.0	4883.9 ± 245.15
	7 d	6.6 ± 0.06	11.0 ± 0.03	1.00 ± 0.002	86.6 ± 0.09	7.6 ± 0.002	1.00 ± 0.003	5775.0 ± 301.4	5861.6 ± 301.49
Cr15	0 d	7.7 ± 0.08	12.5 ± 0.05	1.00 ± 0.003	43.0 ± 0.04	12.0 ± 0.004	0.99 ± 0.003	610.4 ± 33.92	653.4 ± 33.96
	1 d	6.9 ± 0.06	20.4 ± 0.07	0.95 ± 0.002	1388 ± 1.72	7.4 ± 0.004	0.93 ± 0.002	2107.0 ± 105.9	3495.0 ± 107.62
	3 d	7.4 ± 0.08	10.9 ± 0.03	1.00 ± 0.002	91.3 ± 0.12	5.5 ± 0.001	1.00 ± 0.003	2782.0 ± 127.1	2873.3 ± 127.22
	5 d	6.3 ± 0.05	15.4 ± 0.04	0.98 ± 0.003	90.1 ± 0.11	4.3 ± 0.002	0.85 ± 0.002	4129.0 ± 223.2	4219.1 ± 223.31
	7 d	6.5 ± 0.06	7.6 ± 0.03	1.00 ± 0.003	0.01 ± 0.00	11.8 ± 0.004	0.91 ± 0.002	4096.0 ± 211.9	4096.01 ± 211.9

Furthermore, the corrosion morphologies of the alloys after immersion experiments were observed using SEM, and the results are shown in Figure 6. From Figure 6a, after 1 day of immersion, the Cr0 sample displays a distinct pitting pit, and EDS analysis indicates that it is primarily composed of Fe oxides. As the immersion time increased, no new pitting pits were observed until the 7th day of immersion, at which various pit morphologies emerged, as depicted in Figure 6c–f. This result is consistent with the impedance tests conducted after the immersion experiment for Cr0, where impedance decreases after 1 day of immersion, followed by a gradual stabilisation of the surface and an increase in impedance with prolonged immersion time. New pits appeared after the 7th day, leading to a subsequent decrease in impedance. Regarding the Cr-containing HEAs, no pitting phenomenon was observed during the immersion test, as shown in Figure 6g,h. For Cr3 to Cr15, a protective film gradually formed on the alloys' surface with increasing immersion time, as evidenced by the continuous rise in impedance. However, with the increasing Cr content, the impact of immersion time on the samples' impedance diminishes. High Cr content can foster the formation of the protective film on alloy surfaces more rapidly.

3.5. Corrosion Property Comparison

Figure 7 illustrates the comparison of corrosion current density and pitting potential in a 3.5 wt.% NaCl solution between the studied alloys and the reported Cr-containing HEAs, demonstrating the effect of the Cr element on corrosion properties. It is evident that with increasing Cr content, the pitting potential of the alloys continuously rises while the corrosion current density generally declines. A minimal addition of Cr (8.3 at.%) significantly enhances the corrosion resistance of the alloys, which remains relatively stable from Cr3 to Cr10. This indicates that under single-phase conditions, the addition of Cr exerts a positive effect on improving the corrosion resistance of HEAs. As for $FeCoNiCr_x$ (x = 0, 0.5, 1) HEAs, an increase in Cr content initially results in a decrease in corrosion current density, followed by an increase. The pitting potential shows an opposite trend, with excessive Cr addition leading to localised corrosion in FeCoNiCr, thereby deteriorating corrosion resistance [15]. For $AlCoFeNiCr_x$ (x = 0.5, 0.75, 1, 1.5, 2) HEAs, a minimal addition of Cr (11.1 at.%) cannot cause alloy passivation. As Cr content further increases to 15.8 at.%, the corrosion current density significantly decreases. Continuing to increase Cr content to 27.3 at.% reduce the corrosion current density, and the pitting potential continues to rise. However, further increasing Cr content leads to an increase in corrosion current density with no passivation phenomenon. This may be related to changes in the alloy phase structure and alterations in the Al/Cr ratio affecting the formation of the passivation film, resulting in decreased corrosion resistance [17]. These findings indicate that the influence of the Cr element on the corrosion resistance of HEAs correlates with their effects on phase structure changes and interactions with other elements. Only under single-phase conditions does increasing Cr content have a continuous positive effect on improving alloy corrosion performance.

Furthermore, we compare the corrosion properties of our alloys with those of the reported HEAs in the AlFeCoNiCr composition system and the typical stainless steel. From Figure 8, it is evident that the designed alloys exhibit superior corrosion resistance, with lower i_{corr} and higher E_{pit} in a Cl^--containing environment. By adjusting the content of the Al element, the reported HEAs show low corrosion current density ($Al_{0.6}FeCoNiCr$) or high pitting potential ($Al_{0.1}FeCoNiCr$). In contrast, the alloys prepared in this study not only possess a lower corrosion current density, indicative of their superior resistance to general corrosion but also exhibit a higher pitting potential, signifying their notable resistance to localised corrosion in Cl^--containing solution.

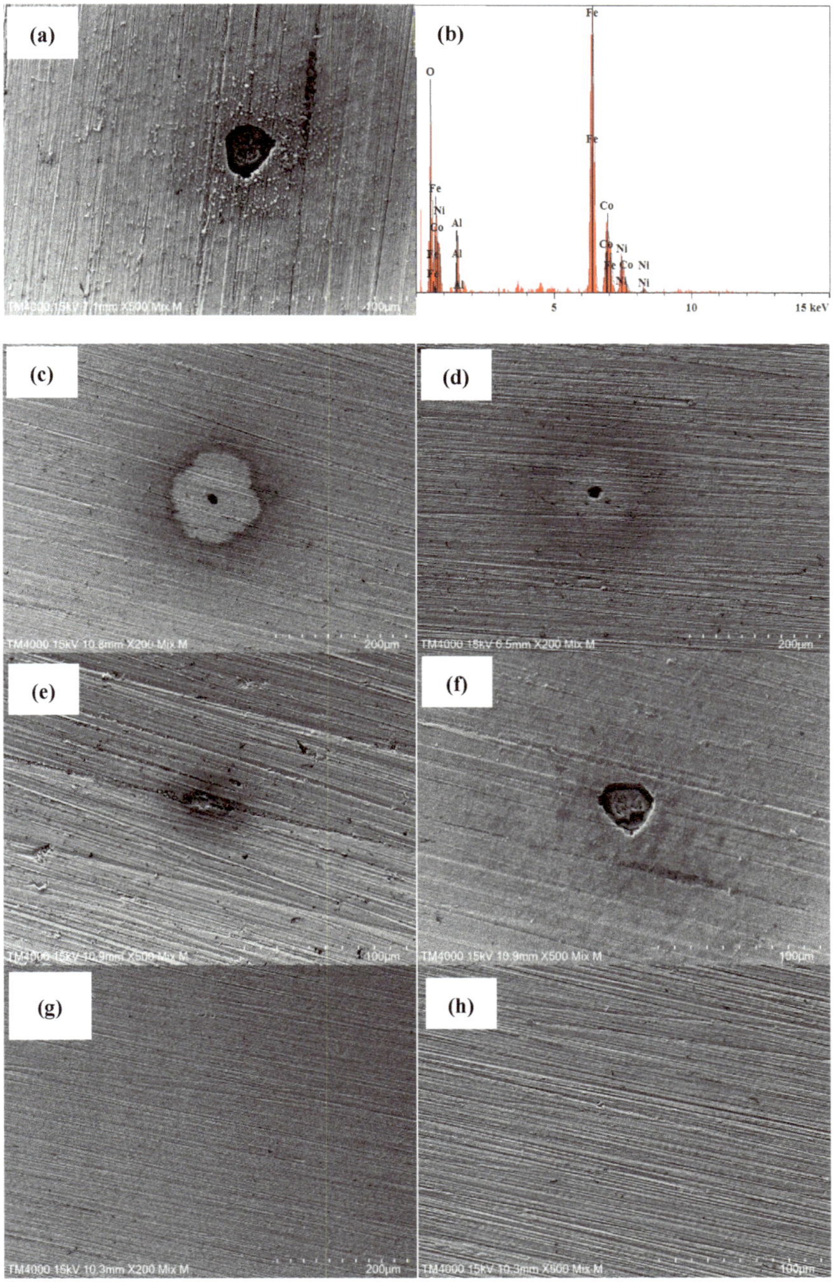

Figure 6. SEM micrographs of the alloy surface after immersion test: (**a**) Cr0 immersion for 1 d and its EDS analysis (**b**); (**c–f**) Cr0 immersion for 7 d; (**g**) Cr3 immersion for 7 d; (**h**) Cr10 immersion for 7 d.

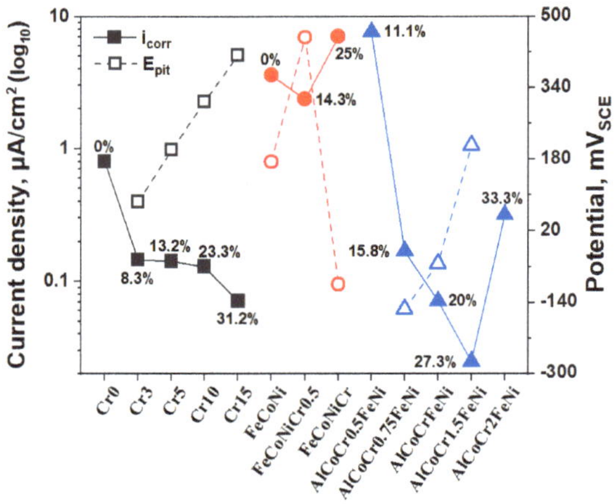

Figure 7. The corrosion properties comparison between the studied alloys and the reported Cr-containing HEAs. The square symbol represents the studied alloys system. The circle symbol represents the FeCoNiCr$_x$ alloys system. The triangle symbol represents the AlCoCr$_x$FeNi alloys system.

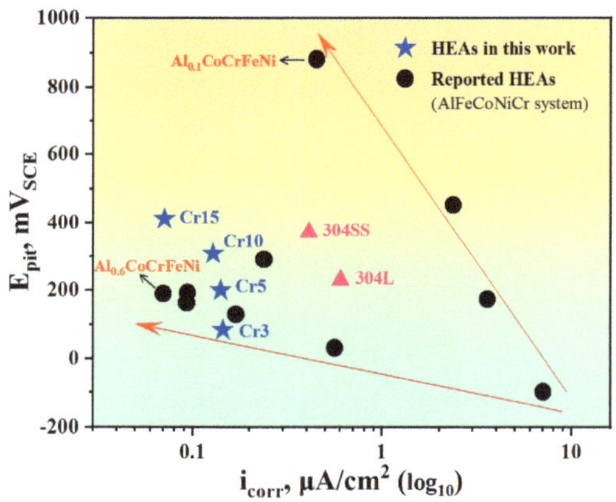

Figure 8. The corrosion properties comparison between the studied alloys and the reported HEAs in the AlFeCoNiCr composition system [14–16,44–46].

Additionally, our developed HEAs show superior corrosion resistance even compared to the two commercial stainless steels of 304 and 304L. The Cr element content of the latter alloys is generally between 18–20 at.%, exceeding the maximum Cr content of the designed alloy. The Cr5 alloys with minor Cr exhibit lower corrosion current density and higher pitting potential than the 304L, indicating its better general corrosion and localised corrosion resistance. It can be reasonably inferred that the corrosion resistance of the investigated HEAs can be further improved after rational plastic deformation and heat treatment to eliminate possible segregation and optimise the microstructure of the as-cast alloys. It is recommended in the future work. However, the phase-formation empirical rule helps to determine the single-phase HEAs, and with the increase in Cr content, the corrosion properties of the designed alloys are continuously enhanced to reach the desired potential and current density direction. It also suggests that by judiciously regulating the composition, maybe the ratio of Cr and Al elements, it is possible to achieve higher corrosion resistance, thereby promoting the potential application of HEAs in marine environments.

4. Conclusions

This study systematically investigates the effect of Cr addition on the phase structure, microstructure, and corrosion performance of the $Al_{0.3}FeCoNiCr_x$ HEAs. The results indicate that the alloys within the designed Cr content range maintain a single-phase FCC structure consistent with most empirical criteria. With the increase in Cr content, the as-cast microstructure of the alloy transitions from cellular structures to typical dendritic–interdendritic morphologies. The results of the electrochemical and immersion test demonstrate that increasing the Cr content has a positive effect on improving the pitting potential of the alloys, reducing the corrosion current density, and enhancing the protective efficacy of the passivation film. The study shows that under single-phase conditions, the Cr element is effective in slowing down the corrosion rate of the $Al_{0.3}FeCoNiCr_x$ HEAs, and this improvement effect continues to strengthen with increasing Cr content, promoting the earlier formation of a protective film on the alloy surface. The designed alloys exhibit significantly superior corrosion performance compared to the current reported HEAs in the AlCoFeNiCr composition system and the commercial stainless steel of 304. In the future, by integrating empirical rules for HEAs phase formation and further synergistic regulation of Al and Cr elements, it is possible to develop HEAs with even better corrosion resistance.

Author Contributions: Conceptualisation, C.W., W.Z. and H.H.; methodology, C.W., M.C. and H.S.; resources and material manufacturing, H.S., Y.T., N.W. and M.C.; investigation, M.C., C.W. and H.S.; writing—original draft preparation, C.W. and H.S.; and data analysis, M.C., H.S., C.W. and Y.T. All the authors contributed to writing—review and editing. All authors have read and agreed to the published version of the manuscript.

Funding: This research was funded by the China National Nuclear Corporation (WDZC-2023-AW-0303), the Program for Innovation Team of Guangdong Ocean University (CXTD2024008).

Institutional Review Board Statement: Not applicable.

Informed Consent Statement: Not applicable.

Data Availability Statement: The raw data required to reproduce these findings can be obtained by contacting the corresponding author.

Conflicts of Interest: The authors declare no conflict of interest.

References

1. Yeh, J.W.; Chen, S.K.; Lin, S.J.; Gan, J.Y.; Chin, T.S.; Shun, T.T.; Tsau, C.H.; Chang, S.Y. Nanostructured high-entropy alloys with multiple principal elements: Novel alloy design concepts and outcomes. *Adv. Eng. Mater.* **2004**, *6*, 299–303. [CrossRef]
2. Cheng, H.; Liu, Z.; Luo, H.; Pan, Z.; Wang, X.; Zhao, Q.; Qi, X.; Li, X. Tuning the microstructure to improve corrosion resistance of additive manufacturing high-entropy alloy in proton exchange membrane fuel cells environment. *Corros. Sci.* **2023**, *213*, 110969. [CrossRef]
3. Cheng, H.; Pan, Z.; Fu, Y.; Wang, X.; Wei, Y.; Luo, H.; Li, X. Corrosion-resistant high-entropy alloy coatings: A review. *J. Electrochem. Soc.* **2021**, *168*, 111502. [CrossRef]
4. Li, W.; Xie, D.; Li, D.; Zhang, Y.; Gao, Y.; Liaw, P.K. Mechanical behavior of high-entropy alloys. *Prog. Mater. Sci.* **2021**, *118*, 100777. [CrossRef]
5. Ma, E.; Wu, X. Tailoring heterogeneities in high-entropy alloys to promote strength–ductility synergy. *Nat. Commun.* **2019**, *10*, 5623. [CrossRef]
6. Mu, Y.; Zhang, L.; Xu, L.; Prashanth, k.; Zhang, N.; Ma, X.; Jia, Y.; Xu, Y.; Jia, Y.; Wang, G. Frictional wear and corrosion behavior of AlCoCrFeNi high-entropy alloy coatings synthesized by atmospheric plasma spraying. *Entropy* **2020**, *22*, 740. [CrossRef]
7. Muangtong, P.; Rodchanarowan, A.; Chaysuwan, D.; Chanlek, N.; Goodall, R. The corrosion behaviour of CoCrFeNi-x (x = Cu, Al, Sn) high entropy alloy systems in chloride solution. *Corros. Sci.* **2020**, *172*, 108740. [CrossRef]
8. Linder, C.; Rao, S.G.; Boyd, R.; Greczynski, G.; Eklund, P.; Munktell, S.; Febvrier, A.; Björk, E.M. Effect of Mo content on the corrosion resistance of $(CoCrFeNi)_{1-x}Mo_x$ thin films in sulfuric acid. *Thin Solid Film.* **2024**, *790*, 140220. [CrossRef]
9. Hsu, Y.J.; Chiang, W.C.; Wu, J.K. Corrosion behavior of $FeCoNiCrCu_x$ high-entropy alloys in 3.5% sodium chloride solution. *Mater. Chem. Phys.* **2005**, *92*, 112–117. [CrossRef]
10. Shi, Y.; Yang, B.; Xie, X.; Brechtl, J.; Dahmen, K.A.; Liaw, P.K. Corrosion of $Al_xCoCrFeNi$ high-entropy alloys: Al-content and potential scan-rate dependent pitting behavior. *Corros. Sci.* **2017**, *119*, 33–45. [CrossRef]

11. Zhou, J.; Cheng, Y.; Wan, Y.; Yang, J. Strengthening by Ti, Nb, and Zr doping on microstructure, mechanical, tribological, and corrosion properties of CoCrFeNi high-entropy alloys. *J. Alloys Compd.* **2024**, *984*, 173819. [CrossRef]
12. Chou, Y.L.; Wang, Y.C.; Yeh, J.W.; Shih, H.C. Pitting corrosion of the high-entropy alloy $Co_{1.5}CrFeNi_{1.5}Ti_{0.5}Mo_{0.1}$ in chloride-containing sulphate solutions. *Corros. Sci.* **2010**, *52*, 3481–3491. [CrossRef]
13. Liu, J.; An, X.; Zhang, J.; Komg, Q.; Li, Q.; Wang, H.; Yao, W.; Wang, Q. Microstructure, mechanical and corrosion properties of Co-and Cu-free $Al_xCrFeNi_{2.5}Mo_{1-x}$ high entropy alloys. *Intermetallics* **2023**, *153*, 107775. [CrossRef]
14. Chai, W.; Lu, T.; Pan, Y. Corrosion behaviors of $FeCoNiCr_x$ (x = 0, 0.5, 1.0) multi-principal element alloys: Role of Cr-induced segregation. *Intermetallics* **2020**, *116*, 106654. [CrossRef]
15. Tsau, C.H.; Lin, S.X.; Fang, C.H. Microstructures and corrosion behaviors of FeCoNi and CrFeCoNi equimolar alloys. *Mater. Chem. Phys.* **2017**, *186*, 534–540. [CrossRef]
16. Jiang, Y.Q.; Li, J.; Juan, Y.F.; Lu, Z.J.; Jia, W.L. Evolution in microstructure and corrosion behavior of $AlCoCr_xFeNi$ high-entropy alloy coatings fabricated by laser cladding. *J. Alloys Compd.* **2019**, *775*, 1–14. [CrossRef]
17. Qiu, Y.; Gibson, M.A.; Fraser, H.L.; Birbilis, N. Corrosion characteristics of high entropy alloys. *Mater. Sci. Technol.* **2015**, *31*, 1235–1243. [CrossRef]
18. Qiu, Y.; Thomas, S.; Gibson, M.A.; Fraser, H.L.; Birbilis, N. Corrosion of high entropy alloys. *Npj Mater. Degrad.* **2017**, *1*, 15. [CrossRef]
19. Taylor, C.D.; Lu, P.; Saal, J.; Frankel, G.S.; Scully, J.R. Integrated computational materials engineering of corrosion resistant alloys. *Npj Mater. Degrad.* **2018**, *2*, 6. [CrossRef]
20. Sheng, G.U.O.; Liu, C.T. Phase stability in high entropy alloys: Formation of solid-solution phase or amorphous phase. *Prog. Nat. Sci. Mater. Int.* **2011**, *21*, 433–446.
21. Guo, S. Phase selection rules for cast high entropy alloys: An overview. *Mater. Sci. Technol.* **2015**, *31*, 1223–1230. [CrossRef]
22. Yang, X.; Zhang, Y. Prediction of high-entropy stabilized solid solution in multi-component alloys. *Mater. Chem. Phys.* **2012**, *132*, 233–238. [CrossRef]
23. Kube, S.A.; Sohn, S.; Uhl, D.; Datye, A.; Mehta, A.; Schroers, J. Phase selection motifs in High Entropy Alloys revealed through combinatorial methods: Large atomic size difference favors BCC over FCC. *Acta Mater.* **2019**, *166*, 677–686. [CrossRef]
24. Guo, S.; Ng, C.; Lu, J.; Liu, C.T. Effect of valence electron concentration on stability of fcc or bcc phase in high entropy alloys. *J. Appl. Phys.* **2011**, *109*, 103505. [CrossRef]
25. Poletti, M.G.; Battezzati, L. Electronic and thermodynamic criteria for the occurrence of high entropy alloys in metallic systems. *Acta Mater.* **2014**, *75*, 297–306. [CrossRef]
26. King, D.J.M.; Middleburgh, S.C.; McGregor, A.G.; Cortie, M.B. Predicting the formation and stability of single phase high-entropy alloys. *Acta Mater.* **2016**, *104*, 172–179. [CrossRef]
27. Singh, A.K.; Kumar, N.; Dwivedi, A.; Subramaniam, A. A geometrical parameter for the formation of disordered solid solutions in multi-component alloys. *Intermetallics* **2014**, *53*, 112–119. [CrossRef]
28. Toda-Caraballo, I.; Rivera-Díaz-del-Castillo, P.E.J. A criterion for the formation of high entropy alloys based on lattice distortion. *Intermetallics* **2016**, *71*, 76–87. [CrossRef]
29. Ke, G.Y. Simple Solid Solution in $Al_xCo_yCr_zCu_{0.5}Fe_vNi_w$ High-Entropy Alloys. Master's thesis, National Tsing Hua University, Hsinchu City, Taiwan, 2005; pp. 22–98.
30. Marcus, P. On some fundamental factors in the effect of alloying elements on passivation of alloys. *Corros. Sci.* **1994**, *36*, 2155–2158. [CrossRef]
31. Lu, P.; Saal, J.E.; Olson, G.B.; Li, T.; Sahu, S.; Swanson, O.J.; Frankel, G.S.; Gerard, A.Y.; Scully, J.R. Computational design and initial corrosion assessment of a series of non-equimolar high entropy alloys. *Scr. Mater.* **2019**, *172*, 12–16. [CrossRef]
32. Burstein, G.T.; Pistorius, P.C.; Mattin, S.P. The nucleation and growth of corrosion pits on stainless steel. *Corros. Sci.* **1993**, *35*, 57–62. [CrossRef]
33. Tang, Y.; Zuo, Y.; Wang, J.; Zhao, X.; Niu, B.; Lin, B. The metastable pitting potential and its relation to the pitting potential for four materials in chloride solutions. *Corros. Sci.* **2014**, *80*, 111–119. [CrossRef]
34. Li, J.; Yang, X.; Zhu, R.; Zhang, Y. Corrosion and serration behaviors of $TiZr_{0.5}NbCr_{0.5}V_xMo_y$ high entropy alloys in aqueous environments. *Metals* **2014**, *4*, 597–608. [CrossRef]
35. Torbati-Sarraf, H.; Shabani, M.; Jablonski, P.D.; Pataky, P.D.; Poursaee, A. The influence of incorporation of Mn on the pitting corrosion performance of CrFeCoNi High Entropy Alloy at different temperatures. *Mater. Des.* **2019**, *184*, 108170. [CrossRef]
36. Luo, H.; Gao, S.; Dong, C.; Li, X. Characterization of electrochemical and passive behaviour of Alloy 59 in acid solution. *Electrochim. Acta* **2014**, *135*, 412–419. [CrossRef]
37. Ai, C.; He, F.; Guo, M.; Zhou, J.; Wang, Z.; Yuan, Z.; Guo, Y.; Liu, L.Y.; Liu, L. Alloy design, micromechanical and macromechanical properties of $CoCrFeNiTa_x$ eutectic high entropy alloys. *J. Alloys Compd.* **2018**, *735*, 2653–2662. [CrossRef]
38. Ezhilselvi, V.; Nithin, J.; Balaraju, J.N.; Subramanian, S. The influence of current density on the morphology and corrosion properties of MAO coatings on AZ31B magnesium alloy. *Surf. Coat. Technol.* **2016**, *288*, 221–229. [CrossRef]
39. Luo, H.; Dong, C.F.; Li, X.G.; Xiao, K. The electrochemical behaviour of 2205 duplex stainless steel in alkaline solutions with different pH in the presence of chloride. *Electrochim. Acta* **2012**, *64*, 211–220. [CrossRef]
40. Wei, L.; Liu, Y.; Li, Q.; Cheng, Y.F. Effect of roughness on general corrosion and pitting of $(FeCoCrNi)_{0.89}(WC)_{0.11}$ high-entropy alloy composite in 3.5 wt.% NaCl solution. *Corros. Sci.* **2019**, *146*, 44–57. [CrossRef]

41. Dominguez-Benetton, X.; Sevda, S.; Vanbroekhoven, K.; Pant, D. The accurate use of impedance analysis for the study of microbial electrochemical systems. *Chem. Soc. Rev.* **2012**, *41*, 7228–7246. [CrossRef]
42. Wei, Y.; Pan, Z.; Fu, Y.; Yu, W.; He, S.; Yuan, Q.; Luo, H.; Li, H. Effect of annealing temperatures on microstructural evolution and corrosion behavior of Ti-Mo titanium alloy in hydrochloric acid. *Corros. Sci.* **2022**, *197*, 110079. [CrossRef]
43. Marcelin, S.; Pébère, N.; Régnier, S. Electrochemical characterisation of a martensitic stainless steel in a neutral chloride solution. *Electrochim. Acta* **2013**, *87*, 32–40. [CrossRef]
44. Qiu, Y.; Thomas, S.; Fabijanic, D.; Barlow, A.J.; Fraser, H.L.; Birbilis, N. Microstructural evolution, electrochemical and corrosion properties of $Al_xCoCrFeNiTi_y$ high entropy alloys. *Mater. Design.* **2019**, *170*, 107698. [CrossRef]
45. Zhao, Y.; Wang, M.; Cui, H.; Song, X.; Zeng, Y.; Gao, X.; Lu, F.; Wang, C.; Song, Q. Effects of Ti-to-Al ratios on the phases, microstructures, mechanical properties, and corrosion resistance of Al2-xCoCrFeNiTix high-entropy alloys. *J. Alloys Compd.* **2019**, *805*, 585–596. [CrossRef]
46. Kumar, N.; Fusco, M.; Komarasamy, M.; Mishra, R.S.; Bourham, M.; Murty, K.L. Understanding effect of 3.5 wt.% NaCl on the corrosion of $Al_{0.1}CoCrFeNi$ high-entropy alloy. *J. Nucl. Mater.* **2017**, *495*, 154–163. [CrossRef]

materials

Article

Effect of Natural Inhibitors on the Corrosion Properties of Grade 2 Titanium Alloy

Mehdrad Faraji [1,2], Luca Pezzato [1,3,*], Arshad Yazdanpanah [1], Giacomo Nardi [1], Mojtaba Esmailzadeh [4] and Irene Calliari [1]

1 Department of Industrial Engineering, University of Padova, Via Marzolo 9, 35131 Padova, Italy; mehrdad.faraji@studenti.unipd.it (M.F.); arshad.yazdanpanah@phd.unipd.it (A.Y.); giacomo.nardi.1@studenti.unipd.it (G.N.); irene.calliari@unipd.it (I.C.)
2 National Institute for Nuclear Physics (INFN), Padova Division, Via Marzolo 8, 35131 Padova, Italy
3 Institute of Condensed Matter Chemistry and Energy Technologies (ICMATE), National Research Council of Italy, C.so Stati Uniti 4, 35127 Padova, Italy
4 Department of Mechanical Engineering, Persian Gulf University, Bushehr 75169-13817, Iran; m.esmaeilzade@pgu.ac.ir
* Correspondence: luca.pezzato@cnr.it; Tel.: +39-3404954870

Abstract: This study investigates the effects of natural inhibitors (pomegranate, algae, and tomato extracts) on the corrosion resistance of titanium (grade 2). To deepen understanding the inhibition mechanism, Molecular Dynamic (MD) and Monte Carlo (MC) simulations were employed to analyze adsorption behaviors and identify optimal adsorption sites on titanium oxide (TiO_2) surfaces for compounds within the inhibitors. Results indicate non-flat adsorption orientations, with pomegranate peel extract components showing superior inhibition capabilities, attributed to the formation of strong O-H chemical bonds with the TiO_2 surface. In the experimental part of the study Electrochemical Impedance Spectroscopy (EIS) and Potentiodynamic Polarization (PDP) were conducted. Two electrolytes were tested: a solution 3.5% NaCl and a solution 0.5 M NaOH. All the tests were performed with 5% of inhibitor and with the reference solution. Also, inhibition efficiency was calculated on the base of PDP tests. The study found that pomegranate extract can act as a good corrosion inhibitor for titanium alloy in aqueous solutions 0.5 M NaOH. This was demonstrated by the increase in the corrosion potential and impedance modulus and decrease in the corrosion current density after the addition of pomegranate extract to the solution. However, in a 3.5% NaCl solution, the efficacy of pomegranate extract was less pronounced, probably due to the high aggressivity of the electrolyte. Tomato and algae extract have instead shown very low inhibition effects in all the tested conditions.

Keywords: natural inhibitor; corrosion; electrochemical study; titanium; molecular dynamic simulation

Citation: Faraji, M.; Pezzato, L.; Yazdanpanah, A.; Nardi, G.; Esmailzadeh, M.; Calliari, I. Effect of Natural Inhibitors on the Corrosion Properties of Grade 2 Titanium Alloy. *Materials* **2024**, *17*, 5202. https://doi.org/10.3390/ma17215202

Academic Editor: Xiaogang Li

Received: 7 October 2024
Revised: 18 October 2024
Accepted: 23 October 2024
Published: 25 October 2024

1. Introduction

Corrosion is one of the most destructive phenomena that affects many industrial sectors, such as the marine construction, oil and gas and automotive sectors [1–4]. Numerous studies have been conducted via different inhibitors to prevent harmful effects of corrosive processes [5,6]. In particular the interest in green inhibitors as eco-friendly materials has currently grown to address corrosion problems [7]. In recent years, several research have been conducted on employing plant and fruit extracts as natural inhibitors, highlighting them as environmentally friendly alternatives to traditional inhibitors for preventing metal corrosion. The advantages of using these green inhibitors are manifold. Firstly, they effectively safeguard metals from corrosion while preserving ecological equilibrium. Secondly, being derived from renewable, biodegradable, and eco-friendly resources, they contribute to minimizing environmental contamination. Compared to traditional

Table 1. *Cont.*

Name	Chemical Structure	Ref.
	Tomato extract	
hydroxyphenyl chroman-4-one	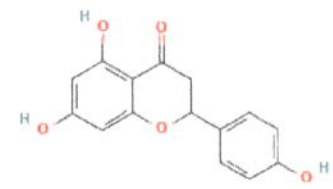	[42]
3,4-Dihydroxy cinnamic acid	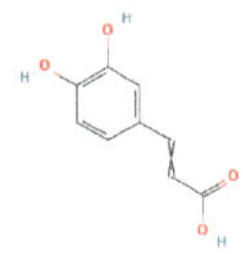	[43]
Quercetin	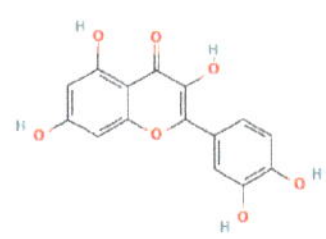	[44]
Kaemferol	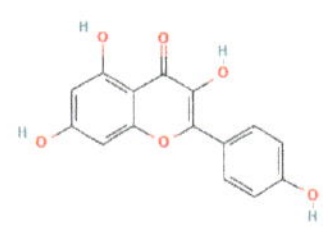	[45]

2.2. Preparation of Inhibitors

The pomegranate, tomato and algae were sourced from Shiraz, Bushehr and the Persian Gulf Sea, Iran. A straightforward, environmentally friendly, and water-based method was used to extract the chemical compounds from the pomegranate peel, tomato peel and algae. The extracts of pomegranate peel, algae and tomato peel were prepared using a similar method. For this purpose, after washing, 100 g of the materials was dried at a temperature of 80 °C for 12 h. The dried materials were then fully ground and pulverized using a laboratory grinder. Next, they were combined with 1 L of deionized water and heated at 70 °C for 8 h to extract the water-soluble compounds. Subsequently, the mixture was centrifuged at 3000 rpm for 20 min to completely eliminate any undissolved particles. Finally, the mixture was heated for 24 h at 80 °C to evaporate the water solvent and obtain the extract in powder form. To assess their inhibitory effects, approximately 700 ppm of each of the prepared materials were mixed in water in order to add this solution to the electrolyte. Therefore, due to the global low amounts of these inhibitors in the electrolyte, the pH of the electrolyte did not change significantly.

2.3. Sample Preparation and Corrosion Tests

An initial ingot of Ti grade 2 alloy was cut transversally with a disc cutter in lubricated conditions. As a result, square samples with a side of 15 mm and a thickness of no more than 3–4 mm were obtained. After cutting, the samples were ground and polished accordingly to a standard metallographic technique.

Then, the samples were washed with ultrasound to obtain an optimal surface finish free of impurities. Both potentiodynamic polarization (PDP) and electrochemical impedance spectroscopy (EIS) tests were conducted on the samples using a Gamry Interface 1010E potentiostat (Gamry Italia, Voghera (PV), Italy) in a three-electrode configuration using a Platinum electrode as a counter electrode, a calomel electrode as a reference electrode and the tested sample (1 cm^2) as a working electrode. PDP tests were conducted between −2 V and +3 V with a scan speed of 1 mV/s. EIS tests were conducted with a sinusoidal perturbation of 10 mV over the frequency range of 100 kHz to 0.1 Hz. All the electrochemical tests were conducted after 30 min of OCP stabilization in two different electrolytes: 0.5 M NaOH and 3.5% NaCl aqueous solutions. Two different corrosive electrolytes were chosen in order to compare the behavior of the inhibitors in a chloride-containing environment and in an alkaline environment. The tests were firstly conducted in the solution without the inhibitor, as reference, and after the addition of 5% of the water-based solution containing the corrosion inhibitor (pomegranate, tomato and algae extract, obtained with the method described in Section 3.1). The results of the EIS were then fitted with a proper equivalent circuit using Z-View software (version 3.3). Corrosion potentials and corrosion currents were graphically extrapolated from PDP tests using the Tafel law, and then the data were employed to calculate the inhibition efficiency. All the corrosion tests were performed in triplicate in order to ensure reproducibility.

3. Results

3.1. Molecular Dynamic Simulation

Monte Carlo (MC) and MD simulations were conducted to investigate the inhibition phenomenon on metal surfaces and to identify low-energy adsorption sites for the various compounds present in the inhibitors. These techniques have gained popularity for providing insights into the preferential adsorption of molecules on metal surfaces [21]. Figures 1 and 2 present simulations of the adsorption orientation of different molecules on a titanium oxide (TiO_2) surface. The results show that molecules exhibit a non-flat adsorption orientation on the TiO_2 surface. After optimization, the estimated adsorption energies of different molecules on the TiO_2 (001) surface and the corresponding adsorption geometries are listed in Table 2.

Table 2. Binding energy and diffusion coefficient of component in a 3.5% NaCl solution.

Components	**Binding Energy (kcal)**	**Diffusion Coefficient D (A^2/ps)**	**R^2**
	Pomegranate peel extract		
Caffeic acid	-4.50×10^3	2.424490×10^{-5}	0.967
Ellagic acid	-1.81×10^4	3.171714×10^{-5}	0.945
Gallic acid	-1.91×10^4	3.499181×10^{-6}	0.988
Urolithine	-1.88×10^4	3.904222×10^{-4}	0921
	Tomato extract		
Hydroxyphenyl chroman-4 one	-3.45×10^3	1.409890×10^{-4}	0.932
3,4-Dihydroxy cinnamic acid	-4.04×10^3	6.265649×10^{-4}	0.941
Quercetin	-4.06×10^3	1.645273×10^{-4}	0.962
Kaempferol	-4.02×10^3	4.103204×10^{-4}	0.933

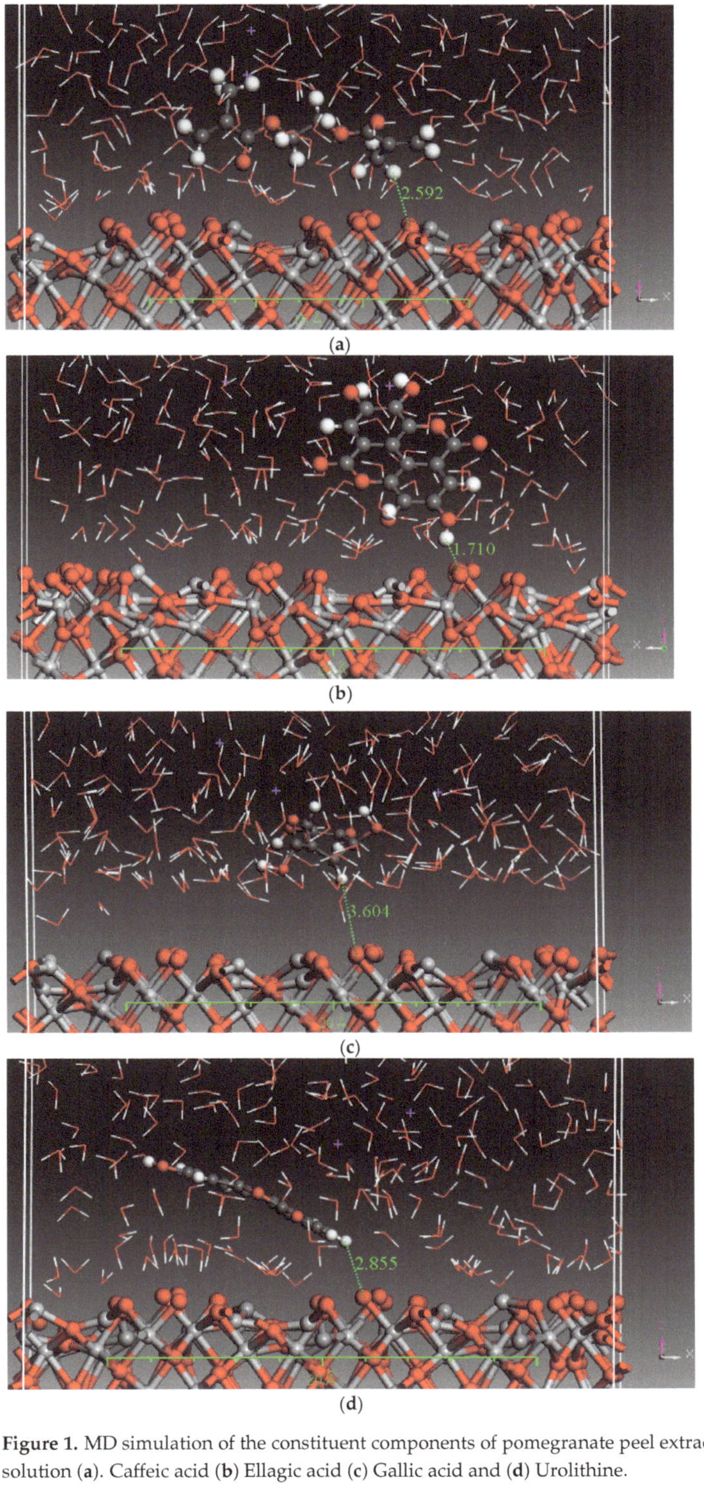

Figure 1. MD simulation of the constituent components of pomegranate peel extract in a 3.5% NaCl solution (**a**). Caffeic acid (**b**) Ellagic acid (**c**) Gallic acid and (**d**) Urolithine.

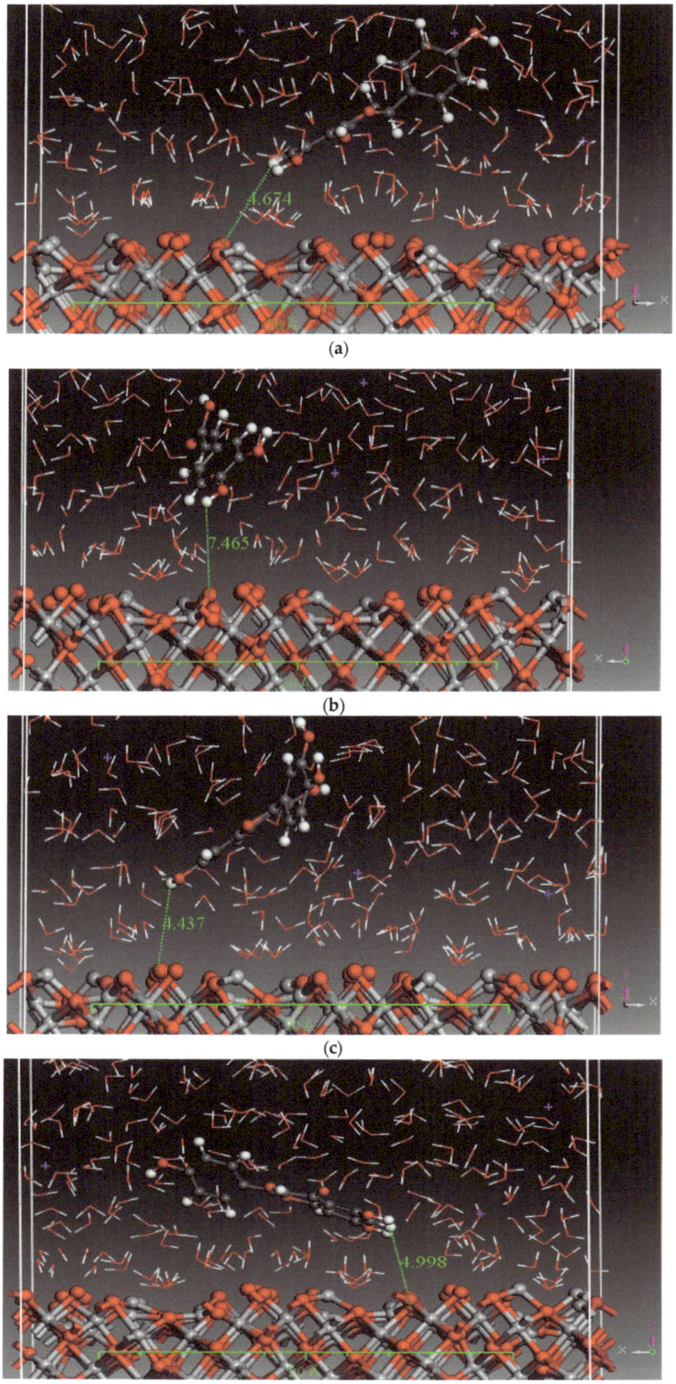

Figure 2. MD simulation of the constituent components of tomato extract in a 3.5% NaCl solution (**a**) hydroxyphenyl chroman-4-one (**b**) 3,4-Dihydroxy cinnamic acid (**c**) Quercetin (**d**) Kaemferol.

Figure 3 shows the changes in the values of MSD for different molecules. The closer the slope is to the value of 1, the greater the ability of molecule movement in the electrolyte will be. Based on this, tomato extract components have the highest mobility capability in the electrolyte. The values of the diffusion coefficients for different molecules of pomegranate peel extract and tomato extract are listed in Table 2.

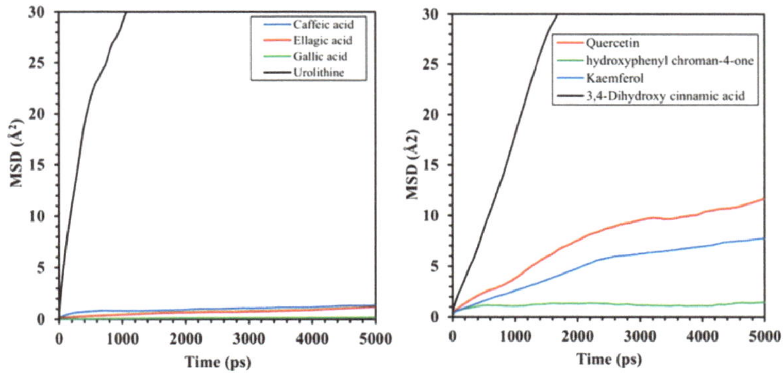

Figure 3. Mean Square Displacement (MSD) analysis.

3.2. Open Circuit Potential (OCP) Measurement

The OCP measurements were conducted to evaluate the thermodynamic stabilities of the grade 2 Ti samples in two environments: with and without inhibitors. Figure 4 shows the OCP measurements of grade 2 Ti samples in 3.5% NaCl (Figure 4a) and 0.5 M NaOH (Figure 4b) solutions with different natural inhibitors. In these figures it can be observed that the OCP values of the Ti samples in both solutions take about 1000 s to reach an indicative steady state. As shown in Figure 4a, by adding pomegranate to a 3.5% NaCl solution, the OCP of the Ti samples is shifted to the higher values. Conversely, the presence of algae in a sodium chloride solution led to a decrease in the OCP value of the Ti. In a 0.5 M NaOH solution, only pomegranate extract increases the OCP value of the Ti sample, from −319 mV to −158 mV. Both the tomato and algae decreased the OCP to −417 mV and −425 mV, respectively.

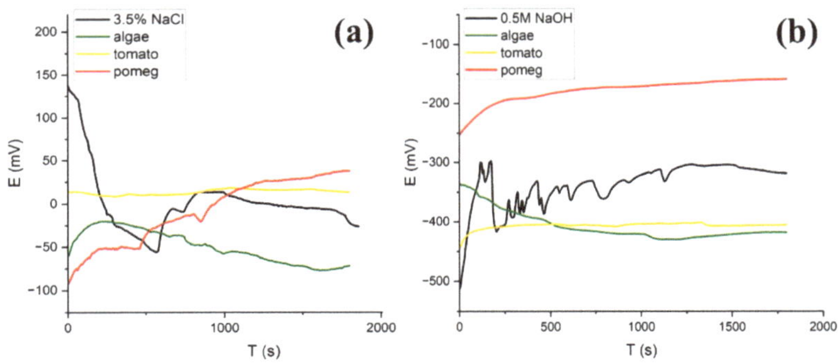

Figure 4. OCP decay analysis of the Ti samples in (**a**) 3.5% NaCl and (**b**) 0.5 M NaOH solution with/without natural inhibitors.

3.3. Electrochemical Impedance Spectroscopy (EIS)

Figures 5 and 6 show the Nyquist and bode plots of the Ti grade 2 samples in 0.5 M NaOH and a 3.5% NaCl solution, with or without the natural inhibitors, whereas Tables 3 and 4 report the results of the fitting of the experimental data. In these figures, the experimental and fitting data can be observed by the scattered dots and straight line, respectively.

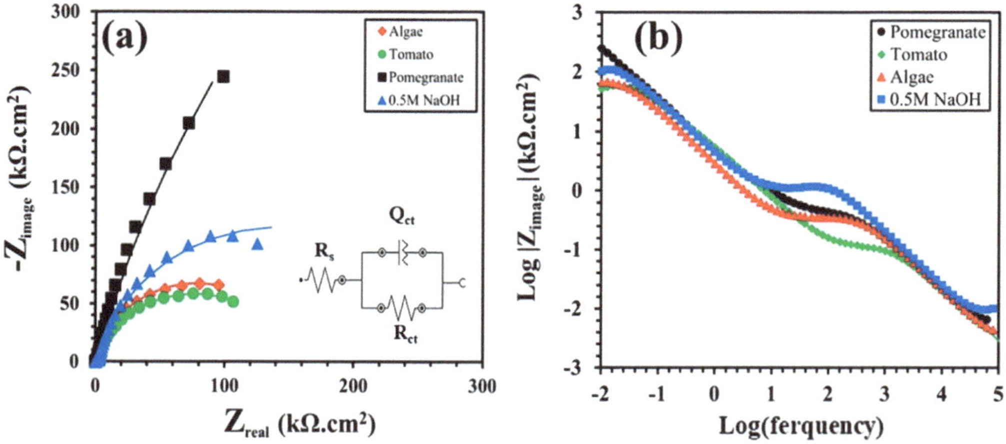

Figure 5. (**a**) Nyquist plot with the equivalent electrical circuit employed to fit the experimental data and (**b**) Bode modulus plot of the Ti grade 2 samples in 0.5 M NaOH solution with/without inhibitors.

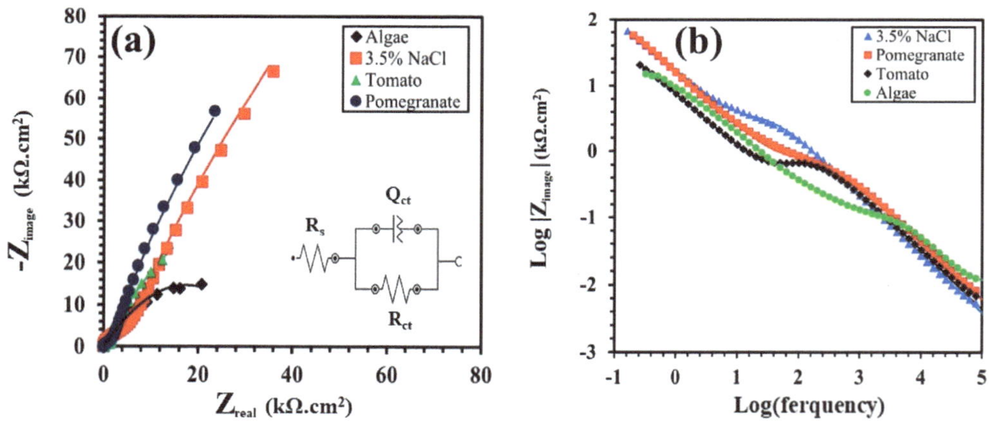

Figure 6. (**a**) Nyquist plot with the equivalent electrical circuit employed to fit the experimental data and (**b**) Bode modulus plot of the Ti samples in 3.5% NaCl solution with/without inhibitors.

Table 3. Results of the fitting of the experimental EIS data for Ti grade 2 in 0.5 M NaOH with/without inhibitors.

Sample	R_{sol} (Ω.cm^{-2})	R_{ct} (Ω.cm^{-2})	Q		χ (%)
			n	Y	
0.5 M NaOH	23.23 ± 0.5	$(2.82 \pm 0.02) \times 10^5$	0.79	1.30×10^{-5}	1.28
Pomegranate	22.72 ± 0.6	$(2.01 \pm 0.01) \times 10^6$	0.85	3.51×10^{-5}	1.93
Tomato	26.70 ± 0.4	$(1.52 \pm 0.03) \times 10^5$	0.83	3.81×10^{-5}	1.18
Algae	24.23 ± 0.8	$(1.56 \pm 0.02) \times 10^5$	0.85	3.53×10^{-6}	1.51

Table 4. Results of the fitting of the experimental EIS data for grade 2 Ti in 3.5% NaCl solution with/without inhibitors.

Sample	R_{sol} (Ω.cm^{-2})	R_{ct} (Ω.cm^{-2})	Q		χ (%)
			n	Y	
3.5% NaCl	46.35 ± 0.4	(8.14 ± 0.01) × 10^5	0.81	1.30 × 10^{-5}	1.39
Pomegranate	36.54 ± 0.5	(1.53 ± 0.02) × 10^6	0.80	1.36 × 10^{-5}	1.57
Tomato	54.62 ± 0.8	92,413 ± 110	0.84	2.61 × 10^{-5}	1.11
Algae	31.14 ± 0.9	39,450 ± 240	0.83	1.53 × 10^{-5}	1.80

Figure 5a displays the equivalent electrical circuit employed to fit the experimental data obtained by the EIS test. The circuit includes a solution resistance (R_s) in series with a parallel capacitive loop charge transfer resistance and constant phase element (CPE) of the oxide layer formed on the surface (R_{ct}/Q_{ct}). The CPE is inserted into the circuit to serve as an alternative to the capacitor, aiming for a closer fitting under conditions where the frequency exponent is below 1.0 [46–50]. According to Figure 5a, the impedance at lower frequencies serves as an indicative measure of both the polarization resistance and the corrosion resistance. Therefore, the corrosion resistance values of the specimens are in the following order: pomegranate > 0.5 NaOH (reference) > tomato > algae. The charge transfer resistance (R_{ct}) of the oxide layer is inversely proportional to the corrosion rate. According to Table 3, the R_{ct} value of the Ti sample in a 0.5 M NaOH solution without any inhibitors is 2.82 × 10^5 Ω.cm^{-2}. By adding the pomegranate extract as an inhibitor, the R_{ct} value was increased to 2.01 × 10^6 Ω.cm^{-2}. In other cases, it was observed that tomato extract and algae have a negative effect on the corrosion resistance of Ti samples in a NaOH solution, so that the R_{ct} values decreased to 1.52 × 10^5 and 1.56 × 10^5 Ω.cm^{-2}, respectively.

Bode impedance plots provide insights into different aspects of corrosion behavior. The high frequency range reflects the impact of surface defects caused by corrosion on the titanium oxide film. The middle frequency range correlates with the corrosion processes occurring within the film itself. Lastly, the low frequency range reveals information about corrosion at the interface between the metal and the oxide film. Referring to Figure 5b, it is observed that the low-frequency range of the algae exhibits significantly lower values in comparison to those of other inhibitors. This observation suggests that the poorly conductive TiO_2 passive layer has been displaced.

Figure 6a displays the results of the EIS tests carried out on a Ti grade 2 sample in a 3.5% NaCl solution. In this case, it does not evidence a significant increase in corrosion resistance using the natural inhibitors. It can be observed that the use of pomegranate extract produces only a slight inhibition effect, especially in comparison with the results obtained in a 0.5 M NaOH solution. According to the equivalent electrical circuit (Figure 6b) and EIS parameters given in Table 4, the R_{ct} value of the Ti sample in a 3.5% NaCl solution without any inhibitors, was increased from 2.82 × 10^5 Ω.cm^{-2} to 2.01 × 10^6 Ω.cm^{-2} by adding pomegranate extract. Inversely the addition of tomato and algae decreased the corrosion resistance of the Ti grade 2 alloy, so that the value of R_{ct} significantly decreased to 92,413 and 39,450 Ω.cm^{-2}, respectively. In Figure 6b, it is observed that the low-frequency range of the algae and tomato exhibit lower values compared to the samples tested with pomegranate extract and to the samples tested without any inhibitors.

3.4. Potentiodynamic Polarization (PDP)

Figure 7 shows the potentiodynamic polarization curves of grade 2 Ti in two different solutions (0.5 M NaOH and 3.5% NaCl) adding different natural inhibitors.

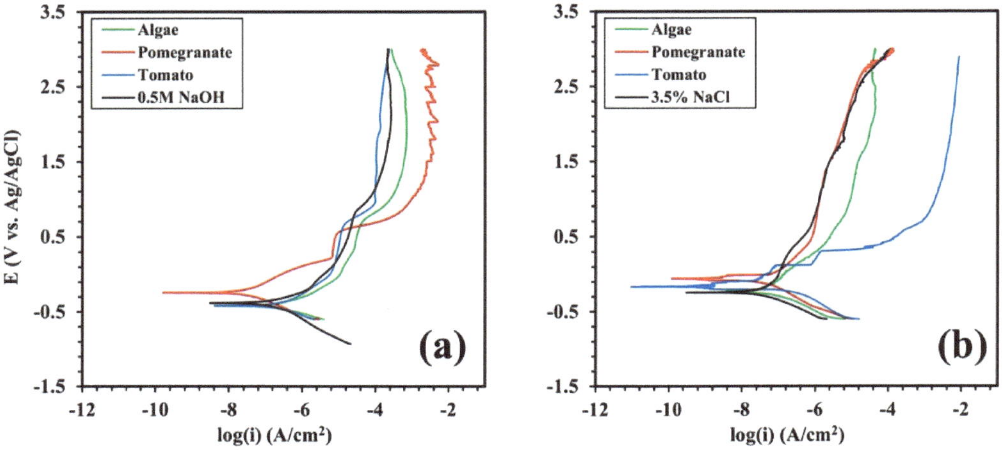

Figure 7. PDP curves of the Ti samples in (**a**) 0.5 M NaOH and (**b**) 3.5% NaCl solution with different inhibitors.

Figure 7a shows the trends in the polarization curves in the case of titanium immersed in an aqueous solution 0.5 M NaOH + natural inhibitors. By considering the graph, it can be observed that the only curve that deviates from the solution without inhibitors (reference solution) is the curve of the pomegranate extract. Table 5 shows the corrosion parameters of the grade 2 Ti samples graphically extrapolated from PDP curves in 0.5 M NaOH, particularly the i_{corr} and E_{corr}. Graphical extrapolation was performed accordingly to the tafel method drawing the tangent to the cathodic and anodic branches of the curves. The presence of pomegranate extract obviously causes a shift in the E_{corr} of NaOH from −383.7 to −244.08 mV. Also, a remarkable decrease in the i_{corr} value of the sample tested in solution containing pomegranate extract can be observed, indicating a high inhibition effect.

Table 5. Corrosion current density, corrosion potential and IE% calculated from PDP performed on grade 2 Ti samples in 0.5 M NaOH.

Sample	E_0 (mV)	I_0 (mA/mm^2)	IE%
0.5 M NaOH	−383.7	$(1.12 \pm 0.01) \times 10^{-6}$	0
Algae	−421.38	$(9.37 \pm 0.01) \times 10^{-7}$	16%
Tomato	−417.28	$(5.71 \pm 0.01) \times 10^{-7}$	49%
Pomegranate	−244.08	$(4.65 \pm 0.01) \times 10^{-7}$	58%

Figure 7b illustrates the behavior of a grade 2 Ti sample in a 3.5% NaCl solution. From the graph, it can be observed that the natural inhibitors under consideration do not significantly increase the resistance of the material, as comparing the anodic branch of the different curves, it is evident that there is no improvement. Considering the corrosion potential there is a slight increase in this parameter employing tomato extract and a more significant increase in the presence of pomegranate extract. Globally, however, it can be stated that from PDP tests the inhibitors do not produce significant positive effects to the corrosion properties of grade 2 Ti in a 3.5% NaCl solution. Table 6 displays data graphically extrapolated from PDP tests in 3.5 NaCl. The presence of pomegranate extract causes a shift in the E_{corr} from −254.57 to −82.40 mV. The i_{corr} value of the sample tested with pomegranate extract in NaCl solution is 9.67×10^{-7} mA/mm^2, which is the lowest between the different samples (and so the best in terms of corrosion resistance), but the differences are very small in comparison to the other tested samples. From the polarization curves it was possible to calculate the inhibition efficiency of the single inhibitors inserted in the different solutions. The efficiencies reported in Tables 5 and 6 have been calculated using the following formula [51]:

$$IE\% = \frac{(i_0 - i)}{i_0} \times 100 \quad (1)$$

where $IE\%$ is the inhibitors percentage efficiency, i_0 is the current density in the case of the solution without inhibitors and i is the current density in the solution with the inhibitor of which the efficiency is to be calculated.

Table 6. Corrosion current density, corrosion potential and IE% calculated from PDP performed on grade 2 Ti samples in 3.5% NaCl.

Sample	E_0 (mV)	I_0 (mA/mm^2)	IE%
3.5% NaCl	−254.57	$(2.67 \pm 0.01) \times 10^{-6}$	0
Algae	−248.99	$(4.04 \pm 0.01) \times 10^{-6}$	−52%
Tomato	−167.65	$(1.01 \pm 0.01) \times 10^{-6}$	62%
Pomegranate	−82.40	$(9.67 \pm 0.01) \times 10^{-7}$	64%

From the tables, it is possible to note that pomegranate extract has a good efficiency in all three different tested solutions whereas the effect of tomato extract and algae extract is controversial and, in any case, lower than the one of pomegranate extract.

4. Discussion

4.1. Simulations

The adsorption energy obtained from Monte Carlo (MC) simulations indicates that components of pomegranate peel extract have higher inhibition performance compared to tomato extract. The adsorption of Gallic acid molecules forms strong O-H chemical bonds with the TiO_2 (001) surface, showcasing their effectiveness in protecting titanium against corrosion.

Additionally, the Mean Square Displacement (MSD) analysis is used to determine diffusion coefficient molecules, identifying different behavior of molecules in diffusion. The mean square displacement quantifies the movement of particles within a system and serves as a metric for characterizing molecular mobility. The formula for calculating MSD is expressed as [52]:

$$\text{MSD} = \left\langle (r_i(t) - r_i(0))^2 \right\rangle \quad (2)$$

where r_i (0) and $r_i(t)$ denote the positions of particle i at the initial time and time t, respectively; the notation $\langle\rangle$ indicates the average value across all particles. The diffusion coefficient D for a particle can be determined by examining the slope of the MSD through the application of the Einstein relationship [52]:

$$\text{D} = \frac{1}{6t} \lim_{t \to \infty} \langle |\text{ri(t)} - \text{ri(0)}|_2 \rangle \quad (3)$$

This relationship enables the calculation of the particle diffusion coefficient, providing insights into the dynamic behavior and mobility of particles within the system over time [52].

An overall investigation into the structural configuration, bonding energies, and permeability characteristics of diverse molecules suggests that within a 3.5% NaCl solution, Urolithin molecules exhibit the greatest efficacy among the components of pomegranate peel extract in obstructing functions. Likewise, within the constituents of the tomato extract, Quercetin molecules assume a pivotal role. In general, the higher bond energies of the constituents within pomegranate peel extract, coupled with their lower permeability coefficient relative to those found in tomato extract, suggest that the formation of a relatively stable molecular film on the titanium oxide surface by pomegranate peel extract may enable it to exert a more potent inhibitory effect.

According to the OCP measurements, the higher OCP values of the Ti sample in both solutions and with the presence of pomegranate extract, indicate the higher thermodynamic stability of the samples. The results from EIS and PDP tests indicate that these natural inhibitors exhibit varying degrees of efficacy, particularly in alkaline and saline solutions.

4.2. *Efficacy of Natural Inhibitors*

Pomegranate extract demonstrated a significant corrosion inhibition effect for Ti grade 2 in a 0.5 M NaOH solution. The increase in charge transfer resistance (R_{ct}) indicates a substantial protective effect, likely due to the formation of a more stable oxide layer on the titanium surface. This finding aligns with previous studies that highlight the role of organic compounds in enhancing the stability of passive films on metals. For instance, it has been shown that polyphenols present in pomegranate can adsorb onto metal surfaces, forming a protective barrier that inhibits corrosion by reducing ion diffusion through the oxide layer [53].

In contrast, the addition of tomato extract and algae resulted in decreased R_{ct} values, suggesting that these extracts may not only lack inhibitory properties but could potentially worsen corrosion processes. Research has indicated that certain organic compounds can disrupt protective oxide films rather than strengthening them [54]. The phytochemicals in tomato and algae may interact unfavorably with the titanium surface, leading to a compromised passive layer and increased corrosion susceptibility.

4.3. *Corrosion Behavior in Saline Environments*

When evaluating titanium's corrosion behavior in a 3.5% NaCl solution, the effectiveness of pomegranate extract as an inhibitor was notably less pronounced compared to its performance in alkaline conditions. The slight increase in R_{ct} with pomegranate extract highlights that while it provides some level of protection, its efficiency is significantly reduced in saline environments. This observation is consistent with the literature indicating that chloride ions can destabilize oxide layers on titanium, leading to increased susceptibility to localized corrosion [55].

The polarization curves further illustrate this trend; while there was a shift in corrosion potential (E_{corr}) with pomegranate extract, the overall improvement in corrosion current density (i_{corr}) was minimal. This suggests that while pomegranate extract has some inhibitive properties, its effectiveness is conditional upon specific environmental conditions, particularly the presence of aggressive ions like chloride. Previous studies have shown that although some natural extracts may enhance corrosion resistance under certain conditions, their performance can be compromised when exposed to aggressive electrolytes [56].

4.4. *Mechanisms of Inhibition*

The mechanisms by which pomegranate extract inhibits corrosion may involve both physical adsorption onto the titanium surface and chemical interactions with the oxide layer. The polyphenolic compounds found in pomegranate are known for their antioxidant properties and their ability to form complexes with metal ions [57]. This interaction may help to stabilize the passive layer on titanium by preventing the dissolution of metal ions into the electrolyte.

Conversely, the negative effects observed with tomato and algae extracts raise concerns regarding their application as corrosion inhibitors. The phytochemicals present in these extracts may not only fail to enhance protective mechanisms but could also introduce additional corrosive species or facilitate localized attack on the titanium surface. Understanding these interactions is crucial for developing effective strategies for enhancing the longevity of titanium implants.

5. Conclusions

In this study, the effects of three natural inhibitors (pomegranate, tomato and algae extract) on the corrosion behavior of grade 2 Ti in two different environments (0.5 M NaOH and 3.5% NaCl) was investigated. The main findings of the work can be summarized as follows:

- MD simulations revealed that components of pomegranate peel extract form strong O-H chemical bonds with the titanium oxide surface, indicating a robust inhibition capability. Additionally, MSD analysis highlighted differences in molecular mobility within the electrolyte, suggesting varied inhibitory potentials among the natural extracts tested.
- OCP decay and EIS tests evidence the promising properties of pomegranate extract as natural corrosion inhibitor for grade 2 titanium in 0.5 M NaOH solution. The presence of the inhibitor in fact produces a shift toward nobler potentials of the OCP and a remarkable increase in the polarization resistance.
- Potentiodynamic polarization tests confirmed the potential of pomegranate extract as a natural inhibitor, even if the effect under polarization is less pronounced than the effect observed at the equilibrium condition with EIS.
- The inhibition mechanism can involve both physical adsorption onto the titanium surface and chemical interactions with the oxide layer.
- In conclusion, it is advisable to choose pomegranate extract as a possible natural inhibitor to conduct further investigations in order to clarify better the mechanism of inhibition.

Author Contributions: Conceptualization, L.P., M.E. and I.C.; methodology, L.P. and A.Y.; software, M.F., G.N. and M.E.; validation, M.F. and M.E.; formal analysis, L.P., G.N. and A.Y.; investigation, G.N., M.F., A.Y., M.E. and L.P.; resources, M.E., L.P. and I.C.; data curation, M.F.; writing—original draft preparation, M.F.; writing—review and editing, M.E., L.P., I.C. and A.Y.; visualization, M.E.; supervision, L.P., M.E., A.Y. and I.C.; project administration, L.P., M.E. and I.C.; funding acquisition, L.P., M.E. and I.C. All authors have read and agreed to the published version of the manuscript.

Funding: The paper was prepared thanks to the financing of the "Shaping a World-class University 2022" program of the University of Padova, project title "Study of the effect of deformation processes on the corrosion properties of different metals" (Scientific Coordinator Luca Pezzato).

Institutional Review Board Statement: Not applicable.

Informed Consent Statement: Not applicable.

Data Availability Statement: The raw data supporting the conclusions of this article will be made available by the authors on request.

Conflicts of Interest: The authors declare no conflicts of interest. The funders had no role in the design of the study; in the collection, analyses, or interpretation of data; in the writing of the manuscript; or in the decision to publish the results.

References

1. Yazdanpanah, A.; Franceschi, M.; Rebesan, P.; Dabalà, M. Correlation of lack of fusion pores with stress corrosion cracking susceptibility of L-PBF 316L: Effect of surface residual stresses. *Materials* **2022**, *15*, 7151. [CrossRef] [PubMed]
2. Yazdanpanah, A.; Pagot, G.; Franceschi, M.; Rebesan, P.; Venturin, M.; Botinha, J.; Gerhmann, B.; De Graeve, I.; Di Noto, V.; Revilla, R.I. Mechanism of alteration in passivity of additively manufactured Ni-Fe-Cr Alloy 718 caused by minor carbon variation. *Electrochim. Acta* **2024**, *503*, 144925. [CrossRef]
3. Yazdanpanah, A.; Revilla, R.I.; Franceschi, M.; Fabrizi, A.; Khademzadeh, S.; Khodabakhshi, M.; De Graeve, I.; Dabala, M. Unveiling the impact of laser power variations on microstructure, corrosion, and stress-assisted surface crack initiation in laser powder bed fusion-processed Ni-Fe-Cr alloy 718. *Electrochim. Acta* **2024**, *476*, 143723. [CrossRef]
4. Abo Nama, H.A.H.; Esen, I.; Karakurt, V.; Ahlatci, H. Effect of the Ti addition on the corrosion behavior of newly developed AA7075-Ti alloys. *J. Alloys Compd.* **2023**, *969*, 172349. [CrossRef]
5. Fayomi, O.; Akande, I.; Odigie, S. Economic impact of corrosion in oil sectors and prevention: An overview. *J. Phys. Conf. Ser.* **2019**, *1378*, 022037. [CrossRef]
6. Panchal, J.; Shah, D.; Patel, R.; Shah, S.; Prajapati, M.; Shah, M. Comprehensive review and critical data analysis on corrosion and emphasizing on green eco-friendly corrosion inhibitors for oil and gas industries. *J. Bio-Tribo-Corros.* **2021**, *7*, 107. [CrossRef]
7. Eddy, N.O.; Ibok, U.J.; Garg, R.; Garg, R.; Iqbal, A.; Amin, M.; Mustafa, F.; Egilmez, M.; Galal, A.M. A brief review on fruit and vegetable extracts as corrosion inhibitors in acidic environments. *Molecules* **2022**, *27*, 2991. [CrossRef]

8. Hossain, N.; Asaduzzaman Chowdhury, M.; Kchaou, M. An overview of green corrosion inhibitors for sustainable and environment friendly industrial development. *J. Adhes. Sci. Technol.* **2021**, *35*, 673–690. [CrossRef]
9. Rani, B.A.; Basu, B.B.J. Green inhibitors for corrosion protection of metals and alloys: An overview. *Int. J. Corros.* **2012**, *2012*, 380217. [CrossRef]
10. Verma, C.; Ebenso, E.E.; Bahadur, I.; Quraishi, M. An overview on plant extracts as environmental sustainable and green corrosion inhibitors for metals and alloys in aggressive corrosive media. *J. Mol. Liq.* **2018**, *266*, 577–590. [CrossRef]
11. Valentini, F.; Pezzato, L.; Dabalà, M.; Brunelli, K. Study of the effect of functionalization with inhibitors on the corrosion properties of PEO-coated additive manufactured AlSi10Mg alloy. *J. Mater. Res. Technol.* **2023**, *27*, 3595–3609. [CrossRef]
12. Okutan, M.; Asan, A.; Singer, H.E. The Effect of the Oğuzlar Walnut Extract as a Green Corrosion Inhibitor on AISI 1010 Mild Steel. *Hittite J. Sci Eng.* **2023**, *10*, 33–41. [CrossRef]
13. Cui, L.; Xiang, T.; Hu, B.; Lv, Y.; Rong, H.; Liu, D.; Zhang, S.; Guo, M.; Lv, Z.; Chen, D. Design of monolithic superhydrophobic concrete with excellent anti-corrosion and self-cleaning properties. *Colloids Surf. A Physicochem. Eng. Asp.* **2024**, *685*, 133345. [CrossRef]
14. Ma, S.Q.; Huang, H.L.; Wan, S.; Hao, L.; Liao, B.K.; Guo, X.P. Experimental, theoretical and computational studies of oxalyl dihydrazide as a novel corrosion inhibitor for mild steel in neutral saline solution. *Colloids Surf. A Physicochem. Eng. Asp.* **2024**, *699*, 134717. [CrossRef]
15. Yazdanpanah, A.; Capuzzo, N.; Khodabakhshi, M.; Dabalà, M. Evaluation of stress-assisted corrosion performance of L-PBF processed Ti6Al4V: A microcapillary electrochemical approach. *Eng. Fail. Anal.* **2024**, *166*, 108891. [CrossRef]
16. Casanova, L.; Gruarin, M.; Pedeferri, M.; Ormellese, M. A comparison between corrosion performances of titanium grade 2 and 7 in strong reducing acids. *Mater. Corros.* **2021**, *72*, 1506–1517. [CrossRef]
17. Ameer, M.; Fekry, A.; Shanab, S. Electrochemical behavior of titanium alloy in 3.5% NaCl containing natural product substances. *Int. J. Electrochem. Sci.* **2011**, *6*, 1572–1585. [CrossRef]
18. M'hiri, N.; Veys-Renaux, D.; Rocca, E.; Ioannou, I.; Boudhrioua, N.M.; Ghoul, M. Corrosion inhibition of carbon steel in acidic medium by orange peel extract and its main antioxidant compounds. *Corros. Sci.* **2016**, *102*, 55–62. [CrossRef]
19. Umoren, S.A.; Solomon, M.M.; Obot, I.B.; Suleiman, R.K. A critical review on the recent studies on plant biomaterials as corrosion inhibitors for industrial metals. *J. Ind. Eng. Chem.* **2019**, *76*, 91–115. [CrossRef]
20. Yaro, A.S.; Khadom, A.A.; Wael, R.K. Apricot juice as green corrosion inhibitor of mild steel in phosphoric acid. *Alex. Eng. J.* **2013**, *52*, 129–135. [CrossRef]
21. Abboud, Y.; Tanane, O.; Bouari, A.E.; Salghi, R.; Hammouti, B.; Chetouani, A.; Jodeh, S. Corrosion inhibition of carbon steel in hydrochloric acid solution using pomegranate leave extracts. *Corros. Eng. Sci. Technol.* **2016**, *51*, 557–565. [CrossRef]
22. Khadom, A.A.; Jassim, S.A.; Kadhim, M.M.; Ali, N.B. Influence of apricot constituents as eco-friendly corrosion inhibitor for mild steel in acidic medium: A theoretical approach. *J. Mol. Liq.* **2022**, *347*, 117984. [CrossRef]
23. Ashassi-Sorkhabi, H.; Mirzaee, S.; Rostamikia, T.; Bagheri, R. Pomegranate (*Punica granatum*) peel extract as a green corrosion inhibitor for mild steel in hydrochloric acid solution. *Int. J. Corros.* **2015**, *2015*, 197587. [CrossRef]
24. Michicotl-Meneses, M.M.; Thompson-Bonilla, M.d.R.; Reyes-López, C.A.; García-Pérez, B.E.; López-Tenorio, I.I.; Ordaz-Pichardo, C.; Jaramillo-Flores, M.E. Inflammation markers in adipose tissue and cardiovascular risk reduction by pomegranate juice in obesity induced by a hypercaloric diet in Wistar rats. *Nutrients* **2021**, *13*, 2577. [CrossRef]
25. Pontonio, E.; Montemurro, M.; Pinto, D.; Trani, A.; Mazzeo, A.; Gobbetti, M.; Rizzello, C. Lactic acid fermentation of pomegranate juice as a tool to improve antioxidant activity. *Front. Microbiol.* **2019**, *10*, 460471. [CrossRef]
26. Abboud, Y.; Chagraoui, A.; Tanane, O.; El Bouari, A.; Hannache, H. Punica granatum leave extract as green corrosion inhibitor for mild steel in Hydrochloric acid. *MATEC Web Conf.* **2013**, *5*, 04029. [CrossRef]
27. Behpour, M.; Ghoreishi, S.; Khayatkashani, M.; Soltani, N. Green approach to corrosion inhibition of mild steel in two acidic solutions by the extract of *Punica granatum* peel and main constituents. *Mater. Chem. Phys.* **2012**, *131*, 621–633. [CrossRef]
28. Chen, G.; Zhang, M.; Pang, M.; Hou, X.-Q.; Su, H.; Zhang, J. Extracts of *Punica granatum* Linne husk as green and eco-friendly corrosion inhibitors for mild steel in oil fields. *Res. Chem. Intermed.* **2013**, *39*, 3545–3552. [CrossRef]
29. Cheyad, M.S.; Salman, T.A. Characterization and study the inhibition activity of pomegranate peel extract for α-brass corrosion in H2SO4 solution. *Orient. J. Chem.* **2017**, *33*, 1241. [CrossRef]
30. Chidiebere, M.A.; Ogukwe, C.E.; Oguzie, K.L.; Eneh, C.N.; Oguzie, E.E. Corrosion inhibition and adsorption behavior of *Punica granatum* extract on mild steel in acidic environments: Experimental and theoretical studies. *Ind. Eng. Chem. Res.* **2012**, *51*, 668–677. [CrossRef]
31. Fouda, A.; Eissa, M.; Fakih, M. Pomegranate aqueous extract (PAE) as an eco-friendly inhibitor for carbon steel used in sanitation plants: Kinetics and bacteria effect. *J. Bio-Tribo-Corros.* **2019**, *5*, 5. [CrossRef]
32. Quraishi, M.; Farooqi, I.; Saini, P. Investigation of some green compounds as corrosion and scale inhibitors for cooling systems. *Corrosion* **1999**, *55*, 493–497. [CrossRef]
33. El-Etre, A.; Abdallah, M.; El-Tantawy, Z. Corrosion inhibition of some metals using lawsonia extract. *Corros. Sci.* **2005**, *47*, 385–395. [CrossRef]
34. Raja, P.B.; Sethuraman, M.G. Natural products as corrosion inhibitor for metals in corrosive media—A review. *Mater. Lett.* **2008**, *62*, 113–116. [CrossRef]

35. Pineda Hernández, D.A.; Restrepo Parra, E.; Arango Arango, P.J.; Segura Giraldo, B.; Acosta Medina, C.D. Innovative method for coating of natural corrosion inhibitor based on Artemisia vulgaris. *Materials* **2021**, *14*, 2234. [CrossRef]
36. Al-Amiery, A.A.; Isahak, W.N.R.W.; Al-Azzawi, W.K. Corrosion inhibitors: Natural and synthetic organic inhibitors. *Lubricants* **2023**, *11*, 174. [CrossRef]
37. Singh, J.; Kaur, H.P.; Verma, A.; Chahal, A.S.; Jajoria, K.; Rasane, P.; Kaur, S.; Kaur, J.; Gunjal, M.; Ercisli, S.; et al. Pomegranate Peel Phytochemistry, Pharmacological Properties, Methods of Extraction, and Its Application: A Comprehensive Review. *ACS Omega* **2023**, *8*, 35452–35469. [CrossRef]
38. Chaudhary, P.; Sharma, A.; Singh, B.; Nagpal, A.K. Bioactivities of phytochemicals present in tomato. *J. Food Sci. Technol.* **2018**, *55*, 2833–2849. [CrossRef]
39. Yu, W.-B.; Hu, Z.-Y.; Jin, J.; Yi, M.; Yan, M.; Li, Y.; Wang, H.-E.; Gao, H.-X.; Mai, L.-Q.; Hasan, T.; et al. Unprecedented and highly stable lithium storage capacity of (001) faceted nanosheet-constructed hierarchically porous TiO_2/rGO hybrid architecture for high-performance Li-ion batteries. *Natl. Sci. Rev.* **2020**, *7*, 1046–1058. [CrossRef]
40. Pang, C.L.; Lindsay, R.; Thornton, G. Structure of Clean and Adsorbate-Covered Single-Crystal Rutile TiO_2 Surfaces. *Chem. Rev.* **2013**, *113*, 3887–3948. [CrossRef]
41. Ong, W.-J.; Tan, L.-L.; Chai, S.-P.; Yong, S.-T.; Mohamed, A.R. Highly reactive {001} facets of TiO_2-based composites: Synthesis, formation mechanism and characterization. *Nanoscale* **2014**, *6*, 1946–2008. [CrossRef] [PubMed]
42. National Center for Biotechnology Information. PubChem Compound Summary for CID 932, 7-Dihydroxy-2-(4-hydroxyphenyl)chroman-4-one. 2024. Available online: https://pubchem.ncbi.nlm.nih.gov/compound/5_7-Dihydroxy-2-_4-hydroxyphenyl_chroman-4-one (accessed on 14 June 2024).
43. National Center for Biotechnology Information. PubChem Compound Summary for CID 2518, 4-Dihydroxy Cinnamic Acid. 2024. Available online: https://pubchem.ncbi.nlm.nih.gov/compound/3_4-Dihydroxy-cinnamic-acid (accessed on 14 June 2024).
44. National Center for Biotechnology Information. PubChem Compound Summary for CID 5280343, Q.R.J. Available online: https://pubchem.ncbi.nlm.nih.gov/compound/Quercetin (accessed on 4 July 2024).
45. National Center for Biotechnology Information. PubChem Compound Summary for CID 5280863, K.R.J. Available online: https://pubchem.ncbi.nlm.nih.gov/compound/Kaempferol (accessed on 4 July 2024).
46. Faraji, M.; Karimi, S.; Esmailzadeh, M.; Pezzato, L. Investigation of the Microstructure and Corrosion Behaviors of Composite Hard-Faced Layers on Al5052 Using SiC and TiB_2 in 3.5% NaCl and 0.5 M H_2SO_4 Solutions. *J. Mater. Eng. Perform.* **2023**, *32*, 9711–9724. [CrossRef]
47. Faraji, M.; Karimi, S.; Esmailzadeh, M.; Pezzato, L.; Calliari, I.; Eskandari, H. The electrochemical and microstructure effects of TiB_2 and SiC addition to AA5052/Al_2O_3 surface composite coatings in 0.5 M H_2SO_4 solution. *J. Bio-Tribo-Corros.* **2022**, *8*, 117. [CrossRef]
48. Faraji, M.; Shabestari, S.; Razavi, S. The effects of cold rolling and stabilization treatment on the improvement of microstructure, mechanical properties, and corrosion behavior of AA5083 sheet. *J. Mater. Eng. Perform.* **2023**, *32*, 2097–2108. [CrossRef]
49. Karimi, S.; Fakhar, N.; Faraji, M.; Fereshteh-Saniee, F. Simultaneous improvement of mechanical strength and corrosion resistance in aluminum alloy 5083 via severe plastic deformation. *Mater. Chem. Phys.* **2024**, *313*, 128755. [CrossRef]
50. Torres, V.; Rayol, V.; Magalhães, M.; Viana, G.; Aguiar, L.; Machado, S.; Orofino, H.; D'elia, E. Study of thioureas derivatives synthesized from a green route as corrosion inhibitors for mild steel in HCl solution. *Corros. Sci.* **2014**, *79*, 108–118. [CrossRef]
51. Zulfareen, N.; Kannan, K.; Venugopal, T.; Gnanavel, S. Synthesis, characterization and corrosion inhibition efficiency of *N*-(4-(Morpholinomethyl Carbamoyl Phenyl) Furan-2-Carboxamide for brass in HCl medium. *Arab. J. Chem.* **2016**, *9*, 121–135. [CrossRef]
52. Zheng, J.; Wang, D.; Zhang, Q.; Song, M.; Jiao, M.; Zhang, Z. Molecular Dynamics Simulation and Structure Changes of Polyester in Water and Non-Aqueous Solvents. *Materials* **2022**, *15*, 2148. [CrossRef]
53. Verma, C.; Ebenso, E.E.; Quraishi, M.; Hussain, C.M. Recent developments in sustainable corrosion inhibitors: Design, performance and industrial scale applications. *Mater. Adv.* **2021**, *2*, 3806–3850. [CrossRef]
54. Zieliński, A.; Sobieszczyk, S. Corrosion of titanium biomaterials, mechanisms, effects and modelisation. *Corros. Rev.* **2008**. [CrossRef]
55. Wang, Z.; Hu, H.; Zheng, Y.; Ke, W.; Qiao, Y. Comparison of the corrosion behavior of pure titanium and its alloys in fluoride-containing sulfuric acid. *Corros. Sci.* **2016**, *103*, 50–65. [CrossRef]
56. Rodrigues, D.C.; Urban, R.M.; Jacobs, J.J.; Gilbert, J.L. In vivo severe corrosion and hydrogen embrittlement of retrieved modular body titanium alloy hip-implants. *J. Biomed. Mater. Res. Part B Appl. Biomater. Off. J. Soc. Biomater. Jpn. Soc. Biomater. Aust. Soc. Biomater. Korean Soc. Biomater.* **2009**, *88*, 206–219. [CrossRef] [PubMed]
57. Mouritz, A.P. *Introduction to Aerospace Materials*; Elsevier: Amsterdam, The Netherlands, 2012. [CrossRef]

materials

MDPI

Article

Effect of Artificial Saliva Modification on Corrosion Resistance of Metal Oxide Coatings on Co-Cr-Mo Dental Alloy

Bożena Łosiewicz [1,*], Patrycja Osak [1], Karolina Górka-Kulikowska [2] and Joanna Maszybrocka [1]

1 Institute of Materials Engineering, Faculty of Science and Technology, University of Silesia in Katowice, 75 Pułku Piechoty 1A, 41-500 Chorzów, Poland

2 Department of Biomaterials and Experimental Dentistry, Poznan University of Medical Sciences, 60-812 Poznań, Poland

* Correspondence: bozena.losiewicz@us.edu.pl; Tel.: +48-32-3497-527

Abstract: Surface modifications not only improve the corrosion resistance of Co-Cr-Mo dental alloys (Bego Wirobond® C) but also ensure their long-term performance and reliability in dental applications. This paper describes the preparation of single-layer TiO_2-ZrO_2 sol–gel coatings on the Co-Cr-Mo dental alloy using the method of dip-coating. The TiO_2-ZrO_2 sol–gel coatings were sintered at 300 and 500 °C. SEM analysis shows that sintering at 300 °C produces a uniform, slightly dense structure without micro-cracks, while sintering at 500 °C results in a denser structure with micro-cracks due to higher stress and shrinkage. EDS confirms that sintering temperature affects the elemental composition of the coating, with higher temperatures causing the volatilization or diffusion of Ti and Zr. Roughness measurements indicate that the Ra value increases with the sintering temperature, meeting dental application requirements. Electrochemical measurements by open-circuit potential, EIS, and cyclic potentiodynamic curves demonstrate that sintering temperature and saliva composition affect corrosion resistance, with NaF and mouthwashes (Listerine Total Care Teeth Protection® and Meridol®) generally increasing charge transfer resistance and double-layer capacitance. The ceramic TiO_2-ZrO_2 coatings significantly reduce pitting corrosion susceptibility at physiological and acidic pH, with the 500 °C sintered coating showing better protective properties. These findings highlight the potential of TiO_2-ZrO_2 coatings in enhancing the performance of Co-Cr-Mo dental alloys.

Keywords: Co-Cr-Mo dental alloy; corrosion resistance; pitting corrosion; saliva; TiO_2-ZrO_2 sol–gel coating

Citation: Łosiewicz, B.; Osak, P.; Górka-Kulikowska, K.; Maszybrocka, J. Effect of Artificial Saliva Modification on Corrosion Resistance of Metal Oxide Coatings on Co-Cr-Mo Dental Alloy. *Materials* **2024**, *17*, 5166. https://doi.org/10.3390/ma17215166

Academic Editor: Xiaogang Li

Received: 27 September 2024
Revised: 18 October 2024
Accepted: 21 October 2024
Published: 23 October 2024

1. Introduction

In modern dentistry, a variety of biomaterials are required to address different dental needs, including restorations, prosthetics, and preventive treatments [1–3]. Base metal alloys are often used in dentistry in many denture infrastructure and tools [4–9]. Co-Cr-Mo alloys belonging to this group are known for their high strength and hardness, which make them suitable for applications requiring durability and resistance to wear [8,10,11]. These alloys exhibit excellent biocompatibility, which is crucial for use in dental and orthopedic applications where they come into direct contact with living tissues [12,13]. Co-Cr-Mo alloys have a high melting point, which is beneficial for their processing and fabrication into various dental and orthopedic components [10]. They have good thermal conductivity, which can be advantageous in certain applications where heat management is important [14]. Co-Cr-Mo alloys exhibit exceptional long-term corrosion resistance in environments with corrosive substances due to their chemical composition and the protective oxide layer formed by chromium [12,13,15]. However, specific conditions such as mechanical stress and exposure to certain ions can influence their long-term performance. These alloys are non-magnetic, ensuring they do not interfere with medical imaging equipment, which makes them suitable for medical implants [16]. They also offer ductility and toughness, which are important for their use in dental prosthetics and other medical devices [17,18]. Co-Cr-Mo alloys are also well suited for various manufacturing processes, including casting and machining, which allows for the precise customization of dental appliances [19,20].

The corrosion resistance of Co-Cr-Mo dental alloys is a critical factor in their use, particularly in dental applications where they are exposed to various corrosive environments [8,12,13,15,21–30]. The alloy exhibits several characteristics that enhance its corrosion resistance. This alloy contains significant amounts of Cr and Mo, which are known for their corrosion-resistant properties. The presence of these elements helps in forming a passive oxide layer on the surface of the alloy, which protects it from further corrosion [8,13,26,27]. The microstructure of Co-Cr-Mo alloys, including the distribution of carbides and the dendritic structure of the solid solution, can influence their corrosion resistance. The proper control of these microstructural features can enhance the alloy's resistance to corrosion [21]. The method of manufacturing can also affect the corrosion resistance of Co-Cr-Mo alloys [12,21–25,28]. Studies have shown that Co-Cr-Mo alloys exhibit capacitive behavior in electrochemical impedance spectroscopy (EIS) tests, indicating a high resistance to corrosion [8,26,27]. The surface roughness of Co-Cr-Mo alloys can influence their corrosion resistance. A smoother surface reduces the contact area with corrosive substances, thereby enhancing the alloy's resistance to corrosion [21,28]. These characteristics collectively contribute to the high corrosion resistance of Co-Cr-Mo dental alloys, making them a preferred material for use in dental and orthopedic applications.

However, the corrosion of Co-Cr-Mo dental alloys during long-term use can be a significant concern due to the potential release of metal ions and corrosion products that may lead to local tissue reactions or systemic effects [15]. The corrosion resistance of the alloy is influenced by several factors, including the alloy composition, surface finish, presence of crevices, and galvanic coupling with other metals [24,26,28]. The results of studies also suggest that modifications of the saliva chemical composition by the addition of fluoride, the pH change, the presence of biofilms, and the use of specific mouthwashes can significantly influence the corrosion resistance of Co-Cr-Mo dental alloys [8,29,30].

Surface modifications of Co-Cr-Mo dental alloys by surface amorphization [31], laser surface modification [32], and surface polishing used to reduce surface roughness [8] can also improve the corrosion resistance of Co-Cr-Mo dental alloys, ensuring their long-term performance and reliability in dental applications.

The sol–gel method is particularly interesting because it offers a controlled and versatile approach for the synthesis of metal oxide coatings on dental alloys, providing enhanced corrosion resistance and improved surface properties [33–37]. This method involves the conversion of a metal alkoxide precursor into a sol, which is a colloidal solution, and then into a gel, which is a solidified form of the sol. The gel is a three-dimensional network of interconnected particles that can be solidified by drying or heat treatment. It can be applied to the metallic surface by techniques such as dipping, spraying, or spin-coating [33,35–37]. The gel adheres to the substrate and can be cured to form a protective solid oxide coating [35–37]. Sol–gel coatings can be modified in many ways to achieve excellent biocompatibility and corrosion resistance, making them an attractive solution in orthodontics and medicine [35,36]. In the sol–gel coating production process, the material passes from the liquid phase (sol) to the solid phase (gel) through the controlled polymerization of particles present in the solution. The result is a thin, uniform coating that can be applied to various surfaces, such as metals used in orthodontic appliances [33,35,36].

To enhance biocompatibility, oxide precursors in the form of various inorganic metal compounds are added to sol–gel coatings [38–40]. Silicates (Ormosils) are precursors for sol–gel-generated SiO_2-based coatings, which are frequently studied for their potential in biocompatible applications [38]. Titania (TiO_2) is used to modify titanium surfaces, enhancing biocompatibility [39]. Zirconia (ZrO_2) is also used to modify titanium surfaces, contributing to biocompatibility [39]. Hybrid Organic–Inorganic (OIH) precursors are a combination of organic and inorganic components, synthesized using sol–gel methods to enhance biocompatibility and corrosion resistance [38,40]. Phosphorus-containing compounds are used to modify sol–gel coatings, aiming to improve both biocompatibility and corrosion protection behavior [40]. These precursors are chosen based on their ability to form stable oxide networks and their compatibility with biological systems, making them suitable for applications requiring enhanced biocompatibility. Recently, TiO_2-ZrO_2 sol–gel coatings were developed to enhance the corrosion resistance of stainless steel for applications in medicine [35,36].

In this work, we continue our interest in the in vitro corrosion resistance of Co-Cr-Mo dental alloys. In a previous study [8], we reported on the results of investigations on the corrosion behavior of the Co-Cr-Mo dental alloy in various artificial saliva environments and a physiological saline solution. The environments were modified with NaF and mouthwashes containing alcoholic (Meridol® by Colgate-Palmolive Company, New York, NY, USA) and aqueous (Listerine Total Care Teeth Protection® by McNeil Consumer Healthcare McNeil-PPC, Inc., Fort Washington, PA, USA) solvents to simulate different oral conditions. It was proven that the corrosion resistance of the Co-Cr-Mo dental alloy can be significantly affected by the composition of the oral environment, particularly with the addition of fluoride ions, which can enhance the protective passive layer on the alloy surface. The aim of this work is to produce for the first time TiO_2-ZrO_2 sol–gel coatings sintered at 300 and 500 °C on the commercial Co-Cr-Mo dental alloy Bego Wirobond® C by BEGO Bremer Goldschlägerei Wilh. Herbst GmbH & Co. KG, Bremen, Germany in order to increase its pitting corrosion resistance in the artificial saliva environment. This study provides new insights into the performance of ceramic TiO_2-ZrO_2 coatings on Co-Cr-Mo substrates in simulated oral environments, which is critical for ensuring the safety, effectiveness, and longevity of dental treatments and devices.

The choice of TiO_2-ZrO_2 sol–gel coatings on Co-Cr-Mo dental alloys is supported by several key factors that enhance their functionality and performance, making them an ideal choice for modern dental implant applications [35,41–48]. TiO_2-ZrO_2 coatings can improve the mechanical properties of Co-Cr-Mo alloys. ZrO_2 is known for its high hardness and toughness, which can enhance the wear resistance and overall durability of dental implants [41]. TiO_2, when doped with certain elements like silver and gallium, exhibits antibacterial properties. This is particularly beneficial in dental applications where preventing bacterial adhesion and biofilm formation is crucial for maintaining oral health and preventing infections [42]. TiO_2-ZrO_2 coatings can be formulated to have high transparency, which is important for aesthetic dental restorations. This property allows the natural color of the underlying tooth structure to show through, enhancing the cosmetic appeal of dental implants [41]. The sol–gel technique used to apply these coatings ensures a strong adhesion between the coating and the substrate, which is essential for maintaining the integrity and functionality of dental implants over time. Electrochemical pretreatment methods can further enhance the adhesion strength of these coatings on Co-Cr-Mo substrates [43]. TiO_2-ZrO_2 coatings are known for their high biocompatibility and can improve the biological performance of dental implants. These coatings can enhance the adhesion of bone to the implant surface, which is crucial for long-term stability and integration of dental implants [41]. ZrO_2 is known for its excellent corrosion resistance, which is particularly important in dental applications where the alloy is exposed to various oral fluids. The addition of TiO_2 further enhances this property, making the coating more durable and resistant to degradation in the oral environment [35]. Corrosion resistance in artificial saliva is vital for long-term dental applications as it affects both biocompatibility and mechanical integrity. It is essential for preventing adverse reactions and ensuring the longevity of dental treatments [29,44]. Understanding corrosion characteristics in artificial saliva helps in selecting materials that are less likely to corrode and more suitable for long-term clinical use [45,46]. Modifying artificial saliva with substances like fluoride can enhance corrosion resistance, supporting oral health and reducing corrosion-related issues [8,47]. This knowledge is crucial for selecting non-precious metal alloys that perform well in the complex and dynamic oral environment and for developing maintenance strategies that enhance clinical outcomes and patient well-being [48].

2. Materials and Methods

2.1. Preparation of Material Surface

The test material is the dental Co-Cr-Mo alloy Bego Wirobond® C (BEGO Bremer Goldschlägerei Wilh. Herbst GmbH & Co. KG, Bremen, Germany), meeting the requirements of the standards ISO 22674 [49] and ISO 9693-1 [50]. The Co-Cr-Mo alloy is an approved class IIa medical device, which is manufactured in accordance with the ISO 13485 standard [51]. Bego Wirobond® C by BEGO Bremer Goldschlägerei Wilh. Herbst GmbH & Co. KG, Bremen, Germany containing Co 59.8(8) wt.%, Cr 31.5(4) wt.%, and Mo 8.8(6) wt.%. is used in dental practice for the production of partial dentures, clasps, veneered crowns, long-span bridgework or bridges with small cross-sections, bars, retainers, and implant-supported superstructures [8].

Co-Cr-Mo Bego Wirobond® C by BEGO Bremer Goldschlägerei Wilh. Herbst GmbH & Co. KG, Bremen, Germany alloy samples were cut into 5 mm thick disks from 15 mm thick cylinders. The samples were ground on #320- to #2500-grit sandpaper and then polished using a colloidal SiO_2 suspension (Buehler Ltd., Lake Bluff, IL, USA). In the next step, the samples were degreased in acetone (Avantor Performance Materials Poland SA, Gliwice, Poland) and then in ultrapure water (Milli-Q Advantage A10 Water Purification System, Millipore SAS, Molsheim, France) in an ultrasonic cleaner for 20 min (VWR International, Radnor, PA, USA).

2.2. Production of TiO_2-ZrO_2 Sol–Gel Coatings

The surface of the Co-Cr-Mo Bego Wirobond® C by BEGO Bremer Goldschlägerei Wilh. Herbst GmbH & Co. KG, Bremen, Germany alloy was modified by depositing metal oxide coatings using the sol–gel method. To visualize the sol–gel process for the dip-coating technique, a flowchart with five steps, including arrows to show the progression from one step to the next, is presented in Figure 1. Each step is detailed with specific conditions or processes. The single-layer TiO_2-ZrO_2 sol–gel coatings were deposited by immersing the samples in a colloidal solution consisting of two precursors—zirconium (1 M ZrBu) and titanium (1 M TiBu). The samples were immersed in the solution prepared in this way at a temperature of 75 °C using the following procedure: immersion for 150 min, rinsing the samples with butanol—$CH_3(CH_2)_3OH$ (Sigma-Aldrich, Saint Louis, MI, USA)—immersion in precursors for 420 min and rinsing in acetic acid CH_3COOH (Sigma-Aldrich, Saint Louis, MI, USA), immersion for 420 min, then rinsing with nitric acid (Avantor Performance Materials Poland SA, Gliwice, Poland), immersion for 420 min and the formation of a sol layer, immersion for 300 min, and then the aging of the sol for 144 h in air. In the next stage, the TiO_2-ZrO_2 sol–gel coatings obtained in this way were sintered at 300 °C and 500 °C for 1 h in an argon atmosphere.

The choice of sintering temperatures of 300 and 500 °C for the TiO_2-ZrO_2 sol–gel coating is justified by several factors related to the material's phase stability and corrosion resistance [35,36,52–54]. According to the literature, TiO_2-ZrO_2 sol–gel coatings remain in the amorphous phase up to 500 °C. This temperature is crucial as it marks the onset of crystallization, which is essential for the development of the desired crystalline structure of the coating. This phase transformation is critical for the material's properties, such as mechanical strength and chemical stability [35]. At temperatures above 500 °C, the presence of oxygen vacancies and lattice distortion in ZrO_2 is confirmed [52]. These defects can influence the electrical properties of the coating. Sintering at 300 and 500 °C ensures that the coating remains in the amorphous phase, which is beneficial for applications requiring specific electrical properties, as amorphous materials do not exhibit long-range order and can have unique properties not found in crystalline materials. Sintering at 500 °C enhances the corrosion resistance of the TiO_2-ZrO_2 coating. This temperature is effective in stabilizing the coating's structure, which in turn improves its resistance to corrosive environments [36].

Annealing at 500 °C is a balance that ensures the structural integrity of the coating without causing excessive thermal damage. It is a temperature that is high enough to induce necessary phase changes and low enough to avoid the formation of unwanted phases or the decomposition of the coating components [53,54]. Additionally, it was found that sol–gel TiO_2-ZrO_2 coatings with an amorphous structure and calcined at 400 °C significantly enhance the corrosion properties of AISI 316 L stainless steel [36]. Although 400 and 500 °C are slightly higher than 300 °C, this indicates that temperatures in this range are effective for improving corrosion resistance, suggesting that 300 °C could also be beneficial for similar applications. Lower sintering temperatures like 300 °C are generally more energy-efficient and safer, as they require less energy input and pose fewer risks of thermal decomposition or phase transformations that could degrade the coating's properties.

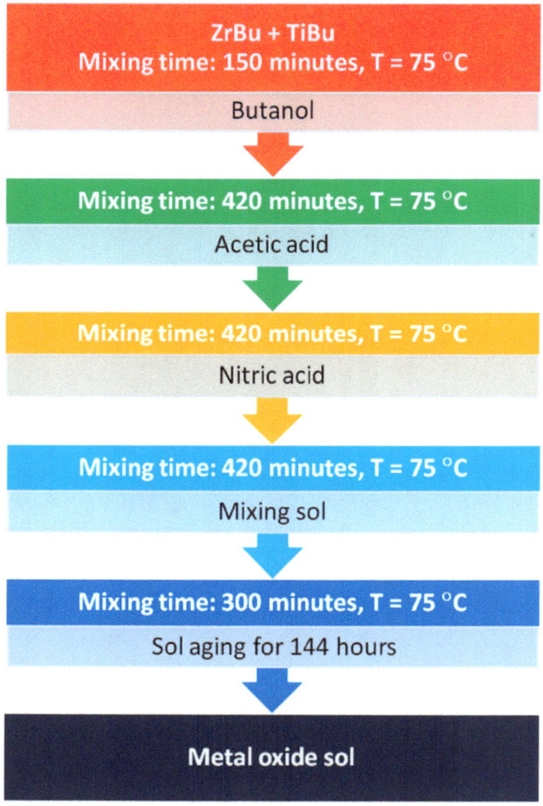

Figure 1. Flowchart of sol–gel process for dip-coating technique.

The dip-coating technique is a method used to deposit a thin film on a substrate by immersing the substrate into a coating solution and then withdrawing it at a controlled rate. This process involves five main steps: immersion, startup, deposition, drainage, and evaporation [55]. Dip-coating is a relatively inexpensive and straightforward process compared to other coating methods. It does not require complex equipment, making it accessible for various applications, including dental alloys [56]. The thickness of the deposited film can be precisely controlled by adjusting the withdrawal speed of the substrate from the coating solution. This control is crucial for applications where uniformity is essential, such as in dental alloys where consistent coating is necessary for performance and longevity [55,57]. Dip-coating can be applied to a wide range of substrates, including dental alloys, which may have complex shapes or sizes. This versatility makes it a suitable method for coating dental implants and other dental devices [57]. It can be combined with other techniques like sol–gel to enhance the properties of the coating. This combination can improve the adhesion, durability, and functionality of the coating on dental alloys [55]. While spin-coating is effective for depositing uniform thin films on flat substrates, it requires precise control over the spinning speed and the amount of coating solution. It may not be as effective for irregularly shaped dental alloys [58]. Sputtering is a high-vacuum process that can also deposit thin films with high purity and good adhesion. However, it is more expensive and requires specialized equipment, which may not be feasible for all dental applications [59].

2.3. Material Characterization

The microstructure of the TiO_2-ZrO_2 sol–gel coatings on the surface of the Co-Cr-Mo substrate was studied using a JEOL JSM-6480 (Peabody, MA, USA) scanning electron microscope (SEM), with a resolution of 3 nm and an accelerating voltage of 20 kV. The local chemical composition of the samples was analyzed using the Energy-Dispersive X-ray Spectroscopy (EDS) technique.

The porosity of TiO_2-ZrO_2 sol–gel coatings on a Co-Cr-Mo substrate, after being sintered at temperatures of 300 and 500 °C, was assessed using a computerized porosity testing method. SEM images, which depicted the surface morphology of the sintered coatings, were analyzed using the ImageJ software v1.54k in conjunction with the JPOR macro [60]. This macro is specifically designed to function with ImageJ and utilizes a thresholding technique to quantify porosity from the microscopic images obtained. To calculate porosity using ImageJ, Equation (1) was used:

$$\text{Porosity (\%)} = (\text{Total Area of Pores}/\text{Total Image Area}) \times 100. \quad (1)$$

The surface roughness of the obtained coatings and the Co-Cr-Mo substrate was determined by contact profilometry using a Mitutoyo Surftest SJ-210 profilometer (Mitutoyo Corporation, Kanagawa, Japan). The measurement was performed in accordance with the ISO 21920-3:2022-06 standard [61]. During the measurement, the speed of the measuring needle was 0.5 mm s^{-1}, and 4 mm long sections were collected. Five measurements were recorded for each sample, and the result presented in the paper is the average of the measurements.

2.4. Corrosion Resistance of TiO_2-ZrO_2 Sol–Gel Coatings

In vitro tests of the pitting corrosion resistance of the Co-Cr-Mo alloy were conducted in an artificial saliva solution prepared in accordance with the AFNOR/NF S90-701 standard [62]. First, 0.1M NaF was added to the artificial saliva solutions with pH = 7.4(1) and pH = 5.5(1) to simulate the active ingredient of toothpastes. A total of 15 mL of commercial antiseptic mouthwashes Listerine Total Care Teeth Protection® (McNeil Consumer Healthcare McNeil-PPC, Inc., Fort Washington, PA, USA), which is an alcohol-based mouthwash (21.6% *v*/*v*), and Meridol® (Colgate-Palmolive Company, New York, NY, USA), which is alcohol-free, was added to the artificial saliva solution with pH 7.4(1). Six types of artificial saliva solution were used in the electrochemical measurements, designated as follows: (1) saliva pH = 7.4, (2) saliva pH = 5.5, (3) saliva pH = 7.4 + 0.1M NaF, (4) saliva pH = 5.5 + 0.1M NaF, (5) saliva pH = 7.4 + 15 mL Listerine Total Care Teeth Protection® by McNeil Consumer Healthcare McNeil-PPC, Inc., Fort Washington, PA, USA, and (6) saliva pH = 7.4 + 15 mL Meridol® by Colgate-Palmolive Company, New York, NY, USA.

A three-electrode system was used for this study, where the working electrode (WE) was a Co-Cr-Mo alloy, the counter electrode (CE) was a Pt foil, and the reference electrode (RE) was a saturated calomel electrode (SCE) using the Autolab/PGSTAT12 (Metrohm Autolab B.V., Utrecht, The Netherlands). Electrochemical measurements were performed at 37(2) °C in solutions degassed with argon. The studies were carried out using classical methods such as the open-circuit potential method, in which the potential was recorded for a time (t) of 6000 s, and using cyclic anodic polarization curves, which were recorded in the potential range from 0.15 V more negative than the open-circuit potential (E_{OC})

value to 0.2 mV at a polarization rate of 1 mV s^{-1}. The studies using the EIS method were performed by recording the EIS spectrum at the E_{OC} potential in the frequency range (f) of 20 kHz–10 mHz with a sinusoidal signal amplitude of 10 mV. EIS data were analyzed using the EQUIVCRT program included in Frequency Response Analyser (FRA) for Windows software v4.9 [63].

3. Results and Discussion

3.1. SEM Observations of TiO_2-ZrO_2 Sol–Gel Coatings

Based on SEM images, the surface morphology of TiO_2-ZrO_2 sol–gel coatings deposited by the dip-coating method under the proposed conditions on the Co-Cr-Mo alloy surface after sintering at different temperatures was presented. Figure 2a,b show SEM images of the surface morphology of a single-layer TiO_2-ZrO_2 sol–gel coating sintered at 300 °C, while Figure 2c,d present SEM images of the surface morphology obtained for the single-layer TiO_2-ZrO_2 sol–gel coating over the Co-Cr-Mo substrate after sintering at 500 °C. Comparative SEM images for the polished surface of the Co-Cr-Mo alloy were presented in an earlier work with the EDS analysis results [8].

One can see that sintering influences the surface morphology of TiO_2-ZrO_2 sol–gel coatings. SEM images for the single-layer TiO_2-ZrO_2 sol–gel coating sintered at 300 °C show a uniform distribution of the obtained ceramic coating on the Co-Cr-Mo alloy substrate (Figure 2a,b). This coating is homogeneous with a slightly dense structure and does not flake off or have visible micro-cracks on the surface. A tight connection of the coating with the Co-Cr-Mo alloy surface is visible. The thickness of the coating was close to 2 μm. SEM images for the single-layer TiO_2-ZrO_2 sol–gel coating sintered at 500 °C also show a complete coverage of the alloy surface with the obtained homogeneous ceramic coating characterized by a more dense structure (Figure 2c,d). However, micro-cracks related to the higher sintering temperature are observed on the surface of this coating with a thickness of about 1 μm. Despite the small micro-cracks formed, the coating does not flake or peel off from the alloy surface and is without any deep cracks. The fact that the TiO_2-ZrO_2 sol–gel coatings crack during sintering primarily is due to two main reasons: stress induced by film shrinkage and the mismatch of thermal expansion [64]. During the sintering process, the sol–gel coatings undergo significant shrinkage, which can induce internal stresses. These stresses can lead to the formation of cracks as the sol–gel coating contracts and the stress exceeds the material's tensile strength. The thermal expansion coefficients of the substrate and the coating may not match, leading to differential thermal expansion during the sintering process. This mismatch can also contribute to the development of internal stresses within the coating, potentially causing it to crack due to the internal pressure mismatch between the thermal expansion coefficients of the Co-Cr-Mo substrate and the TiO_2-ZrO_2 sol–gel coating. The mean coefficient of thermal expansion at 20–500 °C for the Co-Cr-Mo alloy is $14.1 \times 10^{-6}/°C$ [65]. The presence of micro-cracks in the TiO_2-ZrO_2 sol–gel coating sintered at 500 °C could affect long-term mechanical properties and potentially lead to failure under dental application conditions. Micro-cracks can significantly reduce the structural integrity and durability of the coating, making it more susceptible to failure under mechanical stress or other environmental factors commonly encountered in dental applications [66–68]. It was found that UV irradiation on the sol–gel film could effectively eliminate the micro-cracks in all of the TiO_2, ZrO_2, and TiO_2-ZrO_2 sol–gel layers, indicating that micro-cracks can compromise the structural integrity of the coatings [66]. Bu et al. [67] reported on the characterization of ZrO_2–TiO_2 composite coatings on stainless steel obtained using a dip-coating method. It was proven that the presence of micro-cracks can affect the overall performance and durability of the coating [67]. The multifunctional applications of sol–gel-derived TiO_2-ZrO_2 composite coatings were highlighted, noting the importance of minimizing defects like micro-cracks to ensure optimal performance and longevity [68].

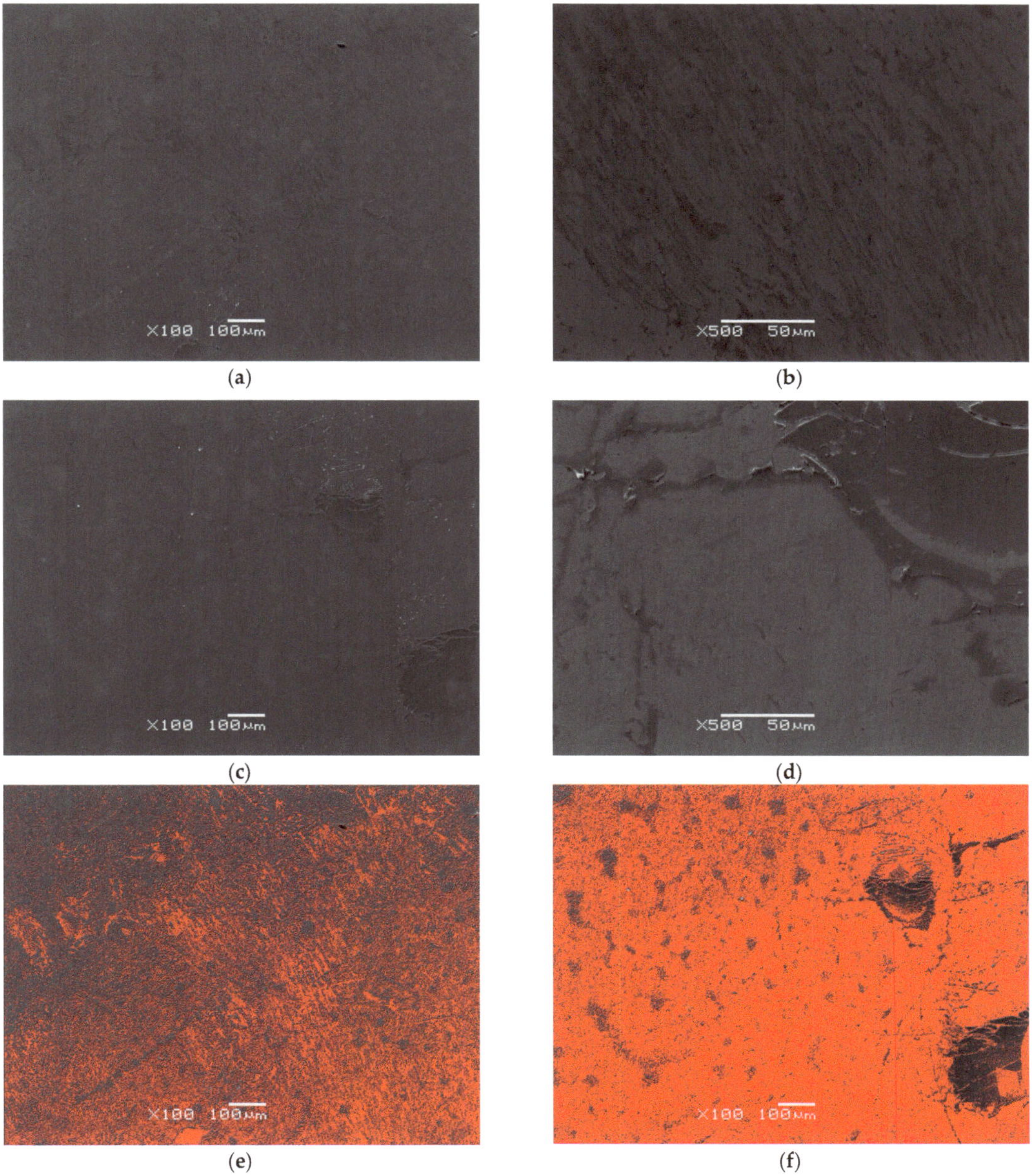

Figure 2. SEM images of the single TiO_2-ZrO_2 sol–gel coating deposited on the surface of a Co-Cr-Mo dental alloy after sintering at the following temperatures: (**a**,**b**) 300 °C; (**c**,**d**) 500 °C; (**e**) 300 °C after a computer analysis of porosity; and (**f**) 500 °C after a computer analysis of porosity. Red indicates the color of the applied filter in the porosity measurement using ImageJ software.

Figure 2e,f include SEM images subjected to a computer analysis of porosity which illustrate the TiO_2-ZrO_2 sol–gel coatings applied to the Co-Cr-Mo dental alloy after sintering at different temperatures of 300 and 500 °C, respectively. The analysis reveals that the porosity of the coatings is 28.84(35)% at 300 °C and increases to 51.57(71)% at 500 °C. The increase in the porosity of the TiO_2-ZrO_2 sol–gel coating with increasing sintering temperature can be attributed to the phase transformation and crystallization processes that occur as the temperature increases. As the temperature reaches 500 °C, the amorphous phase starts to crystallize, which can lead to an increase in porosity due to the rearrangement of atoms and the formation of new crystal structures [35]. This phase transformation and crystallization process can create more void spaces within the material, thus increasing its porosity. It should be added that the porosity of bone tissue varies widely among individuals, with values typically falling between 50% and 90%, reflecting the diverse states of bone health and structure across different people [69]. The TiO_2-ZrO_2 sol–gel coating applied to a Co-Cr-Mo surface and sintered at 500 °C was found to create a biomimetic structure with porosity levels similar to those of human bone tissue, suggesting it could effectively mimic the porous structure of bone in a living organism and potentially promote bone growth.

It should be noted that during sintering, the structure of TiO_2-ZrO_2 sol–gel coatings undergoes several changes that affect their properties. The primary changes include densification, crystallization, grain growth, and phase changes [70–73]. The coatings become denser as the sintering temperature increases. This densification process reduces the porosity of the coatings, which can enhance their mechanical and chemical properties [70,71]. Additionally, during sintering, the precursor used in the initial stage of applying TiO_2-ZrO_2 sol–gel coatings with an amorphous structure undergoes crystallization processes, and the structure changes from amorphous to crystalline, further affecting the surface morphology [70–73]. This change in structure can affect the optical and electrical properties of the coatings [70,71]. With the increasing sintering temperature, the grains within the coating grow larger. This grain growth can influence the mechanical properties and surface roughness of the coatings. Depending on the sintering conditions, the TiO_2 and ZrO_2 phases within the composite may undergo phase changes, which can further affect the overall properties of the coating. These structural changes are crucial for tailoring the properties of TiO_2-ZrO_2 sol–gel coatings to meet specific application requirements [70–73].

3.2. EDS Microanalysis of TiO_2-ZrO_2 Sol–Gel Coatings

EDS microanalysis was used to study the effect of sintering temperature on the elemental composition of the TiO_2-ZrO_2 sol–gel coating on the Co-Cr-Mo alloy. The EDS spectrum, as indicated by Figure 3, was recorded from a specific micro-region of the coating surface. This spectrum shows the distribution of elements within the selected area, providing qualitative and quantitative information about the material's composition. The EDS analysis confirms that Zr and Ti elements are present on the coating surface. Because of the low thickness of the TiO_2-ZrO_2 sol–gel coating, the chemical elements originating from the substrate (Co, Cr, and Mo) are also visible in the exemplary EDS spectra.

Table 1 quantifies the elemental composition, expressed in atomic percentage (at.%), for each sintering temperature of the TiO_2-ZrO_2 sol–gel coating on the Co-Cr-Mo substrate. Local measurements of the chemical composition of the tested materials using the EDS microanalytical method were given as mean values determined from measurements in five micro-areas. For the coating sintered at 300 °C, the elemental content is as follows: Co—33.81(29) at.%, Cr—26.33(71) at.%, Mo—3.55(20) at.%, Ti—20.71(39) at.%, and Zr—15.61(33) at.%. When sintered at 500 °C, the composition changes to the following: Co—36.42(40) at.%, Cr—29.53(69) at.%, Mo—5.64(21) at.%, Ti—18.21(51) at.%, and Zr—10.20(11) at.%. The numbers in parentheses represent the standard deviation associated with the measurement. The comparative EDS spectrum for the metallic substrate without any coating closely matched the expected composition of the Co-Cr-Mo alloy as specified by the manufacturer [8].

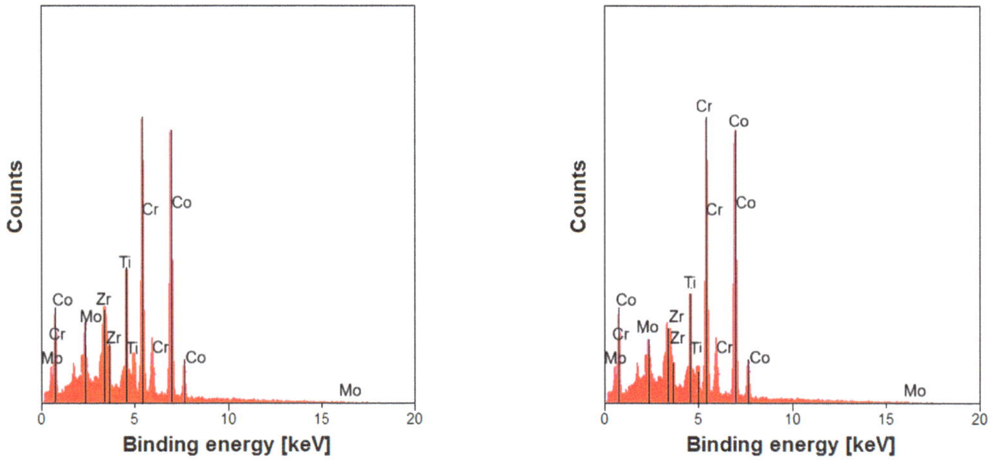

(**a**) (**b**)

Figure 3. The EDS spectrum recorded from selected micro-region for a Co-Cr-Mo alloy with a TiO_2-ZrO_2 sol–gel coating sintered at (**a**) 300 °C and (**b**) 500 °C.

Table 1. Elemental content on surface of Co-Cr-Mo alloy with TiO_2-ZrO_2 sol–gel coating sintered at 300 °C and 500 °C.

Sintering Temperature [°C]	Element Content [at.%]				
	Co	Cr	Mo	Ti	Zr
300	33.81(29)	26.33(71)	3.55(20)	20.71(39)	15.61(33)
500	36.42(40)	29.53(69)	5.64(21)	18.21(51)	10.20(11)

From the data provided, it is evident that the sintering temperature affects the elemental distribution on the surface of the TiO_2-ZrO_2 sol–gel coating. A higher sintering temperature causes less titanium and zirconium to be incorporated into TiO_2-ZrO_2 sol–gel coatings obtained from precursors present in the sol–gel solution. During the sintering process of TiO_2-ZrO_2 sol–gel coatings at high temperatures, the diffusion or volatilization of Ti and Zr can have significant implications on the final properties of the coating [67,70,74]. At high temperatures, the elements within the coating can undergo diffusion, which can affect the homogeneity and properties of the coating. Additionally, the volatilization of certain components can occur, leading to changes in the composition and structure of the coating. The SEM micrographs of ZrO_2–TiO_2 films at different sintering temperatures show that the microstructure of the coating changes with increasing temperature. This indicates that the diffusion of Ti and Zr atoms within the coating can lead to changes in the grain size and morphology, which are crucial for the mechanical and optical properties of the coating [67]. The high-temperature oxidation resistance of the sol–gel-derived $ZrO_2/(Al_2O_3$–$Y_2O_3)$ composite coating on a titanium alloy suggests that the stability of the coating at high temperatures is critical. The diffusion of Ti and Zr can affect the oxidation resistance by altering the protective oxide layer formed on the coating surface [74]. The corrosion behavior of amorphous sol–gel TiO_2–ZrO_2 nano-coatings annealed at 400 °C shows that the coating significantly improves corrosion resistance. This improvement is likely due to the formation of a stable oxide layer on the coating surface, which is influenced by the diffusion and volatilization of Ti and Zr during sintering [36]. The structural and surface chemical properties of sol–gel-derived TiO_2–ZrO_2 oxides with different Zr/Ti ratios indicate that the ratio of Ti to Zr can significantly affect the properties of the coating. Changes in the Zr/Ti ratio during sintering due to diffusion or volatilization can lead to variations in the coating's properties [70]. The increase in the Ti/Zr ratio from 1.33 at 300 °C to 1.78 at 500 °C in the TiO_2-ZrO_2 sol–gel coating indicates that as the sintering temperature increases, there is a greater segregation of titanium into the coating compared to zirconium (Table 1). This suggests that at higher temperatures, the TiO_2 component of the coating becomes more dominant, possibly due to differences in the thermal expansion coefficients or solubility of TiO_2 and ZrO_2, which can lead to the preferential incorporation of titanium into the crystalline structure or the formation of a more Ti-rich phase [75].

The depletion or redistribution of Ti and Zr in a TiO_2-ZrO_2 sol–gel coating sintered at high temperatures can significantly impact its functional properties, particularly corrosion resistance and biocompatibility [36,76–79]. The presence of TiO_2 and ZrO_2 in the coating enhances corrosion resistance due to their chemical stability and ability to form protective oxide layers on the surface of the substrate. The depletion or redistribution of Ti and Zr could lead to a reduction in the thickness or uniformity of these protective layers, thereby decreasing the overall corrosion resistance of the coating [36,76]. The Ti/Zr molar ratio plays a crucial role in determining the structural and chemical properties of the TiO_2-ZrO_2 composite. A change in this ratio due to depletion or redistribution could alter the phase composition and crystallinity of the coating, potentially affecting its corrosion resistance [77]. TiO_2 and ZrO_2 are known for their excellent biocompatibility, which is partly due to their inert nature and ability to form a stable oxide layer that minimizes direct contact with biological fluids. The depletion or redistribution of Ti and Zr could disrupt this protective layer, potentially leading to increased biological reactivity and reduced biocompatibility [78]. The equimolar TiO_2-ZrO_2 nanocomposite, with a predominant tetragonal phase for zirconia, has been shown to exhibit enhanced biocompatibility. Any alteration in the Ti/Zr ratio or phase composition due to depletion or redistribution could compromise these beneficial properties [79].

3.3. *Surface Roughness Study*

The geometric structure of the surface is an important factor determining the quality of a given dental appliance surface, which is the first to come into contact with the host tissues [80,81]. It can vary significantly depending on the specific design and surface modifications [82]. The quality of the surface of the material dedicated to the implant or prosthesis significantly affects the functional properties and durability of the element. Ensuring the appropriate roughness of the dental appliance surface limits the occurrence of health-threatening events during the long-term contact of the material with the aggressive and variable environment of the living organism. In practice, the most commonly used parameter determining the surface roughness of dental materials is Ra, which defines the arithmetical mean deviation of the assessed profile [83]. The Ra parameter provides stable results, not significantly influenced by scratches, contamination, or measurement noise, and is widely recognized for its resilience to these factors, ensuring consistent and reliable measurements in surface roughness analysis [84,85]. From literature data, it is evident that lower Ra values are generally preferred for dental materials to reduce bacterial adhesion and biofilm formation. This is because smoother surfaces are less conducive to bacterial attachment and biofilm accumulation, which can lead to dental caries and other oral health issues [86–88]. Additionally, a low Ra value limits the interaction of prosthetic elements with each other, which leads to longer failure-free use.

The Ra parameter refers to specific profile features and is a measure of surface roughness. Ra belongs to amplitude average parameters representing the arithmetic mean of the absolute ordinate Z(x) within the sampling length (l), as shown in Figure 4, and can be calculated based on Equation (2) [83]:

$$Ra_i = \frac{1}{l}\int_0^l |Z(x)|dx. \tag{2}$$

Since the measured deviations Z_i from the x axis are of the order of a fraction of a millimeter, the Ra parameter is expressed in μm in practice. This parameter denotes the maximum allowable height of a rectangle whose surface area is equal to the surface area formed by the area between the graph of the Z(x) function and the x axis, where the x axis must pass through the Z(x) function so that the surface area of the Z(x) function located below the x axis is equal to the surface area located above the x axis.

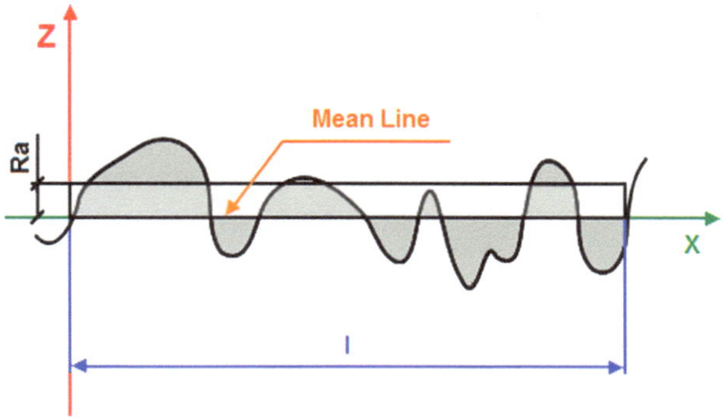

Figure 4. Graphical interpretation of surface roughness parameter Ra.

The Ra measurement for the Co-Cr-Mo alloy was performed in the absence and in the presence of the TiO_2-ZrO_2 sol–gel coating sintered at 300 and 500 °C (Figure 5).

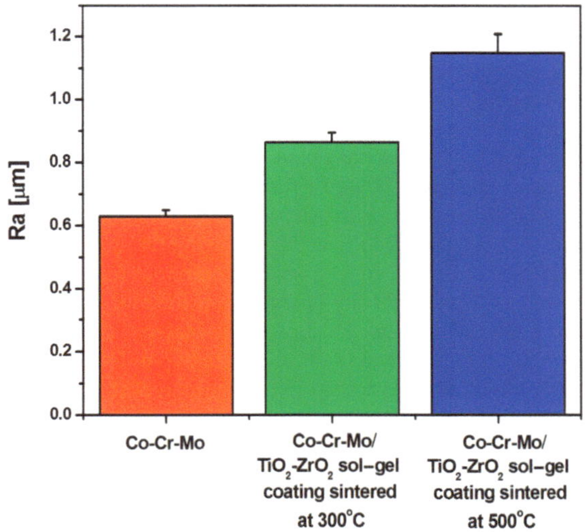

Figure 5. The surface roughness parameter Ra for the Co-Cr-Mo alloy in the absence and in the presence of the TiO_2-ZrO_2 sol–gel coating sintered at 300 °C and 500 °C.

The Ra value for the Co-Cr-Mo alloy in its initial state is 0.630(20) µm, which indicates a smooth surface [89]. The Ra value for the TiO_2-ZrO_2 sol–gel coating sintered at 300 °C and 500 °C is 0.836(30) µm and 1.148(60) µm, respectively. The obtained results indicate that the sintered TiO_2-ZrO_2 coatings obtained on the Co-Cr-Mo alloy meet the requirements for metal alloys in dentistry regarding biocompatibility and the lowest possible Ra parameter value to limit the formation of bacterial biofilm [86]. The lower the surface roughness, the lower the risk of adhesion of *Streptococcus mutans* bacteria to the surface of prosthetic materials, which contribute to the formation of carious lesions [90]. A smooth surface also causes a change in the surface charge of solutions present in the oral cavity, as well as increased hydrophobicity [91].

The increase in surface roughness (Ra value) with the sintering temperature of TiO_2-ZrO_2 sol–gel coatings for dental implants has significant implications for osseointegration, bacterial adhesion, and patient comfort [92–97]. Surface roughness plays a crucial role in the initial bone apposition and subsequent osseointegration of dental implants. Studies have shown that moderately rough surfaces can enhance the mechanical interlocking between the implant and the surrounding bone, promoting better bone growth into surface irregularities [92]. An optimal roughness range is essential for maximizing the surface area available for bone contact without compromising the mechanical stability of the implant–bone interface. This range is typically considered to be between 1 and 2 μm for Ra [92]. Increased surface roughness can also affect bacterial adhesion. Smoother surfaces generally exhibit less bacterial adhesion compared to rougher surfaces, as the latter can provide more niches for bacteria to colonize [93]. Excessive roughness can therefore increase the risk of peri-implantitis, a common complication associated with dental implants [94]. The roughness of the implant surface can also influence patient comfort, particularly if the implant is in contact with soft tissues. A very rough surface may lead to soft tissue irritation and discomfort [95]. In areas where the implant is in close proximity to mucosal tissues, a smoother surface may be preferable to ensure patient comfort [96]. The obtained results indicate that the Ra value for the TiO_2-ZrO_2 sol–gel coating sintered at 500 °C aligns with the optimal range for dental applications (Figure 5).

3.4. In Vitro Electrochemical Tests Using the Open-Circuit Potential Method

The open-circuit potential method is a crucial technique used in in vitro corrosion resistance tests of dental materials [8,12,13,15,21,22,24,26–29,31,32,98–100]. This method is also known as open-circuit voltage, zero-current potential, corrosion potential, equilibrium potential, or rest potential [98]. It is often used to find the E_{OC} of a given electrochemical system, from which other experiments are based. The open-circuit potential method involves measuring the potential difference between a dental material as the WE and a RE in an electrolyte solution under open-circuit conditions. This provides information about the tendency of a material to corrode in the oral environment, as it reflects the approximate corrosion potential (E_{cor}) of the material. Before starting the E_{OC} measurement, it is important to ensure that the surface of the dental material is stable and has reached a steady state. This is typically achieved by allowing the system to be undisturbed until the potential readings stabilize. The open-circuit potential method is often used in conjunction with other electrochemical techniques such as EIS and potentiodynamic polarization to provide a comprehensive understanding of the material's corrosion resistance [5,6,8,99,100].

Experimental data illustrating the dependence of E_{OC} on t for electrodes made of a Co-Cr-Mo alloy with a sol–gel layer sintered at 300 and 500 °C obtained in an artificial saliva solution without and with additives in the form of NaF and Listerine Total Care Teeth Protection® by McNeil Consumer Healthcare McNeil-PPC, Inc., Fort Washington, PA, USA and Meridol® by Colgate-Palmolive Company, New York, NY, USA mouthwashes are presented in Figure 6. Based on the obtained E_{OC} = f(t) curves in the applied corrosive environments, the tendency of each type of tested electrodes to corrode in the environment of unmodified and modified artificial saliva was determined.

The E_{OC} values provided in Table 2 indicate the corrosion behavior of a Co-Cr-Mo electrode with TiO_2-ZrO_2 sol–gel coatings sintered at different temperatures (300 and 500 °C) in various artificial saliva solutions at 37 °C after 6000 s of immersion in the electrolyte. E_{OC} is a measure of the electrode's tendency to corrode in a given environment, with more negative values generally indicating a higher corrosion rate. From the data obtained, several trends can be observed. E_{OC} becomes more negative (indicating higher corrosion rates) when the pH of the saliva decreases from 7.4 to 5.5 for the unmodified saliva solutions—comparing electrolytes (1) and (2). The addition of 0.1M NaF to the saliva solutions significantly increases the corrosion rate (more negative E_{OC} values) for both sintering temperatures—electrolytes (3) and (4). The presence of Listerine Total Care Teeth Protection® by McNeil Consumer Healthcare McNeil-PPC, Inc., Fort Washington, PA, USA and Meridol® by Colgate-Palmolive Company, New York, NY, USA in the saliva solutions at pH 7.4 also increases the corrosion rate, with Listerine Total Care Teeth Protection® by McNeil Consumer Healthcare McNeil-PPC, Inc., Fort Washington, PA, USA showing a stronger effect than Meridol® by Colgate-Palmolive Company, New York, NY, USA for the 500 °C sintered coating—electrolytes (5) and (6). Overall, the coatings sintered at 500 °C tend to have more negative E_{OC} values, suggesting that they may be more susceptible to corrosion in these environments compared to those sintered at 300 °C. These results suggest that the corrosion behavior of the Co-Cr-Mo electrode with TiO_2-ZrO_2 sol–gel coatings is influenced by both the sintering temperature and the composition of the artificial saliva solution. The obtained results also indicate that the surface modification by applying TiO_2-ZrO_2 sol–gel coatings sintered at both 300 and 500 °C increases the corrosion resistance of the dental Co-Cr-Mo alloy in artificial saliva at physiological and acidic pH, which simulates inflammation in the body [8]. The E_{OC} values obtained after 6000 s of stabilization were treated as approximate E_{cor} values in further electrochemical measurements.

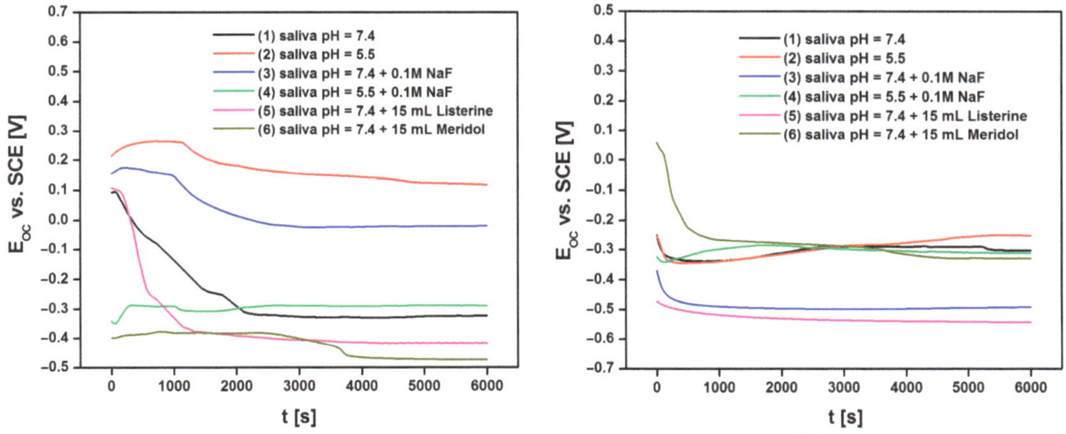

Figure 6. E_{OC}—t dependence in unmodified and modified artificial saliva solutions at 37 °C for Co-Cr-Mo electrode with TiO_2-ZrO_2 sol–gel coating sintered at (**a**) 300 °C and (**b**) 500 °C.

Table 2. E_{OC} for the Co-Cr-Mo electrode with the TiO_2-ZrO_2 sol–gel coating sintered at 300 and 500 °C in unmodified and modified artificial saliva solutions at 37 °C after 6000 s.

Electrolyte Type	E_{OC} [V]	
	TiO_2-ZrO_2 Sol–Gel Coating Sintered at 300 °C	**TiO_2-ZrO_2 Sol–Gel Coating Sintered at 500 °C**
(1) Saliva with pH = 7.4	−0.207(17)	−0.256(13)
(2) Saliva with pH = 5.5	0.023(10)	−0.242(19)
(3) Saliva with pH = 7.4 + 0.1M NaF	−0.020(14)	−0.492(27)
(4) Saliva with pH = 5.5 + 0.1M NaF	−0.291(16)	−0.344(24)
(5) Saliva with pH = 7.4 + 15 mL Listerine Total Care Teeth Protection® by McNeil Consumer Healthcare McNeil-PPC, Inc., Fort Washington, PA, USA	−0.417(23)	−0.543(33)
(6) Saliva with pH = 7.4 + 15 mL Meridol® by Colgate-Palmolive Company, New York, NY, USA	−0.474(23)	−0.328(25)

Various dental materials, including Co-Cr-Mo alloys and titanium-based materials, have been tested using the open-circuit potential method to assess their corrosion resistance in artificial saliva and other simulated oral environments [8,12,13,15,21,22,24,26–29,31,32,99,100]. It was found that E_{OC} is influenced by changes in the surrounding environment, such as temperature, pH, and the presence of different ions. This variability can make it challenging to compare results across different studies or to predict the material's behavior in real-world conditions. While the open-circuit potential method provides information about the corrosion potential of a material, it does not offer detailed insights into the mechanisms of corrosion or the material's resistance to localized corrosion, which are also critical for assessing the material's long-term performance [15,100]. These limitations highlight the need for the careful control of experimental conditions and the use of complementary techniques to fully understand the corrosion behavior of dental materials.

3.5. In Vitro Electrochemical Tests Using Electrochemical Impedance Spectroscopy

Impedance tests were conducted to understand the corrosion mechanisms and kinetics at the electrode–electrolyte interface when the tested dental materials were exposed to unmodified and modified saliva solutions [8,26,27,29,99,100]. The electrode potential was stabilized until a steady E_{OC} was reached, as EIS requires a linear and steady-state system. The electrochemical corrosion resistance of metallic electrodes with a passive layer should be tested by the EIS method in the low-frequency range. This frequency range is particularly sensitive to the properties of the passive layer and the processes occurring at the electrode–electrolyte interface. The low-frequency range allows for the evaluation of the impedance behavior that is influenced by the capacitive properties of the passive layer, the charge transfer resistance across the interface, and the presence of any corrosion products or defects in the passive layer.

The EIS analysis involved the use of Bode diagrams to compare experimental and simulated data for Co-Cr-Mo electrodes coated with TiO_2-ZrO_2 sol–gel coatings sintered at 300 °C (Figure 7) and 500 °C (Figure 8) obtained under potentiostatic control at E_{OC} in artificial saliva solution at different pH levels and in the presence of additives such as NaF, Listerine Total Care Teeth Protection® by McNeil Consumer Healthcare McNeil-PPC, Inc., Fort Washington, PA, USA, and Meridol® by Colgate-Palmolive Company, New York, NY, USA.

The decreasing value of the logarithm of the impedance modulus at the lowest frequency in Figures 7a and 8a is indicative of a decline in the corrosion resistance of the material, potentially due to changes in the passive layer or the electrochemical environment at the interface. The impedance modulus is a measure of how much the system opposes the flow of current, and it is influenced by the properties of the material and the electrolyte solution [8,26,27,29,99,100]. The impedance modulus at low frequencies is particularly sensitive to the capacitive properties of the passive layer formed by the coating and the charge transfer resistance at the electrode–electrolyte interface. When the logarithm of the impedance modulus decreases at low frequencies, it suggests that the passive layer may be less effective in protecting the underlying material from corrosion. This could be due to the presence of defects in the coating, the dissolution of the coating material, or the formation of less protective corrosion products. The decrease in the impedance modulus can also be related to a reduction in the charge transfer resistance, which is a measure of the ease with which ions can be exchanged at the interface, indicating a more active corrosion process.

In the case of the Co-Cr-Mo electrode with a TiO_2-ZrO_2 sol–gel coating sintered at 300 °C, the highest value of the logarithm of the impedance modulus at the lowest tested frequency f = 10 mHz equal to 5.32(79) Ω cm^2 was recorded in the artificial saliva environment with neutral pH and NaF (3) addition, which indicates the highest corrosion resistance among all the electrolytes used from (1) to (6), as shown in Figure 6a. The lowest log$|Z|$ value of 3.82(56) Ω cm^2 for the same electrode type was obtained in artificial saliva solution at pH = 7.4 modified with Meridol® by Colgate-Palmolive Company, New York, NY, USA mouthwash (6), which proves the strong influence of the additive in the form of ready-to-use Meridol by Colgate-Palmolive Company, New York, NY, USA mouthwash on the corrosion resistance of the surface-modified Co-Cr-Mo electrode in artificial saliva.

The increase in the sintering temperature of the TiO_2-ZrO_2 sol–gel coating up to 500 °C affects the corrosion resistance of the Co-Cr-Mo electrode (Figure 8a). In the case of the Co-Cr-Mo electrode with a TiO_2-ZrO_2 sol–gel coating sintered at 500 °C, the highest value of $\log|Z|_{f = 10\ mHz}$ equal to 5.45(71) Ω cm^2 was recorded in the environment of unmodified artificial saliva at pH = 7.4 (1), which indicates the highest corrosion resistance determined in electrolytes from (1) to (6). The lowest corrosion resistance for the same type of electrode was found based on the lowest $\log|Z|$ value of 3.72(47) Ω cm^2 in artificial saliva solution at pH = 7.4 with the addition of 0.1M NaF (3). It should be emphasized that the $\log|Z|$ values obtained at the lowest frequency f = 10 mHz indicate an increase in the corrosion resistance of the Co-Cr-Mo electrode covered with TiO_2-ZrO_2 sol–gel coatings sintered both at 300 and 500 °C in comparison with the $\log|Z|_{f = 10\ mHz}$ values obtained for the Co-Cr-Mo electrode without surface modification in artificial saliva at pHs of 7.4 and 5.5 [8]. The slope of all curves showing the relationship $\log|Z| = \log(f)$ in Figures 7a and 8a in the mid-frequency range is close to —1.

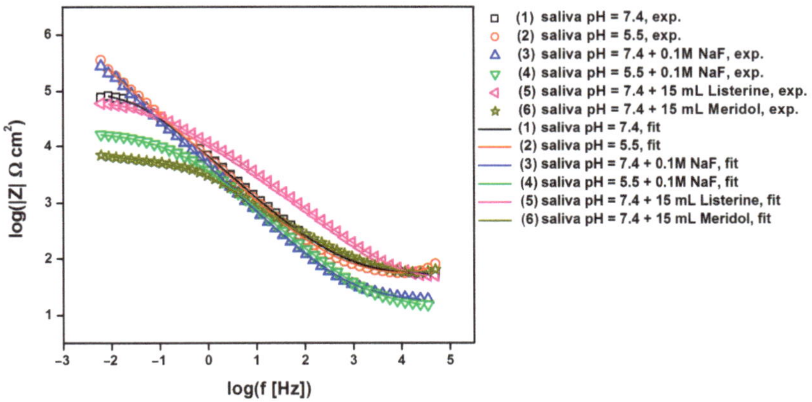

(a)

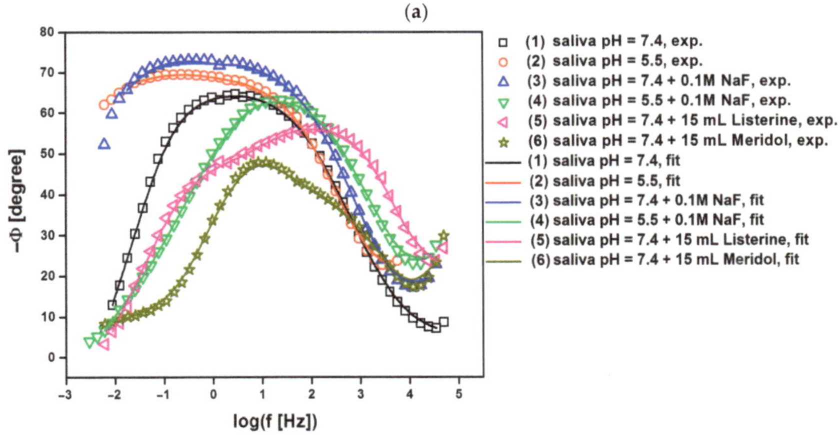

(b)

Figure 7. Experimental (symbols) and simulated (lines) Bode diagrams for a Co-Cr-Mo electrode coated with a TiO_2-ZrO_2 sol–gel coating sintered at 300 °C in artificial saliva solution before and after modification at 37 °C: (**a**) impedance module; (**b**) phase angle shift.

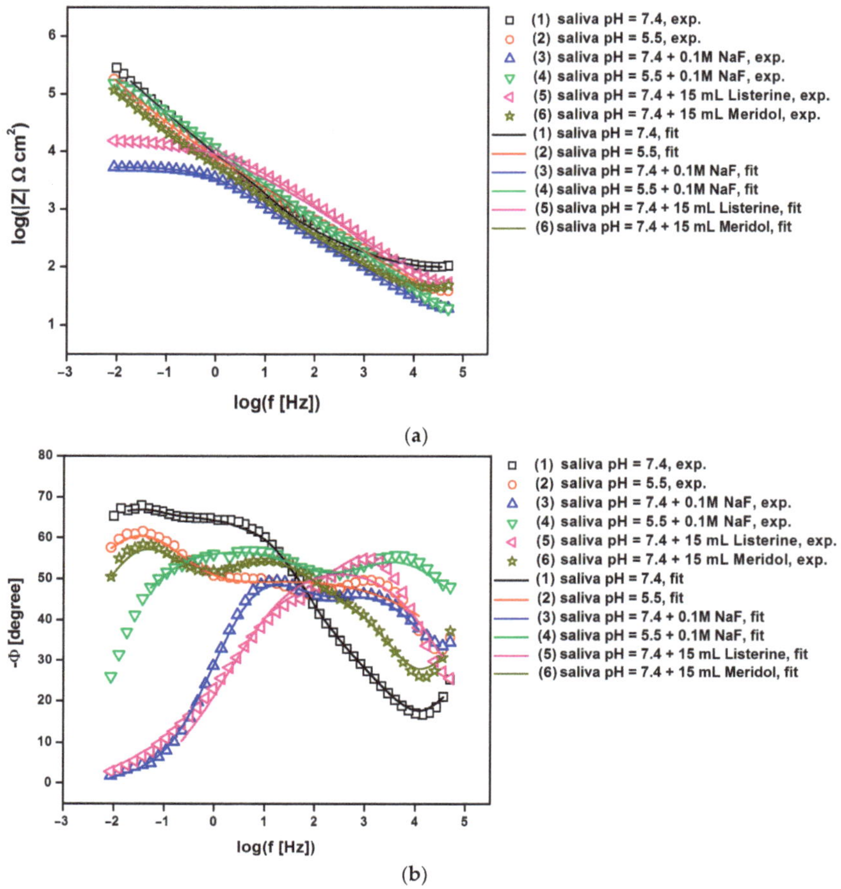

Figure 8. Experimental (symbols) and simulated (lines) Bode diagrams for a Co-Cr-Mo electrode coated with a TiO_2-ZrO_2 sol–gel coating sintered at 500 °C in artificial saliva solution before and after modification at 37 °C: (**a**) impedance module; (**b**) phase angle shift.

The shape of the EIS spectra shown in Figures 7b and 8b is characteristic of passivated metallic materials undergoing pitting corrosion in a biological environment containing aggressive chloride ions [8,26,27,29,99,100]. In the case of the Co-Cr-Mo electrode covered with TiO_2-ZrO_2 sol–gel coatings sintered at 300 °C, one time constant is visible in the circuit in solutions (1) to (6) in Figure 7b. In Figure 8b, a similar impedance behavior is seen for the Co-Cr-Mo electrode coated with a TiO_2-ZrO_2 sol–gel coating sintered at 500 °C in a neutral pH artificial saliva solution (1) and with the addition of ready-to-use Listerine Total Care Teeth Protection® by McNeil Consumer Healthcare McNeil-PPC, Inc., Fort Washington, PA, USA mouthwash (5). The presence of only one time constant in the circuit indicates that there is a single frequency response in the system, suggesting that the corrosion process involves a single dominant mechanism or process under these conditions [8,26,29,99,100]. In the remaining tested corrosive environments (2), (3), (4), and (6), two time constants are observed in the circuit obtained for the Co-Cr-Mo electrode covered with TiO_2-ZrO_2 sol–gel coatings sintered at 500 °C (Figure 8b). The obtained Bode plots show two distinct regions of frequency response, suggesting that the corrosion process involves two distinct mechanisms or processes under these conditions [100].

To understand the detailed mechanism and kinetics of corrosion processes for the Co-Cr-Mo electrode covered with TiO_2-ZrO_2 sol–gel coatings sintered at 300 and 500 °C in the artificial saliva environment before and after the modification of its chemical composition, the obtained EIS experimental data were approximated using appropriate electrical equivalent circuits (Figure 9).

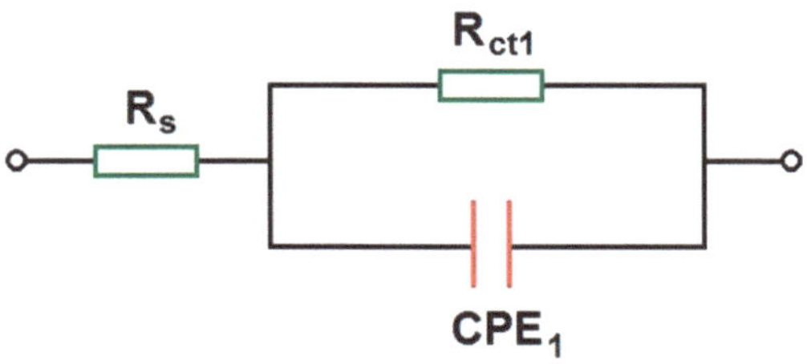

(a)

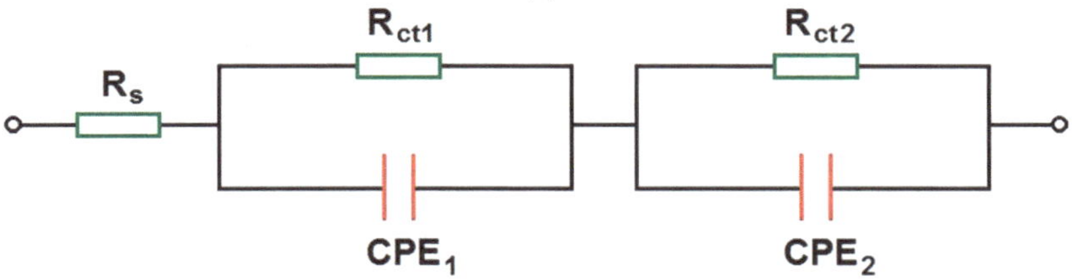

(b)

Figure 9. An equivalent electrical circuit for the Co-Cr-Mo electrode | TiO_2-ZrO_2 sol–gel coating | saliva solution in the pitting corrosion process: (**a**) 1CPE model and (**b**) 2CPE model.

The simplest electrical equivalent circuit in the form of the modified Randles circuit was used to model the experimental impedance spectra with one time constant. This circuit consists of a solution resistance (R_s) in series with a parallel combination of double-layer capacitance (C_{dl}) and charge transfer resistance (R_{ct}). As the capacitance often differs from the ideal due to the roughness of the electrode surface, C_{dl} was represented by a constant phase element (CPE). The CPE's impedance is defined by Equation (3):

$$\hat{Z}_{CPE} = \frac{1}{T(j\omega)^{\phi}} \quad (3)$$

where T is the capacitance parameter, dependent on the electrode potential, and ϕ is related to the deviation from purely capacitive behavior [63]. When ϕ equals 1, the CPE behaves like a pure capacitor, and T equals C_{dl}. Other values of f represent different behaviors, such as Warburg impedance, resistance, or inductance. This so-called 1CPE model includes one semicircle on the complex plane plot, characterized by R_s, T_1, ϕ_1, and R_{ct1} (Figure 9a). This model is commonly used to assess the corrosion resistance of passivated metallic electrodes in simulated body fluids [8,26,29,99,100].

For the experimental impedance spectra with two time constants, a more complex model was used to study the interfacial properties of the electrochemical system (Figure 9b). This so-called 2CPE model consists of two Randles circuits in series, producing two semicircles on the complex plane plot. The high-frequency semicircle is described by R_s, T_1, ϕ_1, and R_{ct1}. The low-frequency semicircle described by T_2, ϕ_2, and R_{ct2}. The use of two time constants in the 2CPE model suggests a more complex electrochemical behavior, possibly due to the presence of multiple layers or processes at the electrode surface [100]. The resistance values (R_s, R_{ct1}, and R_{ct2}) are generally higher in the 2CPE model as compared with those in the 1CPE model, indicating a more detailed representation of the corrosion protection offered by the coating.

The average double-layer capacitance ($\overline{C}_{dl}$) of the corrosion process was determined from the Brug relationship described by Equation (4) [101]:

$$T = \overline{C}_{dl}^{\phi}(R_s^{-1} + R_{ct}^{-1})^{1-\phi} \quad (4)$$

Bode diagrams in Figures 7 and 8 confirm the very good fit of the experimental EIS data to the theoretical data. Table 3 presents the equivalent electrical circuit parameters obtained from EIS experiments for Co-Cr-Mo electrodes coated with TiO_2-ZrO_2 sol–gel coatings sintered at 300 °C and exposed to various artificial saliva environments at 37 °C. The comparative fitting parameters for the Co-Cr-Mo electrodes coated with TiO_2-ZrO_2 sol–gel coatings sintered at 500 °C are presented in Tables 4 and 5. The detailed error in determining the fitting parameters was below 15%.

Table 3. Equivalent electrical circuit parameters obtained by fitting EIS experimental data using the 1CPE model for a Co-Cr-Mo electrode with a TiO_2-ZrO_2 sol–gel coating sintered at 300 °C in an unmodified and modified artificial saliva environment at 37 °C.

Artificial Saliva Solution	R_s [Ω cm^2]	T_1 [F cm^{-2} s$^{\phi-1}$]	ϕ_1	R_{ct1} [Ω cm^2]	$\overline{C}_{dl}$ [F cm^{-2}]
(1) pH = 7.4	55(8)	3.66(55)·10^{-5}	0.74(11)	1.01(15)·10^5	3.88·10^{-3}
(2) pH = 5.5	58(8)	3.79(57)·10^{-5}	0.76(11)	3.21(48)·10^5	2.79·10^{-3}
(3) pH = 7.4 + 0.1M NaF	20(3)	4.70(71)·10^{-5}	0.72(10)	8.21(23)·10^5	1.62·10^{-2}
(4) pH = 5.5 + 0.1M NaF	14(2)	6.63(92)·10^{-5}	0.71(10)	1.64(25)·10^4	1.65·10^{-3}
(5) pH = 7.4 + 15 mL Listerine Total Care Teeth Protection® by McNeil Consumer Healthcare McNeil-PPC, Inc., Fort Washington, PA, USA	31(4)	2.32(35)·10^{-5}	0.63(9)	6.64(96)·10^4	2.92·10^{-3}
(6) pH = 7.4 + 15 mL Meridol® by Colgate-Palmolive Company, New York, NY, USA	47(7)	8.79(99)·10^{-5}	0.84(12)	7.14(99)·10^3	3.50·10^{-2}

Table 4. Equivalent electrical circuit parameters obtained by fitting EIS experimental data using the 1CPE model for a Co-Cr-Mo electrode with a TiO_2-ZrO_2 sol–gel coating sintered at 500 °C in an unmodified and modified artificial saliva environment at 37 °C.

Artificial Saliva Solution	R_s [Ω cm^2]	T_1 [F cm^{-2} s$^{\phi-1}$]	ϕ_1	R_{ct} [Ω cm^2]	$\overline{C}_{dl}$ [F cm^{-2}]
(1) pH = 7.4	45(6)	2.26(34)·10^{-5}	0.74(10)	2.34(35)·10^6	4.08·10^{-3}
(5) pH = 7.4 + 15 mL Listerine Total Care Teeth Protection ® by McNeil Consumer Healthcare McNeil-PPC, Inc., Fort Washington, PA, USA	25(4)	1.71(26)·10^{-5}	0.78(11)	1.50(23)·10^4	1.57·10^{-3}

Table 5. Equivalent electrical circuit parameters obtained by fitting EIS experimental data using the 2CPE model for a Co-Cr-Mo electrode with a TiO_2-ZrO_2 sol–gel coating sintered at 500 °C in an unmodified and modified artificial saliva environment at 37 °C.

Artificial Saliva Solution	R_s [Ω cm^2]	T_1 [F cm^{-2} s$^{\phi-1}$]	ϕ_1	R_{ct1} [Ω cm^2]	T_2 [F cm^{-2} s$^{\phi-1}$]	ϕ_2	R_{ct2} [Ω cm^2]	$\overline{C}_{dl}$ [F cm^{-2}]
(2) pH = 5.5	10(1)	5.1(77)·10^{-5}	0.79(12)	8.82(92)·10^3	5.8(87)·10^{-5}	0.52(8)	191(29)	7.5·10^{-4}
(3) pH = 7.4 + 0.1M NaF	8(1)	4.3(65)·10^{-5}	0.74(11)	5.06(76)·10^3	9.8(97)·10^{-5}	0.56(8)	253(38)	3.9·10^{-4}
(4) pH = 5.5 + 0.1M NaF	4(1)	2.5(38)·10^{-7}	0.66(10)	2.46(37)·10^5	4.6(69)·10^{-5}	0.56(8)	219(33)	1.0·10^{-4}
(6) pH = 7.4 + 15 mL Meridol® by Colgate-Palmolive Company, New York, NY, USA	21(3)	8.7(73)·10^{-5}	0.83(12)	2.87(43)·10^4	7.6(94)·10^{-5}	0.57(9)	115(17)	2.6·10^{-4}

For the Co-Cr-Mo electrode with the TiO_2-ZrO_2 sol–gel coating sintered at 300 °C (Table 3), the data show that R_s varies with the electrolyte type, with the lowest values observed in the presence of 0.1M NaF, regardless of the pH level. The T_1 values are generally similar across different electrolytes, with slightly higher values in the presence of NaF. The $\phi 1$ values are close to 0.75, suggesting a non-ideal capacitive behavior. R_{ct1} is the highest in the presence of 0.1M NaF, indicating a more significant barrier to electron transfer. The $\overline{C}_{dl}$ varies, with the highest value observed in the saliva with added NaF at pH 7.4.

For the Co-Cr-Mo electrode with the TiO_2-ZrO_2 sol–gel coating sintered at 500 °C (Tables 4 and 5), R_s is lower compared to that of the coating sintered at 300 °C, indicating a more conductive electrolyte environment. The T_1 values are lower, and the ϕ_1 values are close to 0.75, similar to the electrode with the coating sintered at 300 °C. R_{ct1} is significantly higher in the unmodified saliva at pH 7.4, suggesting a more protective coating.

When using the 2CPE model (Table 5), which includes two time constants, the data show that the second CPE (T_2, ϕ_2) and the second charge transfer resistance (R_{ct2}) are introduced. The T_2 values are lower than T_1, and the ϕ_2 values are around 0.55–0.57, indicating a different type of non-ideal capacitive behavior. The R_{ct2} values are lower than R_{ct1}, suggesting a less significant barrier to electron transfer associated with the second time constant. The $\overline{C}_{dl}$ values are lower compared to the electrode with the TiO_2-ZrO_2 sol–gel coating sintered at 300 °C, indicating a possible difference in the electrode's surface properties. The obtained EIS data suggest that the sintering temperature and the composition of the artificial saliva environment affect the electrochemical properties of the Co-Cr-Mo electrodes with TiO_2-ZrO_2 sol–gel coatings. The addition of NaF and the use of mouthwashes like Listerine Total Care Teeth Protection® by McNeil Consumer Healthcare McNeil-PPC, Inc., Fort Washington, PA, USA and Meridol® by Colgate-Palmolive Company, New York, NY, USA significantly alter the electrochemical behavior, with NaF generally increasing the charge transfer resistance and the double-layer capacitance, indicating a more protective environment for the electrode. The EIS data for the unmodified Co-Cr-Mo obtained under the same electrochemical conditions also revealed the effect of pH on the uncoated Co-Cr-Mo electrode [8]. It was found that at pH 7.4, the resistance R_1 is higher than that at pH 5.5, indicating better corrosion resistance at a higher pH. The addition of 0.1 M NaF increases R_1 at both pH levels, suggesting improved corrosion resistance in the presence of fluoride ions. The T_1 values are lower at pH 7.4 compared to pH 5.5, which may indicate a more compact or less defective passive layer at higher pH, and the presence of Listerine Total Care Teeth Protection® by McNeil Consumer Healthcare McNeil-PPC, Inc., Fort Washington, PA, USA and Meridol® by Colgate-Palmolive Company, New York, NY, USA mouthwashes significantly increases R_2, suggesting a protective effect on the electrode surface.

R_{ct} is a measure of the electron transfer kinetics at the electrode–electrolyte interface. A higher R_{ct} value indicates slower electron transfer, which is generally associated with better corrosion resistance [5,6,8,26,29,47,63,99–101]. If the R_{ct} values increase for coatings in NaF and mouthwash environments, this suggests that the coatings exhibit better corrosion resistance in these conditions [8,26,45–48]. $\overline{C}_{dl}$ is related to the interfacial properties and the roughness of the surface [5,6,8,26,29,47,63,99–101]. A lower $\overline{C}_{dl}$ value typically indicates a smoother surface, which can be beneficial for corrosion resistance as it reduces the available surface area for corrosion reactions. The structural differences observed at different sintering temperatures of TiO_2-ZrO_2 sol–gel coatings can influence EIS parameters. For instance, higher sintering temperatures may lead to denser coatings with fewer defects, which can result in higher R_{ct} values. Conversely, lower sintering temperatures may produce less dense coatings with lower R_{ct} values (see Figure 2, Tables 3–5). An increase in R_{ct} for coatings in NaF and mouthwash environments suggests a long-term benefit in terms of corrosion resistance. However, it is important to consider other factors such as the mechanical properties of coatings. If the increased R_{ct} is accompanied by a decrease in mechanical strength, this could potentially lead to issues such as delamination or degradation over time. The observed EIS trends are consistent with known models of corrosion mechanisms in metal oxides in which the formation of a protective oxide layer on the metallic surface leads to increased R_{ct} values [5,6,8,47,99–101].

3.6. In Vitro Cyclic Potentiodynamic Polarization Curve Study

The susceptibility to pitting corrosion of the Co-Cr-Mo electrode covered with the TiO_2-ZrO_2 sol–gel coating sintered at 300 and 500 °C in the unmodified and modified saliva solutions at 37 °C was determined by cyclic potentiodynamic polarization curve tests (Figure 10).

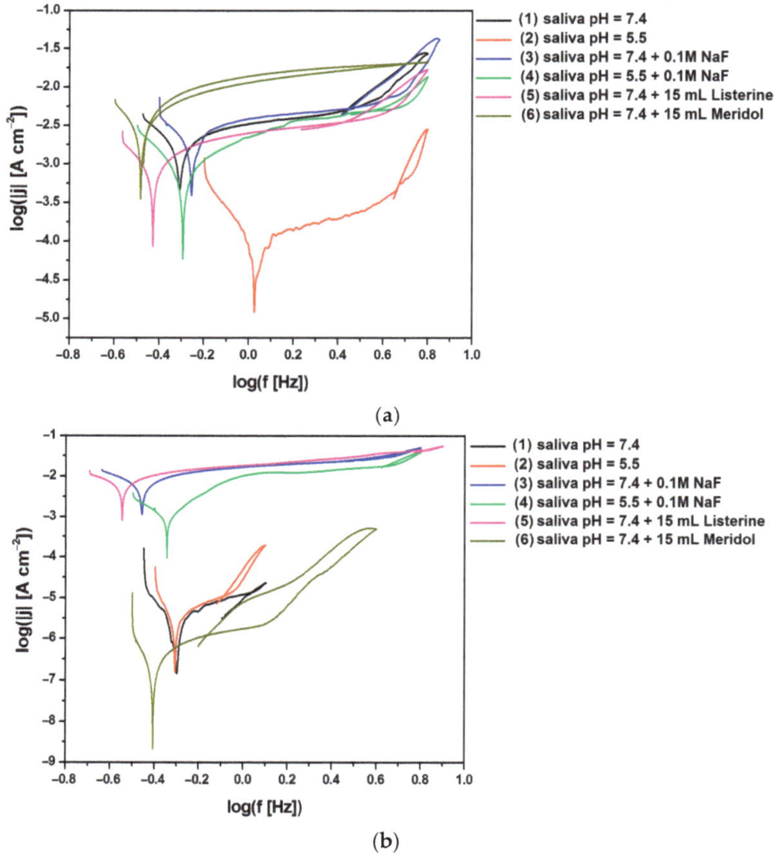

Figure 10. The cyclic potentiodynamic polarization curve at a polarization scan rate of v = 1 mV s^{-1} in the unmodified and modified saliva solutions at 37 °C for the Co-Cr-Mo electrode covered with the TiO_2-ZrO_2 sol–gel coating sintered at (**a**) 300 °C and (**b**) 500 °C.

The key parameters for corrosion resistance include the corrosion potential (E_{cor}), corrosion current density (j_{cor}), protection potential (E_p), and breakdown potential (E_{bd}). The difference between E_{bd} and E_p (E_{bd}—E_p) denoting the width of the hysteresis loop in the anodic curves shown in Figure 10 is also an important indicator of the susceptibility to pitting corrosion.

For the Co-Cr-Mo electrode covered with the TiO_2-ZrO_2 sol–gel coating sintered at 300 °C (Table 6), the lowest corrosion current density (indicating higher corrosion resistance) is observed in the unmodified saliva solution with pH = 7.4, followed by the solution with added NaF. The highest corrosion current density is seen in the saliva solution with pH = 5.5, suggesting that acidic conditions increase the susceptibility to corrosion. The addition of NaF generally improves the corrosion resistance, as evidenced by lower j_{cor} values. The presence of Listerine Total Care Teeth Protection® by McNeil Consumer Healthcare McNeil-PPC, Inc., Fort Washington, PA, USA and Meridol® by Colgate-Palmolive Company, New York, NY, USA also affects the corrosion behavior, with Meridol® by Colgate-Palmolive Company, New York, NY, USA showing a significant negative shift in E_{cor} and a high E_{bd}—E_p value, indicating a protective effect.

Table 6. The key parameters of the corrosion resistance of the Co-Cr-Mo electrode covered with the TiO_2-ZrO_2 sol–gel coating sintered at 300 °C in the unmodified and modified saliva solutions at 37 °C, where E_{cor}—corrosion potential; j_{cor}—corrosion current density; E_p—protection potential; and E_{bd}—breakdown potential (see Figure 10a).

Type of Electrolyte	E_{cor} [V]	j_{cor} [A cm^{-2}]	E_p [V]	E_{bd} [V]	E_{bd}—E_p [V]
(1) Saliva pH = 7.4	−0.309(36)	$-4.88(58)\cdot10^{-8}$	0.438(66)	0.797(86)	0.359
(2) Saliva pH = 5.5	0.027(3)	$-2.71(33)\cdot10^{-7}$	0.670(98)	0.799(98)	0.129
(3) Saliva pH = 7.4 + 0.1M NaF	−0.070(8)	$-1.50(18)\cdot10^{-9}$	0.454(68)	0.849(99)	0.395
(4) Saliva pH = 5.5 + 0.1M NaF	−0.292(34)	$-3.05(36)\cdot10^{-8}$	0.545(80)	0.798(95)	0.253
(5) Saliva pH = 7.4 + 15 mL Listerine Total Care Teeth Protection ® by McNeil Consumer Healthcare McNeil-PPC, Inc., Fort Washington, PA, USA	−0.428(52)	$-6.71(80)\cdot10^{-9}$	0.420(65)	0.793(98)	0.373
(6) Saliva pH = 7.4 + 15 mL Meridol® by Colgate-Palmolive Company, New York, NY, USA	−0.482(58)	$-1.16(14)\cdot10^{-7}$	-0.455(78)	0.623(77)	0.623

In the case of the Co-Cr-Mo electrode covered with the TiO_2-ZrO_2 sol–gel coating sintered at 500 °C (Table 7), the corrosion current densities are generally lower than those for the electrode with the coating sintered at 300 °C, indicating better corrosion resistance. The lowest j_{cor} is observed in the unmodified saliva solution with pH = 7.4, followed by the solution with added NaF. The acidic saliva solution (pH = 5.5) again shows higher corrosion current density, confirming the detrimental effect of acidity. The addition of NaF and Listerine Total Care Teeth Protection® by McNeil Consumer Healthcare McNeil-PPC, Inc., Fort Washington, PA, USA improves corrosion resistance, while that of Meridol® by Colgate-Palmolive Company, New York, NY, USA shows a significant negative shift in E_{cor} and a high E_{bd}—E_p value, similar to the electrode sintered at 300 °C, suggesting a protective effect. Based on the obtained results, it can be concluded that the produced TiO_2-ZrO_2 coatings have an impact on reducing the susceptibility to pitting corrosion of the Co-Cr-Mo dental alloy in the oral cavity environment with physiological and acidic pH [8]. According to the literature, Co-Cr-Mo alloys exhibit varying degrees of corrosion resistance depending on the processing method, surface modification, and the conditions of exposure [8,26,27,29,30,32].

The results of cyclic potentiodynamic polarization curves indicating reduced pitting susceptibility for TiO_2-ZrO_2 sol–gel coatings sintered at 500 °C can be attributed to several mechanistic factors. To delve deeper into these reasons, it is essential to consider microstructural changes, the role of micro-cracks, and the influence of environmental factors such as pH and the composition of artificial saliva. Sintering at higher temperatures can lead to increased densification and grain growth within coatings. This can result in a more homogeneous and compact structure, which is generally more resistant to corrosion (see Figure 2). Micro-cracks can form in coatings due to thermal stresses during sintering, which can actually enhance passivation properties. This is because they can provide pathways for the electrolyte to access the underlying metal, promoting the formation of a more uniform and protective oxide layer. The pH and the composition of artificial saliva can significantly influence the initiation of pitting corrosion [8,26,45–48]. A more neutral or slightly alkaline pH can favor the formation of a stable passivation layer, whereas a more acidic environment can lead to its breakdown. The presence of certain ions (e.g., chloride, fluoride) can also affect the stability of the passivation layer and the propensity for pitting corrosion. The composition of artificial saliva can provide buffering effects that stabilize the pH at the coating surface, which can be beneficial for maintaining the integrity of the passivation layer [102].

Table 7. The key parameters of the corrosion resistance of the Co-Cr-Mo electrode covered with the TiO_2-ZrO_2 sol–gel coating sintered at 500 °C in the modified saliva solutions at 37 °C, where E_{cor}—corrosion potential; j_{cor}—corrosion current density; E_p—protection potential; and E_{bd}—breakdown potential (see Figure 10b).

Type of Electrolyte	E_{cor} [V]	j_{cor} [A cm^{-2}]	E_p [V]	E_{bd} [V]	E_{bd}—E_p [V]
(1) Saliva pH = 7.4	−0.297(39)	$-1.17(13)\cdot10^{-9}$	0.037(6)	0.102(16)	0.065
(2) Saliva pH = 5.5	−0.305(43)	$-1.28(14)\cdot10^{-9}$	−0.111(13)	0.098(14)	0.209
(3) Saliva pH = 7.4 + 0.1M NaF	−0.456(59)	$-3.91(43)\cdot10^{-7}$	0.437(48)	0.794(92)	0.357
(4) Saliva pH = 5.5 + 0.1M NaF	−0.344(44)	$-3.51(39)\cdot10^{-8}$	0.633(72)	0.798(99)	0.165
(5) Saliva pH = 7.4 + 15 mL Listerine Total Care Teeth Protection® by McNeil Consumer Healthcare McNeil-PPC, Inc., Fort Washington, PA, USA	−0.546(73)	$-1.95(22)\cdot10^{-8}$	0.799(88)	0.890(84)	0.091
(6) Saliva pH = 7.4 + 15 mL Meridol® by Colgate-Palmolive Company, New York, NY, USA	−0.406(52)	$-1.22(12)\cdot10^{-9}$	−0.161(19)	0.556(72)	0.717

The comparison between the effects of different artificial saliva components on TiO_2-ZrO_2 coatings is crucial for understanding their performance in real-world dental applications. These components, such as NaF and various mouthwash formulations, can interact with the coatings at the molecular level, influencing their corrosion resistance and overall stability. Fluoride ions can interact with TiO_2-ZrO_2 coatings by forming fluoride-rich oxide layers. These layers are generally more stable and protective than the native oxide layers, providing enhanced corrosion resistance. Fluoride ions can replace hydroxide ions in the oxide layer, forming compounds such as TiF_4 and ZrF_4, which are highly resistant to corrosion [103]. The presence of fluoride ions can lead to the formation of a more stable and uniform passivation layer, reducing the susceptibility to pitting corrosion [8,26,45–48]. Fluoride-rich oxide layers can reduce the release of metal ions from the coatings, further enhancing their stability in corrosive environments [102]. Mouthwashes contain a variety of components, including alcohol, chlorhexidine, essential oils, and other antimicrobial agents. These components can have different effects on TiO_2-ZrO_2 coatings. Alcohol is often used as a solvent and antimicrobial agent in mouthwashes. It can influence the wetting properties of the coatings and potentially affect the stability of the passivation layer. Alcohol can reduce the surface tension of saliva, affecting the adhesion and cohesion of the passivation layer [104]. The effects of mouthwash components on corrosion resistance can be variable, depending on the specific formulation and concentration of the active ingredients [105]. Some mouthwash components, particularly those with high alcohol content or strong antimicrobial agents, may disrupt the passivation layer, potentially increasing the susceptibility to corrosion [105]. NaF generally enhances the stability of the passivation layer, providing better long-term corrosion resistance. Mouthwashes, depending on their composition, may have variable effects, ranging from beneficial to potentially disruptive. For dental implants, the use of NaF-containing products can be beneficial for maintaining the integrity of the coatings. However, the choice of mouthwash should be considered, as some formulations may affect the stability of the passivation layer [8].

The investigated TiO_2-ZrO_2 coatings provide significant protection in both physiological and acidic conditions. The underlying mechanisms can be attributed to the unique properties of the dual composition and its interaction with the environment at different pH levels. In physiological conditions, both TiO_2 and ZrO_2 form stable oxide layers that protect the underlying metal from corrosion. The stability of these layers is due to the formation of strong metal–oxygen bonds [106]. In acidic conditions, the stability of the oxide layers can be challenged due to the increased level of activity of hydrogen ions. However, the dual composition of TiO_2-ZrO_2 can mitigate this effect by forming mixed oxide layers that are more resistant to acidic attack [107]. The combination of TiO_2 and ZrO_2 can lead to a synergistic effect where the formation of a mixed oxide layer provides better protection than either oxide alone. This mixed layer can offer enhanced passivation properties, reducing the susceptibility to corrosion in both physiological and acidic conditions. The dual composition can also improve the mechanical properties of the coatings, making them more resistant to crack propagation and delamination, which are common issues in corrosive environments [108]. At physiological pH, the oxide layers formed by both TiO_2 and ZrO_2 are stable and protective. The formation of a compact and adherent oxide layer prevents the ingress of corrosive species [102]. In acidic conditions, the increased concentration of hydrogen ions can lead to the partial dissolution of the oxide layer. However, the dual composition of TiO_2-ZrO_2 can resist this dissolution better than either oxide alone. The mixed oxide layer can repel hydrogen ions more effectively, maintaining its integrity even in acidic environments [103]. The ability of TiO_2-ZrO_2 coatings to provide significant protection in both physiological and acidic conditions is clinically significant for several reasons. The oral environment can experience fluctuations in pH due to dietary habits, oral hygiene practices, and the presence of certain bacteria [109]. The robustness of these coatings ensures consistent protection against corrosion under varying conditions [105]. The resistance of these coatings to both physiological and acidic conditions can lead to an improved longevity of dental implants, reducing the need for replacement and minimizing patient discomfort [110].

To provide a comprehensive evaluation of TiO_2-ZrO_2 coatings, it is essential to compare them with other commonly used coatings such as pure TiO_2, ZrO_2, or hydroxyapatite. Additionally, discussing the expected long-term performance based on accelerated aging tests or literature evidence will help appreciate the practical applicability of these coatings in dental alloys. The dual composition of TiO_2-ZrO_2 generally exhibits better corrosion resistance compared to pure TiO_2 or ZrO_2 coatings. This is due to the formation of a more stable and protective mixed oxide layer [107]. TiO_2-ZrO_2 coatings often show higher durability and resistance to corrosion compared to hydroxyapatite coatings, which can degrade over time in physiological environments [106]. The combination of TiO_2 and ZrO_2 can lead to improved mechanical properties, including higher toughness and resistance to crack propagation [108]. TiO_2-ZrO_2 coatings typically exhibit better mechanical strength and wear resistance compared to hydroxyapatite coatings [102]. Both TiO_2 and ZrO_2 are known for their excellent biocompatibility. The dual composition maintains this biocompatibility while offering additional benefits in terms of corrosion resistance and mechanical properties [103]. Hydroxyapatite coatings are highly biocompatible and osteoconductive. TiO_2-ZrO_2 coatings, while not osteoconductive, are also biocompatible and can be designed to support osseointegration [110].

Accelerated aging tests can be performed by exposing the obtained TiO_2-ZrO_2 coatings to simulated oral environments for extended periods. These tests can help predict the long-term stability and performance of the coatings [105]. Combining mechanical stress (e.g., cyclic loading) with chemical stress (e.g., acidic and alkaline solutions) can provide a more comprehensive understanding of the coatings' durability [107]. Studies on similar coatings have shown that TiO_2-ZrO_2 coatings can maintain their integrity and protective properties over extended periods in physiological environments [106]. The resistance of TiO_2-ZrO_2 coatings to degradation in both acidic and alkaline conditions suggests that they can withstand the varying pH of the oral environment over time [103].

The obtained TiO_2-ZrO_2 sol–gel coatings show excellent adherence, flexibility, and compactness, which are crucial for effective corrosion protection. They are relatively thin and are capable of forming a protective barrier against corrosive environments. The TiO_2-ZrO_2 sol–gel coatings provide enhanced corrosion resistance by creating a barrier that prevents the ingress of corrosive species. This is particularly important for dental appliances where the alloy is exposed to saliva and other biological fluids. The sol–gel method allows for the incorporation of various metal oxides into the coating, which can further enhance the properties of the coating. Further research should focus on the precise fabrication of nanostructured mixed metal oxides using the sol–gel method, which can further improve the mechanical and protective properties of the coatings.

Translating the findings on TiO_2-ZrO_2 sol–gel coatings into clinical practice involves several critical steps, including scaling up the technology, addressing regulatory hurdles, ensuring biocompatibility, and optimizing manufacturing processes. The sol–gel process used to create these coatings needs to be optimized for large-scale production. This includes standardizing the precursor solutions, coating parameters (e.g., thickness, uniformity), and sintering conditions [111]. Implementing rigorous quality control measures is essential to ensure the consistency and reliability of the coatings. This includes monitoring the coating thickness, adhesion, and corrosion resistance [105]. Before clinical use, the coatings must undergo preclinical testing to evaluate their safety and efficacy. This includes in vitro tests (e.g., cytotoxicity, biocompatibility) and in vivo animal studies [112]. Successful preclinical testing should be followed by clinical trials to assess the performance of the coatings in human subjects. These trials are crucial for gaining regulatory approval [113]. The coatings must be tested for cytotoxicity and biocompatibility using standardized methods. This includes assessing the effects of the coatings on cell viability, proliferation, and differentiation [114]. Evaluating the immunological response to the coatings is also important to ensure they do not induce adverse reactions in the body [115]. The sol–gel process must be optimized for industrial manufacturing, ensuring that the coatings can be produced consistently and efficiently [111]. Automating the coating process can improve efficiency and reduce variability. This includes automating the application of the sol–gel solution, drying, and sintering steps [51]. The coatings must be compatible with the dental alloys commonly used in implants and restorations. This includes ensuring adhesion and mechanical compatibility [102]. Dentists and oral surgeons need to be trained in the application and handling of implants and restorations coated with TiO_2-ZrO_2 [110]. The potential of TiO_2-ZrO_2 sol–gel coatings into clinical practice requires a careful consideration of scaling up the technology, addressing regulatory hurdles, ensuring biocompatibility, and optimizing manufacturing processes. By following standardized testing protocols and adhering to regulatory guidelines, these coatings can be successfully integrated into dental restorations, offering improved longevity and performance.

4. Conclusions

The applied sol–gel method allows us to obtain thin TiO_2-ZrO_2 coatings on the surface of the Co-Cr-Mo dental alloy (Bego Wirobond® C by BEGO Bremer Goldschlägerei Wilh. Herbst GmbH & Co. KG, Bremen, Germany), minimizing material waste due to precise and accurate synthesis, which reduces the amount of raw materials used in the process. SEM analysis shows that sintering temperatures at 300 and 500 °C significantly affect the coatings' morphology and structure, with higher temperatures causing micro-cracks but increasing density. EDS confirms elemental composition changes with temperature, potentially due to volatilization or diffusion. Roughness measurements indicate that the coatings meet dental requirements for biocompatibility and bacterial resistance. Electrochemical tests reveal that the coatings enhance corrosion protection, particularly at higher sintering temperatures and in the presence of NaF and mouthwashes (Listerine Total Care Teeth Protection® by McNeil Consumer Healthcare McNeil-PPC, Inc., Fort Washington, PA, USA, Meridol® by Colgate-Palmolive Company, New York, NY, USA), suggesting their suitability for various oral environments. These coatings are effective in reducing pitting corrosion susceptibility, critical for the longevity and safety of dental devices. The findings underscore the potential of TiO_2-ZrO_2 sol–gel coatings in improving the performance of Co-Cr-Mo dental alloys in various oral environments.

Author Contributions: Conceptualization, B.Ł. and K.G.-K.; methodology, B.Ł., P.O., K.G.-K. and J.M.; software, B.Ł., P.O. and J.M.; validation, K.G.-K., P.O. and J.M.; formal analysis, P.O., K.G.-K. and J.M.; investigation, P.O., K.G.-K. and J.M.; resources, B.Ł. and K.G.-K.; data curation, P.O., K.G.-K. and J.M.; writing—original draft preparation, B.Ł. and P.O.; writing—review and editing, K.G.-K. and J.M.; visualization, K.G.-K., P.O. and J.M. All authors have read and agreed to the published version of the manuscript.

Funding: This research received no external funding.

Institutional Review Board Statement: Not applicable.

Informed Consent Statement: Not applicable.

Data Availability Statement: The original contributions presented in the study are included in the article, further inquiries can be directed to the corresponding author.

Acknowledgments: The authors wish to express their gratitude to Joanna Waś (BOSMAL Automotive Research and Development Institute, Bielsko-Biała, Poland) for technical support with the production of sol–gel coatings.

Conflicts of Interest: The authors declare no conflicts of interest.

References

1. Yu, Y.M.; Lu, Y.P.; Zhang, T.; Zheng, Y.F.; Liu, Y.S.; Xia, D.D. Biomaterials science and surface engineering strategies for dental peri-implantitis management. *Military Med. Res.* **2024**, *11*, 29. [CrossRef]
2. Alsuraifi, A.; Sulaiman, Z.M.; Mohammed, N.A.R.; Mohammed, J.; Ali, S.K.; Abdualihamaid, Y.H.; Husam, F.; Ayad, A. Explore the most recent developments and upcoming outlooks in the field of dental nanomaterials. *Beni-Suef Univ. J. Basic Appl. Sci.* **2024**, *13*, 82. [CrossRef]
3. Shen, C.; Rawls, H.R.; Esquivel-Upshaw, J.F. *Phillips' Science of Dental Materials*, 13th ed.; Elsevier Health Sciences: Amsterdam, The Netherlands, 2021.
4. Hakim, L.K.; Yari, A.; Nikparto, N.; Mehraban, S.H.; Cheperli, S.; Asadi, A.; Darehdor, A.A.; Nezaminia, S.; Dortaj, D.; Nazari, Y.; et al. The current applications of nano and biomaterials in drug delivery of dental implant. *BMC Oral Health* **2024**, *24*, 126. [CrossRef] [PubMed]
5. Smołka, A.; Dercz, G.; Rodak, K.; Łosiewicz, B. Evaluation of corrosion resistance of nanotubular oxide layers on the Ti13Zr13Nb alloy in physiological saline solution. *Arch. Metall. Mater.* **2015**, *60*, 2681–2686. [CrossRef]
6. Szklarska, M.; Dercz, G.; Łosiewicz, B.; Rak, J.; Simka, W. The influence of passivation type on corrosion resistance of Ti15Mo alloy in simulated body fluids. *Arch. Metall. Mater.* **2015**, *60*, 2687–2693. [CrossRef]
7. Łosiewicz, B.; Stróż, A.; Osak, P.; Maszybrocka, J.; Gerle, A.; Dudek, K.; Balin, K.; Łukowiec, D.; Gawlikowski, M.; Bogunia, S. Production, Characterization and Application of Oxide Nanotubes on Ti-6Al-7Nb Alloy as a Potential Drug Carrier. *Materials* **2021**, *14*, 6142. [CrossRef]
8. Łosiewicz, B.; Osak, P.; Kubisztal, J.; Górka-Kulikowska, K. Effect of Artificial Saliva Modification on the Corrosion Resistance and Electronic Properties of Bego Wirobond® C Dental Alloys. *Appl. Sci.* **2023**, *13*, 12185. [CrossRef]
9. Smołka, A.; Rodak, K.; Dercz, G.; Dudek, K.; Łosiewicz, B. Electrochemical Formation of Self-Organized Nanotubular Oxide Layers on Ti13Zr13Nb Alloy for Biomedical Applications. *Acta Phys. Pol. A* **2014**, *125*, 932–935. [CrossRef]
10. Rudolf, R.; Majerič, P.; Lazić, V.; Raić, K.T. Processing of Cobalt-Chrome Dental Alloys. In *Advanced Dental Metallic Materials*; Springer Series in Materials Science; Springer Nature: Cham, Switzerland, 2024; Volume 338. [CrossRef]
11. Vutova, K.; Stefanova, V.; Markov, M.; Vassileva, V. Study on Hardness of Heat-Treated CoCrMo Alloy Recycled by Electron Beam Melting. *Materials* **2023**, *16*, 2634. [CrossRef]
12. Atapour, M.; Sanaei, S.; Wei, Z.; Sheikholeslam, M.; Henderson, J.D.; Eduok, U.; Hosein, Y.K.; Holdsworth, D.W.; Hedberg, Y.S.; Ghorbani, H.R. In vitro corrosion and biocompatibility behavior of CoCrMo alloy manufactured by laser powder bed fusion parallel and perpendicular to the build direction. *Electrochim. Acta* **2023**, *445*, 142059. [CrossRef]
13. Gurel, S.; Nazarahari, A.; Canadinc, D.; Gerstein, G.; Maier, H.J.; Cabuk, H.; Bukulmez, T.; Cananoglu, M.; Yagci, M.B.; Toker, S.M.; et al. From corrosion behavior to radiation response: A comprehensive biocompatibility assessment of a CoCrMo medium entropy alloy for utility in orthopedic and dental implants. *Intermetallics* **2022**, *149*, 107680. [CrossRef]
14. Watanabe, M.; Takahashi, T.; Imaizumi, S.; Zhao, Y.; Adachi, M.; Ohtsuka, M.; Chiba, A.; Koizumi, Y.; Fukuyama, H. Thermophysical properties of liquid Co–Cr–Mo alloys measured by electromagnetic levitation in a static magnetic field. *Thermochim. Acta* **2022**, *708*, 179119. [CrossRef]
15. Qi, J.; Cole, T.; Foster, A.; Rainforth, W.M. Tribocorrosion behaviour of CoCrMo in simulated body fluid under anaerobic conditions. *Wear* **2024**, *556–557*, 205523. [CrossRef]
16. Available online: https://www.vestedmetals.net/featured-metal-products-cobalt-alloys/#:~:text=High/corrosion/resistance, /high/temperature/resistance,/non-magnetic/CoCrMo/(CCM/is (accessed on 26 September 2024).

17. Jiang, W.; An, X.; Ni, S.; Wang, L.; He, J.; Chen, Z.; Huang, Y.; Song, M. Achieving excellent strength-ductility combination through the control of intricate substructures in an additively manufactured Co–Cr–Mo alloy. *Mater. Sci. Eng. A* **2023**, *886*, 145687. [CrossRef]
18. Yamanaka, K.; Mori, M.; Sato, S.; Chiba, A. Stacking-fault strengthening of biomedical Co–Cr–Mo alloy via multipass thermomechanical processing. *Sci. Rep.* **2017**, *7*, 10808. [CrossRef]
19. Roudnicka, M.; Bigas, J.; Molnarova, O.; Palousek, D.; Vojtech, D. Different Response of Cast and 3D-Printed Co-Cr-Mo Alloy to Heat Treatment: A Thorough Microstructure Characterization. *Metals* **2021**, *11*, 687. [CrossRef]
20. Szala, M.; Beer-Lech, K.; Gancarczyk, K.; Kilic, O.B.; Pędrak, P.; Özer, A.; Skic, A. Microstructural Characterisation of Co-Cr-Mo Casting Dental Alloys. *Adv. Sci. Technol. Res. J.* **2017**, *11*, 76–82. [CrossRef] [PubMed]
21. Mace, A.; Khullar, P.; Bouknight, C.; Gilbert, J.L. Corrosion properties of low carbon CoCrMo and additively manufactured CoCr alloys for dental applications. *Dent. Mater.* **2022**, *38*, 1184–1193. [CrossRef]
22. Zhang, M.; Yang, Y.; Song, C.; Bai, Y.; Xiao, Z. Effect of the heat treatment on corrosion and mechanical properties of CoCrMo alloys manufactured by selective laser melting. *Rapid Prototyp. J.* **2018**, *24*, 1235–1244. [CrossRef]
23. Sing, S.L.; Yeong, W.Y. Effect of Heat Treatment on Cobalt-Chromium-Molybdenum Alloy Fabricated by Selective Laser Melting. In Proceedings of the 3rd International Conference on Progress in Additive Manufacturing (Pro-AM 2018), Singapore, 14–17 May 2018; pp. 458–463. [CrossRef]
24. Seyedi, M.; Zanotto, F.; Monticelli, C.; Balbo, A.; Liverani, E.; Fortunato, A. Microstructural characterization and corrosion behaviour of SLM CoCrMo alloy in simulated body fluid. *Metall. Ital.* **2018**, *110*, 45–50.
25. Mirea, R.; Biris, I.M.; Ceatra, L.C.; Ene, R.; Paraschiv, A.; Cucuruz, A.T.; Sbarcea, G.; Popescu, E.; Badea, T. In Vitro Physical-Chemical Behaviour Assessment of 3D-Printed CoCrMo Alloy for Orthopaedic Implants. *Metals* **2021**, *11*, 857. [CrossRef]
26. Pontes, J.R.; Toptan, F.; Pinto, A.M.; Ariza, E.; Alves, A.C. Influence of processing method of CoCrMo dental alloy on the corrosion behavior in artificial saliva. *Mater. Corros.* **2022**, *74*, 576–584. [CrossRef]
27. Kosec, T.; Leban, M.B.; Kurnik, M.; Kopač, I. Comparison of the corrosion properties of CoCrMo dental alloys in artificial saliva. *Mater. Tehnol.* **2021**, *55*, 819–824. [CrossRef]
28. Döring, J.; Basten, S.; Ecke, M.; Herbster, M.; Kirsch, B.; Halle, T.; Lohmann, C.H.; Bertrand, J.; Aurich, J. Surface integrity modification of CoCrMo alloy by deep rolling in combination with sub-zero cooling as potential implant application. *J. Biomed. Mater. Res.* **2023**, *111*, 946–957. [CrossRef]
29. Lu, C.; Zheng, Y.; Zhong, Q. Corrosion of dental alloys in artificial saliva with Streptococcus mutans. *PLoS ONE* **2017**, *12*, e0174440. [CrossRef]
30. Puskar, T.; Jevremovic, D.; Williams, R.J.; Eggbeer, D.; Vukelic, D.; Budak, I. A Comparative Analysis of the Corrosive Effect of Artificial Saliva of Variable pH on DMLS and Cast Co-Cr-Mo Dental Alloy. *Materials* **2014**, *7*, 6486–6501. [CrossRef]
31. Liu, C.; Zhou, Z.; Li, K.Y. Improved corrosion resistance of CoCrMo alloy with self-passivation ability facilitated by carbon ion implantation. *Electrochim. Acta* **2017**, *241*, 331–340. [CrossRef]
32. Ataiwi, A.H.; Majed, R.A.; Muhsin, A.A. Effect of Laser Surface Modification on the Corrosion Resistance of Dental Alloys in Artificial Saliva Containing Alcoholic Beverages. *Iraqi J. Laser Part A* **2013**, *12*, 43–52.
33. Zarkov, A. Sol–Gel Technology Applied to Materials Science: Synthesis, Characterization and Applications. *Materials* **2024**, *17*, 462. [CrossRef]
34. Dell'Era, A.; Catauro, M. Sol–Gel Method Applied to Crystalline Materials. *Crystals* **2021**, *11*, 903. [CrossRef]
35. Stambolova, I.; Dimitrov, O.; Yordanov, S.; Lakov, L.; Jivov, B.; Vassilev, S.; Blaskov, V.; Shipochka, M. Application of ZrO_2 and ZrO_2-TiO_2 Coatings as Corrosion Barriers: Surface Composition and Structure. *Bulg. Soc. NDT Int. J. "NDT Days"* **2019**, *2*, 160–164.
36. Ćurković, L.; Otmačić Ćurković, H.; Žmak, I.; Mustafa, M.K.; Gabelica, I. Corrosion Behavior of Amorphous Sol–Gel TiO_2–ZrO_2 Nano Thickness Film on Stainless Steel. *Coatings* **2021**, *11*, 988. [CrossRef]
37. Navas, D.; Fuentes, S.; Castro-Alvarez, A.; Chavez-Angel, E. Review on Sol-Gel Synthesis of Perovskite and Oxide Nanomaterials. *Gels* **2021**, *7*, 275. [CrossRef] [PubMed]
38. Barrino, F. Hybrid Organic–Inorganic Materials Prepared by Sol–Gel and Sol–Gel-Coating Method for Biomedical Use: Study and Synthetic Review of Synthesis and Properties. *Coatings* **2024**, *14*, 425. [CrossRef]
39. Alcázar, J.C.B.; Lemos, R.M.J.; Conde, M.C.M.; Chisini, L.A.; Salas, M.M.S.; Noremberg, B.S.; da Motta, F.V.; Demarco, F.F.; Tarquinio, S.B.; Carreño, N.L.V. Preparation, characterization, and biocompatibility of different metal oxide/PEG-based hybrid coating synthesized by sol–gel dip coating method for surface modification of titanium. *Prog. Org. Coat.* **2019**, *130*, 206–213. [CrossRef]
40. El hadad, A.A.; García-Galván, F.R.; Mezour, M.A.; Hickman, G.J.; Soliman, I.E.; Jiménez-Morales, A.; Barranco, V.; Galván, J.C.; Perry, C.C. Organic-inorganic hybrid coatings containing phosphorus precursors prepared by sol–gel on Ti6Al4V alloy: Electrochemical and in-vitro biocompatibility evaluation. *Prog. Org. Coat.* **2020**, *148*, 105834. [CrossRef]
41. Bukhari, S.B.; Imran, M.; Bashir, M.; Riaz, S.; Naseem, S. Room temperature stabilized TiO_2 doped ZrO_2 thin films for teeth coatings—A sol-gel approach. *J. Alloys Compd.* **2018**, *767*, 1238–1252. [CrossRef]
42. D'Agostino, A.; Bertolini, M.; Bono, N.; Pavarini, M.; Tarsini, P.; Candiani, G.; De Nardo, L.; Chiesa, R. Antibacterial titanium dioxide coatings for CoCrMo orthopaedic implants. *Appl. Surf. Sci.* **2023**, *609*, 155300. [CrossRef]

43. Zai, W.; Wu, C.L.; Man, H.C. Improving adhesion strength of sol-gel coating on CoCrMo by electrochemical pretreatment. *Mater. Lett.* **2020**, *270*, 127696. [CrossRef]
44. Renita, D.; Rajendran, S.; Chattree, A. Influence of Artificial Saliva on the Corrosion Behavior of Dental Alloys: A review. *Ind. J. Adv. Chem. Sci.* **2016**, *4*, 478–483.
45. Shim, H.M.; Oh, K.T.; Woo, J.Y.; Hwang, C.J.; Kim, K.N. Corrosion resistance of titanium-silver alloys in an artificial saliva containing fluoride ions. *J. Biomed. Mater. Res. B Appl. Biomater.* **2005**, *73B*, 252–259. [CrossRef] [PubMed]
46. Cui, W.F.; Dong, Y.Y.; Bao, Y.C.; Qin, G.W. Improved corrosion resistance of dental Ti50Zr alloy with (TiZr) N coating in fluoridated acidic artificial saliva. *Rare Met.* **2021**, *40*, 2927–2936. [CrossRef]
47. Łosiewicz, B.; Osak, P.; Górka-Kulikowska, K.; Goryczka, T.; Dworak, M.; Maszybrocka, J.; Aniołek, K. Effect of Artificial Saliva Modification on Pitting Corrosion and Mechanical Properties of the Remanium®-Type Orthodontic Archwire. *Materials* **2023**, *16*, 6791. [CrossRef]
48. Roscher, J.; Holze, R. Corrosion of nonprecious metal alloys in dentistry in the presence of common saliva additives. *Mater. Corros.* **2021**, *72*, 1410–1416. [CrossRef]
49. *ISO 22674:2023-03*; Dentistry—Metallic Materials for Fixed and Removable Restorations and Appliances. ISO: Geneva, Switzerland, 2023.
50. *ISO 9693-1:2012*; Dentistry—Compatibility Testing—Part 1: Metal-Ceramic Systems. ISO: Geneva, Switzerland, 2012.
51. *ISO 13485:2016-04*; Medical Devices—Quality Management Systems—Requirements for Regulatory Purposes. ISO: Geneva, Switzerland, 2016.
52. Shishodia, G.; Gupta, S.; Pahwa, N.; Shishodia, P.K. ZrO_2 Nanoparticles Synthesized by the Sol–Gel Method: Dependence of Size on pH and Annealing Temperature. *J. Electron. Mater.* **2024**, *53*, 5159–5168. [CrossRef]
53. Bhattacharyya, S.; Medda, S.K.; Naskar, M.K. A Preparative Approach of TiO_2-ZrO_2 Coating Using Aquo-Based TiO_2 Precursor Useful for Light Reflective Application. *T. Indian Ceram. Soc.* **2021**, *80*, 227–233. [CrossRef]
54. Zhang, W.; Ji, G.; Bu, A.; Zhang, B. Corrosion and Tribological Behavior of ZrO_2 Films Prepared on Stainless Steel Surface by the Sol–Gel Method. *ACS Appl. Mater. Interfaces* **2015**, *7*, 28264–28272. [CrossRef]
55. Tang, W.; Fischer, N.G.; Kong, X.; Sang, T.; Ye, Z. Hybrid coatings on dental and orthopedic titanium implants: Current advances and challenges. *BMEMat* **2024**, e12105. [CrossRef]
56. Butt, M.A. Thin-Film Coating Methods: A Successful Marriage of High-Quality and Cost-Effectiveness—A Brief Exploration. *Coatings* **2022**, *12*, 1115. [CrossRef]
57. Jilani, A.; Abdel-wahab, M.S.; Hammad, A.H. *Advance Deposition Techniques for Thin Film and Coating*; InTechOpen: London, UK, 2017. [CrossRef]
58. Rasool, A.; Kossar, S.; Amiruddin, R. Different Types of Thin Film Deposition Techniques and Application. In *Handbook of Emerging Materials for Semiconductor Industry*; Song, Y.S., Thoutam, L.R., Tayal, S., Rahi, S.B., Samuel, T.S.A., Eds.; Springer: Singapore, 2024. [CrossRef]
59. Available online: https://zh.kindle-tech.com/faqs/what-is-the-advantage-of-sputtering (accessed on 11 October 2024).
60. Grove, C.; Jerram, D.A. jPOR: An ImageJ Macro to Quantify Total Optical Porosity from Blue-Stained Thin Sections. *Comput. Geosci.* **2011**, *37*, 1850–1859. [CrossRef]
61. *ISO 21920-3:2022-06*; Geometrical Product Specifications (GPS)—Surface Texture: Profile—Part 3: Specification Operators. ISO: Geneva, Switzerland, 2022.
62. *AFNOR/NF S90-701*; Matériel Médico-Chirurgical—Biocompatibilité des Matériaux et Dispositifs Médicaux—Méthodes d'Extraction. French Standardization Association: Paris, France, 1988.
63. Eco Chemie, B.V. (Ed.) Kanaalweg. In *User Manual for Frequency Response Analysis (FRA) for Windows Version 4.9*; Kanaalweg: Utrecht, The Netherlands, 2001.
64. Bockmeyer, M.; Löbmann, P. Crack formation in TiO_2 films prepared by sol–gel processing: Quantification and characterization. *Thin Solid Films* **2007**, *515*, 5212–5219. [CrossRef]
65. Li, H.; Liang, K.; Mei, L.; Gu, S.; Wang, S. Corrosion protection of mild steel by zirconia sol-gel coatings. *J. Mater. Sci. Lett.* **2001**, *20*, 1081–1083. [CrossRef]
66. Li, Y.; Yang, H.; He, Y.; Zhang, Q.; Shi, Y.; Chen, Y. Properties of UV-irradiated TiO_2, ZrO_2, and TiO_2-ZrO_2 films as pore-sealing layers on micro-arc-oxidized aluminum alloys. *J. Sol-Gel Sci. Technol.* **2020**, *93*, 70–78. [CrossRef]
67. Bu, A.; Wang, J.; Zhang, J.; Bai, J.; Shi, Z.; Liu, Q.; Ji, G. Corrosion behavior of ZrO_2–TiO_2 nanocomposite thin films coating on stainless steel through sol–gel method. *J. Sol-Gel Sci. Technol.* **2017**, *81*, 633–638. [CrossRef]
68. Simon, S.M.; George, G.; Sajna, M.S.; Prakashan, V.P.; Jose, T.A.; Vasudevan, P.; Saritha, A.C.; Biju, P.R.; Joseph, C.; Unnikrishnan, N.V. Recent advancements in multifunctional applications of sol-gel derived polymer incorporated TiO_2-ZrO_2 composite coatings: A comprehensive review. *Appl. Surf. Sci. Adv.* **2021**, *6*, 100173. [CrossRef]
69. Jiao, J.; Hong, Q.; Zhang, D.; Wang, M.; Tang, H.; Yang, J.; Qu, X.; Yue, B. Influence of Porosity on Osteogenesis, Bone Growth and Osteointegration in Trabecular Tantalum Scaffolds Fabricated by Additive Manufacturing. *Front. Bioeng. Biotechnol.* **2023**, *11*, 1117954. [CrossRef]
70. Zou, H.; Lin, Y. Structural and surface chemical properties of sol–gel derived TiO_2–ZrO_2 oxides. *Appl. Catal. A Gen.* **2004**, *265*, 35–42. [CrossRef]

71. Stambolova, I.; Stoyanova, D.; Shipochka, M.; Boshkova, N.; Eliyas, A.; Simeonova, S.; Grozev, N.; Boshkov, N. Surface Morphological and Chemical Features of Anticorrosion ZrO_2–TiO_2 Coatings: Impact of Zirconium Precursor. *Coatings* **2021**, *11*, 703. [CrossRef]
72. Mansilla, Y.; Arce, M.D.; González-Oliver, C.; Basbus, J.; Troiani, H.; Serquis, A. Characterization of Stabilized ZrO_2 Thin Films Obtained by Sol-Gel Method. *Appl. Surf. Sci.* **2021**, *569*, 150787. [CrossRef]
73. Lin, Y.; Andrews, C.; Sodano, H. Enhanced Piezoelectric Properties of Lead Zirconate Titanate Sol-Gel Derived Ceramics Using Single Crystal PbZr0.52Ti0.48O3 Cubes. *J. Appl. Phys.* **2010**, *108*, 064108. [CrossRef]
74. Yao, Y.B.; Yao, Z.J.; Zhou, Y.J. High-temperature oxidation resistance of sol–gel-derived $ZrO_2/(Al_2O_3$–$Y_2O_3)$ coating on titanium alloy. *J. Sol-Gel Sci. Technol.* **2016**, *80*, 612–618. [CrossRef]
75. Naumenko, A.; Gnatiuk, I.; Smirnova, N.; Eremenko, A. Characterization of sol-gel derived TiO_2/ZrO_2 films and powders by Raman spectroscopy. *Thin Solid Films* **2012**, *520*, 4541–4546. [CrossRef]
76. Li, M.; Jin, Z.X.; Zhang, W.; Bai, Y.H.; Cao, Y.Q.; Li, W.M.; Wu, D.; Li, A.D. Comparison of chemical stability and corrosion resistance of group IV metal oxide films formed by thermal and plasma-enhanced atomic layer deposition. *Sci. Rep.* **2019**, *9*, 10438. [CrossRef] [PubMed]
77. Anisah, S.; Kanezashi, M.; Nagasawa, H.; Tsuru, T. Effect of the Ti/Zr ratio on the hydrothermal and chemical stability of TiO_2-ZrO_2 nanofiltration membranes. *Sep. Purif. Technol.* **2021**, *274*, 119060. [CrossRef]
78. Cui, Z.; Zhu, J.; Jiang, H.; Wu, S.; Zhu, S. Research Progress of the Surface Modification of Titanium and Titanium Alloys for Biomedical Application. *Acta Metall. Sin.* **2022**, *58*, 837–856. [CrossRef]
79. Ramamoorthy, S.; Das, S.; Balan, R.; Lekshmi, I.C. TiO_2-ZrO_2 nanocomposite with tetragonal zirconia phase and photocatalytic degradation of Alizarin Yellow GG azo dye under natural sunlight. *Mater. Today Proc.* **2021**, *47*, 4641–4646. [CrossRef]
80. Heimes, D.; Becker, P.; Pabst, A.; Smeets, R.; Kraus, A.; Hartmann, A.; Sagheb, K.; Kämmerer, P.W. How does dental implant macrogeometry affect primary implant stability? A narrative review. *Int. J. Implant Dent.* **2023**, *9*, 20. [CrossRef]
81. Salerno, M.; Itri, A.; Frezzato, M.; Rebaudi, A. Surface microstructure of dental implants before and after insertion: An in vitro study by means of scanning probe microscopy. *Implant Dent.* **2015**, *24*, 248–255. [CrossRef]
82. Sotova, C.; Yanushevich, O.; Kriheli, N.; Grigoriev, S.; Evdokimov, V.; Kramar, O.; Nozdrina, M.; Peretyagin, N.; Undritsova, N.; Popelyshkin, E.; et al. Dental Implants: Modern Materials and Methods of Their Surface Modification. *Materials* **2023**, *16*, 7383. [CrossRef]
83. *ISO 4287:1997*; Geometrical Product Specifications (GPS)—Surface Texture: Profile Method—Terms, Definitions and Surface Texture Parameters. International Organization for Standardization: Geneva, Switzerland, 1997.
84. Sahlathasneem, K.; Surinder, D. A Comprehensive Review of Noise Measurement, Standards, Assessment, Geospatial Mapping and Public Health. *Ecol. Quest.* **2023**, *34*, 1–26. [CrossRef]
85. Na, G.; Jo, N.H.; Eun, Y. Performance degradation due to measurement noise in control systems with disturbance observers and saturating actuators. *J. Frank. Inst.* **2019**, *356*, 3922–3947. [CrossRef]
86. Schubert, A.; Wassmann, T.; Holtappels, M.; Kurbad, O.; Krohn, S.; Bürgers, R. Predictability of Microbial Adhesion to Dental Materials by Roughness Parameters. *Coatings* **2019**, *9*, 456. [CrossRef]
87. Engel, A.S.; Kranz, H.T.; Schneider, M.; Tietze, J.P.; Piwowarcyk, A.; Kuzius, T.; Arnold, W.; Naumowa, E.A. Biofilm formation on different dental restorative materials in the oral cavity. *BMC Oral Health* **2020**, *20*, 162. [CrossRef] [PubMed]
88. Kreve, S.; Dos Reis, A.C. Bacterial adhesion to biomaterials: What regulates this attachment? A review. *Jpn. Dent. Sci. Rev.* **2021**, *57*, 85–96. [CrossRef]
89. Witzke, K.; Kensbock, R.; Willsch, C.U.; Fricke, K.; Bekeschus, S.; Metelmann, H.-R. Mechanical and Plasma Electrolytic Polishing of Dental Alloys. *Materials* **2023**, *16*, 6222. [CrossRef]
90. Hjerppe, J.; Rodas, S.; Korvala, J.; Pesonen, P.; Kaisanlahti, A.; Özcan, M.; Suojanen, J.; Reunanen, J. Surface Roughness and Streptococcus mutans Adhesion on Metallic and Ceramic Fixed Prosthodontic Materials after Scaling. *Materials* **2021**, *14*, 1027. [CrossRef]
91. Gharechahi, M.; Moosavi, H.; Forghani, M. Effect of Surface Roughness and Materials Composition. *J. Biomater. Nanobiotechnology* **2012**, *3*, 541. [CrossRef]
92. Wennerberg, A.; Albrektsson, T. Effects of titanium surface topography on bone integration: A systematic review. *Clin. Oral Implants Res.* **2009**, *20*, 172–184. [CrossRef]
93. Bollen, C.M.L.; Lambrechts, P.; Quirynen, M. Comparison of surface roughness of oral hard materials to the threshold surface roughness for bacterial plaque retention: A review of the literature. *Dent. Mater.* **1997**, *13*, 258–269. [CrossRef]
94. Rokaya, D.; Srimaneepong, V.; Wisitrasameewon, W.; Humagain, M.; Thunyakitpisal, P. Peri-implantitis Update: Risk Indicators, Diagnosis, and Treatment. *Eur. J. Dent.* **2020**, *14*, 672–682. [CrossRef]
95. Schoberleitner, I.; Augustin, A.; Egle, D.; Brunner, C.; Amort, B.; Zelger, B.; Brunner, A.; Wolfram, D. Is It All about Surface Topography? An Intra-Individual Clinical Outcome Analysis of Two Different Implant Surfaces in Breast Reconstruction. *J. Clin. Med.* **2023**, *12*, 1315. [CrossRef] [PubMed]
96. Scarano, A.; Piattelli, M.; Caputi, S.; Favero, G.A. Bacterial adhesion on commercially pure titanium and zirconium oxide disks: An in vivo human study. *J. Periodontol.* **2004**, *75*, 292–296. [CrossRef] [PubMed]

97. Singh, A.V.; Katz, A.; Maharjan, R.S.; Gadicherla, A.K.; Richter, M.H.; Heyda, J.; Del Pino, P.; Laux, P.; Luch, A. Coronavirus-mimicking nanoparticles (CorNPs) in artificial saliva droplets and nanoaerosols: Influence of shape and environmental factors on particokinetics/particle aerodynamics. *Sci. Total Environ.* **2023**, *860*, 160503. [CrossRef] [PubMed]
98. Motoyoshi, M. (Ed.) *Current Techniques and Materials in Dentistry*; MDPI AG: Basel, Switzerland, 2022; ISBN 3036544135.
99. Łosiewicz, B.; Osak, P.; Maszybrocka, J.; Kubisztal, J.; Stach, S. Effect of Autoclaving Time on Corrosion Resistance of Sandblasted Ti G4 in Artificial Saliva. *Materials* **2020**, *13*, 4154. [CrossRef] [PubMed]
100. Osak, P.; Maszybrocka, J.; Kubisztal, J.; Ratajczak, P.; Łosiewicz, B. Long-Term Assessment of the In Vitro Corrosion Resistance of Biomimetic ACP Coatings Electrodeposited from an Acetate Bath. *J. Funct. Biomater.* **2021**, *12*, 12. [CrossRef]
101. Brug, G.J.; van den Eden, A.L.G.; Sluyters-Rehabach, M.; Sluyters, J.H. The analysis of electrode impedances complicated by the presence of a constant phase element. *J. Electroanal. Chem. Interf. Electrochem.* **1984**, *176*, 275–295. [CrossRef]
102. Geetha, M.; Singh, A.K.; Asokamani, R.; Gogia, A.K. Ti based biomaterials, the ultimate choice for orthopaedic implants—A review. *Prog. Mater. Sci.* **2009**, *54*, 397–425. [CrossRef]
103. Reclaru, L.; Meyer, J.M.; Eschler, P.Y. Influence of pH and fluoride on corrosion of titanium and its alloys. *Biomaterials* **1998**, *19*, 1285–1293.
104. White, D.J. Reactivity of fluoride in plaque, mouthrinse, and dentifrice systems. *Caries Res.* **2001**, *35*, 136–141.
105. *ISO 10271:2021-02*; Dentistry—Corrosion Test Methods for Metallic Materials. ISO: Geneva, Switzerland, 2021.
106. Sun, L.; Berndt, C.C.; Gross, K.A.; Kucuk, A. Properties of nanostructured TiO_2 coatings synthesized by plasma spraying. *J. Am. Ceram. Soc.* **2001**, *84*, 2118–2124.
107. Oh, S.; Bram, M.; Pilliar, R.M. Effects of sintering temperature on pitting corrosion resistance of porous titanium. *J. Biomed. Mater. Res. B Appl. Biomater.* **2003**, *67*, 636–644.
108. Wang, C.A.; Zhou, Y.C.; Huang, Y. Thermal shock resistance of thermal barrier coatings. *Surf. Coat. Technol.* **2004**, *177–178*, 117–121.
109. Singh, A.V.; Vyas, V.; Salve, T.S.; Cortelli, D.; Dellasega, D.; Podestà, A.; Milani, P.; Gade, W.N. Biofilm formation on nanostructured titanium oxide surfaces and a micro/nanofabrication-based preventive strategy using colloidal lithography. *Biofabrication* **2012**, *4*, 025001. [CrossRef] [PubMed]
110. Buser, D.; Schenk, R.K.; Steinemann, S.; Fiorellini, J.P.; Fox, C.H.; Stich, H. Influence of surface characteristics on bone integration of titanium implants. A histomorphometric study in miniature pigs. *J. Biomed. Mater. Res.* **1991**, *25*, 889–902. [CrossRef]
111. Brinker, C.J.; Scherer, G.W. *Sol-Gel Science: The Physics and Chemistry of Sol-Gel Processing*; Academic Press: Cambridge, MA, USA, 1990.
112. *ISO 10993-1:2021-06*; Biological Evaluation of Medical Devices—Part 1: Evaluation and Testing Within a Risk Management Process. ISO: Geneva, Switzerland, 2021.
113. FDA Guidance for Industry and FDA Staff, Factors to Consider When Making Benefit-Risk Determinations in Medical Device Premarket Approval and De Novo Classifications. Available online: https://www.fda.gov/media/99769/download (accessed on 12 October 2024).
114. *ISO 10993-5:2009*; Biological Evaluation of Medical Devices—Part 5: Tests for In Vitro Cytotoxicity. ISO: Geneva, Switzerland, 2009.
115. *ISO 10993-10:2023-07*; Biological Evaluation of Medical Devices—Part 10: Tests for Irritation and Sensitization. ISO: Geneva, Switzerland, 2023.

materials

MDPI

Article

Investigation of Precipitation Behavior of a Novel Ni-Fe-Based Superalloy during High-Temperature Aging Treatment

Jun Cheng [1], Kejian Li [1], Zhengang Yang [1], Xin Huo [2], Manjie Fan [2], Songlin Li [2], Shengzhi Li [2], Qu Liu [1,*], Qingxian Ma [1] and Zhipeng Cai [1]

[1] Department of Mechanical Engineering, Tsinghua University, Beijing 100084, China; chengj22@mails.tsinghua.edu.cn (J.C.); kejianli@mail.tsinghua.edu.cn (K.L.); yzghebut@163.com (Z.Y.); maqxdme@mail.tsinghua.edu.cn (Q.M.); czpdme@mail.tsinghua.edu.cn (Z.C.)

[2] Shanghai Turbine Plant, Shanghai Electric Power Generation Equipment Co., Ltd., Shanghai 200240, China; huoxin@shanghai-electric.com (X.H.); fanmj@shanghai-electric.com (M.F.); lisl3@shanghai-electric.com (S.L.); lishzh@shanghai-electric.com (S.L.)

* Correspondence: liuqu@mail.tsinghua.edu.cn; Tel.: +86-10-62790147

Abstract: The precipitation behavior of a novel Ni-Fe-based superalloy developed for advanced ultra-supercritical (A-USC) coal-fired power plant applications during high-temperature aging treatment was investigated. The results showed that the major precipitates in the novel alloy were randomly distributed MC carbides, $M_{23}C_6$ carbides at grain boundaries, and the γ'-Ni_3 (Al, Ti) phase in grain interiors after aging. MC remained relatively stable during both short-term and long-term aging. $M_{23}C_6$ quickly precipitated and exhibited a discrete distribution at grain boundaries during short-term aging, and partly developed into continuous films during long-term aging. After uniform precipitation, the shape of γ' remained spherical, and the size kept increasing with aging time according to the Lifshitz–Slyozov–Wagner (LSW) model. The hardness of the novel alloy was mainly associated with the precipitation behavior of γ'; as γ' gradually precipitated, the hardness steadily increased; after complete precipitation, as the size of γ' increased, the hardness first increased and then decreased, reaching the peak hardness when the average radius of γ' achieved the critical size. In addition, the novel alloy exhibited abnormal coarsening behavior at grain boundaries during both short-term and long-term aging. The coarsened grain boundaries were actually precipitate-free zones (PFZs) and the coarsened and elongated rod-like particles inside were identified as γ' precipitates. The mechanism of strain-induced grain boundary migration and the discontinuous coarsening reaction is proposed for the formation of PFZs. Furthermore, PFZs were considered to be potential crack sources during the creep rupture test, leading to earlier failure of the material.

Keywords: Ni-Fe-based superalloy; advanced ultra-supercritical; high-temperature aging treatment; precipitation behavior; grain boundary migration

Citation: Cheng, J.; Li, K.; Yang, Z.; Huo, X.; Fan, M.; Li, S.; Li, S.; Liu, Q.; Ma, Q.; Cai, Z. Investigation of Precipitation Behavior of a Novel Ni-Fe-Based Superalloy during High-Temperature Aging Treatment. *Materials* **2024**, *17*, 4875. https://doi.org/10.3390/ma17194875

Academic Editor: Daniela Kovacheva

Received: 14 September 2024
Revised: 30 September 2024
Accepted: 1 October 2024
Published: 4 October 2024

1. Introduction

Developing advanced ultra-supercritical (A-USC) units with high parameters can significantly improve the efficiency of coal-fired power generation and effectively reduce carbon dioxide (CO_2) emissions [1–3]. Conventional ferritic or austenitic heat-resistant steel in current power plants cannot meet the service performance requirements of A-USC units [4,5]. Ni-based superalloys including solution-strengthened, γ'-Ni_3 (Al, Ti) phase precipitation-strengthened, and γ''-Ni_3Nb phase precipitation-strengthened types have become the main candidate materials for A-USC units due to their excellent creep strength and corrosion resistance [6,7]. The candidate materials for the high-temperature components of A-USC units that have been developed include Inconel 740/740H [8], Inconel 617/617B [9], Haynes 230/282 [10], and Nimonic 263 [11], among which Inconel 740H and 617B (also known as CCA 617) have the best comprehensive performance. However, 740H and 617B are prohibitively expensive due to the high content of Co, Mo, and W and have poor workability for large components. Recently, the Xi'an Thermal Power Research Institute Co., Ltd. (Xi'an, China) and China Huaneng Group Co., Ltd. (Beijing, China) jointly developed a novel Ni-Fe-based superalloy [12] with high comprehensive cost-effectiveness by increasing the Fe content and reducing the Co content. The novel alloy belongs to Ni-based superalloys strengthened by the γ' phase, with higher tensile strength than that of the 617B alloy and extremely low Nb content. It can effectively solve the problem of segregation in large thick-walled forgings and is expected to be used for manufacturing high-temperature components of A-USC units.

The γ' precipitation-strengthened Ni-based superalloys usually contain a large amount of alloying elements, so the precipitates often form a metastable microstructure after heat treatment [13]. Therefore, high-temperature aging for these alloys often results in the evolution of precipitate size, shape, and distribution, even undergoing phase reactions to transform into topologically close-packed (TCP) phases such as σ and η, which significantly affect the mechanical properties of the materials [14]. For instance, with the progression of aging, the size of γ' gradually increases [14–20], the shape may undergo a transformation between cubic and spherical [18,19], and TCP phases may be formed in some alloys [14], which in turn degrades the mechanical properties including tensile strength [15,16,18–20], toughness [17], and creep strength [14,19]. The evolution of carbides can also have an impact on the strength, plasticity, and fracture behavior of alloys to some extent [21–23]. In addition, several previous studies have pointed out that the formation of precipitate-free zones (PFZs) at grain boundaries weakens the creep resistance of alloys [24,25]. Undoubtedly, it can be concluded that ideal properties can be achieved by optimizing the size, shape, and distribution of the precipitates, especially γ', and controlling harmful phase transformation reactions [26]. Therefore, it is necessary to study the precipitation behavior of Ni-based superalloys during high-temperature aging and further investigate the correlation between the microstructure and mechanical properties to ensure the reliability and safety of high-temperature components in service.

The objective of this paper is to investigate precipitates' evolution of the novel alloy during high-temperature aging treatment. Precipitation behavior during the precipitation process was mainly studied through the short-term aging treatment of the novel alloy in the solution treatment state, and precipitation behavior during the coarsening process was mainly studied through the long-term aging treatment of the novel alloy in the standard heat treatment state. The precipitation and coarsening of the γ' phase were focused on and the abnormal coarsening behavior at grain boundaries during aging was studied in detail. In addition, the correlation between precipitation behavior and mechanical properties was elucidated. The research results of this paper can provide reference for subsequent welding research and the development and improvement of novel Ni-based superalloys.

2. Materials and Methods

The novel alloy material used in this paper was produced by a Shanghai turbine plant from Shanghai Electric Power Generation Equipment Co., Ltd. using heavy forging technology. After measurement by an electron probe microanalyzer (EPMA, JXA8230, JEOL, Tokyo, Japan), its main chemical composition was determined, as shown in Table 1. To eliminate the inherent dendritic and precipitated phases [12,27], the material underwent solution treatment after forging with parameters of 1080 °C × 3.5 h (air cooling for 0.5 h) + 1000 °C × 2 h (air cooling to room temperature). After two-step aging treatment of the novel alloy in the solution treatment (ST) state at 650 °C × 20 h (cooling rate of 5 °C/h) + 800 °C × 20 h (cooling rate of 5 °C/h), the novel alloy in the standard heat treatment (SHT) state was obtained.

Table 1. Chemical composition of the novel alloy.

Elements	Fe	Ni	Cr	Al + Ti	Mo	C + W + Co + B + Nb + Zr
wt.%	42.0	Bal.	16.0	3.4	0.5	0.7

The aging experiments of the novel alloy were conducted using a chamber furnace. As shown in Figure 1a, considering the temperature range for rapid precipitation of the γ' phase, the specimens (6 mm × 6 mm × 3 mm) in the ST state were subjected to short-term aging treatment at 650 °C, 750 °C, 800 °C, and 850 °C, respectively. After reaching the preset temperatures at a rate of 10 °C/s, each group of specimens was taken out sequentially and quenched with water to room temperature at an aging time of 1/2/10/60 h based on the real-time precipitation of the γ' phase. In view of the actual service conditions of the material, as shown in Figure 1b, the long-term aging specimens were in the SHT state before the experiment, and the conditions were 650 °C/5827 h, 675 °C/1615 h, 700 °C/3804 h, and 700 °C/4368 h. In addition, a creep rupture test was performed at 700 °C/230 MPa to study the relationship between abnormally coarsened behavior at grain boundaries and creep rupture failure.

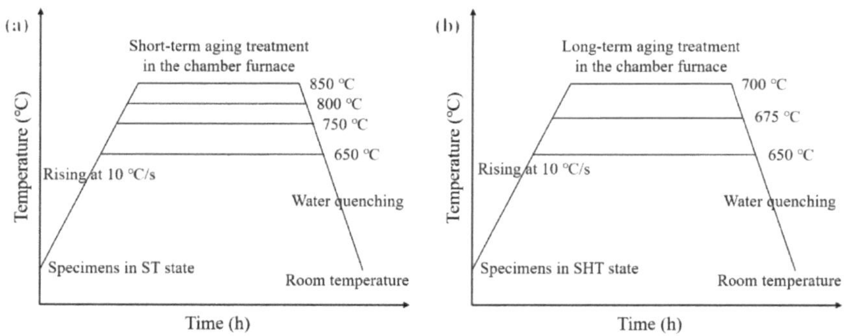

Figure 1. Schematic diagrams of the experimental procedures of (**a**) short-term aging treatment and (**b**) long-term aging treatment.

After the aging experiments and creep rupture test, the metallographic specimens were prepared for subsequent microstructure observation and mechanical property testing. After grinding to 3000 grit and polishing to a 2.5 μm surface finish, metallographic specimens were obtained by immersion in an etching solution (composed of 30 mL glycerol + 25 mL hydrochloric acid + 5 mL nitric acid) for about 7 min at room temperature. An optical microscope (OM, OLYMPUS DP72, Tokyo, Japan) was used to observe metallographic specimens and analyze microscopic features at a relatively macroscopic scale (low magnification). A scanning electron microscope (SEM, GeminiSEM 300, ZEISS, Jena, Germany) was used as the main tool for microstructure observation. The acceleration voltage of the SEM was set to 15 kV, the work distance was adjusted to 8.5–9 mm, and the imaging mode was selected as secondary electron mode with a magnification range of 50–50,000×. The average grain size was determined according to the ASTM E112-13 standard [28], and the size of the γ' phase was calculated by Image J software (version 1.54f). The SEM equipped with energy dispersive spectroscopy (EDS) and EPMA were used to analyze the chemical composition of specific microregions. The crystallographic information of the specimens was characterized by the SEM equipped with an electron backscatter diffraction (EBSD) detector. The specimens for EPMA and EBSD observations were electro-polished using a solution containing 65 mL phosphoric acid + 15 mL sulfuric acid + 12 mL glycerol + 3 mL deionized water + 5 g CrO_3. After the specimens were prepared by a focused ion beam (FIB), transmission electron microscopy (TEM, Tecnai G2 F20 S-TWIN, SemiStar Corp., Morgan Hill, CA, USA) was used for the morphology observation and EDS and selected area electron diffraction (SAED) were used for analysis.

Microhardness testing was conducted based on the ASTM E92-17 standard [29] by using a microhardness tester (FM-810, FT, Tokyo, Japan) with a 200 gf load held for 10 s. The hardness of each specimen was measured at least seven times to obtain the average value, and the distance between the adjacent measurement points was far enough to another measurement. An in situ nanomechanical testing system (FT-NMT04, Femto Tools, Buchs,

Switzerland) installed in the Zeiss SEM was used to test the reduced modulus distribution of a 220 μm × 160 μm region. The continuous stiffness measurement (CSM) method was adopted to measure continuously the reduced modulus as a function of indentation depth with a frequency of 140 Hz and an amplitude of 1 nm. Approximately 4000 CSM indentations with a 3 μm spacing and a 100 nm maximum depth were generated over the entire region, forming a spatial property mapping with high resolution.

3. Results and Discussion

3.1. Initial Microstructure

The overall microstructural characteristics of the novel alloy in the solution treatment (ST) state is shown in Figure 2a. The γ matrix was composed of equiaxed austenite grains with a grain size of approximately 150 μm (ASTM grade 3), containing numerous annealing twins and some randomly distributed blocky particles with a size of 1–10 μm, as shown in Figure 2b. According to the EDS elemental analysis results (Figure 2b) and elemental mapping (Figure 2c), the blocky particles were identified as MC carbides rich in Ti [30] and contained small amounts of carbide-forming elements such as Mo, W, and Nb. MC carbides with a melting point of about 1300 °C mainly formed during solidification and remained in the matrix because of incomplete dissolution and elimination during solution treatment [31]. In addition, the enrichment of Zr, B, and Nb was found in small-sized MC carbides, which may lead to HAZ liquation cracking in the novel alloy during subsequent welding [32,33].

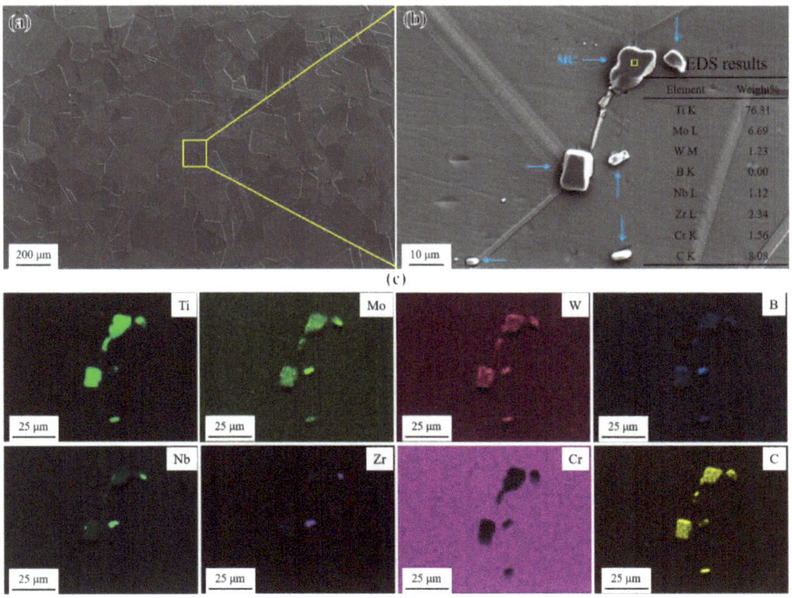

Figure 2. Microstructural characteristics of the novel alloy in the ST state: (**a**) overall microstructure; (**b**) morphology and EDS elemental analysis results of MC; and (**c**) EDS elemental mapping of MC in (**b**).

The overall microstructural characteristics of the novel alloy in the standard heat treatment (SHT) state are shown in Figure 3a. It can be seen that the grain size remained almost unchanged compared to the ST state. And the morphology and composition of blocky MC were also basically stable [34] (Figure 3b). Moreover, the microstructure in the SHT state included discrete small-sized precipitates distributed at grain boundaries (Figure 3c) and a large number of fine spherical γ' particles with an average radius of about 12 nm uniformly distributed in grain interiors [30] (Figure 3d).

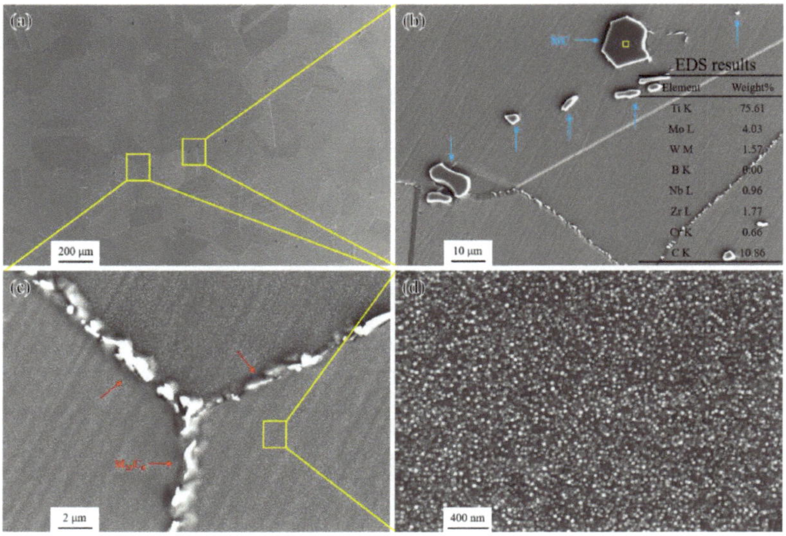

Figure 3. Microstructural characteristics in the SHT state: (**a**) overall microstructure; (**b**) morphology and EDS elemental analysis results of MC; (**c**) morphology of precipitates at grain boundaries; and (**d**) morphology of γ' particles.

According to the TEM analysis results and SAED patterns shown in Figure 4, the discrete precipitates with a size less than 1 um at grain boundaries were identified as $M_{23}C_6$ carbides rich in Cr and Mo [30], which maintained a coherent relationship with the γ matrix and precipitated along grain boundaries during the standard heat treatment. Generally speaking, discrete small-sized $M_{23}C_6$ carbides can hinder dislocation movement and grain boundary sliding, thus improving the creep rupture strength of the novel alloy to some extent [35].

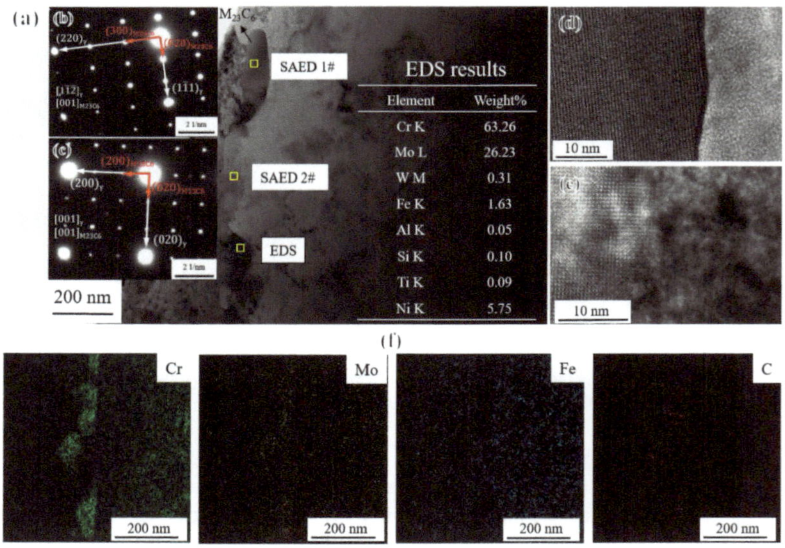

Figure 4. TEM analysis of $M_{23}C_6$: (**a**) bright field image at the grain boundary; SAED results of $M_{23}C_6$ and γ matrix of (**b**) position 1# and (**c**) position 2#; high-resolution image of (**d**) position 1# and (**e**) position 2#; and (**f**) EDS elemental mapping of $M_{23}C_6$ in (**a**).

As a precipitation-strengthened superalloy, $M_{23}C_6$ carbides at grain boundaries have a certain influence on the strength of the novel alloy, but the fine spherical γ' phase is the main strengthening phase [36]. The γ' phase holds a coherent relation with the γ matrix [37] and enhances the strength of the novel alloy by hindering dislocation movement. Its size, content, shape and distribution have a decisive impact on the mechanical properties of the novel alloy [12,26]. As shown in Figure 5, the Vickers hardness of the novel alloy in the ST state was about 164 HV, while it reached about 359 HV in the SHT state, indicating that the strengthening effect of the fine γ' particles precipitating during the standard heat treatment was extremely significant.

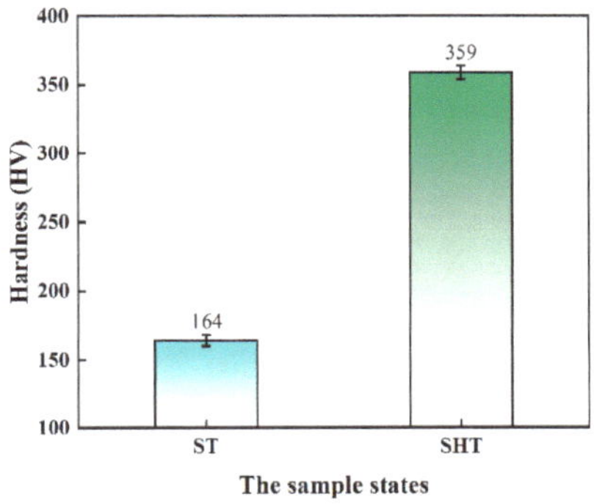

Figure 5. Comparison of hardness between ST and SHT states.

3.2. Precipitation Behavior during Short-Term Aging Treatment

3.2.1. Precipitation Behavior at Grain Boundaries

The microstructure of the novel alloy in the ST state after short-term aging treatment at different temperatures for 60 h is shown in Figure 6. Overall, the grain size remained at ASTM grade 3 at all four temperatures, consistent with the ST state. And there was no obvious change in the quantity and morphology of randomly distributed blocky MC carbides [31]. In addition, with the increase in the aging temperature, the contrast difference between different grains became larger and larger, with almost no difference at 650 °C, a slight difference at 750 °C, an obvious difference at 800 °C, and the most obvious difference at 850 °C, which is speculated to be caused by the precipitation of the γ' phase.

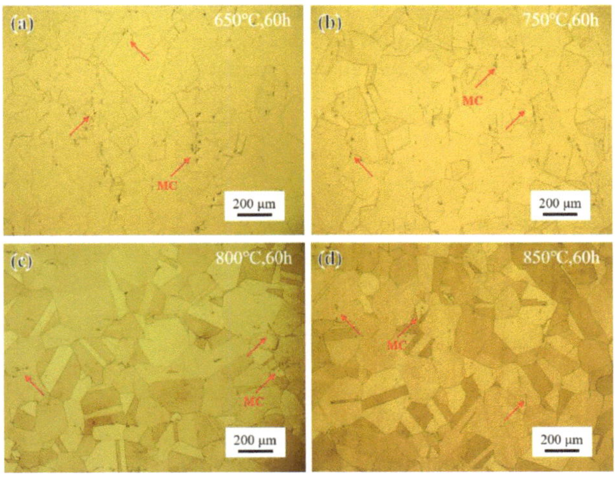

Figure 6. OM images of microstructure after short-term aging for 60 h: (**a**) 650 °C; (**b**) 750 °C; (**c**) 800 °C; and (**d**) 850 °C.

The morphology of $M_{23}C_6$ carbides at grain boundaries of the novel alloy after short-term aging for 60 h at different temperatures is shown in Figure 7. At 650 °C, there was almost no precipitation of $M_{23}C_6$ carbides, and the grain boundaries were very clean. At 750 °C, 800 °C, and 850 °C, as the aging time increased, $M_{23}C_6$ at grain boundaries gradually precipitated, aggregated, connected and grew, and finally formed a discrete chain-like distribution [21]. Furthermore, with the increase in the aging temperature, the precipitation behavior of the γ' phase in grain interiors changed significantly, and the uniform distribution of fine γ' particles could be gradually clearly seen, being completely invisible at 650 °C, blurred at 750 °C, relatively obvious at 800 °C, and most clearly visible at 850 °C, which was consistent with the contrast difference results of the OM images (Figure 6).

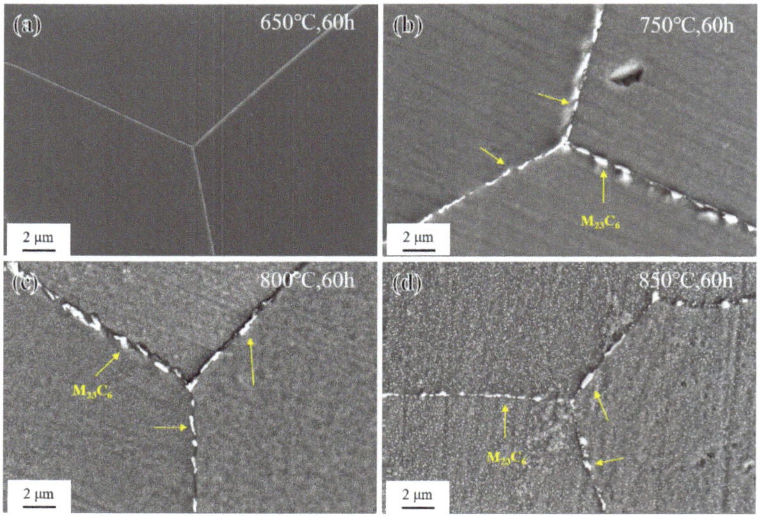

Figure 7. Morphology of $M_{23}C_6$ after short-term aging for 60 h: (**a**) 650 °C; (**b**) 750 °C; (**c**) 800 °C; and (**d**) 850 °C.

It is particularly noteworthy that during short-term aging at 800 °C and 850 °C, as the aging time increased, the grain boundaries of the novel alloy exhibited abnormal coarsening behavior and formed precipitate-free zones (PFZs), as shown in Figure 8. Further related research will be elaborated in Section 3.4.

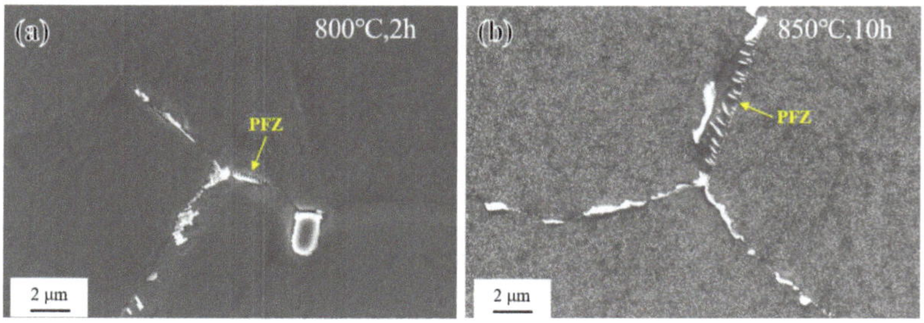

Figure 8. Abnormal coarsening behavior at grain boundaries during short-term aging: (**a**) 800 °C/2 h and (**b**) 850 °C/10 h.

3.2.2. Precipitation Behavior in Grain Interiors

The morphology of the γ' phase in the grain interiors of the novel alloy after short-term aging treatment for 60 h at different temperatures is shown in Figure 9. The size of the γ'

phase (the mean radii were invisible, at 11 nm, 18 nm, and 33 nm, respectively) increased rapidly with the aging temperature, while the volume fraction (invisible, 61%, 50%, and 30%, respectively) and the number density gradually decreased, which was attributed to the higher solubility of Ti and Al in the γ matrix at higher temperatures [38]. It is worth noting that the precipitation rate of the γ' phase was very slow during short-term aging at 650 °C, so its size was extremely small and could not be observed. Therefore, only performing the first step of the aging treatment at 650 °C for 20 h cannot quickly achieve the effect of strengthening the novel alloy. In order to regulate the size and volume fraction of the γ' phase to optimize the strength and toughness matching of the material, a second step of the aging treatment at 800 °C for 20 h was added to the novel alloy. When the novel alloy after the first step of aging was subjected to the second step of aging treatment (i.e., standard heat treatment), the γ' phase could quickly and completely precipitate to reach equilibrium, and then gradually grow to reach a relatively suitable size and volume fraction (Figure 3d), thereby ensuring excellent mechanical properties of the material [26].

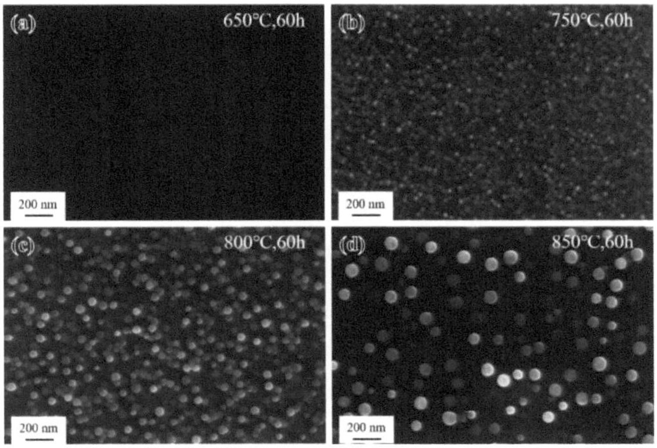

Figure 9. Morphology of the γ' phase after short-term aging for 60 h: (**a**) 650 °C; (**b**) 750 °C; (**c**) 800 °C; and (**d**) 850 °C.

The hardness of the specimens subjected to short-term aging at different temperatures for various time is shown in Figure 10a (the aging time of the specimen in the ST state is 0 h, which is set to 0.01 h for logarithmic plotting). At 650 °C, the hardness increased continuously with aging time; at 750 °C, the hardness continuously increased and gradually reached the peak hardness of 377 HV at 60 h; at 800 °C and 850 °C, the hardness first increased and then decreased, reaching the peak hardness (349 HV and 331 HV, respectively) at 2 h. To explain the differences in peak hardness at different temperatures, the equilibrium phase diagram of the main precipitates of the novel alloy was calculated using the thermodynamic calculation software JMatPro (version 7.0), as shown in Figure 10b. It can be seen that from 650 °C to 850 °C, as the temperature increases, the mass fraction of the γ' phase gradually decreases, indicating a decrease in volume fraction. Therefore, the peak hardness of the novel alloy gradually decreased [39]. The relationship between the measured peak hardness (the measured peak hardness here is the maximum hardness in the measured value, not necessarily the actual peak hardness during aging) and the content of γ' is fitted, as shown in Figure 10c. The fitting determination coefficient (R^2) is 0.97244, indicating a good fitting effect, which once again proves that the variation in hardness was directly related to the precipitation behavior of the γ' phase. Based on the fitted line, the peak hardness at 650 °C is predicted to be approximately 404 HV.

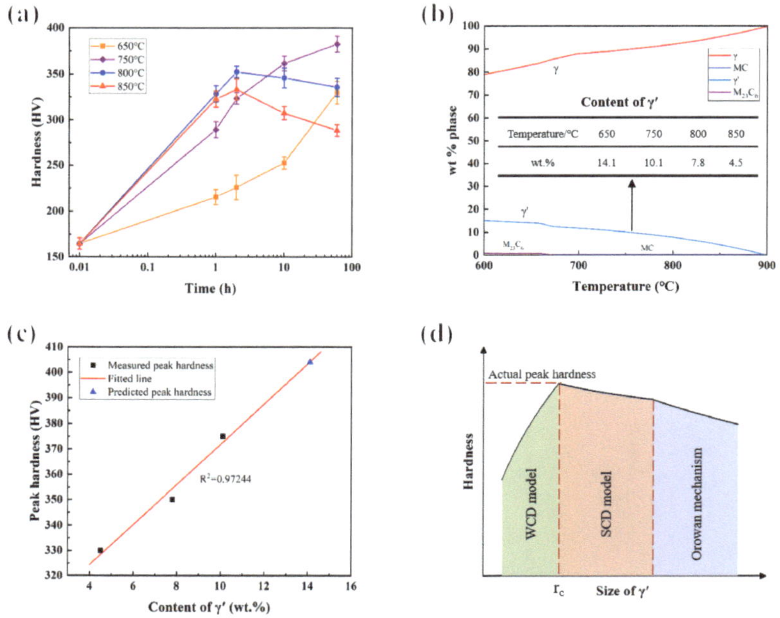

Figure 10. (**a**) A plot of hardness versus short-term aging time at different temperatures; (**b**) the equilibrium phase diagram of the main precipitates; (**c**) a plot of peak hardness versus the content of γ'; and (**d**) schematic diagram of hardness versus the size of the γ' phase under different dislocation strengthening mechanisms.

In γ' precipitation-strengthened Ni-based superalloys, γ' acts as a barrier to the movement of dislocations through a cutting mechanism and the Orowan bypassing mechanism [37] to achieve the strengthening effect. The type of cutting mechanism can be roughly subdivided into two groups, the weakly coupled dislocations (WCD) model [40] and the strongly coupled dislocations (SCD) model [41–43]. Combined with the precipitation behavior of γ', the variation in hardness with time in Figure 10a can be explained through the dislocation strengthening mechanisms of γ', as shown in Figure 10d. In the early stage of short-term aging, a large amount of γ' was precipitated, and its strengthening effect was positively correlated with the volume fraction, so the hardness of the novel alloy increased rapidly. After a certain period of time, the precipitation of γ' was almost over, and its volume fraction remained basically unchanged. At this time, the size of γ' was still small, and the strengthening mechanism was the WCD model. The strengthening effect was enhanced with the increase in the size of γ', and the hardness increased accordingly [44]. As the size of γ' increased to reach the critical size r_c, the strengthening mechanism began to transform into the SCD model. The strengthening effect of γ' corresponding to the transition point was the strongest [45] and the hardness reached the actual peak hardness. With the further increase of the size of γ', the hindering effect on dislocations gradually decreased and the hardness kept decreasing. When the critical resolved shear stress (CRSS) required for dislocations to bypass γ' was smaller than the shear stress of the SCD model, the strengthening mechanism shifted to the Orowan mechanism, the strengthening effect became weaker [39], and the hardness also decreased accordingly.

Conversely, the critical size of γ' of the novel alloy can also be inferred from the change in hardness. The morphology of γ' after short-term aging at 850 °C for 1–60 h is shown in Figure 11. The mean radii were 8 nm, 10 nm, 16 nm, and 33 nm, respectively. At 1 h, according to Figure 10a, the hardness was in the rising stage, indicating that the strengthening mechanism was the WCD model. Therefore, it can be judged that the critical size was larger than the average radius (8 nm) of γ' at this time. When the aging time was 10 h, the hardness was in the decreasing stage, suggesting that the strengthening mechanism was the SCD model or the Orowan mechanism, so the critical size was smaller than the average radius (16 nm) of γ' at this time. Consequently, it can be concluded that the critical size of achieving actual peak hardness during aging was within the range of 8–16 nm.

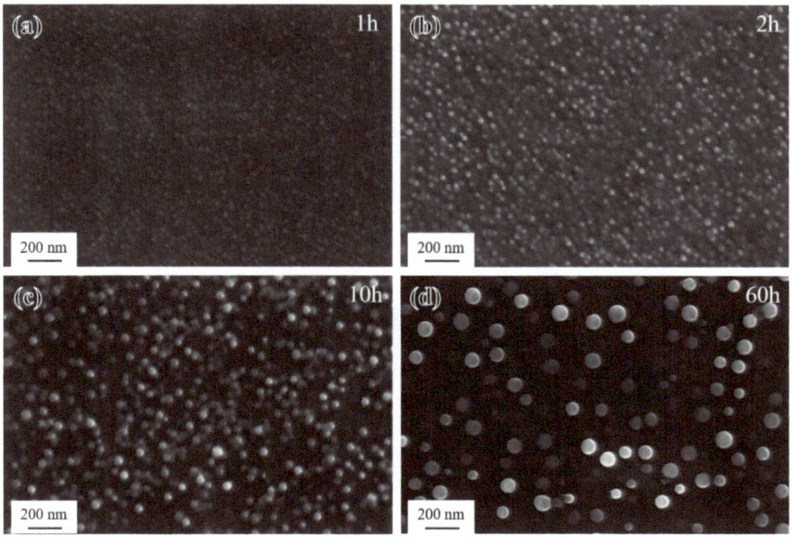

Figure 11. Morphology of γ' after short-term aging at 850 °C: (**a**) 1 h; (**b**) 2 h; (**c**) 10 h; and (**d**) 60 h.

3.3. Precipitation Behavior during Long-Term Aging Treatment

3.3.1. Precipitation Behavior at Grain Boundaries

The microstructure of the novel alloy in the SHT state after long-term aging treatment at different temperatures is shown in Figure 12. Overall, the grain size was relatively unchanged during long-term aging and remained at ASTM grade 3, consistent with the SHT state. In addition, there were no obvious changes in the morphology and distribution of MC carbides [46].

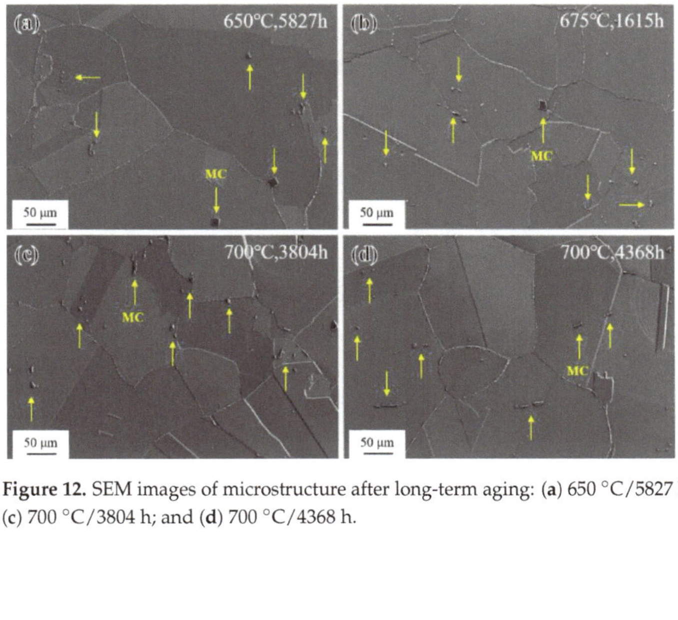

Figure 12. SEM images of microstructure after long-term aging: (**a**) 650 °C/5827 h; (**b**) 675 °C/1615 h; (**c**) 700 °C/3804 h; and (**d**) 700 °C/4368 h.

The morphology of $M_{23}C_6$ at grain boundaries of the novel alloy after long-term aging at different temperatures is shown in Figure 13. Compared with the SHT state (Figure 3c), some discrete $M_{23}C_6$ had developed into a short and thin continuous film (about 10 μm long) at grain boundaries after long-term aging [13], which might reduce the impact toughness and creep rupture life of the novel alloy to some extent [18,47] due to reducing the pinning effect and facilitating crack nucleation and propagation at grain boundaries.

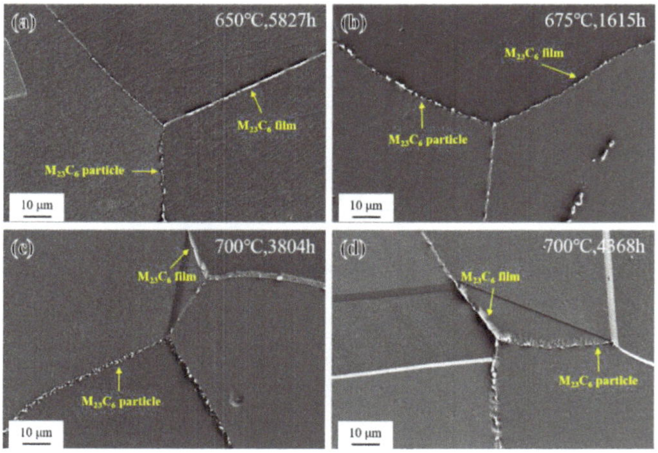

Figure 13. Morphology of $M_{23}C_6$ after long-term aging: (**a**) 650 °C/5827 h; (**b**) 675 °C/1615 h; (**c**) 700 °C/3804 h; and (**d**) 700 °C/4368 h.

As shown in Figure 14, after long-term aging at 650–700 °C, the novel alloy also exhibited abnormal coarsening behavior at grain boundaries and formed PFZs, but it was different from that of short-term aging only occurring at 800 °C and 850 °C. Further related research will be elaborated in Section 3.4.

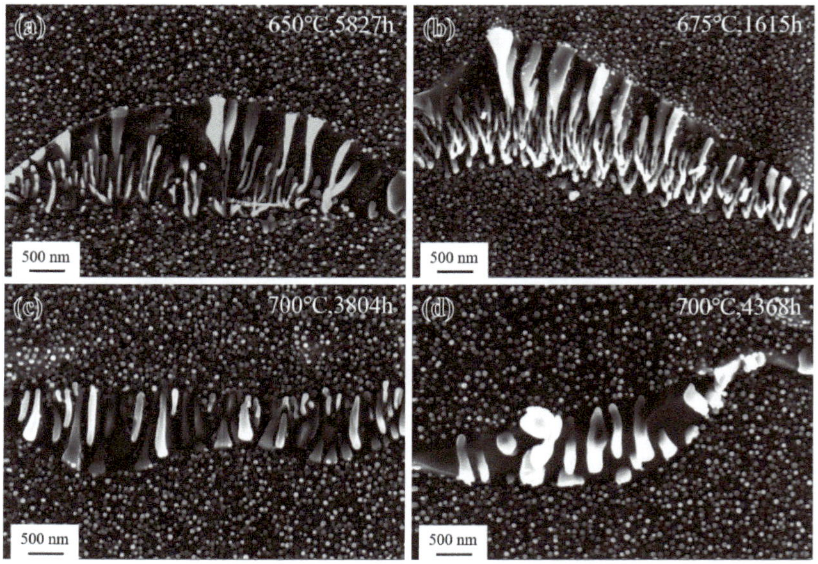

Figure 14. Abnormal coarsening behavior at grain boundaries during long-term aging: (**a**) 650 °C/5827 h; (**b**) 675 °C/1615 h; (**c**) 700 °C/3804 h; and (**d**) 700 °C/4368 h.

3.3.2. Precipitation Behavior in Grain Interiors

The morphology of the γ' phase in the grain interiors of the novel alloy after long-term aging treatment at different temperatures is shown in Figure 15. It can be seen that γ' particles (the mean radii were 21 nm, 18 nm, 29 nm, and 30 nm, respectively) maintained a spherical shape and uniform distribution during long-term aging but underwent a certain degree of coarsening process at all three temperatures compared to the SHT state (Figure 3d). Comparing Figure 15a,b, it can be seen that the average size of γ' particles at 650 °C was larger, which was related to the much longer aging time at this temperature. Figure 15c,d show that the size of γ' particles gradually increased as the aging time increased during long-term aging at 700 °C, while the number density gradually decreased [13]. The aging time at 700 °C was shorter than that at 650 °C, but the temperature was significantly higher, so the size of γ' particles was relatively larger.

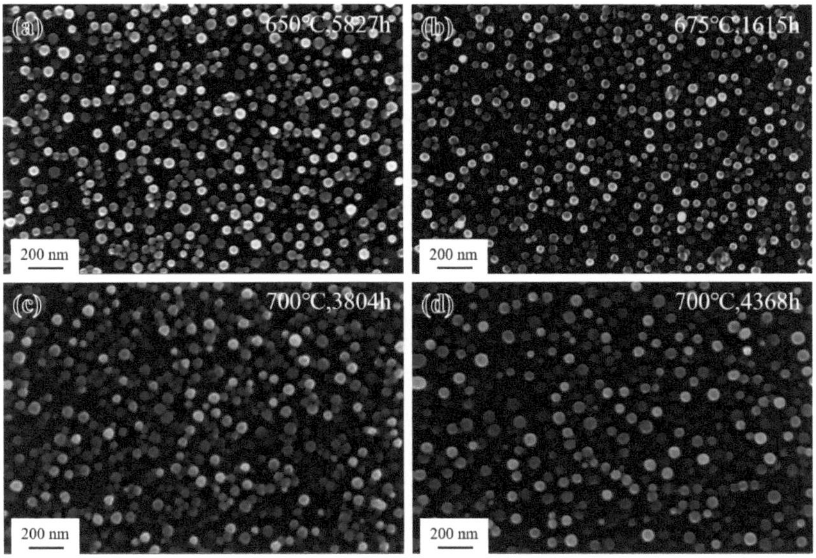

Figure 15. Morphology of γ' phase after long-term aging: (**a**) 650 °C/5827 h; (**b**) 675 °C/1615 h; (**c**) 700 °C/3804 h; and (**d**) 700 °C/4368 h.

Many findings [13,48–52] have confirmed that the coarsening behavior of γ' particles in Ni-based superalloys follows the diffusion-controlled [53] Lifshitz–Slyozov–Wagner (LSW) growth kinetics model. The LSW model states that when the chemical composition and volume fraction of γ' particles remain constant at the same temperature, the size of γ' particles varies according to the formula $r_t^3 - r_0^3 = kt$ [54,55], where r_t and r_0 are the average radius of γ' particles at aging times t and 0, respectively, and k is the coarsening rate [49]. The cubic of the average size of γ' particles after various long-term aging times at three temperatures is plotted as a function of time in Figure 16a. According to the result at 700 °C, it can be seen that the cubic of the average radius of γ' particles increases linearly with the aging time and R^2 of the fitted line is 0.99941, which indicates that the coarsening behavior of γ' particles in the novel alloy during long-term aging also followed the diffusion-controlled LSW model.

During long-term aging, it was mainly the reduction in the specific free energy at the γ'/γ interface [49] that drove the coarsening of γ' particles. As the aging progressed, due to the gradual increase in lattice misfit, the coherency of the γ'/γ interface was disrupted and the interfacial energy increased [56], resulting in an enhanced driving force for the coarsening of γ' particles. The coarsening behavior of γ' particles was then promoted to occur at the expense of interfacial energy reduction. Actually, although immobile in the γ matrix, γ' particles tended to reduce the interfacial energy through the diffusion of the forming elements Al and Ti from smaller particles to larger particles, which meant promoting the coarsening of large particles by dissolving small γ' particles [31]. Therefore, the number density of γ' particles decreased but the average size increased. From the slope k of the line fitted by the LSW model, it can be seen that the higher the temperature, the faster the coarsening of γ' particles.

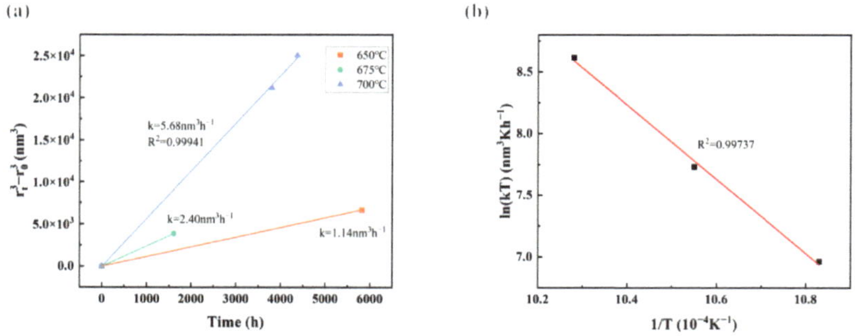

Figure 16. (**a**) A plot of the size of the γ' phase versus long-term aging time at different temperatures and (**b**) the relationship between temperature (T) and the coarsening rate (k).

In addition, after obtaining the coarsening rate k at the three temperatures according to Figure 16a, the coarsening activation energy Q of γ' particles can also be calculated by the formula $ln(kT)$ = constant $-Q/RT$ [57], where T is the absolute temperature and R is the ideal gas constant of 8.314 J/(mol K). The relationship between $ln(kT)$ and $1/T$ is shown in Figure 16b. The slope of the fitted line (R^2 = 0.99737) is -3.01, so the coarsening activation energy of γ' particles is calculated to be 250.3 kJ/mol, which is close to the volume diffusion activation energies of both Ti and Al in Ni (257 and 270 kJ/mol, respectively) [58]. Similar activation energy values for γ' coarsening have also been reported in some Ni-based superalloys [48,49,58,59]. This result again suggests that the coarsening of γ' particles was mainly achieved by the volume diffusion of Al and Ti, and the coarsening activation energy was dominated by the mole fractions of Ti and Al atoms in the γ matrix [12].

The hardness measurement results of the novel alloy after long-term aging at different temperatures are shown in Figure 17. According to the results of short-term aging, the critical size of the novel alloy was within the range of 8–16 nm. As shown in Figure 15, the average radius of γ' particles after long-term aging at 675 °C for 1615 h was the smallest at 18 nm, exceeding the critical size. Therefore, its hardness (381 HV) was in the decreasing stage shown in Figure 10d, and the strengthening mechanism was the SCD model or the Orowan mechanism. Similarly, after long-time aging at 650 °C for 5827 h and 700 °C for 3804 h and 4368 h, the average radii of γ' particles also exceeded the critical size, the hardness was in the decreasing stage, and the larger the average size of γ' particles, the lower the hardness. In addition, it can also be judged that the hardness of the SHT state (359 HV) was in the rising stage, and its strengthening mechanism was the WCD model, so the average radius of γ' particles (12 nm) had not reached the critical size. Consequently, it can be further concluded that the critical size of achieving actual peak hardness was within the range of 12–16 nm.

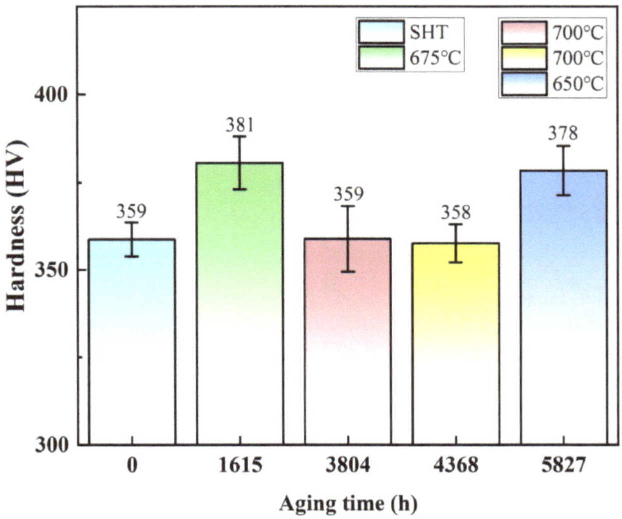

Figure 17. Comparison of hardness before and after long-term aging at different temperatures.

3.4. Abnormal Coarsening Behavior at Grain Boundaries

3.4.1. Characterization of PFZs

In the study of precipitates' evolution during the high-temperature aging treatment of the novel alloy mentioned above, abnormal coarsening behavior at grain boundaries occurred during short-term aging at 800 °C and 850 °C (Figure 8), and during long-term aging at 650 °C, 675 °C, and 700 °C (Figure 14). Partial grain boundaries migrated, and precipitate-free zones (PFZs) [24,25,31,60–63] were formed. Actually, PFZs used to specifically describe the coarsened grain boundary features were not truly free of second phases. On the contrary, while a large number of previously existing fine γ' phases disappeared, some coarsened and elongated rod-like particles were formed in these regions, so PFZs are also known as precipitate-denuded regions or coarsened zones [64]. From the EDS elemental analysis results (Figure 18b) and EPMA elemental mapping (Figure 18c), it can be seen that the rod-like particles were rich in Ni, Al, and Ti elements, suggesting that they were likely abnormally coarsened γ' precipitates.

To confirm the speculation about the rod-like particles, the region containing the PFZ (as indicated by the yellow frame) in Figure 19a was cut and sampled using the FIB technique for TEM observation, and the bright field image was obtained, as shown in Figure 19b. The SAED patterns shown in Figure 19c indicate that the rod-like phase held a coherent relation with the γ matrix and involved the same face-centered cubic (FCC) crystal orientation. Combined with the EDS elemental mapping of Ni, Ti, and Al shown in Figure 19d, it can be confirmed that the rod-like phase was indeed an abnormally coarsened γ' phase [62,63]. In addition, there was no presence of the η phase in PFZs, which could be attributed to the low ratio of Ti to Al (Ti: Al = 1.43) in the novel alloy [13].

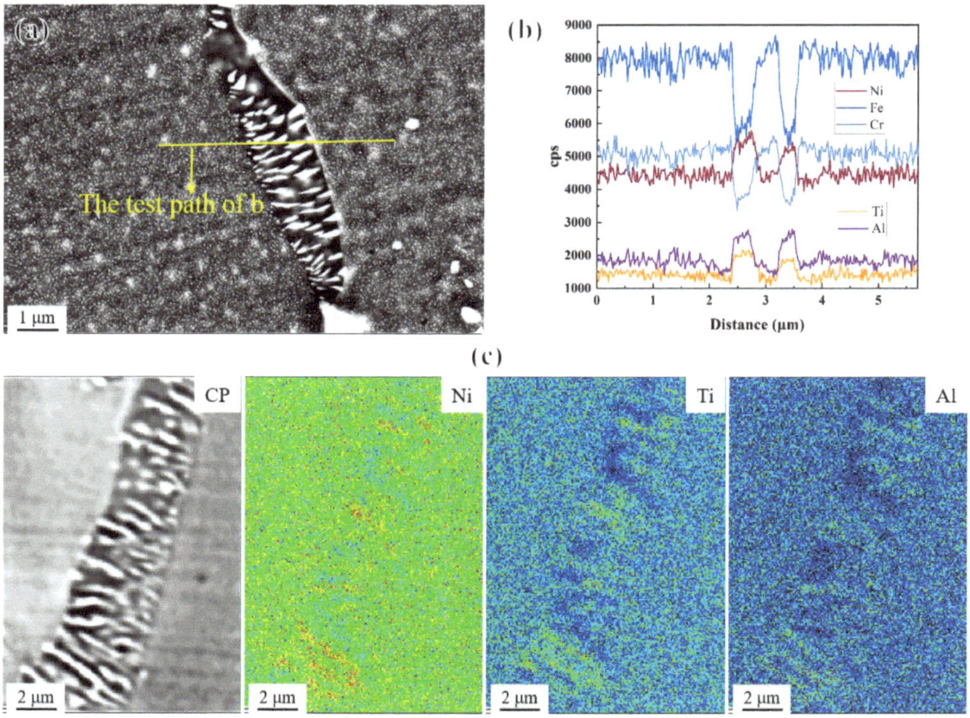

Figure 18. (**a**) SEM image of PFZ; (**b**) EDS elemental analysis along the test path in (**a**); and (**c**) EPMA elemental mapping of PFZ.

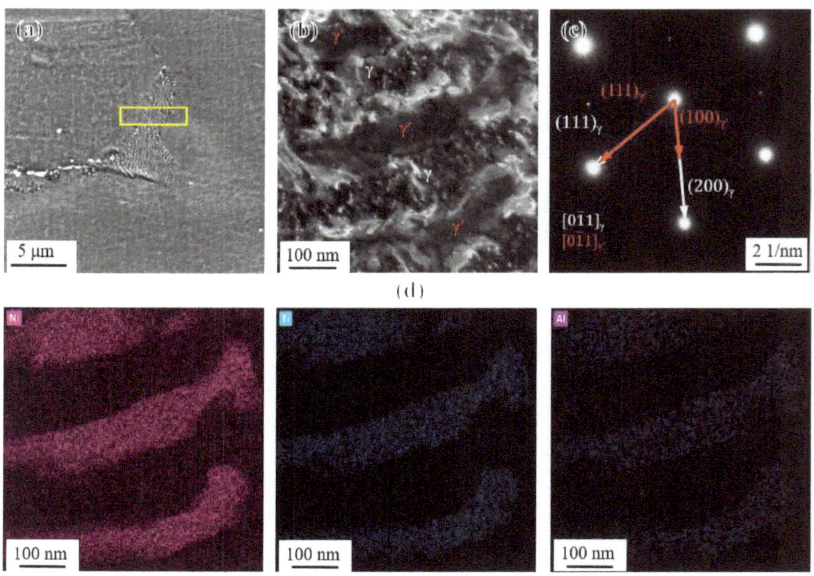

Figure 19. (**a**) Microstructure of PFZ taking the position of the TEM sample prepared by FIB; (**b**) bright field image of the TEM sample; (**c**) SAED patterns of rod-like γ' phase and γ matrix; and (**d**) EDS elemental mapping of rod-like γ' phase.

3.4.2. Formation of PFZs

In order to clarify the formation of the PFZ, the crystal orientation and dislocation density distribution of grains on both sides of the coarsened grain boundary were investigated by EBSD. As shown in Figure 20a,b, the crystal orientation of the PFZ remained consistent with the left grain, indicating that the grain boundary migrated from the left grain to the right grain. Furthermore, as shown in Figure 20c, the kernel average misorientation (KAM) of the left grain was obviously lower than that of the right grain. Generally speaking, the larger the KAM, the higher the geometrically necessary dislocation (GND) density [65] and the higher the strain gradient and stored strain energy inside the crystal [66]. Therefore, it can be inferred that grain boundaries migrated from grains with a lower dislocation density to grains with a higher dislocation density, forming PFZs to reduce the strain energy near the grain boundaries [60,67].

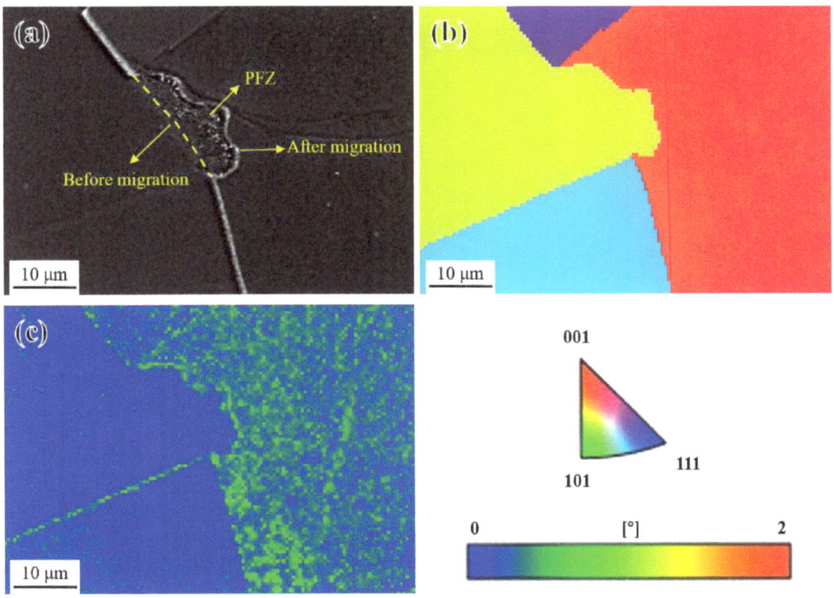

Figure 20. EBSD analysis of PFZ; (**a**) SEM image; (**b**) IPF map of (**a**); and (**c**) KAM map of (**a**).

In fact, similar abnormal coarsening behavior at grain boundaries has also been observed in many Ni-based superalloys, including TG700A [31,60], 617B [61], 740H [62], Nimonic 263 [63], etc. And PFZs were also formed, with the rod-like γ' phase appearing inside. However, most researchers have discovered this phenomenon during high-temperature creep (especially in the third stage of creep), and believe that PFZs only occur when both high-stress or strain and high-temperature conditions are simultaneously present and when the stress or strain is applied for a sufficiently long period of time. Accordingly, several mechanisms including Nabarro–Herring creep, grain boundary carbide formation, and grain boundary sliding or migration have been proposed to explain the formation of PFZs. Comparatively, the occurrence during aging is rarely mentioned. The lack of external stress resulted in the previous mechanisms no longer being fully applicable to the abnormal coarsening behavior at grain boundaries discovered during aging.

Based on the above research results, this paper proposes a new mechanism of grain boundary migration and the discontinuous coarsening reaction to explain the abnormal coarsening behavior at grain boundaries during aging of the novel alloy, as shown in Figure 21. During the forging and heat treatment processes of large thick-walled structural components, high residual stress or strain is easily generated inside the material due to uneven temperature distribution and plastic deformation [68,69]. During aging, the significant decrease in the lattice constant [12] results in strong negative creep behavior, causing high shrinkage stress within the novel alloy. Under the coupling effect of high temperature and high stress or strain, dislocation pile-up occurs due to a significant difference in orientation between adjacent grains, and the grain boundary migrates towards the grain with a higher dislocation density [70]. After grain boundary migration occurs, the rapid diffusion and local segregation of Al and Ti along the moving grain boundaries lower the solidus temperature of the γ' phase, causing the fine spherical γ' particles in the area through which the grain boundary migration passes to dissolve [31]. At the same time, the local solute atoms are in a supersaturated state. Due to the stability of the composition, the new γ' phase directionally precipitates along the grain boundary migration direction and discontinuous coarsening reactions occur under the Oswald-like mechanism, forming rod-like abnormally coarsened and elongated γ' precipitates inside the area passed by the migrating grain boundary [71,72]. Finally, PFZs lacking fine spherical γ' but containing the coarse rod-like γ' phase are formed, manifested as abnormally coarsened grain boundaries through metallographic observation.

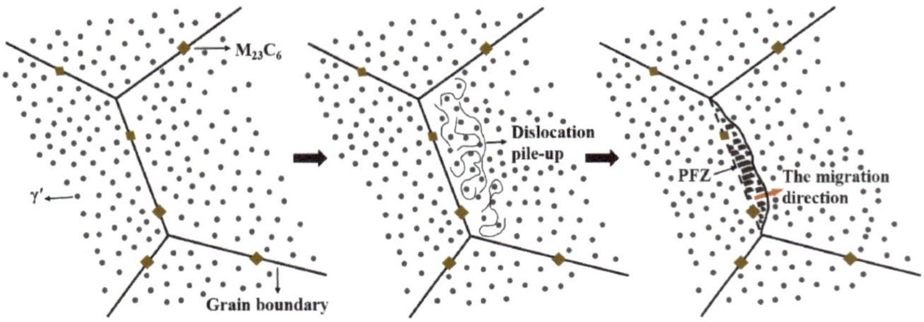

Figure 21. Schematic diagram of the formation of PFZ.

The mechanism proposed above can explain why abnormal coarsening behavior at grain boundaries occurred under different conditions for short-term aging and long-term aging. When the aging temperature reached 800 °C or above, the speed of grain boundary migration and element diffusion was greatly improved, so abnormal coarsening behavior at grain boundaries could occur in just a few hours, as shown in Figure 8. When the temperature dropped to 650 °C, due to the cumulative effect, the behavior could also occur when the aging time reached several thousand hours, as shown in Figure 14.

Overall, there are still some limitations in the research on the mechanism. For instance, residual stress and negative creep shrinkage stress are only qualitatively described and not quantitatively measured. The study on the formation conditions of PFZs is still insufficient, and subsequent creep experiments can be considered to quantitatively study the formation conditions and evolution of PFZs. In addition, most of the key components of the A-USC units need to be welded, and considering that the welding residual stress, welding metallurgical behavior, and post-weld heat treatment (PWHT) may affect the formation of PFZs, it is also necessary to study the weldability of the novel alloy.

3.4.3. Influence of PFZs

After the occurrence of abnormal coarsening behavior at grain boundaries in the novel alloy, PFZs in the coarsened grain boundaries lacked a sufficient amount of uniformly distributed spherical γ' particles for strengthening, and the coarsened and elongated rod-like γ' phase may even have had a weakening effect. Therefore, it is speculated that the mechanical properties of PFZs were weaker compared to the intragranular regions. A high-resolution nanoindentation test in a 220 μm × 160 μm region was conducted to compare the mechanical properties of PFZs and near grains. The SEM image and inverse pole figure (IPF) of the selected area are shown in Figure 22a,b, respectively. Figure 22c is a local nanoindentation image of the area after the completion of the test. It can be seen from Figure 22d corresponding to Figure 22a that the reduced modulus in PFZs was significantly lower than that in the intragranular regions on both sides, which indicated that PFZs of the novel alloy indeed had poor mechanical properties and might fail first during service [24,25,31,62].

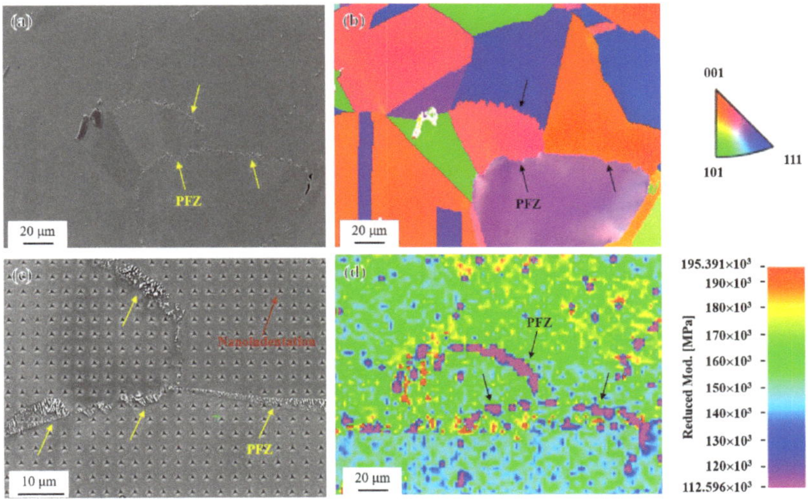

Figure 22. In situ nanomechanical test of PFZs: (**a**) SEM image; (**b**) IPF map of (**a**); (**c**) local nanoindentation image of (**a**); and (**d**) reduced modulus map of (**a**).

Observation of the creep rupture specimen at 700 °C/230 MPa confirms that abnormal coarsening behavior at grain boundaries had a significant impact on the creep rupture performance of the novel alloy. From Figure 23a, it can be seen that creep micro-voids originated from the PFZ [31], gradually grew and expanded, connected to form cracks, and finally promoted the rupture, as shown in Figure 23b. Consequently, combined with the analysis of the weaker properties of PFZs mentioned above, it can be inferred that the abnormal coarsening behavior at grain boundaries was an important reason for the earlier failure of the novel alloy during the high-temperature creep rupture test. Inhibiting this behavior will greatly improve the creep rupture performance.

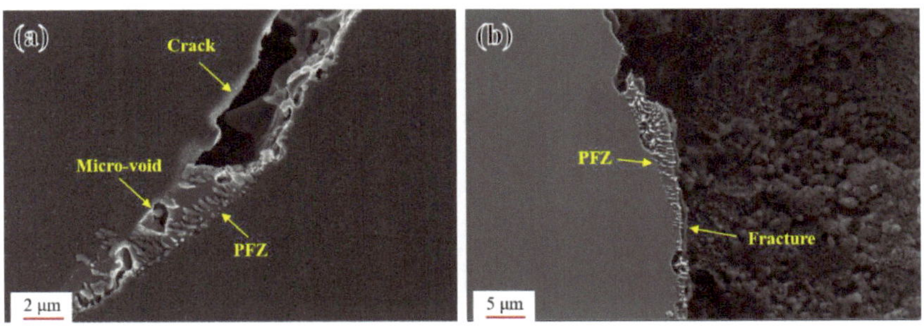

Figure 23. Cross-section SEM images of the creep rupture specimen at 700 °C/230 MPa: (**a**) the relationship between the PFZ and the crack. and (**b**) the relationship between the PFZ and the fracture.

4. Conclusions

In this paper, the precipitation behavior of a novel Ni-Fe-based superalloy in the solution treatment (ST) state and the standard heat treatment (SHT) state during aging treatment was studied through microstructure characterization and mechanical property testing, and a detailed analysis of abnormal coarsening behavior at gain boundaries was provided. The following conclusions are made based on the research results above:

(1) When the novel alloy was in the ST state, the grain size was at ASTM grade 3, and blocky MC carbides rich in Ti were randomly distributed in grain interiors and at grain boundaries. When the novel alloy was in the SHT state, the grain size remained almost unchanged, and MC was also stable. Discrete $M_{23}C_6$ carbides rich in Cr precipitated at grain boundaries, and numerous uniformly distributed fine spherical γ' particles precipitated in grain interiors, which directly led to a significant increase in hardness.

(2) After short-term aging of the novel alloy in the ST state, there was no obvious change in the grain size and MC. At 650 °C, almost no precipitates were observed at grain boundaries, and the γ' phase precipitated slowly, resulting in a small quantity and size. At 750 °C, 800 °C, and 800 °C, discrete $M_{23}C_6$ and the fine γ' phase rapidly precipitated and gradually grew. After the complete precipitation of the γ' phase, the hardness reached the peak hardness when the average radius of the γ' phase increased to the critical size, and then gradually decreased with the increase in the size of the γ' phase.

(3) After long-term aging of the novel alloy in the SHT state, the grain size and MC remained almost unchanged. Discrete $M_{23}C_6$ at grain boundaries underwent a transition to continuous films. The size of the γ' phase in grain interiors significantly increased, and the coarsening behavior followed the Lifshitz–Slyozov–Wagner (LSW) model with a coarsening activation energy of 250.3 kJ/mol. The hardness increased at 650 °C and 675 °C and slightly decreased at 700 °C, while all were below the peak value, which was related to the fact that the average radii of γ' exceeded the critical size (12–16 nm).

(4) The novel alloy exhibited abnormal coarsening behavior at grain boundaries during both short-term and long-term aging. Partial grain boundaries migrated, and precipitate-free zones (PFZs) were formed. It was confirmed that the grain boundary migrated from the grain with a lower KAM to the grain with a higher KAM, forming abnormally coarsened and elongated rod-like γ' particles along the migration direction. It is proposed that the formation mechanism of PFZs is strain-induced grain boundary migration and a discontinuous coarsening reaction.

(5) PFZs had weaker mechanical properties compared to nearby grains and were considered to be potential crack sources during the creep rupture test, leading to an earlier failure of the novel alloy. To ensure the service performance of the novel alloy, it is necessary to inhibit the abnormal coarsening behavior at grain boundaries. Based on the study results of the formation mechanism of PFZs, relevant measures should consider optimizing grain boundary character distribution (GBCD), improving composition segregation, and reducing residual stress through appropriate heat treatment. Once the abnormal coarsening behavior is controlled, the performance will be greatly improved, and it is predicted that the much cheaper novel alloy may be able to match the durability of Ni-based superalloys, such as 617B and 740H, and further promote the development of practical applications for thermal power units above 650 °C.

Author Contributions: Conceptualization, J.C. and Q.L.; methodology, J.C., K.L., Z.Y., X.H., M.F., S.L. (Songlin Li), S.L. (Shengzhi Li), Q.L. and Z.C.; validation, K.L., Z.Y., M.F. and S.L. (Shengzhi Li); formal analysis, J.C., K.L. and Z.Y.; investigation, J.C., K.L., Z.Y., X.H., M.F. and Q.L.; resources, X.H., M.F., S.L. (Songlin Li) and S.L. (Shengzhi Li); data curation, J.C. and Z.Y.; writing—original draft preparation, J.C.; writing—review and editing, J.C., K.L. and Q.L.; visualization, Q.M.; supervision, X.H., S.L. (Songlin Li) and Q.L.; project administration, K.L., Q.L., Q.M. and Z.C.; funding acquisition, Q.L., Q.M. and Z.C. All authors have read and agreed to the published version of the manuscript.

Funding: This research was funded by the National Key R&D Program of China, grant number 2023YFB3405400.

Institutional Review Board Statement: Not applicable.

Informed Consent Statement: Not applicable.

Data Availability Statement: The raw data supporting the conclusions of this article will be made available by the authors upon request.

Conflicts of Interest: Author Mr. Xin Huo, Ms. Manjie Fan, Mr. Songlin Li and Mr. Shengzhi Li were employed by Shanghai Turbine Plant, Shanghai Electric Power Generation Equipment Co., Ltd. The remaining authors declare that the research was conducted in the absence of any commercial or financial relationships that could be construed as a potential conflict of interest.

References

1. Bugge, J.; Kjær, S.; Blum, R. High-efficiency coal-fired power plants development and perspectives. *Energy* **2006**, *31*, 1437–1445. [CrossRef]
2. Yu, K.; Zhou, Z.; Zhou, G.Y.; Cao, G.; Yan, J.; Zhang, P.; Li, P.; Liu, P.; Yuan, Y. Microstructural evolution and creep behavior of a novel Ni-Fe-based superalloy GH2070P for 700 °C class A-USC steam pipes. *J. Alloys Compd.* **2024**, *1004*, 175843. [CrossRef]
3. Cheng, S.H.; Qin, X.Z.; Wen, B.; Wang, J.Q.; Wu, Y.S.; Zhou, L.Z. Effects of Al and Ti on thermal expansion behavior of a Ni-Fe-Cr-based superalloy: A combined experimental and first-principle study. *J. Mater. Res. Technol.-JMR&T* **2024**, *29*, 3805–3812.
4. Zhao, Z.; Li, S.; Ma, X.; Yang, X.; Wen, Z.; Li, Y.; Yan, J.; Zhang, C. Research status of main steam pipe candidate material characteristics and service performance for 700 °C ultra supercritical unit. *Therm. Power Gener.* **2021**, *50*, 1–12.
5. Viswanathan, R.; Coleman, K.; Rao, U. Materials for ultra-supercritical coal-fired power plant boilers. *Int. J. Press. Vessel. Pip.* **2006**, *83*, 778–783. [CrossRef]
6. Liu, D.Y.; Yan, L.Y.; Hou, S.X. The evaluation of high temperature corrosion of the Inconel 740H in simulated coal-fired environments. *Corros. Sci.* **2019**, *161*, 108150. [CrossRef]
7. Jiang, H.; Liu, C.; Dong, J.; Zhang, M. The effect of Mo and Ti elements on long-term microstructure stability of 617B nickel-base superalloy. *J. Alloys Compd.* **2020**, *821*, 153217. [CrossRef]
8. Chong, Y.; Liu, Z.D.; Godfrey, A.; Liu, W.; Weng, Y.Q. Microstructure evolution and mechanical properties of Inconel 740H during aging at 750 °C. *Mater. Sci. Eng. A-Struct. Mater. Prop. Microstruct. Process.* **2014**, *589*, 153–164.
9. Joseph, C.; Persson, C.; Colliander, M.H. Influence of heat treatment on the microstructure and tensile properties of Ni-base superalloy Haynes 282. *Mater. Sci. Eng. A-Struct. Mater. Prop. Microstruct. Process.* **2017**, *679*, 520–530. [CrossRef]
10. Tytko, D.; Choi, P.P.; Kloewer, J.; Kostka, A.; Inden, G.; Raabe, D. Microstructural evolution of a Ni-based superalloy (617B) at 700 °C studied by electron microscopy and atom probe tomography. *Acta Mater.* **2012**, *60*, 1731–1740. [CrossRef]
11. Luo, Q.S.; Chi, K.N.; Li, S.X.; Barnard, P. Microstructural Stability and Lattice Misfit Characterisations of Nimonic 263. In Proceedings of the ASME 2012 Pressure Vessels and Piping Conference, Toronto, ON, Canada, 15–19 July 2012; pp. 197–206.
12. Cai, J.; Qian, C.; Huo, X.; Liu, Q.; Li, K.; Ji, W.; Li, Z.; Yang, Z.; Cheng, J.; Fan, M.; et al. Characterizing Short-Time Aging Precipitation Behavior of a Novel Nickel–Iron-Based Alloy via Electrical Performance. *Materials* **2024**, *17*, 4143. [CrossRef] [PubMed]
13. Zhong, Z.H.; Gu, Y.F.; Yuan, Y. Microstructural stability and mechanical properties of a newly developed Ni-Fe-base superalloy. *Mater. Sci. Eng. A-Struct. Mater. Prop. Microstruct. Process.* **2015**, *622*, 101–107. [CrossRef]
14. Kim, I.S.; Choi, B.G.; Hong, H.U.; Do, J.; Jo, C.Y. Influence of thermal exposure on the microstructural evolution and mechanical properties of a wrought Ni-base superalloy. *Mater. Sci. Eng. A* **2014**, *593*, 55–63. [CrossRef]
15. Huang, Y.; Zhang, R.; Zhou, Z.; Zhang, P.; Yan, J.; Yuan, Y.; Gu, Y.; Cui, C.; Zhou, Y.; Sun, X. Microstructure optimization for higher strength of a new Fe–Ni-based superalloy. *Mater. Sci. Eng. A* **2023**, *865*, 144632. [CrossRef]
16. Wang, Z.T.; Ning, Y.Q.; Yu, H.; Zhang, B.Y.; Xie, B.C.; Huang, S.; Zhang, W.Y.; Zhang, B.J. Mechanical degradation behavior and γ' coarsening mechanism of a Ni-based superalloy during long-term high-temperature thermal exposure. *J. Mater. Res. Technol.-JMR&T* **2024**, *30*, 9510–9520.
17. Fan, X.K.; Li, F.Q.; Liu, L.; Cui, H.C.; Lu, F.G.; Tang, X.H. Evolution of γ' Particles in Ni-Based Superalloy Weld Joint and Its Effect on Impact Toughness During Long-Term Thermal Exposure. *Acta Metall. Sin.-Engl. Lett.* **2020**, *33*, 561–572. [CrossRef]
18. Hou, K.L.; Wang, M.; Ou, M.Q.; Li, H.Z.; Hao, X.C.; Ma, Y.C.; Liu, K. Effects of microstructure evolution on the deformation mechanisms and tensile properties of a new Ni-base superalloy during aging at 800 °C. *J. Mater. Sci. Technol.* **2021**, *68*, 40–52. [CrossRef]
19. Zhang, L.L.; Chen, J.Y.; Tang, X.; Xiao, C.B.; Zhang, M.J.; Yang, Q. Evolution of Microstructures and Mechanical Properties of K439B Superalloy During Long-Term Aging at 800 °C. *Acta Metall. Sin.* **2023**, *59*, 1253–1264.
20. Li, Y.S.; Zhao, W.J.; Chen, X.; Sang, P.; Ju, Y.F. Phase-field simulation of evolution kinetics of second γ' phase in Ni-Al alloy under tensile stress. *J. Mater. Res. Technol.-JMR&T* **2022**, *17*, 1450–1458.
21. Luo, R.; Liu, Q.; Gao, P.; Liu, T.; Ding, H.; Zhang, D.; Cao, Y.; Cheng, X. Effect of microstructural evolution and mechanical properties of Haynes 230 superalloy during long-term aging at 700 °C. *J. Alloys Compd.* **2023**, *947*, 169495. [CrossRef]
22. Li, J.; Tang, L.; Li, J.; Zhou, R.; Zhong, L.; Liu, W.; Li, X. Study on the Performance Evolution of Inconel 740H During Operation on the Key Component Test Facility at 700 °C. *Rare Met. Mater. Eng.* **2021**, *50*, 3304–3312.
23. Gao, S.; Hou, J.; Guo, Y.; Zhou, L. Microstructural evolution and mechanical properties of solution annealed cast IN617B alloy during long-term thermal exposure. *Mater. Sci. Eng. A-Struct. Mater. Prop. Microstruct. Process.* **2017**, *704*, 302–310. [CrossRef]

24. Peterson, A.; Baker, I. The formation mechanism, growth, and effect on the mechanical properties of precipitate free zones in the alumina-forming austenitic stainless steel Fe-20Cr-30Ni-2Nb-5Al during creep. *Mater. Sci. Eng. A-Struct. Mater. Prop. Microstruct. Process.* **2021**, *820*, 141561. [CrossRef]
25. Masoumi, F.; Thélaud, L.; Shahriari, D.; Jahazi, M.; Cormier, J.; Devaux, A.; Flipo, B.C.D. High temperature creep properties of a linear friction welded newly developed wrought Ni-based superalloy. *Mater. Sci. Eng. A-Struct. Mater. Prop. Microstruct. Process.* **2018**, *710*, 214–226. [CrossRef]
26. Joseph, J.; Annasamy, M.; Kada, S.R.; Hodgson, P.D.; Barnett, M.R.; Fabijanic, D.M. Optimising the Al and Ti compositional window for the design of γ' (L12)-strengthened Al–Co–Cr–Fe–Ni–Ti high entropy alloys. *Mater. Sci. Eng. A* **2022**, *835*, 142620. [CrossRef]
27. Yang, J.; Heogh, W.; Ju, H.; Kang, S.; Jang, T.S.; Jung, H.D.; Jahazi, M.; Han, S.C.; Park, S.J.; Kim, H.S.; et al. Functionally graded structure of a nitride-strengthened Mg 2 Si-based hybrid composite. *J. Magnes. Alloys* **2024**, *12*, 1239–1256. [CrossRef]
28. *ASTM E112-13*; Standard Test Methods for Determining Average Grain Size. ASTM International: West Conshohocken, PA, USA, 2013.
29. *ASTM E92-17*; Standard Test Methods for Vickers Hardness and Knoop Hardness of Metallic Materials. ASTM International: West Conshohocken, PA, USA, 2013.
30. Huang, J.; Yan, J.B.; Lei, D.J.; Yi, D.W.; Zhu, X.Y.; Zhou, J.B.; He, Y.H.; Li, P.; Huang, L.H.; Liu, D.X. Tensile properties and deformation characteristics of nickel-iron-based superalloys for steam turbine rotors. *Emergent Mater.* **2024**. [CrossRef]
31. Xu, Q.S.; Cao, T.S.; Ye, F.; Xu, F.H.; Li, H.F.; Fang, X.D.; Zhao, J. Creep-induced microstructural evolution in a nickel-based superalloy designed for advanced ultra-supercritical boilers. *Mater. Charact.* **2018**, *139*, 311–318. [CrossRef]
32. Donth, B.; Blaes, N.; Diwo, A.; Bottcher, D. Large-Scale Manufacture of Nickel Alloy Turbine Rotor Forgings for A-USC Power Plants. *Trans. Indian Natl. Acad. Eng.* **2020**, *5*, 75–82. [CrossRef]
33. Zhang, X.Q.; Chen, H.B.; Xu, L.M.; Xu, J.J.; Ren, X.K.; Chen, X.Q. Cracking mechanism and susceptibility of laser melting deposited Inconel 738 superalloy. *Mater. Des.* **2019**, *183*, 108105. [CrossRef]
34. Sun, W.; Qin, X.; Guo, J.; Lou, L.; Zhou, L. Thermal stability of primary MC carbide and its influence on the performance of cast Ni-base superalloys. *Mater. Des.* **2015**, *69*, 81–88. [CrossRef]
35. Wang, C.S.; Guo, Y.A.; Guo, J.T.; Zhou, L.Z. Microstructural stability and mechanical properties of a boron modified Ni-Fe based superalloy for steam boiler applications. *Mater. Sci. Eng. A-Struct. Mater. Prop. Microstruct. Process.* **2015**, *639*, 380–388. [CrossRef]
36. Wang, C.; Guo, Y.; Guo, J.; Zhou, L. Microstructural changes and their effect on tensile properties of a Ni-Fe based alloy during long-term thermal exposure. *Mater. Sci. Eng. A* **2016**, *670*, 178–187. [CrossRef]
37. Wang, W.Z.; Hong, H.U.; Kim, I.S.; Choi, B.G.; Jeong, H.W.; Kim, M.Y.; Jo, C.Y. Influence of γ' and grain boundary carbide on tensile fracture behaviors of Nimonic 263. *Mater. Sci. Eng. A* **2009**, *523*, 242–245. [CrossRef]
38. Martens, V.; Nembach, E. Strengthening of the nimonic alloy PE 16 by ordered particles of Ni3(Al, Ti). *Acta Metall.* **1975**, *23*, 149–153. [CrossRef]
39. Zhu, J.; An, C.; Lu, Y.; Zhu, M.; Xuan, F. Research Progress on Effect of gamma' Phase on Strength, Fatigue and Creep Properties of Nickel-Based Superalloys. *Mater. Mech. Eng.* **2023**, *47*, 71.
40. Kozar, R.W.; Suzuki, A.; Milligan, W.W.; Schirra, J.J.; Savage, M.F.; Pollock, T.M. Strengthening Mechanisms in Polycrystalline Multimodal Nickel-Base Superalloys. *Metall. Mater. Trans. A-Phys. Metall. Mater. Sci.* **2009**, *40A*, 1588–1603. [CrossRef]
41. Sun, F.; Tong, J.Y.; Feng, Q.; Zhang, J.X. Microstructural evolution and deformation features in gas turbine blades operated in-service. *J. Alloys Compd.* **2015**, *618*, 728–733. [CrossRef]
42. Du, B.N.; Yang, J.X.; Cui, C.Y.; Sun, X.F. Effects of grain refinement on the microstructure and tensile behavior of K417G superalloy. *Mater. Sci. Eng. A-Struct. Mater. Prop. Microstruct. Process.* **2015**, *623*, 59–67. [CrossRef]
43. Huther, W.; Reppich, B. Interaction of dislocations with coherent, stress-free, ordered particles. *Z. Met.* **1978**, *69*, 628–634. [CrossRef]
44. Xu, Q.S.; Cao, T.S.; Xu, F.H.; Fang, X.D.; Cheng, C.Q.; Zhao, J. Precipitation behavior and mechanical properties of TG700A alloy during aging. *Chin. J. Nonferrous Met.* **2018**, *28*, 1559–1567.
45. Peng, Z.C.; Tian, G.F.; Jiang, J.; Li, M.Z.; Chen, Y.; Zou, J.W.; Dunne, F.P.E. Mechanistic behaviour and modelling of creep in powder metallurgy FGH96 nickel superalloy. *Mater. Sci. Eng. A-Struct. Mater. Prop. Microstruct. Process.* **2016**, *676*, 441–449. [CrossRef]
46. Jiang, S.B.; Tian, Z.L.; Liu, W.B.; Chen, H.; Yang, Z.G.; Liu, Z.D.; Zhang, C.; Weng, Y.Q. Microstructural evolution and hardness of a heat resistant alloy during long term aging at 700 °C. *J. Alloys Compd.* **2018**, *765*, 1267–1274. [CrossRef]
47. Wang, C.; Zhao, H.; Guo, Y.; Guo, J.; Zhou, L. Structural stability and mechanical properties of phosphorus modified Ni-Fe based superalloy GH984. *Mater. Res. Innov.* **2014**, *18*, 324–330. [CrossRef]
48. Zhao, S.Q.; Xie, X.S.; Smith, G.D.; Patel, S.J. Gamma prime coarsening and age-hardening behaviors in a new nickel base superalloy. *Mater. Lett.* **2004**, *58*, 1784–1787. [CrossRef]
49. Baldan, A. Review Progress in Ostwald ripening theories and their applications to the γ'-precipitates in nickel-base superalloys: Part II: Nickel-base superalloys. *J. Mater. Sci.* **2002**, *37*, 2379–2405. [CrossRef]
50. Wang, C.S.; Guo, Y.G.; Guo, J.T.; Zhou, L.Z. Microstructural characteristics and mechanical properties of a Mo modified Ni-Fe-Cr based alloy. *Mater. Sci. Eng. A-Struct. Mater. Prop. Microstruct. Process.* **2016**, *675*, 314–322. [CrossRef]
51. Zhao, X.B.; Dang, Y.Y.; Yin, H.F.; Yuan, Y.; Lu, J.T.; Yang, Z.; Gu, Y.F. Evolution of the microstructure and microhardness of a new wrought Ni-Fe based superalloy during high temperature aging. *J. Alloys Compd.* **2015**, *644*, 66–70. [CrossRef]

found. Firstly, when the Li^+ concentration was only a few mg/kg, the corrosion behavior of Zr-4 alloy is almost the same as that in pure water. Secondly, when the Li^+ concentration was tens of mg/kg, the corrosion transition was earlier, and the corrosion after transition was accelerated. Thirdly, when the Li^+ concentration increased to 350 mg/kg, the stage before corrosion transition was no longer measurable, and the corrosion rate continued to increase after corrosion transition. Injecting boric acid could mitigate the accelerated corrosion effect of LiOH on Ziraloys [9]. Although autoclave corrosion tests have shown that the corrosion resistance of the Zr-Sn-Nb alloy in LiOH solutions was better than Zircaloys [10], there was still a lack of research on the effect of injecting boron on the Zr-Sn-Nb alloy.

There are three understandings of the deterioration of corrosion resistance induced by Li^+. Firstly, Li^+ replaces the position of Zr^{4+} in ZrO_2 lattice, resulting in the generation of additional oxygen vacancies and promoting oxygen diffusion [11]. Secondly, Li^+ promotes the recrystallization of ZrO_2, leading to an increase in grain boundary and the formation of pores and cracks in oxides [12]. Thirdly, Li^+ increases the porosity of oxide films, leading to the preferential dissolution of connection positions in the ZrO_2 lattice, causing the penetration of protective oxide films, ultimately leading to the rapid diffusion of the oxygen-to-oxide/metal (O/M) interface [13]. Xie et al. [14] found the aggregation of Li at the ZrO_2 grain boundary, and they believed it reduced the grain boundary energy of oxide film, resulting in defects and cracking.

As for the corrosion inhibition mechanism of injecting boron, there are also three understandings. Firstly, the reaction between boric acid and the -OLi group on the surface may alleviate the corrosion problem caused by Li^+ concentration [15]. Secondly, boron promotes the formation of oxides from dissolved alloy elements, covering the pores in ZrO_2 and hindering corrosion [16]. Thirdly, Boron changes the chemical environment in the pores of ZrO_2, reducing the solubility of oxides [17]. However, there is still a lack of clear conclusions about the corrosion inhibition mechanism of injecting boron.

To sum up, LiOH is a commonly used non-volatile alkaline agent in PWRs. It poses a risk of local concentration on the surface of fuel cladding, which can lead to the accelerated corrosion of zirconium alloy. Therefore, it is considered to inject a certain concentration of boric acid to mitigate the effects of Li^+. However, the mechanism by which zirconium alloy accelerated the corrosion caused by the local concentration of LiOH is still unclear, and the laws of boric acid alleviating Zr-Sn-Nb alloy corrosion are also unknown. In order to clarify the influence of injecting boron on the corrosion resistance of Zr-Sn-Nb alloy, this paper studied the corrosion behavior of the Zr-Sn-Nb alloy in concentrated LiOH solutions with different boron concentrations, and an attempt has been made to explain the corrosion inhibition mechanism of zirconium alloys by injecting boron to the aqueous solutions of the primary loop of PWRs.

2. Materials and Methods

2.1. Alloy and Materials

In this study, we selected a Zr-Sn-Nb alloy plate with a thickness of (2.1 $\pm$ 0.1) mm as a material. We obtain Zr-Sn-Nb plate from State Nuclear Bao Ti Zirconium Industry Co., Ltd., Baoji, China and the composition is shown in Table 1.

Table 1. Range of Zr-Sn-Nb alloy composition (wt%).

Element	Sn	Nb	Fe	Cr	Zr
Content	0.9–1.2	0.25–0.35	0.3–0.4	0.05–0.10	Bal.

2.2. Corrosion Experiment

The corrosion experiment was conducted using an autoclave. Plates were cut into (30 $\pm$ 1) mm $\times$ (20 $\pm$ 1) mm $\times$ (1.4 $\pm$ 0.1) mm sheets. Before corrosion, specimens were pickled in the standard metallographic etching solution for about 1 min to obtain a shiny surface. These specimens were moved immediately into the flowing water after pickling. Rinsing in the flowing water lasted for about 10 min and then specimens were transferred into the boiling deionized water for another 10 min to eliminate the influence of fluorine. The corrosion test was carried out in a static autoclave at 360 ± 3 °C/18.6 ± 1.4 MPa following the ASTM-G2/G2M-06 standard [18]. The autoclave was deaerated each time during heating-up at 140–150 °C. After deaeration, the autoclave was heated to 360 ± 3 °C, and then the corrosion test started. The aqueous solution compared in this study is shown in Table 2. By injecting 50 mg/kg and 200 mg/kg of boron to the concentrated LiOH solutions, the effects of different boron concentrations on the corrosion resistance of the Zr-Sn-Nb alloy were compared. Specimens were taken and weighed every 30 days, and corrosion resistance was evaluated by weight gain. In this work, at least three parallel samples were left in the autoclave to ensure an average measuring of weight gain.

Table 2. Aqueous solutions compared in this study (mg/kg).

Type of Solutions	Li Concentration	B Concentration	Dissolved Oxygen
B = 0 mg/kg	100 ± 5	0	≤0.1
B = 50 mg/kg	100 ± 5	50 ± 5	≤0.1
B = 200 mg/kg	100 ± 5	200 ± 10	≤0.1

2.3. Microscopic Analysis after Corrosion

Scanning electron microscope (SEM) was used to observe the surface and cross-sectional morphology of oxide films. Transmission electron microscope (TEM) was used to obtain a more detailed local microstructure of the oxide films. The thin foils of cross-sections of the oxide films were prepared using focused ion beam (FIB). X-ray diffraction (XRD) was used to analyze the crystal structure of the sample after corrosion testing. Atom probe tomography (APT) was used to analyze the aggregation of elements in the microzone of the oxide film. Glow discharge optical emission spectrometry (GDOES) was conducted to measure the distributions of elements in oxide films along the depth direction. The oxygen/nitrogen/hydrogen determinator was used to analyze the hydrogen contents of the corroded specimens, and the hydrides were observed in the matrix of the corroded specimens using an optical microscope.

3. Results

3.1. Effect of Boron Injection on Corrosion Behavior of Zr-Sn-Nb Alloy

3.1.1. Corrosion Kinetics

The weight gain profile of Zr-Sn-Nb alloy is shown in Figure 1. The injection of boron significantly reduced the corrosion rate of Zr-Sn-Nb alloy in aqueous solutions with a Li concentration of 100 mg/kg. In B = 0 mg/kg solutions, the corrosion transition of Zr-Sn-Nb alloy was occurred in 60 days, after that, the corrosion rate increased sharply, and the weight gain was more than 10,000 mg/dm^2 in 420 days. In B = 50 mg/kg and B = 200 mg/kg solutions, the corrosion rates were much lower than that in B = 0 mg/kg solutions, and after 510 days, the weight gain was 202.38 mg/dm^2 and 184.77 mg/dm^2, respectively, which was only 1/50 of that in B = 0 mg/kg solutions. It is clear that the corrosion rate of Zr-Sn-Nb alloy decreases significantly after increasing the B concentration from 0 to 50 mg/kg. When the B content was increased from 50 mg/kg to 200 mg/kg, the corrosion inhibition effect was significantly reduced. After 510 days, the weight gain of the Zr-Sn-Nb alloy in B = 200 mg/kg solutions was only about 18 mg/dm^2 lower compared to that in B = 50 mg/kg solution.

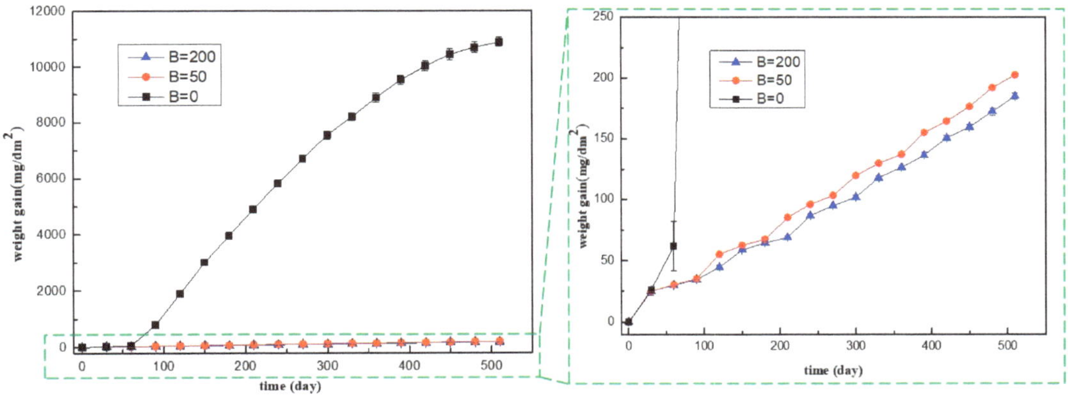

Figure 1. Corrosion kinetics of Zr-Sn-Nb alloy corroded in Li = 100 mg/kg solutions with, 50, 200 mg/kg boron, black square data represent the weight gain of samples corroded in B = 0 mg/kg solutions, red circular data represent the weight gain of samples corroded in B = 50 mg/kg solutions, blue triangle data represent weight gain of samples corroded in B = 200 mg/kg solutions.

3.1.2. Top Appearance

The photograph of corrosion specimens are shown in Figure 2. In B = 200 mg/kg and B = 50 mg/kg solutions, the surface of Zr-Sn-Nb alloy remained uniformly grey-black throughout the corrosion test, and there were no obvious defects on surface. In B = 0 mg/kg solutions, the specimen only showed uniform grey-black color in the first 30 days. After 60 days, a small number of white spots began to appear, and gradually increased and covered the whole surface with the prolongation of the test time. After 210 days, cracks appeared on the surface of specimen. After 480 days, severe cracking occurred in the oxide films.

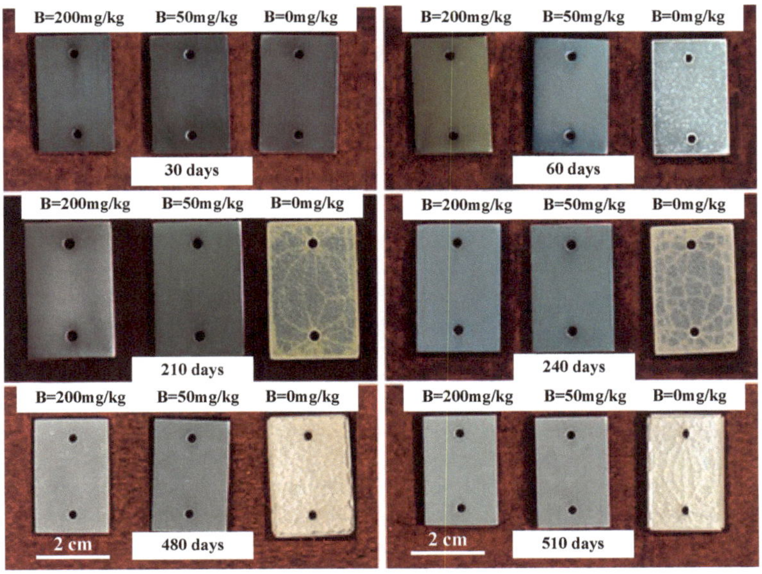

Figure 2. Top appearance of the Zr-Sn-Nb alloy in different solutions at different corrosion times, the solutions samples corroded in are labeled in the upper part of sub-figures, and the corrosion times are labeled in the bottom part of sub-figures.

3.1.3. SEM Morphology of Oxide Films

The SEM surface morphology of oxide films are shown in Figure 3. In B = 200 mg/kg and B = 50 mg/kg solutions, the oxide films were uniform and dense. In B = 0 mg/kg solutions, the oxide films present a large number of groove corrosion morphology after 30 days, which may be related to plate textile [19]. After 180 days, cracks began to appear. Comparing the surface morphologies of oxide films in different solutions after the same cycle test, it can be found that there are a large number of oxide particles and a small number of corrosion pits on the surface in B = 0 mg/kg solution.

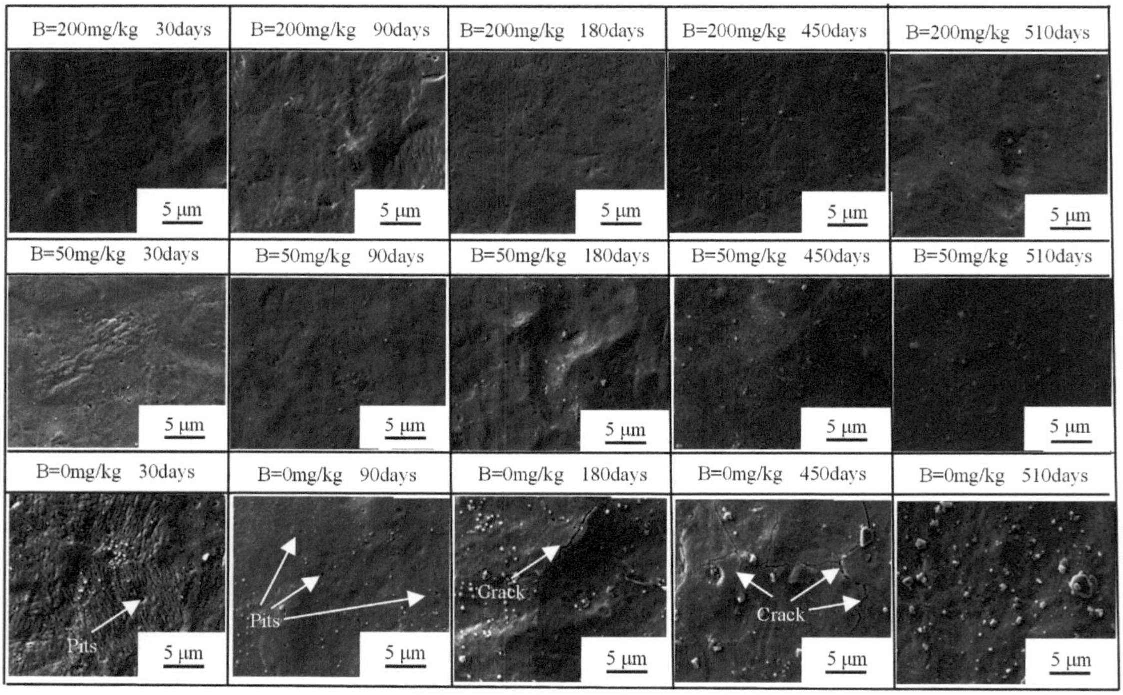

Figure 3. SEM surface morphology of Zr-Sn-Nb alloy in different solutions at different corrosion times, the solutions samples corroded in and corrosion time are labeled in the upper part of sub-figures.

The SEM cross-sectional morphology of Zr-Sn-Nb alloy corroded for several days (containing Zr matrix and oxide films) are shown in Figure 4. In B = 0 mg/kg solutions, a large number of cracks appeared in oxide films after 90 days. In B = 200 mg/kg and B = 50 mg/kg solutions, cracks appeared in oxide films after 360 and 450 days, respectively. Inhomogeneous oxidation of Zr-Sn-Nb alloys leads to the accumulation of stress in oxide films, which may destroy oxide films and cause cracks [20]. Inhomogeneous oxidation was more severe in B = 0 mg/kg solution, resulting in cracking of oxide films after 90 days corrosion, which was earlier than that in B = 200 mg/kg and B = 50 mg/kg solutions. In addition, after 180 days, cracks perpendicular to surface had appeared in oxide films in B = 0 mg/kg solution, indicating that stress generated by the growth of oxide films led to brittle cracking, and such cracks directly promoted the mass transfer process from surface to O/M interface, accelerating corrosion. As can be seen in Figure 4, with the extension of the test time, the cracks continued to expand to O/M interface, while no cracks appeared until the end of the 510-day test in B = 200 mg/kg and B = 50 mg/kg solutions.

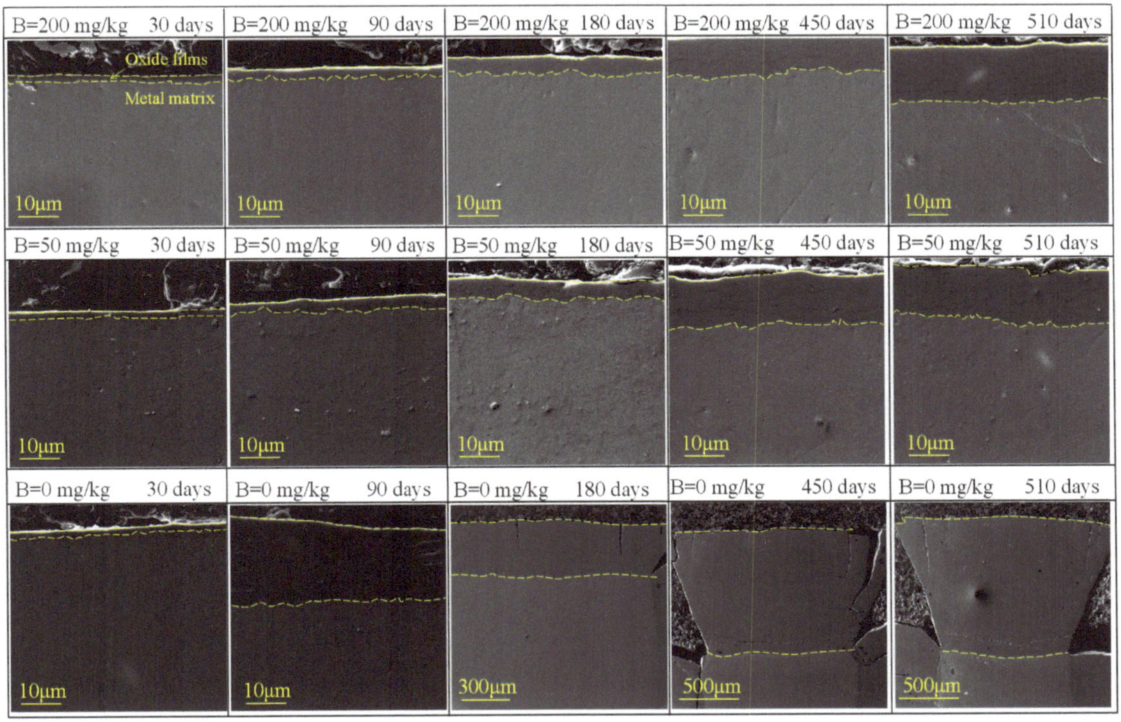

Figure 4. SEM cross-sectional morphology of Zr-Sn-Nb alloy in different solutions at different corrosion times, the solutions samples corroded in and corrosion time are labeled in the upper part of sub-figures.

3.1.4. Hydrogen Absorption Concentrations

The hydrogen absorption concentrations of Zr-Sn-Nb alloy in different solutions are shown in Table 3. In B = 0 mg/kg solutions, hydrogen absorption concentrations increased sharply to 400 mg/kg after 90 days, and to 3900 mg/kg after 510 days. In B = 200 mg/kg and B = 50 mg/kg solutions, the hydrogen absorption concentrations was lower than that in B = 0 mg/kg solution. When boron concentration was increased from 50 mg/kg to 200 mg/kg, the hydrogen absorption concentration was slightly decreased.

Table 3. Hydrogen absorption concentration of Zr-Sn-Nb alloy corroded in different solutions at different times (mg/kg).

Solutions \ Times	30 Days	90 Days	180 Days	270 Days	360 Days	450 Days	510 Days
B = 0 mg/kg	18	400	2400	3700	3500	3800	3900
B = 50 mg/kg	14	18	20	34	44	58	74
B = 200 mg/kg	20	16	14	28	40	46	49

3.1.5. Hydrides Morphology

After chemical etching, hydrides morphology is shown in Figure 5. At the beginning of the corrosion test, the hydrides length and density in B = 200 mg/kg solutions was lower than that in B = 50 mg/kg and B = 0 mg/kg solutions. As time went on, the hydrides length and density was increased, which tended to be similar in the B = 200 mg/kg and B = 50 mg/kg solutions. In B = 0 mg/kg solutions, hydrides had a tendency to interconnect after 90 days, and then hydrides' density increased sharply until it covered the whole observation field of view.

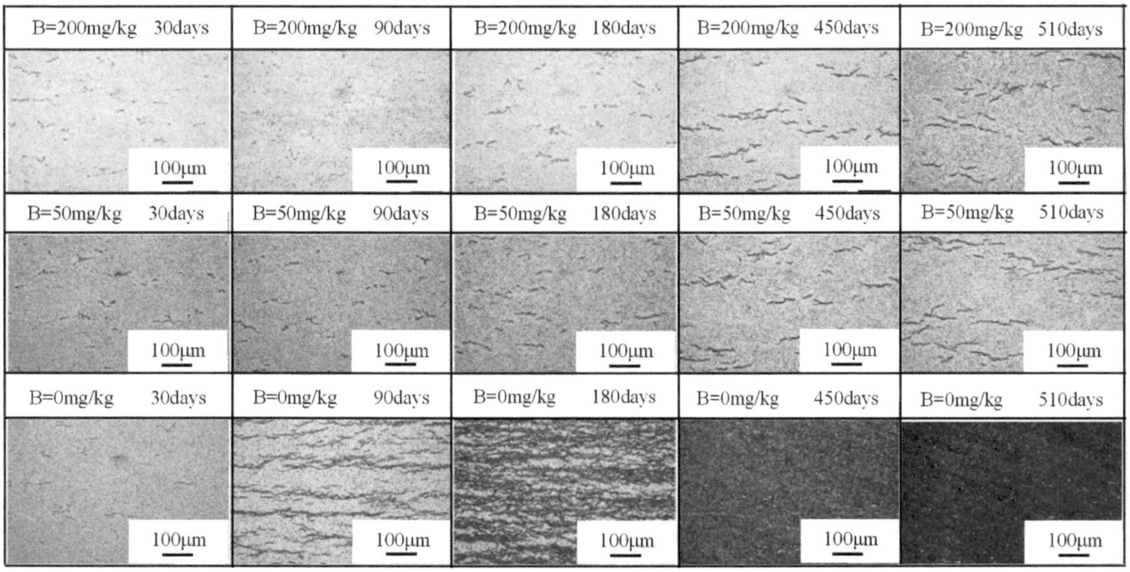

Figure 5. Hydride morphology of Zr-Sn-Nb alloy in different solutions at different corrosion times, the solutions samples corroded in and corrosion time are labeled in the upper part of sub-figures.

3.2. Effect of Boron Injection on Microstructure of Oxide Films of Zr-Sn-Nb Alloys

3.2.1. Cross-Sectional Microstructures of Oxide Films

The oxide films microstructure of Zr-Sn-Nb alloy corroded in B = 200 mg/kg solutions for 30 days (studied by TEM) are shown in Figure 6. From Figure 6a, oxide films were composed of an equiaxed crystal layer, a columnar crystal layer and an oxygen-enriched layer. The outer films were porous, and the thickness was about 200 nm, which consisted of equiaxed crystals with the size of 20–50 nm. The presence of pores between equiaxed crystals indicates that the outer films were not protective, as shown in Figure 6b. Below the equiaxed crystal layer was a columnar crystal layer with a thickness of 1.2 μm. The columnar crystal grew perpendicular to surface with a width of 50–100 nm. To reduce stress accumulation, zirconium alloys preferentially form columnar crystal oxides, and this oxide films can prevent the flow of corrosive media to O/M interface [21]. Near O/M interface, some island-shaped second phase particles (SPPs) with sizes of 50–200 nm could be observed. They are $Zr(Nb,Fe,Cr)_2$, and all of them cracked on the side near surface, as shown in Figure 6c. Zr in $Zr(Nb,Fe,Cr)_2$ were preferentially oxidized, followed by Fe, Cr, and Nb. The Pilling–Bedworth ratio (PB ratio) of Fe is 2.05, Cr is 2.00, Nb is 2.67. They are larger than that of Zr. So in the subsequent oxidation, the volume expansion of SPPs is inconsistent with that of Zr, generating additional stress and leading to cracking around SPPs [22]. An oxygen-enriched layer with a thickness of 200 nm also existed near O/M interface, and the morphology was shown in Figure 6d,e. Performing electron spectroscopy point scanning at position 1, 2 and 3 in Figure 6e, the results are shown in Table 4. Oxygen-enriched layer can effectively prevent the further diffusion of oxygen into Zr matrix by continuously dissolving oxygen atoms.

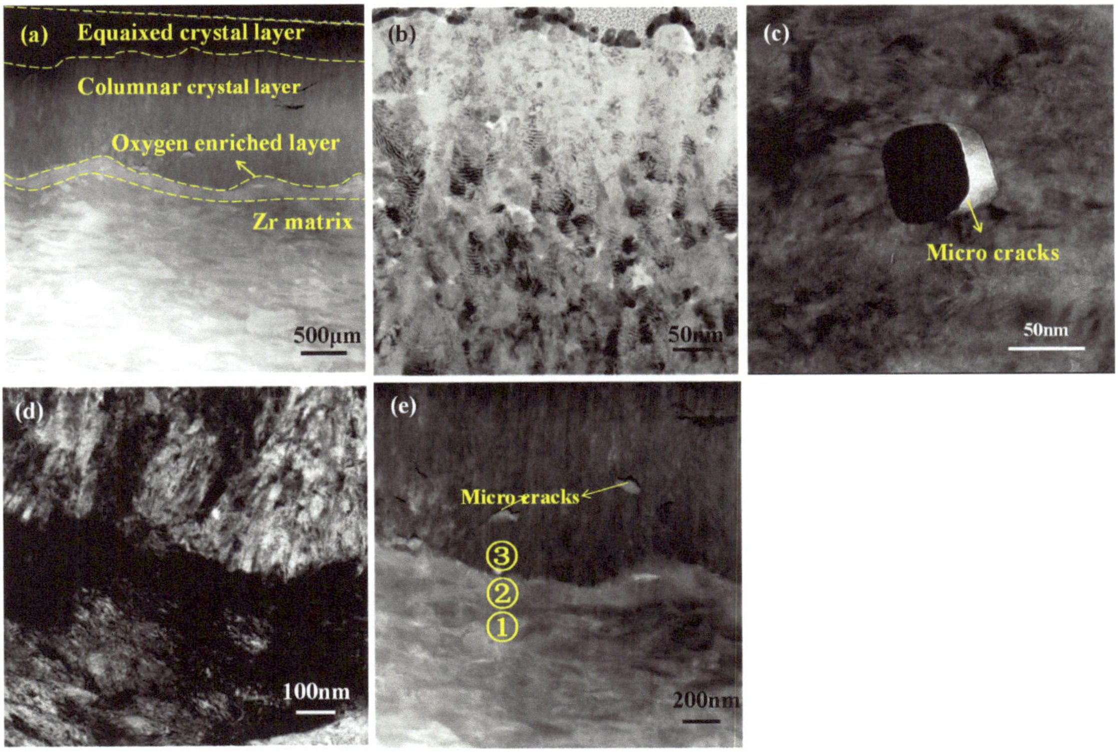

Figure 6. Cross-sectional microstructure of oxide films corroded in B = 200 mg/kg solutions for 30 days: (**a**) cross-section morphology of oxide film; (**b**) pores between equiaxed crystals; (**c**) SPPs observed near O/M interface; (**d**) bright field image of oxygen enriched layer; and (**e**) HAADF image of oxygen-enriched layer.

Table 4. The electron spectroscopy results of positions 1, 2, and 3 in Figure 6e (wt%).

Position	O	Cr	Fe	Zr	Nb	Sn
1	2.21	0.00	0.07	96.84	0.41	0.46
2	11.77	0.08	0.06	86.77	0.07	1.26
3	36.83	0.00	0.04	61.94	0.22	0.97

Figure 7 shows the cross-sectional microstructure of oxide films corroded in B = 0 mg/kg solutions for 30 days. The oxide films consisted of an equiaxed crystal layer, a columnar crystal layer, and an oxygen enriched layer, which were consistent with the cross-sectional microstructure corroded in B = 200 mg/kg solutions. This indicates that the corrosion mechanism has not changed in different solutions. From Figure 7a, it can be seen that cracks appeared in oxide films. This type of crack was usually caused by stress. When the local stress in oxide films exceeds the limit, cracks parallel to the surface will occur. The local stress was usually caused by uneven oxidation. In B = 0 mg/kg solutions, the Zr-Sn-Nb alloy exhibits significant uneven oxidation leading to oxide cracking, which means that the concentrated Li aggravates the inhomogeneous oxidation of Zr-Sn-Nb alloys. Table 5 shows electron spectroscopy results of positions 1, 2, and 3 in Figure 7b. There is also a oxygen-enriched layer nearby the O/M interface, which is coincided with the result of Figure 6.

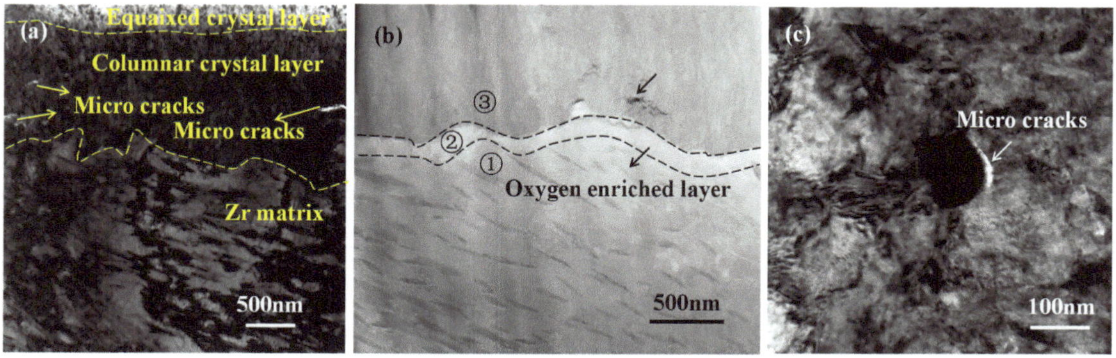

Figure 7. Cross-sectional microstructure of oxide films corroded in B = 0 mg/kg solutions for 30 days: (**a**) cross-section morphology of oxide film; (**b**) morphology of O/M interface; and (**c**) SPPs observed near O/M interface.

Table 5. The electron spectroscopy results of positions 1, 2, and 3 in Figure 7b (wt%).

Position	O	Cr	Fe	Zr	Nb	Sn
1	2.59	0.43	2.10	93.11	1.77	0.00
2	5.78	0.03	0.03	91.34	0.39	2.43
3	23.39	0.00	0.03	75.48	0.19	0.92

3.2.2. Crystal Structure of Oxide Films

XRD was performed to obtain the crystal structure of oxide films, and the results are shown in Figure 8. All the XRD results showed diffraction peaks of Zr [23]. After 30-day corrosion, the oxide films in B = 0 mg/kg solutions was mainly monoclinic ZrO_2 (m-ZrO_2) [24], and the oxide films in B = 50 mg/kg solutions were composed of m-ZrO_2 and tetragonal ZrO_2 (t-ZrO_2) [25]. In B = 200 mg/kg solutions, the oxide films were mainly composed of t-ZrO_2. After being corroded for 180 days and 360 days, the oxide films in B = 0 mg/kg and B = 50 mg/kg solutions mainly consisted of m-ZrO_2 and contained a small amount of t-ZrO_2, and oxide films in B = 200 mg/kg solutions were made up of m-ZrO_2 and t-ZrO_2. After being corroded for 510 days, the oxide films in B = 0 mg/kg, B = 50 mg/kg, and B = 200 mg/kg solutions were mainly made up of m-ZrO_2. However, the number of m-ZrO_2 diffraction peaks in B = 200 mg/kg and B = 50 mg/kg solutions is less than that in B = 0 mg/kg solutions.

During the oxidation of zirconium alloys, t-ZrO_2 was formed firstly, and then t-ZrO_2 is converted into m-ZrO_2 [3]. Therefore, the XRD results show that the oxidation of the Zr-Sn-Nb alloy is effectively inhibited in B = 200 mg/kg solutions. However, in all solutions, the oxidation process of Zr-Sn-Nb did not show significant changes, with the difference being the change in corrosion rate.

3.2.3. Distribution of Elements in Oxide Films

GDOES was used to obtain the element distribution along the depth direction in oxide films, and the results are shown in Figure 9. In B = 0 mg/kg solutions, Li infiltrated into oxide films and Li concentration decreased with depth. In B = 50 mg/kg and B = 200 mg/kg solutions, B and Li infiltrated into oxide films, and the infiltration of B did not affect the infiltration of Li. The variation of B concentration showed a periodic wave shape law. The variation pattern of Li concentration is different from B. These indicate that a corrosion transition may have an impact on B distribution in oxide films, while it has no effect on Li distribution.

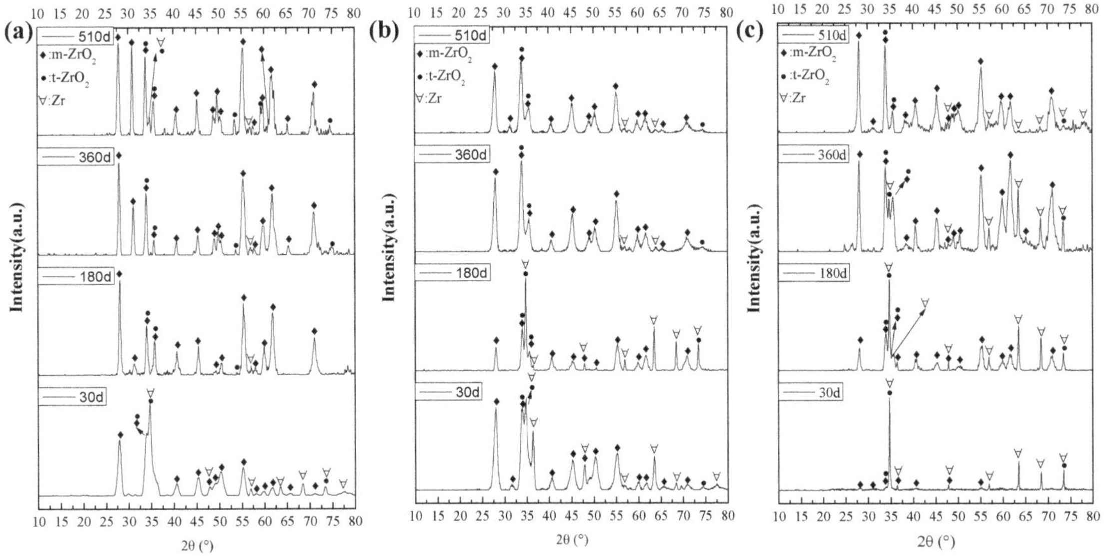

Figure 8. XRD results of Zr-Sn-Nb alloys immersed in different solutions at different times: (**a**) B = 200 mg/kg, (**b**) B = 50 mg/kg, (**c**) B = 0 mg/kg, the corrosion times are labeled in the upper left corner of sub-figures.

APT was conducted to study the local element segregation in oxide films, and the element distribution of oxide films corroded in B = 200 mg/kg solutions for 510 days are shown in Figure 10. Fe, Li, and O accumulated in a certain area of oxide films, while Zr and Sn were uniformly distributed in the entire oxide. B was not detected, which may be related to the low B concentration. Taking a cylindrical area (with a diameter of 20 nm) along the direction indicated by the arrow in Figure 10a, and the concentration distribution of Fe and Li are shown in Figure 10b. Li and Fe gathered at the same location.

Fe usually exists in the form of $Zr(Nb,Fe,Cr)_2$ in Zr-Sn-Nb alloys [2]. This kind of SPPs has good corrosion resistance and will oxidize later than the Zr matrix. Due to the extremely low solid solubility of Fe in ZrO_2, when Zr matrix around $Zr(Nb,Fe,Cr)_2$ is oxidized, Fe will be discharged and aggregated at SPPs boundaries [26]. Meanwhile, others have also found the phenomenon of Fe segregation at oxide grain boundaries [27]. There are different opinions on how Li and O enter oxide films and how they affect the corrosion resistance of zirconium alloys. Research has confirmed that Li in oxides films exists in the form of LiOH • H_2O or LiOH, and most of Li exists in porous oxides [28]. Xie et al. [14] found that Li will aggregate at grain boundaries during corrosion. Based on the research results mentioned above, the process of Li accelerated corrosion is as follows: Li^+ and OH^- diffuse inward along pores or grain boundaries of oxides, and continuously aggregate at grain boundaries and pores. The size of Li^+ is 76 pm and the size of Zr^{4+} is 72 pm. When a large amount of Li^+ aggregate at grain boundary, it is easy to replace the Zr^{4+} in ZrO_2, leading to an increase in oxygen vacancy. The oxygen vacancies promote the diffusion of H, O, and another corrosion medium along grain boundaries, accelerating corrosion.

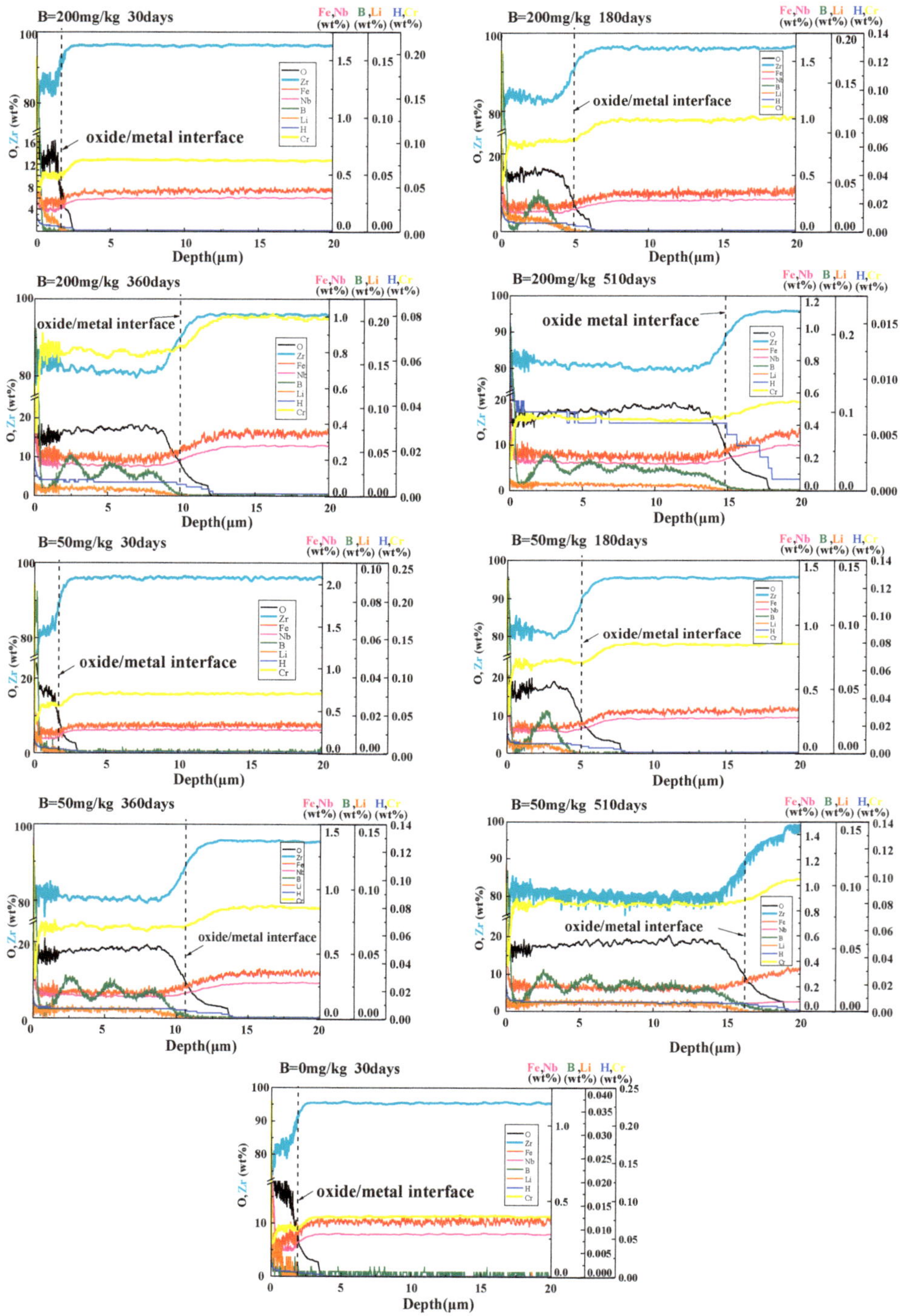

Figure 9. Element distribution along depth direction in oxide films in different solutions at different times, the solutions' samples corroded in and the corrosion times are labeled in the upper left corner of sub-figures.

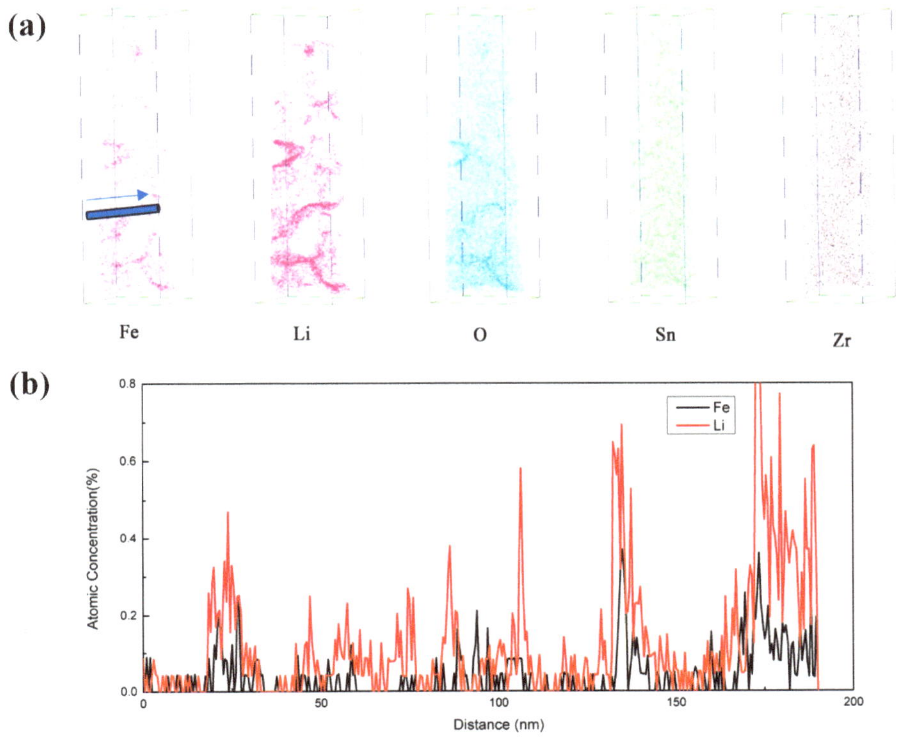

Figure 10. APT results of oxide films corroded in B = 200 mg/kg solutions for 510 days: (**a**) APT three-dimensional reconstruction of the distribution of elements in oxide films, (**b**) distribution of Fe and Li concentrations along the arrow direction in the sub-figure (**a**).

In order to analyze the influence of boron on the segregation behavior of Li and Fe, the APT analysis of oxide films corroded in B = 200 mg/kg solutions was compared to that in LiOH = 1 mol/L solutions (no boron injected) conducted by Xie et al. [14]. It was found that the injection of boron did not slow down the segregation of Li and Fe at grain boundaries. In contrast, the peak concentration of Li at grain boundaries (0.6%~0.8%) was higher than that in B = 0 mg/kg solutions (0.1%), which may be related to the longer corrosion time. The peak concentration of Fe at the grain boundary (0.4%) is approximately equal to that in B = 0 mg/kg solutions (0.4%). Based on the above analysis, it can be concluded that the injection of boron will not affect the segregation behavior of Li and Fe.

3.3. Accelerated Corrosion Mechanism Induced by Li and Corrosion Inhibition Mechanism of Injecting Boron

The weight gain, the length and concentration of hydrides in B = 50 mg/kg and B = 200 mg/kg solutions was much lower than that in B = 0 mg/kg solutions. These indicate that the injection of boron inhibits the corrosion of Zr-Sn-Nb alloys in concentrated Li solutions. The oxide morphology, composition, and structure of Zr-Sn-Nb alloys in B = 0 mg/kg, B = 50 mg/kg and B = 200 mg/kg solutions showed no significant differences, which means injecting boron did not change corrosion process. It was found that B and Li can infiltrate the oxide films, and Li exhibits an aggregation effect in oxides. This indicates that boron injection does not affect the infiltration and aggregation behavior of Li.

In order to clarify the effects of pH, undissolved LiOH and Li^+ on corrosion of Zr-Sn-Nb alloy, we calculate $pH_{360°C}$, undissolved LiOH and Li^+ concentration in different solutions (aqueous solutions in this paper and in other literature) and compare corrosion resistance of zirconium alloys, as shown in Table 6. When the Li concentration exceeds 70 mg/kg, if boron is not injected to the solution, Li will significantly accelerate the corrosion of zirconium alloys. When the B concentration exceeds 1/3 of Li concentration, injecting boron will have a good corrosion inhibition effect. However, increasing B concentration after exceeding the critical concentration cannot enhance the corrosion inhibition effect. In addition, When the Li concentration exceeds 700 mg/kg, injecting boron cannot slow down the accelerated corrosion effect. It also can be found that the Li-induced accelerated corrosion is not related to $pH_{360°C}$, undissolved LiOH, and Li^+ concentration.

Table 6. $PH_{360°C}$, undissolved LiOH, Li^+, and corrosion resistance of Zr-Sn-Nb alloys indifferent solutions.

Serial Number	Li Concentration (mg/kg)	B Concentration (mg/kg)	$pH_{360°C}$	Undissolved LiOH (mol/L)	Li^+ Concentration (mg/kg)	Corrosion Accelerated?
1	70	0	10.56	4.53×10^{-7}	≈56	yes
2	70	100	10.44	3.41×10^{-7}	≈70	no
3	70	1000	9.83	8.30×10^{-8}	≈70	no
4	100	0	10.76	1.13×10^{-6}	≈100	yes
5	100	50	10.71	1.00×10^{-6}	≈100	no
6	100	200	10.56	7.15×10^{-7}	≈100	no
7	300	0	11.05	4.22×10^{-6}	≈300	yes
8	300	100	10.98	3.64×10^{-6}	≈300	no
9	300	1000	10.44	1.05×10^{-6}	≈300	no
10	700	0	11.30	1.33×10^{-5}	≈700	yes
11	700	100	11.26	1.23×10^{-5}	≈700	yes
12	700	200	11.23	1.13×10^{-5}	≈700	yes
13	700	1000	10.90	5.27×10^{-6}	≈700	no
14	700	2000	10.51	2.17×10^{-6}	≈700	no

In the three solutions studied in this paper, the Li concentration was basically the same, indicating that the accelerated corrosion is not the sole effect of Li. Although zirconium alloys undergo accelerated corrosion in LiOH solutions, accelerated corrosion does not occur in $LiNO_3$, NaOH, KOH, RbOH, and CsOH solutions [7], indicating that single OH^- cannot accelerate corrosion. Therefore, the accelerated corrosion in LiOH solution is the result of the cooperative effect of Li^+ and OH^-.

According to the Macdonald model [29], the oxide film exhibits semiconductor properties in aqueous solutions. Anionic vacancies in oxides conduct oxygen from aqueous solutions to O/M interface, leading to the oxidation of the matrix, where the concentration of oxygen vacancies determines the rate of oxidation. According to the cross-sectional microstructure, oxide films of Zr-Sn-Nb alloys are consist of equiaxed crystal oxide (outermost), columnar crystal oxide (middle), and oxygen-enriched layer (bottom).

Based on the discussions mentioned above, we propose the accelerated corrosion mechanism induced by the Li and corrosion inhibition mechanism of injecting boron, as shown in Figure 11. In B = 0 mg/kg solutions, Li^+ and OH^- accelerate corrosion through the six following steps:

1. Li^+ and OH^- diffuse from aqueous solutions to the top of columnar crystal through pores of equiaxed crystal. Moreover, the concentration of corrosive medium is prone to occur in these pores, leading to the enrichment of Li^+ and OH^- at the top of columnar crystals.
2. Li^+ and OH^- diffuse along the grain boundaries of columnar crystal towards by solid-state diffusion, resulting in a uniform distribution of Li in the depth direction. In oxide films, columnar crystals also contain micropores [30]. Therefore, Li^+ and OH^- also tend to be enriched near the micropores.
3. Oxides exhibit acidity or neutrality in solutions, while their surfaces exhibit electronegativity in alkaline solutions [31], requiring the neutralization of cation in solutions. OH^- enriched at the top of columnar crystal makes the local environment appear

alkaline, promoting negative charge accumulate on the surface of the nearby columnar crystal. At this time, Li^+ adsorbs on the surface of columnar crystal, which is conducive to its diffusion to the interior of columnar crystal.

4. When Li^+ is adsorbed on the surface of columnar crystal, it can easily replace Zr^{4+} in ZrO_2 due to the very close size of Li^+ and Zr^{4+}. In addition, Li^+ continuously diffuse along grain boundaries to the O/M interface.
5. When Li^+ replaces Zr^{4+}, 1.5 oxygen vacancies form near it [32]. Therefore, due to the presence of a large amount of Li^+ at the top of columnar crystals, grain boundaries and micropores, a large number of oxygen vacancies will also be generated nearby.
6. Oxygen in solutions can occupy these oxygen vacancies easily through short-range diffusion, resulting in an excess of oxygen in ZrO_2 lattice. These oxygen diffuse rapidly along grain boundaries to O/M interface through solid-state diffusion, accelerating corrosion [33].

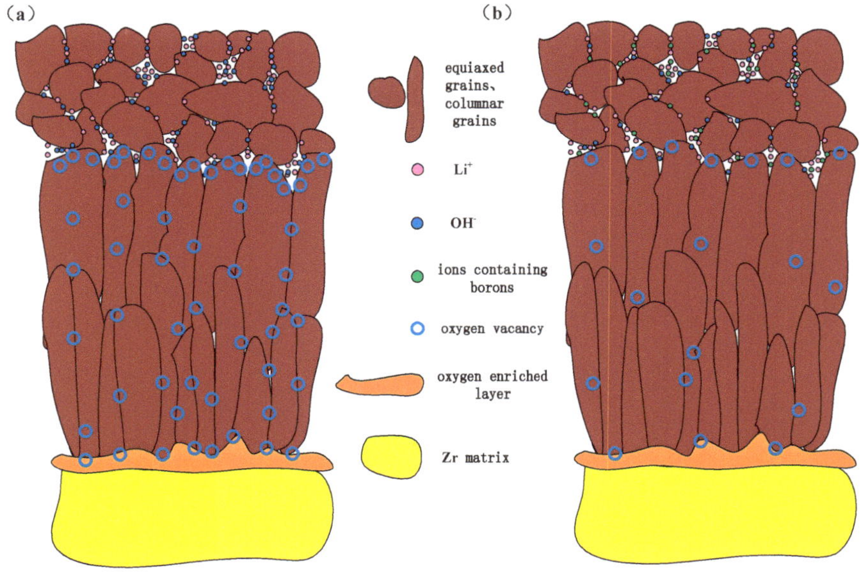

Figure 11. Schematic diagram of mechanism: (**a**) LiOH accelerating corrosion and (**b**) corrosion inhibition of injecting boron.

After injecting boron to solutions, the corrosion rate decreased significantly. According to the calculation results in Table 4, the injection of boric acid did not change pH, that is, the alkaline environment did not change, and did not affect the Li^+ concentration in solutions. Li still exists stably in oxide films. The APT results also showed that the aggregation of Li at grain boundaries was not affected by injecting boron. Calculations and experimental results suggest that the injection of boric acid did not change the steps 1–4 mentioned above. In addition, B infiltrated into oxide films and migrated to O/M interface, without reducing the aggregation of Li in columnar crystals. The size of B^{3+} is 27 pm, which is much smaller than that of Li^+, Zr^{4+}, and O^{2-}. So, B^{3+} can be easily incorporated in the ZrO_2 lattice, affecting steps 5–6. The corrosion inhibition mechanism of injecting boron is shown in Figure 10b: Li^+, OH^- and B^{3+} diffuse from aqueous solutions to the top of columnar crystal. When Li^+ replaces Zr^{4+} of the ZrO_2 lattice, B^{3+} also incorporated in the ZrO_2 lattice, and its three positive charges balance the loss of positive charges caused by Li^+ replacing Zr^{4+}. It maintains the electrical neutrality of ZrO_2. Therefore, the generation of oxygen vacancies is suppressed, resulting in insufficiency of oxygen vacancies in oxides, which can transport oxygen to O/M interface, thereby slowing down corrosion. Due to the limited dissociation of boric acid in high-temperature water, it is necessary to inject sufficient boric acid to provide sufficient B^{3+}. When the boric acid concentration exceeds a certain degree, the B^{3+} in the ZrO_2 lattice will reach saturation, and corrosion inhibition effect will reach the upper limit. Therefore, when the B concentration increases from 50 mg/kg to 200 mg/kg, the improvement in the corrosion inhibition effect is much smaller than that when B concentration increases from 0 to 50 mg/kg.

4. Conclusions

To clarify the influence of injecting boron on the corrosion resistance of the Zr-Sn-Nb alloy, the 360 °C/18.6 MPa autoclave corrosion test was performed in concentrated LiOH solutions with different boron concentrations. Corrosion resistance and the hydrides' absorption behavior were obtained. SEM, TEM, GDOES, and APT were used to characterize the oxide films in different solutions. An attempt has been made to explain the accelerated corrosion mechanism induced by the Li mechanism and the corrosion inhibition mechanism of injecting boron.

1. In B = 0 mg/kg solutions, Zr-Sn-Nb alloys were corroded severely, and oxide films showed significant cracking after 180 days. After 510 days, the weight gain was 10879.01 mg/dm^2. After injecting 50 mg/kg and 200 mg/kg boron in LiOH solutions, the surface of oxide films maintained uniform and dense throughout the entire test. After 510 days, the weight gain increased to 202.38 mg/dm^2 in B = 50 mg/kg solutions, and 184.77 mg/dm^2 in B = 200 mg/kg solutions. Injecting boron significantly reduced the corrosion rate, hydrogen concentration, and length of Zr-Sn-Nb alloys.
2. The accelerated corrosion mechanism induced by the Li is as follows: Li^+ tends to be incorporated in oxides in an alkaline environment, leading to the generation of a large number of oxygen vacancies. Oxygen vacancies carry oxygen from the solutions to the O/M interface, accelerating corrosion.
3. The corrosion inhibition mechanism of injecting boron is as follows: after B^{3+} incorporated in oxides films, the generation of oxygen vacancies is inhibited. This leads to insufficiency of oxygen vacancies, thereby slowing down corrosion. However, when boron concentration exceeds the critical value, the B^{3+} incorporated in the ZrO_2 lattice will reach saturation, and corrosion inhibition effect will reach the upper limit. Continuing to increase the boron concentration cannot significantly improve corrosion inhibition effect.

The results of this work can do some help to clarify the influence of the boron content in LiOH solution on corrosion resistance of Zr-Sn-Nb alloy, and can provide reference for the optimization of primary coolant in PWRs.

Author Contributions: Conceptualization, Y.Z. and Z.Y.; Methodology, J.X., P.D. and M.T.; Corrosion Test, Z.Y. and M.T.; Preparation and Analysis of Solutions, M.T.; Microscopic Analysis, Z.C., J.X. and P.D.; Data Process and Mechanism analysis, Y.Z. and Z.W.; Writing—original draft, Z.W. and Z.C.; Writing—review and editing, Y.Z., J.X., P.D., Z.Y. and M.T. All authors have read and agreed to the published version of the manuscript.

Funding: This research received no external funding.

Institutional Review Board Statement: Not applicable.

Informed Consent Statement: Not applicable.

Data Availability Statement: The data presented in this study are available on request from the corresponding author due to confidentiality rules.

Conflicts of Interest: The authors declare no conflicts of interest.

References

1. Duan, Z.; Yang, H.; Satoh, Y.; Murakami, K.; Kano, S.; Zhao, S.; Shen, J.; Abe, H. Current status of materials development of nuclear fuel cladding tubes for light water reactors. *Nucl. Eng. Des.* **2017**, *316*, 131–150. [CrossRef]

2. Wu, Z.; Jia, Y.; Dai, X.; Yi, W. Effect of Reprocess on Microstructure and Corrosion Resistance of Zr-Sn-Nb Alloy. *Met.* **2022**, *12*, 1822.
3. Motta, A.; Couet, A.; Comstock, R. Corrosion of zirconium alloys used for nuclear fuel cladding. *Annu. Rev. Mater. Res.* **2015**, *45*, 311–343. [CrossRef]
4. Yu, Z.; Werden, J.W.; Capps, N.A.; Linton, K.D.; Couet, A. (S)TEM/EDS study of native precipitates and irradiation induced Nb-rich platelets in high-burnup M5. *J. Nucl. Mater.* **2021**, *554*, 152667. [CrossRef]
5. Zhou, B.; Yao, M.; Li, Z.; Wang, X.; Zhou, J.; Long, C.; Liu, Q.; Luan, B. Optimization of N18 zirconium alloy for fuel cladding of water reactors. *J. Mater. Sci. Technol.* **2012**, *28*, 606–613. [CrossRef]
6. Chen, Z.; Zhao, Y.; Gong, B.; Tang, M.; Wu, Z. Influence of Ammonia on the Corrosion Behavior of a Zr-Sn-Nb Alloy in High Temperature Water. *Front. Mater.* **2022**, *9*, 910186. [CrossRef]
7. Jeong, Y.; Baek, J.; Kim, S. Corrosion characteristics and oxide microstructure of Zircaloy-4 in aqueous alkali hydroxide solution. *J. Nucl. Mater.* **1999**, *270*, 322–333. [CrossRef]
8. Han, J.; Rheem, K. The corrosion characteristics of Zircaloy-4 fuel cladding in LiOH-H_3BO_3 solutions. *J. Nucl. Mater.* **1994**, *217*, 197–199. [CrossRef]
9. Cox, B.; Wu, C. Transient effects of lithium hydroxide and boric acid on zircaloy corrosion. *J. Nucl. Mater.* **1995**, *22*, 169–178. [CrossRef]
10. Park, J.; Seung, J.; Choi, B.; Yong, H. Corrosion and oxide characteristics of Zr-1.5Nb-0.4Sn-0.2Fe-0.1Cr alloys in 360 °C pure water and LiOH solution. *J. Nucl. Mater.* **2008**, *373*, 343–350. [CrossRef]
11. Yilmazbayhan, A.; Breval, E.; Motta, A.T.; Comstock, R.J. Transmission electron microscopy examination of oxide layers formed on Zr alloys. *J. Nucl. Mater.* **2006**, *349*, 265–281. [CrossRef]
12. Yao, M.; Zhou, B.; Li, Q.; Liu, W.; Geng, X.; Lu, Y. A superior corrosion behavior of zircaloy-4 in lithiated water at 360 °C/18.6 mpa by beta-quenching. *J. Nucl. Mater.* **2008**, *374*, 197–203. [CrossRef]
13. Oskarsson, M.; Ahlberg, E.; Pettersson, K. Oxidation of Zircaloy-2 and Zircaloy-4 in water and lithiated water at 360 °C. *J. Nucl. Mater.* **2001**, *295*, 97–108. [CrossRef]
14. Xie, S.; Zhou, B.; Liang, X.; Li, Q.; Zhang, J. The distribution of Li ions in the oxide film formed on Zircaloy-4 corroded in lithiated water at 633 k. *Materials* **2020**, *13*, 873. [CrossRef]
15. Wei, K.; Wang, X.; Zhu, M.; Guan, H.; Zhang, J. Effects of Li, B and H elements on corrosion property of oxide films on ZIRLO alloy in 300°C/14MPa lithium borate buffer solutions. *Corros. Sci.* **2020**, *181*, 109216. [CrossRef]
16. Kim, T.; Choi, K.; Yoo, S.; Lee, Y. Influence of dissolved hydrogen on the early stage corrosion behavior of zirconium alloys in simulated light water reactor coolant conditions. *Corros Sci.* **2018**, *131*, 235–244. [CrossRef]
17. Cox, B.; Wu, C. Dissolution of zirconium oxide films in 300 °C LiOH. *J. Nucl. Mater.* **1993**, *199*, 272–284. [CrossRef]
18. *ASTM G2/G2M-06(2011)e1.2011*; Standard Test Method for Corrosion Testing of Products of Zirconium, Hafnium, and Their Alloys in Water at 680°F(360°C) or in Steam at 750°F (400°C). ASTM: West Conshohocken, PA, USA, 2011.
19. Liu, J.; Yu, H.; Karamched, P.; Hu, J.; Grovenor, C. Mechanism of the α-Zr to hexagonal-ZrO transformation and its impact on the corrosion performance of nuclear Zr alloys. *Acta Mater.* **2019**, *179*, 328–341. [CrossRef]
20. Polatidis, E.; Frankel, P.; Wei, J.; Klaus, M.; Comstock, R.; Ambard, A.; Lyon, S.; Cottis, R.; Preuss, M. Residual stresses and tetragonal phase fraction characterisation of corrosion tested zircaloy-4 using energy dispersive synchrotron x-ray diffraction. *J. Nucl. Mater.* **2013**, *432*, 102–112. [CrossRef]
21. Pecheur, D.; Godlewski, J.; Billot, P.; Thomazet, J. Microstructure of oxide films formed during the waterside corrosion of the zircaloy-4 cladding in lithiated environment. In *Zirconium in the Nuclear Industry: 15th International Symposium*; ASTM STP 1505; Sunriver: Philadelphia, OR, USA, 1996.
22. Proff, C.; Abolhassani, S.; Lemaignan, C. Oxidation behaviour of zirconium alloys and their precipitates—A mechanistic study. *J. Nucl. Mater.* **2013**, *432*, 222–238. [CrossRef]
23. Ni, N. Study of Oxidation Mechanisms of Zirconium Alloys by Electron Microscopy. Ph.D. Thesis, Oxford University, Oxford, UK, 2011.
24. Igawa, N.; Ishii, Y. Crystal structure of metastable tetragonal zirconia up to 1473 K. *J. Am. Ceram. Soc.* **2001**, *84*, 1169–1171. [CrossRef]
25. Katz, G. X-ray diffraction powder pattern of metastable cubic ZrO_2. *J. Am. Ceram. Soc.* **1971**, *54*, 531. [CrossRef]
26. Cao, X.; Yao, M.; Peng, J.; Zhou, B. Corrosion behavior of $Zr(Fe_x,Cr_{1-x})_2$ alloys in 400 °C superheated steam. *Acta Metall. Sin. (Engl. Lett.)* **2011**, *47*, 882–886.
27. Hu, J.; Aarholt, T.; Setiadinata, B.; Li, K.; Grovenor, C. A multi-technique study of "barrier layer" nano-porosity in Zr oxides during corrosion and hydrogen pickup using (S)TEM, TKD, APT and Nano SIMS. *Corros. Sci.* **2019**, *158*, 108109. [CrossRef]
28. Xie, S.; Zhou, B.; Liang, X.; Liu, W.; Li, H.; Li, Q.; Yao, M.; Zhang, J. A novel mechanism for nodular corrosion of zircaloy-4 corroded in 773 K superheated steam. *Corros. Sci.* **2017**, *126*, 44–54. [CrossRef]
29. Macdonald, D. Some personal adventures in passivity—A review of the point defect model for film growth. *Russ. J. Electrochem.* **2012**, *48*, 235–258. [CrossRef]
30. Hu, J.; Liu, J.; Lozano-Perez, S.; Grovenor, C.; Mader, E. Hydrogen pickup during oxidation in aqueous environments: The role of nano-pores and nano-pipes in zirconium oxide films. *Acta Mater.* **2019**, *180*, 105–115. [CrossRef]

31. Wei, K.; Chen, L.; Qu, Y.; Zhang, Y.; Zhang, J. Zeta potential of microarc oxidation film on ZIRLO alloy in different aqueous solutions. *Corros. Sci.* **2018**, *143*, 129–135. [CrossRef]
32. Krausová, A.; Macák, J.; Sajdl, P.; Novotny, R.; Reniuková, V.; Vrtílková, V. In-situ electrochemical study of Zr-1Nb alloy corrosion in high temperature Li+ containing water. *J. Nucl. Mater.* **2015**, *467*, 302–310. [CrossRef]
33. Liu, W.; Zhou, B.; Li, Q.; Yao, M. Detrimental role of LiOH on the oxide film formed on Zircaloy-4. *Corros. Sci.* **2005**, *47*, 1855–1860. [CrossRef]

materials

MDPI

Article

Enhancing the Tensile Properties and Ductile-Brittle Transition Behavior of the EN S355 Grade Rolled Steel via Cost-Saving Processing Routes

Vadym Zurnadzhy [1,2], Vera Stavrovskaia [1], Yuliia Chabak [1,2], Ivan Petryshynets [2], Bohdan Efremenko [1], Kaiming Wu [3], Vasily Efremenko [1,2,3,*] and Michail Brykov [4]

1 Physics Department, Pryazovskyi State Technical University, 49000 Dnipro, Ukraine; zurnadzhy_v_i@pstu.edu (V.Z.); verochkastavrovska@gmail.com (V.S.); chabak_y_g@pstu.edu (Y.C.); efremenko_b_v@pstu.edu (B.E.)
2 Institute of Materials Research, Slovak Academy of Sciences, 04001 Kosice, Slovakia; ipetryshynets@saske.sk
3 International Research Institute for Steel Technology, Collaborative Innovation Center for Advanced Steels, Wuhan University of Science and Technology, 430081 Wuhan, China; wukaiming@wust.edu.cn
4 Faculty of Engineering and Physics, National University Zaporizhzhia Polytechnic, 69063 Zaporizhzhia, Ukraine; brykov@zp.edu.ua
* Correspondence: vgefremenko@gmail.com or efremenko_v_g@pstu.edu

Abstract: Structural rolled steels are the primary products of modern ferrous metallurgy. Consequently, enhancing the mechanical properties of rolled steel using energy-saving processing routes without furnace heating for additional heat treatment is advisable. This study compared the effect on the mechanical properties of structural steel for different processing routes, like conventional hot rolling, normalizing rolling, thermo-mechanically controlled processing (TMCP), and TMCP with accelerating cooling (AC) to 550 °C or 460 °C. The material studied was a 20 mm-thick sheet of S355N grade (EN 10025) made of low-carbon (V+Nb+Al)-micro-alloyed steel. The research methodology included standard mechanical testing and microstructure characterization using optical microscopy, scanning and transmission electronic microscopies, energy dispersive X-ray spectrometry, and X-ray diffraction. It was found that using different processing routes could increase the mechanical properties of the steel sheets from S355N to S550QL1 grade without additional heat treatment costs. TMCP followed by AC to 550 °C ensured the best combination of strength and cold-temperature resistance due to formation of a quasi-polygonal/acicular ferrite structure with minor fractions of dispersed pearlite and martensite/austenite islands. The contribution of different structural factors to the yield tensile strength and ductile–brittle transition temperature of steel was analyzed using theoretical calculations. The calculated results complied well with the experimental data. The effectiveness of the cost-saving processing routes which may bring definite economic benefits is concluded.

Keywords: structural steel; EN S355; cost-saving; hot rolling; normalizing rolling; TMCP; accelerated cooling; microstructure; mechanical properties

Citation: Zurnadzhy, V.; Stavrovskaia, V.; Chabak, Y.; Petryshynets, I.; Efremenko, B.; Wu, K.; Efremenko, V.; Brykov, M. Enhancing the Tensile Properties and Ductile-Brittle Transition Behavior of the EN S355 Grade Rolled Steel via Cost-Saving Processing Routes. *Materials* **2024**, *17*, 1958. https://doi.org/10.3390/ma17091958

Academic Editor: Xiaogang Li

Received: 28 March 2024
Revised: 14 April 2024
Accepted: 17 April 2024
Published: 23 April 2024

1. Introduction

S355 grade rolled steel plate (Euronorm EN 10025) is a widely used metallurgical product [1–3]. S355 grade specifies that steel sheets with 16–40 mm thickness should have a yield tensile strength (YTS) of not less than 345 MPa and an ultimate tensile strength (UTS) within the 470–630 MPa interval [4]. Consequently, the total elongation (TEL) (ductility) should not be lower than 22%. This combination of mechanical properties meets the requirements in many structural applications, including civil and infrastructural construction, heavy machinery, shipbuilding, offshore, and wind farms, predetermining the widespread use of this type of flat steel products [5–7]. The S355 grade sheets are produced using low-carbon Mn-Si steel with micro-additions of carbide/nitride-forming elements (Nb, Ti, V, and Al).

Due to lower carbon content and limited amounts of alloying elements, S355 grade steel is suitable for welded structures where the properties of the heat-affected zone are important for the integrity of the structures [8–10]. The EN 10025-3 [11] and EN 10025-6 [12] standards propose using a heat treatment, specifically normalization or quenching/tempering, to ensure the desired mechanical behavior in the steel of the S355–S690 grades. However, heat treatment is an energy-consumable process with significant production costs and greenhouse gas emissions.

The required steel properties must be achieved based on energy (cost)-saving solutions, which is challenging for the metallurgical industry. According to this approach, heating should be used only before rolling to form the desired final structure in the rolled steel, and no additional furnace heating should be applied for the heat treatment. The corresponding cost-saving technologies are associated with the final stages of steel sheet manufacturing, specifically with finish rolling and subsequent accelerated cooling (AC) to ensure an appropriate mechanism of the γFe $\rightarrow$ αFe phase transformation [13]. These technological processes include (a) normalizing rolling (NR) [14], (b) thermo-mechanically controlled processing (TMCP) [15], and (c) thermo-mechanically controlled processing followed by accelerated cooling (TMCP/AC) [16–18]. NR is a thermo-mechanical treatment with rolling completion in the lower part of the single-phase (austenite) temperature region, which inhibits the austenite grain growth and contributes to the total grain refinement. Eventually, the size of the ferrite grain after NR was close to that of furnace normalization, allowing the exclusion of this heat treatment from the technological route [19,20]. Accordingly, wide usage of NR was reported, even for manufacturing thick plates intended for offshore platform applications [21]. TMCP differs from NR by a lower completion temperature of finish rolling and a greater deformation (reduction) in the last rolling passes [22–24]. This effectively suppresses the austenite recrystallization, refining the ferrite grain and maintaining the strain-hardening effect ("Dislocation Engineering" [25]) with remarkably enhanced mechanical properties [26,27]. TMCP is most often used for producing high-strength steel for oil/gas pipelines [28] but is also applied while processing structural steel (S355J2 and S355G8) for more general applications [6]. Conventional hot rolling (HR), NR, and TMCP form a "ferrite + pearlite" structure with a high grade of structural banding due to a relatively slow cooling rate in still air.

A more advanced combination of properties is associated with a pearlite-free structure comprising a nonequilibrium ferrite with irregular (quasi-polygonal) [29] or acicular [30] morphologies, even lath-like bainite [31]. Bhadeshia et al. considered that the quasi-polygonal ferrite results from a diffusional phase transition having a bulk chemical composition similar to the parent austenite [32], whereas the acicular ferrite is identical to bainite regarding the transformation mechanism [33]. The acicular ferrite is characterized by high-angle boundaries, which is beneficial for the hindering of crack propagation [34]. Under a certain cooling rate, the martensite/austenite (M/A) conglomerates also appeared in the structure contributing to steel strength [35]. The pearlite-free structures mentioned earlier could be formed in low-carbon structural steel using post-rolling AC, which suppresses the austenite$\rightarrow$pearlite transformation [36,37]. It is particularly important that this structure could be formed using heating before HR, not requiring the additional costs for furnace heating of the rolled sheets [38]. The combination of TMCP and AC led to an advanced steel strength that meets the requirements of the X80–X120 API 5L grades (Misra et al. [17,18], Ramirez et al. [39]).

Furthermore, ensuring the appropriate low-temperature impact behavior of S355 steel (especially in the transverse direction) is important since it has a wide range of applications. The absorbed impact energy (E) measured using the Charpy test controls this behavior. According to EN 10025-3, the minimum E values for longitudinal (L) and transverse (T) specimens are defined as 40 and 20 J for S355N grade, respectively, and the minimum E values for L and T specimens are defined as 47 and 27 J for S355NL grade, respectively, with the testing conducted at -20 °C in both cases. For the S355NL grade, additional testing performed at -50 °C obtained minimum E values of 27 and 16 J for L and T specimens, respectively [4]. Usually, an increased strength is associated with an increased temperature of the ductile-to-brittle transition (T_{DBT}), though the "TMCP + AC" process can improve the strength and cold resistance of the rolled steel, including due to the "ferrite/pearlite" banding elimination [39]. The effectiveness of the accelerated controlled cooling after the hot rolling of high-strength structural steel was proven by Fan et al. [16], who revealed the decrease in effective grain size and increase in the dislocations density under intensification of the steel cooling. Thus, NR, TMCP, and AC in different combinations are prospective energy-saving processing routes in the final stage of steel sheet manufacturing, promoting the S355 grade steel to a much higher property level without significant additional costs. The optimal route selection should imply the comparison of different technological schemes applied to the steel sheets of the same thickness and similar chemical composition, preferably processed using the steel of the same heat. However, such comprehensive studies are rarely reported in the literature. Consequently, this study aimed to compare various technological schemes of processing the sheets made of (V+Nb+Al)-micro-alloyed structural steel, focusing on the compliance of mechanical properties with the requirements of Euronorm EN 10025. The research task was to determine the target microstructure that can be formed in the steel during its rolling and sequential cooling, in terms of gaining the optimal combination of strength and low-temperature ductile–brittle behavior.

2. Materials and Methods

The study material was S355N grade (EN 10025) rolled steel sheet. The steel contained (in wt.%) 0.12, 1.52, 0.19, 0.010, 0.010, 0.052, 0.033, 0.033, and 0.0025 of C, Mn, Si, S, P, V, Nb, Al, and Ca, respectively (Cr, Ni, and Cu were less than 0.05 wt.% each and Mo was less than 0.01 wt.%). The chemical composition and carbon equivalent of 0.39 complied with the EN 10025 requirements. The steel was smelted in a converter and cast into 220 mm thick slabs. After heating at 1180–1200 °C, the slabs were reduced (rolled) to a 20 mm thickness according to different technological routes discerned using the finish rolling temperature (FRT) and the afterward sheet cooling rate as follows (Figure 1):

- conventional HR with FRT of 980–1000 °C (which was above the non-recrystallization temperature, T_{nr}) and air-cooling (route "a");
- NR with FRT of 800–830 °C and air-cooling (route "b");
- TMCP with FRT of 700–720 °C (in two-phase ($\alpha + \gamma$) interval) and air-cooling (route "c");
- TMCP/AC with FRT of 790–810 °C (in the single-phase (γFe) interval) and an accelerated cooling by water to 550 °C (route "d") or 460 °C (route "e").

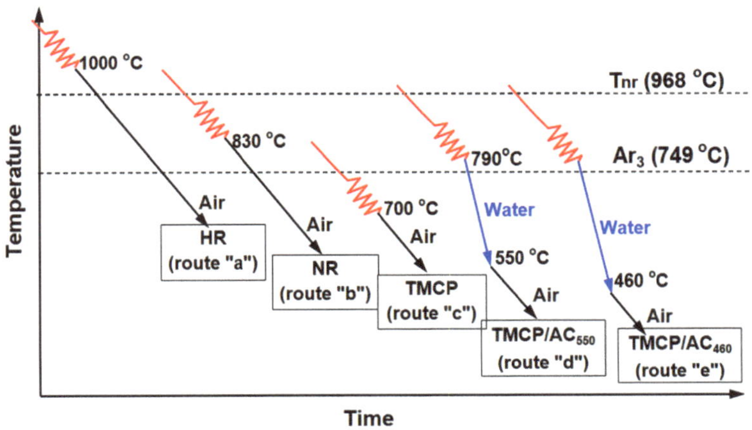

Figure 1. The schemes of the technological routes of S355N steel sheets processing.

The processing parameters were selected based on the austenite → ferrite transformation A_{r3} temperature of 748.7 °C and T_{nr} of 968.1 °C, which were calculated as follows [40]:

$$A_{r3}\ (°C) = 910 - 310C - 80Mn - 20Cu - 15Cr - 55Ni - 80Mo, \quad (1)$$

$$T_{nr}(°C) = 887 + 464C + \left(6445Nb - 644\sqrt{Nb}\right) + \left(732V - 230\sqrt{V}\right) + 890Ti + 363Al - 357Si, \quad (2)$$

where the chemical elements were taken in wt.%.

Routes from "c" to "e" included FRT below T_{nr} to retain the strain-strengthening effect. Routes "d" and "e" had a higher FRT than route "c" to decrease the proeutectoid (polygonal) ferrite amount and increase the fraction of hard phases (acicular ferrite, bainite). Under TMCP/AC, water cooling was used to decrease the temperature of the sheet to 550 ± 20 °C (TMCP/AC_{550}) or 460 ± 20 °C (TMCP/AC_{460}) with a cooling intensity of 15 ± 4 and 18 ± 4 °C·s^{-1}, respectively. After the completion of each processing route, the treated sheets were gathered in a stack and held for 40 h for slow cooling to 100 °C.

Mechanical properties were determined according to EN 10025 using tensile and impact tests. The longitudinal tensile specimens (5 mm in diameter and 30 mm in length gauge) were tested at room temperature with a tensile speed of 6 mm·min^{-1}. The absorbed E value was measured using a Charpy pendulum-type tester on V-notched specimens of 10 mm × 10 mm × 55 mm size. The impact testing temperatures varied as 0, −20, −40, and −60 °C to define the ductile–brittle transition temperature (T_{DBT}) of steel. The results were averaged using three tensile and three Charpy specimens for each regime (testing temperature). The anisotropy index (*Ai*) was calculated as follows:

$$A_i = \frac{E_L}{E_T}, \quad (3)$$

where, E_L and E_T represent the mean absorbed energy values in the longitudinal and transverse directions, respectively.

Optical microscopy (OM, Axiovert 40 MAT, Carl Zeiss, Jena, Germany) and scanning electronic microscopy (SEM, JSM-7000F, JEOL, Tokyo, Japan) were utilized for the microstructure characterization of the mirror-polished specimens etched in the 4 vol.% Nital reagent. The ferrite grain size was measured using the intercept method according to ASTM E112. The effective grain size was measured for the acicular ferrite structures. Additionally, the same SEM was used to observe the rupture surface of the impact specimen. The fine structure of the steel was characterized using transmission electron microscopy (TEM, JEM-100-C-XII, JEOL, Tokyo, Japan) equipped with energy dispersive X-ray spectroscopy (EDX, Inca-sight, Oxford Instruments, Abingdon, UK) at 100 kV acceleration voltage. For TEM, thin foils were mechanically polished to 0.1–0.15 mm thickness, followed by electropolishing in 6-vol% perchloric acid solution using a fluid-jet polishing machine. X-ray diffraction (XRD) was performed using the diffractometer (X'Pert PRO, PANalytical, Malvern, UK) with Cu-K_α radiation under 40 kV voltage, 50 mA tube current, 0.033° scan step, and 0.069°·s^{-1} scan speed. The broadening of ferrite peaks in XRD was used to measure the dislocation density under the assumption that dislocations caused the strain broadening in ferrite [41].

3. Results

3.1. Mechanical Properties Assessment

The tensile properties of the experimental steel, depending on the processing route, are presented in Table 1. For all cases, the strength indicators (YTS and UTS) met the requirements of EN 10025-3 for the S355N grade. After conventional HR (scheme HR), the steel had YTS, UTS, and TEL of 390 ± 6 MPa, 556 ± 8 MPa, and 27 ± 1%, respectively, fully complying with the requirements of the S355 grade. NR improved YTS by 55 MPa (to 445 ± 8 MPa), improved UTS by 14 MPa (to 570 ± 9 MPa), and increased TEL by two points (up to 29 ± 2%). Consequently, this combination of strength/ductility increased the grade of the steel to S420N. A more significant advancing of strength was achieved using TMCP with finish rolling in a two-phase temperature interval of 700–720 °C: YTS increased by 86 MPa (22.0%) and UTS increased by 20 MPa (3.6%) compared to HR. Despite the slight decrease in TEL to 25%, the mechanical properties of the TMCP-treated steel complied with the requirements of the high S460N grade. Notably, NR and TMCP had a greater impact on YTS rather than UTS. Thus, under the HR, NR, and TMCP routes with the rolled steel cooling in still air, the steel grade could be maximally increased to S460N.

Table 1. Effect of the processing route on the mechanical properties of 20 mm-thick sheets of a S355N grade. L: longitudinal specimens, T: transversal specimens. The scatter of experimental values of absorbed impact energy and its mean value (in parenthesis) are given depending on the testing temperature.

Processing Rout, Corresponding EN 10025 Grades	Direction	Tensile Testing Properties			Absorbed Impact Energy (J) under Testing at				
		YTS (MPa)	UTS (MPa)	TEL (%)	0 °C	−20 °C	−40 °C	−50 °C (Interpolation)	−60 °C
Hot rolling S355N, S355NL	L	390 ± 6	556 ± 8	27 ± 2	145–222 (189 ± 38)	96–196 (150 ± 30)	62–152 (109 ± 7)	(87)	22–114 (65 ± 29)
	T	–	–	–	35–82 (60 ± 11)	30–72 (45 ± 9)	10–65 (30 ± 7)	(27)	4–52 (24 ± 7)
Normalizing rolling S(355,420)N, S(355,420)NL	L	445 ± 8	570 ± 9	29 ± 2	151–194 (180 ± 10)	80–202 (165 ± 9)	33–194 (121 ± 14)	(103)	20–120 (84 ± 15)
	T	–	–	–	66–77 (70 ± 8)	52–61 (58 ± 10)	46–56 (52 ± 9)	(52)	36–70 (52 ± 10)
TMCP S(355-460)N, S(355-460)NL	L	476 ± 5	576 ± 10	25 ± 2	152–203 (179.0 ± 23)	131–190 (160 ± 27)	46–129 (128 ± 37)	(87)	21–129 (45 ± 20)
	T	–	–	–	32–82 (58 ± 12)	20–75 (47 ± 9)	10–60 (34 ± 9)	(34)	5–61 (34 ± 8)
TMCP/AC_{550} S(355-460)N, S(355-460)NL, S(460,500)Q, S(460,500)QL, S(460,500)QL1	L	525 ± 9	640 ± 10	22 ± 2	152–200 (171 ± 21)	57–183 (134 ± 8)	71–189 (115 ± 30)	(83)	48–61 (53 ± 8)
	T	–	–	–	71–96 (81 ± 2)	67–71 (71 ± 4)	36–71 (62 ± 2)	(50)	36–59 (47 ± 2)
TMCP/AC_{460} S(355-460)N, S(355-460)NL, S(460-550)Q, S(460-550)QL S(460-550)QL1	L	554 ± 8	660 ± 9	17 ± 2	113–203 (174 ± 37)	95–191 (146 ± 24)	115–181 (150 ± 5)	(117)	41–144 (83 ± 13)
	T	–	–	–	41–90 (64 ± 7)	34–80 (56 ± 3)	7–54 (38 ± 3)	(41)	5–71 (44 ± 5)

The steel strength was further improved using processing schemes that combined TMCP and AC. Under TMCP/AC_{550}, YTS increased by 35%, reaching 525 ± 9 MPa compared to the HR route. UTS increased by 15%, reaching 640 ± 10 MPa compared to HR, and it was a much higher increment compared to NR and TMCP. A decreased ductility (TEL of 22%) accompanied the increased strength. However, in general, the steel quality level increased to the S500Q grade, which refers to the steel subjected to the quenching-and-tempering heat treatment according to EN 10025-6. Moreover, higher strength properties were obtained (YTS of 544 ± 8 MPa and UTS of 660 ± 9 MPa) matching the level of the S550Q grade (the TEL value of 16.5% also complied with the S550Q grade) when AC reached 460 °C (TMCP/AC_{460}).

These data on the variation in the steel quality (grade number) depending on the processing route was based on the analysis of strength properties and ductility. A more precise analysis of the data of the variation of the absorbed E depending on the testing temperatures could derive a complete picture, as graphically illustrated in Figure 2a,b. The E values gradually decreased with the decreasing testing temperature for the longitudinal and transverse specimens for each processing route, which is characteristic of the body-centered cubic (BCC) lattice [42]. The steel retained a mean absorbed energy at a high level (>100 J) in the longitudinal direction up to cooling at −40 °C, irrespective of the processing route. At −60 °C, the mean E value varied from 47 J for TMCP to 83–84 J for NR and

$TMCP/AC_{460}$. Accordingly, the absorbed impact energy at the test temperatures controlled by EN 10025-3 and EN 10025-6 (−20, −40, −50, and −60 °C) was several times higher than the minimum standard values. Under the testing at 0 °C, all processing routes ensured approximately the same E values, with some advantage for the HR specimens (Figure 2a). The NR specimens had an advantage at −20 °C. The $TMCP/AC_{460}$ specimens had the best impact toughness at −40 and −60 °C (together with NR specimens in the latter case).

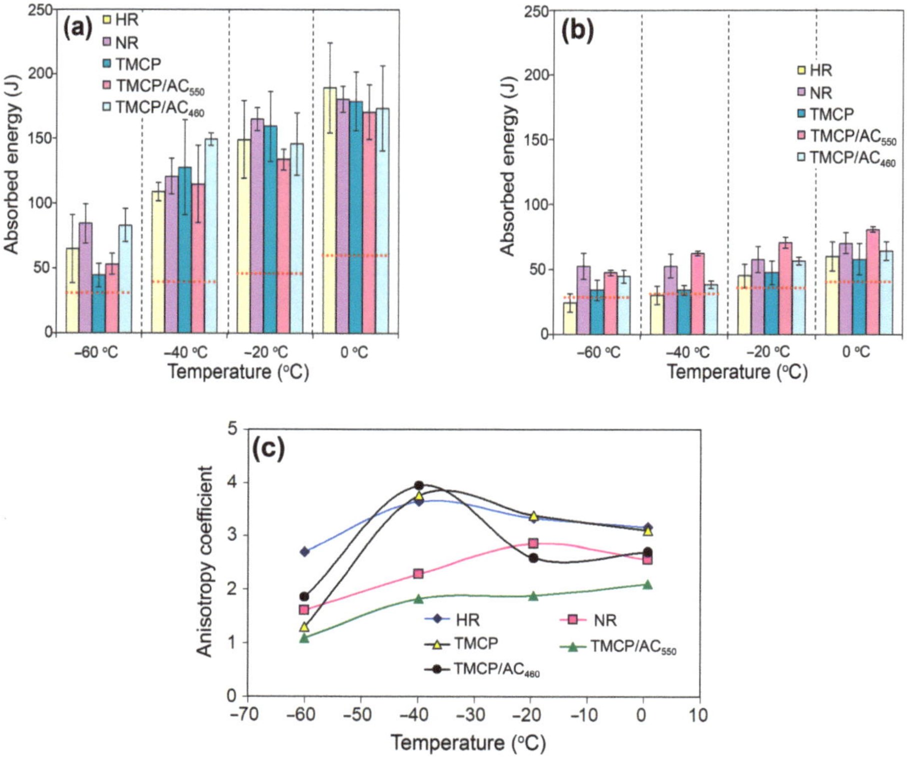

Figure 2. The temperature-wise evolution of the absorbed impact energy for (**a**) longitudinal and (**b**) transverse specimens. (**c**) The temperature-wise ranging of the processing routes by the anisotropy index. (Red dotted lines in (**a**,**b**) show the minimum E level corresponding to the category "QL1").

Furthermore, the transverse specimens absorbed two to three times less energy than the longitudinal specimens (Figure 2b). For different processing routes, an average E value varied from 58–81 J at 0 °C to 29–62 J at −40 °C and 24–52 J at −60 °C. At all testing temperatures, with the exception of −60 °C, the $TMCP/AC_{550}$ specimens showed the highest mean E values. At −60 °C, $TMCP/AC_{550}$ and NR provided the best impact behavior. In contrast, the HR and TMCP specimens absorbed the lowest energy at any testing temperature. The data showed that $TMCP/AC_{550}$ provided the best low-temperature impact toughness in the transverse direction.

The difference in low-temperature impact behavior in longitudinal and transverse directions was assessed using *Ai*, which is the ratio of the mean absorbed energy values in the two directions (Figure 2c). Under the testing temperatures up to −40 °C, the highest *Ai* values were attributed to HR (*Ai* = 3.2–3.6), TMCP (*Ai* = 3.1–3.7), and $TMCP/AC_{460}$ at -40 °C (*Ai* = 3.9). The lowest *Ai* value of 1.8–2.1 was attributed to the $TMCP/AC_{550}$ specimens. Under the testing temperature of −60 °C, *Ai* reached its minimum for each processing route, and the lowest *Ai* value of 1.1 was ascribed to $TMCP/AC_{550}$.

The distinction in the direction-wise impact behavior was closely related to the specific features of the fracture of the specimen. The rapture surface of the Charpy specimens tested at −40 °C (steel was subjected to NR) is illustrated in Figure 3. The fracture of the longitudinal specimen mainly consisted of dimples and tear ridges, manifesting the ductile trans-granular mechanism of the crack propagation (Figure 3a). The different sizes of the dimples indicated the micro-voids merging at higher strains, which were beneficial for the impact toughness [43]. The transverse specimen had minor dimple areas on the surface. In contrast, the inter-granular pattern combined with the quasi-cleavage areas was dominant, implying mostly brittle fracture type (Figure 3b). Notably, the inter-granular facets were not smooth and were covered with the micro-relief left after the crack branched along the grain boundaries (inset in Figure 3b). The dimples and micro-relief proved that rupture proceeded through increased absorption of energy, maintaining the E values at the acceptable level of 45–56 J.

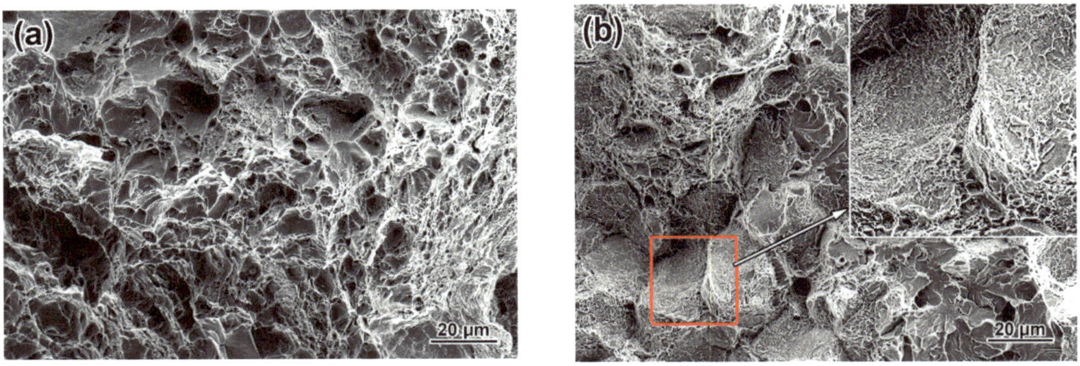

Figure 3. The fracture patterns of the V-notched specimens after the normalizing rolling (testing at −40 °C): (**a**) longitudinal, (**b**) transversal.

The scatter of experimental values of absorbed energy and its mean value depending on the test temperature (the values for −50 °C were obtained using interpolation) are presented in Table 1. These data allowed for the specification of the EN 10025 grades that complied with the experimental results, specifically with respect to impact toughness. A comparison of the results with the standard norms showed that all applied processing routes provided high impact toughness (absorbed impact energy). Therefore, the steel sheet fully met the requirements of the following grades: (a) S355N and S355NL after HR, (b) S(355,420)N and S(355,420)NL after NR, (c) S(355-460)N and S(355-460)NL after TMCP, (d) S(355-460)N, S(355-460)NL, S(460,500)Q, S(460,500)QL, and S(460,500)QL1 after TMCP/AC_{550}, and (e) S(355-460)N, S(355-460)NL, S(460-550)Q, S(460-550)QL, and S(460-550)QL1 after TMCP/AC_{460}. Compliance of the steel with the grades marked "L" and "L1" indicated that its impact toughness met the standard over the entire controlled temperature range, including low temperatures from −40 to −60 °C, emphasizing the high quality of the steel and its compliance with the northern operational requirements.

The presented data showed that the steel strength progressively increased as the processing route changed from HR to TMCP/AC, corresponding to an increased steel grade from S355 for HR to S460 for TMCP and then to S500 and S550 for TMCP/AC_{550}, TMCP/AC_{460}. Although the last two grades refer to the quenched-and-tempered steel (furnace heating), this study obtained the same high properties without the heat treatment. The ductility of steel (TEL) increased to 29% after NR, followed by a gradual decrease to 17% according to the sequence: NR → TMCP → TMCP/AC_{550} → TMCP/AC_{460}. Although TEL decreased, it retained its values matching the corresponding grades that increased in the same sequence. TMCP/AC_{550} should be considered an optimal processing route for the steel studied with respect to the complex properties, like strength, ductility, low-temperature impact toughness, and anisotropy. YTS and TEL of the steel exceeded 500 MPa and 20%, respectively, for implementing TMCP/AC_{550}. Additionally, the best cold resistance in the transverse direction and minimal anisotropy in impact toughness were ensured.

3.2. Microstructure Characterization

The advancement in the mechanical properties of steel described earlier was due to the evolution of its microstructure on changing the processing route (Figure 4). The HR specimens had a classic "ferrite + pearlite" structure consisting of a major fraction of polygonal ferrite and 16 vol. % of pearlite (Figure 4a). Pearlite colonies were stretched along the rolling direction to form the "ferrite/pearlite" banding pattern. The ferrite grains had a near-polyhedral equiaxed shape, a non-uniform size varying from 2–3 μm to 30–36 μm, and an average diameter of 24.1 ± 1.1 μm. The grains had high-angle boundaries and a minor amount of dislocations, while the cellular substructure was not observed (Figure 5a).

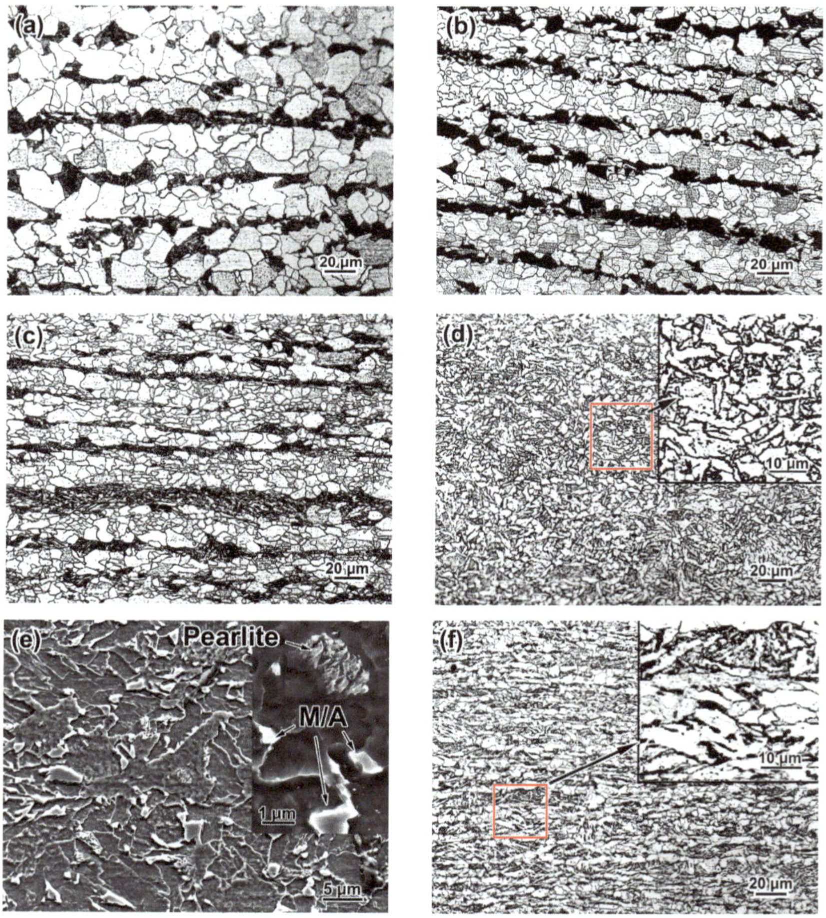

Figure 4. Microstructure of steel after the processing routes: (**a**) HR, (**b**) NR, (**c**) TMCP, (**d**,**e**) TMCP/AC_{550}, (**f**) TMCP/AC_{460}. ((**a**–**d**,**f**) are OM images, (**e**) is SEM image).

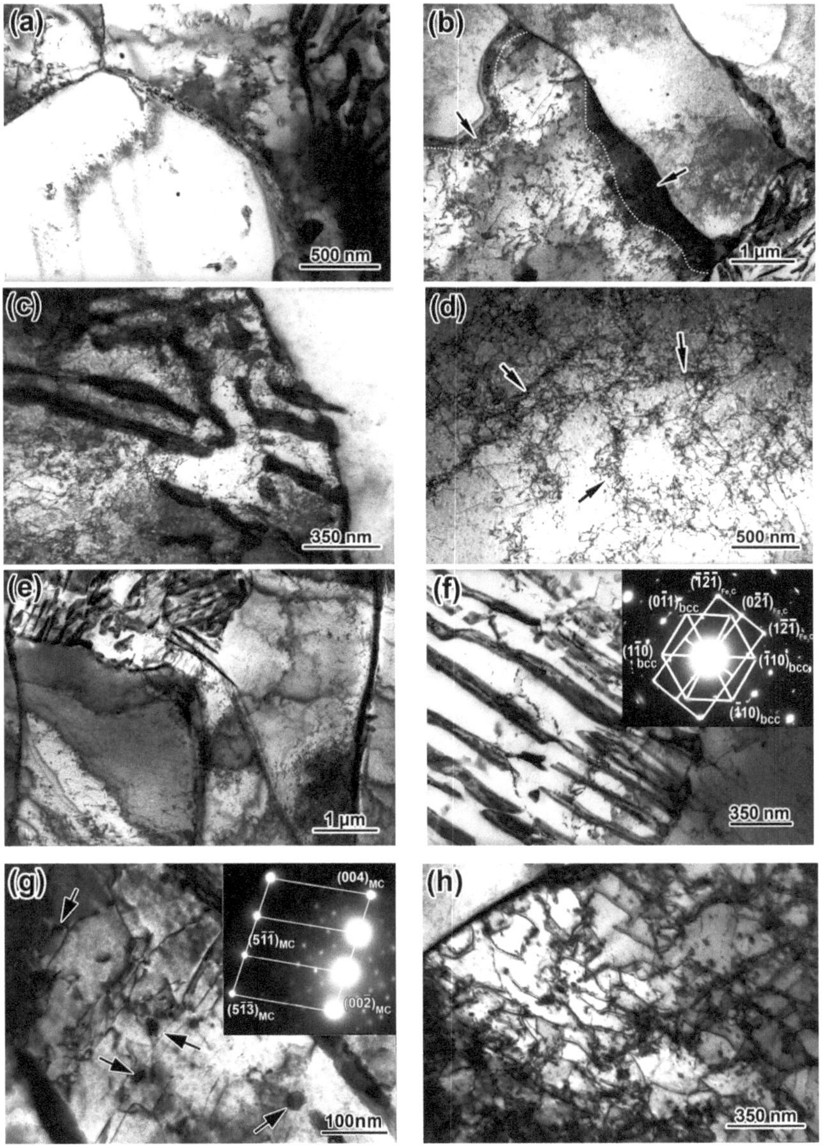

Figure 5. TEM images of steel after HR, NR, and TMCP treatments: (**a**,**b**,**e**) ferrite grains with a nearby pearlite colony, (**c**) pearlite colony with high dislocation density, (**d**) dislocation "walls" in ferrite grains (shown by the arrows), (**f**) cementite lamellae and selected area electron diffraction (SAED) showing the reflection on the [111] zone axis of ferrite and the reflection on the [01$\bar{2}$] zone axis of cementite, (**g**) precipitates (Nb,V)C (shown by the arrows) and SAED showing the reflection on the [111] zone axis of carbide, (**h**) dislocation clots around the nano-precipitates inside ferrite grain. ((**a**)—HR, (**b**–**d**)—NR, (**e**–**h**)—TMCP).

NR refined the ferrite grain with the average size decreasing to 15.0 ± 0.85 μm and the grains becoming slightly elongated (Figure 4b). In addition to the fine grains, there were occasional coarse grains with up to 40 μm length and 15–17 μm width. The NR specimens retained the banding pattern, which was more pronounced compared to the HR specimens. Visually, the grains had a higher density of lattice defects with a non-uniform distribution of dislocations within the grains. Additionally, grains with a dislocation sink to the boundaries were revealed (shown using dotted lines and arrows in Figure 5b). An increased density of dislocations was also found in the ferrite layers within the pearlite colonies (Figure 5c). Many ferrite grains showed cellular substructures with the dislocation "walls" dividing the grain to the smaller sub-grains of 0.13–0.45 μm in size (0.24 ± 0.05 μm in average) (shown using arrows in Figure 5d). The heavier dislocation pattern of the NR specimens was due to finishing the deformation close to T_{nr}. The dispersed precipitates of Nb and V carbides were not observed in the NR structure.

TMCP further stretched and refined the ferrite grains to 12.1 ± 0.85 μm in average diameter. The grain dimension became more uniform, and the size of the coarse grains showed a maximum decrease to 25 μm length and 15–17 μm width (Figure 4c). A pronounced "ferrite/pearlite" banding characterized the TMCP specimens. The last rolling passes were performed at low temperatures of 700–720 °C under TMCP, inhibiting the dynamic and metadynamic recrystallization [44]. Additionally, this temperature range refers to the precipitation of fine V/Nb carbides [45], contributing to recrystallization suppression [46]. Subsequently, the structure acquired a heavily deformed pattern with a high dislocation density. As shown in Figure 5f, the dislocations inside the carbide plates were identified as cementite belonging to the pearlite colony, as depicted in the selected area electron diffraction (SAED) pattern in the inset to Figure 5f. As shown by the arrows in Figure 5g, the ferrite grains had fine grainy precipitates of (Nb,V)C carbide apart from the plate particles. The SAED analysis shown in the inset of Figure 5g confirmed the cubic crystal lattice (of NaCl-type) characteristic for MC carbide [47]. According to the EDX spectra in Figure 6a, the precipitates contained 16.82 wt.% of Nb, 2.12 wt.% of V, and nearly 80 wt.% of iron. The low content of strong carbide-forming elements in (Nb,V)C carbide was the artifact caused by the small size of the precipitate. Consequently, the EDX results overestimated the iron content due to the contribution of the surrounding ferric matrix [48]. The (Nb,V)C precipitates were nano-sized with an 8.1 ± 0.3 nm average value and a diameter varying in the 2–14 nm range, and 63% of the precipitate did not exceed 10 nm in size (Figure 6b). A comparison of the structures of TMCP and NR specimens showed the latter had no nano-particles and allowed to the presumption that the progressive accumulation of crystal defects due to a lower FRT under TMCP facilitated the (Nb,V)C carbide precipitation. As shown in Figure 5h, the formed precipitates interacted with dislocations using the Orowan mechanism in the grains, preferably oriented relative to the deformation direction, forming a dislocation "forest" that contributed to the steel strength [49,50].

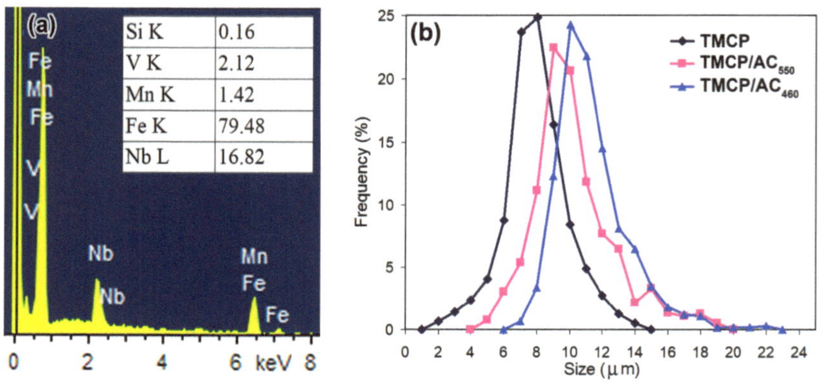

Figure 6. (**a**) EDS spectra and chemical composition of MC precipitate. (**b**) Size distribution of MC carbide.

The application of AC just after HR changed the structure significantly. Mostly, the TMCP/AC_{550} processing route suppressed the pearlite transformation, sharply decreasing the pearlite fraction and eliminating the "ferrite/pearlite" banding (Figure 4d). The phase transformation temperature decreased due to fast cooling. Consequently, a quasi-polygonal ferrite (irregular shape) or acicular ferrite replaced the polyhedral-shaped ferrite, with the former predominant in the structure (the area of acicular ferrite is shown in the inset in Figure 4d). The ferrite grain was additionally refined compared to TMCP, with the average size reduced to 8.9 ± 0.4 μm. The length and width of the coarsest grains did not exceed 10 and 4 μm, respectively. The colonies of fine grains of pearlite with up to 2–3 μm size were seen inside the ferrite grains. Additionally, the small grainy "islands" (0.2–2 μm) were dispersed along the grain boundaries, presumably being the martensite/austenite (M/A) conglomerates [51], as illustrated in the inset in Figure 4e. The TEM image showed the M/A "islands" as the dark-contrast bulky inclusions (denoted by single arrows in Figure 6a). SAED revealed austenite retained the in M/A constituent (inset in Figure 7a). Additionally, austenite was retained as thin films (20–95 nm) between the ferrite grains (shown by doubled arrows in Figure 7a). The SAED analysis in the inset in Figure 7b confirmed the occasional cementite films along the ferrite grains (Figure 7b). The ferrite grains had a sub-grain pattern since dislocation "walls" divided them. As shown in the left part of Figure 6c, the dislocations interacted with the nano-sized precipitates of (Nb,V)C carbide, and their dispersion is illustrated by the dark-field image in the right part of Figure 6c. The size range of the (Nb,V)C precipitates was 5–19 nm, and almost 64% of the precipitates were smaller than 10 nm (Figure 6b).

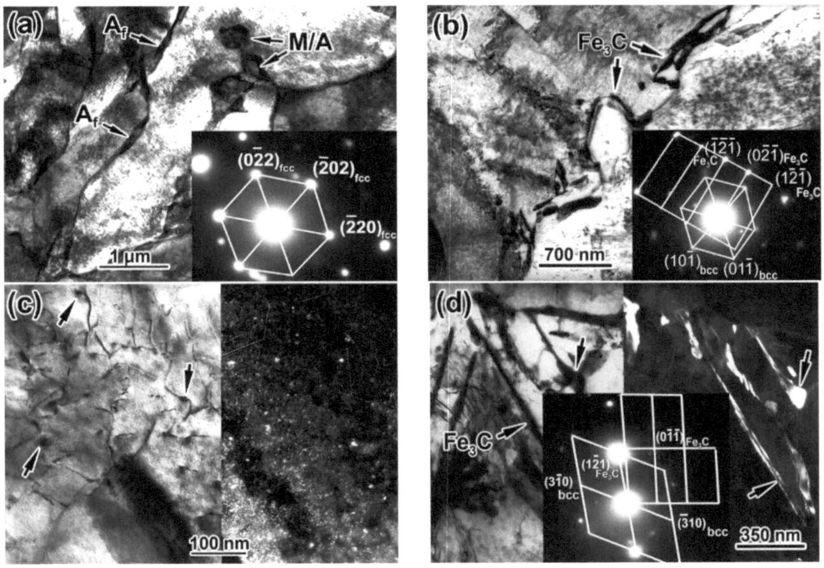

Figure 7. Fine microstructure (TEM images) of steel after TMCP/AC_{550} and TMCP/AC_{460} treatments: (**a**) M/A "islands" (M/A), austenite films (A_f) in acicular ferrite grains and the SAED of M/A showing the reflection on the [111] zone axis of austenite, (**b**) cementite films and the SADEs of the reflections on the zone axes of [111] of ferrite and $[01\bar{2}]$ of cementite, (**c**) the bright-field and dark-field images of (Nb,V)C precipitates (shown by the arrows), (**d**) cementite lamellae (shown by the arrows) in bainite and the SAEDs of the reflections on the zone axes of [135] of ferrite and $[\bar{1}00]$ of cementite. ((**a**,**b**)—TMCP/AC_{550}; (**c**)—TMCP/AC_{460}).

Compared to TMCP/AC_{550}, the phase transformation proceeded to a lower temperature during TMCP/AC_{460}, as evidenced by the appearance of approximately 25 vol. % of bainite along with irregular/acicular ferrite (Figure 4f). As shown in the TEM image in Figure 7d, thin carbide lamellae oriented from the boundaries inward of the ferrite grains characterized bainite. The dark-field observation in the cementite reflection and the corresponding SAED analysis in the inset of Figure 7d confirmed that the lamellae were cementite carbide. This observation explained the decrease in TEL to 17% compared to 22% in TMCP/AC_{550} since the borderline cementite precipitates were considered a negative factor affecting steel ductility [52]. Additionally, nano-sized (Nb,V)C precipitates were found in the structure (Figure 7c) with an approximately similar size distribution as the TMCP/AC_{550} structure (Figure 6b). The mean size of the ferrite grains after TMCP/AC_{460} was 9.9 ± 0.9 μm.

4. Discussion

4.1. The Contribution of Structural Factors to the Yield Strength of S355N Steel

The results mentioned earlier showed that using only a specific cost-saving finish processing route could significantly improve the mechanical properties of low-carbon (Nb, V, Al)-micro-alloyed steel from S355 to S550 grade. This was possible due to the inhibited grain growth and recrystallization processes, along with the decreased phase transformation temperature. The best combination of strength, ductility, and impact toughness was attributed to TMCP/AC_{550}, which provided a structure mainly comprised of fine irregular-shaped ferrite grains strengthened using nano-sized carbides and M/A islands. According to Wang et al., fine and evenly dispersed M/A islands could contribute to the mechanical properties of low-carbon steel [51]. Other benefiting features of the TMCP/AC_{550} structure are (a) a predominant cementite-free state of the grain boundaries and (b) the retained austenite (as boundary-allocated films and M/A islands) that can hamper crack propagation [53] and enhance the properties due to the TRIP-effect [54,55]. These features were especially conducive to the low-temperature impact behavior of steel. Compared to TMCP/AC_{550}, the precipitation of cementite carbides between the ferrite laths during bainite transformation at 460 °C (TMCP/AC_{460}) decreased ductility and low-temperature toughness in the transverse direction.

The analysis of the contribution of a particular structural factor to the mechanical properties can assess its importance for steel quality. It is well known that the yield strength of low-carbon welded steels obeys the additive approach expressed as follows [28,56,57]:

$$YTS = \sigma_o + \Delta\sigma_{SS} + \Delta\sigma_D + \Delta\sigma_{GB} + \Delta\sigma_{DP} + \Delta\sigma_P + \Delta\sigma_{M/A}, \quad (4)$$

where σ_o represents lattice friction stress (Pierls–Nabarro stress). $\Delta\sigma_{SS}$ represents solid solution strengthening due to interstitial and substitutional atoms. $\Delta\sigma_{GB}$, $\Delta\sigma_D$, and $\Delta\sigma_P$ represent grain boundary strengthening, dislocation strengthening, and pearlite strengthening, respectively. $\Delta\sigma_{DP}$ and $\Delta\sigma_{M/A}$ represent the strengthening due to dispersed precipitates and M/A islands, respectively.

The frictional stress of the α-Fe lattice (Pierls–Nabarro stress) can be roughly assessed as 2 G × 10^{-4} MPa [58]. Since the shear modulus of iron (G) is equal to 84,000 MPa, the frictional stress of the steel lattice adopted in this study was 17 MPa regardless of the steel structure.

Solid solution strengthening, which occurs due to strengthening when atoms of alloying and impurity elements are dissolved in ferrite, is calculated as follows [58]:

$$\Delta\sigma_{SS} = \sum_{i=1}^{n} k_i c_i, \quad (5)$$

where k_i represents the coefficient of the *i*-element and was equal to 5440, 690, 83, 32, 30, and 30 MPa·wt%$^{-1}$ for (C+N), P, Si, Mn, Ni, and Cr, respectively [57]. c_i represents the *i*-element content in ferrite (in wt. %). Nb and V were not considered since it was presumed that they were completely bound in (Nb,V)C carbide.

Dislocation strengthening was calculated using the Taylor equation as follows [57]:

$$\Delta\sigma_D = \alpha M G b \rho^{1/2}, \tag{6}$$

where, α represents a coefficient that depends on the nature of the interaction of the dislocation during strain hardening and is equal to 0.25 [59]. M represents the Taylor factor that is equal to 2.73 for ferrite [57]. G represents the shear modulus that is equal to 81.6 GPa for ferrite [28]. b represents the magnitude of the Burgers vector that is equal to 0.25 nm for iron. ρ represents the dislocations density.

Grain boundary strengthening was estimated according to the Hall–Petch law as follows:

$$\Delta\sigma_{GB} = k_y d^{-1/2}, \tag{7}$$

where, k_y: the Hall–Patch slope coefficient that was adopted as 0.63 MPa·m$^{-1/2}$ for ferrite–pearlite steels [58]. d: average grain size.

In the case of the formation of sub-grains, sub-structural strengthening is calculated instead of grain boundary strengthening as follows [58]:

$$\Delta\sigma_S = k_C l^{-m}, \tag{8}$$

where kc represents the coefficient characterizing the sub-grain boundaries structure and is taken as 1.5×10^{-4} MPa·m. l represents the average sub-grain size. m represents the coefficient adopted as 0.5 [23].

The strengthening due to the precipitates ($\Delta\sigma_{DP}$) can be modelled using the Ashby–Orowan equation as follows [60]:

$$\Delta\sigma_{DP} = \frac{6.66}{L} \ln \frac{D}{4.96 \times 10^{-4}}, \tag{9}$$

where D represents the mean planar intercept diameter of a precipitate. L represents the surface-to-surface precipitate spacing and is calculated as follows [61]:

$$L = D\left[\left(\frac{\pi}{4f}\right)^{0.5} - 1\right], \tag{10}$$

where f represents the volume fraction of the precipitates derived from TEM images.

The pearlite contribution to the strength of steel depends on its volume fraction (P, vol. %) as follows [58]:

$$\Delta\sigma_P = 2.4P. \tag{11}$$

The contribution of the M/A islands was calculated as follows:

$$\Delta\sigma_{M/A} = \sigma_{M/A} f_{M/A}. \tag{12}$$

where $\sigma_{M/A}$ is taken as 600 MPa [62]. $f_{M/A}$ represents the volume fraction of the M/A islands.

The theoretical yield strength for different processing routes was calculated using the values of the structural parameters presented in Table 2. The results of the calculations are depicted in Table 3 and Figure 8. Their analysis shows satisfactory compliance between theoretical results and experimental YTS values. This generally confirms the reliability of the adopted approach in understanding and assessing the factors determining the strength of structural steel. The biggest difference in results relates to the TMCP route: the calculation for TMCP gives an excess over the experimental value by 38.5 MPa (which however does not exceed 10% of the latter). A similar deviation (28.2 MPa), but with the opposite sign, refers to the HR route. These differences may be caused by an error in determining the average grain size, which depends on the location of sampling for the study. Also, in the case of TMCP, the difference may be affected by an error in the calculation of the parameters of dispersed precipitates, which were observed in the local micro-areas of the specimen.

For other processing routes, the difference between the calculated data and experimental values did not exceed 15 MPa (2.8%), while the best match in the results was ascribed to TMCP/AC_{550}.

Table 2. Structural parameters used for the theoretical yield strength calculation of the studied steel.

Processing Route	ρ, cm^{-2}	d, µm	D, nm	f, vol.%	P, vol.%	$f_{M/A}$, vol.%
HR	4.2×10^8	24.1	–	–	16.9	–
NR	2.1×10^9	15.0	–	–	15.1	–
TMCP	4.4×10^9	12.1	8.1	0.56	15.9	–
TMCP/AC_{550}	8.5×10^9	8.9	10.1	0.51	1.0	3.5
TMCP/AC_{460}	1.5×10^{10}	9.9	11.2	0.50	–	–

Table 3. The contribution of structural factors to the yield strength of experimental steel (MPa). In the last column, in parentheses, the difference in YTS is shown in % to the experimental YTS.

Processing Route	σ_o	$\Delta\sigma_{SS}$	$\Delta\sigma_D$	$\Delta\sigma_{GB}$	$\Delta\sigma_{DP}$	$\Delta\sigma_P$	$\Delta\sigma_{M/A}$	YTS (Calculation)	YTS (Experiment)	$YTS_{calc} - YTS_{exp}$
HR	17	154.1	28.5	128.3	–	40.8	–	361.7	390.0 ± 6	−28.2 (7.2%)
NR	17	154.1	63.8	162.7	–	36.0	–	433.1	445.0 ± 8	−11.9 (2.7%)
TMCP	17	154.1	92.4	181.8	39.0	38.4	–	514.5	476.0 ± 5	38.5 (8.1%)
TMCP/AC_{550}	17	154.1	128.4	211.2	31.0	2.4	21.0	536.0	525.0 ± 9	11.0 (2.1%)
TMCP/AC_{460}	17	154.1	170.5	200.2	28.4	–	–	559.1	544.0 ± 8	−15.1 (2.8%)

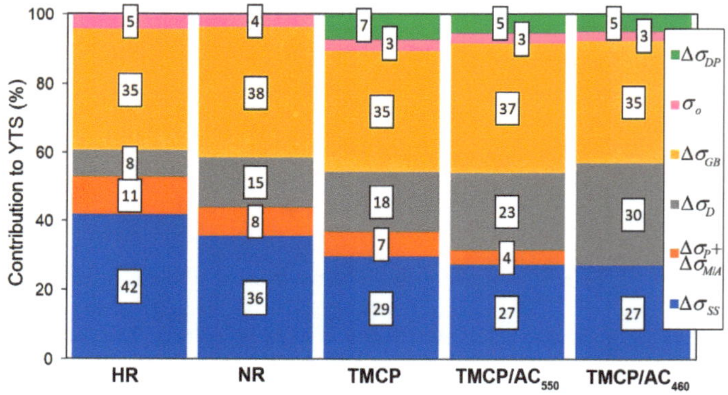

Figure 8. Contribution (%) of structural strengthening factors to theoretic YTS.

As shown in Figure 8, the contribution percentage of various structural factors to the yield strength changed as the processing route changed. The solid-solution strengthening had the most significant impact of 42% in the hot-rolled state, and its contribution gradually decreased to 27% for TMCP/AC. Instead, the contribution of dislocation strengthening increased sharply from 8% for HR to 30% for TMCP/AC_{460}. The latter resulted from the increased dislocation density due to the inhibited recrystallization because of the decreased FRT and the shear component of phase transition under M/A and bainite formation [63]. The "acicular ferrite + M/A + bainite" structure replaced "ferrite + pearlite", decreasing the pearlite strengthening effect from 11% for HR to 0% for TMCP/AC_{460}. The impact of grain boundaries remained stable at a high level of 35–38%, i.e., Hall–Petch strengthening was crucial for steel strength regardless of the processing mode. Although the dispersed (Nb,V)C carbides added just 28–39 MPa, they contributed indirectly through grain refinement when acting as the nuclei for the ferrite grains [64,65].

4.2. Variation of the Ductile–Brittle Transition Temperature Depending on the Processing Route

The evolution of a steel structure usually affects the strength and brittle fracture resistance. The latter is often assessed using T_{DBT} [64]. The change in the ductile–brittle transition temperature (ΔT_{DBT}) of pure iron under the effect of different structural factors can be evaluated as follows [58]:

$$\Delta T_{DBT} = \Delta T_{SS} + \Delta T_D + \Delta T_{GB} + \Delta T_{DP} + \Delta T_P + \Delta T_{M/A}, \quad (13)$$

where ΔT_{SS}, ΔT_D, ΔT_{GB}, ΔT_{DP}, ΔT_P, and $\Delta T_{M/A}$ are the variations of T_{DBT} due to the influence of different strengthening factors such as solid solution, dislocation, dispersed precipitates, pearlite, grain boundary, and M/A islands, respectively.

It was proposed that the effect of the *i*-structural factor on ΔT_{DBT} could be calculated by multiplying its contribution to YTS (i.e., $\Delta\sigma_i$) and the coefficient of embrittlement (K_i) as follows [58]:

$$\Delta T_{DBT_i} = \sum_{i=1}^{n} K_i \cdot \Delta\sigma_i, \quad (14)$$

Taking the K_i values from [58], the Equation (13) can be presented as follows:

$$\Delta T_{DBT} = 0.5 \cdot \Delta\sigma_{SS} + 0.4 \cdot \Delta\sigma_D + 0.3 \cdot \Delta\sigma_{DP} + 0.9 \cdot \Delta\sigma_P + 0.5 \cdot \Delta\sigma_{M/A} - 0.7 \cdot \Delta\sigma_{GB}. \quad (15)$$

The calculated results of ΔT_{DBT} are gathered in Table 4 and shown in Figure 9 as a vector diagram. It was obvious that the majority of strengthening factors increased T_{DBT}, implying steel embrittlement. However, the most deteriorating effect was associated with a solid-solution strengthening (ΔT_P of 77 °C). The grain boundary strengthening was the only factor that was accompanied by a decreased T_{DBT} (ΔT_{GB} of −90–−147 °C). As shown in the last column of Table 4 where the total ΔT_{DBT} values are present, each of the applied processing routes should increase the cold-resistance threshold compared to pure iron due to the complex strengthening effect. In other words, the used processing routes should result in pure iron embrittlement. Consequently, the most significant increase of 35.4 and 33.0 °C in T_{DBT} corresponds to HR and TMCP, respectively, mostly due to the presence of pearlite and increased dislocation density in the case of TMCP. Despite the presence of pearlite, T_{DBT} had a smaller increase of 21 °C after NR due to the positive effect of grain refinement. The most remarkable result implied that TMCP/AC_{550} should provide a close-to-zero change in the ductile–brittle threshold (ΔT_{DBT} of 2.5 °C) due to further grain refinement and the replacement of pearlite by quasi-polygonal/acicular ferrite. Thus, despite a significantly increased strength compared to HR, TMCP/AC_{550} should not affect the low-temperature impact behavior of steel. The improved low-temperature behavior of the TMCP/AC_{550}-treated steel in a transverse direction confirmed this conclusion. A comparison of the absorbed E for transverse specimens after HR and TMCP/AC_{550} showed that the TMCP/AC_{550} specimens had 1.5–2.0 times higher E values than HR specimens at any testing temperature (Table 1). However, TMCP/AC_{460} should somewhat worsen the low-temperature behavior of steel since T_{DBT} increased by 13.6 °C (Table 4), which is explained by an increased dislocation density due to bainite transformation. In the case of TMCP/AC_{460}, grain refinement and pearlite elimination could not compensate for the dislocation-induced embrittlement. In conclusion, TMCP/AC_{550} is considered the most favorable processing route for the steel studied.

Table 4. The contribution of structural factors to the temperature of ductile–brittle transition (°C).

Processing Route	ΔT_{SS}	ΔT_D	ΔT_{DP}	ΔT_P	$\Delta T_{M/A}$	ΔT_{GB}	ΔT_{DBT}
HR	77.1	11.4	-	36.7	-	−89.8	35.4
NR	77.1	25.5	-	32.4	-	−113.9	21.1
TMCP	77.1	32.5	11.7	34.6	-	−127.3	33.0
TMCP/AC_{550}	77.1	51.3	9.3	2.2	10.5	−147.8	2.5
TMCP/AC_{460}	77.1	68.2	8.5	-	-	−140.2	13.6

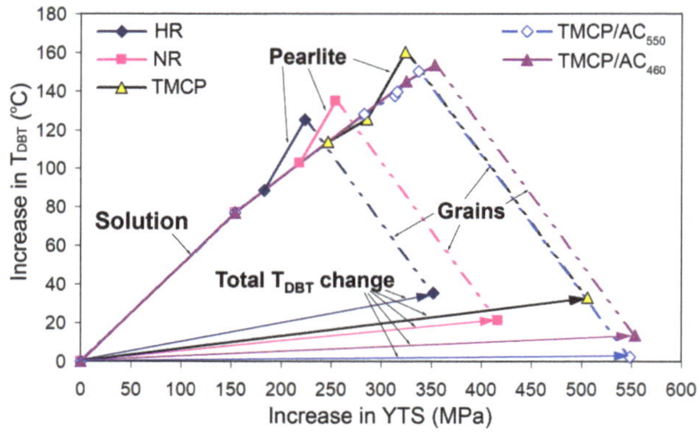

Figure 9. Vector diagram of the change in T_{DBT} caused by strengthening effects of different structural factors depending on the processing scheme.

There are different approaches to evaluating T_{DBT} for steel. For T_{DBT}, the temperature T_{50} is often used, which corresponds to 50% ductile area on the surface of the broken specimen [66]. Additionally, T_{DBT} is estimated regarding the specified minimum value of absorbed energy [67] or T_{DBT} is taken as a temperature of the double decrease in the impact toughness relative to testing at room temperature [68]. This study followed the latter approach using the temperature dependencies of absorbed energy, as presented in Figure 3. The temperature that refers to a double decrease in the absorbed energy relative to testing at 0 °C was assessed using interpolation. The calculated values of T_{DBT} are depicted in Figure 10. T_{DBT} consistently decreased in the longitudinal direction as finish temperature decreased and post-rolling cooling intensified from –30.8 °C in the HR specimens to –48.5 °C in the TMCP/AC_{460} specimens. Although the processing route effect was less noticeable in the transverse direction, TMCP/AC_{550} stood out (T_{DBT} = −36.5 °C) and complied with the high cold resistance assessed earlier (Table 4 and Figure 9). These results proved that the combination of TMCP and AC significantly advanced the ductile-to-brittle behavior of the studied steel.

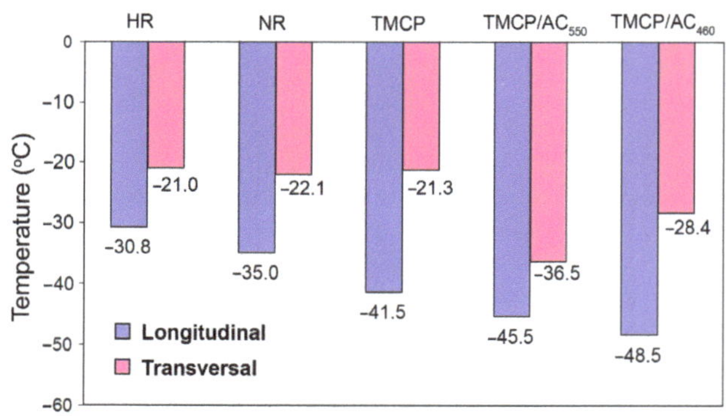

Figure 10. The temperature corresponding to the double reduction of the absorbed energy relative to testing at 0 °C.

The obtained data showed that varying only the finish rolling temperature and increasing the post-rolling cooling rate could increase the yield strength of (Nb+V+Al)-microalloyed steel of S355 grade by about 30% from 390 to 544 MPa. The increased YTS was accompanied by the persistent or increased low-temperature impact toughness in longitudinal and transverse directions (at −20–−60 °C). Moreover, TMCP/AC decreased the rolling anisotropy, which was vital for large welded structures experiencing fatigue issues [69,70].

The results presented in this paper are based on the theoretical and experimental studies on processing routes for structural high-strength low-alloy (HSLA) steel previously fulfilled by Ishikawa et al. [71], Wang et al. [35], Misra et al. [17,18], Bhadeshia et al. [32,33], and many other researchers. The novelty of this work is that it compared different treatment modes performed on one batch of rolled steel, which neutralizes the influence of the chemical composition and makes it possible to clearly assess the influence of parameters of the technological process. This allowed us to more accurately determine the grade of rolled steel (according to EN 10025) that can be achieved depending on the processing route.

Significantly, the mechanical properties of the S355 grade steel were enhanced through a cost-saving approach that may bring definite economic benefits. This statement is supported by the rough estimation of the costs connected with steel sheet manufacturing. If we assume the hot rolling as a baseline, then NR and low-temperature TMCP (FRT of 700–720 °C) increase the production costs approximately by USD 5–7 per ton and USD 15–20 per ton, respectively, due to an increased electricity consumption. Similar cost increases can be caused by the combination of TMCP (with FRT of 790–810 °C) and accelerated water cooling. Alternatively, heat treatment with additional furnace heating (normalization) can be used to ensure the required mechanical properties. It is a much more costly operation since natural gas is consumed in large amounts (80–110 m^3 per ton), which accounts for approximately USD 120–150 per ton. Moreover, to produce the steel sheets of S460Q-S550Q grades, the two-step "quenching-and-tempering" heat treatment is required, resulting in even higher production costs. Therefore, it is obvious that replacing heat treatment with other processing routes not associated with additional heating of the metal is economically feasible, such that it can result in savings of at least USD 100 per ton of rolled steel. Another aspect that is not mentioned yet is the possibility of further cost reduction by reducing the content of alloying elements. As can be seen from the presented data, accelerated cooling at the last stage of processing gives a significant increase in strength, which makes it possible to reduce the content of alloying and micro-alloying additives in the production of steel sheets of S355 and S410 categories. Despite the attractiveness of the cost-saving approaches, it is associated with corresponding investments. The use of the described processing routes calls for the availability of appropriate technological equipment, including a powerful rolling mill for the low-temperature deformation of steel (under NR and TMCP routes), as well as the presence of an accelerated cooling device (for TMCP/AC routes).

A limitation of the present research was studying steel sheets of only one thickness (20 mm). This does not allow evaluation of the potential capabilities of the processing routes used to improve the quality of steel-rolled products of a wider thickness range. Also, the economic benefits of using the energy-saving approaches should be assessed more precisely considering the technological peculiarities of each operation. Therefore, similar studies need to be performed for structural steel sheets of various thicknesses, including the heavy plates. This could be a possible avenue for future research on this topic.

5. Conclusions

This study investigated the microstructure and mechanical properties of 20 mm-thick sheets of low-carbon (V+Nb+Al)-micro-alloyed rolled steel, depending on the final manufacturing stage. Based on the obtained results, the major conclusions are drawn as follows:

1. Several processing routes without furnace heating were used at the final stage production of 20 mm-thick sheets of S355N grade steel (EN 10025), specifically HR, NR, TCMP, and AC. The variation in the processing route could increase the mechanical properties of steel from S355N to S550QL and S550QL1 grades without the additional heat treatment costs. The order of the used processing routes with the increased strength was as follows: conventional HR $\rightarrow$ NR $\rightarrow$ TMCP $\rightarrow$ TMCP/AC_{550} $\rightarrow$ TMCP/AC_{460}.

2. NR improved the properties mainly through grain refinement. A further decrease in the FRT under the TMCP-included processes was accompanied by a progressive grain refinement of up to 10.5–11 numbers of ASTM E112 and an increased dislocation density that formed a sub-grained structure. The accumulation of lattice crystal defects stimulated the precipitation of nano-sized particles (2–20 nm) of (Nb,V)C carbide, which further interacted with dislocations using the Orowan mechanism.
3. The application of AC with a cooling intensity of 15–18 $°C·s^{-1}$ after TMCP suppressed the formation of pearlite and eliminated the ferrite + pearlite structural banding. TMCP/AC_{550} formed a structure consisting of quasi-polygonal and acicular ferrite with minor fractions of dispersed pearlite and M/A islands. It ensured an optimal combination of strength (YTS of 525 MPa), ductility (TEL of 22%), and sub-zero absorbed impact energy of 115 and 62 J in the longitudinal and transversal specimens at −40 °C, respectively. The appearance of bainite under TMCP/AC_{460} led to a moderately decreased ductility (TEL of 17%) and transversal sub-zero impact toughness, which was associated with the appearance of cementite lamellae at the grain boundaries.
4. The strengthening mechanism contribution to the yield strength was defined analytically. The solid solution and grain boundary were the primary contributors to strength, irrespective of the processing route. For TMCP, the strengthening due to dispersed precipitates accounted for less than 10% of YTS. With the decreased finish rolling temperature and the involvement of water cooling, the contribution of the dislocation mechanism increased significantly, approaching 30% after TMCP/AC_{460}.
5. The effects of strengthening mechanisms on ΔT_{DBT} for steel were calculated. TMCP/AC routes minimally affected the low-temperature impact toughness of steel compared with HR, NR, and TMCP. This output complied with the experimental data showing that TMCP/AC routes ensured the lowest values of the ductile–brittle transition threshold in the longitudinal direction (−48.5 °C, TMCP/AC_{460}) and in the transversal direction (−36.5 °C, TMCP/AC_{550}).

Author Contributions: Conceptualization, V.Z., V.S. and V.E.; methodology, V.E., K.W., I.P. and M.B.; software, B.E.; validation, Y.C. and B.E.; formal analysis, Y.C. and M.B.; investigation, V.S., V.Z., I.P., B.E. and Y.C.; resources, V.S., I.P. and K.W.; data curation, Y.C.; writing—original draft preparation, V.Z., V.E. and V.S.; writing—review and editing, V.Z., V.E., V.S., K.W. and M.B.; visualization, V.Z., B.E. and I.P.; supervision, V.E.; project administration, Y.C.; funding acquisition, I.P., V.E. and K.W. All authors have read and agreed to the published version of the manuscript.

Funding: The work was supported by the National Research Foundation of Ukraine (Project Number 2021.01/0189), the Ministry of Education and Science of Ukraine (Project Number 0123U100374), and "EU Next Generation EU through the Recovery and Resilience Plan for Slovakia" (Project Numbers 09I03-03-V01-00061 and 09I03-03-V01-00099).

Institutional Review Board Statement: Not applicable.

Informed Consent Statement: Not applicable.

Data Availability Statement: Data are contained within the article.

Conflicts of Interest: The authors declare no conflicts of interest.

Nomenclature

AC	accelerated cooling
E	absorbed impact energy
FRT	finish rolling temperature
HR	hot rolling
K_i	coefficient of embrittlement
L	longitudinal direction
NR	normalizing rolling

OM	optical microscopy
SAED	selected area electron diffraction
SEM	scanning electronic microscopy
T	transverse direction
T_{DBT}	ductile–brittle transition temperature
TEL	total elongation
TEM	transmission electron microscopy
TMCP	thermo-mechanically controlled processing
TMCP/AC	thermo-mechanically controlled processing followed by accelerated cooling
T_{nr}	non-recrystallization temperature
UTS	ultimate tensile strength
XRD	X-ray diffraction
YTS	yield tensile strength

References

1. Sun, J.; Hensel, J.; Klassen, J.; Nitschke-Pagel, T.; Dilger, K. Solid-state phase transformation and strain hardening on the residual stresses in S355 steel weldments. *J. Mater. Process. Technol.* **2019**, *265*, 173–184. [CrossRef]
2. Yu, B.; Chen, Z.; Wang, P.; Song, X. A comparative study on the mechanical behavior of S355f steel repair-welded joints. *J. Constr. Steel Res.* **2023**, *205*, 107878. [CrossRef]
3. Karnaukh, S.G.; Markov, O.E.; Kukhar, V.V.; Shapoval, A.A. Classification of steels according to their sensitivity to fracture using a synergetic model. *Int. J. Adv. Manuf. Technol.* **2022**, *119*, 5277–5287. [CrossRef]
4. *EN 10025-1:2004*; Hot Rolled Products of Structural Steels—Part 1: General Technical Delivery Conditions. CEN: Brussels, Belgium, 2004.
5. Kala, Z.; Valeš, J. Imperfection sensitivity analysis of steel columns at ultimate limit state. *Arch. Civ. Mech. Eng.* **2018**, *18*, 1207–1218. [CrossRef]
6. Igwemezie, V.; Mehmanparast, A.; Kolios, A. Materials selection for XL wind turbine support structures: A corrosion-fatigue perspective. *Mar. Struct.* **2018**, *61*, 381–397. [CrossRef]
7. Fu, L.; Xu, G.; Yan, Y.; Yang, J.; Xie, J. The application and research progress of high strength and high performance steel in building structure. *IOP Conf. Ser. Mater. Sci. Eng.* **2018**, *392*, 022008. [CrossRef]
8. Rozumek, D.; Lewandowski, J.; Lesiuk, G.; Correia, J.A. The influence of heat treatment on the behavior of fatigue crack growth in welded joints made of S355 under bending loading. *Int. J. Fatigue* **2020**, *131*, 105328. [CrossRef]
9. Brykov, M.N.; Petryshynets, I.; Džupon, M.; Kalinin, Y.A.; Efremenko, V.G.; Makarenko, N.A.; Pimenov, D.Y.; Kováč, F. Microstructure and properties of heat affected zone in high-carbon steel after welding with fast cooling in water. *Materials* **2020**, *13*, 5059. [CrossRef]
10. Yan, R.; Mela, K.; Yang, F.; Bamby, H.E.; Veljkovic, M. Equivalent material properties of the heat-affected zone in welded cold-formed rectangular hollow section connections. *Thin-Walled Struct.* **2023**, *184*, 110479. [CrossRef]
11. *EN 10025-3:2004*; Hot Rolled Products of Structural Steels—Part 3: Technical Delivery Conditions for Normalized/Normalized Rolled Weldable Fine Grain Structural Steel. CEN: Brussels, Belgium, 2004.
12. *EN 10025-6:2004*; Hot Rolled Products of Structural Steels—Part 6: Technical Delivery Conditions for Flat Products of High Yield Strength Structural Steels in the Quenched and Tempered Condition. CEN: Brussels, Belgium, 2004.
13. González, R.; García, J.O.; Barbés, M.A.; Quintana, M.J.; Verdeja, L.F.; Verdeja, J.I. Structural ultrafine grained steels obtained by advanced controlled rolling. *J. Iron Steel Res. Int.* **2013**, *20*, 62–70. [CrossRef]
14. Gorni, A.A.; Da Soares, M.R.S. Microstructural evolution of the normalizing plate rolling of niobium microalloyed steels. *Tecnol. Metal. Mater. Min.* **2015**, *12*, 72–76. [CrossRef]
15. Buchmayr, B. Thermomechanical treatment of steels—A real disruptive technology since decades. *Steel Res. Int.* **2017**, *88*, 1700182. [CrossRef]
16. Fan, Y.; Wang, Q.; Liu, H.; Wang, T.; Wang, Q.; Zhang, F. Effect of controlled cooling on microstructure and tensile properties of low C Nb-Ti-containing HSLA steel for construction. *Metals* **2017**, *7*, 23. [CrossRef]
17. Zhao, J.; Hu, W.; Wang, X.; Kang, J.; Yuan, G.; Di, H.; Misra, R.D.K. Effect of microstructure on the crack propagation behavior of microalloyed 560 MPa (X80) strip during ultra-fast cooling. *Mater. Sci. Eng. A* **2016**, *666*, 214–224. [CrossRef]
18. Tang, S.; Liu, Z.Y.; Wang, G.D.; Misra, R.D.K. Microstructural evolution and mechanical properties of high strength microalloyed steels: Ultra Fast Cooling (UFC) versus Accelerated Cooling (ACC). *Mater. Sci. Eng. A* **2013**, *580*, 257–265. [CrossRef]
19. Laber, K.B.; Dyja, H. The effect of the normalizing rolling of S355J2G3 steel round bars on the selected mechanical properties of finished product. *Solid State Phenom.* **2010**, *165*, 294–299. [CrossRef]
20. Ilievski, R.; Martinova, Z.; Krstevski, B.; Maddeski, J. Determination of the restart temperature for normalized rolling of C-Mn steel in plate mill. In Proceedings of the VIII International Congress Machines, Technologies, Materials'2011, Varna, Bulgaria, 19–21 September 2011.

21. Banasiak, M.; Hornik, A.; Szczęch, S.; Majta, J.; Kwiecień, M.; Cebo-Rudnicka, A.; Rywotycki, M.; Muszka, K. Effect of hot-rolled heavy section bars post-deformation cooling on the microstructure refinement and mechanical properties of microalloyed steels. *Metals* **2021**, *11*, 1284. [CrossRef]
22. Sami, Z.; Rayane, K.; Alaeddine, K. Effects of thermo-mechanical parameters on microstructural and mechanical properties of API X70 steel. *JOM* **2024**. [CrossRef]
23. Costa, L.; Melo, G.; Castro, N.; Buschinelli, A. Microstructural characterization of API 5L X65 and X70 steels manufactured by TMCP process. *Tecnol. Metal. Mater. Min.* **2022**, *19*, e2511. [CrossRef]
24. Roccisano, A.; Nafisi, S.; Stalheim, D.; Ghomashchi, R. Effect of TMCP rolling schedules on the microstructure and performance of X70 steel. *Mater. Charact.* **2021**, *178*, 111207. [CrossRef]
25. Huang, M.X.; He, B.B. Alloy design by dislocation engineering. *J. Mater. Sci. Technol.* **2018**, *34*, 417–420. [CrossRef]
26. Hu, J.; Du, L.X.; Xie, H.; Gao, X.H.; Misra, R.D.K. Microstructure and mechanical properties of TMCP heavy plate microalloyed steel. *Mater. Sci. Eng. A* **2014**, *607*, 122–131. [CrossRef]
27. Singh, P.; Mula, S.; Ghosh, S.A. Grain refinement, strain hardening and fracture in thermomechanically processed ultra-strong microalloyed steel. *Mater. Today Commun.* **2023**, *38*, 107582. [CrossRef]
28. Guo, B.; Fan, L.; Wang, Q.; Fu, Z.; Wang, Q.; Zhang, F. Effect of finish rolling temperature on the microstructure and tensile properties of Nb-Ti microalloyed X90 pipeline steel. *Metals* **2016**, *6*, 323. [CrossRef]
29. Krauss, G.; Thompson, S.W. Ferritic microstructures in continuously cooled low- and ultralow-carbon steels. *ISIJ Int.* **1995**, *35*, 937–945. [CrossRef]
30. Cho, L.; Tselikova, A.; Holtgrewe, K.; De Moor, E.; Schmidt, R.; Findley, K.O. Critical assessment 42: Acicular ferrite formation and its influence on weld metal and heat-affected zone properties of steels. *Mater. Sci. Technol.* **2022**, *38*, 1425–1433. [CrossRef]
31. De-Castro, D.; Eres-Castellanos, A.; Vivas, J.; Caballero, F.G.; San-Martín, D.; Capdevila, C. Morphological and crystallographic features of granular and lath-like bainite in a low carbon microalloyed steel. *Mater. Charact.* **2022**, *184*, 111703. [CrossRef]
32. Bhadeshia, H.K.D.H. Diffusional formation of ferrite in iron and its alloys. *Prog. Mater. Sci.* **1985**, *29*, 321–386. [CrossRef]
33. Babu, S.S.; Bhadeshia, H.K.D.H. Stress and the acicular ferrite transformation. *Mater. Sci. Eng. A* **1992**, *156*, 1–9. [CrossRef]
34. Sun, L.; Liu, X.; Xu, X.; Lei, S.; Li, H.; Zhai, Q. Review on niobium application in microalloyed steel. *J. Iron Steel Res. Int.* **2022**, *29*, 1513–1525. [CrossRef]
35. Wang, Y.; Wang, Q.; Liu, L.; Xu, W. Fracture mode of martensite-austenite constituents containing multiphase steel controlled by microstructural and micromechanical aspects. *Mech. Adv. Mater. Struct.* **2015**, *22*, 591–596. [CrossRef]
36. Efremenko, V.G.; Popov, E.S.; Kuz'min, S.O.; Trufanova, O.I.; Efremenko, A.V. Introduction of three-stage thermal hardening technology for large diameter grinding balls. *Metallurgist* **2014**, *57*, 849–854. [CrossRef]
37. Kong, X.; Lan, L. Optimization of mechanical properties of low carbon bainitic steel using TMCP and accelerated cooling. *Procedia Eng.* **2014**, *81*, 114–119. [CrossRef]
38. Efremenko, V.G.; Zotov, D.S.; Zurnadzhy, V.I.; Kussa, R.A.; Savenko, V.I.; Sagirov, R.I.; Bocharova, O.A.; Efremenko, A.V. Computer modelling-based selection of accelerated cooling parameters for advanced high-strength structural steel. *IOP Conf. Ser. Mater. Sci. Eng.* **2021**, *1037*, 012030. [CrossRef]
39. Ramirez, M.F.G.; Hernández, J.W.C.; Ladino, D.H.; Masoumi, M.; Goldenstein, H. Effects of different cooling rates on the microstructure, crystallographic features, and hydrogen induced cracking of API X80 pipeline steel. *J. Mater. Res. Technol.* **2021**, *14*, 1848–1861. [CrossRef]
40. Karjalainen, L.P.; Maccagno, T.M.; Jonas, J.J. Softening and flow stress behaviour of Nb microalloyed steels during hot rolling simulation. *ISIJ Int.* **1995**, *35*, 1523–1531. [CrossRef]
41. Ungár, T. Microstructural parameters from X-ray diffraction peak broadening. *Scr. Mater.* **2004**, *51*, 777–781. [CrossRef]
42. Zhang, Y.H.; Ma, E.; Sun, J.; Han, W.Z. A unified model for ductile-to-brittle transition in body-centered cubic metals. *J. Mater. Sci. Technol.* **2023**, *141*, 193–198. [CrossRef]
43. Zurnadzhy, V.I.; Efremenko, V.G.; Wu, K.M.; Petryshynets, I.; Shimizu, K.; Zusin, A.M.; Brykov, M.N.; Andilakhai, V.A. Tailoring strength/ductility combination in 2.5 wt% Si-alloyed middle carbon steel produced by the two-step Q-P treatment with a prolonged partitioning stage. *Mater. Sci. Eng. A* **2020**, *791*, 139721. [CrossRef]
44. Ding, S.; Taylor, T.; Khan, S.A.; Sato, Y.; Yanagimoto, J. Further understanding of metadynamic recrystallization through thermomechanical tests and EBSD characterization. *J. Mater. Process. Technol.* **2022**, *299*, 117359. [CrossRef]
45. Zheng, Y.; Wang, Q.; Zhu, L.; Han, B.; Guo, Z.; Wang, B.; Feng, J.; Lu, S.; Shen, W.; Cao, R. Microstructure evolution and carbide precipitation behavior of microalloyed TS800TB steel during hot rolling and coiling processes. *Mater. Sci. Eng. A* **2022**, *840*, 142902. [CrossRef]
46. Foder, J.; Burja, J.; Klančnik, G. Grain size evolution and mechanical properties of Nb, V–Nb, and Ti–Nb boron type S1100QL steels. *Metals* **2021**, *11*, 492. [CrossRef]
47. Chabak, Y.; Efremenko, V.; Zurnadzhy, V.; Puchý, V.; Petryshynets, I.; Efremenko, B.; Fedun, V.; Shimizu, K.; Bogomol, I.; Kulyk, V.; et al. Structural and tribological studies of "(TiC + WC)/Hardened Steel" PMMC coating deposited by air pulsed plasma. *Metals* **2022**, *12*, 218. [CrossRef]
48. Efremenko, B.V.; Shimizu, K.; Espallargas, N.; Efremenko, V.G.; Kusumoto, K.; Chabak, Y.G.; Belik, A.G.; Chigarev, V.V.; Zurnadzhy, V.I. High-temperature solid particle erosion of Cr-Ni-Fe-C arc cladded coatings. *Wear* **2020**, *460–461*, 203439. [CrossRef]

49. Singh, P.P.; Ghosh, S.; Mula, S. Strengthening behaviour and failure analysis of hot-rolled Nb+V microalloyed steel processed at various coiling temperatures. *Mater. Sci. Eng. A* **2022**, *859*, 144210. [CrossRef]
50. Chabak, Y.; Efremenko, B.; Petryshynets, I.; Efremenko, V.; Lekatou, A.G.; Zurnadzhy, V.; Bogomol, I.; Fedun, V.; Koval', K.; Pastukhova, T. Structural and tribological assessment of biomedical 316 stainless steel subjected to pulsed-plasma surface modification: Comparison of LPBF 3D printing and conventional fabrication. *Materials* **2021**, *14*, 7671. [CrossRef] [PubMed]
51. Wang, C.; Wu, X.; Liu, J.; Xu, N. Transmission electron microscopy of martensite/austenite islands in pipeline steel X70. *Mater. Sci. Eng. A* **2006**, *438–440*, 267–271. [CrossRef]
52. Zurnadzhy, V.; Efremenko, V.; Petryshynets, I.; Dabalà, M.; Franceschi, M.; Wu, K.; Kováč, F.; Chabak, Y.; Puchy, V.; Brykov, M. Alternative approach for the intercritical annealing of (Cr, Mo, V)-alloyed TRIP-assisted steel before austempering. *Metals* **2022**, *12*, 1814. [CrossRef]
53. Koval', A.D.; Efremenko, V.G.; Brykov, M.N.; Andrushchenko, M.I.; Kulikovskii, R.A.; Efremenko, A.V. Principles for developing grinding media with increased wear resistance. Part 1. Abrasive wear resistance of iron-based alloys. *J. Frict. Wear* **2012**, *33*, 39–46. [CrossRef]
54. Rykavets, Z.M.; Bouquerel, J.; Vogt, J.-B.; Duriagina, Z.A.; Kulyk, V.V.; Tepla, T.L.; Bohun, L.I.; Kovbasyuk, T.M. Investigation of the microstructure and properties of TRIP 800 steel subjected to low-cycle fatigue. *Prog. Phys. Met.* **2019**, *20*, 620–633. [CrossRef]
55. Hesse, O.; Merker, J.; Brykov, M.; Efremenko, V. Zur Festigkeit niedriglegierter Stäble mit erhöhtem Kohlenstoffgehalt gegen abrasiven Verschleiß [On the strength of low-alloy steels with increased carbon content against abrasive wear]. *Tribol. Schmierungstech.* **2013**, *60*, 37–43.
56. Hajy Akbary, F.; Sietsma, J.; Miyamoto, G.; Kamikawa, N.; Petrov, R.H.; Furuhara, T.; Santofimia, M.J. Analysis of the mechanical behavior of a 0.3C-1.6Si-3.5Mn(wt%) quenching and partitioning steel. *Mater. Sci. Eng. A* **2016**, *677*, 505–514. [CrossRef]
57. Kostryzhev, A.; Marenych, O.; Killmore, C.; Pereloma, E. Strengthening mechanisms in thermomechanically processed NbTi-microalloyed steel. *Metall. Mater. Trans. A* **2015**, *46*, 3470–3480. [CrossRef]
58. Gol'dshtejn, M.I. *Metal Physics of High-Strength Alloys*; Metallurgy: Moscow, Russia, 1986; 312p.
59. Huang, M.; Rivera-Díaz-del-Castillo, P.E.J.; Bouaziz, O.; Van der Zwaag, S. Modelling strength and ductility of ultrafine grained BCC and FCC alloys using in reversible thermodynamics. *Mater. Sci. Technol.* **2009**, *25*, 833–839. [CrossRef]
60. Maropoulos, S.; Paul, J.D.H.; Ridley, N. Microstructure–property relationships in tempered low alloy Cr–Mo–3·5Ni–V steel. *Mater. Sci. Technol.* **1993**, *9*, 1014–1020. [CrossRef]
61. Gladman, T.; Mcivor, I.D.; Dulieu, D. Microalloying 75. In Proceedings of the International Symposium on High Strength Low Alloy Steels, New York, NY, USA, 1–3 October 1977.
62. Fan, L.; Zhou, D.; Wang, T.; Li, S.; Wang, Q. Tensile properties of an acicular ferrite and martensite/austenite constituent steel with varying cooling rates. *Mater. Sci. Eng. A* **2014**, *590*, 224–231. [CrossRef]
63. Yakubtsov, I.; Zhang, R.; Boyd, D. Particularities of the formations of bainite and martensite/austenite phase in low carbon low alloy steels during continuous cooling. *Int. J. Mater. Res.* **2011**, *102*, 504–512. [CrossRef]
64. Sahay, S.K.; Bhadeshia, H.K.D.H.; Honeycombe, R.W.K. Carbide precipitation and the nucleation of allotriomorphic ferrite in an Fe-W-C steel. *Mater. Sci. Eng. A* **1992**, *157*, 101–105. [CrossRef]
65. Zaichuk, N.; Shymchuk, S.; Tkachuk, A.; Feshchuk, Y.; Szczot, J. Structure and properties of surface bandage shelves for the gas turbine engine's blades. In *Advances in Design, Simulation and Manufacturing IV, Proceedings of the 4th International Conference on Design, Simulation, Manufacturing: The Innovation Exchange, DSMIE-2021, Lviv, Ukraine, 8–11 June 2021*; Lecture Notes in Mechanical Engineering; Springer: Cham, Switzerland, 2021; pp. 602–612. [CrossRef]
66. Moura, C.; Vilela, J.J.; Rabell, E.G.; Martins, G.P.; Gonçalves Carneiro, J.R. Evaluation of the ductile-to-brittle transition temperature in steel low carbon. In Proceedings of the International Nuclear Atlantic Conference—INAC 2009, Rio de Janeiro, Brazil, 27 September 2009.
67. Chao, Y.J.; Ward, J.D.; Sands, R.G. Charpy impact energy, fracture toughness and ductile–brittle transition temperature of dual-phase 590 steel. *Mater. Des.* **2007**, *28*, 551–557. [CrossRef]
68. Chatterjee, A.; Chakrabarti, D.; Moitra, A.; Mitra, R.; Bhaduri, A.K. Effect of normalization temperatures on ductile–brittle transition temperature of a modified 9Cr–1Mo steel. *Mater. Sci. Eng. A* **2014**, *618*, 219–231. [CrossRef]
69. Ostash, O.P.; Kulyk, V.V.; Poznyakov, V.D.; Gaivorons'kyi, O.A.; Vira, V.V. Influence of the modes of heat treatment on the strength and cyclic crack-growth resistance of 65G steel. *Mater. Sci.* **2019**, *54*, 776–782. [CrossRef]
70. Mao, L.; Wang, W.; Liu, Z.; Sha, M.; Zhang, D. Investigation of the fatigue crack growth behavior of S355 steel weldments of motor hangers of high-speed trains. *Eng. Fail. Anal.* **2022**, *140*, 106425. [CrossRef]
71. Ishikawa, N.; Endo, S.; Kondo, J. High performance UOE linepipes. *JFE Tech. Rep.* **2006**, *1*, 20–26.

materials

Article

Accelerated Formation of Oxide Layers on Zircaloy-4 Utilizing Air Oxidation and Comparison with Water-Corroded Oxide Layers

Shanmugam Mannan Muthu [1], Hyeon-Bae Lee [1], Bright O. Okonkwo [2], Dong Wang [1,3], Changheui Jang [1,*] and Taehyung Na [4]

1 Department of Nuclear and Quantum Engineering, Korea Advanced Institute of Science and Technology, 291 Daehak-ro, Yuseong-gu, Daejeon 34141, Republic of Korea; muthumech1992@kaist.ac.kr (S.M.M.); lhdory@kaist.ac.kr (H.-B.L.); wangdong@kaist.ac.kr or w854187495@stu.xjtu.edu.cn (D.W.)

2 Institute of Corrosion Science and Technology, Guangzhou 510530, China; boo@icost.ac.cn

3 Shaanxi Key Laboratory of Advanced Nuclear Energy and Technology, State Key Laboratory of Multiphase Flow in Power Engineering, School of Nuclear Science and Technology, Xi'an Jiaotong University, Xi'an 710049, China

4 Central Research Institute, Korea Hydro & Nuclear Power Co., Ltd., Daejeon 34101, Republic of Korea; taehyung.na@khnp.co.kr

* Correspondence: chjang@kaist.ac.kr

Abstract: For the dry storage of Canada Deuterium Uranium (CANDU) spent nuclear fuels, the integrity of Zircaloy-4 fuel cladding has to be verified. However, the formation of ~10 μm-thick oxide layers in typical CANDU reactor operating conditions takes several years, which makes sample preparation a slow process. To overcome such limitations, in this study, an accelerated formation of an oxide layer on Zircaloy-4 cladding tube was developed with a combination of high-temperature water corrosion (HT-WC) and air oxidation (AO). First, Zircaloy-4 tubes were corroded in oxygenated (2 ppm dissolved oxygen) high-temperature water (360 °C/19.5 MPa) for 500 h. Then, the tubes were air-oxidized at 500 °C for 30 h. Finally, the tubes were corroded again in HT-WC for 500 h to produce ~10 μm-thick oxide layers. The morphology and characteristics of the oxide layer in each step were analyzed using X-ray diffraction, scanning and transmission electron microscopy. The results showed that the oxide layer formed in the accelerated method was comparable to that formed in HT-WC in terms of morphology and oxide phases. Thus, the accelerated oxide formation method can be used to prepare an oxidized Zircaloy-4 cladding tube for CANDU fuel integrity analysis.

Keywords: Zircaloy-4; spent fuel cladding; oxide layer; corrosion; oxidation

Citation: Muthu, S.M.; Lee, H.-B.; Okonkwo, B.O.; Wang, D.; Jang, C.; Na, T. Accelerated Formation of Oxide Layers on Zircaloy-4 Utilizing Air Oxidation and Comparison with Water-Corroded Oxide Layers. *Materials* **2023**, *16*, 7589. https://doi.org/10.3390/ma16247589

Academic Editor: Xiaogang Li

Received: 2 November 2023
Revised: 29 November 2023
Accepted: 6 December 2023
Published: 11 December 2023

1. Introduction

Zirconium-based alloys have been extensively used as fuel cladding materials in water-cooled nuclear reactors like pressurized water reactors (PWR), boiling water reactors (BWR), and CANDU reactors, mostly owing to their low neutron absorption cross-section and, in part, their high strength and good corrosion resistance [1,2]. Nonetheless, during the normal operation of water reactors, the Zr-based claddings are exposed to high-temperature and high-pressure water (300–360 °C and 8–15.5 MPa), and oxide layers are formed on the surface. The thickness of the oxide layer formed on the spent nuclear fuels varies depending on water reactor types and fuel cladding materials [3,4]. For example, the oxide layer thickness could be 100–150 μm for fuel cladding of Zircaloy-4 and Nb-containing advanced Zr-based cladding in typical PWR spent fuel. Meanwhile, for CANDU reactors where Zircaloy-4 is used as a fuel cladding material, the thickness of the oxide layer is 10 μm at the time of refueling [3–5].

When removed from the reactors, spent nuclear fuels are stored in the spent nuclear fuel pool filled with highly borated water for several years until the level of decay heat drops sufficiently. Then, the spent nuclear fuels are transported to a dry storage facility, where they will be stored until permanent disposal. In the case of CANDU reactors, for dry storage, spent nuclear fuels are sealed within a container filled with inert gas, so the further corrosion of fuel cladding could be minimized during the storage period [5–7]. However, there is some concern that air and moisture could be trapped during the sealing process, which could cause additional corrosion of CANDU spent fuel cladding with an oxide layer, aided by the high temperature caused by the remaining decay heat from spent fuels. To assess this additional oxidation, simulated CANDU spent fuel cladding tubes with ~10 μm-thick oxide layers are needed.

2. Research Background

It is well known that the Zr-based cladding would be corroded in high-temperature water and form the ZrO_2 layer. Several researchers have studied the corrosion behavior of the Zr-based cladding materials in simulated high-temperature water. Diana et al. [7] studied the long-term corrosion behavior of Zircaloy-4 in PWR primary water at 310 °C and 10 MPa for 3024 h, followed by electrochemical measurement using EIS impedance spectra. It is elucidated from the EIS plot results that the inner oxide layer acts as a barrier and the outer layer is more porous. Kim et al. [8] performed the corrosion test and analysis on the Zr-1.5Nb alloy in distilled water with a static autoclave condition at 360 °C and 18.9 MPa following the corrosion test per ASTM G2-88 [9] procedures. The authors further studied the effect of Nb-rich precipitates on the oxidation rate of the alloy. It is clearly understood from the results that the zirconium matrix oxidized faster than the β-Nb phase. Jiang et al. [10] compared the corrosion behavior of Zr-1Nb and Zircaloy-4 in LiOH-H_3BO_3 and pure water conditions at 330 °C and 14 MPa for 200 h. The authors observed weight loss in both LiOH-H_3BO_3 and pure water conditions. More weight loss was found on the Zircaloy-4 in Li/B water conditions. Also, the mechanical properties of the corroded samples were reduced as compared to the as-received samples. Kim et al. [11] performed the corrosion test on the zirconium alloy in a hydrogenated water environment at 325 °C. The oxide layer formed on the alloy surface is mainly comprised of monoclinic ZrO_2. The authors observed the tetragonal to monoclinic ZrO_2 phase transformation. The cracks were observed on the 200 nm-size oxide layer near the oxide–matrix interface.

Meanwhile, studies on the oxidation of Zr-based alloys have also been performed to understand the oxidation behavior during the loss of coolant accident (LOCA) scenario, which would cause degradation of mechanical properties and the release of heat and hydrogen. When the cladding materials are exposed to air, the oxide thickness reaches a critical value, and cracks may form, which would provide the pathway for air to make direct contact with the metal surface, causing faster oxidation [12–15]. Beak et al. [16] investigated the oxidation characteristics of the Zircaloy-4 under steam conditions for 3600 s in the temperature range of 700–1200 °C. The authors observed grain coarsening, phase changes, and increasing oxide scale thickness as the temperature increased. Tung et al. [17] characterized the oxidation performance of the Zircaloy-4 cladding in an air environment at 500–800 °C for 150 h. The authors found that the oxidation rate increased with increasing the test temperature and the exposure time. Some discontinuous microcracks were noticed on the scale formed on the alloy above 650 °C. Jordan et al. [13] compared the oxidation behavior of the pure Zr, Zr-3, Zircaloy-4, Zr-1Nb, and Zr-2.5 Nb alloy samples in an air and oxygen environment from 400–800 °C. The authors reported that all the samples experienced to faster breakaway oxidation and formed a thick scale. Less protective oxide scales formed on the alloys after oxidation in air environment as compared to the oxygen.

The main purpose of this study is to produce a 10 μm-thick oxide layer on the Zircaloy-4 cladding to simulate the CANDU spent fuel cladding condition by combining high-temperature water corrosion and air oxidation to accelerate the growth of the oxide layer. To check the validity of this approach, the characteristics of the oxide layer formed by the accelerating process were compared with those formed in high-temperature water. For the characterization, X-ray diffraction (XRD), scanning electron microscope (SEM), and transmission electron microscope (TEM) analyses were performed. The results were compared and discussed in view of the applicability of the accelerated process to produce the simulated CANDU spent fuel cladding.

3. Experimental Methods

3.1. Materials

Zircaloy-4 is extensively used as a fuel cladding tube in the nuclear industry. In this research work, a tube of Zircaloy-4 was chosen as the substrate material, provided by the KEPCO Nuclear Fuel Co., Ltd. (Daejeon, Republic of Korea). The elemental analysis was performed on the substrate using inductively coupled plasma optical emission spectroscopy (ICP-OES), and the compositions were Zr-1.45%Sn-0.2%Fe-0.1%Cr.

3.2. High-Temperature DO Water Corrosion and the Oxidation Test

For the corrosion test, the specimens were prepared in a cylindrical tube with dimensions of 10 mm in length and 0.4 mm in thickness, and the outer diameter of the tube was 10 mm. Prior to the test, the specimens were polished with SiC sheets with 600 grit size followed by rinsing with acetone to remove the oxides and other impurities. The average surface roughness of the substrate material is 0.12 µm. The high-temperature water corrosion (HT-WC) test in a oxygenated HTW environment and the air oxidation (AO) test were carried out on Zircaloy-4 in three different conditions and the study diagram is shown in Figure 1. These were the following:

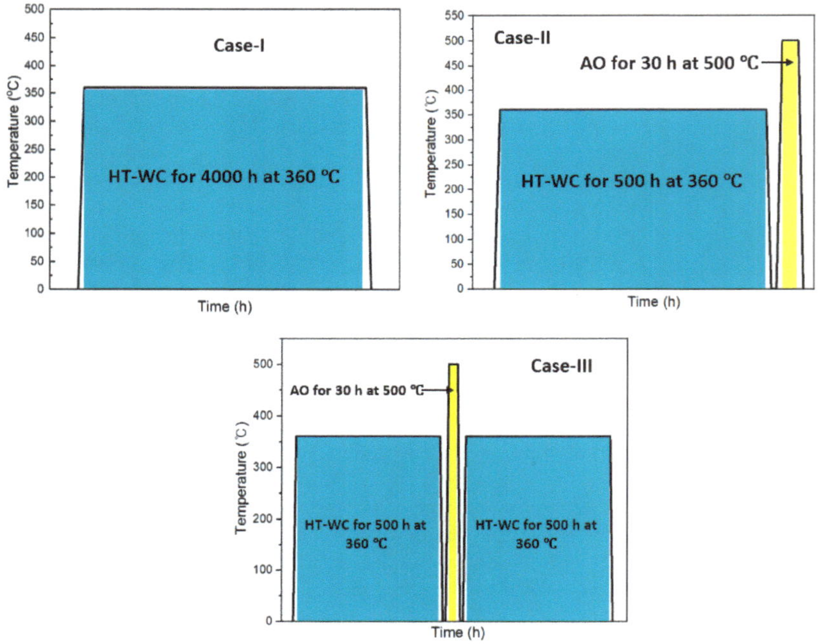

Figure 1. The study diagram of the Zircaloy-4 after subjected to different cases.

Case I: Corrosion in a oxygenated high-temperature water (HT-WC) for 4000 h.

Prior to the corrosion test, the samples were hanged in the autoclave using stainless steel wires. The high-temperature water corrosion (HT-WC) test was conducted on the Zircaloy-4 samples in a simulated primary water reactor environment containing lithium hydroxide (LiOH-2.2 ppm, Junsei Chemical Co., Ltd., Tokyo, Japan) and boric acid (H_3BO_3-1200 ppm, Daejung Chemicals and Metals Co., Ltd., Siheung, Republic of Korea) at a temperature of 360 °C and 19.5 MPa. To accelerate the oxidation rate of Zircaloy-4, a dissolved oxygen (DO) content of 2000 ppb was used. The test was performed up to 4000 h with an interval of 24, 100, 300, 500, 1000, 2000, 3000, and 4000 h. The specimen weight was measured at the end of each exposure time to determine the oxidation rate by the weight change method. Initially, 24 specimens were loaded, then 3 specimens were removed at each interval and measured, and the remaining specimens were further exposed till next interval.

Case II: HT-WC for 500 h and air oxidation (AO) in an air environment at 500 °C for 30 h.

In Case II, the samples were subjected to HT-WC for 500 h, followed by air oxidation at 500 °C (AO) in an air environment for 30 h. The weight change was monitored regularly at 10, 20, and 30 h. Initially, 10 specimens were exposed to HT-WC and subjected to AO. One specimen was removed after 10 and 20 h AO. All remaining 8 specimens were exposed to 30 h AO, of which 3 were used for measurement and analysis. The remaining 5 specimens were used for Case III test.

Case III: HT-WC for 500 h, AO at 500 °C for 30 h, and HT-WC for 500 h. For this test, 5 specimens from Case II were further exposed to HT-WC for 500 h. In Case III, the samples were subjected to HT-WC for 500 h and AO at 500 °C for 30 h. Then, the samples were again exposed to high-temperature water for 500 h.

3.3. Analysis Methods

The change in weight of the samples was measured using an electronic weighing balance (Toledo, OH, USA), with an accuracy of 0.0001 g. The surface morphology and elemental analysis of the oxide scale was characterized by a scanning electron microscope (FE-SEM: Hitachi SU-5000, Hitachi Hi-Tech, Tokyo, Japan) equipped with energy dispersive spectroscopy (EDS). The phase analysis of the oxide layer was performed by X-ray diffraction analysis (XRD: RIGAKU D/MAX-2500, RIGAKU Corp, Tokyo, Japan) ranging from 20° to 90° with a step size of 0.01° and a scan speed of 4° min^{-1}. To determine the thickness of the oxide layer, cross-sectional analysis was performed on the corroded samples using a Focused Ion Beam (DB-FIB: FEI Helios G4 UX, Thermo Fisher Scientific solutions Co., Ltd., Seoul, Republic of Korea) and a transmission electron microscope (TEM: FEI Talos F200X, Thermo Fisher Scientific solutions Co., Ltd., Republic of Korea). The FIB milling process was employed to prepare the TEM sampling, and the sputtering was obtained with a Ga^+ ion beam. The elemental distribution on the oxide layer was examined using EDS mapping and line scan analysis. High-resolution TEM (HR-TEM) images and the Fast Fourier Transform (FFT) diffraction (SAED) pattern were obtained using HR-TEM images.

4. Results

4.1. Weight Gain Measurement

The weight change in the samples per unit area was calculated using the following Formula (1) [18].

$$\Delta W = \frac{W_f - W_i}{A} \tag{1}$$

where ΔW is the weight change in the samples per unit area; W_i and W_f are the initial and final weight of the samples and A is the surface area of the samples. The weight gain measurement results are shown in Figure 1. For Case I (Figure 2a), the weight gain of the samples increased rapidly until 500 h, reaching 25.24 mg/dm^2 with parabolic behavior. During the initial stage, the weight gain of the sample is high due to the formation of a new oxide layer on the surface. Then, the rate slowed down until 2000 h, followed by the linear weight gain after 2000 h. This may be attributed to the formation of cracks, resulting in oxide growth rate was increased slightly. Finally, the weight gain of the samples was 64.17 mg/dm^2 after being exposed for 4000 h for Case I.

In Case II, the sample was initially subjected to HT-WC for 500 h, followed by air oxidation at 500 °C for 30 h. Figure 2b illustrates that the weight gain of the sample was 125.10 mg/dm^2. In Case III, the samples were exposed to 500 h in the HT-WC condition, 30 h in air oxidation, and again 500 h in the HT-WC condition. The weight gain of sample was 153.97 mg/dm^2. It is clear from the weight gain plot that the samples experienced a larger weight gain during the air oxidation test. The weight gain results of Zircaloy-4 after corrosion in three different conditions are listed in Table 1.

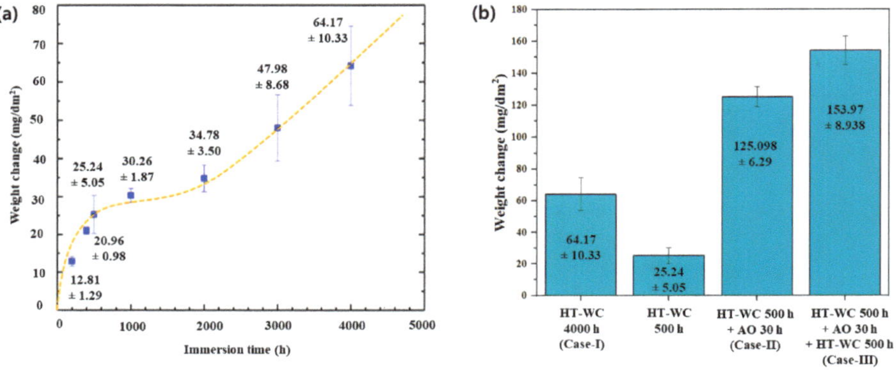

Figure 2. Weight gain plot of the Zircaloy-4 after exposure to (**a**) high-temperature water corrosion (HT-WC) condition and (**b**) a combination of HT-WC and air oxidation (AO) conditions.

Table 1. Weight gain of the Zircaloy-4 after exposure to different conditions.

Cases	Condition	Weight Gain (mg/dm^2)
I	HT-WC for 4000 h	64.17
II	HT-WC for 500 h + AO for 30 h	125.10
III	HT-WC for 500 h + AO for 30 h + HT-WC for 500 h	153.97

4.2. XRD Analysis

To determine the oxide phases, XRD analysis was performed on the bare and corrosion tested Zircaloy-4 samples and the results are shown in Figure 3. Zr phase was found on the bare Zircaloy-4 sample. There are no oxide phases were observed on the bare sample. For all cases, strong peaks of Zr matrix (PDF #05-0665) and m-ZrO_2 (PDF #37-1484) were observed. In addition, a very small peak (at 30.1°) of t-ZrO_2 (PDF #50-1089) was observed. The relative intensities of m-ZrO_2 peaks were greater for Case III, which is consistent with the higher weight gain of Case III. Generally, t-ZrO_2 would appear when the zirconium-based alloys are exposed to temperatures above 1070 °C; however, it could be formed under high stress of the oxide scale [19,20].

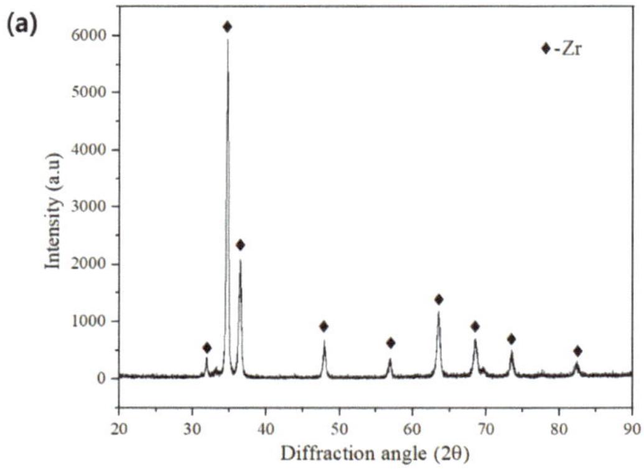

Figure 3. *Cont.*

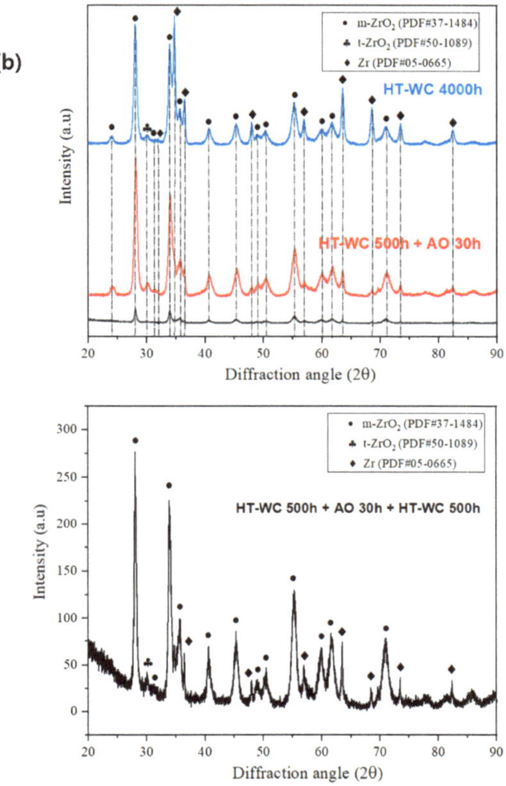

Figure 3. XRD analysis of the (**a**) bare and (**b**) corroded Zircaloy-4 after exposure to high-temperature water corrosion (HT-WC) and an air oxidation environment.

4.3. Characterization of the Oxide Layer

4.3.1. Case I: HT-WC for 4000 h

The surface morphology of the oxide layer developed on the sample after exposure to a HT-WC for 4000 h is displayed in Figure 4a,b. It can be seen from the SEM images with different magnifications that the oxide scales covered the entire surface of the sample. Along with the fine particle shapes, nodule oxides are found on the scales, which are spread on the entire sample surface. The XRD results (Figure 3) suggest that the oxide scales are ZrO_2. The cross-section SEM image of the Zircaloy-4 after HT-WC for 4000 h is displayed in Figure 4c. The average thickness of the oxide layer was 5.12 μm for the outer sides of the Zircaloy-4 tubes. For detailed oxide scale analysis, the TEM sampling was prepared using FIB equipped with SEM. The SEM image of FIB fabricated area of the corroded Zircaloy-4 sample after HT-WC for 4000 h is shown in Figure 4d. Some microcracks, pores, and a large bubble were found formed on the scale of Zircaloy-4 after subjected to HT-WC for 4000 h.

Figure 5 shows the results of a TEM-EDS cross-sectional analysis results of FIB sample of the oxide layer that formed on Zircaloy-4 exposed to HT-WC for 4000 h. It can be seen clearly from the TEM image that there is a bubble in the top part of the ZrO_2 oxide layer. In addition, several horizontal cracks were noticed on the scale. The EDS mapping results show that bubble is formed between the inner ZrO_2 and outer Fe-rich oxides. A vertical crack was present at the top surface of the ZrO_2 scale in the bubble, which is clearly highlighted. The formation of bubbles has been well reported in the literature [20–22]. The main reasons behind the bubble formation on the top region of the oxide layer are (i) the large variation in the thermal expansion co-efficient between the

ZrO_2 matrix and the Fe-rich oxide in the water; and (ii) longer exposure time of the oxide layer in a high-temperature water medium.

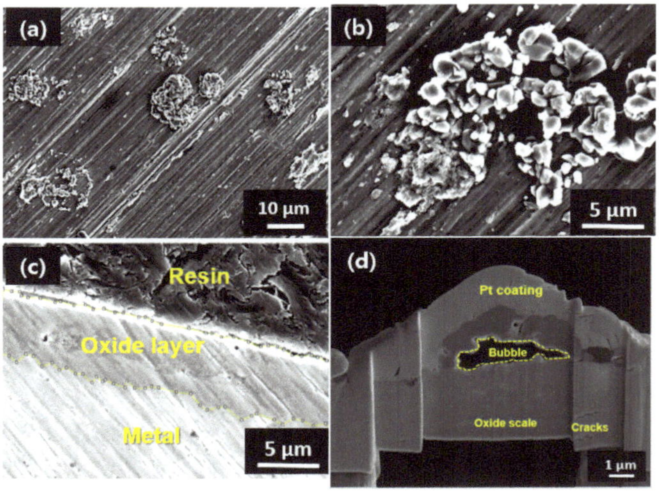

Figure 4. SEM (**a**,**b**) surface, (**c**) cross-sectional images, and (**d**) FIB image for the TEM sample of Zircaloy-4 exposed to high-temperature water (HT-WC) for 4000 h (Case I).

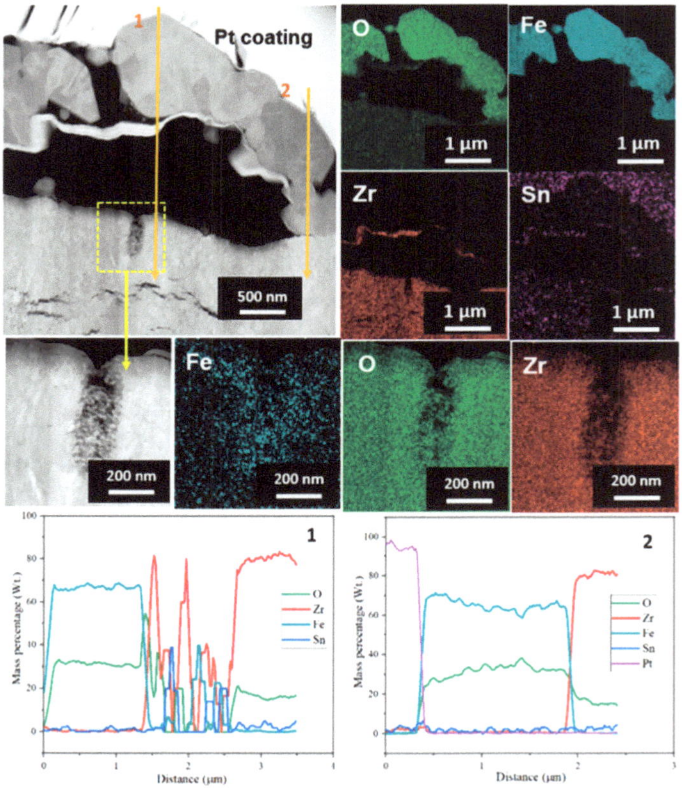

Figure 5. TEM image and EDS analysis results of the oxide layer on Zircaloy-4 exposed to the HT-WC for 4000 h (Case I).

The EDS line scanning analysis was carried out on the oxide scale to study the elemental distribution. The results of the EDS line scan were in good agreement with the EDS mapping results, showing the presence of outer Fe-rich oxide and inner ZrO_2. The HR-TEM and FFT pattern results (Figure 6) illustrate that the oxide scale present above the bubble region is mainly composed of Zr, Fe, and O-rich phases. The FFT pattern of region i indicates the presence of the hexagonal closed pack (hcp-$ZrFe_2$) phase, which is in agreement with the literature [23,24]. Meanwhile, in region ii, near a crack, the amorphous m-ZrO_2 was identified.

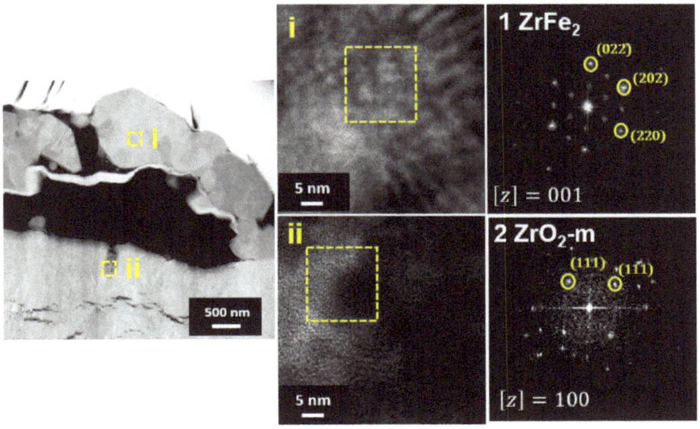

Figure 6. HR-TEM micrograph and FFT patterns of the oxide layer on the Zircaloy-4 exposed to the HT-WC for 4000 h (Case I).

The formation of Fe-rich oxide and $ZrFe_2$ phase on Zircaloy-4 (Fe content of 0.2%) during exposed HT-WC conditions could be attributed to the diffusion of Fe from the autoclave made of stainless steel. Similarly, Jeong et al. [25] found the Zr_3Fe precipitate after the corrosion test on the Zr-Nb alloy in high-temperature water conditions at 360 °C and supercritical water conditions at 500 °C.

4.3.2. Case II: HT-WC for 500 h + AO for 30 h

The surface morphology of the oxide layer developed on the sample after exposure to a HT-WC for 500 h followed by AO for 30 h (Case II) is displayed in Figure 7a,b. It is clear that, compared to the Case I sample (Figure 4), more and larger oxide nodules were found on the air-oxidized surface of the Case II sample. Figure 7c,d illustrates the SEM and FIB-EDS results of the oxide layer of the Zircaloy-4 exposed to HT-WC for 500 h followed by AO for 30 h. FIB image clearly show that the oxide layer contains several cracks, horizontal to the interface of oxide and metal. Ishii et al. [26] observed the cracks on the oxide scales during the oxidation studies on the Zircaloy-2 samples in air environment at 450 °C. The main reason for the formation of cracks in the oxide scale is generation of compressive stresses during the oxidation [26,27]. The scale thickness was measured ~10 μm. The EDS line scanning result (Figure 7e) shows the scale mainly consisted of Zr and O. The O content was very low (10%) in the transition region, which is found between the scale and matrix regions.

TEM and EDS mapping analysis were performed on the oxide layer formed on the sample exposed to a HT-WC for 500 h followed by AO for 30 h (Case II) at different regions, such as the top region of the oxide layer and the oxide–matrix interface. TEM images (Figure 8) clearly display the presence of several micro- and nano-sized cracks at the top region of oxide layers. These kinds of cracks were not identified by the SEM analysis. The scales are mainly rich in Zr, O, and Sn. In addition, a Fe-rich oxide precipitate was observed in the top region of the scale, which was formed during exposure to high-temperature water prior to air oxidation. TEM analysis at the oxide–matrix interface region is shown in

Figure 9. The cracks were found in the oxide layer close to the oxide–matrix interface. The transition region is clearly visible by TEM analysis, and the thickness of the transition region is ~200 nm. In the transition region, the oxygen was partially distributed, which suggests that the transition region was formed by the diffusion of oxygen into the matrix [28,29]. HR-TEM analysis was carried out at the top and oxide–matrix regions of oxide layers and the results are shown in Figure 10a,b. The FFT pattern result demonstrates that the oxide layer is mainly composed of m-ZrO_2.

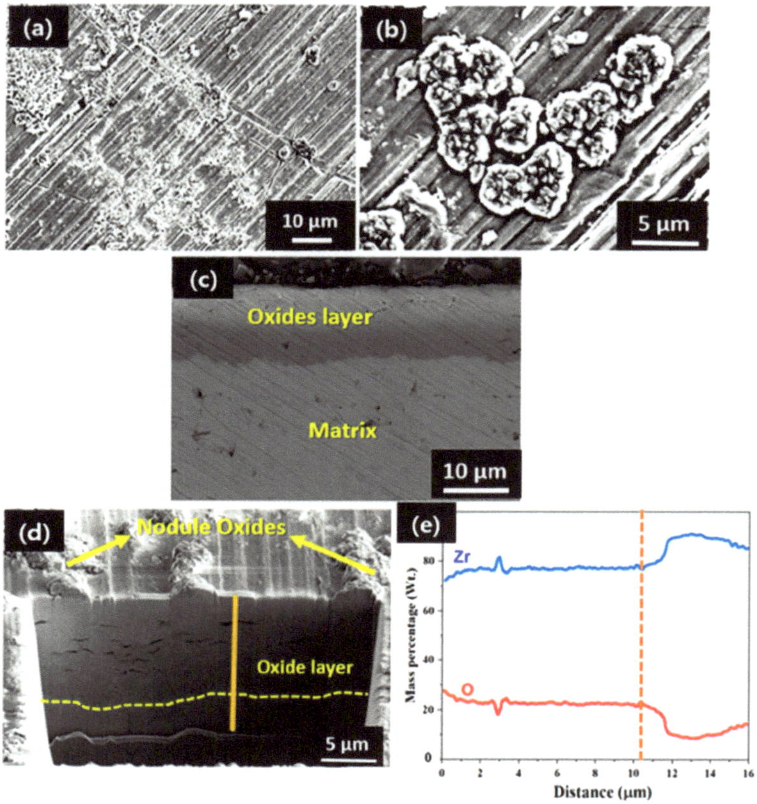

Figure 7. (**a–c**) SEM analysis at the surface and cross-section and (**d**,**e**) FIB-EDS analysis of the Zircaloy-4 exposed to HT-WC for 500 h followed by AO for 30 h (Case II).

4.3.3. Case III: HT-WC for 500 h + AO for 30 h + HT-WC for 500 h

The surface morphology of the oxide layer developed on the sample exposed to a HT-WC for 500 h, followed by AO for 30 h, and again HT-WC for 500h (Case III) is shown in Figure 11. Several small and nodular oxides were evenly distributed on the entire surface of the oxide scales. In addition, some microcracks were clearly seen on the oxide layer. The overall surface morphology and distribution of nodule oxides are similar to those for Case I and Case II. The cross-sectional SEM image (Figure 11c) shows that the average scale thickness is ~12.8 μm and the EDS line scanning result (Figure 11d) indicates that the scale is ZrO_2. The FIB image (Figure 11e) reveals the existence of several larger cracks in the oxide layer.

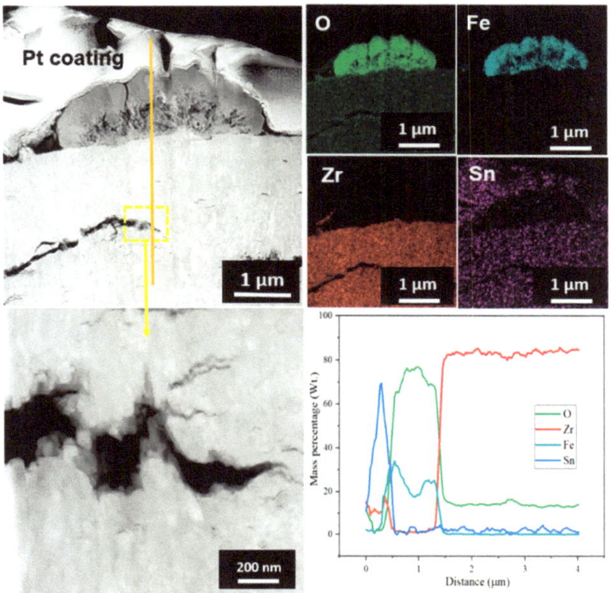

Figure 8. TEM-EDS analysis results at the top region of the oxide layer of Zircaloy-4 exposed to HT-WC for 500 h followed by AO for 30 h (Case II).

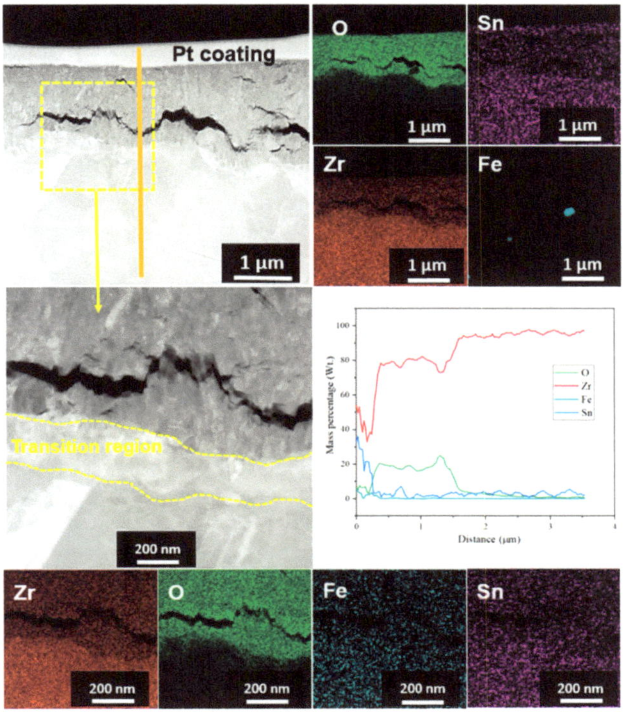

Figure 9. TEM-EDS analysis results at the metal–oxide interface region of the oxide layer of Zircaloy-4 exposed to HT-WC for 500 h followed by AO for 30 h (Case II).

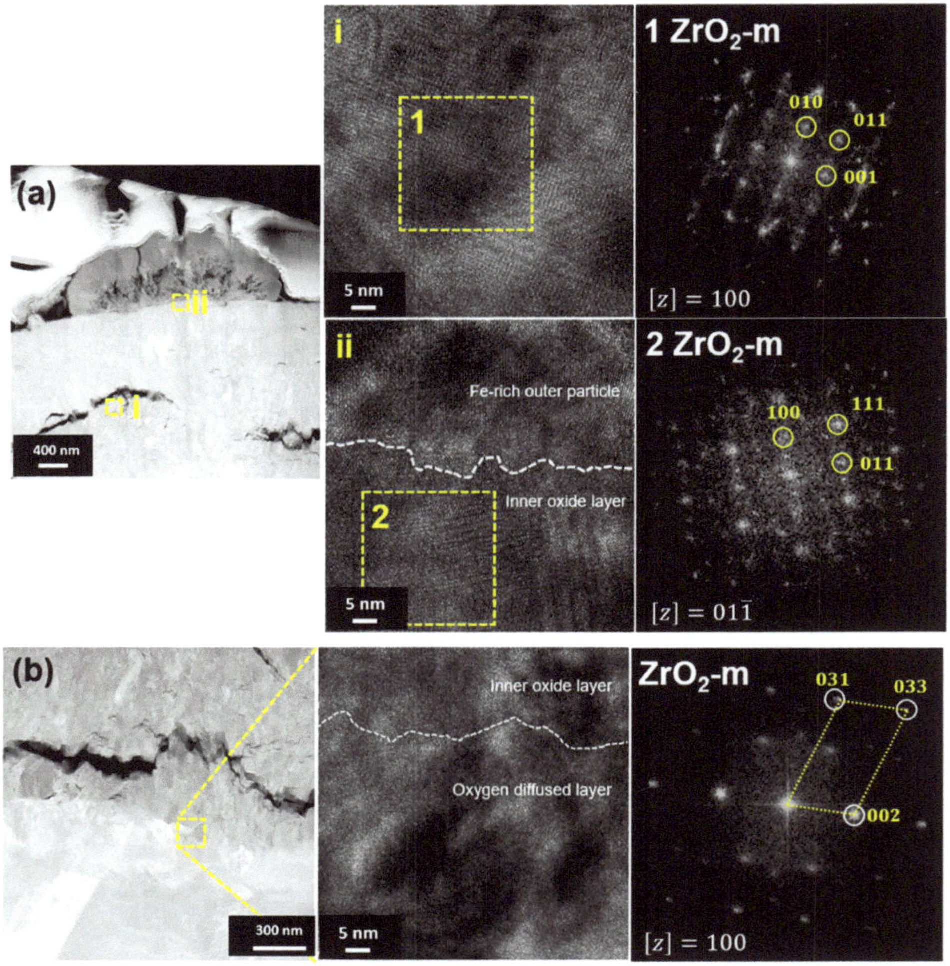

Figure 10. HR-TEM micrographs and FFT patterns at the (**a**) top and (**b**) metal–oxide interface regions of the oxide layer of Zircaloy-4 exposed to HT-WC for 500 h followed by AO for 30 h (Case II).

TEM and EDS mapping analysis were performed on the oxide layer formed on the sample exposed to a HT-WC for 500 h followed by AO for 30 h (AO) and HT-WC for 500 h (Case III) at different regions, such as the top region of the oxide layer, and the oxide–matrix interface. The TEM image (Figure 12) shows that the top region of the oxide layer contains the mixture of equiaxed and columnar grains. Similar to Case I, a large bubble was present at the top region of the oxide layer, but the presence of Fe-rich oxide was not clear above the bubble. The EDS mapping results show that Zr and O were mainly distributed in the top region of the oxide layer while Fe content was almost negligible. Cracks were not found in the observed region, but large cracks appeared somewhat below the top region (~5 μm from top surface). The results of TEM-EDS analysis at the oxide–matrix interface are shown in Figure 13. The cracks were found in the oxide layer close to the oxide–matrix interface. The transition region observed for Case II sample was not clear for Case III sample. HR-TEM analysis was carried out at the top and oxide–matrix regions of the oxide layer and the results are shown in Figure 14a,b. The FFT pattern result demonstrates that the oxide layer is mainly composed of m-ZrO_2. It should be noted that, though Fe-rich oxides and intermetallics were not observed in the region where TEM analysis was performed, their presence could be anticipated due to the similarity in corrosion environment to Case I. In addition, the presence of hydrides was observed at the oxide–matrix interface.

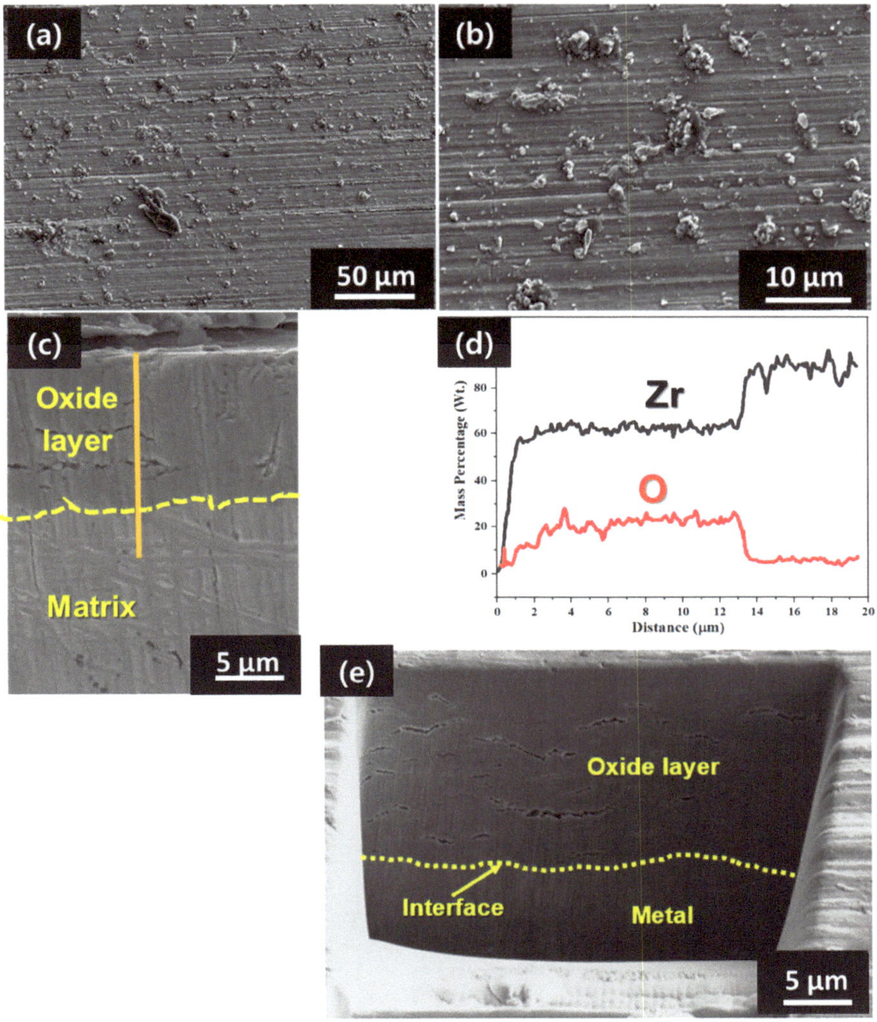

Figure 11. (**a**,**b**) SEM-EDS at the surface and (**c**,**d**) cross-section and (**e**) FIB analysis of the oxide layer of Zircaloy-4 exposed to HT-WC for 500 h followed by AO for 30 h and HT-WC for 500 h (Case III).

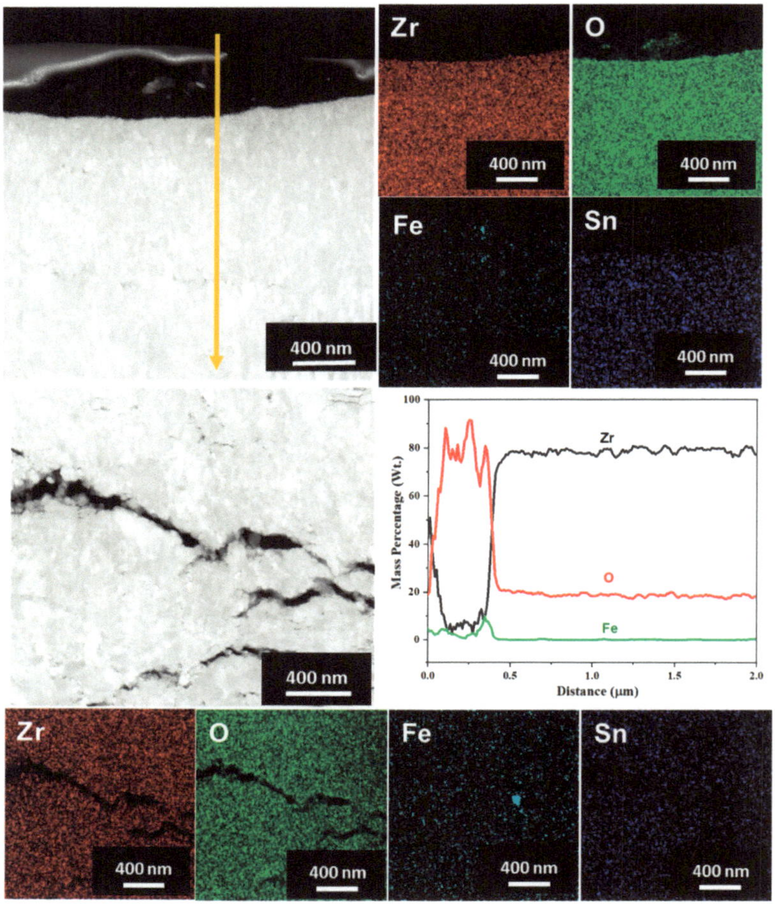

Figure 12. TEM and EDS mapping results at the top region of the oxide layer formed on the Zircaloy-4 exposed to HT-WC for 500 h followed by AO for 30 h and HT-WC for 500 h (Case III).

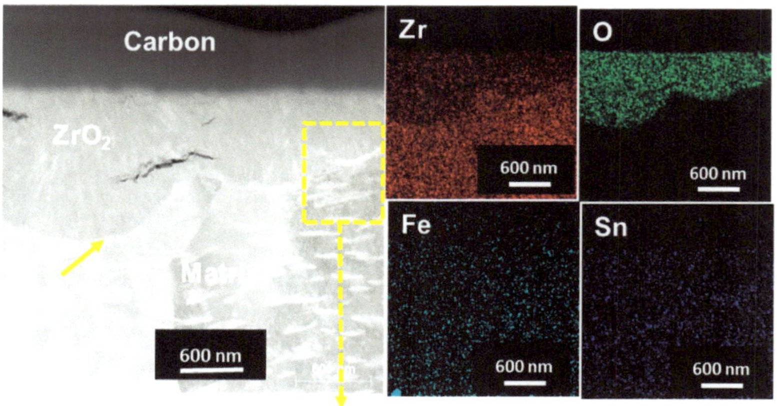

Figure 13. TEM and EDS mapping results at the oxide–matrix interface of the Zircaloy-4 exposed to HT-WC for 500 h followed by AO for 30 h and HT-WC for 500 h (Case III).

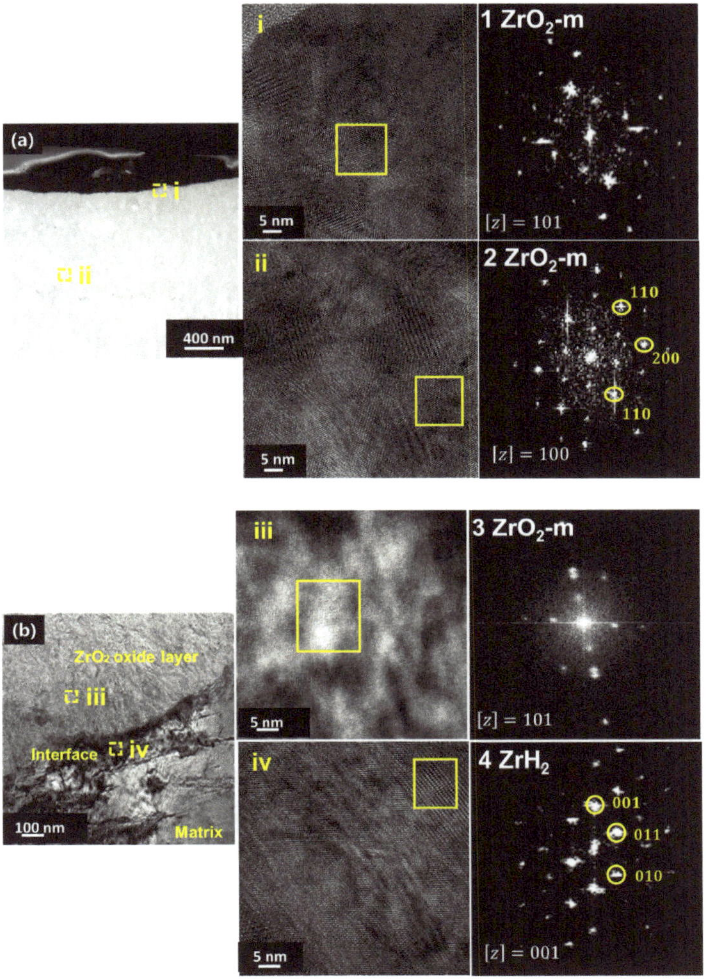

Figure 14. HR-TEM micrographs and FFT patterns at the (**a**) top and (**b**) metal–oxide interface regions of the oxide layer of Zircaloy-4 exposed to HT-WC for 500 h followed by AO for 30 h and HT-WC for 500 h (Case II).

5. Discussion

5.1. Oxide Layer Formation on the Zircaloy-4 in HT-WC Environment

In this study, the weight gain (mg/dm^2) was found on the Zircaloy-4 after being exposed to a HT-WC environment for 4000 h due to the growth of a new oxide layer. The sample weight was increased with respect to exposure time. The XRD results confirmed the Zr-matrix and m-ZrO_2 phases present on the oxide scale. The formation of the ZrO_2 oxide phase on Zicaloy-4 is due to the diffusion of oxygen ions into the Zr matrix during exposure to a water medium at 360 °C. The oxide layer formation mechanism was clearly explained by the following Equation (2).

$$Zr^{4+} + 2O^{2-} \rightarrow ZrO_2 \quad (2)$$

Ni et al. [28] observed the distribution of ZrO_2 at the oxide–matrix interface of PWR-corroded Zircaloy-4 at 330 °C. But the ZrO_2 layer is not continuous due to the short immersion time and low water temperature. Along with this, transparent layer grains were developed with a size of 30–500 nm. It is well known that the corrosion behavior of zirconium materials depends mainly on the oxide characteristics, such as grain morphology and crystal structure [25]. In addition, other elements present in the zirconium alloys also lead to the formation of oxide precipitates, which also affect the oxidation properties. From the XRD and EDS results, no other precipitates were found on the ZrO_2 layer due to the occurrence of low Sn (1.45%) in the Zircaloy-4. But Hong et al. [30] found that increasing the amount of Sn in Zircaloy-4 made it less resistant to oxidation in the PWR at 360 °C and 180 bars. Jeong et al. [31] investigated the influence of Sn addition with Zr on the oxidation performance of the Zr-alloy in LiOH conditions. The authors concluded that the addition of Sn content (1%) to the Zr alloy would resist Li penetration into the oxide scale and further reduce the oxidation rate.

Based on the crystal structure, ZrO_2 can be classified as monoclinic and tetragonal. According to the grain shape, it is divided into columnar and equiaxed crystals. It is well known that the existence of columnar and tetragonal phases on the oxide layer could enhance the oxidation resistance. In this current study, Zr-matrix and m-ZrO_2 phases are the predominant oxide phases. Generally, the t-ZrO_2 phase will appear above the oxidation temperature of 1070 °C. However, a minor amount of t-ZrO_2 phase was observed due to high compressive stress, which could stabilize the tetragonal phase. At the initial stage, a tetragonal phase was formed, which was changed into a monoclinic When the Zircaloy-4 is subjected to oxidation, volume expansion may occur on the scale, resulting in compressive stresses. During the relaxation of the compressive stress, the tetragonal oxide phase destabilized and transformed into the monoclinic phase. Due to stress reduction in the oxide scale, the tetragonal phase was changed to the monoclinic phase, which leads to crack development.

5.2. The Bubble Formation Mechanism

The FIB image (Figure 4d) clearly indicates the existence of bubbles on the top surface of the oxide scales that formed Zircaloy-4 after 3000 h and 4000 h of exposure to HT-WC. The bubble formation may cause the oxide layer to fail during HT-WC at elevated temperatures. Therefore, it is significant to study the bubble formation mechanism. For the detailed discussion, the images with higher magnification were obtained from the FIB images of the HT-WC sample. The length of the bubble was found to be in the range of 3–4 μm.

Three different stages were involved in the bubble formation process. These are the initial bubble formation stage, the bubble accumulation stage, and the bubble connection stage. Generally, the Kirkendall mechanism was used to analyze the initial bubble formation mechanism, which has been stated by several researchers during the oxidization test on the Zircaloy-4. The main reasons behind the bubble formation on the top region of the oxide layer are (i) the large variation in the thermal expansion coefficient between the ZrO_2 matrix and the Fe-rich oxide in the water medium; and (ii) rapid oxide growth by oxygen diffusion resulting in compressive stress. Therefore, several cracks were formed on the scales, which affect the oxidation properties at high pressure and temperature [21,32,33].

5.3. Air Oxidation Mechanism

The air oxidation test was conducted on the HTW-corroded Zircaloy-4 samples at 500 °C for 30 h to study the oxidation characteristics of the HTW-corroded surface. It can be noticed from the weight gain plot that the oxidation rate is high in the air oxidation (AO) condition as compared to the HT-WC condition. This may be attributed to the higher test temperature (500 °C) and large amount of oxygen in the air oxidation test as compared to the HT-WC test. The ZrO_2 oxides were developed on the Zircaloy-4 surface due to the reaction of oxygen ions from the atmospheric air and the Zircaloy-4 matrix. The oxide growth occurred continuously due to the penetration of oxygen through the oxide layer.

Therefore, the volume of the oxides formed from the metal is higher than the metal (the Pilling–Bedworth ratio (PBR) of zirconium and zirconium oxide is 1.56) [29,34].

It is essential to consider that several factors, such as morphology, crystal structure, porosity, and cracks, are present on the scales when examining the oxidation characteristics of the Zr-based alloys. The surface morphology of the oxidized surface (Figure 7) illustrated more oxide growth on the surface than on the HTW-corroded surfaces. Along with that, more nodule-rich oxide phases were noticed. When the HTW-corroded Zircaloy-4 during oxidation in air, a compressive stress was induced on the oxide scale, causing a tetragonal phase was developed at the initial stage. During the cooling period in the atmospheric air, outward oxide growth occurred on the corroded surface, causing the compressive stress to be relaxed. Therefore, the tetragonal phase was changed to the monoclinic phase. The monoclinic phase is less protective as compared to the tetragonal phase. A thick scale was formed on the HTW-corroded surface after air oxidation. Several cracks were developed on the scale, which are parallel to the metal–oxide interface. These cracks were mainly attributed to the cyclic growth of stress on the oxide scale during heating and cooling. In addition, the Pilling–Bedworth ratio (PBR) of ZrO_2/Zr is 1.56 because high-volume expansion may induce more stress on the scale [24,27]. TEM results show the existence of several cracks on the scale of the oxidized sample, which are not able to be detected by SEM cross-section analysis. Moreover, Fe- and O-rich precipitate were found on the top surface of the scale, and Sn was partially distributed on the scale. The oxide–matrix interface region was analyzed by TEM to study the adhesion behavior. The thickness of the scale was found in the range of 1.3–1.5 μm after HT-WC for 500 h. Cracks are clearly seen by TEM. These cracks may increase the oxide growth rate. This may also be one of the reasons for higher oxidation when exposed to air. The white region was noticed in the matrix region below the oxide scale for 200 nm. This region is called the oxygen transition region, which indicates that the oxygen was distributed into the Zr matrix region.

5.4. Comparison of Oxide Layer Characteristics

The characteristics of oxide layers formed in three cases of corrosion and oxidation conditions are summarized in Table 2. It can be said that the characteristics of oxide layers formed in Case I and III differ only in terms of thickness and the presence of bubbles on the top region of the oxide layer. However, it is anticipated that the presence of bubbles would not affect the overall integrity of the oxide layer during long-term storage in a dry storage facility because of the location of bubbles and the relatively smaller size compared to oxide layer thickness.

Table 2. Comparison of the characteristics of the oxide layer formed on Zircaloy-4 exposed to three cases of corrosion and oxidation conditions.

Cases	Oxide Thickness	Oxide Phase	Location of CRACKS	Bubbles
I	~5 μm	Mostly m-ZrO_2 + little t-ZrO_2	Top layer to interface	Yes
II	~10 μm	Mostly m-ZrO_2 + little t-ZrO_2	Top layer to interface	No
III	~13 μm	Mostly m-ZrO_2 + little t-ZrO_2	Top layer to interface	No

6. Conclusions

In this work, the Zircaloy-4 fuel cladding tubes were subjected to a combination of corrosion in oxygenated high-temperature water at 360 °C/19.5 MPa (HT-WC) and air oxidation at 500 °C (AO) to form a ~10 μm-thick oxide layer on the surface. The following three cases of corrosion and oxidation conditions were used and the resulting oxide layers were compared:

Case I: HT-WC for 4000 h,

Case II: HT-WC for 500 h + AO for 30 h, and

Case III: HT-WC for 500 h + AO at 500 °C for 30 h + HT-WC for 500 h.

Based on the tests and analysis, the following conclusions were drawn:

1. The weight gain was 64.17 mg/dm^2 after HT-WC for 4000 h (Case I) and the scale thickness was ~5 μm. Meanwhile, the weight gain was 153.97 mg/dm^2 after HT-WC for 500 h followed by AO for 30 h and HT-WC for 500 h (Case I) and the scale thickness was ~13 μm. Thus, air oxidation showed accelerated weight gain and oxide layer growth compared to high-temperature water corrosion.
2. In all three cases, the oxide layer mainly consisted of m-ZrO_2 with little t-ZrO_2 confirmed by the XRD analysis and TEM analysis. Also, horizontal cracks were widely present from the top of the oxide layer to the oxide–matrix interface.
3. The oxide layer formed in the accelerated method (Case III) was comparable to that formed in HT-WC in terms of morphology and oxide phases. Thus, the accelerated oxide formation method can be used to prepare an oxidized Zircaloy-4 cladding tube for CANDU fuel integrity analysis.

It should be noted that, even though the accelerated formation of the oxide layer by adopting air oxidation was successful, the microstructure change and resulting mechanical property change in Zr matrix should be investigated in the future.

Author Contributions: S.M.M., investigation and writing—original draft; H.-B.L., experimentation and investigation; B.O.O., experimentation and investigation; D.W., investigation; C.J., conceptualization, supervision, and writing—review and editing; T.N., formal analysis and resources. All authors have read and agreed to the published version of the manuscript.

Funding: This work was supported by the Institute of Korea Spent Nuclear Fuel grant funded by the Korean government, the Ministry of Trade, Industry and Energy (No. 2021040101002C).

Institutional Review Board Statement: Not applicable.

Informed Consent Statement: Not applicable.

Data Availability Statement: The raw/processed data required to reproduce these findings cannot be shared at this time due to technical or time limitations. They will be shared upon request.

Conflicts of Interest: Author Taehyung Na was employed by the company Korea Hydro & Nuclear Power Co., Ltd. The remaining authors declare that the research was conducted in the absence of any commercial or financial relationships that could be construed as a potential conflict of interest.

References

1. Wang, P.; Was, G.S. Oxidation of Zircaloy-4 during in situ proton irradiation and corrosion in PWR primary water. *J. Mater. Res.* **2015**, *30*, 1335–1348. [CrossRef]
2. Jiang, G.; Xu, D.; Yang, W.; Liu, L.; Zhi, Y.; Yang, J. High-temperature corrosion of Zr–Nb alloy for nuclear structural materials. *Prog. Nucl. Energy* **2022**, *154*, 104490. [CrossRef]
3. Sun, C.; Hui, R.; Qu, W.; Yick, S. Progress in corrosion resistant materials for supercritical water reactors. *Corros. Sci.* **2009**, *51*, 2508–2523. [CrossRef]
4. Wang, Y.M.; Feng, W.; Xing, Y.R.; Ge, Y.L.; Guo, L.X.; Ouyang, J.H.; Jia, D.C.; Zhou, Y. Degradation and structure evolution in corrosive LiOH solution of microarc oxidation coated Zircaloy-4 alloy in silicate and phosphate electrolytes. *Appl. Surf. Sci.* **2018**, *431*, 2–12. [CrossRef]
5. Kim, S.; Kang, J.H.; Lee, Y. Hydride embrittlement resistance of Zircaloy-4 and Zr-Nb alloy cladding tubes and its implications on spent fuel management. *J. Nucl. Mater.* **2022**, *559*, 153393. [CrossRef]
6. Naceur, A.; Marleau, G. CANDU-6 operation simulations using accident tolerant cladding candidates. *Ann. Nucl. Energy* **2019**, *124*, 472–489. [CrossRef]
7. Diniasi, D.; Golgovici, F.; Marin, A.H.; Negrea, A.D.; Fulger, M.; Demetrescu, I. Long-term corrosion testing of Zy-4 in a LiOH solution under high pressure and temperature conditions. *Materials* **2021**, *14*, 4586. [CrossRef]
8. Kim, H.G.; Choi, B.K.; Park, J.Y.; Cho, H.D.; Jeong, Y.H. Analysis of oxidation behavior of the β-Nb phase formed in Zr–1.5Nb alloy by using the HVEM. *J. Alloys Compd.* **2009**, *481*, 867–871. [CrossRef]
9. *ASTMG2-88*; Standard Test Method for Corrosion Testing of Products of Zirconium, Hafnium, and Their Alloys in Water at 680 °F or in Steam at 750 °F. ASTM International: West Conshohocken, PA, USA, 2002.
10. Jiang, G.; Xu, D.; Wei, Y.; Liu, H.; Liu, L.; Kuang, W. Corrosion of Zircaloy-4, Zr-1Nb and FeCrAl alloys in deaerated water at 330 °C and 14 MPa: Superior corrosion resistance and corrosion mechanism of FeCrAl alloy by experiments and molecular dynamics simulations. *J. Chem. Eng.* **2023**, *462*, 142143. [CrossRef]
11. Kim, T.; Kim, J.; Choi, K.J.; Yoo, S.C.; Kim, S.; Kim, J.H. Phase transformation of oxide film in zirconium alloy in high temperature hydrogenated water. *Corros. Sci.* **2015**, *99*, 134–144. [CrossRef]

12. De Gabory, B.; Motta, A.T.; Wang, K. Transmission electron microscopy characterization of Zircaloy-4 and ZIRLO™ oxide layers. *J. Nucl. Mater.* **2015**, *456*, 272–280. [CrossRef]
13. Vandegrift, J.L.; Price, P.M.; Stroud, J.P.; Parga, C.J.; Van Rooyen, I.J.; Jaques, B.J.; Butt, D.P. Oxidation behavior of Zirconium, Zircaloy-3, Zircaloy-4, Zr-1Nb, and Zr-2.5 Nb in air and oxygen. *Nucl. Mater. Energy* **2019**, *20*, 100692. [CrossRef]
14. Geng, D.; Deng, J.; Liu, L.; Sun, Q. Characterization of surface coarse-grained structure and improved corrosion resistance for Zr-4 alloy at high temperature. *J. Nucl. Mater.* **2023**, *583*, 154550. [CrossRef]
15. Lee, H.B.; Chen, J.; Meng, S.; Jang, C.; Park, S.H.; Xiao, Q.; Lee, H.J.; Na, K.H. Characterization of oxide layers formed on type 316 stainless steel exposed to the simulated PWR primary water environment with varying dissolved hydrogen and zinc concentrations. *J. Nucl. Mater.* **2021**, *556*, 153193. [CrossRef]
16. Baek, J.H.; Park, K.B.; Jeong, Y.H. Oxidation kinetics of Zircaloy-4 and Zr–1Nb–1Sn–0.1 Fe at temperatures of 700–1200 °C. *J. Nucl. Mater.* **2004**, *335*, 443–456. [CrossRef]
17. Yung, T.Y.; Lu, W.F.; Tsai, K.C.; Chuang, C.H.; Chen, P.T. Oxidation Behavior Characterization of Zircaloy-4 Cladding with Different Hydrogen Concentrations at 500–800 °C in an Ambient Atmosphere. *Materials* **2022**, *15*, 2997. [CrossRef] [PubMed]
18. Zhong, X.; Li, M.; Zhou, B.; Yao, M.; Shen, J. In-situ electrochemical impedance spectroscopy investigations on the early stage corrosion behavior of Zircaloy-4 in high-temperature lithiated water. *Corros. Sci.* **2021**, *182*, 109223. [CrossRef]
19. Ni, N.; Lozano-Perez, S.; Sykes, J.M.; Smith, G.D.W.; Grovenor, C.R.M. Focussed ion beam sectioning for the 3D characterisation of cracking in oxide scales formed on commercial ZIRLO™ alloys during corrosion in high temperature pressurised water. *Corros. Sci.* **2011**, *53*, 4073–4083. [CrossRef]
20. Steinbrueck, M.; da Silva, F.O.; Grosse, M. Oxidation of Zircaloy-4 in steam-nitrogen mixtures at 600–1200 °C. *J. Nucl. Mater.* **2017**, *490*, 226–237. [CrossRef]
21. Wei, T.; Zhang, R.; Yang, H.; Liu, H.; Qiu, S.; Wang, Y.; Du, P.; He, K.; Hu, X.; Dong, C. Microstructure, corrosion resistance and oxidation behavior of Cr-coatings on Zircaloy-4 prepared by vacuum arc plasma deposition. *Corros. Sci.* **2019**, *158*, 108077. [CrossRef]
22. Li, Q.; Wang, Y.; Du, P.; Song, P.; Zhang, R.; Li, Z.; Lu, J. Oxidation properties and microstructure of a chromium coating on zircaloy-4 fuel cladding material applied by atmospheric plasma spraying. *J. Nucl. Mater.* **2022**, *560*, 153496. [CrossRef]
23. Park, J.Y.; Kim, H.G.; Jeong, Y.H. Corrosion behavior of Zr– Nb alloys in 360 °C water and 500 °C supercritical water. *Met. Mater. Int.* **2006**, *12*, 497–503. [CrossRef]
24. Annand, K.; Nord, M.; MacLaren, I.; Gass, M. The corrosion of Zr (Fe, Cr)$_2$ and Zr$_2$Fe secondary phase particles in Zircaloy-4 under 350 °C pressurized water conditions. *Corros. Sci.* **2017**, *128*, 213–223. [CrossRef]
25. Jeong, Y.H.; Baek, J.H.; Kim, S.J.; Kim, H.G.; Ruhmann, H. Corrosion characteristics and oxide microstructures of Zircaloy-4 in aqueous alkali hydroxide solutions. *J. Nucl. Mater.* **1999**, *270*, 322–333. [CrossRef]
26. Ishii, Y.; Sykes, J.M. Microstructure of oxide layers formed on Zircaloy-2 in air at 450 °C. *Mater. High Temp.* **2000**, *17*, 23–28. [CrossRef]
27. Xie, S.; Zhou, B.; Liang, X.; Liu, W.; Li, H.; Li, Q.; Yao, M.; Zhang, J. A novel mechanism for nodular corrosion of Zircaloy-4 corroded in 773 K superheated steam. *Corros. Sci.* **2017**, *126*, 44–54. [CrossRef]
28. Ni, N.; Hudson, D.; Wei, J.; Wang, P.; Lozano-Perez, S.; Smith, G.D.W.; Sykes, J.M.; Yardley, S.S.; Moore, K.L.; Lyon, S.; et al. How the crystallography and nanoscale chemistry of the metal/oxide interface develops during the aqueous oxidation of zirconium cladding alloys. *Acta. Mater.* **2012**, *60*, 7132–7149. [CrossRef]
29. Steinbrück, M.; Böttcher, M. Air oxidation of Zircaloy-4, M5®and ZIRLO™ cladding alloys at high temperatures. *J. Nucl. Mater.* **2011**, *414*, 276–285. [CrossRef]
30. Hong, H.S.; Kim, S.J.; Lee, K.S. Long-term oxidation characteristics of oxygen-added modified Zircaloy-4 in 360 °C water. *J. Nucl. Mater.* **1999**, *273*, 177–181. [CrossRef]
31. Jeong, Y.H.; Kim, J.H.; Kim, H.G. Effects of Sn on the oxidation and oxide characteristics of Zr alloys in water and a LiOH solution. *Met. Mater. Int.* **2004**, *10*, 453–459. [CrossRef]
32. Hong, H.S.; Moon, J.S.; Kim, S.J.; Lee, K.S. Investigation on the oxidation characteristics of copper-added modified Zircaloy-4 alloys in pressurized water at 360 °C. *J. Nucl. Mater.* **2001**, *297*, 113–119. [CrossRef]
33. Wei, K.; Wang, X.; Zhu, M.; Guan, H.; Xu, C.; Xue, W.; Zhang, J. Effects of Li, B and H elements on corrosion property of oxide films on ZIRLO alloy in 300 °C/14 MPa lithium borate buffer solutions. *Corros. Sci.* **2021**, *181*, 109216. [CrossRef]
34. Lai, P.; Lu, J.; Zhang, H.; Liu, Q.; Zeng, Q.; Guo, X.; Zhang, L. The corrosion behavior of M5 (Zr–1Nb-0.12 O) alloy in 360 °C water with dissolved oxygen. *J. Nucl. Mater.* **2020**, *532*, 152079. [CrossRef]

materials

Article

Molecular Dynamics Simulations of Displacement Cascades in BCC-Fe: Effects of Dislocation, Dislocation Loop and Grain Boundary

Pandong Lin [1,2], Shugang Cui [1,2], Junfeng Nie [3,*], Lei He [3] and Wendong Cui [3]

[1] Key Laboratory for Liquid-Solid Structural Evolution and Processing of Materials, Ministry of Education, Shandong University, Jinan 250061, China
[2] School of Materials Science and Engineering, Shandong University, Jinan 250061, China
[3] Key Laboratory of Advanced Reactor Engineering and Safety, Ministry of Education, Institute of Nuclear and New Energy Technology, Tsinghua University, Beijing 100084, China
* Correspondence: niejf@tsinghua.edu.cn

Abstract: The interactions between displacement cascades and three types of structures, dislocations, dislocation loops and grain boundaries, in BCC-Fe are investigated through molecular dynamics simulations. Wigner–Seitz analysis is used to calculate the number of point defects induced in order to illustrate the effects of three special structures on the displacement cascade. The displacement cascades in systems interacting with all three types of structure tend to generate more total defects compared to bulk Fe. The surviving number of point defects in the grain boundary case is the largest of the three types of structures. The changes in the atomic structures of dislocations, dislocation loops and grain boundaries after displacement cascades are analyzed to understand how irradiation damage affects them. These results could reveal irradiation damage at the microscale. Varied defect production numbers and efficiencies are investigated, which could be used as the input parameters for higher scale simulation.

Keywords: molecular dynamics; displacement cascade; dislocation; dislocation loop; grain boundary

Citation: Lin, P.; Cui, S.; Nie, J.; He, L.; Cui, W. Molecular Dynamics Simulations of Displacement Cascades in BCC-Fe: Effects of Dislocation, Dislocation Loop and Grain Boundary. *Materials* **2023**, *16*, 7497. https://doi.org/10.3390/ma16237497

Academic Editor: Franz Saija

Received: 16 October 2023
Revised: 27 October 2023
Accepted: 30 October 2023
Published: 4 December 2023

1. Introduction

In nuclear power plants, many structural materials such as A508-III steel and tungsten are exposed to large amounts of neutron irradiation, which could result in irradiation hardening, embrittlement and creep [1–3]. For example, irradiation results in increased hardness, yield stress and ductile–brittle transition temperature. Molecular dynamics (MD) simulation has been extensively used to study defect production from displacement cascades in bulk body-centered cubic (BCC), face-centered cubic (FCC) and close-packed hexagonal (HCP) metals [4–10]. For example, Sahi et al. [7] conducted molecular dynamics simulations to investigate the effect of applied strain and temperature on a displacement cascade in Zr. Their results demonstrated that the number of defects induced by the displacement cascade increased under tensile uniaxial hydrostatic strain. Many irradiation defects are generated during the displacement cascade process. Results show that the number of point defects after a displacement cascade depends on varied factors such as energy of primary knocked-on atoms (PKA) and temperature.

As the microstructure evolves, the dislocations, dislocation loops and grain boundaries could become the dominant defects in materials [11,12]. Indeed, dislocations and grain boundaries are inherent defects in materials while dislocation loops are attributed to irradiation. Therefore, the three types are the typical structures when considering both irradiated and unirradiated factors. Apart from this, displacement cascade processes may interact with these defects, leading to varied defect production numbers and efficiencies, which could be used as the input parameters for higher scale simulations such as dislocation dynamics (DD) simulation and crystal plasticity (CP) theory. For example, Roman. E [13] studied the collision cascade process when interacting with 1/2<110> screw dislocations in aluminum. The number of residual point defects was evaluated and compared with that in the pristine material under the same irradiation conditions. Dai et al. [14] used molecular dynamics simulations to reveal that the primary damage production was reduced due to the presence of *a*-type dislocation loops, *c*-component dislocation loops and a tilt grain boundary. Fu et al. [15] found that the residual vacancy number usually exceeded the residual interstitial number when the displacement cascades interacted with straight edge dislocation in tungsten via a molecular dynamics simulation. However, the influence of these defects on the displacement cascade in BCC-Fe is not extensively studied, especially the understanding of surviving defects induced by the displacement cascade. In addition, a comparison of varied interactions due to different structures also needs to be studied. The purposes are (1) clarifying the interaction mechanism between the defect structure and the displacement cascade and (2) obtaining varied defect production numbers and efficiencies, in order to provide input parameters for different structures in the mesoscale model.

In addition, MD simulations suggest that the point defects induced via the displacement cascade may enter the structure of dislocation segments or grain boundaries. For example, S. Heredia-Avalos et al. [16] found that a migration of the 1/2<111> edge dislocation by gliding along its slip plane could occur, which was attributed to a collision cascade through an atomic simulation. Fu et al. [15] found that the cascade near a dislocation promoted climb and some of dipoles were sheared off by the cascade. Lu et al. [17] stated that the dislocation climb in edge dislocations and cross-slip in screw dislocations are caused by displacement cascades in Fe-20Cr-25Ni alloys, observed through molecular dynamics. It could be concluded that these structures are changed deeply in the presence of a displacement cascade. Note that clarifying the influence of the displacement cascade process on defect structures is also of concern in this work. At present, there are few comparison studies on the interaction between the three types of defect structures and displacement cascades.

In this study, we investigate the interactions between displacement cascades and three types of structures, dislocations, dislocation loops and grain boundaries, using MD simulations in BCC-Fe, which is basic components of structural materials [18]. The effects of three types of structures on the total and surviving defect number are determined. The atomic structural changes in dislocations, dislocation loops and grain boundaries induced by displacement cascades are analyzed.

2. Methods

Large-scale atomic/molecular massively parallel simulator (LAMMPS) code [19,20] is used to launch all MD simulations. The simulation system contains around 3.3 million atoms, as shown in Table 1, and periodic boundary conditions are applied in all three directions. The interatomic potentials calculated using the embedded atom method (EAM) developed by Mendelev et al. [21] were applied to describe the inter-atomic Fe interactions, and were successfully applied to simulate irradiation damage in BCC Fe [22–24].

Table 1. Type of defects and size and atoms in the simulation cells.

Structure	x	y	z	Number of Atoms
BCC 1/2[1–11] edge dislocation	35.0 nm	37.2 nm	30.3 nm	3,390,600
BCC 1/2[1–11] screw dislocation	32.0 nm	37.1 nm	32.3 nm	3,233,622
BCC [001] dislocation loop	34.3 nm	34.3 nm	34.3 nm	3,456,177
BCC $\sum$5(013) grain boundary	27.1 nm	27.1 nm	27.1 nm	1,710,000

2.1. Generation of Edge Dislocation, Screw Dislocation, Dislocation Loop and Grain Boundary

Before the displacement cascade simulation, the accurate models of edge dislocation, screw dislocation, the dislocation loop and the grain boundary should be introduced, respectively. 1/2[1–11] edge dislocation with the x slip direction and screw dislocation with the y slip direction were established using Atomsk software (version beta-0.12.1) [25]. The BCC crystal containing dislocation was oriented along the x [112], y [11–1] and z [−110] directions. When it comes to the dislocation loop, the model consisting of a [001] dislocation loop whose diameter is 3.5 nm was oriented along the x [100], y [010] and z [001] directions. Moreover, the habit plane of the interstitial-type dislocation loop is perpendicular to their Burgers vectors. Note that the [001] dislocation loop is square-shaped. The ∑5(013) grain boundary was also built using Atomsk. Figure 1d shows the configuration of the ∑5(013) grain boundary where the tilt axis is aligned along the x direction, and the grain boundary planes are normal to the y direction. The schematic of simulation cells is shown in Figure 1 and the details of the simulation cells used in this work are listed in Table 1. Furthermore, the conjugate gradient method was applied to all cases to determine the global minimum energy configurations. Such being the case, the relaxed configuration could be obtained.

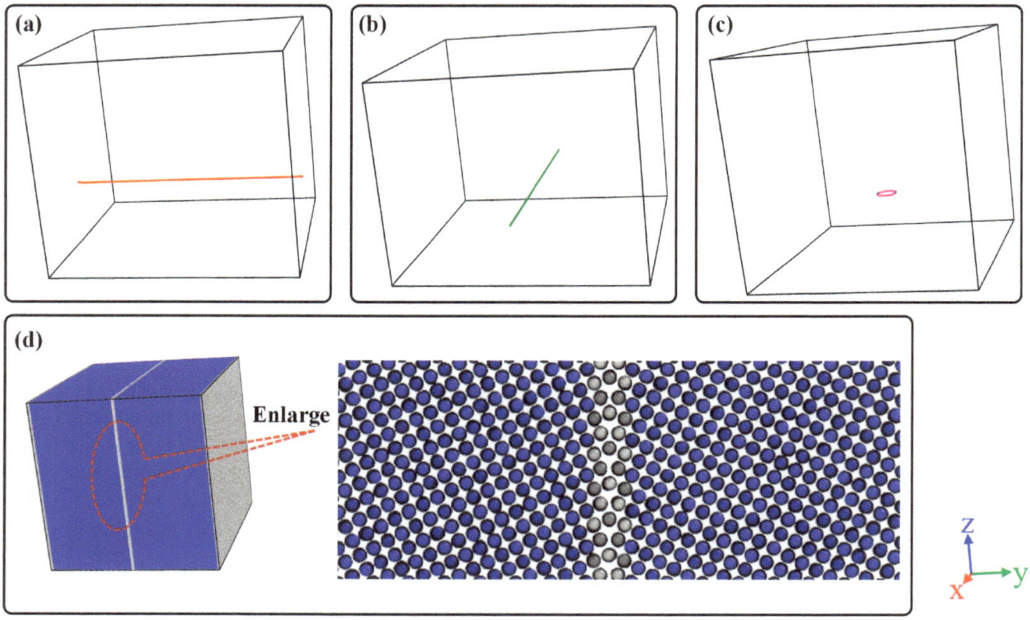

Figure 1. Schematic of defects in a simulation box: (**a**) 1/2[1–11] screw dislocation, (**b**) 1/2[1–11] edge dislocation, (**c**) [001] dislocation loop and (**d**) ∑5(013) grain boundary. BLue balls represent atoms in bcc structure while white balls denote atoms of grain boundary.

2.2. Displacement Cascade Simulations

The schematic of the displacement cascade simulation box is shown in Figure 1. The PKA energy was 10 keV and the temperature was set as 300 K. The PKA energy is given in the form of atomic velocity and the PKA was placed close to the dislocation, dislocation loop and grain boundary within 3.0 nm with the aim of investigating the interaction between cascade events and special structures. For each case, 8 independent cases were simulated with random seeds. In all cases, it should be guaranteed that the displacement cascades evolved and relaxed for several picoseconds until the temperature and the generated number of defect displacements stabilized. Note that the constant-energy constant-volume ensemble (NVE) was used in all cascade simulations. Before initial displacement damage simulations, all simulation cells of Fe crystals were relaxed for 50 ps in the NPT ensemble to ensure that the system reached thermal equilibrium. The cell dimensions were held constant for further displacement cascade simulations. An adaptive time step method [19,20] is used during displacement cascade simulations to ensure the appropriate displacement of atoms per time step. Here, the maximum distance for an atom to move in one timestep is 0.00125 nm.

2.3. Analysis and Visualization Tools

Common Neighbor Analysis (CNA) [26,27] is used to distinguish the structure of atoms (BCC, FCC and HCP and so on). The number of interstitials and vacancies is counted using the Wigner–Seitz analysis [28]. A vacancy is represented by an empty W-S cell with no Fe atoms, whereas a W-S cell with two Fe atoms corresponds to an interstitial. The dislocation extraction method (DXA) [29] is used to examine the dislocation and its Burgers vector. All analyses were completed using the Open Visualization Tool (OVITO) [30] code.

2.4. Defect Counting Method

Below is described the defect analysis method. At first, Wigner–Seitz analysis was used to calculate the total number of interstitials (I_t) and vacancies (V_t). Then, it could be found that some portion of the point defects (interstitials or vacancies) moved into dislocations, dislocation loops and grain boundaries. This portion of the point defects are defined as I_d and V_d. The other part resides in the matrix of material, which are named as I_m and V_m. Generally, it could be obtained that $I_t = I_d + I_m$ and $V_t = V_d + V_m$.

3. Results and Discussion

3.1. Cascade Simulation of BCC-Fe with Dislocations

In this section, the cascade simulations of BCC-Fe with 1/2<111> edge and screw dislocations are analyzed from two perspectives. The first part is the defect production. The other part is the displacement-cascade-induced atomic structural changes in the edge and screw dislocations, respectively.

3.1.1. The Influence of Dislocations on Defect Production

Figure 2 represents the number of point defects induced by the displacement cascade for BCC-Fe. The simulated data points are averaged from the eight samples generated from the random seeds, which is used in Sections 3.2 and 3.3. The standard deviation is also listed in Figure 2. It could be seen that displacement cascade tends to increase the number of total defects when interacting with both screw and edge dislocations as shown in Figure 2a. Whether edge dislocation or screw dislocation, there are more than twice as many point defects in the presence of the dislocation as in the absence of a dislocation. As has been mentioned in Section 2, the point defects may go into dislocations or be still left in the materials after the displacement cascade. If we focus on the surviving point defects left in the matrix, the number of defects (I_m and V_m in Figure 2b,c) for the cases of both edge and screw dislocations is more than that in the case without dislocation, which means that the defect production efficiency in the materials with high dislocation density is higher. Apart from that, it could be concluded that (a) more interstitials as opposed to vacancies move into dislocation for both screw and edge dislocations, which could suppress the migration of interstitials and their recombination with vacancies, leading to more point defects left in the simulation system, and (b) compared with edge dislocations, screw dislocations are more likely to absorb point defects. More discussions about structural changes are listed in the next part.

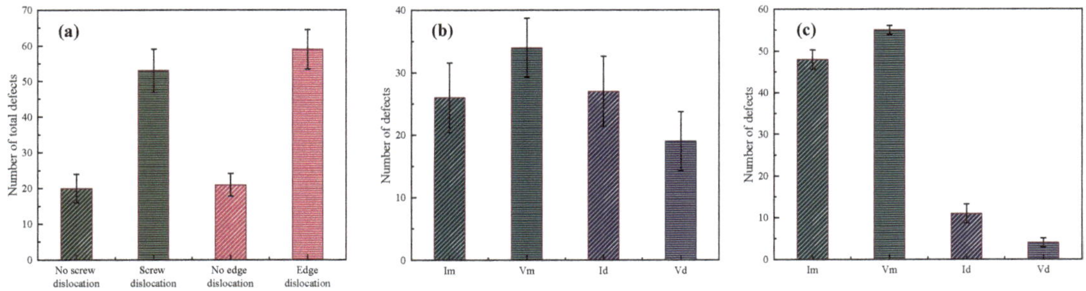

Figure 2. The number of point defects induced by displacement cascade for BCC-Fe with or without the dislocations: (**a**) The effect of screw and edge dislocation on the total number of point defects. (**b**) The effects of screw dislocation on I_m, V_m, I_d, V_d. (**c**) The effects of edge dislocation on I_m, V_m, I_d, V_d.

3.1.2. Atomic Structural Changes in Dislocations

Figure 3 shows the distribution of interstitials and vacancies during the displacement cascade simulation at 0.25 ps. It could be seen that the interstitials basically surround the vacancies for the cases with both edge and screw dislocations. High-energy particles knock the struck atom off its lattice site, creating a vacancy. The dislocated atoms continue to emit outwards, hitting other lattice atoms; such being the case, the interstitials are at the periphery of the displacement cascade region.

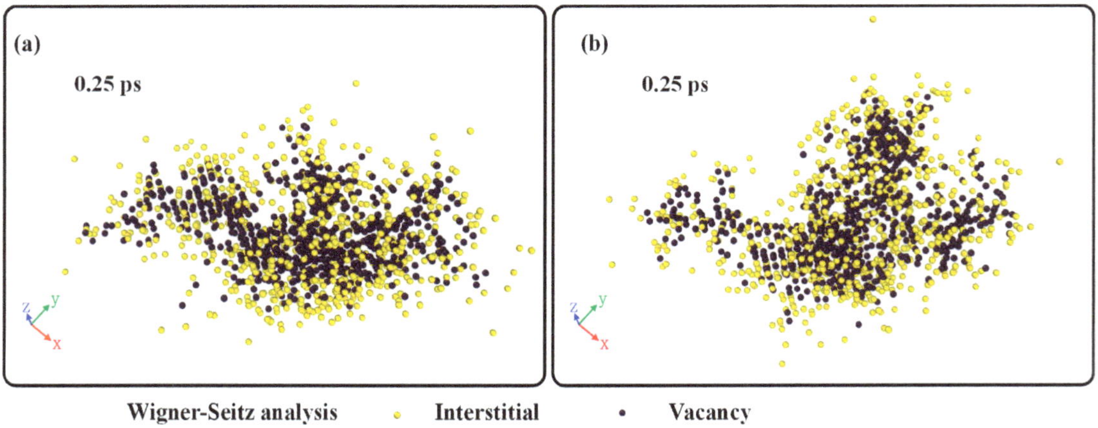

Figure 3. The distribution of interstitials and vacancies during displacement cascade simulation: (**a**) Edge dislocation and (**b**) Screw dislocation.

Figures 4 and 5 show the atomic structures of the edge and screw dislocations before and after the displacement cascade for BCC-Fe, using the CNA and DXA analysis. The Burgers vector of both the edge and screw dislocations is set as 1/2[1–11].

For the edge dislocation, the following could be concluded: (1) The migration and bending of dislocation along the y direction are clearly discernible. In terms of dislocation under a fixed temperature, no motion of dislocation occurs. Therefore, the motion of dislocation during the displacement cascade is attributed to the temperature gradient. Apart from that, with the arrival of the irradiation-induced point defects, dislocation segments in the displacement cascade region actually climb, leading to the bending of the dislocation. (2) On the other hand, the displacement cascade would eliminate some atoms of dislocation as shown in Figure 4f. Generally, the extra-high energy is introduced by the PKA, as a consequence of which some atoms of the dislocation that are in the displacement cascade region become amorphous during the displacement cascade. The dislocation is divided into two small dislocation segments when it cools down. (3) The displacement cascade could facilitate dislocation climb, which can be seen in Figure 4d,f. Note that dislocation climb could be divided into negative and positive [17]. The preference for positive or negative dislocation climb depends on the type of defects that go into dislocations. For a positive climb to happen, the edge dislocation needs to absorb vacancies, while the edge dislocation needs to absorb interstitials to have a negative climb. In other words, the availability of interstitials and vacancies determines whether a negative climb or positive climb occurs. (4) A point defect cluster is generated in the dislocation core, as shown in Figure 4d. Except for a few interstitial atoms and vacancy-entry dislocations, most point defects exist in the form of clusters.

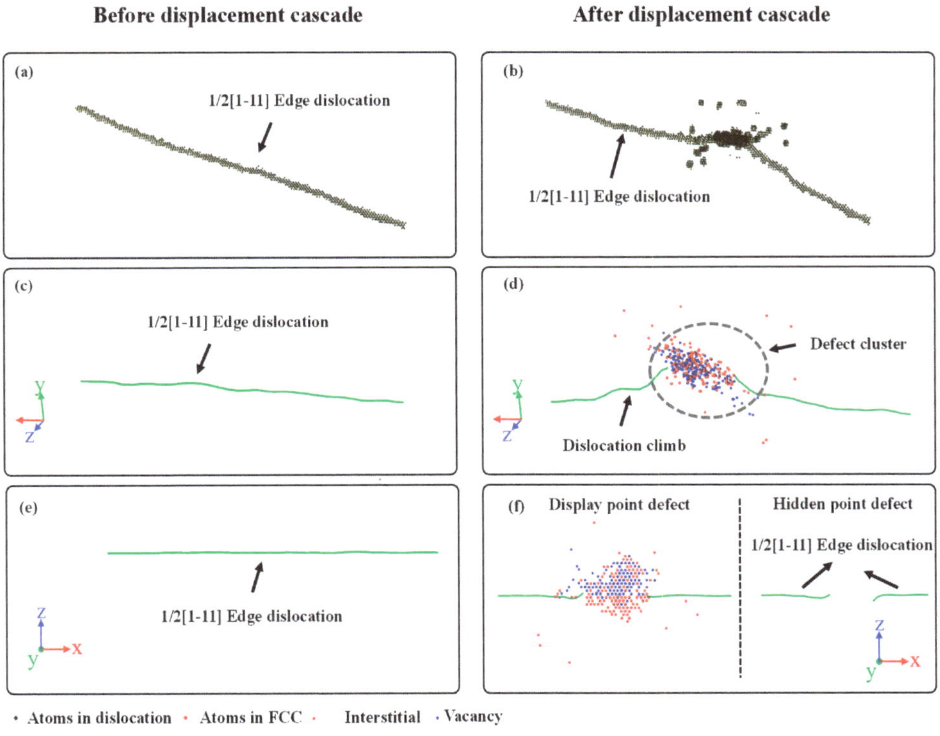

Figure 4. The atomic structure of an edge dislocation. (**a**,**c**,**e**) are the structures before displacement cascade. (**b**,**d**,**f**) are the structures after displacement cascade.

For screw dislocations, there are many similarities with edge dislocations. It can be seen that some of the atoms are also eliminated by the cascade as shown in Figure 5f. Moreover, a point defect cluster also forms due to the interaction between the screw dislocation and the cascade, as shown in Figure 5d. In contrast to the dislocation climb of edge dislocations, the screw dislocation may cross-slip during the interaction. Furthermore, note that the number of point defects entering the screw dislocation is significantly higher than that entering the edge dislocation, which explains the fewer surviving interstitials and vacancies in the matrix of BCC-Fe. Last but not least, two small dislocation segments are also observed in the finial atomic configuration.

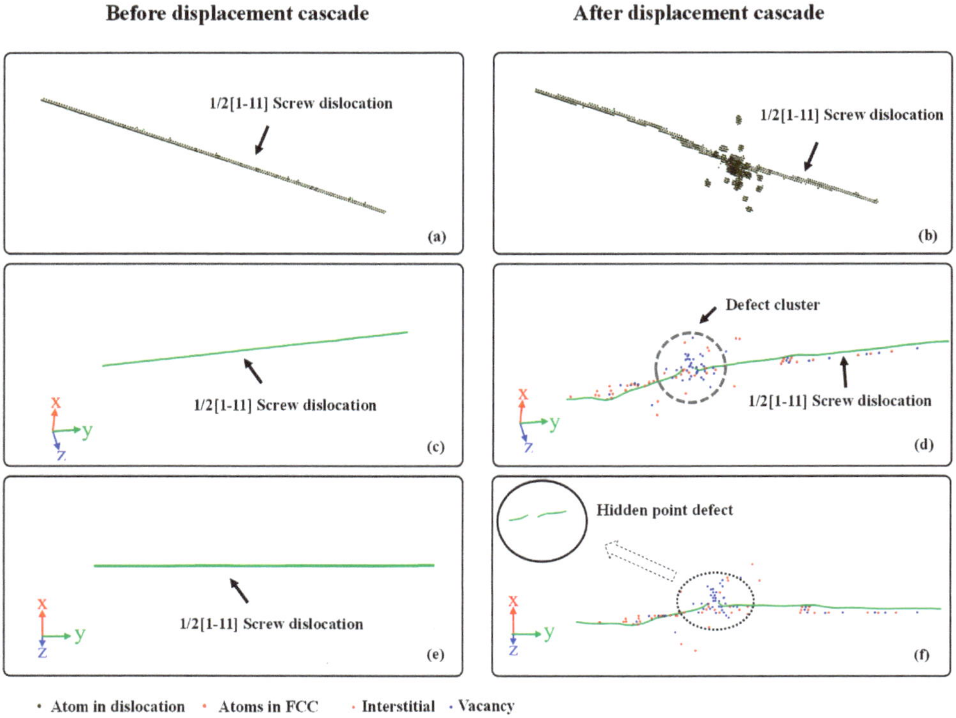

Figure 5. The atomic structure of a screw dislocation. (**a**,**c**,**e**) are the structures before displacement cascade. (**b**,**d**,**f**) are the structures after displacement cascade.

3.2. Cascade Simulation of BCC-Fe with Dislocation Loop

In this section, the effect of the [001] dislocation loop on the displacement cascade simulation of BCC-Fe is discussed. Figure 6 shows the number of point defects induced by the displacement cascade for BCC-Fe with or without the dislocation loops. It could be concluded that (1) more total point defects are generated when displacement cascades interact with the [001] dislocation loop, as shown in Figure 6a, and (2) fewer surviving defects are left in the matrix of BCC-Fe when compared to the case without a dislocation loop. Such being the case, the defect production efficiency in the BCC-Fe with a high density of [001] dislocation loops is lower than in the bulk case.

Figure 7 shows the interaction evolution between the dislocation loop and the displacement cascade. The following could be found: (1) At first, the cascade eliminates some of the atoms in the dislocation loop as shown in Figure 7b, which is similar to the case of 1/2<111> dislocation. (2) By interacting with the displacement cascade, dislocation dissociation occurs, which can be described as [001] $\rightarrow$ 1/2[111] + 1/2[$-$1$-$11], as shown in Figure 7c. Due to the instability of the 1/2[111] and 1/2 [$-$1$-$11] dislocation segments, dislocation dissociation disappears in the interaction with the displacement cascade. (3) In the later part of the displacement cascade process, the [001] dislocation segment is regenerated due to the recovery of interstitials and vacancies, and finally a complete [001] dislocation loop is formed. Note that the interstitials and vacancies induced by the displacement cascade enter the dislocation loop, which can be seen in Figure 7e. Unlike the 1/2<111> dislocation loop, the <100> dislocation segment is more likely to adsorb vacancies rather than interstitials.

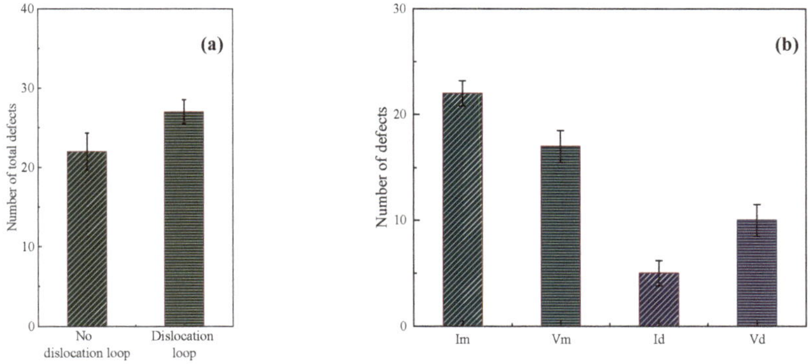

Figure 6. The number of point defects induced by the displacement cascade for BCC-Fe with or without the dislocation loops: (**a**) The effect of [001] dislocation loop on the total number of irradiation-induced point defects. (**b**) The effects of [001] dislocation loop on I_m, V_m, I_d, V_d.

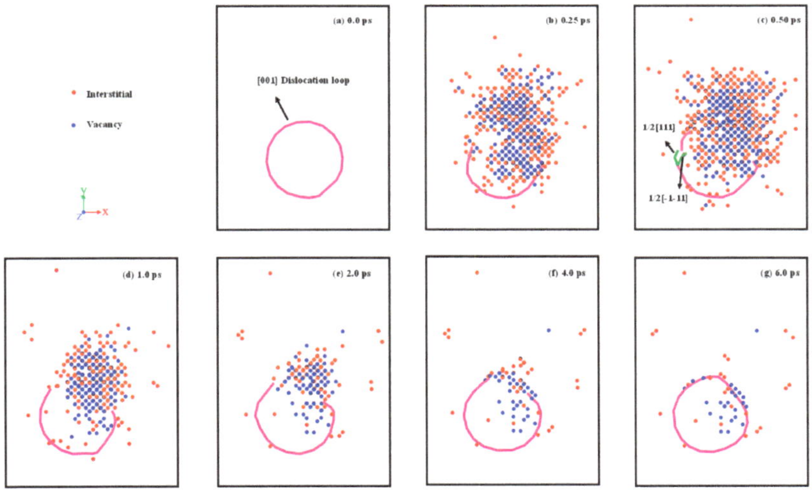

Figure 7. The interaction evolution between dislocation loop and displacement cascade.

3.3. Cascade Simulation of BCC-Fe with Grain Boundary

In this part, the displacement cascade interacting with the grain boundary is investigated. Figure 8 shows the effects of the ∑5(013) grain boundary on I_m, V_m, I_d and V_d. It can be seen that a displacement cascade that overlaps a grain boundary feeds the grain boundary with excess defects, leaving few defects in the matrix of BCC-Fe. In other words, the ∑5(013) grain boundary could reduce the surviving interstitials in bulk Fe by 70% and vacancies by 55%, and from this the grain boundary could be seen as having weak biased sinks for the annihilation of irradiation-induced defects. This is echoed by the fact that the grain boundary could trap defects but preferentially trap interstitials [31–33]. This phenomenon has also been observed in W [34], Mo [35], Fe-Cr alloys [36] and Fe [37]. Moreover, the time for stabilization of the irradiation damage in the grain boundary system is longer than in bulk Fe, which is attributed to the preferential sink property of the grain boundary over point defects. In terms of the structure of the grain boundary, the initial structure is 'kite'-shaped [38], as shown in Figure 8b. When interacting with the displacement cascade, the trapped interstitials fill the largest free volume within the grain boundary, retaining the 'kite' structure, which is consistent with previous work [39]. The difference lies in the fact that the 'kite' shape has an extra atom in its center, as shown in Figure 8c. Moreover, the reversible grain boundary phase transition, reported in FCC metals [38], is not found in this work.

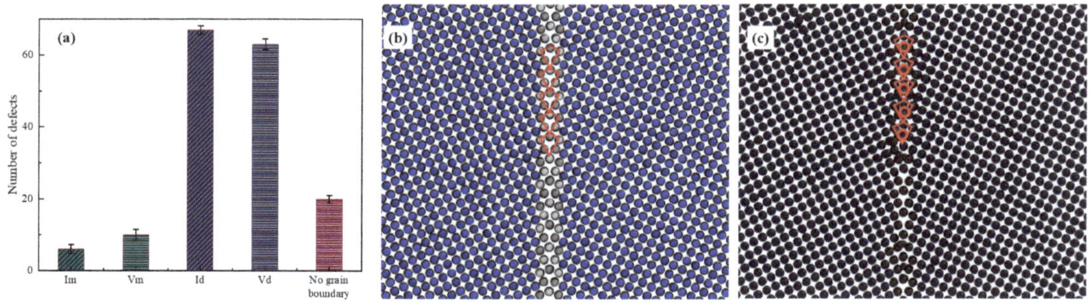

Figure 8. (**a**) The effects of $\sum5(013)$ grain boundary on I_m, V_m, I_d and V_d. Projection from $\langle100\rangle$ axis on GBs, showing the change in structures: (**b**) the grain boundary is 'kite'-shaped without the cascade and (**c**) the grain boundary remains 'kite'-shaped with the cascade. Bule balls represent atoms in bcc structure while white balls denote atoms of grain boundary.

Figure 9 shows the evolution of the displacement cascade with the simulation time in the presence of the $\sum5(013)$ grain boundary. Note that the grain boundaries seem to migrate slightly towards the side of the incoming PKA, which means the displacement cascade could damage the grain boundary structure. Furthermore, this could be attributed to both the temperature gradient during the cascade evolution as well as the excess point defect produced in the displacement cascade [39].

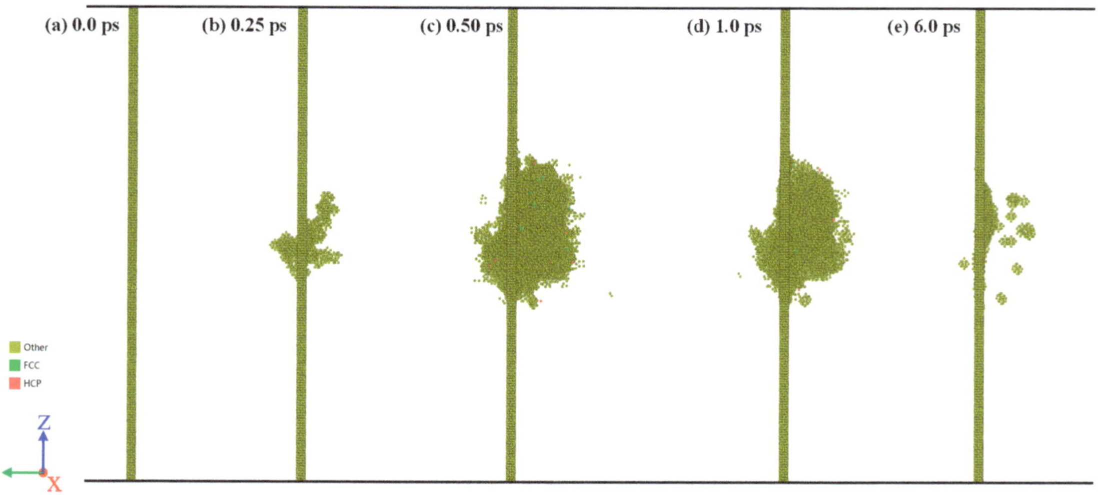

Figure 9. Evolution of the displacement cascade with the simulation time in the $\sum5(013)$ grain boundary. The atoms of BCC structure are hidden, as identified via the CNA method.

4. Conclusions

In this work, a molecular dynamics method is used to investigate the interaction between a displacement cascade and three types of structures: dislocations, dislocation loops and grain boundaries. The defect production and the displacement cascade induced atomic structural changes are discussed briefly. Below are listed the conclusions.

- Due to the presence of dislocations, dislocation loops and grain boundaries, the defect production efficiency changes. All three types of structures can cause the cascade to generate more total point defects. Moreover, a portion of these defects enter into the dislocation, dislocation loop and grain boundary. During the interaction, 1/2[1–11] edge dislocation climb occurs while screw dislocation may cross-slip. Compared with edge dislocations, screw dislocations are more likely to absorb point defects.
- As for the [001] dislocation loop, a lower number of surviving defects is left in the matrix of BCC-Fe, which means that the defect production efficiency in the BCC-Fe with a high density of [001] dislocation loops is lower than in the bulk case. Compared to the 1/2<111> dislocation segment, the <100> dislocation segment is more likely to adsorb vacancies rather than interstitials.
- The number of point defects in the grain boundary case is the largest of the three types of structures. Grain boundaries could trap defects but preferentially trap interstitials. Furthermore, grain boundaries seem to migrate slightly towards the side of the incoming PKA, but retain their 'kite' structure.

This paper explores the effects of three types of defects: dislocations, dislocation loops and grain boundaries, on the displacement cascade process, which can reveal irradiation damage at the microscale. Varied defect production numbers and efficiencies are investigated, which could be used as the input parameters for higher scale simulations. Future research can be carried out based on parameter transformation between microscale and higher scale simulations.

Author Contributions: P.L.: Data curation, Formal analysis, Investigation, Methodology, Project administration, Resources, Validation, Software, Visualization, Writing—original draft. J.N.: Supervision, Funding acquisition, Writing—review & editing, Data curation, Formal analysis, Investigation, Methodology, Project administration, Resources, Validation, Software, Visualization. S.C., W.C. and L.H.: Writing—review & editing, Data curation. All authors have read and agreed to the published version of the manuscript.

Funding: This work was supported by the Modular HTGR Super-critical Power Generation Technologycollaborative project between CNNC and Tsinghua University Project (grant number ZHJTIZYFGWD20201), the National Key R&D Plan of China (grant number 2020YFB1901600), the National Science Technology Major Project of China (grant numbers 2017ZX06902012 and 2017ZX06901024), the CNNC Youth Elite Scientific Research Project the Key Research and Development Program of Shandong Province (grant Numbers 2020CXGC010305 and 2021ZLGX01).

Institutional Review Board Statement: Not applicable.

Informed Consent Statement: Not applicable.

Data Availability Statement: All the data that support the findings of this study are available upon reasonable request.

Conflicts of Interest: The authors declare no conflict of interest.

References

1. Li, C.L.; Shu, G.G.; Yan, J.L.; Liu, W.; Duan, Y.G. Effects of Neutron, Ion and Proton Irradiation on Nano-Indentation Hardness of RPV Steels. *Mater. Sci. Forum* **2020**, *999*, 39–46. [CrossRef]
2. Xu, C.L.; Liu, X.B.; Wang, H.K.; Li, Y.F.; Jia, W.Q.; Qian, W.J.; Quan, Q.W.; Zhang, H.J.; Xue, F. A study of predicting irradiation-induced transition temperature shift for RPV steels with XGBoost modeling. *Nucl. Eng. Technol.* **2021**, *53*, 2610–2615. [CrossRef]
3. Toyama, T.; Nagai, Y.; Al Mazouzi, A.; Hatakeyama, M.; Hasegawa, M.; Ohkubo, T.; Van Walle, E.; Gerard, R. Intergranular Segregation in the Pressure Vessel Steel of a Commercial Nuclear Reactor Studied by Atom Probe Tomography. *Mater. Trans.* **2013**, *54*, 2119–2124. [CrossRef]
4. Voskoboinikov, R.; Osetsky, Y.; Bacon, D. Computer simulation of primary damage creation in displacement cascades in copper. I. Defect creation and cluster statistics. *J. Nucl. Mater.* **2008**, *377*, 385–395. [CrossRef]
5. Yu, G.; Ma, Y.; Cai, J.; Lu, D.-G. Molecular dynamics simulations of displacement cascades in Fe-10%Cr systems. *Chin. Phys. B* **2012**, *21*, 036101. [CrossRef]

6. Tikhonchev, M.Y.; Svetukhin, V.V.; Kadochkin, A.S.; Gaganidze, E. Molecular dynamics simulation of atomic displacement cascades in Fe-9 at % Cr and Fe-9 at % Cr-0.1 at % C alloys. *Russ. Met.* **2011**, *2011*, 415–422. [CrossRef]
7. Sahi, Q.-U.; Kim, Y.-S. Molecular dynamics simulations of the coupled effects of strain and temperature on displacement cascades in α-zirconium. *Nucl. Eng. Technol.* **2018**, *50*, 907–914. [CrossRef]
8. Pan, R.; Wu, L.; Wu, X.; Tang, A.; Zhang, W.; He, W.; Wen, B.; Ovcharenko, Y.; Kharchenko, D.O. Microstructure evolution of cascade annealing in irradiated pure α-Zr: A molecular dynamics study. *Radiat. Eff. Defects Solids* **2019**, *174*, 752–764. [CrossRef]
9. Fu, J.; Chen, Y.; Fang, J.; Gao, N.; Hu, W.; Jiang, C.; Zhou, H.-B.; Lu, G.-H.; Gao, F.; Deng, H. Molecular dynamics simulations of high-energy radiation damage in W and W–Re alloys. *J. Nucl. Mater.* **2019**, *524*, 9–20. [CrossRef]
10. Korchuganov, A.V.; Zolnikov, K.P.; Kryzhevich, D.S. Atomistic simulation of structural damage during ion irradiation of iron single crystals. *J. Phys. Conf. Ser.* **2017**, *830*, 012067. [CrossRef]
11. Choi, S.I.; Kim, J.H. Radiation-induced dislocation and growth behavior of zirconium and zirconium alloys—A review. *Nucl. Eng. Technol.* **2013**, *45*, 385–392. [CrossRef]
12. Wang, X.; Gao, N.; Setyawan, W.; Xu, B.; Liu, W.; Wang, Z. Effect of irradiation on mechanical properties of symmetrical grain boundaries investigated by atomic simulations. *J. Nucl. Mater.* **2017**, *491*, 154–161. [CrossRef]
13. Voskoboinikov, R.E. MD simulations of collision cascades in the vicinity of a screw dislocation in aluminium. *Nucl. Instrum. Methods Phys. Res. Sect. B Beam Interact. Mater. At.* **2013**, *303*, 104–107. [CrossRef]
14. Dai, C.; Long, F.; Saidi, P.; Béland, L.K.; Yao, Z.; Daymond, M.R. Primary damage production in the presence of extended defects and growth of vacancy-type dislocation loops in hcp zirconium. *Phys. Rev. Mater.* **2019**, *3*, 043602. [CrossRef]
15. Fu, B.Q.; Fitzgerald, S.P.; Hou, Q.; Wang, J.; Li, M. Effect of collision cascades on dislocations in tungsten: A molecular dy-namics study. *Nucl. Instrum. Methods Phys. Res. Sect. B Beam Interact. Mater. At.* **2017**, *393*, 169–173. [CrossRef]
16. Heredia-Avalos, S.; Denton, C.; Moreno-Marín, J.; Martinez, E.; Caturla, M. Collision cascade effects near an edge dislocation dipole in alpha-Fe: Induced dislocation mobility and enhanced defect clustering. *J. Nucl. Mater.* **2021**, *543*, 7. [CrossRef]
17. Lu, Z.; Xu, L.; Chen, T.; Tan, L.; Xu, H. Interactions between displacement cascade and dislocation and their influences on Peierls stress in Fe-20Cr-25Ni alloys. *Comput. Mater. Sci.* **2019**, *160*, 279–286. [CrossRef]
18. Zinkle, S.; Singh, B. Microstructure of neutron-irradiated iron before and after tensile deformation. *J. Nucl. Mater.* **2006**, *351*, 269–284. [CrossRef]
19. Plimpton, S. Fast Parallel Algorithms for Short-Range Molecular Dynamics. *J. Comput. Phys.* **1995**, *117*, 1–19. [CrossRef]
20. Thompson, A.P.; Aktulga, H.M.; Berger, R.; Bolintineanu, D.S.; Brown, W.M.; Crozier, P.S.; Veld, P.J.I.; Kohlmeyer, A.; Moore, S.G.; Nguyen, T.D.; et al. LAMMPS-a flexible simulation tool for parti-cle-based materials modeling at the atomic, meso, and continuum scales. *Comput. Phys. Commun.* **2022**, *271*, 34. [CrossRef]
21. Mendelev, M.I.; Han, S.; Srolovitz, D.J.; Ackland, G.J.; Sun, D.Y.; Asta, M. Development of new interatomic potentials ap-propriate for crystalline and liquid iron. *Philos. Mag.* **2003**, *83*, 3977–3994. [CrossRef]
22. Stimac, J.; Serrao, C.; Mason, J. Dependence of simulated radiation damage on crystal structure and atomic misfit in metals. *J. Nucl. Mater.* **2023**, *585*, 13. [CrossRef]
23. Lin, P.; Nie, J.; Liu, M. Investigation of the effects of point defects on the tensile strength of BCC-Fe using molecular dynamics. *Appl. Phys. A* **2021**, *127*, 16. [CrossRef]
24. Wang, M.; Wang, F.; Wang, H.; Zhang, J.; Zhao, X.; Wu, H. Effects of bismuth nanoparticles on the nano-cutting properties of single-crystal iron materials: A molecular dynamics study. *Appl. Phys. A* **2023**, *129*, 12. [CrossRef]
25. Hirel, P. Atomsk: A tool for manipulating and converting atomic data files. *Comput. Phys. Commun.* **2015**, *197*, 212–219. [CrossRef]
26. Clarke, A.S.; Jónsson, H. Structural changes accompanying densification of random hard-sphere packings. *Phys. Rev. E* **1993**, *47*, 3975–3984. [CrossRef]
27. Honeycutt, J.D.; Andersen, H.C. Molecular dynamics study of melting and freezing of small Lennard-Jones clusters. *J. Phys. Chem.* **1987**, *91*, 4950–4963. [CrossRef]
28. Nordlund, K.; Ghaly, M.; Averback, R.S.; Caturla, M.; de la Rubia, T.D.; Tarus, J. Defect production in collision cascades in elemental semiconductors and fcc metals. *Phys. Rev. B* **1998**, *57*, 7556–7570. [CrossRef]
29. Stukowski, A.; Bulatov, V.V.; Arsenlis, A. Automated identification and indexing of dislocations in crystal interfaces. *Model. Simul. Mater. Sci. Eng.* **2012**, *20*, 085007. [CrossRef]
30. Stukowski, A. Visualization and analysis of atomistic simulation data with OVITO—The Open Visualization Tool. *Model. Simul. Mater. Sci. Eng.* **2010**, *18*, 015012. [CrossRef]
31. LaGrange, T.; Arakawa, K.; Yasuda, H.; Kumar, M. Preferential void formation at crystallographically ordered grain boundaries in nanotwinned copper thin films. *Acta Mater.* **2015**, *96*, 284–291. [CrossRef]
32. Samaras, M.; Derlet, P.M.; Van Swygenhoven, H.; Victoria, M. Computer Simulation of Displacement Cascades in Nano-crystalline Ni. *Phys. Rev. Lett.* **2002**, *88*, 12550. [CrossRef]
33. Bai, X.M.; Voter, A.F.; Hoagland, R.G.; Nastasi, M.; Uberuaga, B.P. Efficient annealing of radiation damage near grain boundaries via interstitial emission. *Science* **2010**, *327*, 1631–1634. [CrossRef] [PubMed]
34. Li, X.; Duan, G.; Xu, Y.; Zhang, Y.; Liu, W.; Liu, C.; Liang, Y.; Chen, J.-L.; Luo, G.-N. Annihilating vacancies via dynamic reflection and emission of interstitials in nano-crystal tungsten. *Nucl. Fusion* **2017**, *57*, 116055. [CrossRef]
35. Novoselov, I.I.; Kuksin, A.Y.; Yanilkin, A.V. Energies of formation and structures of point defects at tilt grain boundaries in molybdenum. *Phys. Solid State* **2014**, *56*, 1401–1407. [CrossRef]

36. Zhang, J.; He, H.; Liu, W.; Kang, L.; Yun, D.; Chen, P. Effects of grain boundaries on the radiation-induced defects evolution in BCC Fe–Cr alloy: A molecular dynamics study. *Nucl. Mater. Energy* **2020**, *22*, 100726. [CrossRef]
37. Zolnikov, K.P.; Korchuganov, A.V.; Kryzhevich, D.S.; Chernov, V.M.; Psakhie, S.G. Formation of Point Defect Clusters in Metals with Grain Boundaries under Irradiation. *Phys. Mesomech.* **2019**, *22*, 355–364. [CrossRef]
38. Frolov, T.; Olmsted, D.L.; Asta, M.; Mishin, Y. Structural phase transformations in metallic grain boundaries. *Nat. Commun.* **2013**, *4*, 1899. [CrossRef]
39. Wang, X.Y.; Gao, N.; Xu, B.; Wang, Y.N.; Shu, G.G.; Li, C.L.; Liu, W. Effect of irradiation and irradiation defects on the mobility of Σ5 symmetric tilt grain boundaries in iron: An atomistic study. *J. Nucl. Mater.* **2018**, *510*, 568–574. [CrossRef]

MDPI AG
Grosspeteranlage 5
4052 Basel
Switzerland
Tel.: +41 61 683 77 34

Materials Editorial Office
E-mail: materials@mdpi.com
www.mdpi.com/journal/materials

www.ingramcontent.com/pod-product-compliance
Lightning Source LLC
Chambersburg PA
CBHW041428210926
51850CB00013B/3377